Asymptotic and Computational Analysis

PURE AND APPLIED MATHEMATICS

A Program of Monographs, Textbooks, and Lecture Notes

LECTURE NOTES

IN PURE AND APPLIED MATHEMATICS

1. *N. Jacobson,* Exceptional Lie Algebras
2. *L. -Å. Lindahl and F. Poulsen,* Thin Sets in Harmonic Analysis
3. *I. Satake,* Classification Theory of Semi-Simple Algebraic Groups
4. *F. Hirzebruch, W. D. Newmann, and S. S. Koh,* Differentiable Manifolds and Quadratic Forms (out of print)
5. *I. Chavel,* Riemannian Symmetric Spaces of Rank One (out of print)
6. *R. B. Burckel,* Characterization of C(X) Among Its Subalgebras
7. *B. R. McDonald, A. R. Magid, and K. C. Smith,* Ring Theory: Proceedings of the Oklahoma Conference
8. *Y.-T. Siu,* Techniques of Extension on Analytic Objects
9. *S. R. Caradus, W. E. Pfaffenberger, and B. Yood,* Calkin Algebras and Algebras of Operators on Banach Spaces
10. *E. O. Roxin, P.-T. Liu, and R. L. Sternberg,* Differential Games and Control Theory
11. *M. Orzech and C. Small,* The Brauer Group of Commutative Rings
12. *S. Thomeier,* Topology and Its Applications
13. *J. M. Lopez and K. A. Ross,* Sidon Sets
14. *W. W. Comfort and S. Negrepontis,* Continuous Pseudometrics
15. *K. McKennon and J. M. Robertson,* Locally Convex Spaces
16. *M. Carmeli and S. Malin,* Representations of the Rotation and Lorentz Groups: An Introduction
17. *G. B. Seligman,* Rational Methods in Lie Algebras
18. *D. G. de Figueiredo,* Functional Analysis: Proceedings of the Brazilian Mathematical Society Symposium
19. *L. Cesari, R. Kannan, and J. D. Schuur,* Nonlinear Functional Analysis and Differential Equations: Proceedings of the Michigan State University Conference
20. *J. J. Schäffer,* Geometry of Spheres in Normed Spaces
21. *K. Yano and M. Kon,* Anti-Invariant Submanifolds
22. *W. V. Vasconcelos,* The Rings of Dimension Two
23. *R. E. Chandler,* Hausdorff Compactifications
24. *S. P. Franklin and B. V. S. Thomas,* Topology: Proceedings of the Memphis State University Conference
25. *S. K. Jain,* Ring Theory: Proceedings of the Ohio University Conference
26. *B. R. McDonald and R. A. Morris,* Ring Theory II: Proceedings of the Second Oklahoma Conference
27. *R. B. Mura and A. Rhemtulla,* Orderable Groups
28. *J. R. Graef,* Stability of Dynamical Systems: Theory and Applications
29. *H.-C. Wang,* Homogeneous Branch Algebras
30. *E. O. Roxin, P.-T. Liu, and R. L. Sternberg,* Differential Games and Control Theory II
31. *R. D. Porter,* Introduction to Fibre Bundles
32. *M. Altman,* Contractors and Contractor Directions Theory and Applications
33. *J. S. Golan,* Decomposition and Dimension in Module Categories
34. *G. Fairweather,* Finite Element Galerkin Methods for Differential Equations
35. *J. D. Sally,* Numbers of Generators of Ideals in Local Rings
36. *S. S. Miller,* Complex Analysis: Proceedings of the S.U.N.Y. Brockport Conference
37. *R. Gordon,* Representation Theory of Algebras: Proceedings of the Philadelphia Conference
38. *M. Goto and F. D. Grosshans,* Semisimple Lie Algebras
39. *A. I. Arruda, N. C. A. da Costa, and R. Chuaqui,* Mathematical Logic: Proceedings of the First Brazilian Conference

40. *F. Van Oystaeyen,* Ring Theory: Proceedings of the 1977 Antwerp Conference
41. *F. Van Oystaeyen and A. Verschoren,* Reflectors and Localization: Application to Sheaf Theory
42. *M. Satyanarayana,* Positively Ordered Semigroups
43. *D. L. Russell,* Mathematics of Finite-Dimensional Control Systems
44. *P.-T. Liu and E. Roxin,* Differential Games and Control Theory III: Proceedings of the Third Kingston Conference, Part A
45. *A. Geramita and J. Seberry,* Orthogonal Designs: Quadratic Forms and Hadamard Matrices
46. *J. Cigler, V. Losert, and P. Michor,* Banach Modules and Functors on Categories of Banach Spaces
47. *P.-T. Liu and J. G. Sutinen,* Control Theory in Mathematical Economics: Proceedings of the Third Kingston Conference, Part B
48. *C. Byrnes,* Partial Differential Equations and Geometry
49. *G. Klambauer,* Problems and Propositions in Analysis
50. *J. Knopfmacher,* Analytic Arithmetic of Algebraic Function Fields
51. *F. Van Oystaeyen,* Ring Theory: Proceedings of the 1978 Antwerp Conference
52. *B. Kedem,* Binary Time Series
53. *J. Barros-Neto and R. A. Artino,* Hypoelliptic Boundary-Value Problems
54. *R. L. Sternberg, A. J. Kalinowski, and J. S. Papadakis,* Nonlinear Partial Differential Equations in Engineering and Applied Science
55. *B. R. McDonald,* Ring Theory and Algebra III: Proceedings of the Third Oklahoma Conference
56. *J. S. Golan,* Structure Sheaves over a Noncommutative Ring
57. *T. V. Narayana, J. G. Williams, and R. M. Mathsen,* Combinatorics, Representation Theory and Statistical Methods in Groups: YOUNG DAY Proceedings
58. *T. A. Burton,* Modeling and Differential Equations in Biology
59. *K. H. Kim and F. W. Roush,* Introduction to Mathematical Consensus Theory
60. *J. Banas and K. Goebel,* Measures of Noncompactness in Banach Spaces
61. *O. A. Nielson,* Direct Integral Theory
62. *J. E. Smith, G. O. Kenny, and R. N. Ball,* Ordered Groups: Proceedings of the Boise State Conference
63. *J. Cronin,* Mathematics of Cell Electrophysiology
64. *J. W. Brewer,* Power Series Over Commutative Rings
65. *P. K. Kamthan and M. Gupta,* Sequence Spaces and Series
66. *T. G. McLaughlin,* Regressive Sets and the Theory of Isols
67. *T. L. Herdman, S. M. Rankin, III, and H. W. Stech,* Integral and Functional Differential Equations
68. *R. Draper,* Commutative Algebra: Analytic Methods
69. *W. G. McKay and J. Patera,* Tables of Dimensions, Indices, and Branching Rules for Representations of Simple Lie Algebras
70. *R. L. Devaney and Z. H. Nitecki,* Classical Mechanics and Dynamical Systems
71. *J. Van Geel,* Places and Valuations in Noncommutative Ring Theory
72. *C. Faith,* Injective Modules and Injective Quotient Rings
73. *A. Fiacco,* Mathematical Programming with Data Perturbations I
74. *P. Schultz, C. Praeger, and R. Sullivan,* Algebraic Structures and Applications Proceedings of the First Western Australian Conference ón Algebra
75. *L. Bican, T. Kepka, and P. Nemec,* Rings, Modules, and Preradicals
76. *D. C. Kay and M. Breen,* Convexity and Related Combinatorial Geometry: Proceedings of the Second University of Oklahoma Conference
77. *P. Fletcher and W. F. Lindgren,* Quasi-Uniform Spaces
78. *C.-C. Yang,* Factorization Theory of Meromorphic Functions
79. *O. Taussky,* Ternary Quadratic Forms and Norms
80. *S. P. Singh and J. H. Burry,* Nonlinear Analysis and Applications
81. *K. B. Hannsgen, T. L. Herdman, H. W. Stech, and R. L. Wheeler,* Volterra and Functional Differential Equations

82. *N. L. Johnson, M. J. Kallaher, and C. T. Long,* Finite Geometries: Proceedings of a Conference in Honor of T. G. Ostrom
83. *G. I. Zapata,* Functional Analysis, Holomorphy, and Approximation Theory
84. *S. Greco and G. Valla,* Commutative Algebra: Proceedings of the Trento Conference
85. *A. V. Fiacco,* Mathematical Programming with Data Perturbations II
86. *J.-B. Hiriart-Urruty, W. Oettli, and J. Stoer,* Optimization: Theory and Algorithms
87. *A. Figa Talamanca and M. A. Picardello,* Harmonic Analysis on Free Groups
88. *M. Harada,* Factor Categories with Applications to Direct Decomposition of Modules
89. *V. I. Istrăţescu,* Strict Convexity and Complex Strict Convexity: Theory and Applications
90. *V. Lakshmikantham,* Trends in Theory and Practice of Nonlinear Differential Equations
91. *H. L. Manocha and J. B. Srivastava,* Algebra and Its Applications
92. *D. V. Chudnovsky and G. V. Chudnovsky,* Classical and Quantum Models and Arithmetic Problems
93. *J. W. Longley,* Least Squares Computations Using Orthogonalization Methods
94. *L. P. de Alcantara,* Mathematical Logic and Formal Systems
95. *C. E. Aull,* Rings of Continuous Functions
96. *R. Chuaqui,* Analysis, Geometry, and Probability
97. *L. Fuchs and L. Salce,* Modules Over Valuation Domains
98. *P. Fischer and W. R. Smith,* Chaos, Fractals, and Dynamics
99. *W. B. Powell and C. Tsinakis,* Ordered Algebraic Structures
100. *G. M. Rassias and T. M. Rassias,* Differential Geometry, Calculus of Variations, and Their Applications
101. *R.-E. Hoffmann and K. H. Hofmann,* Continuous Lattices and Their Applications
102. *J. H. Lightbourne, III, and S. M. Rankin, III,* Physical Mathematics and Nonlinear Partial Differential Equations
103. *C. A. Baker and L. M. Batten,* Finite Geometries
104. *J. W. Brewer, J. W. Bunce, and F. S. Van Vleck,* Linear Systems Over Commutative Rings
105. *C. McCrory and T. Shifrin,* Geometry and Topology: Manifolds, Varieties, and Knots
106. *D. W. Kueker, E. G. K. Lopez-Escobar, and C. H. Smith,* Mathematical Logic and Theoretical Computer Science
107. *B.-L. Lin and S. Simons,* Nonlinear and Convex Analysis: Proceedings in Honor of Ky Fan
108. *S. J. Lee,* Operator Methods for Optimal Control Problems
109. *V. Lakshmikantham,* Nonlinear Analysis and Applications
110. *S. F. McCormick,* Multigrid Methods: Theory, Applications, and Supercomputing
111. *M. C. Tangora,* Computers in Algebra
112. *D. V. Chudnovsky and G. V. Chudnovsky,* Search Theory: Some Recent Developments
113. *D. V. Chudnovsky and R. D. Jenks,* Computer Algebra
114. *M. C. Tangora,* Computers in Geometry and Topology
115. *P. Nelson, V. Faber, T. A. Manteuffel, D. L. Seth, and A. B. White, Jr.* Transport Theory, Invariant Imbedding, and Integral Equations: Proceedings in Honor of G. M. Wing's 65th Birthday
116. *P. Clément, S. Invernizzi, E. Mitidieri, and I. I. Vrabie,* Semigroup Theory and Applications
117. *J. Vinuesa,* Orthogonal Polynomials and Their Applications: Proceedings of the International Congress
118. *C. M. Dafermos, G. Ladas, and G. Papanicolaou,* Differential Equations: Proceedings of the EQUADIFF Conference
119. *E. O. Roxin,* Modern Optimal Control: A Conference in Honor of Solomon Lefschetz and Joseph P. Lasalle
120. *J. C. Díaz,* Mathematics for Large Scale Computing
121. *P. S. Milojević,* Nonlinear Functional Analysis

122. *C. Sadosky,* Analysis and Partial Differential Equations: A Collection of Papers Dedicated to Mischa Cotlar
123. *R. M. Shortt,* General Topology and Applications: Proceedings of the 1988 Northeast Conference
124. *R. Wong,* Asymptotic and Computational Analysis: Conference in Honor of Frank W. J. Olver's 65th Birthday

Other Volumes in Preparation

Asymptotic and Computational Analysis

Conference in Honor of Frank W. J. Olver's 65th Birthday

edited by

R. Wong
The University of Manitoba
Winnipeg, Manitoba, Canada

Marcel Dekker, Inc. New York and Basel

Library of Congress Cataloging-in-Publication Data

Asymptotic and computational analysis : conference in honor of Frank W. J. Olver's 65th birthday / [edited by] R. Wong.
p. cm. -- (Lecture notes in pure and applied mathematics ; 124)
Summary: Papers presented at the International Symposium on Asymptotic and Computational Analysis, held June 1989, Winnipeg, Man., sponsored by the Dept. of Applied Mathematics, University of Manitoba and the Canadian Applied Mathematics Society.
Includes bibliographical references.
ISBN 0-8247-8347-6 (alk. paper)
1. Numerical analysis--Congresses. 2. Asymptotic expansions--Congresses. 3. Olver, Frank W. J., 1924- --Congresses.
I. Olver, Frank W. J., 1924- . II. Wong, R. (Roderick),
III. International Symposium on Symptotic and Computational Analysis (1989 : Winnipeg, Man.) IV. University of Manitoba. Dept. of Applied Mathematics. V. Canadian Applied Mathematics Society.
VI. Series: Lecture notes in pure and applied mathematics ; v. 124.
QA299.6.A88 1990
519.4--dc20 90-2810
CIP

ISBN 0-8247-8347-6

This book is printed on acid-free paper.

MARCEL DEKKER, INC.
270 Madison Avenue, New York, New York 10016

Current printing (last digit):
10 9 8 7 6 5 4 3 2 1

PRINTED IN THE UNITED STATES OF AMERICA

Preface

These proceedings consist of a collection of papers presented at the International Symposium on Asymptotic and Computational Analysis. The symposium was held in honor of Professor Frank W. J. Olver on the occasion of his 65th birthday, and took place at the Delta Winnipeg Hotel in Winnipeg, Canada. It was sponsored by the Department of Applied Mathematics at the University of Manitoba and the Canadian Applied Mathematics Society.

Professor Olver obtained his B.Sc. in 1945 with first class honors in mathematics from the University of London, and was awarded a D.Sc. degree by the same university in 1961 for the work he had done up to that time in the field of Mathematical Analysis. He spent 15 years at the National Physical Laboratory in England, where he worked with some top-notch numerical analysts, and served for two years as head of the Numerical Methods Section. In 1961 Professor Olver left England for the U.S., and worked at the National Bureau of Standards from 1961 - 1986. Since 1969 he has been a research professor in the Institute for Physical Science and Technology at the University of Maryland.

Professor Olver has made many important contributions to asymptotics, numerical analysis and special functions. These include an algorithm for second-order recurrence relations, the computation of zeros of second-order differential equations, uniform asymptotic solutions of differential equations with turning point and simple pole, and estimates of remainders in asymptotic expansions.

His book "Asymptotics and Special Functions" has been the standard reference in these fields for fifteen years, and is now considered a classic. During the conference, Robert E. O'Malley, SIAM's vice president for publication, cited the high standards set by Olver as founding editor of the SIAM Journal on Mathematical Analysis for the period 1969 - 1974. He has continued to serve on the editorial board of this journal since that time, and is also an associate editor for Mathematics Computation.

The symposium was attended by 120 people, representing more than ten countries spanning three continents. There were eight one-hour speakers, viz: R. A. Askey, J. A. Cochran, W. Gautschi, D. S. Jones, F. W. J. Olver, R. E. O'Malley, Jr., W. H. Reid, and F. Ursell. In addition, there were 17 invited half-hour talks and 23 contributed talks. There are a total of 39 papers in these proceedings. Their topics cover the fields of special functions, asymptotics of integrals, asymptotic solutions to differential equations, numerical analysis, singular perturbations and linear difference equations.

We wish to thank all those who participated for their contribution, and all those who helped to make this symposium possible. We gratefully acknowledge the funding provided by the Natural Sciences and Engineering Council of Canada, and the University of Manitoba. Thanks are also due to Marcel Dekker, Inc., for including the Proceedings in their Lecture Notes in Pure and Applied Mathematics Series.

R. Wong

Contents

Underlined name indicates the presenter of the paper.

CONTRIBUTED PAPERS

Contributors

INVITED PAPERS

J. D. Allen Professor, Department of Mathematical Sciences, Rensselaer Polytechnic Institute, Troy, New York

Richard A. Askey Professor, Department of Mathematics, University of Wisconsin–Madison, Madison, Wisconsin

L. Berg Professor, Department of Mathematics, Wilhelm-Pieck Universitat, Rostock, East Germany

W. G. C. Boyd Professor, School of Mathematics, University of Bristol, Bristol, England

B. C. Carlson Professor, Ames Laboratory, Department of Mathematics, Iowa State University of Science and Technology, Ames, Iowa

L. M. Ciasullo Professor, Department of Pure and Applied Mathematics, Washington State University, Pullman, Washington

James A. Cochran Professor, Department of Pure and Applied Mathematics, Washington State University, Pullman, Washington

J. N. L. Connor Professor, Department of Chemistry, The University of Manchester, Manchester, England

Guangwei Fan Professor, Department of Mathematical Sciences, University of Cincinnati, Ohio

Luigi Gatteschi Professor, University of Torino, Department of Mathematics, Torino, Italy

Walter Gautschi Professor, Department of Computer Sciences, Purdue University, West Lafayette, Indiana

Floyd B. Hanson Professor, Department of Mathematics, Statistics, and Computer Science, University of Illinois at Chicago, Chicago, Illinois

Yasuhiko Ikebe Professor, Department of Mathematics, University of Utah, Salt Lake City, Utah

D. S. Jones Professor, Department of Mathematical Sciences, University of Dundee, Dundee, Scotland

Nicholas D. Kazarinoff Professor, Department of Mathematics, State University of New York at Buffalo, Buffalo, New York

Marek Kowalski Professor, Department of Mathematics, University of Utah, Salt Lake City, Utah

Anthony W. Leung Professor, Department of Mathematical Sciences, University of Cincinnati, Cincinnati, Ohio

R. E. Meyer Professor, Department of Mathematics, University of Wisconsin–Madison, Madison, Wisconsin

Frank W. J. Olver Professor, Institute for Physical Science and Technology, University of Maryland, College Park, Maryland

Robert E. O'Malley, Jr. Professor, Department of Mathematical Sciences, Rensselaer Polytechnic Institute, Troy, New York

W. H. Reid Professor, Department of Mathematics, University of Chicago, Chicago, Illinois

Yasutaka Sibuya Professor, Department of Mathematics, University of Minnesota, Minneapolis, Minnesota

Donald R. Smith Professor, Department of Mathematics, University of California, San Diego, La Jolla, California

Kusum Soni Professor, Department of Mathematics, University of Tennessee, Knoxville, Tennessee

R. P. Soni Professor, Department of Mathematics, University of Tennessee, Knoxville, Tennessee

Frank Stenger Professor, Department of Mathematics, University of Utah, Salt Lake City, Utah

Nico M. Temme Professor, Center for Mathematics and Computer Science, Amsterdam, The Netherlands

F. Ursell Professor, Department of Mathematics, University of Manchester, Manchester, England

Frederic Y. M. Wan Professor, Department of Applied Mathematics, University of Washington, Seattle, Washington

Joseph S. Wilkowski Professor, Department of Mathematics, Manhattan College, Bronx, New York

Jet Wimp Professor, Department of Mathematics and Computer Sciences, Drexel University, Philadelphia, Pennsylvania

R. V. M. Zahar Professor, Professeur Titulaire, University of Montreal, Department of Information and Operations Research, Montreal, Quebec, Canada

CONTRIBUTED PAPERS

R. M. Barron Professor, Department of Mathematics and Statistics, and Fluid Dynamics Research Institute, University of Windsor, Windsor, Ontario, Canada

J. F. Bird Professor, The Johns Hopkins University, Applied Physics Laboratory, Laurel, Maryland

P. A. Carinhas Professor, Department of Physics, Texas A & M University, College Station, Texas

O. P. Chandna Professor, Department of Mathematics and Statistics, and Fluid Dynamics Research Institute, University of Windsor, Windsor, Ontario, Canada

Anthony M. J. Davis Professor, Department of Mathematics, The University of Alabama, Tuscaloosa, Alabama

Luca Dieci Professor, School of Mathematics, Georgia Institute of Technology, Atlanta, Georgia

R. A. Farrell Professor, The Johns Hopkins University, Applied Physics Laboratory, Laurel, Maryland

C. L. Frenzen Professor, Mathematics Department, Southern Methodist University, Dallas, Texas

S. A. Fulling Professor, Department of Mathematics, Texas A & M University, College Station, Texas

William M. Y. Goh Professor, Drexel University, Department of Mathematics and Computer Science, Philadelphia, Pennsylvania

D. Kaminski Professor, Weapons Effectiveness Group, Defence Research Establishment, Valcartier, Courcelette, Quebec, Canada

Francesco Mainardi Professor, Department of Physics, University of Bologna, Bologna, Italy

R. B. Paris Professor, Department of Mathematical and Computer Sciences, Dundee Institute of Technology, Dundee, Scotland

N. Rudraiah Professor, Department of Mathematics and Statistics, and Fluid Dynamics Research Institute, University of Windsor, Windsor, Ontario, Canada

Robert D. Russell Professor, Department of Mathematics and Statistics, Simon Fraser University, Burnaby, British Columbia, Canada

Eric Schmutz Professor, Drexel University, Department of Mathematics and Computer Science, Philadelphia, Pennsylvania

Shagi-Di Shih Professor, Department of Mathematics, University of Wyoming, Laramie, Wyoming

D. Siegel Professor, Department of Applied Mathematics, University of Waterloo, Waterloo, Ontario, Canada

R. Spigler Professor, Department of Mathematical Methods and Models for Applied Science, University of Padova, Padova, Italy

B. J. Stoyanov Professor, The Johns Hopkins University, Applied Physics Laboratory, Laurel, Maryland

M. Vianello Professor, Department of Pure and Applied Mathematics, University of Padova, Padova, Italy

Giuliano Vitali Professor, Institute of Agronomy, University of Bologna, Bologna, Italy

A. D. Wood Professor, School of Mathematical Sciences, Dublin City University, Glasnevin, Dublin, Ireland

Asymptotic and Computational Analysis

Part I

Invited Papers

Graphs as an Aid to Understanding Special Functions

Richard A. Askey Professor, Department of Mathematics, University of Wisconsin–Madison, Madison, Wisconsin

1. INTRODUCTION

Graphs of special functions can be very useful, both in teaching and in discovering problems to work on. The use of graphs to help understand properties of functions goes back a long way. In 1848, Wilbraham [17] discovered the nonuniform convergence of some Fourier series by looking at graphs of the partial sums of the expansion

$$\sum_{k=0}^{\infty} (-1)^k \frac{\cos(2k+1)x}{2k+1} = \begin{cases} \frac{\pi}{4}, & |x| < \frac{\pi}{2} \\ \frac{-\pi}{4}, & \frac{\pi}{2} < |x| < \pi \end{cases}.$$

Later, graphs of this expansion and of

$$\sum_{k=1}^{\infty} \frac{\sin\ kx}{k} = \frac{\pi - x}{2}, \qquad 0 < x < 2\pi$$

by Michelson and Stratton [10] led to the rediscovery of this, which now goes under the name of Gibbs phenomena. Hewitt and Hewitt [6] gave a nice historical survey of this work, and provided more detailed graphs which help explain some of the more subtle aspects of Gibbs phenomena.

One book that contained many graphs is Jahnke and Emde [7], [8], [9]. In the preface to the 1933 edition, Emde wrote the following:

> "Many new tables have been added and old ones extended. If every new table had been included the price of the book would have become excessive, but where it has not been possible to include a table, curves have been given. These curves have been plotted to such a scale and with such fine subdivision that the numerical values can be read off with some accuracy."

This is quoted from the 1945 edition [8].

While this may have been a reasonable goal in 1909 or 1933 it no longer is. Graphs are a very good way to display qualitative information, but not a good way to store accurate quantitative data. This means that some of the graphs in handbooks of special functions are drawn to the wrong scale. However some were just right. Consider the following graph.

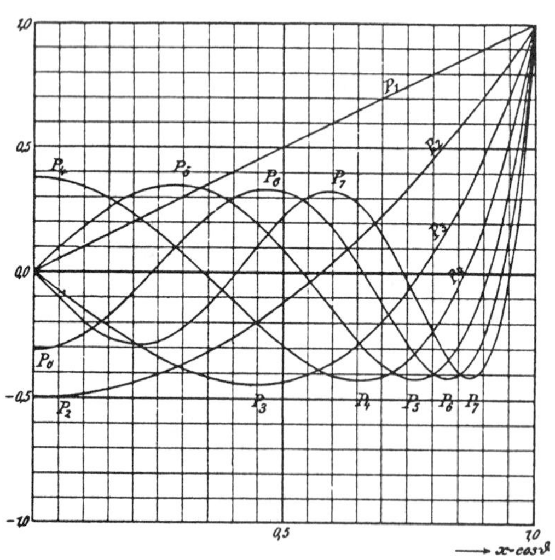

Figure 1. Legendre polynomial $P_n(x)$

This graph is excellent, and provides some important information about Legendre polynomials that had not been observed before. However it took almost 40 years before anyone looked carefully at the graph and made the right observation. This observation, of J. Todd, will be mentioned later. First, some notation and some other graphs.

$$P_n(x) = \frac{1}{2^n n!}\frac{d^n}{dx^n}(x^2-1)^n . \tag{1.1}$$

Legendre polynomials are one instance of orthogonal polynomials. For general orthogonal polynomials, set

$$p_n(x) = k_n x^n + \dots , \tag{1.2}$$

and assume that

$$\int_{-\infty}^{\infty} p_n(x)p_m(x)d\alpha(x) = h_n\delta_{m,n} , \qquad m,n = 0,1,\dots, \tag{1.3}$$

with $d\alpha(x) \geqq 0$. In the general theory it is often assumed that the polynomials are orthonormal, so $h_n = 1$. However for specific sets of polynomials this introduces unwanted square roots, and there are some important facts which are only true, or only useful, with a different normalization.

The polynomials treated are hypergeometric functions. Here is the standard notation.

$$(a)_n = \Gamma(n+a)/\Gamma(a), \qquad n = 0,1,\dots, \tag{1.4}$$

$${}_pF_q\left(\begin{matrix}a_1,\dots,a_p\\ b_1,\dots,b_q\end{matrix};t\right) = \sum_{k=0}^{\infty}\frac{(a_1)_k\cdots(a_p)_k t^k}{(b_1)_k\cdots(b_q)_k k!} \tag{1.5}$$

2. LEGENDRE POLYNOMIALS

Legendre polynomials arise in many places, so many people have to try to find the right formulas to help solve their problems. There are many known formulas about Legendre polynomials, and it is not easy to separate them into formulas which are specific to Legendre polynomials, those which have extensions to some other sets of orthogonal polynomials, and those that hold for general orthogonal polynomials. Also the first look at a long list of formulas is intimidating. As an aid to students who are seeing this material for the first time, it is useful to either show them graphs of Legendre polynomials, or have them compute and draw such graphs, and then try to prove the facts which seem to be true from looking at the graphs. Here are three graphs of Legendre polynomials.

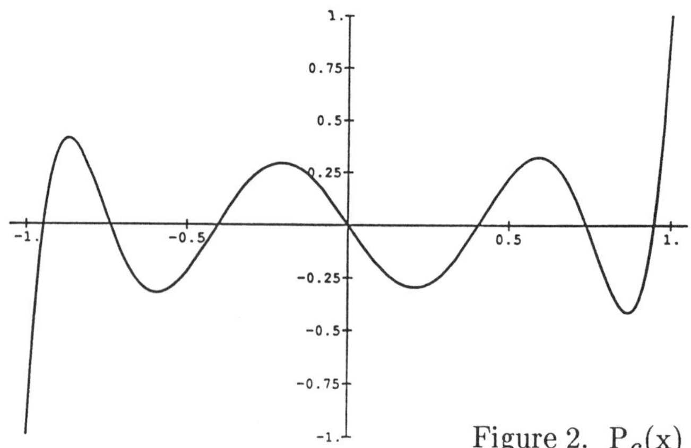

Figure 2. $P_6(x)$

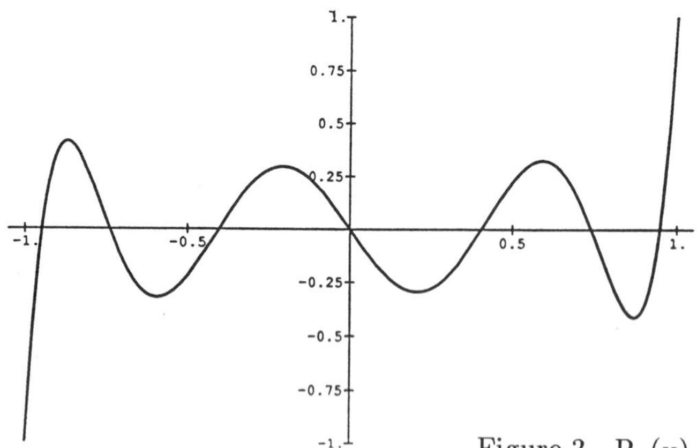

Figure 3. $P_7(x)$

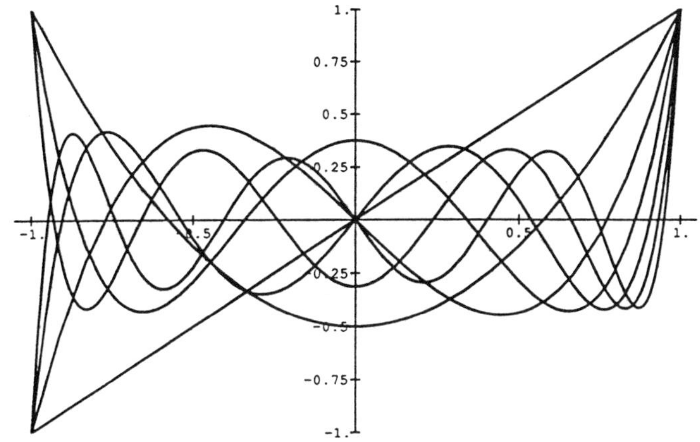

Figure 4. $P_n(x)$ $n = 1,\ldots,7$

These graphs are given on $[-1,1]$, since most of the interest in Legendre polynomials is for these values of x. For $x > 1$ they grow monotonically to infinity as $x \to \infty$ for $n \geq 1$, and as $n \to \infty$ for $x > 1$ fixed. Before continuing the reader should look at the graphs and see what facts they suggest.

Here are some useful formulas involving Legendre polynomials.

$$P_n(x) = \frac{(-1)^n}{2^n n!} \frac{d^n}{dx^n} (1-x^2)^n \tag{2.1}$$

$$P_n(x) = \frac{(-1)^n}{2^n n!} \sum_{k=0}^{\lfloor n/2 \rfloor} \binom{n}{k} \frac{(2n-2k)!}{(n-2k)!} (-1)^k x^{n-2k} \tag{2.2}$$

$$P_n(x) = {}_2F_1\left(\begin{matrix}-n, n+1\\ 1\end{matrix}; \frac{1-x}{2}\right) \tag{2.3}$$

$$(1-2xr+r^2)^{-1/2} = \sum_{n=0}^{\infty} P_n(x) r^n \tag{2.4}$$

$$P_n(\cos\theta) = \sum_{k=0}^{n} \frac{(\frac{1}{2})_k(\frac{1}{2})_{n-k}}{k!(n-k)!} \cos(n-2k)\theta \tag{2.5}$$

$$(2n+1)xP_n(x) = (n+1)P_{n+1}(x) + nP_{n-1}(x) \tag{2.6}$$

$$\sum_{k=0}^{n} (2k+1)P_k(x)P_k(y) = (n+1)\frac{[P_{n+1}(x)P_n(y)-P_n(x)P_{n+1}(y)]}{x-y} \tag{2.7}$$

$$\sum_{k=0}^{n} (2k+1)[P_k(x)]^2 = (n+1)[P'_{n+1}(x)P_n(x)-P'_n(x)P_{n+1}(x)] \tag{2.8}$$

$$\int_{-1}^{1} P_m(x)P_n(x)dx = \frac{2\delta_{m,n}}{2n+1}, \qquad m,n = 0,1,\ldots \tag{2.9}$$

where

$$\delta_{m,n} = \begin{matrix} 0 & \text{if} & m \neq n \\ 1 & \text{if} & m = n \end{matrix}$$

$y = P_n(x)$ satisfies

$$(1-x^2)y'' - 2xy' + n(n+1)y = 0\,. \tag{2.10}$$

In his proof of Todd's conjecture, which will be stated later, Szegö used the following formulas.

$$(1-x)[P'_{n+1}(x)+P'_n(x)] = (n+1)[P_n(x)-P_{n+1}(x)] \tag{2.11}$$

$$(1+x)[P'_{n+1}(x)-P'_n(x)] = (n+1)[P_n(x)+P_{n+1}(x)] \tag{2.12}$$

$$xP'_n(x) + (n+1)P_n(x) = P'_{n+1}(x) \tag{2.13}$$

The formulas (2.1) – (2.10) are all reasonable formulas, and if I do not remember every constant in them, I know the general form and with a little work can compute the constants I do not remember. However formulas like (2.11) – (2.13) are of a different type, and I find them much harder to remember. There are substitute formulas, especially for (2.11) and (2.12), but to state them a more general set of orthogonal polynomials needs to be introduced. These are the Jacobi polynomials $P_n^{(\alpha,\beta)}(x)$. I will state analogues of some of (2.1) – (2.10) for Jacobi polynomials. Further formulas can be found in Szegö [15, Chapter IV] and Erdélyi [5, Chapter 10].

$$(1-x)^{\alpha}(1+x)^{\beta}P_n^{(\alpha,\beta)}(x) = \frac{(-1)^n}{2^n n!}\frac{d^n}{dx^n}[(1-x)^{n+\alpha}(1+x)^{n+\beta}] \tag{2.14}$$

$$P_n^{(\alpha,\beta)}(x) = \frac{(\alpha+1)_n}{n!}\,{}_2F_1\left(\begin{matrix}-n, n+\alpha+\beta+1\\ \alpha+1\end{matrix}; \frac{1-x}{2}\right) \tag{2.15}$$

$$\int_{-1}^{1} P_n^{(\alpha,\beta)}(x)P_m^{(\alpha,\beta)}(x)(1-x)^{\alpha}(1+x)^{\beta}dx$$

$$= \frac{2^{\alpha+\beta+1}\Gamma(n+\alpha+1)\Gamma(n+\beta+1)}{(2n+\alpha+\beta+1)\Gamma(n+1)\Gamma(n+\alpha+\beta+1)}\delta_{m,n} \tag{2.16}$$

when $\alpha,\beta > -1$, and for $n = 0$ the product $(2n+\alpha+\beta+1)\Gamma(n+\alpha+\beta+1)$ is interpreted as $\Gamma(\alpha+\beta+2)$ so there is no indeterminancy when $\alpha + \beta + 1 = 0$.

$y = P_n^{(\alpha,\beta)}(x)$ satisfies

$$(1-x^2)y'' + [\beta-\alpha-(\alpha+\beta+2)x]y' + n(n+\alpha+\beta+1)y = 0 \tag{2.17}$$

Formula (2.15) needs to be interpreted as a limit when $\alpha = -1,\ldots,-n$, and $\alpha,\beta > -1$ is necessary in (2.16), but the other formulas do not need this restriction. Here are some other useful formulas.

$$\frac{d}{dx} P_n^{(\alpha,\beta)}(x) = \frac{(n+\alpha+\beta+1)}{2} P_{n-1}^{(\alpha+1,\beta+1)}(x) \tag{2.18}$$

$$(n+\alpha+1)P_n^{(\alpha,\beta)}(x) - (n+1)P_{n+1}^{(\alpha,\beta)}(x)$$
$$= \frac{(2n+\alpha+\beta+2)(1-x)}{2} P_n^{(\alpha+1,\beta)}(x) \tag{2.19}$$

$$P_n^{(\alpha,\beta)}(-x) = (-1)^n P_n^{(\beta,\alpha)}(x) \tag{2.20}$$

$$(2n+\alpha+\beta+1)P_n^{(\alpha,\beta)}(x)$$
$$= (n+\alpha+\beta+1)P_n^{(\alpha+1,\beta)}(x) - (n+\beta)P_{n-1}^{(\alpha+1,\beta)}(x) . \tag{2.21}$$

The three term recurrence relation (2.6) and its consequence, the Christoffel–Darboux formula (2.7), are actually facts about general orthogonal polynomials, as are the formulas (2.19) and (2.21). Formula (2.20) also extends to general orthogonal polynomials. Formulas (2.14), (2.17) and (2.18) only hold for Jacobi polynomials and their limiting cases. There is another way to state (2.19) that makes it easier to understand. This will be given below.

General orthogonal polynomials satisfy a three term recurrence relation that extends (2.6). It is

$$xp_n(x) = A_n p_{n+1}(x) + B_n p_n(x) + C_n p_{n-1}(x) \tag{2.22}$$

with A_n, B_n, C_n real, $A_n C_{n+1} > 0$, $n = 0,1,\ldots$. There is a converse. If (2.22) is satisfied, along with the required conditions in the next line, then there is a positive measure $d\alpha(x)$ so that (1.3) holds.

From (2.22) it is easy to show that

$$\sum_{k=0}^{n} \frac{p_k(x)p_k(y)}{h_k} = \frac{A_n[p_{n+1}(x)p_n(y)-p_n(x)p_{n+1}(y)]}{h_n(x-y)} . \tag{2.23}$$

Then

$$\sum_{k=0}^{n} \frac{[p_k(x)]^2}{h_k} = \frac{A_n[p'_{n+1}(x)p_n(x)-p'_n(x)p_{n+1}(x)]}{h_n} . \tag{2.24}$$

If $\{p_n(x)\}$ is orthogonal on $a \leqq x \leqq b$, then

$$\frac{p_n(x)}{p_n(a)} - \frac{p_{n+1}(x)}{p_{n+1}(a)} = \lambda_n q_n(x)\,(x-a) \tag{2.25}$$

where $\{q_n(x)\}$ is orthogonal with respect

$$(x-a)d\alpha(x)$$

when (1.3) holds. This is the general fact that gives (2.19). To find the general result that gives (2.21), set $y = a$ in (2.23) and rewrite it as

$$\sum_{k=0}^{n} \frac{p_k(x)p_k(a)}{h_k} = \frac{A_n p_n(a)p_{n+1}(a)}{(x-a)h_n} \left[\frac{p_{n+1}(x)}{p_{n+1}(a)} - \frac{p_n(x)}{p_n(a)} \right] = \mu_n q_n(x) .$$

Then subtracting this formula with n and $n-1$ gives

$$\mu_n q_n(x) - \mu_{n-1} q_{n-1}(x) = \frac{p_n(x)p_n(a)}{h_n} . \tag{2.26}$$

This implies (2.21). The polynomials $q_n(x)$ satisfy

$$\int_a^b q_n(x)q_m(x)(x-a)d\alpha(x) = 0, \quad m \neq n . \tag{2.27}$$

To obtain an analogue of (2.20), if $\{p_n(x)\}$ is orthogonal with respect to $w(x)$ and $\{q_n(x)\}$ is orthogonal with respect to $w(-x)$, then

$$q_n(x) = \lambda_n p_n(-x)$$

for a constant λ_n.

The constants in (2.19) – (2.21) can be found by looking at the highest coefficient, and the value when $x = 1$ in (2.19) and (2.21) to determine the second unknown coefficient. There are only two unknown constants in the general form of both of these, for any one of the three apparent constants can be taken to be one.

There are many other important formulas, but these will suffice for our current goal. This is to state and prove facts about $P_n(x)$ which are suggested by the graphs. Listing these, we have:

(a) $P_n(x)$ is even or odd depending on whether n is even or odd.

(b) $P_n(1) = 1$ and $|P_n(x)| \leqq P_n(1) = 1$.

(c) The zeros of $P_n(x)$ are real, simple and all lie in the interval $-1 < x < 1$.

(d) More generally than (b), if $P_n'(y_{k,n}) = 0$, $y_{n-1,n} < \dots < y_{1,n}$, if $y_{0,n} = 1$, $y_{n,n} = -1$, and if

$$\mu_{k,n} = |P_n(y_{k,n})|, \quad k = 0,1,\dots,n,$$

then

$$\mu_{k,n} < \mu_{k-1,n}, \quad k = 1,2,\dots,\lfloor n/2 \rfloor . \tag{2.28}$$

All of these are suggested by the separate graphs of $P_6(x)$ and $P_7(x)$. Looking at the graphs of $P_n(x)$ together, as in Figures 1 or 4, it seems likely that:

(e) The zeros of $P_n(x)$ and $P_{n+1}(x)$ separate each other.

(f) The location of the maxima of $|P_n(x)|$ and $|P_{n+1}(x)|$ separate each other.

(g) $\mu_{k,n} < \mu_{k,n-1}$, $k = 1,2,\dots,n-1$.

The inequality in (g) is the observation Todd made.

Most of these are well known, and proofs can be found in a number of different books. For example Szegö [15] contains proofs of all except (g). For most a short comment will suffice.

(a) follows from (2.1) or (2.2), or from the fact that the weight function is even.

(b) follows from (2.5), which is an immediate corollary of (2.4) after factoring the left hand side of (2.4) as

$$(1-re^{i\theta})^{-1/2}(1-re^{-i\theta})^{-1/2}, \qquad x = \cos\theta,$$

and taking the Cauchy product of the series. Observe that (b) does not follow from either (2.2) or (2.3). As a result these are probably poor ways to compute $P_n(x)$ for large n. They are.

(c) follows from (2.1) and Rolle's theorem. For if $n \geqq 2$,

$$\frac{d}{dx}(1-x^2)^n$$

vanishes to order $n-1$ at 1, at -1, and at one point between them. Then apply Rolle's theorem $n-1$ more times. (c) is also a general fact about all sets of orthogonal polynomials. For if

$$\int_a^b p_n(x)p_m(x)d\alpha(x) = 0, \quad m \neq n, \ d\alpha(x) \geqq 0,$$

and if $p_n(x)$ changes sign at $x_1,\dots,x_r$, $a < x_k < b$, $k = 1,\dots,r$ then

$$\varepsilon \int_a^b p_n(x) \prod_{k=1}^{r} (x-x_k)d\alpha(x) > 0$$

for either $\varepsilon = 1$ or $\varepsilon = -1$, and

$$\int_a^b p_n(x) \prod_{k=1}^{r} (x-x_k)d\alpha(x) = 0$$

if $r < n$, by orthogonality. Therefore $p_n(x)$ has n sign changes in $a < x < b$, and since it can have at most n zeros, they are all real, simple and on the smallest interval containing the support of $d\alpha(x)$.

(d) is harder to prove, and is not true in general. The only known proof uses an idea of Sonine, and monotonicity can be proven for general Jacobi polynomials by this method. Here is the argument for Legendre polynomials. See Szegö [15, Chapter VII], or better, work it out yourself if you want to see how it works for Jacobi polynomials.

Since we are interested in the maximum points of $|P_n(x)|$, and these occur when the derivative is zero, it is natural to consider $[P_n(x)]^2$ rather than $|P_n(x)|$. Here are two graphs of $[P_8(x)]^2$. The first is on $[-1,1]$. Clearly the most interesting part of the graph is for x^2 somewhat smaller than 1, so the next graph gives more detail when $[P_8(x)]^2$ is small.

What would be nice is to find a function that increases for $0 \leqq x \leqq 1$ and interpolates the maximum points of $[P_n(x)]^2$. Sonine's idea is to consider

$$f(x) = [P_n(x)]^2 + \frac{(1-x^2)[P_n'(x)]^2}{n(n+1)}$$

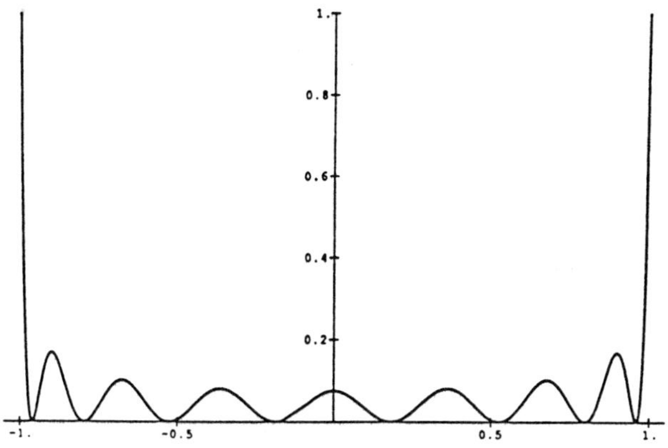

Figure 5. $[P_8(x)]^2$

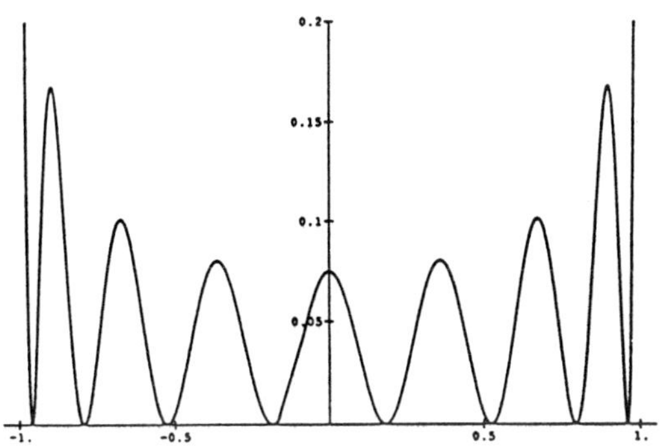

Figure 6. $[p_8(x)]^2$

which interpolates the values at $P_n'(x) = 0$ and at $x = 1$. Differentiating gives

$$f'(x) = 2P_n(x)P_n'(x) + \frac{2(1-x^2)P_n'(x)\ P_n''(x)}{n(n+1)} - \frac{2x[P_n'(x)]^2}{n(n+1)}$$

$$= \frac{2P_n'(x)}{n(n+1)}[(1-x^2)P_n''(x) - xP_n'(x) + n(n+1)P_n(x)]$$

$$= \frac{2x[P_n'(x)]^2}{n(n+1)} .$$

This implies $f'(x) \geqq 0, 0 \leqq x \leqq 1$, and $f'(x) \leqq 0, -1 \leqq x \leqq 0$, so $f(x)$ is increasing for $0 < x \leqq 1$ and decreasing for $-1 \leqq x < 0$. This implies (2.27).

The graph of $f(x)$ for $n = 8$ is given next. Notice not only the monotonicity, but the lack of convexity.

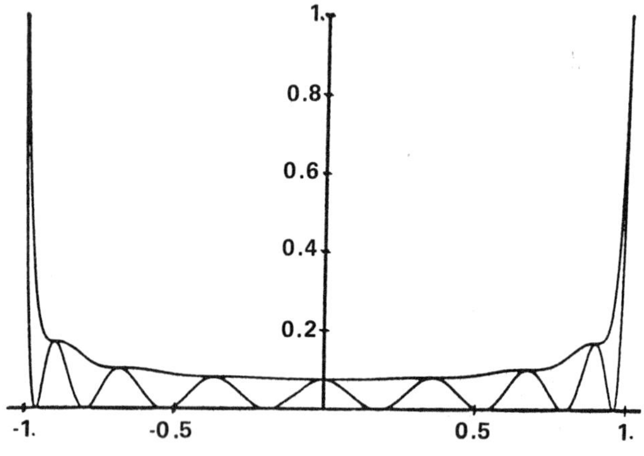

Figure 7. $[P_8(x)]^2$ and an upper envelope

The separation of the zeros of $P_n(x)$ and $P_{n+1}(x)$ is a fact about general orthogonal polynomials. To prove it, let $x_{k,n}$ denote the zeros of $P_n(x)$ ordered by $x_{k+1,n} < x_{k,n}$, $k = 1,2,\ldots,n-1$. Then putting $x = x_{k,n+1}$, and $x = x_{k+1,n+1}$ in (2.24) gives

$$p_{n+1}'(x_{k,n+1})p_n(x_{k,n+1})p_{n+1}'(x_{k+1,n+1})p_n(x_{k+1,n+1}) > 0 . \qquad (2.29)$$

For the left hand side of (2.24) is positive, and (2.29) contains all the terms that are not necessarily positive when (2.24) is multiplied by itself for $x = x_{k,n+1}$ and $x = x_{k+1,n+1}$. Since the zeros $x_{k,n+1}$ are simple, $p_{n+1}(x)$ has constant sign for $x_{k+1,n+1} < x < x_{k,n+1}$, and since it changes sign at the end points of this interval,

$$p'_{n+1}(x_{k,n+1})p'_{n+1}(x_{k+1,n+1}) < 0\,. \tag{2.30}$$

Then (2.29) and (2.30) imply that

$$p_n(x_{k,n+1})p_n(x_{k+1,n+1}) < 0\,,$$

so $p_n(x)$ vanishes in the interval $x_{k+1,n+1} < x < x_{k,n+1}$. This gives n zeros for $p_n(x)$, so

$$x_{k+1,n} < x_{k+1,n+1} < x_{k,n} < x_{k,n+1}\,. \tag{2.31}$$

To prove the separation of the points where the maxima of $P_n(x)$ occur, one can just remark that

$$\{P'_{n+1}(x)\}$$

is a set of orthogonal polynomials. However this separation is also a fact about general orthogonal polynomials. If

$$p'_n(y_{k,n}) = 0, \quad k = 1,2,\ldots,n-1$$

and

$$y_{k+1,n} < y_{k,n}$$

then the same argument gives

$$p'_n(y_{k,n+1})p_{n+1}(y_{k,n+1})p'_n(y_{k+1,n+1})p_{n+1}(y_{k+1,n+1}) > 0\,.$$

But $p_n(x)$ changes sign at its zeros, so

$$p_{n+1}(y_{k,n+1})p_{n+1}(y_{k+1,n+1}) < 0 .$$

Therefore

$$p_n'(y_{k,n+1})p_n'(y_{k+1,n+1}) < 0$$

and so $p_n'(x)$ has a change of sign in $y_{k+1,n+1} < x < y_{k,n+1}$. Therefore

$$y_{k+1,n} < y_{k+1,n+1} < y_{k,n} < y_{k,n+1} \cdot$$

Finally consider Todd's conjecture (g),

$$\mu_{k,n} < \mu_{k,n-1}, \ k = 1,2,\ldots,n-1 . \tag{2.32}$$

Here we will prove monotonicity for the maxima of

$$\left|\frac{P_n^{(\alpha,\alpha)}(x)}{P_n^{(\alpha,\alpha)}(1)}\right| .$$

When it is useful we will denote the maxima of $|P_n^{(\alpha,\alpha)}(x)/P_n^{(\alpha,\alpha)}(1)|$ by $\mu_{k,n}(\alpha)$. Consider

$$\left[\frac{P_n^{(\alpha,\alpha)}(x)}{P_n^{(\alpha,\alpha)}(1)}\right]^2 - \left[\frac{P_{n+1}^{(\alpha,\alpha)}(x)}{P_{n+1}^{(\alpha,\alpha)}(1)}\right]^2$$

$$= \left[\frac{P_n^{(\alpha,\alpha)}(x)}{P_n^{(\alpha,\alpha)}(1)} - \frac{P_{n+1}^{(\alpha,\alpha)}(x)}{P_{n+1}^{(\alpha,\alpha)}(1)}\right]\left[\frac{P_n^{(\alpha,\alpha)}(x)}{P_n^{(\alpha,\alpha)}(1)} + \frac{P_{n+1}^{(\alpha,\alpha)}(x)}{P_{n+1}^{(\alpha,\alpha)}(1)}\right]$$

$$= \frac{n!}{(\alpha+1)_n}\,(1-x)P_n^{(\alpha+1,\alpha)}(x)\;\frac{n!}{(\alpha+1)_n}\,(1+x)P_n^{(\alpha,\alpha+1)}(x)$$

(by (2.19), (2.20) and (2.15))

$$= (1-x^2)\frac{\left[(n+2\alpha+2)P_n^{(\alpha+1,\alpha+1)}(x)+(n+\alpha+1)P_{n-1}^{(\alpha+1,\alpha+1)}(x)\right]}{2(n+\alpha+1)}$$

$$\left[\frac{n!}{(\alpha+1)_n}\right]^2 \cdot \frac{\left[(n+2\alpha+2)P_n^{(\alpha+1,\alpha+1)}(x)-(n+\alpha+1)P_{n-1}^{(\alpha+1,\alpha+1)}(x)\right]}{2(n+\alpha+1)}$$

(by (2.21) and (2.20))

$$= \frac{(1-x^2)}{(n+1)^2}\left[\frac{d}{dx}\,\frac{P_{n+1}^{(\alpha,\alpha)}(x)}{P_{n+1}^{(\alpha,\alpha)}(1)}\right]^2 - \frac{(1-x^2)}{(n+2\alpha+1)^2}\left[\frac{d}{dx}\,\frac{P_n^{(\alpha,\alpha)}(x)}{P_n^{(\alpha,\alpha)}(1)}\right]^2$$

(by (2.18)) .

Denote

$$R_n^{(\alpha,\alpha)}(x) = \frac{P_n^{(\alpha,\alpha)}(x)}{P_n^{(\alpha,\alpha)}(1)}$$

for simplicity. Then

$$f(x) := \left[R_n^{(\alpha,\alpha)}(x)\right]^2 + \frac{(1-x^2)}{(n+2\alpha+1)^2}\left[\frac{d}{dx}R_n^{(\alpha,\alpha)}(x)\right]^2$$

$$= \left[R_{n+1}^{(\alpha,\alpha)}(x)\right]^2 + \frac{(1-x^2)}{(n+1)^2}\left[\frac{d}{dx}R_{n+1}^{(\alpha,\alpha)}(x)\right]^2 \tag{2.33}$$

Differentiating gives

$$(n+2\alpha+1)^2 f'(x) = 2\frac{d}{dx} R_n^{(\alpha,\alpha)}(x) .$$

$$\left[(n+2\alpha+1)^2 R_n^{(\alpha,\alpha)}(x) - x\frac{d}{dx} R_n^{(\alpha,\alpha)}(x) + (1-x^2)\frac{d^2}{dx^2} R_n^{(\alpha,\alpha)}(x)\right]$$

and

$$(n+1)^2 f'(x) = \frac{2d}{dx} R_{n+1}^{(\alpha,\alpha)}(x)$$

$$\left[(n+1)^2 R_{n+1}^{(\alpha,\alpha)}(x) - x\frac{d}{dx} R_{n+1}^{(\alpha,\alpha)}(x) + (1-x^2)\frac{d^2}{dx^2} R_{n+1}^{(\alpha,\alpha)}(x)\right] .$$

Thus the zeros of $f'(x)$ contain the zeros of $\frac{d}{dx} R_n^{(\alpha,\alpha)}(x)$ and those of $\frac{d}{dx} R_{n+1}^{(\alpha,\alpha)}(x)$. But these zeros are distinct, so this gives $2n-1$ zeros of $f'(x)$. However, from the first equality in (2.33) $f(x)$ is a polynomial of degree $2n$, so all the zeros of $f'(x)$ are accounted for. Therefore

$$f'(x) = \lambda_n \frac{d}{dx} R_n^{(\alpha,\alpha)}(x) \frac{d}{dx} R_{n+1}^{(\alpha,\alpha)}(x)$$

for a constant λ_n. Comparing coefficients of the highest powers of x gives

$$f'(x) = \frac{(2\alpha+1)}{(n+2\alpha+1)} \frac{d}{dx} R_n^{(\alpha,\alpha)}(x) \frac{d}{dx} R_{n+1}^{(\alpha,\alpha)}(x) . \tag{2.34}$$

Because of the separation of the zeros of $\frac{d}{dx} R_n^{(\alpha,\alpha)}(x)$ and $\frac{d}{dx} R_{n+1}^{(\alpha,\alpha)}(x)$, they take on opposite signs for $y_{k,n} < x < y_{k,n+1}$ when these zeros are ordered by $y_{k+1,n} < y_{k,n}$. Thus if $\alpha > -\frac{1}{2}$, $f(x)$ decreases for $y_{k,n} < x < y_{k,n+1}$, and when $-1 < \alpha < -\frac{1}{2}$, $f(x)$ increases on this interval. This implies

$$\mu_{k,n+1}(\alpha) < \mu_{k,n}(\alpha),\ \alpha > -\frac{1}{2},\ k = 1,2,\ldots,n,\ n = 1,2,\ldots, \tag{2.35}$$

and

$$\mu_{k,n+1}(\alpha) > \mu_{k,n}(\alpha),\ -1 < \alpha < -\frac{1}{2},\ k = 1,2,\ldots,n,\ n = 1,2,\ldots . \qquad (2.36)$$

This result was proven for $\alpha = 0$ by Szegö [16], and for $\alpha > -\frac{1}{2}$ by Szász [12]. Actually his proof also works for $-1 < \alpha < -\frac{1}{2}$, but he did not make this observation. It is a useful observation, since the inequalities change at $\alpha = -\frac{1}{2}$. When $\alpha = -\frac{1}{2}$, $\mu_{k,n} = 1$ so there is equality in both (2.35) and (2.36).

As I remarked earlier, Szegö used (2.11) – (2.13). If one has these formulas then his proof is shorter than the one I gave. However I like the version given here since it uses more general facts. In particular, I can remember all the facts I used, at least up to constants, and I can never remember (2.11) – (2.13). On the other hand, my argument is probably too tight, and does not have the flexibility one needs when trying to extend these results.

The Todd–Szegö inequality was rediscovered and proved by Cornille and Martin [3], and then used to estimate

$$\sum_{k=1}^{N} (-1)^k P_n(x_k) \qquad (2.37)$$

when $1 \geqq x_1 \geqq \cdots \geqq x_N \geqq 0$ and when $1 \geqq x_1 \geqq \cdots \geqq x_N \geqq -1$. In a sequel [4] they considered analogous problems for

$$\left[\frac{(n+2\lambda)!}{n!}\right]^{1/2} \left[\frac{1-x}{2}\right]^{\lambda} \left[\frac{1+x}{2}\right]^{\lambda} \frac{P_n^{(\lambda,\lambda)}(x)}{P_n^{(\lambda,\lambda)}(1)} \qquad (2.38)$$

when $\lambda = 1,2,\ldots$.

3. JACOBI POLYNOMIALS

Some of the facts in (a) to (g) hold for general orthogonal polynomials. Others do not, but can be considered for Jacobi polynomials. Inequality (b) extends to

$$|P_n^{(\alpha,\beta)}(x)| \leqq P_n^{(\alpha,\beta)}(1), \ -1 \leqq x \leqq 1, \ \alpha \geqq \beta > -1, \ \alpha \geqq -\tfrac{1}{2}. \tag{3.1}$$

There are many proofs which work for a restricted range of α and β, but the only proof that works for the full set $\alpha \geqq \beta > -1, \alpha \geqq -\frac{1}{2}$ as far as I know, is the proof used to prove (d). As mentioned above, this is given in Szegö [15, Chapter VII].

The remaining one is (g). Here there are some interesting open problems. First, there is the quadratic transformation

$$\frac{P_{2n}^{(\alpha,\alpha)}(x)}{P_{2n}^{(\alpha,\alpha)}(1)} = \frac{P_n^{(\alpha,-1/2)}(2x^2-1)}{P_n^{(\alpha,-1/2)}(1)}. \tag{3.2}$$

If $y_{k,n}$ are the location of the relative maxima of $\left|\frac{P_n^{(\alpha,\beta)}(x)}{P_n^{(\alpha,\beta)}(1)}\right|$ ordered by $1 = y_{0,n} > y_{1,n} > \dots > y_{n,n} = -1$, set

$$\mu_{k,n}(\alpha,\beta) = \left|\frac{P_n^{(\alpha,\beta)}(y_{k,n})}{P_n^{(\alpha,\beta)}(1)}\right|. \tag{3.3}$$

The quadratic transformation (3.2) and (2.35) imply

$$\mu_{k,n+1}(\alpha,\beta) < \mu_{k,n}(\alpha,\beta), \quad k = 1,2,\dots,n; \ n = 1,2,\dots, \tag{3.4}$$

when $\alpha = \beta > -\frac{1}{2}$ and when $\alpha > \beta = -\frac{1}{2}$. This suggests the following conjecture.

Conjecture 1. If $\alpha > \beta > -\frac{1}{2}$ then

$$\mu_{k,n+1}(\alpha,\beta) < \mu_{k,n}(\alpha,\beta),\ k = 1,2,\ldots,n;\ n = 1,2,\ldots\,. \tag{3.5}$$

I have tried to prove (3.5) for twenty years, without success. One reason it is hard is that the inequality (3.5) seems to reverse when $\alpha = 0,\ \beta = -1$. Observe that

$$P_n^{(0,-1)}(x) = \frac{P_n(x)+P_{n-1}(x)}{2} \tag{3.6}$$

follows from (2.21) with $\alpha = -1,\ \beta = 0$ after use of (2.20). Here is a graph of $P_n^{(0,-1)}(x)$ for $n = 2,3,\ldots,7$. Notice the monotonicity seems to go the other way.

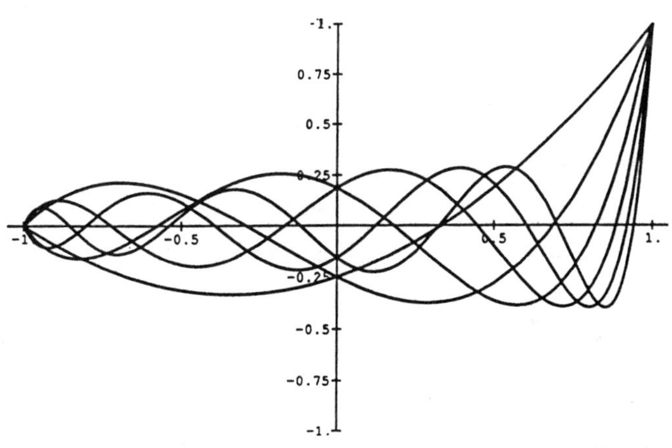

Figure 8. $P_n^{(0,-1)}(x),\ n = 2,\ldots,7$

If this can be proven then one gets the inequality

$$\frac{P_n(x)+P_{n-1}(x)}{2} = P_n^{(0,-1)}(x) > J_0(j_{1,1}) \tag{3.7}$$

where the right hand side is the first minimum of the Bessel function $J_0(x)$ for $x > 0$.

Since

$$\lim_{n\to\infty} \mu_{k,n}(0,\beta) = |J_0(j_{k,1})| \tag{3.8}$$

(from Mehler's formula when $\beta = 0$ and its easy extension for general β, see Szegö [15, Chapter 8]) and the right hand side is the k'th relative extrema of $J_0(x)$ for $x > 0$, Szegö's inequality (2.31) and the conjectured inequality

$$\mu_{k,n+1}(0,-1) > \mu_{k,n}(0,-1) \tag{3.9}$$

would give two sided bounds for the k'th relative extrema of $J_0(x)$.

The real reason it is so easy to prove (3.5) when $\alpha = \beta$ but not for $\alpha \neq \beta$ is that when $\alpha = \beta$ the case $\alpha = \beta = -\frac{1}{2}$ has equality instead of an inequality. For $\alpha > \beta$ there is is no longer such a case. Thus the cases where the inequality reverses are not definite, but depend on the particular maximum points. However it should be possible to prove Conjecture 1 and also settle the question of whether the inequality (3.5) reverses when $\beta = -1, \alpha > -1$ as it does when $\alpha = -1$, when $\alpha = -1/2$ (from a quadratic transformation) and as it seems to when $\alpha = 0$.

Observe that the normalization of $P_n^{(\alpha,\beta)}(x)$ played no role in (a) – (f), but was essential in (g). There are other normalizations possible, but this is the most interesting one if the normalization only uses a function of n. For example,

$$\frac{P_n^{(\frac{1}{2},\frac{1}{2})}(\cos\theta)}{P_n^{(\frac{1}{2},\frac{1}{2})}(1)} = \frac{\sin(n+1)\theta}{(n+1)\sin\theta} \tag{3.10}$$

is the function we have considered when $\alpha = \beta = \frac{1}{2}$. It is also possible to consider

$$\frac{\sin(n+1)\theta}{\sin\theta} = U_n(x), \quad x = \cos\theta . \tag{3.11}$$

Here is the corresponding graph from [1, Figure 22.7]

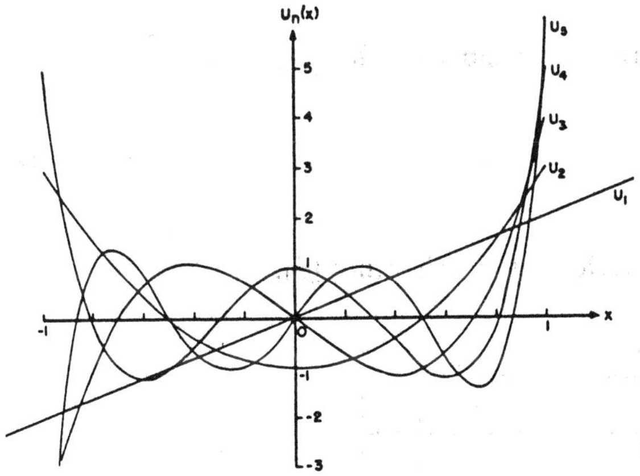

Figure 9. $U_n(x)$ $n = 1,\ldots,5$

The monotonicity is now lost, and the successive extreme points go to infinity as $n \to \infty$. These questions are most interesting when the functions are normalized so that the extrema have a limit when $n \to \infty$.

There are other ways to normalize the functions if one allows a factor of x as well. For example, the Laguerre polynomials $L_n^0(x)$ have the following graph.

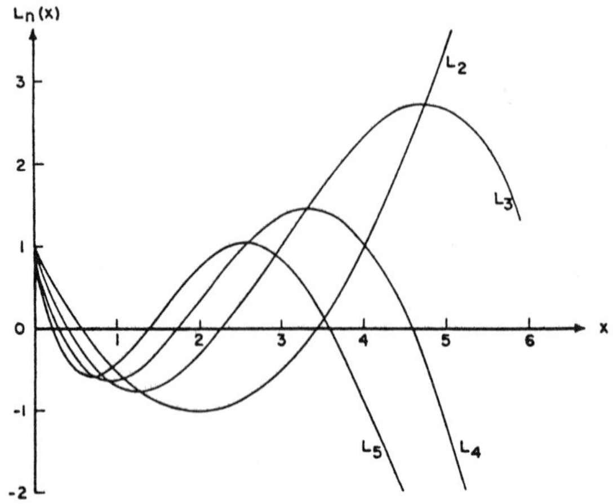

Figure 10. $L_n^0(x)$ n = 2,...,5

However the right functions to consider for this problem are

$$\varphi_n(x) = L_n^0(x)e^{-x/2} \tag{3.12}$$

as Todd [17] observed. See the following graphs from [9].

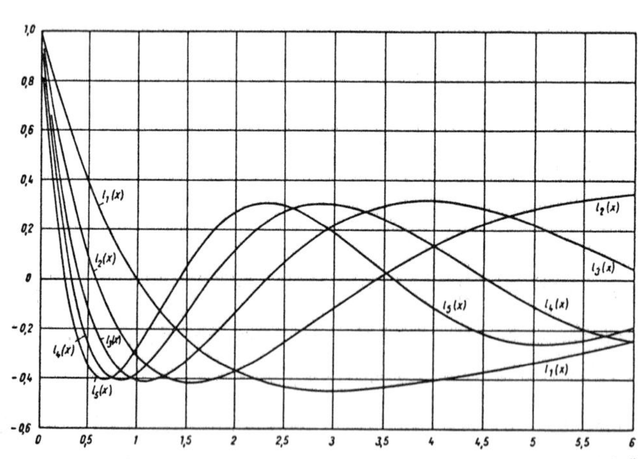

Figure 11. $L_n^0(x)e^{-x/2}$ n = 1,...,5

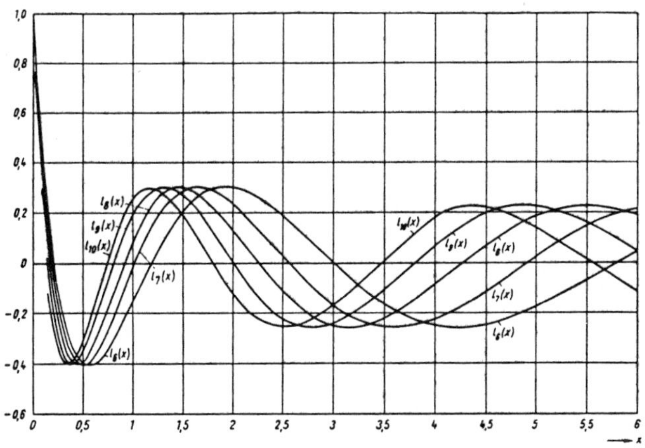

Figure 12. $L_n^0(x)e^{-x/2}$ $n = 6,\ldots,10$

Many more examples could be given, but some pictures should be left to the reader.

As a last example for Jacobi polynomials, consider the zeros of $P_n^{(\alpha,\beta)}(x)$. Denote these by

$$x_{k,n}(\alpha,\beta) \tag{3.13}$$

with $x_{k,n}(\alpha,\beta) > x_{k+1,n}(\alpha,\beta)$. Markoff and Stieltjes showed that this is a decreasing function of α for fixed β,k,n. See Szegö [15, Chapter VI]. Here is a picture of $P_8^{(\alpha,0)}(x)$, $\alpha = 0,1,2,3$ which illustrates this monotonicity. The first graph gives the general picture, the second rescales the values of the functions so sufficient detail can be seen about the zeros.

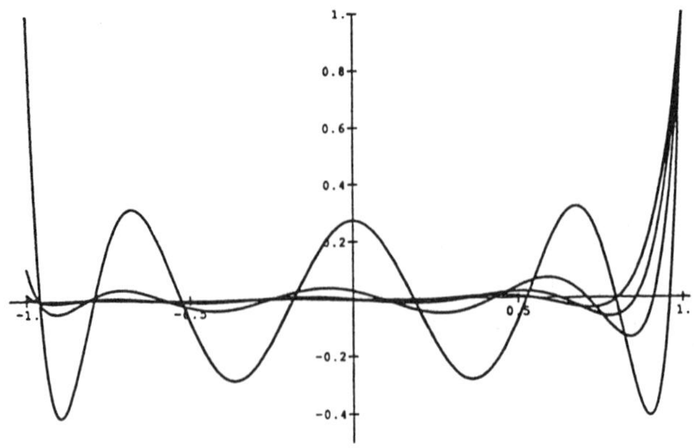

Figure 13. $P_8^{(\alpha,0)}(x)$, $\alpha = 0,1,2,3$

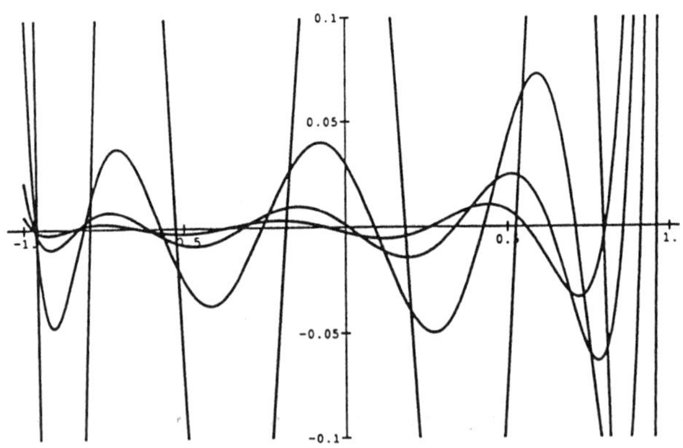

Figure 14. Details from Figure 13.

The monotonicity is illustrated, but the graph suggests more. It suggests that

$$x_{k+1,n}(\alpha,\beta) < x_{k,n}(\gamma,\beta) \tag{3.14}$$

for $\alpha < \gamma \leqq \alpha + 2$ and that (3.14) might hold for $\alpha + 2 < \gamma \leqq \alpha + 3$. The last is probably too much to hope for, but I am willing to conjecture that

$$x_{k+1,n}(\alpha,\beta) < x_{k,n}(\alpha+2,\beta)\ . \tag{3.15}$$

Here is a proof of (3.14) for $\gamma = \alpha + 1$, and so by monotonicity for $\alpha < \gamma \leqq \alpha + 1$. Recall (2.19), which we rewrite as

$$\frac{P_n^{(\alpha,\beta)}(x)}{P_n^{(\alpha,\beta)}(1)} - \frac{P_{n+1}^{(\alpha,\beta)}(x)}{P_{n+1}^{(\alpha,\beta)}(1)} = \frac{(2n+\alpha+\beta+2)(1-x)}{2(\alpha+1)} \frac{P_n^{(\alpha+1,\beta)}(x)}{P_n^{(\alpha+1,\beta)}(1)} \tag{3.16}$$

Recall that the zeros of $P_n^{(\alpha,\beta)}(x)$ and $P_{n+1}^{(\alpha,\beta)}(x)$ and also the extrema of these functions separate each other. The graphs of $P_n^{(\alpha,\beta)}(x)/P_n^{(\alpha,\beta)}(1)$ and $P_{n+1}^{(\alpha,\beta)}(x)/P_{n+1}^{(\alpha,\beta)}(1)$ intersect between their zeros, so we have more than (3.14). For

$$x_{k+1,n}(\alpha,\beta) < x_{k+1,n+1}(\alpha,\beta) < x_{k,n}(\alpha+1,\beta) < x_{k,n}(\alpha,\beta) \tag{3.17}$$

4. OTHER FUNCTIONS

Here are two final pictures. The first gives a graph of $Q_n(x)$, a second solution of the differential equation (2.10). It is defined by

$$Q_n(x) = \frac{1}{2}\int_{-1}^{1} \frac{P_n(t)}{x-t}\,dt \tag{4.1}$$

for x not in $-1 \leqq x \leqq 1$ and by the average of the values across the cut for $-1 < x < 1$. Another formula is

$$Q_n(x) = \frac{1}{2} P_n(x)\log\frac{1+x}{1-x} - W_{n-1}(x) \tag{4.2}$$

where

$$W_{n-1}(x) = \sum_{k=1}^{n} \frac{1}{k} P_{k-1}(x)P_{n-k}(x) .$$

A third expression is

$$Q_n(\cos\theta) = 2\,\frac{n!}{(\frac{3}{2})_n} \sum_{k=0}^{\infty} \frac{(\frac{1}{2})_k (n+1)_k}{k!\,(n+\frac{3}{2})_k} \cos(n+2k)\theta . \tag{4.3}$$

This corresponds to one of the many Legendre polynomial formulas not given in section 2. It is analogous to the series

$$P_n(\cos\theta) = \frac{4}{\pi}\,\frac{n!}{(\frac{3}{2})_n} \sum_{k=0}^{\infty} \frac{(\frac{1}{2})_k (n+1)_k}{k!\,(n+\frac{3}{2})_k} \sin(n+2k+1)\theta . \tag{4.4}$$

See Szegö [15, Chapter IV] and [8].

Jahnke and Emde [8] contains the following graph. Clearly many of the same types of facts continue to hold. They should be proven.

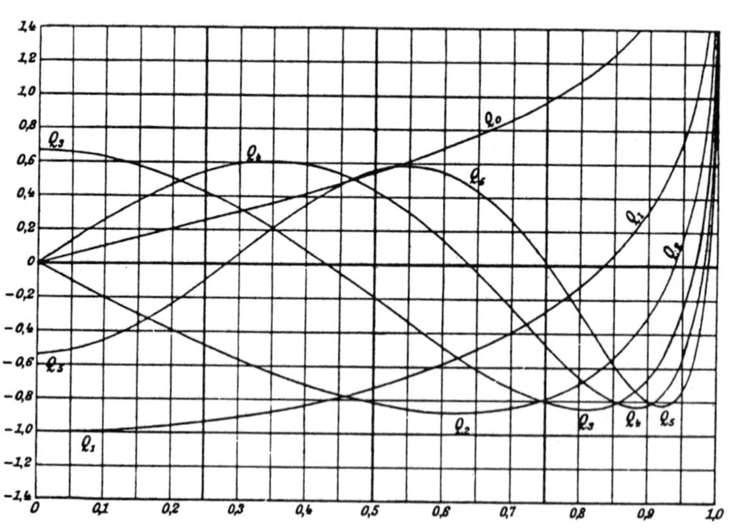

Figure 15. $Q_n(x)$ $n = 0,\ldots,5$

The second graph comes from [1, Figure 9.1]. Here it deals with the Bessel functions $J_0(x)$, $J_1(x)$, $Y_0(x)$ and $Y_1(x)$. The Y's are a second solution of the Bessel equation

$$x^2y'' + xy' + (x^2-\alpha^2)y = 0$$

which are normalized appropriately.

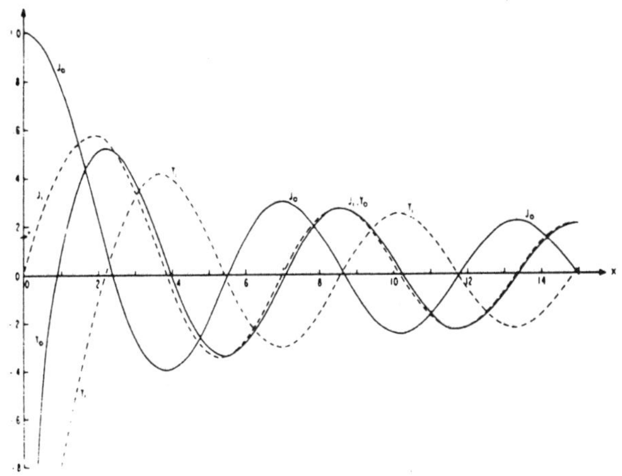

Figure 16. $J_0(x)$, $Y_0(x)$, $J_1(x)$, $Y_1(x)$

Notice how closely $J_1(x)$ and $Y_0(x)$ agree, once $x \geqq 3$. This is clear from their asymptotic expansions, but the graph makes it clear. There are two ways to consider conjugate functions. One is to consider two appropriate solutions of a second order differential equation, as $\cos n\theta$ and $\sin n\theta$ satisfy

$$y'' + n^2y = 0 .$$

For Bessel functions, Muckenhoupt and Stein [11] used a different method, where they used $J_{\alpha+1}(x)$ instead of $Y_\alpha(x)$. The graphs show why this is a very reasonable thing to try. Of course there are deeper reasons coming from singular Cauchy–Riemann type equations, but it is nice to also have an elementary way of thinking about these questions.

It may be time to reconsider Conjecture 1 from an asymptotic point of view. It was first studied this way by Cooper [2], before Szegö found his proof of (g). However Cooper's claims for Jacobi polynomials are clearly incorrect, since they disagree with the case $\alpha = \beta$, where there are theorems. I want complete answers, since they are often more useful than results which only hold for n sufficiently large. However not every problem has a solution from a beautiful formula, and this may be an instance of this. So, while I will work with formulas to try to find simple proofs, I encourage others to consider these problems from an asymptotic perspective. It is with this that I close a paper dedicated to Frank Olver, whose work on asymptotic methods has all of us in his debt. He is also responsible for the last graph.

After this paper was finished, I proved (g) for $Q_n(x)$. Details will appear elsewhere.

Acknowledgement: Paul Nevai has supplied the new graphs. This made it possible for me to finish the paper on time for this volume, so I thank him for this aid, but also for all the work he has done in trying to convince me to become a late twentieth century mathematician. Graphs have played a central role in some of his work, but their effect has been hidden.

The graphs in Figures 1 and 15 are reprinted from [8]. Those in Figures 11 and 12 are reprinted from [9]. They are reprinted with permission.

REFERENCES

1. M. Abramowicz and I. Stegun, ed., *Handbook of Mathematical Functions with Formulas, Graphs and Mathematical Tables*, Nat. Bureau of Standards Appl. Math. Ser. 55 (1964), Wash. D.C., Paperback reprint, Dover, New York, 1965.

2. R. Cooper, The extremal values of Legendre polynomials and of certain related functions, *Proc. Camb. Phil. Soc. 46* (1950), 549–554.

3. H. Cornille and A. Martin, Constraints on the phase of scattering amplitudes due to positivity, *Nuclear Physics B49* (1972), 413–440.

4. H. Cornille and A. Martin, Constraints on the phases of helicity amplitudes due to positivity, *Nuclear Physics B77* (1974), 141–162.

5. A. Erdélyi et al, *Higher Transcendental Functions*, vol. 2, McGraw-Hill, New York, 1953.

6. E. Hewitt and R. E. Hewitt, The Gibbs–Wilbraham Phenomenon: an episode in Fourier analysis, *Archive for History of Exact Sciences 21* (1980), 129–160.

7. E. Jahnke and F. Emde, *Funktionentafeln mit Formeln und Kurven*, Teubner, Leipzig, 1909.

8. E. Jahnke and F. Emde, *Tables of Functions*, fourth edition, Dover, New York, 1945.

9. E. Jahnke, F. Emde and F. Lösch, *Tables of Higher Functions*, sixth edition, 1960 McGraw Hill, New York.

10. A. A. Michelson and S. W. Stratton, A new harmonic analyser, *Phil. Mag. (5) 45* (1898), 85–91.

11. B. Muckenhoupt and E. M. Stein, Classical expansions and their relation to conjugate harmonic functions, *Trans. Amer. Math. Soc. 118* (1965), 17–92.

12. O. Szász, On the relative extrema of ultraspherical polynomials, *B.U.M.I. (3) 5* (1950), 125–127. *Collected Papers* 1297–1299.

13. O. Szász, On the relative extrema of Bessel functions, *B.U.M.I. (3) 5* (1950), 225–229, *Collected Papers* 1300–1305.

14. O. Szász, On the relative extrema of the Hermite orthogonal functions, *J. Indian Math. Soc. 25* (1951), 1340–1345, *Collected Papers* 1340–1345.

15. G. Szegö, *Orthogonal Polynomials*, Amer. Math. Soc. Colloq. Publ. 23, Amer. Math. Soc., Providence, 1939, fourth edition 1975.

16. G. Szegö, On the relative extrema of Legendre polynomials, *B.U.M.I. (3) 5* (1950), 120–121. *Collected Papers*, vol. 3, 219–220.

17. J. Todd, On the relative extrema of the Laguerre orthogonal functions, *B.U.M.I. (3) 5* (1950), 122–125.

18. H. Wilbraham, On a certain periodic function, *Cambridge and Dublin Math. J. 3* (1848), 198–201.

Asymptotics of Integrals, Series, and Operators

L. Berg Professor, Department of Mathematics, Wilhelm-Pieck Universitat, Rostock, East Germany

I. PRELIMINARIES

The asymptotic formula

$$\int_a^b \exp[-g(s,t)]dt \sim [2\pi/g_{tt}(s,x)]^{1/2}\exp[-g(s,x)] \quad (1)$$

for $s \to s_0$ with $s \in S$ dates back to Laplace (1812) and can be considered to be the basic result in asymptotics of parameter integrals. At first we assume all entries to be real, later on we also speak about the complex case, where the integral in question can be a contour integral. The boundaries a,b can depend on the variable s, but they are also allowed to be infinite. The parameter x can depend on s too, it is exactly or approximately the maximum point of the integrand and at first it is an inner point of the interval (a,b). The set S can be continuous or discrete and also higher-dimensional, it always shall possess s_0 as a limit point not belonging to S.

In many applications the function g(s,t) is linear in s, cf. Berg (1967), Olver (1974) and Temme (1987). However, this is not necessary, cf. Laplace (1812), Berg (1958) and Evgrafov (1961). In what follows we begin with a proof of formula (1), afterwards we survey twelve modifications of it, in particular series instead of integrals. These topics are partly already considered in the German book Berg (1968) and in §§ 17 and 18 of the Russian book Riekstiņš (1977), but there are added also some later results, in particular of my former collaborators R. Riedel, H.-J. Schell and E. Wagner, without obtaining completeness. The survey ends with a functional analytic version of asymptotics of operators following Berg (1983).

We exclude here the extensive theory of the asymptotic behaviour of solutions of different types of equations and refer for differential equations only to Olver (1974), Goering (1977) and for difference equations to Olver and Sookne (1972) and to Berg (1986).

II. THE BASIC FORMULA OF LAPLACE

In the following we use the abbreviation that for functions of s,t we drop these two letters when t = x, for example we write g = g(s,x). We also drop the special movement $s \to s_0$ and the restriction $s \in S$. It is useful to split the proof of (1) into two steps. The original proof of (1) is known as the method of Laplace, cf. Olver (1974), where are contained also further literature and historical remarks.

THEOREM 1: If we have

$$g_t = o(g_{tt}^{1/2}), \tag{2}$$

and if there exists a positive function $w = w(s)$ with

$$w^2 g_{tt} \to +\infty, \tag{3}$$

as well as

$$g_{tt}(s,u) \sim g_{tt} \tag{4}$$

uniformly for all u with $|u-x| \leq w$, then (1) is valid with $a = x-w$ and $b = x+w$.

PROOF: By means of Taylor's formula

$$g(s,t) = g + g_t(t-x) + (1/2)g_{tt}(s,u)(t-x)^2 \tag{5}$$

with $|u-x| < w$ and the substitution $t-x = g_{tt}^{-1/2}v$ we obtain

$$\int_{x-w}^{x+w} \exp[-g(s,t)]dt \tag{6}$$

$$= g_{tt}^{-1/2}e^{-g}\int_{-wg_{tt}^{1/2}}^{wg_{tt}^{1/2}} \exp[-g_t g_{tt}^{-1/2}v - g_{tt}(s,u)v^2/(2g_{tt})]dv.$$

Since the integrand at the right-hand side possesses the majorant $\exp[|v|-v^2/4]$, the corresponding integral converges in view of (2), (3), (4) and the theorem of Lebesgue to

$$\int_{-\infty}^{+\infty} \exp[-v^2/2]dv = (2\pi)^{1/2}. \tag{7}$$

This proves the theorem.

COROLLARY: If the assumptions of Theorem 1 are satisfied, and if additionally we have

$$\int_a^{x-w} \exp[-g(s,t)]dt = o(g_{tt}^{-1/2}e^{-g}), \tag{8a}$$

$$\int_{x+w}^{b} \exp[-g(s,t)]dt = o(g_{tt}^{-1/2}e^{-g}), \tag{8b}$$

then (1) is valid with a and b from (8).

The hypotheses (8) require further efforts to check them. The simplest sufficient condition for them stems from H. Schubert, generalizations of it are contained in Berg (1959).

CRITERION: If the assumptions of Theorem 1 are satisfied, if $a \leq x-w$, $x+w \leq b$, and if $g_t(s,t)$ is monotonously increasing for $a \leq t \leq b$, then (8) is valid [and (1) too].

PROOF: From (2), (3), (4) and Taylor's formula we see that

$$g_t(s,x+w) = g_t + g_{tt}(s,u)w \sim g_{tt}w \tag{9}$$

is positive. Hence, using the notation $y = x+w$, we have

$$\int_y^b \exp[-g(s,t)]dt \leq g_t^{-1}(s,y)\int_y^b \exp[-g(s,t)]g_t(s,t)dt$$
$$\leq g_t^{-1}(s,y)\exp[-g(s,y)].$$

But this together with (5) for $t = y$ and (9) implies the statement (8b). Formula (8a) is proven analogously.

REMARKS: (i) Condition (2) is satisfied when

$$g_t(s,x) = 0. \tag{10}$$

This in conjunction with $g_{tt} > 0$ means that $\mathbf{x}$ is the minimum point of g(s,t), i.e. the maximum point of the integrand in (1). For applications it is very important that it is not necessary to know this point exactly, it suffices to know an approximation $\mathbf{x}$ satisfying (2), which can be found by the powerful method of asymptotic iteration, cf. Berg (1965) and in the higher-dimensional case Krätzel (1977).

(ii) In order to find the function w = w(s) of Theorem 1 choose an elementary function with a parameter p and try to determine p satisfying (3) and (4). Usually a function like $w = s^p$ will do. When $g_{tt} \to \infty$ choose w to be a small constant or $w = (g_{tt})^p$ with $-1/2 < p < 0$.

(iii) It is possible to weaken the uniformity of (4), since only the existence of a convergent majorant is necessary. The other hypotheses of Theorem 1 are quite natural.

(iv) There is a close connection between (1) and the cental limit theorem of probability theory, because the integrand in (7) is the standardized normal distribution, disregarding a constant factor. This leads later on to the possibilities of choosing other distribution functions as limit functions.

(v) For the case of Laplace integrals and more generally the case g(s,t) = k(st) + h(t) it is also possible to invert formula (1) under suitable Tauberian conditions. Concerning this general Tauberian theory we refer to Wagner (1968) and (1970). The results can be used to determine the asymptotic behaviour of the solutions of linear convolution equations of the second kind, cf. Wagner (1978), and for systems of such equations, Wagner (1979).

(vi) Let us mention that for the numerical inversion of the Laplace transform a combination of asymptotic and Fourier methods was worked out by Belov and Cirulis (1985).

III. MODIFICATIONS OF THE BASIC FORMULA

Now we want to indicate several modifications of the Laplace formula (1) without going into too many details. We begin with very simple cases, where at first the assumptions of Theorem 1 with regard to g(s,t) are preserved. The later more complicated cases are only sketched.

1. A further factor Formula (1) generalizes to

$$\int_a^b \exp[-g(s,t)]h(s,t)dt \sim (2\pi/g_{tt})^{1/2}e^{-g}h \tag{11}$$

with h = h(s,x) as before, if h(s,t) satisfies

$$h(s,t) \sim h$$

uniformly for $|t-x| \leq w$. This modification gives us the possibility to determine x not as the maximum point of the whole integrand, but as the maximum point of the main part of it. Moreover, formula (11) can be used if the whole integrand does not have a second derivative with respect to t.

2. A factor with singularity In the case that h is infinite or zero formula (1) generalizes e. g. to

$$\int_a^b \exp[-g(s,t)]|t-x|^p dt \sim (2g_{tt})^q\Gamma(q)e^{-g} \tag{12}$$

with q = (p+1)/2, as long as p > -1. The proof can be done as before, only (7) goes over into an integral, which can be expressed by the Γ-function. When the additional factor changes its sign at the point t = x the proof is more difficult and requires assumptions on the third derivative of g(s,t). Of course, the case of a logarithmic singularity can be treated too, cf. Riedel and Treutler (1975).

3. *Asymptotic bisection* If the approximation **x** to the maximum point is at the same time a boundary point of the integration, then we have

$$\int_a^x \exp[-g(s,t)]dt \sim [\pi/(2g_{tt})]^{1/2}e^{-g}, \tag{13a}$$

$$\int_x^b \exp[-g(s,t)]dt \sim [\pi/(2g_{tt})]^{1/2}e^{-g}. \tag{13b}$$

If we only need one of these formulas, then we also need the assumptions only with respect to the corresponding interval of integration. Hence we can use also two different funktions w, putting $a = x-w_1$ and $b = x+w_2$ in Theorem 1.

4. *Finite boundaries* If (3) is replaced by $w^2g_{tt} \to 2p^2$ with a certain constant p, then (1) changes to

$$\int_{a-w}^{a+w} \exp[-g(s,t)]dt \sim [2\pi/g_{tt}]^{1/2}e^{-g}\mathrm{erf}(p), \tag{14}$$

because in this case at the end of the proof we arrive instead of (7) at an integral, which can be expressed by the error function normalized by $\mathrm{erf}(\infty) = 1$.

5. *Non vanishing derivative* When (2) is replaced by

$$g_t \sim p\ g_{tt}^{1/2} \tag{15}$$

for a certain constant p, then (1) becomes

$$\int_a^b \exp[-g(s,t)]dt \sim [2\pi/g_{tt}]^{1/2}\exp(-g+p^2/2), \tag{16}$$

because in this case, at the end of the proof we arrive at

$$\int_{-\infty}^{+\infty} \exp(-pv-v^2/2)dv = (2\pi)^{1/2}\exp(p^2/2).$$

Assumption (15) means that x is already too far away from the maximum point of the integrand.

6. Maxima of another order Using the abbreviation

$$g_n(s,t) = (\partial/\partial t)^n g(s,t), \tag{17}$$

formula (13b) generalizes to

$$\int_x^b \exp[-g(s,t)]dt \sim (1/n)\Gamma(1/n)(n!/g_n)^{1/n}e^{-g}, \tag{18}$$

if we replace (5) by Taylor's formula of order n with

$$g_k = o(g_n^{k/n}) \tag{2'}$$

for k = 1,2,...,n-1,

$$w^n g_n \to +\infty, \tag{3'}$$

$$g_n(s,u) \sim g_n \tag{4'}$$

uniformly for all u with $x \leq u \leq x+w$, and if we modify (8b) in an obvious way. The case that (2') is generalized as in (15) was considered by Riedel (1967), the case n = 4 having been already considered by Wagner (1962). In general, now it is not possible to express the final integrals by well known functions. The special case n = 1 from (18) reads

$$\int_x^b \exp[-g(s,t)]dt \sim g_t^{-1}e^{-g}. \tag{19}$$

Changing the direction of t we obtain the analogous formula

$$\int_a^x \exp[-g(s,t)]dt \sim -g_t^{-1}e^{-g}, \tag{20}$$

and the same change is possible in (18). The asypmtotic representations (19) and (20) sharpen the estimates (8).

7. A mixed formula Riedel (1968) has constructed a general formula containing the special case

$$\int_x^b \exp[-g(s,t)]dt \tag{21}$$
$$\sim [\pi/(2g_{tt})]^{1/2}\exp[-g+g_t^2/(2g_{tt})]\operatorname{erfc}[g_t(2g_{tt})^{-1/2}]$$

under the assumptions

$$w(g_t+pg_{tt}^{1/2}) \to +\infty \tag{22}$$

for a certain positive constant p, (4) and a suitable modification of (8b). The condition (22) is a linear combination of (3′) with n = 1 and the square root of (3). In case of (2) formula (22) becomes (13b), in case of $g_t g_{tt}^{-1/2} \to +\infty$ it becomes (19). In general $g_t g_{tt}^{-1/2}$ can have strong oscillations, and (21) expresses a uniform asymptotic behaviour. This result has been generalized by Schell (1978), (1979) to asymptotic representations of real integrals, which are uniformly valid with respect to a second parameter varying in a certain domain, the latter one can depend on s.

8. The higher-dimensional case According to Hsu (1948) and Riedel (1965) formula (1) generalizes to

$$\iint \exp[-g(s,t,r)]dtdr \sim 2\pi H^{-1/2}\exp[-g(s,x,y)] \tag{23}$$

in the two-dimensional case with the Hessian

$$H = \begin{vmatrix} g_{tt} & g_{tr} \\ g_{rt} & g_{rr} \end{vmatrix}$$

at the inner point t = x, r = y. For simplicity this shall be the maximum point of the integrand, so that

$$g_t(s,x,y) = g_r(s,x,y) = 0. \tag{24}$$

The proof of (23) can be done either by means of the two-dimensional Taylor formula or by iterated application of formula (1). Similar modifications as before are possible, however, in the case that (x,y) is a boundary point of the domain of integration the result depends essentially on the angle β of the boundary at this point. Only in the smooth case with $\beta = \pi$ we have asymptotic bisection. Moreover, the generalization to the higher-dimensional case is possible, cf. Bleistein and Handelsman (1986) and Fedoryuk (1987).

9. Asymptotic expansions Formula (11) can be considered in view of (1) as the first term of an asymptotic expansion

$$\int_a^b \exp[-g(s,t)]h(s,t)dt \approx \sum_{n=0}^{\infty} h_n(s,x)Q_n(s) \tag{25}$$

with $h_n(s,t)$ as in (17),

$$Q_n(s) = (1/n!)\int_c^d \exp[-g(s,t)](t-x)^n dt \tag{26}$$

and suitable boundaries c,d. For the proof we have to apply Taylor's formula at the point $t = x$ to the function h(s,t) and to use suitable remainder estimates. The expansion (25) is meant in the generalized sense that only

$$h_{2n+1}Q_{2n+1}(s) = o[h_{2n}Q_{2n}(s)],$$
$$h_{2n+2}Q_{2n+2}(s) = o[h_{2n}Q_{2n}(s)]$$

for all n. The approximation by the first term of (25) is optimal if we choose x in such a way that $Q_1(s) = 0$. Formula (25) is particularly useful if the coefficients (26) can be calculated explicitly, for instance if the generating function

$$F(s,z) = \sum_{n=0}^{\infty} Q_n(s)z^n = \int_c^d \exp[z(t-x)-g(s,t)]dt$$

is well known. Several investigations concern explicit remainder estimates, cf. Schell (1970), (1972), Olver (1974), (1980), Berg (1977), Wong (1984) and Riekstiņš (1986), which for numerical calculations can be improved by converging factors. Of great importance are also uniformity problems, cf. Bleistein and Handelsman (1986), Fedoryuk (1987), Schell (1983) and Temme (1987), where very often preliminary substitutions are useful, cf. the already mentioned method of Laplace and, moreover, the case of Chester, Friedman and Ursell (1957). The best known example of (25) is the case $g(s,t) = st^p - q \ln t$ of Watson's lemma with $h(s,t) = h(t)$, $a = c = x = 0$, $b = d = \infty$ and $p > 0$, $q > -1$. Here, Taylor's series for $h(t)$ can be replaced by an asymptotic expansion for $t \to +0$. Some new examples for asymptotic expansions of special functions are worked out by Schell (1987), Wong (1988) and Berg (1989).

10. Infinite series By means of Euler's sum formula the asymptotic representation (1) can be transferred to series

$$\sum_{n=0}^{\infty} \exp[-g(s,n)] \sim (2\pi/g_{tt})^{1/2}e^{-g}. \tag{27}$$

However, the additional assumptions that $g_{tt} \to 0$ and $x \to \infty$ are needed. When $g_{tt} \to 2k \neq 0$ we find that

$$\sum_{n=0}^{\infty} \exp[-g(s,n)] \sim T_k(x)e^{-g} \tag{28}$$

with the 1-periodic function

$$T_k(x) = \sum_{n=-\infty}^{+\infty} \exp[-k(n-x)^2],$$

whereas in case of $g_{tt} \to +\infty$ we have

$$\sum_{n=0}^{\infty} \exp[-g(s,n)] \tag{29}$$

$$\sim \{\exp[-g_{tt}r^2/2] + \exp[-g_{tt}(1-r)^2/2]\}e^{-\varphi}$$

with $r = x-[x]$. Here the term in the braces is oscillating between 0 and 1, but this does not matter since $e^{-\varphi}$ is increasing rapidly. Examples are the entire functions of order zero with $g(s,t) = t^p - t \ln s$ and $2 < p < 3$. According to Riedel (1969) integrals and series can be treated in the same way by using Stieltjes integrals. In Riedel (1983) are found asymptotic expansions for the remainders of alternating series, which can be used to accelerate the convergence.

11. Complex integrals Formula (1) is also valid in the complex case, if we replace (3) by

$$|w^2 g_{tt}| \to +\infty, \tag{30}$$

$$|\arg(w^2 g_{tt})| \leq \pi/2 - c \tag{31}$$

with a certain constant $c > 0$. In the proof instead of (5) Taylor's formula with an integral representation for the remainder term must be used. It is not necessary for $g(s,t)$ to be a holomorphic function. It suffices that the required derivatives exist with respect to the contour of integration. However, in case of holomorphy Cauchy's integral theorem can be used and the contour of integration can be replaced in a suitable way as in the method of steepest descent, cf. Bleistein and Handelsman (1986), or in the method of gradient lines, cf. Cirulis (1973). Now the maximum point of the integral becomes a saddle point, cf. Fedoryuk (1987). Many concrete investigations concern the special functions of mathematical physics, cf. Olver (1974). There are new asymptotic representations for hypergeometric functions in Wagner (1982), which correct old ones, and the corresponding asymptotics of several parameters are to be found in Wagner (1986).

12. Oscillating kernels Under additional assumptions it is also possible to consider the case $c = 0$ in (31). Replacing $g(s,t)$ by $-ig(s,t)$ with real $g(s,t)$, formula (1) becomes

$$\int_a^b \exp[ig(s,t)]dt \sim (2\pi/g_{tt})^{1/2}\exp[i(g+\pi/4)], \qquad (32)$$

if we sharpen (4) to

$$g_{tt}(s,u) - g_{tt} = o(w^{-3}g_{tt}^{-1/2}) \qquad (33)$$

uniformly for all u with $|u-x| \leq w$, and if we once more assume (10). In the proof we now have to use

$$\int_{-\infty}^{+\infty} \exp(it^2)dt = \pi^{1/2}\exp(i\pi/4)$$

instead of (7). The application of (32) is known as the method of stationary phase, cf. Olver (1974) and Fedoryuk (1987). Taking real and imaginary part of (32) we arrive at the formulas

$$\int_a^b \cos[g(s,t)]dt = (2\pi/g_{tt})^{1/2}[\cos(g+\pi/4)+o(1)], \qquad (34a)$$

$$\int_a^b \sin[g(s,t)]dt = (2\pi/g_{tt})^{1/2}[\sin(g+\pi/4)+o(1)]. \qquad (34b)$$

In view of the zeros these asymptotic equations can not be written as asymptotic representations, as it is done in other publications, cf. e. g. Murray (1984), formula (3.57). As before, several modifications are possible, cf. Riedel (1981), (1982) and for the higher-dimensional case Wong and McClure (1981) and once more Fedoryuk (1987).

IV. ASYMPTOTICS OF OPERATORS

The foregoing considerations were done completely in the style of classical analysis. Hence there arises the question of functional analytic methods in asymptotics. The advantage of Fréche spaces for asymptotic expansions was shown by Pittnauer (1972). For the connection between asymptotics and neutrix calculus we refer to van der Corput (1962), for asymptotics and nonstandard analysis to van den Berg (1987), and for the asymptotic properties of distributions to Müller (1982) and Stanković (1988). Here we consider asymptotic representations of operators, which generalize integrals.

Let D be the set of positive functions defined on S, let R be a set of positive functions f(t) defined for $t > c$, and let $K: D \to R$ be a positive and positive homogeneous operator, i.e. it transforms $f(t) \leq g(t)$ for $t > c$ into $(Kf)(s) \leq (Kg)(s)$ for $s \in S$, and satisfies $K(pf) = pKf$ for every constant $p \geq 0$. For simplicity we assume that the carrier of K is the infinite interval $[\mathbf{x},\infty)$ with $c < \mathbf{x} = \mathbf{x}(s)$ and that $\mathbf{x}(s) \to +\infty$ for $s \to s_0$, so that the foregoing inequality $f(t) \leq g(t)$ is only required for $t \geq \mathbf{x}$. We use the method of test-functions y(t) similar to Korovkin's theorem in approximation theory.

THEOREM 2. Let y(t) be a function from D with $y'(t) > 0$ and

$$Ky^r(t) \sim h(r)y^r(x), \quad s \to s_0, \tag{35}$$

for $-r_0 < r < r_0$, and let h(r) be a continuous function with $h(0) = 1$. Then for all functions $f \in D$ with

$$f'(t)/f(t) = o[y'(t)/y(t)], \quad t \to +\infty \tag{36}$$

we have

$$Kf(t) \sim f(x), \quad s \to s_o. \tag{37}$$

PROOF. For every $r > 0$ there exists according to (36) a constant t_1 with

$$-r\, y'(t)/y(t) < f'(t)/f(t) < r\, y'(t)/y(t) \tag{38}$$

for $t > t_1$. By integration we find

$$[y(t)/y(x)]^{-r} < f(t)/f(x) < [y(t)/y(x)]^{r}$$

for $x < t$ and, applying the operator K,

$$y^{r}(x)Ky^{-r} \le f^{-1}(x)Kf \le y^{-r}(x)Ky^{r}. \tag{39}$$

The right-hand side is asymptotically equal to $h(r)$ and the left-hand side to $h(-r)$. Hence the inequality (39) implies (37), since $h(r)$ is continuous and $h(0) = 1$.

REMARKS. (i) The test-function $y(t)$ is optimal, if $h(r) \neq 1$ for $r \neq 0$, because in this case it does not have the property (37), but all smaller functions in the sense of (36) have it.

(ii) If $f(t)$ is increasing, we only need (35) for $r \geq 0$, because we can put $r = 0$ on the left-hand side of (38). If $f(t)$ is decreasing, we analogously need (35) only for $r \leq 0$ with $r = 0$ on the right-hand side of (38).

(iii) In Berg (1983) are found the cases that **x** is an inner point of the carrier, and that for **K** is constructed an asymptotic expansion. Note that the paper just quoted contains some misprints. At the beginning of the proof in p. 251 replace $\varepsilon \neq 0$ by $\varepsilon > 0$, and at the last two equations in p. 255 add the factor $\pi^{-1/2}$ on the right-hand sides.

REFERENCES

1. Belov, M. A., and Cirulis, T. T. (1985). Asymptotic methods for the inversion of integral transforms (Russ.), Zinatne, Riga.
2. van den Berg, I. (1987). Nonstandard asymptotic analysis, Lect. Notes Math. 1249, Springer, Berlin.
3. Berg, L. (1958). Math. Nachr., 17: 101-135.
4. Berg, L. (1959). Math. Nachr., 20: 159-165.
5. Berg, L. (1965). Z. Angew. Math. Mech. 45: 333-352.
6. Berg, L. (1967). Introduction to the operational calculus (Transl. f. Germ.), North-Holland, Amsterdam.
7. Berg, L. (1968). Asymptotische Darstellungen und Entwicklungen, VEB DVW, Berlin.
8. Berg, L. (1977). Computing, 18: 361-363.
9. Berg, L. (1983). Coll. Math. János. Bolyai, 35: Proc. Conf. Functions, Series, Operators 1980, Budapest, 249-256.
10. Berg, L, (1986). Lineare Gleichungssysteme mit Bandstruktur und ihr asymptotisches Verhalten, VEB DVW, Berlin; Hanser, München.
11. Berg, L. (1989). Rostock. Math. Colloq., 37: 4-10.
12. Bleistein,N., and Handelsman, R. A. (1986). Asymptotic expansions of integrals, Dover, New York.
13. Cirulis, T. T. (1973). The method of gradient lines for the asymptotic representation of contour integrals (Russ.), Zinatne, Riga.
14. Chester, C., Friedman, B., and Ursell, F. (1957). Proc. Cambridge Philos. Soc., 53: 599-611.
15. van der Corput, J. G. (1962). J. Analyse. Math., 9: 205-345.
16. Evgrafov, M. A. (1961). Asymptotic estimates and entire functions (Transl. f. Russ.), Gordon & Breach, New York.

17. Fedoryuk, M. V. (1987). Asymptotics: Integrals and series (Russ.), Nauka, Moskow.
18. Goering, H. (1977). Asymptotische Methoden zur Lösung von Differentialgleichungen, WTB 144, Akademie-V., Berlin.
19. Hsu, L. C. (1948). Duke. Math. J., 15: 623-632.
20. Krätzel, E. (1977). Math. Nachr. 78: 249-261.
21. Laplace, P. S. (1812). Theorie analytique des probabilités, Courcier, Paris.
22. Müller, D. (1982). Abelsche und Taubersche Sätze für einige Integraltransformationen von Distributionen, Diss. (A) Friedrich-Schiller-Univ. Jena.
23. Murray, J. D. (1984). Asymptotic Analysis, Springer, New York.
24. Olver, F. W. J. (1974). Asymptotics and special functions, Academic Press, New York.
25. Olver, F. W. J. (1980). SIAM Rev. 22: 188-203.
26. Olver, F. W. J., and Sookne, D. J. (1972). Math. Comp., 26: 941-947.
27. Pittnauer, F. (1972). Vorlesungen über asymptotische Reihen, Lect. Notes Math. 301, Springer, Berlin.
28. Riedel, R. (1965a). Z. Angew. Math. Mech., 45: 323-332.
29. Riedel, R. (1965b). Math. Nach., 30: 295-317.
30. Riedel, R. (1967). Wiss. Z. Univ. Halle, 16 M: 109-123.
31. Riedel, R. (1968). Math. Nachr., 36: 79-92.
32. Riedel, R. (1969). Z. Angew. Math. Mech., 49: 63-75.
33. Riedel, R. (1981). Beiträge Anal. 15: 109-116.
34. Riedel, R. (1982). Z. Anal. Anwendungen 1 (1): 75-82.
35. Riedel, R. (1983). Wiss. Z. Univ. Halle, 32 M: 133-146.
36. Riedel, R., and Treutler, H. (1975). Wiss. Z. Univ. Halle, 24 M: 43-55.
37. Riekstiņš, E. Ya. (1977). Asymptotic expansions of integrals 2 (Russ.), Zinatne, Riga.
38. Riekstiņš, E. Ya. (1986). Remainder estimates in asymptotic expansions (Russ.), Zinatne, Riga.

39. Schell, H.-J. (1970). Math. Nachr. 44: 157-164.
40. Schell, H.-J. (1972). Beiträge Anal., 3: 97-108.
41. Schell, H.-J. (1978). Math. Nachr. 82: 309-323.
42. Schell, H.-J. (1979). Math. Nachr., 88: 391-408.
43. Schell, H.-J. (1983). Z. Anal. Anwendungen 2: 159-173, 427-442.
44. Schell, H.-J. (1987). Z. Anal. Anwendungen 6: 421-438.
45. Stanković, B. (1988). Generalized functions, convergence structures, and their applications, Proc. Conf. 1987, Dubrovnik (B. Stanković, E. Pap, St. Pilipović, V. S. Vladimirov, eds.), Plenum Press, New York.
46. Temme, N. M. (1987). CWI Tract 36: Proc. ICIAM 1987, Paris-La Villette (A. H. P. van der Burgh, R. M. M. Mattheij, eds.), Amsterdam, 117-128.
47. Wagner, E. (1962). Math. Nachr., 24: 371-383.
48. Wagner, E. (1968). Math. Nachr., 36: 323-331.
49. Wagner, E. (1970). Math. Nachr., 47: 303-317.
50. Wagner, E. (1978). Beiträge Anal., 11: 165-183.
51. Wagner, E. (1979). Math. Nachr. 90: 173-187.
52. Wagner, E. (1982). Z. Anal. Anwendungen, 1 (3): 1-11.
53. Wagner, E. (1986). Z. Anal. Anwendungen, 5: 265-276.
54. Wong, R., (1984). SIAM Review 22: 401-435.
55. Wong, R., (1988). Math. Comp. 50: 229 - 234.
56. Wong, R., and McClure (1981). Math. Comp. 37: 509-521.

Asymptotic Expansions for the Coefficient Functions Associated with Linear Second-Order Differential Equations: The Simple Pole Case

W. G. C. Boyd Professor, School of Mathematics,
University of Bristol, Bristol, England

I. INTRODUCTION

In this paper we extend previous work of ours (Boyd [2]) on uniform asymptotic behaviour of solutions of second order linear differential equations. That work was concerned with the uniform asymptotic behaviour of solutions of the differential equation

$$\frac{d^2W}{d\zeta^2}(u,\zeta) = \{u^2\zeta + \psi(\zeta)\}W(u,\zeta) \qquad (1.1)$$

in certain domains Δ which included the turning point $\zeta = 0$. We showed that for sufficiently large u, $W(u,\zeta)$ was a solution of (1.1) in Δ if and only if for some solution $\mathscr{A}i$ of Airy's differential equation,

$$W(u,\zeta) = \mathcal{A}i(u^{2/3}\zeta)\mathcal{A}(u,\zeta) + u^{-4/3}\,\mathcal{A}i\,'(u^{2/3}\zeta)\,\mathcal{B}(u,\zeta), \qquad (1.2)$$

where $\mathcal{A}$, $\mathcal{B}$ were slowly-varying and were holomorphic functions of ζ in Δ. By "slowly-varying" we mean that we were able to show that

$$\mathcal{A}(u,\zeta) = \sum_{s=0}^{n} \frac{A_s(\zeta)}{u^{2s}} + \frac{\mathcal{A}_{n+1}(u,\zeta)}{u^{2n+2}}$$
$$\mathcal{B}(u,\zeta) = \sum_{s=0}^{n-1} \frac{B_s(\zeta)}{u^{2s}} + \frac{\mathcal{B}_n(u,\zeta)}{u^{2n}}, \qquad (1.3)$$

with $\mathcal{A}_{n+1}$, $\mathcal{B}_n = O(1)$ as $u \to \infty$. The central result of the paper, which we have just stated, was given by Theorem 3.2.

We shall arrive at a similar result for the following problem. We shall suppose that we have an equation of the form

$$\frac{d^2W}{d\zeta^2}(u,\nu,\zeta) = \left\{-\frac{u^2}{4\zeta} + \frac{\nu^2-1}{4\zeta^2} + \frac{\psi(\zeta)}{\zeta}\right\} W(u,\nu,\zeta), \qquad (1.4)$$

where the prescribed function $\psi(\zeta)$ is holomorphic in a certain simply-connected domain Δ which includes the regular singular point $\zeta = 0$: the domain Δ will be specified more precisely in §2. (The title of the paper uses the phrase "simple pole" to describe this problem because the large parameter u multiplies only the simple pole term in (1.4).) Such an equation arises when a differential equation with a regular singularity - such as Legendre's equation - is reduced to normal form. Indeed we apply our results to Legendre's equation in §4. We propose to find uniform asymptotic expansions to solutions of (1.4), valid in a domain including $\zeta = 0$, as the real parameter $u \to \infty$. We shall assume that the parameter ν is real and, without loss of generality, that $\nu \geq 0$.

The standard theory for this problem is largely due to Olver ([4], [5], [6]), and is described by him in chapter 12 of his book [6]. The appropriate approximants are of the form $\zeta^{\frac{1}{2}} \mathscr{C}_\nu(u\zeta^{\frac{1}{2}})$ where $\mathscr{C}$ denotes any cylinder function. Thus we seek to show that $W(u,\nu,\zeta)$ is a solution of (1.4) if and only if for some $\mathscr{C}_\nu$,

$$W(u,\nu,\zeta) = \zeta^{\frac{1}{2}} \mathscr{C}_\nu(u\zeta^{\frac{1}{2}}) \mathscr{A}(u,\nu,\zeta) + \frac{\zeta}{u} \mathscr{C}_{\nu+1}(u\zeta^{\frac{1}{2}}) \mathscr{B}(u,\nu,\zeta), \qquad (1.5)$$

where $\mathscr{A}$, $\mathscr{B}$ are slowly-varying, that is to say, of the form (1.3).

Ursell [8] has shown that (1.5) does indeed hold for the Legendre equation. He arrived at this result through an asymptotic analysis of the integral representations of the Legendre functions P and Q. We shall demonstrate the result directly from the differential equation (1.4).

In §2 it is shown how $\mathscr{B}(u,\nu,\zeta)$ may be defined as the solution of a certain Volterra integral equation, and thence how $\mathscr{A}(u,\nu,\zeta)$ may be defined also. In §3, the Volterra equation is used to bound the remainders $\mathscr{B}_n$ and $\mathscr{A}_{n+1}$. In §4 we show how our analysis may be applied to Legendre functions. Appendices A and B are concerned with the definitions of the suprema we shall need in §2, 3.

We do not supply all the details of the analysis in this paper. A fuller exposition will be provided elsewhere.

2. DEFINITION OF THE COEFFICIENT FUNCTIONS $\mathscr{A}$, $\mathscr{B}$

We shall find coupled differential equations satisfied by $\mathscr{A}$ and $\mathscr{B}$, and eliminate $\mathscr{A}$ between them. The resulting equation for $\mathscr{B}$ will turn out to have only one slowly-varying solution: we shall find an integral equation which is satisfied by this solution.

The expression (1.5) will be a solution of (1.4) if and only if

$$\zeta^{\frac{1}{2}} \mathscr{C}_\nu(u\zeta^{\frac{1}{2}})\left[\mathscr{A}'' + \frac{1+\nu}{\zeta}\mathscr{A}' - \frac{\psi}{\zeta}\mathscr{A} + \mathscr{B}' + \frac{1}{2\zeta}\mathscr{B}\right]$$

$$+ \frac{\zeta}{u}\mathscr{C}_{\nu+1}(u\zeta^{\frac{1}{2}})\left[B'' + \frac{1-\nu}{\zeta}\mathscr{B}' - \frac{\psi}{\zeta}\mathscr{B} - \frac{u^2}{\zeta}\mathscr{A}'\right] = 0, \quad (2.1)$$

where the derivatives of $\mathscr{A}$, $\mathscr{B}$ are with respect to ζ. Equation (2.1) will be satisfied by any cylinder function $\mathscr{C}_\nu(u\zeta^{\frac{1}{2}})$ if and only if

$$\zeta\mathscr{A}'' + (1+\nu)\mathscr{A}' - \psi\mathscr{A} + \zeta\mathscr{B}' + \tfrac{1}{2}\mathscr{B} = 0 \quad (2.2a)$$

and

$$u^2\mathscr{A}' - \zeta\mathscr{B}'' - (1-\nu)\mathscr{B}' + \psi\mathscr{B} = 0. \quad (2.2b)$$

Integration of (2.2b) yields

$$\mathscr{A}(u,\nu,\zeta) = -u^{-2}\nu\,\mathscr{B}(u,\nu,\zeta) + u^{-2}\zeta\,\mathscr{B}'(u,\nu,\zeta)$$

$$- u^{-2}\int_0^\zeta \psi(\eta)\,\mathscr{B}(u,\nu,\eta)d\eta + \Lambda(u,\nu), \quad (2.3)$$

where $\Lambda(u,\nu)$ is an arbitrary constant of integration. For simplicity, we shall take $\Lambda(u,\nu) = 1$ in this paper. When (2.3) is used in (2.2a) to eliminate $\mathscr{A}$ one arrives at the equation

$$\zeta^2\mathscr{B}''' + 3\zeta\mathscr{B}''(u^2\zeta + 1 - \nu^2)\mathscr{B}' + \tfrac{1}{2}u^2\mathscr{B}$$

$$= u^2\Lambda\psi + \psi\mathscr{B} - \psi\int_0^\zeta \psi\mathscr{B}\,d\eta + 2\zeta\psi\mathscr{B}'. \quad (2.4)$$

We next wish to find the slowly-varying solution of the differential equation which results on replacing the right hand side of (2.4) by some given function. We shall first find the slowly-varying solution in Δ of the equation

$$\zeta^2\Phi'''(u,\nu,\zeta)+3\zeta\Phi''(u,\nu,\zeta)+(u^2\zeta+1-\nu^2)\Phi'(u,\nu,\zeta)+\tfrac{1}{2}u^2\Phi(u,\nu,\zeta)=\phi(\zeta) \tag{2.5}$$

where the function $\phi(\zeta)$, supposed given, is holomorphic in Δ and, for some positive constant σ, is $O(\zeta^{-\frac{1}{2}-\sigma})$ as $|\zeta| \to \infty$ in Δ.

To construct the slowly-varying solution of (2.5), we note first that the homogeneous form of the differential equation is satisfied by the product of any two cylinder functions of the form $\mathscr{C}_\nu(u\zeta^{\frac{1}{2}})$. We shall adopt the convention that $\zeta^{\frac{1}{2}}$ is defined in the ζ-plane cut along the negative real axis, with that branch being taken for which $\zeta^{\frac{1}{2}}$ is positive when ζ is real and positive; we shall continue to use this convention without further comment until the statement of Theorem 3.1 in §3. It is convenient to introduce the notation

$$\mathscr{C}_\nu^{(0)}(z) = -2J_\nu(z), \quad \mathscr{C}_\nu^{(1)}(z) = H_\nu^{(1)}(z), \quad \mathscr{C}_\nu^{(2)}(z) = H_\nu^{(2)}(z).$$

The factor of -2 in the first of these definitions is chosen in order that we have the symmetric identity

$$\mathscr{C}_\nu^{(0)} + \mathscr{C}_\nu^{(1)} + \mathscr{C}_\nu^{(2)} = 0 . \tag{2.6}$$

We remark that the three solutions $\mathscr{C}_\nu^{(j)}(z)$ $(j = 0,1,2)$ of Bessel's equation may be distinguished by their recessive properties: $\mathscr{C}_\nu^{(0)}(z)$ is recessive as $z \to 0$ (recall that $\nu \geq 0$), while $\mathscr{C}_\nu^{(1)}(z)$ and $\mathscr{C}_\nu^{(2)}(z)$ are recessive as $|z| \to \infty$ in $0 < \arg z < \pi$ and $-\pi < \arg z < 0$ respectively.

By analogy with the result (2.10) of [2], we might expect that the slowly-varying solution of (2.5) for $\zeta \in \Delta$ would be

$$\Phi(u,\nu,\zeta) = -\frac{\pi^2}{8}\left[\mathscr{C}_\nu^{(1)}(u\zeta^{\frac{1}{2}})\mathscr{C}_\nu^{(2)}(u\zeta^{\frac{1}{2}})\int_{C_0(\zeta)}\{\mathscr{C}_\nu^{(0)}(u\xi^{\frac{1}{2}})\}^2\phi(\xi)d\xi\right.$$

$$+\ \mathscr{C}_\nu^{(0)}(u\zeta^{\frac{1}{2}})\mathscr{C}_\nu^{(2)}(u\zeta^{\frac{1}{2}})\int_{C_1(\zeta)}\{\mathscr{C}_\nu^{(1)}(u\xi^{\frac{1}{2}})\}^2\phi(\xi)d\xi$$

$$\left.+\ \mathscr{C}_\nu^{(0)}(u\zeta^{\frac{1}{2}})\mathscr{C}_\nu^{(1)}(u\zeta^{\frac{1}{2}})\int_{C_2(\zeta)}\{\mathscr{C}_\nu^{(2)}(u\xi^{\frac{1}{2}})\}^2\phi(\xi)d\xi\right], \quad (2.7)$$

where the $C_j(\zeta)$ are contours of integration in the ξ-plane cut along the negative real axis starting at $\xi = \zeta$ and running to 0 when $j = 0$, to infinity in the upper half-plane when $j = 1$, or to infinity in the lower half-plane when $j = 2$ (Fig. 1). The contours are to lie entirely within Δ, which will be taken to be the domain $\operatorname{Re}\zeta^{\frac{1}{2}} < d^{\frac{1}{2}}$, for some positive constant d. The boundary $\operatorname{Re}\zeta^{\frac{1}{2}} = d^{\frac{1}{2}}$ we shall denote by $\partial\Delta$. Under the appropriate conditions (see Lemma 2.2), (2.7) is indeed the solution we seek.

We shall find it convenient to write an expression such as (2.7) in the form

$$\Phi(u,\nu,\zeta) = \int_C K(u,\nu,\zeta,\xi)\phi(\xi)d\xi\ . \quad (2.8)$$

The form (2.7) is not convenient for our purposes. By appealing to the identity (2.6), one finds

$$\Phi(u,\nu,\zeta) = -\frac{\pi^2}{8}\left[-\left\{\mathscr{C}_\nu^{(1)}(u\zeta^{\frac{1}{2}})\right\}^2\int_{C_0(\zeta)}\left\{\mathscr{C}_\nu^{(0)}(u\xi^{\frac{1}{2}})\right\}^2\phi(\xi)d\xi\right.$$

$$-\left\{\mathscr{C}_\nu^{(0)}(u\zeta^{\frac{1}{2}})\right\}^2\int_{C_1(\zeta)}\left\{\mathscr{C}_\nu^{(1)}(u\xi^{\frac{1}{2}})\right\}^2\phi(\xi)d\xi$$

$$+\ \mathscr{C}_\nu^{(0)}(u\zeta^{\frac{1}{2}})\mathscr{C}_\nu^{(1)}(u\zeta^{\frac{1}{2}})\left\{\int_{C_0(\zeta)}\mathscr{C}_\nu^{(0)}(u\xi^{\frac{1}{2}})\mathscr{C}_\nu^{(1)}(u\xi^{\frac{1}{2}})\phi(\xi)d\xi\right.$$

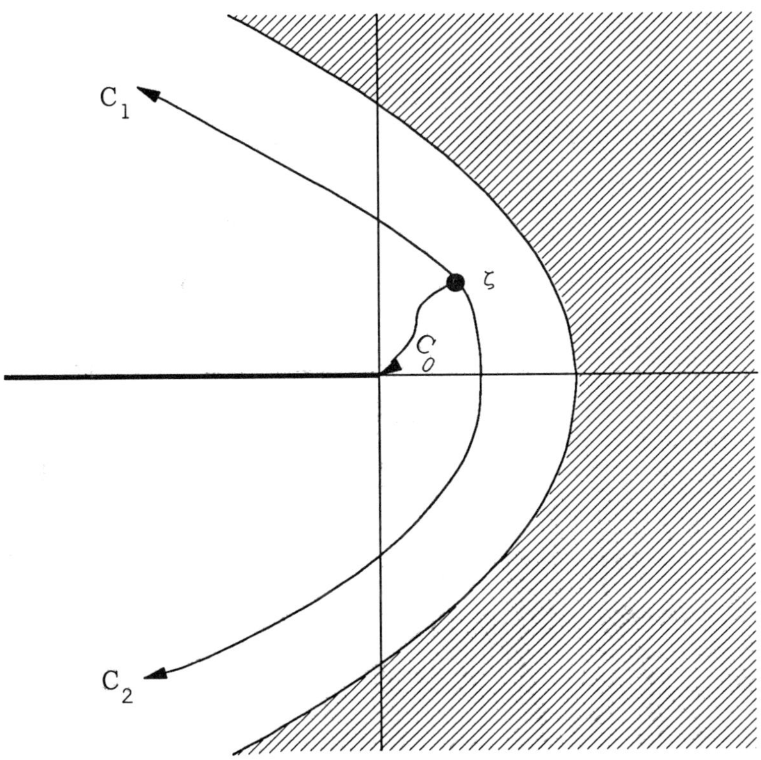

FIG. 1 *The domain* Δ *and the contours* $C_j(\zeta)$.

$$+ \int_{C_{1}(\zeta)} \mathscr{C}_\nu^{(0)}(u\xi^{\frac{1}{2}})\mathscr{C}_\nu^{(1)}(u\xi^{\frac{1}{2}})\phi(\xi)d\xi$$

$$+ \int_{C_{21}} \mathscr{C}_\nu^{(1)}(u\xi^{\frac{1}{2}})\mathscr{C}_\nu^{(2)}(u\xi^{\frac{1}{2}})\phi(\xi)d\xi$$

$$+ \int_{C_{22}} \mathscr{C}_\nu^{(0)}(u\xi^{\frac{1}{2}})\mathscr{C}_\nu^{(2)}(u\xi^{\frac{1}{2}})\phi(\xi)d\xi\Big\}\Big]. \tag{2.9}$$

Here, C_{ij} denotes a contour which is defined as follows. For $i = 0,1,2$, let Ξ_i be defined to be the origin, or infinity in the upper half-plane, or infinity in the lower half-plane respectively: then C_{ij} runs from Ξ_i to Ξ_j. There are a further two representations for $\Phi(u,\nu,\zeta)$, which may be obtained from (2.9) by cyclic permutation of $\{0,1,2\}$.

One may prove from the representation (2.9), and its analogue with $\{0,1,2\}$ replaced by $\{2,0,1\}$, the following remarkable lemma: despite the functions $\mathscr{C}_\nu^{(j)}(u\zeta^{\frac{1}{2}})$ having a branch point at $\zeta = 0$, the solution $\Phi(u,\nu,\zeta)$ given by (2.7) is holomorphic provided $\phi(\zeta)$ is. This result is analogous to those given by Ursell [8] in his Lemmas A and B.

Lemma 2.1. Let Δ denote the domain which comprises the set of points for which $\mathrm{Re}\zeta^{\frac{1}{2}} < d^{\frac{1}{2}}$ for some positive constant d, where $\zeta^{\frac{1}{2}}$ takes its principal values. Suppose the function $\phi(\zeta)$ is holomorphic in Δ and, for some positive constant σ, is $O(\zeta^{-\frac{1}{2}-\sigma})$ as $|\zeta| \to \infty$ in Δ.

Then the function $\Phi(u,\nu,\zeta)$ defined by (2.7) is holomorphic throughout Δ.

Next, we bound $\Phi(u,\nu,\zeta)$ and its first and second derivatives. The version of the identity (2.9) in which $\{0,1,2\}$

is replaced by $\{1,2,0\}$, in conjunction with the maximum modulus theorem, may be used to show that

$$|\zeta^{\frac{1}{2}}\Phi(u,\nu,\zeta| \le \frac{k^2}{2u^2}\int_{\partial\Delta}\left|\frac{\phi(\xi)}{\xi^{\frac{1}{2}}}\,d\xi\right| + \frac{\kappa^2}{u^3}\sup_{\Delta}|\phi(\zeta)| \qquad (2.10)$$

for all $\zeta \in \Xi$ and all $u > 0$. The constants k and κ - suprema involving the cylinder functions - are defined in Appendix A, together with $k', \kappa', k'', \kappa''$ which we need below. We further find that for all $\zeta \in \Delta$ and all $u > 0$,

$$|\zeta\Phi'(u,\nu,\zeta)| \le \frac{k'k}{4u^2d^{\frac{1}{2}}}\int_{\partial\Delta}\left|\frac{\phi(\xi)}{\xi^{\frac{1}{2}}}\,d\xi\right| + \frac{\kappa'\kappa}{2u^2}\sup_{\Delta}|\phi(\zeta)| \qquad (2.11)$$

$$|\zeta^{3/2}\Phi''(u,\nu,\zeta)| \le \frac{3k''k}{8u^2d}\int_{\partial\Delta}\left|\frac{\phi(\xi)}{\xi^{\frac{1}{2}}}\,d\xi\right| + \frac{\kappa''\kappa}{2u}\sup_{\Delta}|\phi(\zeta)|. \qquad (2.12)$$

We may combine (2.10, 11) in the single vector inequality for Φ and Φ'

$$\begin{bmatrix} |\zeta^{\frac{1}{2}}\Phi(u,\nu,\zeta)| \\ |\zeta\Phi'(u,\nu,\zeta)| \end{bmatrix} \le u^{-2}\underset{\sim}{K}\begin{bmatrix} \int_{\partial\Delta}\left|\frac{\phi(\xi)}{\xi^{\frac{1}{2}}}\,d\xi\right| \\ \sup_{\Delta}|\phi(\xi)| \end{bmatrix}, \qquad (2.13)$$

where

$$\underset{\sim}{K} = \begin{bmatrix} \frac{1}{2}k^2 & \kappa^2/u \\ \frac{1}{4}d^{-\frac{1}{2}}k'k & \frac{1}{2}\kappa'\kappa \end{bmatrix}. \qquad (2.14)$$

We shall make direct use of the inequality (2.13) in §3.

We may now assert the following.

Lemma 2.2. With the same conditions as in Lemma 2.1, the function $\Phi(u,\nu,\zeta)$ defined by (2.7) is, for each real non-negative value of ν, the only solution of the differential equation (2.5) which is uniformly bounded for all $\zeta \in \Delta$ and all sufficiently large positive u. Explicit bounds for $\Phi(u,\nu,\zeta)$, $\Phi'(u,\nu,\zeta)$, $\Phi''(u,\nu,\zeta)$ are available from (2.10, 11, 12).

We return now to the equations (2.4) for $\mathfrak{B}(u,\nu,\zeta)$ and suppose that $\psi(\zeta)$ satisfies the same conditions as does $\phi(\zeta)$ in Lemma 2.1, and further that $\psi^{(m)}(\zeta) = O(\zeta^{-\frac{1}{2}-m-\sigma})$ as $|\zeta| \longrightarrow \infty$ in Δ. Then it is reasonable to suppose that the slowly-varying solution of (2.4) that we seek satisfies the integral equation

$$\mathfrak{B} = \int_c K(u,\nu,\zeta,\xi)\{u^2\psi + \psi\mathfrak{B} - \psi\int_0^\xi \psi\mathfrak{B}d\eta + 2\zeta\psi\mathfrak{B}'\}d\xi, \quad (2.15)$$

and indeed we shall show that this is so in §3.

We shall express $\mathfrak{A}(u,\nu,\zeta)$ and $\mathfrak{B}(u,\nu,\zeta)$ as descending power series in u^2 with remainder terms:

$$\mathfrak{A}(u,\nu,\zeta) = \sum_{s=0}^{n-1} \frac{A_s(\nu,\zeta)}{u^{2s}} + \frac{\mathfrak{A}_n(u,\nu,\zeta)}{u^{2n}} \qquad n = 0,1,2,\ldots, \quad (2.16a)$$

$$\mathfrak{B}(u,\nu,\zeta) = \sum_{s=0}^{n-1} \frac{B_s(\nu,\zeta)}{u^{2s}} + \frac{\mathfrak{B}_n(u,\nu,\zeta)}{u^{2n}} \qquad n = 0,1,2,\ldots\ . \quad (2.16b)$$

The coefficients $A_s(\nu,\zeta)$, $B_s(\nu,\zeta)$ in (2.16) are determined by substituting (2.16a,b) into (2.2a,b) and equating corresponding powers of u^2. Integration then yields $A_0(\nu,\zeta) = 1$ and

$$B_s(\nu,\zeta) = -A_s'(\nu,\zeta) + \zeta^{-\frac{1}{2}}\int_0^\zeta \{\psi(\eta)A_s(\nu,\eta) - (\nu+\tfrac{1}{2})A_s'(\nu,\eta)\}\eta^{-\frac{1}{2}}d\eta \quad (2.17a)$$

$$A_s(\nu,\zeta) = -\nu B_{s-1}(\nu,\zeta) + \zeta B_{s-1}'(\nu,\zeta) - \int_0^\zeta \psi(\eta)B_{s-1}(\nu,\eta)d\eta. \quad (2.17b)$$

(Quantities with negative indices are taken to be zero.)

One may show (cf. Olver [6], p. 445):

Lemma 2.3. Let $\psi(\zeta)$ satisfy the same conditions as does $\phi(\zeta)$ in Lemma 2.1, and that in addition its mth derivative is $O(\zeta^{-\frac{1}{2}-m-\sigma})$ as $|\zeta| \to \infty$ in Δ. Then for each non-negative integer s, the function $B_s(\nu,\zeta)$ is holomorphic within Δ; as $|\zeta| \to \infty$ in Δ, $\frac{d^m B_s}{d\zeta^m}(\nu,\zeta) = O(\zeta^{-\frac{1}{2}-m})$ for each non-negative integer m and $\zeta^{\frac{1}{2}}(\zeta^{\frac{1}{2}}B_s(\nu,\zeta))' = O(\zeta^{-\frac{1}{2}-\rho})$, where $\rho = \min(\sigma,1)$.

With these results one readily arrives at the equations

$$\mathcal{B}_n(u,\nu,\zeta) = \int_C K(u,\nu,\zeta,\xi)\{u^2\xi^{\frac{1}{2}}(\xi^{\frac{1}{2}}B_n)' + \psi\mathcal{B}_n - \psi\int_0^\xi \psi\mathcal{B}_n d\eta + 2\xi\psi\mathcal{B}_n'\}d\xi \tag{2.18}$$

$$\mathcal{A}_{n+1}(u,\nu.\zeta) = -\nu\mathcal{B}_n(u,\nu,\zeta) + \zeta\mathcal{B}_n'(u,\nu,\zeta) - \int_0^\zeta \psi(\eta)\mathcal{B}_n(u,\nu,\eta)d\eta. \tag{2.19}$$

In the next section, we shall obtain a bound for the solution $\mathcal{B}_n(u,\nu,\zeta)$ of the integral equation (2.18), and thence, from (2.19), a bound for $\mathcal{A}_{n+1}(u,\nu,\zeta)$.

3. ERROR BOUNDS

To bound $\mathcal{B}_n(u,\nu,\zeta)$, the solution of (2.18), we define a sequence of iterates $\mathcal{B}_n^{(0)}(u,\nu,\zeta)$, $\mathcal{B}_n^{(1)}(u,\nu,\zeta)$, $\mathcal{B}_n^{(2)}(u,\nu,\zeta),\ldots,$ where

$$\mathscr{B}_n^{(0)} = u^2 \int_c K(u,\nu,\zeta,\xi)\xi^{\frac{1}{2}}(\xi^{\frac{1}{2}}B_n)'d\xi + \Lambda_{n+1} \int_c K(u,\nu,\zeta,\xi)\psi d\xi \tag{3.1}$$

$$\mathscr{B}_n^{(r+1)} = \int_c K(u,\nu,\zeta,\xi\xi)\{\psi'\mathscr{B}_n^{(r)} - \psi \int_0^\xi \psi\mathscr{B}_n^{(r)}d\eta + 2\psi\mathscr{B}_n^{(r)\prime}\}d\xi, \quad r = 0,1,2,\ldots \tag{3.2}$$

We define

$$\underset{\sim}{\mathscr{B}}_n^{(r)}(u,\nu,\zeta) = \begin{bmatrix} |\zeta^{\frac{1}{2}}\mathscr{B}_n^{(r)}(u,\nu,\zeta)| \\ |\zeta\mathscr{B}_n^{(r)\prime}(u,\nu,\zeta)| \end{bmatrix}, \quad \underset{\sim}{\mathscr{B}}_n(u,\nu,\zeta) = \begin{bmatrix} |\zeta^{\frac{1}{2}}\mathscr{B}_n(u,\nu,\zeta)| \\ |\zeta\mathscr{B}_n'(u,\nu,\zeta)| \end{bmatrix}, \tag{3.3}$$

$$\underset{\sim}{\psi} = \begin{bmatrix} d^{-\frac{1}{2}}\beta_0' + \beta_0^2 & 2d^{-1}\beta_0 \\ d^{-\frac{1}{2}}b_0' + b_0\beta_0 & 2d^{-2}b_0 \end{bmatrix}, \tag{3.4}$$

$$\underset{\sim}{B}_n(u,\nu) = (\underset{\sim}{I} - u^{-2}\underset{\sim}{K}\underset{\sim}{\Psi})^{-1}\underset{\sim}{K}\begin{bmatrix} \beta_n \\ b_n \end{bmatrix} \tag{3.5}$$

for $u > u_0$, where u_0 is defined to be that value of u for which the spectral radius $\rho(u^{-2}\underset{\sim}{K}\underset{\sim}{\Psi}) = 1$. Note that for fixed ν, $\underset{\sim}{B}_n(u,\nu)$ is uniformly bounded for $u > u_0$.

With the notation we have just introduced, we may assert the truth of the following lemmas, in both of which the conditions of Lemma 2.3 are supposed to hold.

Lemma 3.1. Suppose $\zeta \in \Delta$ and $u > u_0$. Then there exists a function $\mathscr{B}_n(u,\nu,\zeta)$ holomorphic in ζ, which satisfies the integral equation (2.18) and the inequality $\underset{\sim}{\mathscr{B}}_n(u,\nu,\zeta) \le \underset{\sim}{B}_n(u,\nu)$.

Lemma 3.2. Suppose $\zeta \in \Delta$ and $u > u_0$. Then there exists a function $\mathscr{A}_{n+1}(u,\nu,\zeta)$, holomorphic in ζ, which satisfies the relation (2.19) and the inequalities

$$|\mathscr{A}_{n+1}(u,\nu,\zeta)| \leq \underset{\sim}{a}\ \underset{\sim}{B} \qquad |\mathscr{A}'_{n+1}(u,\nu,\zeta)| \leq \underset{\sim}{a}'\ \underset{\sim}{B}_n,$$

where the row vectors $\underset{\sim}{a}$ and $\underset{\sim}{a}'$ are defined by

$$\underset{\sim}{a} = (d^{-\frac{1}{2}}\nu + \beta_0 \qquad 1)$$

$$\underset{\sim}{a}' = (d^{-\frac{1}{2}}b_0 \qquad d^{-1}(1+\nu)) + \left[\frac{3}{8}\, d^{-3/2}k''k \qquad \tfrac{1}{2}d^{-\frac{1}{2}}u\kappa''\kappa\right]\underset{\sim}{K}^{-1},$$

except for those values of u,ν for which $\underset{\sim}{K}$ is singular.

We may thus prove the following theorem.

Theorem 3.1. Suppose that the function $\psi(\zeta)$ is holomorphic in the domain $\mathrm{Re}\ \zeta^{\frac{1}{2}} < d^{\frac{1}{2}}$, for some positive constant d, and that there exists a positive number σ, independent of ζ, such that for each non-negative integer m, its mth derivative is $O(\zeta^{-\frac{1}{2}-m-\sigma})$ as $|\zeta| \longrightarrow \infty$ in Δ.

Then, for fixed non-negative ν, there exists a positive constant u_0 and functions $\mathscr{A}(u,\nu,\zeta)$, $\mathscr{B}(u,\nu,\zeta)$ such that for $u > u_0$ and $\zeta \in \Delta$, the following hold.

(i) $\mathscr{A}(u,\nu,\zeta)$ and $\mathscr{B}(u,\nu,\zeta)$ are holomorphic in ζ.

(ii) For $n = 0,1,2,\ldots,$

$$\mathscr{A}(u,\nu,\zeta) = \sum_{s=0}^{n} \frac{A_s(\nu,\zeta)}{u^{2s}} + O(u^{-2n-2})$$

$$\mathscr{B}(u,\nu,\zeta) = \sum_{s=0}^{n-1} \frac{B_s(\nu,\zeta)}{u^{2s}} + O(u^{-2n})$$

uniformly as $u \longrightarrow \infty$ within Δ. Explicit bounds for the O terms are available from Lemmas 3.1, 3.2.

(iii) A necessary and sufficient condition for $W(u,\nu,\zeta)$ to be a solution of the differential equation

$$\frac{d^2W}{d\zeta^2}(u,\nu,\zeta) = \left\{-\frac{u^2}{4\zeta} + \frac{\nu^2-1}{4\zeta^2} + \frac{\psi(\zeta)}{\zeta}\right\}W(u,\nu,\zeta)$$

is that

$$W(u,\nu,\zeta) = \zeta^{\frac{1}{2}}\mathscr{C}_\nu(u\zeta^{\frac{1}{2}})\mathscr{A}(u,\nu,\zeta) + \frac{\zeta}{u}\mathscr{C}_{\nu+1}(u\zeta^{\frac{1}{2}})\mathscr{B}(u,\nu,\zeta)$$

for some solution $\mathscr{C}_\nu$ of Bessel's differential equation.

It should be noted that we have *not* specified $\zeta^{\frac{1}{2}}$ in the statement of the theorem - indeed $\zeta^{\frac{1}{2}}$ may be specified arbitrarily, either in a cut plane or on a Riemann sheet, and the theorem remains valid.

4. AN APPLICATION: LEGENDRE FUNCTIONS OF LARGE DEGREE

We may use the results of §3 to bound the errors in the uniform asymptotic expansions of the Legendre functions $P^{-\nu}_{u-\frac{1}{2}}(z)$, $Q^{\nu}_{u-\frac{1}{2}}(z)$ as the real parameter $u \to \infty$ for fixed real non-negative values of ν and for z lying in the right hand half-plane, $\text{Re } z \geq 0$. (We eschew the conventional notation $P^{-\mu}_{\nu}(z)$, $Q^{\mu}_{\nu}(z)$ to avoid a clash of notation.) Corresponding results for negative values of u, ν, $\text{Re } z$ may be obtained by the use of appropriate connection formulae.

The equation

$$\frac{d^2w}{dz^2}(u,\nu,z) = \left\{-\frac{u^2-\frac{1}{4}}{1-z^2} - \frac{\nu^2-1}{(1-z^2)^2}\right\}w(u,\nu,z) \tag{4.1}$$

is satisfied by functions of the form $(z^2-1)^{1/4}\mathscr{P}^{\nu}_{u-\frac{1}{2}}(z)$, where $\mathscr{P}^{\nu}_{u-\frac{1}{2}}(z)$ denotes any solution of the associated Legendre equation:

$$(1-z^2)\frac{d^2\mathscr{P}^{\nu}_{u-\frac{1}{2}}}{dz^2} - 2z\frac{d\,\mathscr{P}^{\nu}_{u-\frac{1}{2}}}{dz} + \left\{u^2 - \frac{1}{4} - \frac{\nu^2}{1-z^2}\right\}\mathscr{P}^{\nu}_{u-\frac{1}{2}} = 0. \tag{4.2}$$

The transformations

$$\zeta^{\frac{1}{2}} = \int_1^z \frac{dt}{(1-t^2)^{\frac{1}{2}}} = \cos^{-1} z \tag{4.3a}$$

$$W(u,\nu,\zeta) = \left[\frac{\zeta}{1-z^2}\right]^{1/4} w(u,\nu,z) = \{\zeta(1-z^2)\}^{1/4}\mathscr{P}^{\nu}_{u-\frac{1}{2}}(z) \tag{4.3b}$$

yield an equation of the form (1.4), namely

$$\frac{d^2W}{d\zeta^2}(u,\nu,\zeta) = \left\{-\frac{u^2}{4\zeta} + \frac{\nu^2-1}{4\zeta^2} + \frac{\psi(\zeta)}{\zeta}\right\} W(u,\nu,\zeta), \tag{4.4}$$

where

$$\psi(\zeta) = \frac{4\nu^2-1}{16}\left[\frac{1}{\zeta} - \frac{1}{1-z^2}\right] \tag{4.5}$$

(Olver [6], §12.1). The branches in (4.3a) are defined so that the function $\zeta(z)$ is continuous for $\text{Re } z \geq 0$ and maps the interval of the real axis $0 < z < 1$ to $\frac{1}{4}\pi^2 > \zeta > 0$. (We shall specify $\zeta^{\frac{1}{2}}$ more precisely below.) We note that the function ψ defined in (4.5) actually depends on ν as well as ζ: however, since ν is assumed fixed, Theorem 3.1 applies in this case also.

We shall take the domain Δ to be the image of $\text{Re } z > 0$ under the transformation (4.3a), so that Δ is the set of points satisfying $|\text{Re}\zeta^{\frac{1}{2}}| < \frac{1}{2}\pi$. The function $\psi(\zeta)$ is holomorphic within Δ (Olver [6], p. 470) and, as $|\zeta| \to \infty$ in Δ, its mth derivative is $o(\zeta^{-1-m})$ (Olver [6], p. 464). THe conditions of Theorem 3.2 are thus satisfied. Therefore we may claim the existence of functions $\mathscr{A}(u,\nu,\zeta)$, $\mathscr{B}(u,\nu,\zeta)$ which are slowly-varying and holomorphic within Δ, and are such that for any solution $\mathscr{P}^{\nu}_{u-\frac{1}{2}}(z)$ of the associated Legendre equation (4.2),

$$\{\zeta(1-z^2)\}^{1/4}\mathscr{P}^{\nu}_{u-\frac{1}{2}}(z) = \zeta^{\frac{1}{2}}\mathscr{C}_\nu(u\zeta^{\frac{1}{2}})\mathscr{A}(u,\nu,\zeta) + \frac{\zeta}{u}\mathscr{C}_{\nu+1}(u\zeta^{\frac{1}{2}})\mathscr{B}(u,\nu,\zeta) \tag{4.6}$$

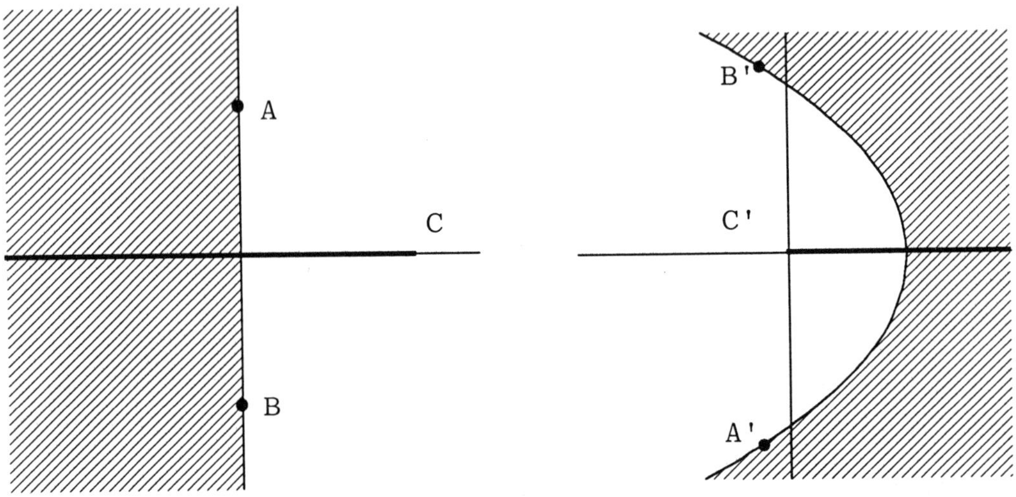

FIG. 2 *The domain* $\mathrm{Re}\, z > 0$ *and its image* Δ *(shown unshaded) under the transformation (4.3a). The images of* A, B, C *are* A′, B′, C′ *respectively. The* z*-plane is cut as indicated for the Legendre functions* $\mathcal{P}^{\nu}_{u-\frac{1}{2}}(z)$, *while the* ζ*-plane is cut for the Bessel functions* $\mathcal{C}_{\nu}(u\zeta^{\frac{1}{2}})$.

for some solution $\mathscr{C}_\nu(u\zeta^{\frac{1}{2}})$ of Bessel's equation: (4.6) will be valid for $\mathrm{Re}\, z > 0$, for ν fixed real and non-negative, and for $u > u_0$, where u_0 is fixed and positive.

It is necessary to specify the branches of the Legendre and Bessel functions in (4.6). Following the usual convention, we shall suppose the Legendre functions are defined in the z-plane cut along the real axis from $-\infty$ to $+1$. We therefore suppose the ζ-plane is cut along the positive real axis, and choose that branch for which $0 \leq \arg\zeta^{\frac{1}{2}} \leq \pi$ (Fig. 2).

Consider first the solution $\{\zeta(1-z^2)\}^{1/4}P^{-\nu}_{u-\frac{1}{2}}(z)$: since the only solutions of the associated Legendre equation which are recessive as $z \rightarrow 1$ are multiples of $P^{-\nu}_{u-\frac{1}{2}}(z)$, we deduce that the corresponding Bessel function in (4.6) must be a multiple of J_ν. That is, we may assert that there exists a constant C^ν_u, independent of ζ, such that

$$P^{-\nu}_{u-\frac{1}{2}}(z) = C^\nu_u\left[\frac{\zeta}{1-z^2}\right]^{1/4}\{J_\nu(u\zeta^{\frac{1}{2}})\mathscr{A}(u,\nu,\zeta) + \frac{1}{u}\, J_{\nu+1}(u\zeta^{\frac{1}{2}})\zeta^{\frac{1}{2}}\mathscr{B}(u,\nu,\zeta)\} \tag{4.7}$$

for $\mathrm{Re}\, z \geq 0$, ν fixed and non-negative, and $u > u_0$.

We shall identify C^ν_u in (4.7) by considering the behaviour of both sides of the equation as $z \rightarrow 1$, or $\zeta \rightarrow 0$. Since $J_\nu(x) \sim (\frac{1}{2}x)^\nu/\Gamma(\nu+1)$ as $x \rightarrow 0$, and $\zeta \sim -2(z-1)$ as $z \rightarrow 1$, we find

$$P^{\nu}_{u-\frac{1}{2}}(z) = C^\nu_u\,\frac{u^\nu e^{\frac{1}{2}\nu\pi i}}{2^{\frac{1}{2}\nu}\,\Gamma(\nu+1)}\,(z-1)^{\frac{1}{2}\nu}\{\mathscr{A}(u,\nu,0) + o(1)\} \quad \text{as} \quad z \rightarrow 1. \tag{4.8}$$

To proceed further we need to evaluate $\mathscr{A}(u,\nu,0)$.

To ensure that the constant C_u^ν is as simple as possible, we wish to have $\mathscr{A}(u,\nu,0) = 1$. This, however, is not consistent with our choice of integration constant, $\Lambda(u,\nu) = 1$, in (2.3). It is not difficult to extend the theory to treat the case when $\Lambda(u,\nu) \neq 1$ by using the results when $\Lambda(u,\nu) = 1$. We do not present the details here: it suffices to note that the form of the results given in Theorem 3.1 remain essentially unaffected.

Returning now to (4.8), since $P^{\nu}_{u-\frac{1}{2}}(z) \sim (z-1)^{\frac{1}{2}\nu}/(2^{\frac{1}{2}\nu}\Gamma(\nu+1))\}$ as $z \to 1$ (Olver [6], p. 171), we find $C_u^\nu = u^{-\nu}e^{-\frac{1}{2}\nu\pi i}$. We therefore deduce that for $\mathrm{Re}\, z \geq 0$, ν fixed and non-negative, and sufficiently large u

$$P^{-\nu}_{u-\frac{1}{2}}(z) = u^{-\nu}e^{-\nu\pi i/2}\left[\frac{\zeta}{1-z^2}\right]^{1/4}\{J_\nu(u\zeta^{\frac{1}{2}})\mathscr{A}(u,\nu,\zeta) + \frac{1}{u}\, J_{\nu+1}(u\zeta^{\frac{1}{2}})\zeta^{\frac{1}{2}}\mathscr{B}(u,\nu,\zeta)\}. \tag{4.9}$$

Consider next the solution $\{\zeta(1-z^2)\}^{1/4}Q^{\nu}_{u-\frac{1}{2}}(z)$. Since it is recessive as $|z| \to \infty$, we deduce that the appropriate Bessel function in (4.6) must be a multiple of $H_\nu^{(1)}$. Arguments similar to those used in arriving at (4.19) yield

$$Q^{\nu}_{u-\frac{1}{2}}(z) = \frac{i\pi}{2(1+2u^{-2}\Lambda_1(u,\nu))}\, u^{\nu}e^{3\nu\pi i/2}\left[\frac{\zeta}{1-z^2}\right]^{1/4} \left\{H_\nu^{(1)}(u\zeta^{\frac{1}{2}})\mathscr{A}(u,\nu,\zeta) + \frac{1}{u}\, H_{\nu+1}^{(1)}(u\zeta^{\frac{1}{2}})\zeta^{\frac{1}{2}}\mathscr{B}(u,\rho,\zeta)\right\} \tag{4.10}$$

for $\mathrm{Re}\, z \geq 0$, ν fixed and non-negative, and sufficiently large u. The coefficient functions $\mathscr{A}$ and $\mathscr{B}$ in (4.10) are the same as those in (4.9).

Once these two results - (4.9) and (4.10) are established, we can at once derive the relation between any solution of Legendre's equation and the corresponding Bessel function. This is in contrast to the method of Olver ([6], chapter 12).

By using Campbell's algorithm [3] for calculating Bessel functions, one may compute the error bounds given by the theory: these will be given elsewhere.

APPENDIX A. SUPREMA OF HANKEL FUNCTIONS

The suprema which we define below in (A.2), (A.3) are evaluated over the domain $Z(\nu)$ in the complex z-plane, defined by

$$Z(\nu) = \{z: \text{Re } z^{\frac{1}{2}} > \nu\}. \tag{A.1}$$

In each case the supremum is normalised so as to ensure that the limit at infinity is 1. The existence of the suprema follow from standard properties of the Hankel functions.

Our definitions of the k-suprema are

$$\begin{aligned} k(\nu) &= \sup|\tfrac{1}{2}\pi z^{\frac{1}{2}} H_\nu^{(1)}(z^{\frac{1}{2}}) H_\nu^{(2)}(z^{\frac{1}{2}})| \\ k'(\nu) &= \sup|\pi z^{3/2} \frac{d}{dz} \{H_\nu^{(1)}(z^{\frac{1}{2}}) H_\nu^{(2)}(z^{\frac{1}{2}})\}| \\ k''(\nu) &= \sup|\tfrac{2}{3}\pi z^{5/2} \frac{d^2}{dz^2} \{H_\nu^{(1)}(z^{\frac{1}{2}}) H_\nu^{(2)}(z^{\frac{1}{2}})\}|. \end{aligned} \tag{A.2}$$

The κ-suprema are defined by

$$\kappa(\nu) = \sup|\tfrac{1}{2}\pi z^{\frac{1}{2}} e^{\mp 2iz^{\frac{1}{2}}} \{H_\nu^{(j)}(z^{\frac{1}{2}})\}^2|$$

$$\kappa'(\nu) = \sup|\tfrac{1}{2}\pi z \; e^{\mp 2iz^{\frac{1}{2}}} \frac{d}{dz} \{H_\nu^{(j)}(z^{\frac{1}{2}})\}^2| \qquad (A.3)$$

$$\kappa''(\nu) = \sup|\tfrac{1}{2}\pi z^{3/2} e^{\mp 2iz^{\frac{1}{2}}} \frac{d^2}{dz^2} \{H_\nu^{(j)}(z^{\frac{1}{2}})\}^2|,$$

where the lower and upper signs apply for $j = 1,2$ respectively.

APPENDIX B. SUPREMA AND INTEGRALS OF THE COEFFICIENT FUNCTIONS

For $n = 0,1,2,\ldots,$ we define

$$b_n(\nu) = \sup_{\Delta}|\zeta^{\frac{1}{2}}(\zeta^{\frac{1}{2}}B_n(\nu,\zeta))'|$$

$$\beta_n(\nu) = \int_{\partial\Delta} |\xi^{\frac{1}{2}}B_n(\xi))' d\xi|. \qquad (B.1)$$

The existence of $b_n(\nu)$ and $\beta_n(\nu)$ follows from Lemma 2.3.

When $n = 0$, (B.1) reduce to

$$b_0 = \sup_{\Delta}|\psi(\zeta)|$$

$$\beta_0 = \int_{\partial\Delta} \left|\frac{\psi(\xi)}{\xi^{\frac{1}{2}}} d\xi\right|. \qquad (B.2)$$

Both constants are used in the definition of $\underset{\sim}{\Psi}$ in (3.4), where we also require

$$b_0' = \sup_{\Delta}|\psi'(\zeta)|$$

$$\beta_0' = \int_{\partial\Delta} \left|\frac{\psi'(\xi)}{\xi^{\frac{1}{2}}} d\xi\right|. \qquad (B.2)$$

ACKNOWLEDGEMENTS

Much of this work was done while the author was visiting the Department of Mathematics at the University of Maryland, College Park. I extend my thanks to the University, and to Professor F.W.J. Olver for many stimulating discussions.

REFERENCES

1. M. Abramowitz, and I.A. Stegun. *Handbook of mathematical functions*, Dover, New York, 1964.
2. W.G.C. Boyd. *Proc. Roy. Soc. (Lond.)* A410: 35-60 (1987).
3. J.B. Campbell. *Comput. Phys. Commun.* *24* 97-105 (1981).
4. F.W.J. Olver. *Philos. Trans. Roy. Soc. (Lond.)* A249: 65-97 (1956).
5. F.W.J. Olver. *Philos. Trans. Roy. Soc. (Lond.)* A250: 479-517 (1958).
6. F.W.J. Olver. *Asymptotics and special functions*, Academic, New York, 1974.
7. E.C. Titchmarsh. *The theory of functions*, Oxford University Press, 2nd edition, 1939.
8. F. Ursell. *Math. Proc. Camb. Phil. Soc.* 95: 367-380 (1984).

Landen Transformations of Integrals

B. C. Carlson Professor, Ames Laboratory, Department of Mathematics, Iowa State University of Science and Technology, Ames, Iowa

Abstract. Landen's transformation changes an integral $I(\zeta)$, depending on parameters $\zeta = (\zeta_1,\ldots,\zeta_k)$, into $I(\eta)$. This and several related transformations are derived by a unified approach, and some applications to numerical computation are discussed.

I. INTRODUCTION

This paper surveys several transformations having the form

$$I(\zeta) = \int_0^\infty \prod_{i=1}^{k} (t+\zeta_i)^{p_i} dt = \int_0^\infty \prod_{i=1}^{k} (t+\eta_i)^{p_i} dt = I(\eta), \tag{1.1}$$

Ames Laboratory is operated for the U. S. Department of Energy by Iowa State University under contract no. W-7405-ENG-82. This work was supported by the Director of Energy Research, Office of Basic Energy Sciences.

where the two integrals are the same except for replacement of the k-tuple ζ by η. An invariance of this kind with $k = 3$ and $p_i = -1/2$, $1 \leq i \leq 3$, was discovered (in a different form) by John Landen in 1775. Its practical importance lies in repeating the transformation $T: \zeta \to \eta = T\zeta$ to get $I(\zeta) = I(T^n\zeta)$ for $n = 1,2,3,\ldots.$ As $n \to \infty$ two of the variables become equal and the integral becomes elementary. This is the method (except for change of variables) that Legendre used to compute his numerical tables of elliptic integrals, and it remained the method of choice until recently. Another application comes from the fact that the η's may be real when two of the ζ's are conjugate complex, and the use of complex arithmetic in the computer can thereby be avoided.

We shall present a unified procedure for deriving several invariances of the type (1.1), including those associated with the arithmetic-geometric mean of Gauss and the Schwab-Borchardt algorithm for computing inverse circular and inverse hyperbolic functions. Other invariances are new in form though not in fundamental content. Some related transformations involving additional terms will be discussed also.

II. CHANGE OF INTEGRATION VARIABLE

Consider the substitution

$$t = s(s+f)/(s+g), \quad dt/ds = (s^2+2gs+fg)/(s+g)^2, \tag{2.1}$$

where f and g ($g \neq 0$) remain to be determined but are independent of s. The first integral in (1.1) is transformed into

$$I(\zeta) = \int_0^\infty (s+g)^{-2-\Sigma p_i}(s^2+2gs+fg)\prod_{i=1}^{k}[s^2+(f+\zeta_i)s+g\zeta_i]^{p_i}\,ds.$$

Since each quadratic polynomial has two linear factors, the

integrand contains 2k+3 linear functions, and this number must somehow be reduced to k if we are to have an invariance of the form (1.1). A way of reducing the number of distinct linear factors is to choose f and g so that two of the quadratic factors, say those for i = 1 and 2, are perfect squares:

$$(f+\zeta_i)/2 = (g\zeta_i)^{1/2}, \qquad i = 1,2.$$

Solving these two equations for f and g, we find

$$f = (\zeta_1\zeta_2)^{1/2}, \qquad g = [(\zeta_1^{1/2}+\zeta_2^{1/2})/2]^2.$$

Already arithmetic and geometric means have appeared on the scene, and they will be prominent in the sequel. The square roots suggest a change of notation:

$$\begin{aligned} &\zeta_1 = x^2, \quad \zeta_2 = y^2, \quad \alpha = (x+y)/2, \quad \gamma = (xy)^{1/2}, \\ &f = \gamma^2, \quad g = \alpha^2, \quad t = s(s+\gamma^2)/(s+\alpha^2). \end{aligned} \tag{2.2}$$

The two perfect squares are

$$s^2+(f+\zeta_1)s+g\zeta_1 = [s+(g\zeta_1)^{1/2}]^2 = (s+\alpha x)^2,$$

$$s^2+(f+\zeta_2)s+g\zeta_2 = [s+(g\zeta_2)^{1/2}]^2 = (s+\alpha y)^2,$$

and the quadratic factor that came from dt/ds is

$$s^2+2gs+fg = s^2+(x+y)\alpha s+xy\alpha^2 = (s+\alpha x)(s+\alpha y).$$

The integrand now contains 2k−1 distinct linear functions of s:

$$\int_0^\infty (t+x^2)^{p_1}(t+y^2)^{p_2}\prod_{i=3}^{k}(t+\zeta_i)^{p_i}\,dt$$
$$= \int_0^\infty (s+\alpha^2)^{-2-\Sigma p_i}(s+\alpha x)^{1+2p_1}(s+\alpha y)^{1+2p_2}\prod_{i=3}^{k}[s^2+(\gamma^2+\zeta_i)s+\alpha^2\zeta_i]^{p_i}\,ds. \tag{2.3}$$

III. THE SCHWAB-BORCHARDT TRANSFORMATION

If $k = 2$ then (2.3) is

$$\int_0^\infty (t+x^2)^{p_1}(t+y^2)^{p_2}\,dt = \int_0^\infty (s+\alpha^2)^{-2-p_1-p_2}(s+\alpha x)^{1+2p_1}(s+\alpha y)^{1+2p_2}\,ds. \tag{3.1}$$

Any one of the three factors on the right side can be eliminated by choosing its exponent to be zero, and there are three invariances with $(p_1,p_2) = (-1,-1)$, $(-1/2,-1)$, or $(-3/4,-1/2)$. Two more with $(-1,-1/2)$ or $(-1/2,-3/4)$ give nothing new beyond a trivial interchange of x and y. In the cases $(-1,-1)$ and $(-3/4,-1/2)$ the integral is respectively a logarithm and an inverse lemniscate function (Carlson, 1971). In the lemniscate case, complex values of $T^n\zeta$ resulting from the choice of negative square roots have been studied by Todd (1988).

Here we shall discuss only the case $(-1/2,-1)$,

$$\int_0^\infty (t+x^2)^{-1/2}(t+y^2)^{-1}\,dt = \int_0^\infty (s+\alpha^2)^{-1/2}(s+\alpha y)^{-1}\,ds, \tag{3.2}$$

which we shall call the Schwab-Borchardt transformation. In the notation of (1.1) this takes the form

$$\int_0^\infty (t+\zeta_1)^{-1/2}(t+\zeta_2)^{-1}dt = \int_0^\infty (t+\eta_1)^{-1/2}(t+\eta_2)^{-1}dt, \tag{3.3}$$

$$\eta_i = (\zeta_i+\lambda)/4, \quad i = 1,2; \quad \lambda = 2(\zeta_1\zeta_2)^{1/2} + \zeta_2. \tag{3.4}$$

In Section VI we shall recover this value of λ by putting $\zeta_3 = \zeta_2$ in $(\zeta_1\zeta_2)^{1/2}+(\zeta_1\zeta_3)^{1/2}+(\zeta_2\zeta_3)^{1/2}$. Because $\eta_1-\eta_2 = (\zeta_1-\zeta_2)/4$, it is easy to see that $T^n\zeta$ converges linearly as $n \to \infty$ to a limit (L,L). Since the integral (3.3) is continuous in ζ, its value is

$$I(L,L) = \int_0^\infty (t+L)^{-3/2}dt = 2L^{-1/2}.$$

Any inverse circular function can be expressed in terms of the integral (3.3) with $\zeta_1 < \zeta_2$, and any logarithmic or inverse hyperbolic function can be similarly expressed with $\zeta_1 > \zeta_2$. For some purposes, especially in connection with elliptic integrals and integral tables, it is very useful to unite these elementary functions in the single function

$$R_C(\zeta_1,\zeta_2) = \frac{1}{2}\int_0^\infty (t+\zeta_1)^{-1/2}(t+\zeta_2)^{-1}dt, \tag{3.5}$$

which is homogeneous in ζ of degree $-1/2$ and has the value $L^{-1/2}$ when $\zeta_1 = \zeta_2 = L$. For example,

$$\begin{aligned}
\arctan x &= xR_C(1,1+x^2), \quad -\infty < x < \infty,\\
\ln x &= (x-1)R_C((1+x)^2/4,x), \quad x > 0,\\
\arccos x &= (1-x^2)^{1/2}\, R_C(x^2,1), \quad 0 \le x \le 1,\\
\operatorname{arccosh} x &= (x^2-1)^{1/2}\, R_C(x^2,1), \quad x \ge 1.
\end{aligned} \tag{3.6}$$

In each case the function multiplying R_C shows the asymptotic behavior as the left side tends to zero. A function equivalent to the reciprocal of R_C was defined by Fubini (1897).

Iteration of (3.4) to compute L and thus R_C from ζ_1 and ζ_2 is a unified form of the algorithm of Schwab (1813) (rediscovered by Borchardt in 1881) for computing inverse circular and inverse hyperbolic functions by iteration of their duplication theorems. Indeed, the homogeneity of R_C recasts the invariance (3.3) as a duplication theorem:

$$R_C(\zeta_1,\zeta_2) = R_C((\zeta_1+\lambda)/4,(\zeta_2+\lambda)/4) = 2R_C(\zeta_1+\lambda,\zeta_2+\lambda). \tag{3.7}$$

To relate the duplication theorems, we note that the equation

$$\arccos(\zeta_1/\zeta_2)^{1/2} = 2\arccos[(\zeta_1+\lambda)/(\zeta_2+\lambda)]^{1/2}$$

is equivalent to $\cos 2\theta = 2\cos^2\theta - 1$ and also, by (3.6), to (3.7). The same remark still holds when cos is replaced by cosh.

Although the convergence of $T^n\zeta$ as $n \to \infty$ is only linear, the computation can be speeded up by a Taylor expansion of $R_C(\zeta_1,\zeta_2)$ about the point (μ,μ), $\mu = (\zeta_1+2\zeta_2)/3$, to fifth order in $\zeta_1-\zeta_2$. The error is then of order $4^{-6n} = (4096)^{-n}$ instead of 4^{-n}. A Fortran code for computing R_C, including its Cauchy principal value when $\zeta_2 < 0$, can be found in (Carlson, 1988, Supplement).

Interesting discussions of the Schwab–Borchardt algorithm are given by Schoenberg (1982), Miel (1983), and Wimp (1984). Choice of negative square roots has been studied by Todd (1978).

IV. THE ARITHMETIC-GEOMETRIC MEAN

If $k = 3$ the integrand on the right side of (2.3) contains two linear factors raised to the power $p_3 \neq 0$. For invariance p_3 must therefore equal p_1 (or p_2 by a trivial exchange of x and y). Choosing $\zeta_3 = 0$ (see Section V for $\zeta_3 \neq 0$), we find

$$\int_0^\infty (t+x^2)^{p_1}(t+y^2)^{p_2}t^{p_1}dt$$
$$= \int_0^\infty (s+\alpha^2)^{-2-2p_1-p_2}(s+\alpha x)^{1+2p_1}(s+\alpha y)^{1+2p_2}(s+\gamma^2)^{p_1}s^{p_1}ds.$$

Only one of the first three factors on the right side may have a nonvanishing exponent, and there are three invariances with $(p_1,p_2) = (-1/2,-1/2)$, $(-3/4,-1/2)$, and $(-1/2,-1)$. The last two are not essentially different from invariances in Section III, for the substitutions $t = \gamma^4/u$ and $s = \alpha^2\gamma^2/v$ transform the cases $(-1/2,-1)$ and $(-3/4,-1/2)$ into the Schwab-Borchardt invariance (3.2) and the lemniscate invariance, respectively, except for interchange of x and y.

However, the case $(p_1,p_2) = (-1/2,-1/2)$ is the Gauss invariance,

$$\int_0^\infty [(t+x^2)(t+y^2)t]^{-1/2}dt = \int_0^\infty [(s+\alpha^2)(s+\gamma^2)s]^{-1/2}ds. \tag{4.1}$$

Since both limits of integration are branch points of the integrand, the integral is the complete elliptic integral of the first kind. The sequence $T^n\zeta$ converges quadratically to (M^2,M^2), where M is Gauss's arithmetic-geometric mean of x and y. Hence the value of the integral is

$$\int_0^\infty (t+M^2)^{-1}t^{-1/2}dt = \pi/M.$$

Extensive discussions of the arithmetic-geometric mean can be found in Cox (1984,1985) and Borweins (1987).

The three invariances listed in Section III and the Gauss invariance (4.1) are four of the twelve in Table 1 of (Carlson,

1971). The remaining eight differ from these only by exchange of x and y or by the substitutions specified earlier in this Section.

V. LANDEN'S TRANSFORMATION

If $k = 3$ and $\zeta_3 = z^2$, the discussion of exponents at the beginning of Section IV is unchanged, but the quadratic polynomial with exponent $p_3 = p_1$ is no longer $(s+\gamma^2)s$ unless $z = 0$:

$$\int_0^\infty (t+x^2)^{p_1}(t+y^2)^{p_2}(t+z^2)^{p_1}dt =$$

$$\int_0^\infty (s+\alpha^2)^{-2-2p_1-p_2}(s+\alpha x)^{1+2p_1}(s+\alpha y)^{1+2p_2}(s+z_+^2)^{p_1}(s+z_-^2)^{p_1}ds,$$

$$2z_\pm = [(x+z)(y+z)]^{1/2} \pm [(x-z)(y-z)]^{1/2}.$$

Just as in Section IV, there are three invariances with $(p_1,p_2) = (-1/2,-1/2)$, $(-3/4,-1/2)$, and $(-1/2,-1)$, and the last two are more complicated versions of invariances in Section III. The case $(-1/2,-1)$ is reduced to the Schwab-Borchardt invariance by a relation that will be proved later in (8.1):

$$\int_0^\infty (t+x^2)^{-1/2}(t+y^2)^{-1}(t+z^2)^{-1/2}dt = 2R_C((xz+y^2)^2,(xy+yz)^2). \tag{5.1}$$

The case $(-3/4,-1/2)$ is reduced to the lemniscate invariance by the relation

$$\int_0^\infty [(t+x^2)(t+y^2)]^{-3/4}(t+z^2)^{-1/2}dt$$

$$= (x+y)^{-1/2} \int_0^\infty (t+u^2)^{-3/4}(t+v^2)^{-1/2}dt, \qquad (5.2)$$

$$u = (x+y)[(xy)^{1/2}+z]^2, \qquad v = (x+y)^2z + 2(xy)^{1/2}(xy+z^2).$$

This formula can be proved by successive use of (Carlson, 1976, (4.12)) and (Carlson, 1977, (9.5-1), (5.9-21), and (6.9-7)).

The case (-1/2,-1/2) is economically expressed in terms of the function

$$R_F(\zeta_1,\zeta_2,\zeta_3) = \frac{1}{2}\int_0^\infty \prod_{i=1}^{3} (t+\zeta_i)^{-1/2}dt, \qquad (5.3)$$

which is homogeneous of degree -1/2 and symmetric in its variables. It has the properties

$$R_F(\zeta_1,\zeta_2,\zeta_2) = R_C(\zeta_1,\zeta_2), \qquad R_F(L,L,L) = L^{-1/2}. \qquad (5.4)$$

The invariance is

$$R_F(x^2,y^2,z^2) = R_F(\alpha^2,z_+^2,z_-^2),$$
$$\alpha = (x+y)/2, \quad 2z_\pm = [(x+z)(y+z)]^{1/2} \pm [(x-z)(y-z)]^{1/2}, \qquad (5.5)$$

which we shall call the Landen invariance because it contains Landen's transformation of the first incomplete elliptic integral. The Gauss invariance (4.1) is the case $z = 0$:

$$R_F(x^2,y^2,0) = R_F(\alpha^2,\gamma^2,0). \qquad (5.6)$$

If $z = y$ the Landen invariance reduces by (5.4) to the Schwab–Borchardt invariance (3.2).

Because the left side of (5.5) is symmetric in x,y,z while

the right side is symmetric only in x and y, the Landen invariance generates three distinct transformations. For example, if x,y,z is a permutation of 1,2,3, the three possible choices of z correspond to the transformations

(x,y,z)	(α,z_+,z_-)
(1,2,3)	(1.500,2.943,1.529)
(2,3,1)	(2.500,2.439,1.025)
(1,3,2)	(2.000,1.936+0.500i,1.936−0.500i).

(When z_+ and z_- are real, we choose the square roots so that $z_+ \geq z_- \geq 0$.) In the first transformation, where x and y are smaller than z, the two smaller variables after transformation (α and z_-) have values that are not far apart (1.500 and 1.529). This means that the modulus (usually denoted by k) of the elliptic integral has increased, for the square of the modulus is the difference between the two larger arguments of R_F divided by the difference between the largest and smallest arguments. Because the modulus has increased, the transformation is called the ascending Landen (AL) transformation, and the inverse transformation from (α,z_+,z_-) to (x,y,z) is the descending Landen (DL) transformation. In the second case, where x and y are larger than z, the two larger variables after transformation (α and z_+) are close together (2.500 and 2.439), and the modulus has decreased. This transformation is called the descending Gauss (DG) transformation, and its inverse is the ascending Gauss (AG) transformation. The third case leads to complex conjugate values of z_+ amd z_- and is less useful than its inverse (see the end of this Section).

The geometric mean that occurs in the Gauss invariance (5.6) is not conspicuous in the Landen invariance (5.5), but it is still there under the surface, not only in the direct but also in the inverse transformation,

$$y+x = 2\alpha, \qquad y-x = (2/\alpha)[(z_+^2-\alpha^2)(z_-^2-\alpha^2)]^{1/2},$$
$$z = z_+z_-/\alpha. \tag{5.7}$$

We shall choose the square roots so that $x \leq y$. To show the formation of arithmetic and geometric means, we define a and b $(a \geq b \geq 0)$ for each transformation as shown below. The third and fourth columns contain the squares of the arithmetic and geometric means of a and b, and the last column gives the difference between those squares:

	a^2	b^2	$(a+b)^2/4$	ab	$(a-b)^2/4$
DG	y^2-z^2	x^2-z^2	$\alpha^2-z_-^2$	$z_+^2-z_-^2$	$\alpha^2-z_+^2$
AL	z^2-x^2	z^2-y^2	$z_+^2-\alpha^2$	$z_+^2-z_-^2$	$z_-^2-\alpha^2$
AG	$\alpha^2-z_-^2$	$\alpha^2-z_+^2$	$(y^2-z^2)/4$	$(y^2-x^2)/4$	$(x^2-z^2)/4$
DL	$z_+^2-\alpha^2$	$z_-^2-\alpha^2$	$(z^2-x^2)/4$	$(y^2-x^2)/4$	$(z^2-y^2)/4$

For example, consider the descending Gauss transformation, which reduces to (5.6) when $z = 0$. If y^2-x^2 is a small quantity, say ε, then $a^2-b^2 = \varepsilon$. If y^2-z^2 is not small, then a is not small, a+b is not small, and a-b must be of order ε. Then the last column shows that $\alpha^2-z_+^2$ is of order ε^2. If the transformation is iterated, with z taken each time to be the smallest initial variable, the two larger final variables (α and z_+) converge quadratically to a common limit L, and the integral becomes the inverse circular function

$$R_F(L^2,L^2,L^2-M^2) = R_C(L^2-M^2,L^2),$$

where M is the arithmetic-geometric mean of a and b. Similarly, if the ascending Landen transformation is iterated, with z taken each time to be the largest initial variable, the two smaller final variables (α and z_-) converge quadratically to a common limit S, and the integral becomes the inverse hyperbolic function

$$R_F(S^2,S^2+M^2,S^2) = R_C(S^2+M^2,S^2).$$

Formal algorithms for iterating DG or AL transformations can be found in (Carlson, 1977, Sec. 9.5); the former or the latter is faster according as the largest or smallest variable of R_F is closer to the middle variable. It is also feasible, though a little less straightforward, to iterate AG or DL transformations.

When an integrand contains the square root of a cubic or quartic polynomial with conjugate complex roots, evaluation of the integral may lead to R_F with one positive and two conjugate complex variables. Taking the complex variables to be z_+^2 and z_-^2 in the inverse transformation (5.7) leads to R_F with real positive variables. However, the same end is achieved by taking the complex variables to be x^2 and y^2 in the direct transformation (5.5), with the advantage that the resulting real variables are closer together in value than they were in the inverse transformation.

VI. THE DUPLICATION THEOREM

Despite their quadratic convergence, Landen transformations have been superseded for many computational purposes by the duplication theorem, which is simpler to use. If an ascending Landen transformation either precedes or follows a descending Gauss transformation, the modulus of the elliptic integral R_F returns to its initial value and we obtain the duplication theorem. One may say loosely that duplication is a square of Landen's transformation. To verify this, we write (5.5) a second time with (x,y,z) replaced by (X,Y,Z) and (α,z_+,z_-) replaced by (A,Z_+,Z_-). If the two transformations are connected by the identification $(z_-,z_+,\alpha) = (X,Y,Z)$, then an AL transformation will automatically be followed by a DG transformation, DG by AL, and a transformation to complex arguments by a transformation back to real arguments. With the help of the relations (implied by (5.5))

$$z_+z_- = \alpha z, \qquad z_+^2+z_-^2 = xy+z^2, \qquad (z_+^2-\alpha^2)(z_-^2-\alpha^2) = (y^2-x^2)^2/16,$$

the result of the two transformations in sequence is found to be

$$4A^2 = (z+x)(z+y), \quad 4Z_+^2 = (y+x)(y+z), \quad 4Z_-^2 = (x+y)(x+z).$$

Using the symmetry and homogeneity of R_F, we get the duplication theorem,

$$R_F(x^2,y^2,z^2) = 2R_F((x+y)(x+z),(y+x)(y+z),(z+x)(z+y)). \qquad (6.1)$$

An equivalent form, showing clearly that the modulus is the same on both sides, is

$$\begin{aligned} R_F(\zeta_1,\zeta_2,\zeta_3) &= R_F((\zeta_1+\lambda)/4,(\zeta_2+\lambda)/4,(\zeta_3+\lambda)/4) \\ &= 2R_F(\zeta_1+\lambda,\zeta_2+\lambda,\zeta_3+\lambda), \end{aligned} \qquad (6.2)$$

$$\lambda = (\zeta_1\zeta_2)^{1/2}+(\zeta_1\zeta_3)^{1/2}+(\zeta_2\zeta_3)^{1/2}.$$

Although the first equality is an invariance of the form (1.1), it does not follow directly from the substitution (2.1) but rather from a substitution derived by Carlson (1977, (9.6-3)). The form of the invariance had been found earlier by W. H. Greiman using successive Landen transformations.

Since $(T\zeta)_i-(T\zeta)_j = (\zeta_i-\zeta_j)/4$, it is easy to see that $T^n\zeta$ converges linearly as $n \to \infty$ to a limit (L,L,L), and the value of $R_F(\zeta_1,\zeta_2,\zeta_3)$ is $L^{-1/2}$ by (5.4).

If $\zeta_2 = \zeta_3$ the duplication theorem for R_F reduces by (5.4) to the duplication theorem (3.7) for the elementary function R_C. Just as for R_C, the computation can be speeded up by a Taylor expansion of $R_F(\zeta_1,\zeta_2,\zeta_3)$ about the point (μ,μ,μ), $\mu = (\zeta_1+\zeta_2+\zeta_3)/3$, to fifth order in the variables $\zeta_i-\mu$, $1 \le i \le 3$. By the symmetry of R_F the expansion can be expressed in terms of the three elementary symmetric functions of these variables, and the vanishing of the symmetric function of degree one greatly simplifies the expansion. The error is then of order $4^{-6n} = (4096)^{-n}$ instead of 4^{-n}. A Fortran code for computing R_F can be found in (Carlson, 1987, Supplement) and in several major software libraries.

VII. A QUARTIC ELLIPTIC INTEGRAL OF THE FIRST KIND

If k = 4 we can write (2.3) in the form

$$\int_0^\infty (t+x^2)^{p_1}(t+y^2)^{p_2}(t+z^2)^{p_3}(t+w^2)^{p_4}dt$$
$$= \int_0^\infty (s+\alpha^2)^{-2-\Sigma p_i}(s+\alpha x)^{1+2p_1}(s+\alpha y)^{1+2p_2}(s+z_+^2)^{p_3}(s+z_-^2)^{p_3} \cdot (s+w_+^2)^{p_4}(s+w_-^2)^{p_4}ds, \tag{7.1}$$

where α and $z_\pm$ are the same as in (5.5) and $w_\pm$ differs from $z_\pm$ only by substitution of w for z:

$$2w_\pm = [(x+w)(y+w)]^{1/2} \pm [(x-w)(y-w)]^{1/2}. \tag{7.2}$$

If $p_3p_4 \neq 0$ invariance requires $p_1 = p_2 = p_3 = p_4 = -1/2$, and the resulting formula is

$$\int_0^\infty [(t+x^2)(t+y^2)(t+z^2)(t+w^2)]^{-1/2}dt$$
$$= \int_0^\infty [s+z_+^2)(s+z_-^2)(s+w_+^2)(s+w_-^2)]^{-1/2}ds. \tag{7.3}$$

The invariant integral is an elliptic integral of the first kind with integrand containing the square root of a quartic polynomial.

By symmetry we may assume that the values of x and y are not separated by either z or w. According to Section V, if the values of x and y are close together, then z_+ or z_- is quadratically close to α according as x and y are larger or smaller than z. Also, w_+ or w_- is quadratically close to α according as x and y are larger or smaller than w. Thus two of the variables z_+, z_-, w_+, w_- are quadratically close to α and so to each other. If the transformation is iterated, with x and y taken each time to be the variables that are nearly equal, the convergence is quadratic and the integral approaches an elementary integral having the form of the left side of (5.1).

Another way of evaluating the left side of (7.3) is to express it in terms of R_F by the reduction theorem (Carlson, 1977, Sec. 9.8), of which (5.1) is a special case, and then use the code for R_F mentioned at the end of Section VI. Although (7.3) appears to be new, the reduction theorem shows that it is a more complicated version of (5.5).

VIII. AN INTEGRAL OF THE THIRD KIND

Because (2.3) yields no invariances if $p_1 \cdots p_5 \neq 0$, it seems unlikely that hyperelliptic integrals have Landen transformations. However, an elliptic integral of the third kind, which may be regarded as a degenerate hyperelliptic integral with two equal variables, has a Landen transformation that can be derived by choosing $p_4 = -1$ in (7.1) and decomposing into partial fractions. As a simple illustration of the procedure, we first prove (5.1).

Choosing $(p_1,p_2,p_3,p_4) = (-1/2,-1/2,-1,0)$ in (7.1), we find

$$\begin{aligned}\int_0^\infty [(t+x^2)(t+y^2)]^{-1/2}(t+z^2)^{-1}dt &= \int_0^\infty [(s+z_+^2)(s+z_-^2)]^{-1}ds \\ &= (z_+^2-z_-^2)^{-1}\ln(z_+^2/z_-^2) \\ &= 2R_C((z_+^2+z_-^2)^2, 4z_+^2z_-^2) \\ &= 2R_C((xy+z^2)^2,(xz+yz)^2),\end{aligned} \tag{8.1}$$

where we have used partial fractions in the second equality and (3.6) in the third. Interchange of y and z yields (5.1).

Next we apply a similar procedure to an elliptic integral of the third kind,

$$R_J(\zeta_1,\zeta_2,\zeta_3,\zeta_4) = \frac{3}{2}\int_0^\infty [(t+\zeta_1)(t+\zeta_2)(t+\zeta_3)]^{-1/2}(t+\zeta_4)^{-1}dt. \quad (8.2)$$

Choosing $(p_1,p_2,p_3,p_4) = (-1/2,-1/2,-1/2,-1)$ in (7.1), we see that

$$R_J(x^2,y^2,z^2,w^2) = \int_0^\infty [(s+\alpha^2)(s+z_+^2)(s+z_-^2)]^{-1/2} \cdot[(s+\alpha^2)/(s+w_+^2)(s+w_-^2)]ds$$

Partial-fraction decomposition of the quantity in the last square brackets leads to Landen's transformation (Zill, 1970,(5.6)) in the form

$$(w_+^2-w_-^2)R_J(x^2,y^2,z^2,w^2) = (w_+^2-\alpha^2)R_J(\alpha^2,z_+^2,z_-^2,w_+^2) + (\alpha^2-w_-^2)R_J(\alpha^2,z_+^2,z_-^2,w_-^2). \quad (8.3)$$

This is not convenient for iteration because the number of R_J functions doubles in each step. However, the difference of the two terms on the right-hand side involves only the simpler functions R_F and R_C. To show this we put $(p_1,p_2,p_3,p_4) = (0,0,-1/2,-1)$ in (7.1):

$$\int_0^\infty (t+z^2)^{-1/2}(t+w^2)^{-1}dt = \int_0^\infty [(s+\alpha^2)(s+z_+^2)(s+z_-^2)]^{-1/2}f(s)ds,$$

$$f(s) = \frac{(s+\alpha x)(s+\alpha y)}{(s+w_+^2)(s+w_-^2)} = 1 + \frac{C}{s+w_+^2} + \frac{D}{s+w_-^2} .$$

In simplifying the coefficients C and D, it is helpful to note that $w_+w_- = \alpha w$ and $w_+^2+w_-^2 = \gamma^2+w^2$, whence

$$(w_+^2-\alpha^2)(w_-^2-\alpha^2) = \alpha^2(\alpha^2-\gamma^2).$$

Thus we see that

$$(-w_+{}^2+w_-{}^2)C = (-w_+{}^2+\alpha x)(-w_+^2+\alpha y)$$

$$= w_+^4-2\alpha^2 w_+^2+\alpha^2\gamma^2$$

$$= (w_+^2-\alpha^2)^2 - \alpha^2(\alpha^2-\gamma^2)$$

$$= (w_+^2-\alpha^2)[(w_+^2-\alpha^2)-(w_-^2-\alpha^2)].$$

It follows that $C = \alpha^2-w_+^2$ and similarly $D = \alpha^2-w_-^2$. We have finally

$$(w_+{}^2-\alpha^2)R_J(\alpha^2,z_+^2,z_-^2,w_+^2) + (w_-^2-\alpha^2)R_J(\alpha^2,z_+^2,z_-^2,w_-^2)$$
$$= 3R_F(\alpha^2,z_+^2,z_-^2) - 3R_C(z^2,w^2). \tag{8.4}$$

This is called a change-of-parameter relation (Zill, 1970,(4.6)) for R_J.

Elimination of one R_J between (8.3) and (8.4) leads to a form of Landen's transformation more suitable for iteration:

$$(w_+^2-w_-^2)R_J(x^2,y^2,z^2,w^2) = 2(w_+^2-\alpha^2)R_J(\alpha^2,z_+^2,z_-^2,w_+^2)$$
$$- 3R_F(x^2,y^2,z^2) + 3R_C(z^2,w^2), \tag{8.5}$$

as well as the same equation with w_+ and w_- interchanged (Zill, 1970, (5.7)). On the right-hand side we have used (5.5).

A special case of R_J is an elliptic integral of the second kind,

$$R_D(\zeta_1,\zeta_2,\zeta_3) = R_J(\zeta_1,\zeta_2,\zeta_3,\zeta_3),$$

with Landen transformation obtained by putting $w = z$ in (8.3) or (8.5). The same procedure as in Section VI yields the duplication theorem,

$$R_D(x^2,y^2,z^2) = 2R_D(x^2+\lambda,y^2+\lambda,z^2+\lambda) + 3/z(z^2+\lambda)$$
$$\lambda = xy+xz+yz, \qquad (8.6)$$

although considerable algebra and a relation between $R_D(\rho,\sigma,\tau)$ and $R_D(\rho,\tau,\sigma)$ are needed along the way. One might expect that the duplication theorem for R_J could be similarly derived from its Landen transformation, but attempts to do this have failed. It is instead obtained as a special case of the addition theorem for R_J (Zill, 1970, (8.11)) and has the form

$$R_J(x^2,y^2,z^2,w^2) = 2R_J(x^2+\lambda,y^2+\lambda,z^2+\lambda,w^2+\lambda) + 3R_C(a^2,b^2),$$
$$\lambda = xy+xz+yz, \qquad a = w^2(x+y+z)+xyz, \qquad b = w(w^2+\lambda). \qquad (8.7)$$

Fortran codes for computing R_D and R_J, including the Cauchy principal value when the last argument of R_J is negative, are given by Carlson (1987, Supplement) and (1988, Supplement). These codes use the duplication theorems in the same way as the codes for R_C and R_F mentioned in Sections III and VI.

IX. A QUARTIC ELLIPTIC INTEGRAL OF THE THIRD KIND

The Landen transformation and change of parameter for R_J shown in (8.3), (8.4), and (8.5) have analogues for an elliptic integral of the third kind with integrand containing the square root of a quartic polynomial,

$$G(\zeta_1,\ldots,\zeta_5) = \int_0^\infty \left[\prod_{i=1}^{4} (t+\zeta_i)\right]^{-1/2} (t+\zeta_5)^{-1}dt. \qquad (9.1)$$

Choosing $p_i = -1/2$, $1 \le i \le 4$, and $p_5 = -1$ in (2.3) and making a partial-fraction decomposition just like the one that led to (8.3), we find the Landen transformation of G:

$$(q_+^2-q_-^2)G(x^2,y^2,z^2,w^2,q^2) = (q_+^2-\alpha^2)G(z_+^2,z_-^2,w_+^2,w_-^2,q_+^2) \\ + (\alpha^2-q_-^2)G(z_+^2,z_-^2,w_+^2,w_-^2,q_-^2), \tag{9.2}$$

where α, $z_\pm$, and $w_\pm$ are defined in (5.5) and (7.2), and $q_\pm$ differs from $z_\pm$ only by substitution of q for z:

$$2q_\pm = [(x+q)(y+q)]^{1/2} \pm [(x-q)(y-q)]^{1/2}. \tag{9.3}$$

Just as in (8.3), the difference of the two terms on the right-hand side of (9.2) involves only simpler functions. To show this we put $(p_1,\ldots,p_5) = (0,0,-1/2,-1/2,-1)$ in (2.3) and make a partial-fraction decomposition just like the one that led to (8.4). The resulting change-of-parameter relation for G is

$$(q_+^2-\alpha^2)G(z_+^2,z_-^2,w_+^2,w_-^2,q_+^2) + (q_-^2-\alpha^2)G(z_+^2,z_-^2,w_+^2,w_-^2,q_-^2) \\ = \int_0^\infty [(s+z_+^2)(s+z_-^2)(s+w_+^2)(s+w_-^2)]^{-1/2}ds \\ - \int_0^\infty [(t+z^2)(t+w^2)]^{-1/2}(t+\zeta^2)^{-1}dt. \tag{9.4}$$

The first term on the right-hand side is the invariant integral in (7.3), and the second term occurs on the left-hand side of (8.1). One of the G's can be eliminated between (9.2) and (9.4) to get the Landen transformation in a form suitable for iteration.

Another way of evaluating G is to express it in terms of R_J, R_F, and R_C by using (Carlson, 1988, (2.35)):

$$(x^2-q^2)G(x^2,y^2,z^2,w^2,q^2)/2 \\ = (x^2-y^2)(x^2-z^2)(x^2-w^2)R_J(X^2,Y^2,Z^2,W^2)/3(x^2-q^2) \\ - R_F(X^2,Y^2,Z^2) + R_C(P^2,Q^2), \tag{9.5}$$

where

$$X = xy+zw, \quad Y = xz+yw, \quad Z = xw+yz,$$

$$W^2 = X^2 - (x^2-z^2)(x^2-w^2)(y^2-q^2)/(x^2-q^2), \tag{9.6}$$

$$P = q^2 + yzw/x, \quad Q = qW/x.$$

REFERENCES

1. Borwein, J. M., and Borwein, P. B. (1987): Pi and the AGM, Wiley, New York.
2. Carlson, B. C. (1971): Algorithms involving arithmetic and geometric means, Amer. Math. Monthly, 78, pp. 496-505.
3. Carlson, B. C. (1976): Quadratic transformations of Appell functions, SIAM J. Math. Anal., 7, pp. 291-304.
4. Carlson, B. C. (1977): Special Functions of Applied Mathematics, Academic Press, New York.
5. Carlson, B. C. (1987): A table of elliptic integrals of the second kind, Math. Comp., 49, pp. 595-606 and S13-S17.
6. Carlson, B. C. (1988): A table of elliptic integrals of the third kind, Math. Comp., 51, pp. 267-280 and S1-S5.
7. Cox, D. A. (1984): The arithmetic-geometric mean of Gauss, Enseign. Math., II Sér., 30, pp. 275-330.
8. Cox, D. A. (1985): Gauss and the arithmetic-geometric mean, Notices Amer. Math. Soc., 32, pp. 147-151.
9. Fubini, G. (1897): Nuovo metodo per lo studio e per il calculo delle funzioni trascendenti elementari, Period. Mat. 12, pp. 169-178.
10. Miel, G. (1983): Of calculations past and present: the Archimedean algorithm, Amer. Math. Monthly, 90, pp. 17-35.
11. Schoenberg, I. J. (1982): Mathematical Time Exposures, The Mathematical Association of America, Chap. 12.
12. Schwab, J. (1813): Elémens de géométrie, vol. 1, C.-J. Hissette, Nancy.
13. Todd, J. (1978): The many limits of mixed means, I, In: General Inequalities,I.(≡ ISNM 41) Beckenbach, E. F. (ed.), Birkhäuser, Basel.
14. Todd, J. (1988): The many limits of mixed means, II, Numer. Math., 54, pp. 1-18.
15. Wimp, J. (1984): Computation with Recurrence Relations, Pitman, Boston, Chap. 13.
16. Zill, D. G., and Carlson, B. C. (1970): Symmetric elliptic integrals of the third kind, Math. Comp., 24, pp. 199-214.

Accelerating the Convergence of Chebyshev Series

L. M. Ciasullo Professor, Department of Pure and Applied Mathematics, Washington State University, Pullman, Washington

James A. Cochran[1] Professor, Department of Pure and Applied Mathematics, Washington State University, Pullman, Washington

I. INTRODUCTION

The numerical approximation of mathematical functions in a manner which is simple, accurate and efficient from a computational point of view is of fundamental importance for solving problems in applied mathematics, engineering and the physical sciences. For example, describing the vibrations of a circular membrane requires the evaluation of Bessel functions, while knowledge of the values of Airy functions is important in optics and light scattering investigations. Extensive mathematical tables for the values of such functions exist and have proved an invaluable aid to computation. For function evaluation on a computer, however, the storage space and search

[1] *This investigation was supported in part by funds provided by Washington State University*

time required in using such tables is often prohibitive. Thus, tables are usually rejected in favor of other approaches.

Since polynomials are quickly and easily evaluated, a popular class of procedures utilizes polynomial representations for function evaluation purposes. Of course, in these methods we need to choose polynomials which approximate the functions under consideration as economically as possible. A frequent choice in this regard is to represent the given function f by a truncation of the Chebyshev series

$$f(x) = \frac{1}{2}a_0 + a_1 T_1(x) + a_2 T_2(x) + \cdots \equiv \sum_{m=0}^{\infty}{}' a_m T_m(x) \tag{1.1}$$

where the $T_m(x)$ are the well-known Chebyshev polynomials

$$T_0(x) = 1, \quad T_1(x) = x, \quad T_2(x) = 2x^2 - 1, \quad T_3(x) = 4x^3 - 3x, \cdots.$$

The importance of such expansions devolves from the fact that, for the fundamental interval [-1,1], they generally differ very little from the best uniform (minimax) polynomial approximation to f of given degree (see, for example, Davis, 1963 and Rivlin, 1974.) Moreover, for smooth f and large enough N, the maximum absolute error in truncating the series (1.1) at the $N+1$-st term is approximately $|a_{N+1}|$, so the accuracy of such a Chebyshev approximation may be ascertained at a glance.

Clearly then the rate at which the Chebyshev coefficients of the given function tend to zero is an important indicator of just how rapidly its Chebyshev series approximation (1.1) is converging and hence of the overall accuracy to be expected with a specific truncation. And any easily applicable technique for improving or accelerating this convergence is obviously of considerable interest. Not unexpectedly, Scraton (1970) found that the application of a *linear fractional* transformation to the argument of $f(x)$ before expanding it in a Chebyshev series can often be used to dramatically improve the state of affairs. One of his examples is the following:

EXAMPLE 1.1 Let us approximate $f(x) = e^{\frac{1}{x}}$ for $1 \leq x \leq 5$ via Chebyshev expansion. Notice that $f(x)$ has an essential singularity at $x = 0$. Traditionally, one begins by transforming the interval [1,5] to the standard interval [-1,1] by setting $t = \frac{1}{2}(x-3)$. This has the effect of sending

$$\begin{aligned} x = 1 &\longmapsto t = -1 \\ x = 5 &\longmapsto t = 1 \\ x = 0 &\longmapsto t = -\frac{3}{2}. \end{aligned}$$

The resulting Chebyshev series is

$$e^{\frac{1}{x}} = 1.622244 - 0.588640\,T_1(t) + 0.274772\,T_2(t) + \cdots + 0.000017\,T_{14}(t) + \cdots.$$

Fifteen terms of this series are necessary to approximate f with an absolute error less than $\varepsilon = 1 \times 10^{-5}$. Observe that the singularity $x = 0$ is "close" to the original interval [1,5], as is the transformed singularity, $t = -\frac{3}{2}$, to the new interval [-1,1].

If instead the linear fractional transformation $y = \frac{3x-5}{2x}$ is used, we are led to the Chebyshev representation

$$e^{\frac{1}{x}} = 1.895736 - 0.743522\,T_1(y) + 0.073861\,T_2(y) + \cdots + 0.000010\,T_5(y) + \cdots.$$

Here, only six terms are necessary to achieve the same accuracy. Note that this new transformation moved the singularity $x = 0$ to the point at infinity and hence as far away as possible from [-1,1].

The efficacy of Scraton's approach is based upon the knowledge that the behavior of the given function f *outside* as well as inside the fundamental interval of representation affects the rate of decay of the Chebyshev coefficients a_m in (1.1) (see Gottlieb and Orszag, 1977, Miller, 1966, and Rivlin, 1974). The Chebyshev coefficients of a function with singularities "near" the interval decay more slowly than those of a function whose singularities are "far away" from the interval of representation. Moving the singularities as far away as possible *before* initiating the expansion process, therefore, minimizes any potential negative effects they may have on the convergence of the series.

If the function of interest has no finite singularities (i.e., is *entire*), the only possible singularity is already at infinity. Thus, accelerating the convergence of Chebyshev expansions for these functions, that is, making a change of variable which causes the coefficients in the series to go to zero faster than they would ordinarily, requires a new approach. In this paper we investigate the effect of the a priori application of *non-linear* fractional changes of scale in such cases. Indeed, we assume that the interval of representation is the standard interval [-1,1] and examine how convergence is affected by the change of variable $z = x^\alpha$, where $\alpha > 0$. Our goal is to determine α so that the Chebyshev series of the given entire function converges as rapidly as possible. For specificity we concentrate on accelerating the Chebyshev series convergence of those functions which are solutions of the differential equation

$$f''(x) - x^n f(x) = 0, \qquad n = 0, 1, 2, \cdots. \tag{1.2}$$

When $n = 0$, these are the familiar exponential functions; when $n = 1$, the solutions are the Airy functions mentioned earlier; when $n = 2$, the solutions can be easily related to Weber or parabolic cylinder functions, etc.

In the next section we show how to obtain asymptotic expansions for the solutions of (1.2) as $x \to \infty$. These expansions allow us to get an asymptotic representation for the Chebyshev coefficients a_m of each f, when m is large. When the non-linear change of scale $x = z^{\frac{1}{\alpha}}$ is incorporated, this representation for a_m leads to the determination of an "optimal" value of α for convergence purposes, and we discuss these considerations in sections III and IV.

In an undertaking of this character numerical examples play an important role in both motivating and then subsequently validating the theoretical results which are obtained. In the final sections of this paper we explore the numerical aspect of our work, illustrating with an array of examples the utility of this new approach to accelerating the convergence of Chebyshev series.

II. ASYMPTOTIC EXPANSIONS FOR ENTIRE FUNCTIONS

Every finite point is an ordinary point of the differential equation (1.2). The point at infinity, however, is an irregular singularity of the finite rank $n+1$ (Ince, 1956). Thomé (1883) first discovered that for a class of differential equations with an irregular singularity of finite rank at $x = \infty$, there exist *formal* solutions of the form

$$e^{r(x)} x^{-\rho} \sum_{j=0}^{\infty} c_j x^{-j}, \; c_0 \neq 0 \tag{2.1}$$

where $r\,(x)$ is a polynomial and ρ is a constant. A classic procedure of Wasow (1965) can be employed to show that in the case of (1.2), these so-called *normal solutions* are in fact asymptotic expansions for the entire functions defined by the original differential equation.

In order to apply the Wasow approach we set $y_1 \equiv f$, $y_2 \equiv f'$ and convert (1.2) into the equivalent first order system

$$\mathbf{y}' = \begin{bmatrix} 0 & 1 \\ x^n & 0 \end{bmatrix} \mathbf{y}. \tag{2.2}$$

If we define

$$Y\,(x) \equiv \begin{bmatrix} \mathbf{y}_1 & \mathbf{y}_2 \end{bmatrix}$$

to be the 2×2 matrix whose columns are solutions of this new *vector* differential equation, then $Y(x)$ is easily seen to satisfy the *matrix* differential equation

$$x^{-n}\frac{dY}{dx} = \begin{bmatrix} 0 & x^{-n} \\ 1 & 0 \end{bmatrix} Y \equiv A(x)\,Y \tag{2.3}$$

where

$$A(x) = \begin{bmatrix} 0 & 0 \\ 1 & 0 \end{bmatrix} + \frac{1}{x^n}\begin{bmatrix} 0 & 1 \\ 0 & 0 \end{bmatrix}.$$

Note that Y is a 2×2 matrix satisfying (2.3) only if each element in the first row of Y is a solution of (1.2). It will be a *fundamental matrix solution* only if these two row elements are linearly independent.

The applicable result of Wasow (1965) is the following:

THEOREM 2.1 Consider the matrix differential equation

$$x^{-q}\frac{dY}{dx} = A(x)Y \tag{2.4}$$

where $A(x)$ is analytic at $x = \infty$ and $q \geq 0$ is integral. Let S be an open sector of the complex x−plane with vertex at the origin and a positive central angle no larger than $\frac{\pi}{q+1}$ (the vertex is not considered a point of the sector). Assume that $A(x)$ is a 2×2 matrix function analytic in S for $|x| \geq x_0$ for some x_0, and having in S a uniform asymptotic expansion

$$A(x) = \sum_{j=0}^{\infty} A_j x^{-j}, \quad x \to \infty, \quad x \in S.$$

If the eigenvalues λ_1, λ_2 of A_0 are distinct, then the differential equation (2.4) has a fundamental matrix solution of the form

$$Y(x) = \hat{Y}(x)\, x^D e^{Q(x)}.$$

Here, $Q(x)$ is a diagonal matrix whose entries are polynomials of degree $q+1$ and which has lead term

$$\frac{x^{q+1}}{q+1}\begin{bmatrix} \lambda_1 & 0 \\ 0 & \lambda_2 \end{bmatrix},$$

D is a constant, diagonal matrix and $\hat{Y}(x)$ has an asymptotic expansion

$$\hat{Y}(x) \sim \sum_{j=0}^{\infty} \hat{Y}_j x^{-j}, \qquad x \to \infty \text{ (in } S\text{)}, \qquad \det \hat{Y}_0 \neq 0.$$

COROLLARY 2.1 If the matrix $A(x)$ in the differential equation (2.4) is analytic in a full neighborhood of $x = \infty$, then a solution of the form described in Theorem 2.1 exists for *every* sector S of angular width $\frac{\pi}{q+1}$.

The existence theorems which lie behind Theorem 2.1 (see Erdelyi 1956, Wasow, 1965) assert that if $Y(x)$ formally satisfies the matrix differential equation (2.4), then there exists a fundamental matrix solution of the equation, call it $\tilde{Y}(x)$, such that

$$Y(x) \sim \tilde{Y}(x), \quad x \to \infty, \; x \in S.$$

Thus, if we can determine a fundamental matrix solution Y to equation (2.3), then the columns of Y will give the desired asymptotic expansions for the linearly independent solutions f_1 and f_2 of our equation of interest, (1.2), and their derivatives.

Unfortunately, we cannot directly apply Theorem 2.1 to equation (2.3) because the lead matrix in this case

$$A_0 = \begin{bmatrix} 0 & 0 \\ 1 & 0 \end{bmatrix}$$

does not have distinct eigenvalues. However, it *is* possible, through a series of transformations, to recast (2.3) so that Theorem 2.1 is applicable. The necessary steps are found in the following lemma:

LEMMA 2.2 The differential equation

$$x^{-n}\frac{dY}{dx} = \begin{bmatrix} 0 & x^{-n} \\ 1 & 0 \end{bmatrix} Y = A(x)\,Y$$

is equivalent to

$$t^{-(n+1)}\frac{dZ}{dt} = \begin{bmatrix} 0 & 1 \\ 1 & nt^{-(n+2)} \end{bmatrix} Z = C(t)\,Z \tag{2.5}$$

where $x = 2^{-\left(\frac{2}{n+2}\right)}t^2$.

PROOF: First, for simplification purposes, interchange the roles of $y_1 = f$ and $y_2 = f'$ by setting

$$Y = TW = \begin{bmatrix} 0 & 1 \\ 1 & 0 \end{bmatrix} W.$$

Substitution into (2.3) then gives

$$x^{-n} T \frac{dW}{dx} = A(x)\, TW$$

or

$$x^{-n} \frac{dW}{dx} = \begin{bmatrix} 0 & 1 \\ x^{-n} & 0 \end{bmatrix} W \equiv J(x)\, W. \tag{2.6}$$

Now consider the change of "dependent" variable

$$W(x) = \begin{bmatrix} 1 & 0 \\ 0 & x^{-\frac{n}{2}} \end{bmatrix} Z \equiv S(x)\, Z.$$

Replacing W by $S(x)\,Z$ and $\frac{dW}{dx}$ by $S'(x)\,Z + S(x)\,Z'$ in equation (2.6) yields, after appropriate manipulation,

$$x^{-\frac{n}{2}} \frac{dZ}{dx} = \begin{bmatrix} 0 & 1 \\ 1 & \frac{n}{2} x^{-(1+\frac{n}{2})} \end{bmatrix} Z = B(x)\, Z \tag{2.7}$$

where

$$B(x) = \begin{bmatrix} 0 & 1 \\ 1 & 0 \end{bmatrix} + x^{-(1+\frac{n}{2})} \begin{bmatrix} 0 & 0 \\ 0 & \frac{n}{2} \end{bmatrix}.$$

Observe that the lead matrix

$$B_0 = \begin{bmatrix} 0 & 1 \\ 1 & 0 \end{bmatrix}$$

of $B(x)$ now has *two* distinct eigenvalues, $\lambda = -1, 1$. In order to avoid working with fractional powers of the independent variable x, we set

$$x = 2^{-\left(\frac{2}{n+2}\right)} t^2.$$

Then

$$\frac{dZ}{dx} = \left[2^{-\left(\frac{n}{n+2}\right)} \frac{1}{t}\right] \frac{dZ}{dt}$$

and so (2.7) becomes

$$2^{\frac{n}{n+2}} t^{-n} \left[2^{-\left(\frac{n}{n+2}\right)} \frac{1}{t}\right] \frac{dZ}{dt} = \begin{bmatrix} 0 & 1 \\ 1 & n\, t^{-(n+2)} \end{bmatrix} Z$$

which, when simplified, is of the desired form (2.5). Since all of these steps may be reversed, the equivalence is proved. □

The transformed differential system (2.5) has a lead matrix C_0 with distinct eigenvalues. Moreover, $C(t)$ is analytic at $t = \infty$, and so now by Theorem 2.1, the system has a fundamental matrix solution of the form

$$Z(t) = \hat{Z}(t)\, t^D e^{Q(t)}$$

where $Q(t)$, D and $\hat{Z}(t)$ have the features described previously. It remains to compute these quantities explicitly. Changing the independent variable from t back to x will then yield a fundamental matrix solution for (2.3), from which we may extract asymptotic expansions for the solutions f_1 and f_2 of (1.2). The final results of such a series of calculations are contained in the following:

THEOREM 2.2 For integer $n \geq 0$, let S be a sector of the complex x-plane with vertex at the origin and a positive central angle no larger than $\frac{2\pi}{n+2}$. Then there exist two linearly independent solutions f_1 and f_2 of the differential equation

$$f'' - x^n f = 0$$

which, along with their derivatives, have the asymptotic expansions

$$\begin{aligned} f_1(x) &= x^{-\frac{n}{4}} \exp\left(\frac{2}{n+2} x^{\frac{n+2}{2}}\right) \left\{1 + \left[\frac{n(n+4)}{16(n+2)}\right] x^{-\frac{(n+2)}{2}}\right. \\ &+ \left.\left[\frac{n(n+4)(3n+4)(3n+8)}{512(n+2)^2}\right] x^{-(n+2)} + O\left(x^{-\frac{3(n+2)}{2}}\right)\right\} \end{aligned} \tag{2.8}$$

$$\begin{aligned} f_2(x) &= x^{-\frac{n}{4}} \exp\left(-\frac{2}{n+2} x^{\frac{n+2}{2}}\right) \left\{1 - \left[\frac{n(n+4)}{16(n+2)}\right] x^{-\frac{(n+2)}{2}}\right. \\ &+ \left.\left[\frac{n(n+4)(3n+4)(3n+8)}{512(n+2)^2}\right] x^{-(n+2)} + O\left(x^{-\frac{3(n+2)}{2}}\right)\right\} \end{aligned} \tag{2.9}$$

$$\begin{aligned} \frac{df_1}{dx} &= x^{\frac{n}{4}} \exp\left(\frac{2}{n+2} x^{\frac{n+2}{2}}\right) \left\{1 - \left[\frac{n(3n+4)}{16(n+2)}\right] x^{-\frac{(n+2)}{2}}\right. \\ &- \left.\left[\frac{n(n+4)(3n+4)(5n+8)}{512(n+2)^2}\right] x^{-(n+2)} + O\left(x^{-\frac{(3n+2)}{2}}\right)\right\} \end{aligned} \tag{2.10}$$

$$\begin{aligned} \frac{df_2}{dx} &= x^{\frac{n}{4}} \exp\left(-\frac{2}{n+2} x^{\frac{n+2}{2}}\right) \left\{1 + \left[\frac{n(3n+4)}{16(n+2)}\right] x^{-\frac{(n+2)}{2}}\right. \\ &- \left.\left[\frac{n(n+4)(3n+4)(5n+8)}{512(n+2)^2}\right] x^{-(n+2)} + O\left(x^{-\frac{(3n+2)}{2}}\right)\right\} \end{aligned} \tag{2.11}$$

as $x \to \infty$ in S.

PROOF: See Appendix A.

III. THE ASYMPTOTIC REPRESENTATION FOR a_m $m \to \infty$

We are now ready to develop an asymptotic representation for the Chebyshev coefficients associated with our functions of interest. As we shall soon see, such a representation will allow us to assess the effect of various nonlinear changes of scale on the efficiency of our Chebyshev expansion.

A. *THE CHEBYSHEV COEFFICIENTS PRIOR TO ANY CHANGE OF SCALE*

As Elliott (1964) has shown, the inner product definition for the Chebyshev coefficient a_m of a given function $f(x)$,

$$a_m = \frac{2}{\pi} \int_{-1}^{1} \frac{f(x) T_m(x)}{\sqrt{1-x^2}} dx$$

can be combined with Cauchy's integral formula

$$f(x) = \frac{1}{2\pi i} \int_C \frac{f(z)}{z-x} dz$$

to yield the contour integral representation

$$a_m = \frac{1}{\pi i} \int_C \frac{f(x)}{\sqrt{x^2-1}\left(x+\sqrt{x^2-1}\right)^m} dx, \quad m = 0, 1, 2, \cdots. \tag{3.1}$$

Here C is a closed contour in the complex x-plane surrounding the interval $-1 \le x \le 1$, on and inside of which $f(x)$ is analytic. Now, if f is *entire*, then *any* closed contour containing the line segment $-1 \le x \le 1$ will suffice.

In another paper, Elliott and Szekeres (1965) have used the integral representation (3.1) in order to derive an asymptotic formula for a_m as $m \to \infty$. The integrand of (3.1) is written as $e^{\psi(x)}$, where

$$\psi(x) \equiv \ln f(x) - \frac{1}{2} \ln\left(x^2 - 1\right) - m \ln\left(x + \sqrt{x^2-1}\right). \tag{3.2}$$

This gives

$$a_m = \frac{1}{\pi i} \int_C e^{\psi(x)} dx = \frac{1}{\pi i} I(m). \tag{3.3}$$

The method of steepest descents (see deBruijn, 1981, Erdelyi, 1956, or Sirovich, 1971, for example) is then used to get an asymptotic representation for $I(m)$, when m is large. First, the saddle points of $\psi(x)$ are found by solving $\psi'(x) = 0$. It is assumed that each saddle point $x = \varsigma$ is *simple*, so $\psi''(\varsigma) \neq 0$. By Cauchy's theorem, the contour C is next deformed into a steepest descent curve in the complex plane through the saddle points. In the neighborhood of each one we have

$$\psi(x) = \psi(\varsigma) + \frac{\psi''(\varsigma)}{2}(x-\varsigma)^2 + O\left[(x-\varsigma)^3\right], \quad x \to \varsigma.$$

Moreover, the equation

$$\psi''(\varsigma)(x-\varsigma)^2 \leq 0$$

defines a straight line in the complex plane through the point $x = \varsigma$. If this line through each saddle point is parameterized as

$$x = \varsigma + \delta t, \qquad -\infty < t < \infty, \qquad \delta = e^{i\left(\frac{\pi}{2} - \frac{1}{2}\arg\psi''(\varsigma)\right)},$$

then, assuming the higher order terms contribute negligibly to the integral along the shifted contour,

$$I(m) = e^{\psi(\varsigma)} \int_C e^{\frac{1}{2}\psi''(\varsigma)(x-\varsigma)^2} dx = \delta\, e^{\psi(\varsigma)} \int_{-\infty}^{\infty} e^{\frac{1}{2}\psi''(\varsigma)(\delta t)^2} dt. \tag{3.4}$$

Now make the substitution

$$u = \sqrt{\frac{1}{2}|\psi''(\varsigma)|}\, t$$

so (3.4) is equal to

$$\frac{\delta e^{\psi(\varsigma)}}{\sqrt{\frac{1}{2}|\psi''(\varsigma)|}} \int_{-\infty}^{\infty} e^{-u^2} du = \sqrt{2\pi}\, \delta |\psi''(\varsigma)|^{-\frac{1}{2}} e^{\psi(\varsigma)}.$$

Finally, remembering (3.2) and folding this all into (3.3) gives

$$a_m \sim \frac{1}{i}\sqrt{\frac{2}{\pi}} \frac{\delta f(\varsigma)}{\sqrt{|\psi''(\varsigma)|}\,(\varsigma^2 - 1)\left(\varsigma + \sqrt{\varsigma^2 - 1}\right)^m}, \qquad m \to \infty \tag{3.5}$$

or a sum of such terms, one for each ς_i, in the event of multiple saddle points.

The representation (3.5) is the desired asymptotic expansion. In order to make use of it, however, we must first be able to find the saddle point(s), ς. Recall that ς is given by the equation $\psi'(\varsigma) = 0$. From (3.2) this means that

$$\frac{f'(\varsigma)}{f(\varsigma)} = \frac{\varsigma}{\varsigma^2 - 1} + \frac{m}{\sqrt{\varsigma^2 - 1}}. \tag{3.6}$$

We recognize this as an equation implicitly defining ς. At this same time we note that differentiating (3.2) twice gives

$$\psi''(\varsigma) = \frac{f''(\varsigma)}{f(\varsigma)} - \left[\frac{f'(\varsigma)}{f(\varsigma)}\right]^2 + \frac{\varsigma^2+1}{(\varsigma^2-1)^2} + \frac{m\varsigma}{(\varsigma^2-1)^{\frac{3}{2}}}. \tag{3.7}$$

It will seldom be possible to find ς exactly using (3.6). It turns out, however, for the functions of interest, that m large implies ς large, and thus we can employ asymptotic approximations for ς and hence $\psi''(\varsigma)$ in (3.5). The functions $f(x)$ to which these results will be applied are the solutions f_1 and f_2 of the differential equation (1.2). In this case, we know

$$\frac{f''(\varsigma)}{f(\varsigma)} = \varsigma^n$$

directly from the differential equation. Moreover, an asymptotic representation for $\frac{f'(\varsigma)}{f(\varsigma)}$ is available from (2.8) to (2.11). Both of these expressions may then be substituted into the formula (3.7) for $\psi''(\varsigma)$.

B. THE CHEBYSHEV COEFFICIENTS AFTER THE NONLINEAR VARIABLE CHANGE

We now introduce the non-linear change of scale $z = x^\alpha$, where $\alpha > 0$. Our goal is to see how this transformation is manifested in the asymptotic formula (3.5) for a_m. We begin this task by writing all relevant expressions in terms of the new independent variable, z. Under the change of variable $x = z^{\frac{1}{\alpha}}$ and letting $u(z) = f(x)$, the original differential equation (1.2) becomes

$$\alpha^2 z^2 \frac{d^2u}{dz^2} + \alpha(\alpha-1)z\frac{du}{dz} - z^{\frac{n+2}{\alpha}}u = 0. \tag{3.8}$$

Since our interest is in entire functions, we perforce concentrate on the case in which

$$\frac{n+2}{\alpha} = k$$

with integral k. In other words, the values of the parameter α are restricted to

$$\alpha = \frac{n+2}{k}, \quad k = 1, 2, 3, \cdots.$$

Without loss of generality we work with $u = f_1$ as given by (2.8). Then

$$\begin{aligned} u &= z^{-\frac{n}{4\alpha}} \exp\left(\frac{2}{n+2} z^{\frac{n+2}{2\alpha}}\right) \left\{1 + \left[\frac{n(n+4)}{16(n+2)}\right] z^{-\frac{(n+2)}{2\alpha}} \right. \\ &+ \left. \left[\frac{n(n+4)(3n+4)(3n+8)}{512(n+2)^2}\right] z^{-\frac{(n+2)}{\alpha}} + O\left(z^{-\frac{3(n+2)}{2\alpha}}\right)\right\} \end{aligned} \tag{3.9}$$

and

$$\begin{aligned} \frac{du}{dz} &= \frac{1}{\alpha} z^{\frac{n+4(1-\alpha)}{4\alpha}} \exp\left(\frac{2}{n+2} z^{\frac{n+2}{2\alpha}}\right) \left\{1 - \left[\frac{n(3n+4)}{16(n+2)}\right] z^{-\frac{(n+2)}{2\alpha}} \right. \\ &- \left. \left[\frac{n(n+4)(3n+4)(5n+8)}{512(n+2)^2}\right] z^{-\frac{(n+2)}{\alpha}} + O\left(z^{-\frac{3(n+2)}{2\alpha}}\right)\right\} \end{aligned} \tag{3.10}$$

as $z \to \infty$. If we denote $\frac{d}{dz} \equiv {}'$, it follows from properties of asymptotic power series that we may divide (3.10) by (3.9), giving

$$\frac{u'}{u} = \frac{1}{\alpha} z^{\frac{n+2(1-\alpha)}{2\alpha}} \left\{1 - \frac{n}{4} z^{-\frac{(n+2)}{2\alpha}} - \frac{n(n+4)}{32} z^{-\frac{(n+2)}{\alpha}} + O\left(z^{-\frac{3(n+2)}{2\alpha}}\right)\right\} \tag{3.11}$$

and using (3.8) and (3.11)

$$\frac{u''}{u} = \frac{1}{\alpha^2} z^{\frac{n+2(1-\alpha)}{\alpha}} + \frac{1-\alpha}{\alpha^2} z^{\frac{n+2(1-2\alpha)}{2\alpha}} \left\{1 - \frac{n}{4} z^{-\frac{(n+2)}{2\alpha}} + O\left(z^{-\frac{(n+2)}{\alpha}}\right)\right\} \tag{3.12}$$

as $z \to \infty$. Employing (3.11) in (3.6) we are now in a position to derive an asymptotic representation for the saddle point(s) ς; namely,

$$\frac{1}{\alpha} \varsigma^{\frac{n+2(1-\alpha)}{2\alpha}} \left\{1 - \frac{n}{4} \varsigma^{-\frac{(n+2)}{2\alpha}} + O\left(\varsigma^{-\frac{(n+2)}{\alpha}}\right)\right\} = \frac{m}{\varsigma} + \frac{1}{\varsigma} + \frac{m}{2\varsigma^3} + \frac{1}{\varsigma^3} + O\left(\varsigma^{-4}\right) \tag{3.13}$$

for $\varsigma \to \infty$ and hence,

$$\varsigma = (\alpha m)^{\frac{2\alpha}{n+2}} (1 + \varepsilon), \qquad \varepsilon = o(1) \text{ as } m \to \infty.$$

Substituting this back into (3.13) yields

$$\varepsilon \sim \left[\frac{n+4\alpha}{2(n+2)}\right] m^{-1} + \frac{\alpha}{n+2} (\alpha m)^{-\frac{4\alpha}{n+2}}, \qquad m \to \infty.$$

Thus,

$$\varsigma = (\alpha m)^{\frac{2\alpha}{n+2}} \left\{1 + \left[\frac{n+4\alpha}{2(n+2)}\right] m^{-1} + \frac{\alpha}{n+2}(\alpha m)^{-\frac{4\alpha}{n+2}} + o\left(m^{-\gamma}\right)\right\} \tag{3.14}$$

where $\gamma = \min\left[1, \frac{4\alpha}{n+2}\right]$. We include both correction terms since it is not clear, for given α and n, which term, $\left[\frac{n+4\alpha}{2(n+2)}\right] m^{-1}$ or $\frac{\alpha}{n+2}(\alpha m)^{-\frac{4\alpha}{n+2}}$, dominates. Also, combining (3.11) and (3.12) with (3.14) gives

$$\psi'' \sim \left[\frac{n+2(1-\alpha)}{2\alpha^2}\right] \varsigma^{\frac{n+2(1-2\alpha)}{2\alpha}} + \left[\frac{n+4\alpha}{4\alpha}\right] \varsigma^{-2} + \frac{m}{\varsigma^2} + \frac{3m}{2\varsigma^4}$$

and so

$$\begin{aligned}\psi'' &= \left[\frac{n+2}{2\alpha^2}\right] (\alpha m)^{1-\frac{4\alpha}{n+2}} \left\{1 + \left[\frac{(n+4\alpha)(n+2-4\alpha)}{4\alpha(n+2)}\right] m^{-1}\right. \\ &+ \left.\frac{1}{2}(\alpha m)^{-\frac{4\alpha}{n+2}} + o\left(m^{-\gamma}\right)\right\}\end{aligned}$$

Lastly, $u(\varsigma)$ must be expressed in terms of α and m. To the indicated order in m, we find

$$u = (\alpha m)^{-\frac{n}{2(n+2)}} \exp\left[\frac{2\alpha m}{n+2} + \frac{n+4\alpha}{2(n+2)} + \frac{(\alpha m)^{1-\frac{4\alpha}{n+2}}}{n+2} + o\left(m^{1-\gamma}\right)\right].$$

Finally, we have the desired representation for a_m as $m \to \infty$:

$$a_m = \frac{\delta\, \alpha\, 2^{1-m}}{i\sqrt{\pi(n+2)}} (\alpha m)^{-\frac{(2\alpha m+n+1)}{n+2}} \exp\left[\frac{2\alpha m}{n+2} + \frac{(\alpha m)^{1-\frac{4\alpha}{n+2}}}{4\alpha} + o\left(m^{1-\gamma}\right)\right] \tag{3.15}$$

where $\alpha = \frac{n+2}{k}$, $k = 1, 2, 3, \ldots$ and $\gamma = \min\left[1, \frac{4\alpha}{n+2}\right]$. Note that in this final step we have made use of the fact that for small x,

$$(1+x)^m = \exp\left[mx + m\, o(x)\right]$$

asymptotically.

IV. THEORETICAL IMPLICATIONS

In the previous section, we found an asymptotic representation for a_m which incorporated the non-linear change of variable $x = z^{\frac{1}{\alpha}}, \alpha > 0$. We now want to examine formula (3.15) in a more theoretical light. We begin by determining α such that, for fixed m and n, the magnitude of a_m is minimized.

A. THE "OPTIMAL" CHOICE OF α

The starting point of our investigation is the asymptotic formula (3.15) which we rewrite as

$$|a_m| \sim C(m,n)\,\alpha^{\frac{1}{n+2}}\left(\frac{e}{\alpha m}\right)^{\frac{2\alpha m}{n+2}}\exp\left[\frac{(\alpha m)^{1-\frac{4\alpha}{n+2}}}{4\alpha}\right],\quad m\to\infty \tag{4.1}$$

where

$$C(m,n)\equiv\frac{2^{1-m}m^{-\left(\frac{n+1}{n+2}\right)}}{\sqrt{\pi(n+2)}}$$

is a positive function depending only on the integral values of m and n. Before proceeding, however, we note the following useful lemma:

LEMMA 4.1 If $f(x)$ is a solution of (1.2), $x = z^{\frac{1}{\alpha}}$, and

$$a_m = \frac{2}{\pi}\int_{-1}^{1}\frac{f(z^{\frac{1}{\alpha}})\,T_m(z)}{\sqrt{1-z^2}}dz,\qquad m = 0,1,2,\cdots$$

is the mth Chebyshev coefficient of $f(z^{\frac{1}{\alpha}})$, then for each fixed m, a_m is an analytic function of α, for $\alpha > 0$.

PROOF: See Appendix B.

With this result, we may now prove

THEOREM 4.1 If $|a_m|$ has the representation (4.1) as $m\to\infty$, m integral and $n = 0,1,2,\cdots$, then $|a_m|$ is asymptotically a monotonically decreasing function of α, for all $\alpha > 0$.

PROOF: Define

$$F(\alpha)\equiv\alpha^{\frac{1}{n+2}}\left(\frac{e}{\alpha m}\right)^{\frac{2\alpha m}{n+2}}\exp\left[\frac{(\alpha m)^{1-\frac{4\alpha}{n+2}}}{4\alpha}\right].$$

It will be enough to show that F is a strictly decreasing function of α, for sufficiently large m. In fact, since the natural logarithm is an increasing function of its argument, we know F decreases if and only if $G(\alpha)\equiv\ln F(\alpha)$ decreases. Now,

$$G(\alpha)\equiv\ln F(\alpha) = \frac{1}{n+2}\{\ln\alpha + 2\alpha m\,[1-\ln(\alpha m)]\} + \frac{m}{4}(\alpha m)^{-\frac{4\alpha}{n+2}}$$

so that

$$G'(\alpha) = \frac{1}{n+2} \cdot \frac{1}{\alpha} \{1 - 2\alpha m \ln(\alpha m)\} - \frac{m}{n+2} \cdot (\alpha m)^{-\frac{4\alpha}{n+2}} \{1 + \ln(\alpha m)\} .$$

Hence, $G'(\alpha) < 0$ if and only if

$$(\alpha m)^{1-\frac{4\alpha}{n+2}} \{1 + \ln(\alpha m)\} + 2\alpha m \ln(\alpha m) > 1$$

and given $\alpha > 0$, this latter inequality is satisfied for m large enough. Owing to Lemma 4.1, $|a_m|$ is thus asymptotically a strictly decreasing function of α. □

COROLLARY 4.1 If $\alpha = \frac{n+2}{k}$, $k = 1, 2, \ldots$, in the previous theorem, then for fixed n, the value $k = 1$ asymptotically minimizes $|a_m|$ as a function of k.

Although, in the strict sense an asymptotic result, this corollary at least suggests that the "optimal" value of α occurs when $k = 1$, namely, when $\alpha = n + 2$. The following numerical examples give empirical evidence that m need not be too large before the desired monotonicity occurs. We restrict attention to $k \leq n + 2$.

EXAMPLE 4.1

(i) If $n = 0$, we have $\alpha = \frac{2}{k}$, $k = 1, 2$. Then from (4.1)

$$\begin{aligned} |a_8| &\doteq 7.51 \times 10^{-16} \; (k = 1) \\ |a_8| &\doteq 4.04 \times 10^{-7} \; (k = 2) \end{aligned}$$

(ii) If $n = 1$, we have $\alpha = \frac{3}{k}$, $k = 1, 2, 3$. Thus, (4.1) implies

$$\begin{aligned} |a_8| &\doteq 6.73 \times 10^{-19} \; (k = 1) \\ |a_8| &\doteq 1.02 \times 10^{-8} \; (k = 2) \\ |a_8| &\doteq 2.28 \times 10^{-6} \; (k = 3) \end{aligned}$$

For all of the above calculations, the presence of multiple saddle points has been taken into account. Note that in each of these examples, the magnitude of the Chebyshev coefficient a_8 when $k = 1$ actually turns out to be smaller than that of a_8 for any other value of k.

B. EVENNESS OF u AND THE "OPTIMAL" α

It would appear at this stage that the "optimal" value of α in the change of variable $x = z^{\frac{1}{\alpha}}$ has been found, namely, $\alpha = n + 2$. However, numerical work (see Section VI) suggests that this question is more subtle and merits deeper consideration.

To begin with, it is necessary to be more specific as to what we mean by "optimal". Given our function $f(x)$, a meaningful goal would be to determine $\alpha \left(= \frac{n+2}{k}\right)$ so that for a given accuracy criterion, the number of non-zero terms needed in the Chebyshev expansion of $u(z) = f\left(z^{\frac{1}{\alpha}}\right)$ is minimized. The following example is enlightening in this regard:

EXAMPLE 4.2 For $n = 0$ and $k = 2$, the function under consideration is $u(z) = \cosh z$. If u is approximated by the first eight non-zero terms in its Chebyshev expansion, the truncation error is approximately

$$|a_{16}| \doteq 1.48 \times 10^{-18}.$$

When $k = 1$, we have $u(z) = \cosh\left(\sqrt{z}\right)$. In this case, the error is about

$$|a_8| \doteq 3.73 \times 10^{-16}.$$

Notice that $k = 1$ did *not* yield the smaller error and hence the more economical representation for u. The reason is as follows:
Since

$$\cosh z = 1 + \frac{z^2}{2!} + \frac{z^4}{4!} + \cdots$$

is even, its Chebyshev series

$$\cosh z = \frac{1}{2}a_0 + a_2T_2 + a_4T_4 + \cdots \tag{4.2}$$

consists only of even terms. On the other hand,

$$\cosh\left(z^{\frac{1}{2}}\right) = 1 + \frac{z}{2!} + \frac{z^2}{4!} + \frac{z^3}{6!} + \cdots$$

has *both* even and odd terms in its Chebyshev series

$$\cosh\left(z^{\frac{1}{2}}\right) = \frac{1}{2}a_0 + a_1T_1 + a_2T_2 + \cdots. \tag{4.3}$$

Now certainly $|a_8|\,(k = 1) < |a_8|\,(k = 2)$, as asserted by Corollary 4.1, but the ninth *coefficient*, a_{16} in the series (4.2) (which corresponds to $k = 2$) is smaller in magnitude than the ninth coefficient, a_8 in (4.3).

Observe that in this example, there is a loss of both evenness and economy in the Chebyshev series when going from $k = 2$ to $k = 1$. Loosely speaking, the even u only needs "half as many" terms in its Chebyshev series. This suggests that there is a definite relationship between the evenness of the transformed u and its "best" Chebyshev series representation. Indeed, after replacing α by $\frac{n+2}{k}$ in (4.1) we have the following

THEOREM 4.2 If

$$|a_m| \sim \frac{1}{2^{m-1}\, m\sqrt{\pi(n+2)}} \left[\frac{m(n+2)}{k}\right]^{\frac{1}{n+2}} \left[\frac{k\,e}{m(n+2)}\right]^{\frac{2m}{k}} \exp\left\{\frac{m}{4}\left[\frac{k}{m(n+2)}\right]^{\frac{4}{k}}\right\}$$

as $m \to \infty$ and n is fixed, then the "optimal" value of $\alpha = \frac{n+2}{k}$ in the change of variable $x = z^{\frac{1}{\alpha}}$ is $\alpha = \frac{n+2}{2}$. (Here, "optimal" means that the m-th non-zero Chebyshev coefficient when $k = 2$ is asymptotically smaller in magnitude than the corresponding coefficient when $k = 1$.)

PROOF: Consider the differential equation (3.8) with $\alpha = \frac{n+2}{k}$:

$$\left(\frac{n+2}{k}\right)^2 z^2 u'' + \left(\frac{n+2}{k}\right)\left(\frac{n+2}{k} - 1\right) z u' - z^k u = 0. \tag{4.4}$$

When $k = 1$, this becomes

$$(n+2)^2 z\, u'' + (n+2)(n+1)\, u' - u = 0. \tag{4.5}$$

Note that this equation has a regular singular point at $z = 0$. The indicial equation

$$r^2 - \frac{1}{n+2} r = 0$$

yields the roots $r = 0, \frac{1}{n+2}$, and standard theory then tells us that there are two linearly independent solutions of the form

$$u(z) = \sum_{j=0}^{\infty} b_j z^j$$

and

$$v(z) = z^{\frac{1}{n+2}} \sum_{j=0}^{\infty} d_j z^j$$

where $b_0,\ d_0 \neq 0$. Our interest is in the solution which is non-zero at the origin, namely the entire function u. Observe that the equation (4.5) cannot be satisfied by a function which is solely even or odd. This means that u has both even and odd terms in its power series, and hence in its Chebyshev series representation.

We also note that equation (4.5) has an irregular singular point at $x = \infty$, and by (3.9) [see also (2.8), (2.9)]

$$f_1\left(z^{\frac{1}{n+2}}\right) \sim z^{-\frac{n}{4(n+2)}} \exp\left\{\frac{2}{n+2} z^{\frac{1}{2}}\right\}$$

$$f_2\left(z^{\frac{1}{n+2}}\right) \sim z^{-\frac{n}{4(n+2)}} \exp\left\{-\frac{2}{n+2} z^{\frac{1}{2}}\right\}$$

as $z \to \infty$, are likewise two linearly independent solutions. Therefore, there exist constants σ and τ such that

$$u(z) = \sigma f_1 + \tau f_2.$$

Because f_2 is recessive as $z \to \infty$, we have

$$u(z) \sim \sigma f_1, \qquad z \to \infty. \tag{4.6}$$

It follows immediately then from (3.5) that the asymptotic form of the $m+1$-st Chebyshev coefficient of u is given by

$$|a_m| \sim \sigma \frac{[m(n+2)]^{\frac{1}{n+2}}}{2^{m-1}\, m \sqrt{\pi(n+2)}} \left[\frac{e}{m(n+2)}\right]^{2m} \exp\left\{\frac{1}{4m^3(n+2)^4}\right\}, \quad m \to \infty. \tag{4.7}$$

The analysis when $k = 2$ is carried out in analogous fashion. (4.4) becomes

$$\left(\frac{n+2}{2}\right)^2 z u'' + \frac{n}{2}\left(\frac{n+2}{2}\right) u' - zu = 0 \tag{4.8}$$

which still has a regular singular point at $z = 0$. But now the indicial roots are $r = 0,\ \frac{2}{n+2}$. Interestingly this time, when we concentrate on the solution u which is non-zero at the origin, classical Frobenius theory shows that it is even. There are only even terms, therefore, in the Chebyshev series representation for u.

As before, the asymptotic relationship (4.6) between u and f_1 holds. In view of the above, however, it follows that the $p+1$-st non-zero Chebyshev coefficient satisfies

$$|a_{2p}| \sim \sigma \frac{2[p(n+2)]^{\frac{1}{n+2}}}{2^{2p-1}\, 2p \sqrt{\pi(n+2)}} \left[\frac{e}{p(n+2)}\right]^{2p} \exp\left\{\frac{1}{2p(n+2)^2}\right\}, \quad p \to \infty. \tag{4.9}$$

The extra factor of two is needed since there are two saddle points which contribute identically in this case. (Recall the discussion surrounding the original formula (3.5)). When $p = m$, (4.7) and (4.9) are asymptotic representations for the $m + 1$-st non-zero Chebyshev coefficient in the cases $k = 1$ and $k = 2$, respectively.

Notice that the only difference between the coefficients in these two cases is in the powers of two and in the powers of the exponential: in (4.7) we have

$$\exp\left\{\frac{1}{4m^3(n+2)^4} - m\ln 2\right\}$$

and in (4.9)

$$\exp\left\{\frac{1}{2m(n+2)^2} - 2m\ln 2\right\}. \tag{4.10}$$

For large m, (4.10) is the smaller of the two factors and so we conclude that the $m + 1$-st non-zero coefficient a_{2m} when $k = 2$ is asymptotically smaller in magnitude than the corresponding coefficient a_m when $k = 1$. □

Remark: For odd values of n the "optimal" change of variable $z = x^{\frac{n+2}{2}}$ (which occurs when $k = 2$) is not defined for $-1 \leq x < 0$. For these n, the transformation may only be used for $0 \leq x \leq 1$.

V. NUMERICAL PROCEDURES FOR COEFFICIENT EVALUATION

In this and the next section we explore the numerical aspect of our work. Here we provide a brief overview of general three-term recurrence relations and then remind readers of the backward recurrence technique of Clenshaw. We have used the Clenshaw method to compute all of our Chebyshev coefficients, and we illustrate the applicability of the method with two examples.

A. *THREE-TERM RECURRENCE RELATIONS*

Consider the second-order, linear, homogeneous three-term recurrence relation

$$y_{n+1} + a_n y_n + b_n y_{n-1} = 0, \qquad n = 1, 2, 3, \cdots, \tag{5.1}$$

where a_n, b_n are complex sequences such that $b_n \neq 0$ for all n. This difference equation generally has two linearly independent solutions, say f_n and g_n, called a *fundamental set.* In the special case in which f_n, g_n are such that

$$f_n = o(g_n) \qquad \text{as} \qquad n \to \infty,$$

we say that f_n is *minimal at infinity* or simply *minimal*; g_n is then called *dominant*. It is well-known that the recurrence relation (5.1) is unstable for the computation of minimal f_n in the direction of increasing n. (See Wimp (1984), for example).

Rayleigh (1910) originally appears to have suggested the idea of using recurrence relations such as (5.1) in the direction of decreasing, rather than increasing n whenever one is interested in minimal solutions. J.C.P. Miller (B.A.A.S., 1952) subsequently codified the approach of *backward* (or *downward*) recurrence in his attempts to compute modified Bessel functions. Miller's algorithm can be described as follows. To compute minimal f_n from the recurrence relation (5.1), it is assumed there is a normalizing condition of the type

$$\lambda_0 f_0 + \lambda_1 f_1 + \lambda_2 f_2 + \cdots = \gamma \neq 0 \tag{5.2}$$

where the λ_i constitute a bounded sequence of real numbers. Then for arbitrary (usually large) N the starting values

$$y_N = 1, \qquad y_{N+1} = y_{N+2} = \cdots = 0$$

are assumed and one recurs downward using (5.1) in the form

$$y_{n-1} = -\frac{1}{b_n}(y_{n+1} + a_n y_n), \qquad n = N, N-1, N-2, \cdots, 1.$$

The normalizing condition (5.2) is truncated to

$$\lambda_0 y_0 + \lambda_1 y_1 + \cdots + \lambda_N y_N = \gamma_N$$

and the trial solution is computed as

$$\varphi_n = \frac{\gamma}{\gamma_N} y_n, \qquad n = 0, 1, 2, \ldots, N.$$

Since f_n is a minimal solution of (5.1) then for fixed n,

$$\varphi_n \to f_n \qquad \text{as} \qquad N \to \infty.$$

Both Miller (B.A.A.S., 1952) and Gautschi (1967) show this algorithm *converges* (see also Olver, 1964). Appropriate values of N are usually determined empirically in each individual instance.

B. CLENSHAW'S METHOD OF BACKWARD RECURRENCE

In an important 1957 paper, C.W. Clenshaw used a variant of the Miller Algorithm to show how backward recurrence could be used to calculate the Chebyshev coefficients of a function satisfying a linear, ordinary differential equation with polynomial coefficients. His procedure is as follows.

Assume we want to obtain the Chebyshev coefficients for a function of the type described above. Denote the order of the differential equation by q, and suppose that $f, f', \ldots, f^{(q)}$ have uniformly convergent Chebyshev series

$$f^{(s)}(x) = \sum_{m=0}^{\infty}{}' a_m^{(s)} T_m(x), \qquad s = 0, 1, \cdots, q. \tag{5.3}$$

(Recall the $\prime$ notation, signifying that only $\frac{1}{2}$ of the zeroth terms are included in the summations.) Now define $C_m(g)$ to be the coefficient of $T_m(x)$ in the expansion of g for (integral) $m > 0$ and *twice* this coefficient when $m = 0$. The following easily derivable relations play an important role in the recurrence process:

$$2m\, a_m^{(s)} = a_{|m-1|}^{(s+1)} - a_{m+1}^{(s+1)}, \; s = 0, 1, \cdots, q-1 \tag{5.4}$$

and

$$C_m\left(x^p f^{(s)}\right) = \frac{1}{2^p} \sum_{j=0}^{p} \binom{p}{j} a_{|m-p+2j|}^{(s)}. \tag{5.5}$$

After substituting the series (5.3) into the differential equation, (5.5) may be used to obtain a recurrence relation for the coefficients. Employing this relation in conjunction with (5.4) we then recur downwards. For an appropriately large N, we assume

$$a_m^{(s)} = 0, \qquad m > N \qquad \text{and} \qquad s = 0, 1, \ldots, q.$$

Starting values are assigned to a_N, $a_N^{(s)}$, $s = 1, 2, \ldots, q$ so that the lower order $a_m^{(s)}$ may be computed recursively. These trial values, of course, must then be normalized or scaled. This is usually accomplished by evaluating the series (5.3) when $s = 0$ at a convenient point. Here it is helpful to note that, owing to the nature of the Chebyshev polynomials,

$$\begin{aligned}
f(-1) &= \frac{1}{2}a_0 - a_1 + a_2 - a_3 + \cdots \\
f(0) &= \frac{1}{2}a_0 - a_2 + a_4 - a_6 + \cdots \\
f(1) &= \frac{1}{2}a_0 + a_1 + a_2 + a_3 + \cdots
\end{aligned}$$

It is also important to observe that complications arise in the application of this procedure when there is more than one linearly independent solution of the differential equation of interest, which, along with all appropriate derivatives, possesses a convergent Chebyshev expansion.

EXAMPLE 5.1 (Recall Example 4.2) Consider the differential equation (1.2) for $n = 0$ under the change of variable $x = z^{\frac{1}{2}}$:

$$4zu''(z) + 2u'(z) - u = 0.$$

This has solutions $u_1 = \cosh\left(z^{\frac{1}{2}}\right)$ and $u_2 = \sinh\left(z^{\frac{1}{2}}\right)$, the first of which is entire, while the second one has a branch singularity at the origin. In order to compute the Chebyshev coefficients for u_1 by backward recurrence, we employ

$$\begin{aligned} a_m^{(2)} &= a_{m+2}^{(2)} + 2(m+1)\, a_{m+1}^{(1)} \\ a_m^{(1)} &= a_{m+2}^{(1)} + 2(m+1)\, a_{m+1} \end{aligned}$$

from (5.4) along with

$$a_m = 4a_{m+1}^{(2)} + 2(2m+1)\, a_m^{(1)}, \qquad m = N-1, N-2, \cdots, 1, 0,$$

which we derived directly from the differential equation. The results given in Table 5.1 have been determined using $N = 16$ and the normalization

$$u_1(0) = \frac{1}{2}a_0 - a_2 + a_4 - a_6 + \cdots = 1.$$

In some instances it may turn out that all of the a_m are determined before the last step of the recurrence process at $m = 0$ has been reached. In such cases some equations will not have been used in the initial application of the procedure. These remaining or "residual" equations, however, must still be satisfied. Occasionally, they may automatically be satisfied by the computed a_m. More commonly it occurs that the one trial solution already obtained does *not* satisfy these residual equations, and the entire recurrence process must be repeated with different starting values in order to determine other linearly independent trial solutions. The final solution will then be a linear combination of these several solutions.

REMARK: If the function under examination is one of a number of solutions (to the given differential equation) which, along with all appropriate derivatives, have convergent Chebyshev expansions, then additional trial solutions

TABLE 5.1: Chebyshev Coefficients for $\cosh(z^{\frac{1}{2}})$

m	a_m	$a_m^{(1)}$	$a_m^{(2)}$	Normalized a_m
0	1.76039×10^{37}	8.65817×10^{36}	1.43961×10^{36}	2.04169
1	4.32010×10^{36}	7.19161×10^{35}	7.18876×10^{34}	.501042
2	1.79737×10^{35}	1.79689×10^{34}	1.28334×10^{33}	2.08457×10^{-2}
3	2.99457×10^{33}	2.13879×10^{32}	1.18816×10^{31}	3.47308×10^{-4}
4	2.67340×10^{31}	1.48516×10^{30}	6.75059×10^{28}	3.10059×10^{-6}
5	1.48514×10^{29}	6.75052×10^{27}	2.59632×10^{26}	1.72246×10^{-8}
6	5.62539×10^{26}	2.16359×10^{25}	7.21191×10^{23}	6.52429×10^{-11}
7	1.54541×10^{24}	5.15135×10^{22}	1.51509×10^{21}	1.79236×10^{-13}
8	3.21958×10^{21}	9.46932×10^{19}	2.49192×10^{18}	3.73405×10^{-16}
9	5.26072×10^{18}	1.38440×10^{17}	3.29618×10^{15}	6.101348×10^{-19}
10	6.92198×10^{15}	1.64809×10^{14}	3.58280×10^{12}	8.02806×10^{-22}
11	7.49130×10^{12}	1.62854×10^{11}	3.25708×10^{9}	8.68835×10^{-25}
12	6.78558×10^{9}	1.35712×10^{8}	2.51317×10^{6}	7.86987×10^{-28}
13	5.21967×10^{6}	9.66605×10^{5}	1.66656×10^{3}	6.05374×10^{-31}
14	3.45216×10^{3}	5.95200×10^{1}	9.60000×10^{-1}	4.00379×10^{-34}
15	1.98400	3.20000×10^{-2}	0.00000	2.30103×10^{-37}
16	1.00000×10^{-3}	0.00000	0.00000	1.15997×10^{-40}

will *always* be necessary.

EXAMPLE 5.2 For $n = 1$, the differential equation (1.2) is

$$y''(x) - x\,y(x) = 0.$$

A fundamental set of solutions to this well-known equation consists of the Airy functions

$$\begin{aligned} y_1(x) &= Ai(x) = c_1\,f(x) - c_2\,g(x) \\ y_2(x) &= Bi(x) = \sqrt{3}\,[c_1\,f(x) + c_2\,g(x)] \end{aligned}$$

where

$$\begin{aligned} c_1 &= Ai(0) \doteq 0.35502805 \\ c_2 &= -Ai'(0) \doteq 0.25881940 \end{aligned}$$

Here, $f(x)$, $g(x)$ are the auxiliary functions (see Abramowitz and Stegun (1972), p. 446):

$$\begin{aligned} f(x) &= 1 + \frac{1}{3!}\,x^3 + \frac{1\cdot 4}{6!}\,x^6 + \frac{1\cdot 4\cdot 7}{9!}\,x^9 + \cdots \\ &= \sum_{j=0}^{\infty} 3^j \left(\frac{1}{3}\right)_j \frac{x^{3j}}{(3j)!} \end{aligned}$$

and

$$\begin{aligned} g(x) &= x + \frac{2}{4!}\,x^4 + \frac{2\cdot 5}{7!}\,x^7 + \frac{2\cdot 5\cdot 8}{10!}\,x^{10} + \cdots \\ &= \sum_{j=0}^{\infty} 3^j \left(\frac{2}{3}\right)_j \frac{x^{3j+1}}{(3j+1)!}. \end{aligned}$$

Suppose we are interested in obtaining the Chebyshev coefficients of f. Two boundary conditions, such as $f(0) = 1$, $f'(0) = 0$ will be necessary to specify f. Moreover, employing the recurrence relations

$$\begin{aligned} a_{m-1}^{(2)} &= a_{m+1}^{(2)} + 2m\,a_m^{(1)} \\ a_{m-1}^{(1)} &= a_{m+1}^{(1)} + 2m\,a_m \\ a_{m-1} &= 2\,a_m^{(2)} - a_{m+1} \end{aligned}$$

TABLE 5.2: Chebyshev Coefficients for $f(x)$

m	a_m	m	a_m
0	2.00347	13	2.54856×10^{-12}
1	0.12504	14	1.06187×10^{-14}
2	2.60440×10^{-3}	15	1.69923×10^{-13}
3	4.16920×10^{-2}	16	1.24928×10^{-15}
4	1.04181×10^{-3}	17	4.33770×10^{-18}
5	1.08512×10^{-5}	18	6.94112×10^{-17}
6	1.73674×10^{-4}	19	4.33776×10^{-19}
7	2.71291×10^{-6}	20	1.29097×10^{-21}
8	1.88386×10^{-8}	21	2.06577×10^{-20}
9	3.01486×10^{-7}	22	1.12260×10^{-22}
10	3.42531×10^{-9}	23	2.91279×10^{-25}
11	1.78395×10^{-11}	24	4.67793×10^{-24}
12	2.85481×10^{-10}		

for $m = N, N-1, \cdots, 1$, we are led to a residual equation

$$R(a) = a_1 - a_3 - 4\,a_1^{(1)} = 0$$

which must also be satisfied. Thus, we will actually need *three* linearly independent trial solutions, β_m, γ_m, σ_m. The final coefficients of f will then be the linear combination

$$a_m = \beta_m\, X + \gamma_m\, Y + \sigma_m\, Z$$

where X, Y, Z are solutions of

$$\begin{aligned} \beta(0)\, X + \gamma(0)\, Y + \sigma(0)\, Z &= 1 \\ \beta'(0)\, X + \gamma'(0)\, Y + \sigma'(0)\, Z &= 0 \\ R(\beta)\, X + R(\gamma)\, Y + R(\sigma)\, Z &= 0. \end{aligned}$$

Choosing $N = 24$ and the starting conditions

$$\begin{array}{lll} \beta_{24} = 1 & \beta_{24}^{(1)} = 0 & \beta_{24}^{(2)} = 0 \\ \gamma_{24} = 0 & \gamma_{24}^{(1)} = 1 & \gamma_{24}^{(2)} = 0 \\ \sigma_{24} = 0 & \sigma_{24}^{(1)} = 0 & \sigma_{24}^{(2)} = 1 \end{array}$$

we obtain the coefficients listed in Table 5.2.

VI. NUMERICAL RESULTS—VALIDATION OF THE THEORY

Clenshaw's Method of Backward Recurrence, as described in the previous section, has been employed to compute the Chebyshev coefficients of various solutions of (1.2) under the change of variable $x = z^{\frac{1}{\alpha}}$ with $\alpha = \frac{n+2}{k}$. Comparison of the coefficients in each of the cases $k = 1, 2, \ldots, n+2$ confirms the significant result of Theorem 4.2, namely, that $\alpha = \frac{n+2}{2}$ $(k = 2)$ yields the most economical change of variable of the type considered. That is, the choice $\alpha = \frac{n+2}{2}$ minimizes the magnitude of the m-th non-zero Chebyshev coefficient of $u(z) = f(z^{\frac{1}{\alpha}})$, for fixed (large) m and n. We consider in detail the cases $n = 0, 1, 2$. Other cases can be analyzed in comparable fashion.

A. THE CASE n=0

Here, $\alpha = \frac{2}{k}$ for $k = 1, 2$. The case in which $k = 1$ $(u(z) = \cosh(z^{\frac{1}{2}}))$ has been thoroughly discussed in Examples 4.2 and 5.1 and Table 5.1 gives the resulting Chebyshev coefficients. When $k = 2$, we have $x = z$, so there is no

TABLE 6.1: Comparison of Chebyshev Coefficients for the Case $n = 0$

m	coefficients of $\cosh(z^{\frac{1}{2}})$ $(k = 1)$	m	coefficients of $\cosh z$ $(k = 2)$
0	2.04169	0	2.53213
1	0.50104	2	0.27150
2	2.08457×10^{-2}	4	5.47424×10^{-3}
3	3.47308×10^{-4}	6	4.49773×10^{-5}
4	3.10058×10^{-6}	8	1.99212×10^{-7}
5	1.72246×10^{-8}	10	5.50590×10^{-10}
6	6.52429×10^{-11}	12	1.03915×10^{-12}
7	1.79236×10^{-13}	14	1.42376×10^{-15}
8	3.73405×10^{-16}	16	1.48018×10^{-18}
9	6.10135×10^{-19}	18	1.20743×10^{-21}
10	8.02806×10^{-22}	20	7.93367×10^{-25}
11	8.68835×10^{-25}	22	4.29079×10^{-28}

change of variable. The associated differential equation is $u'' - u = 0$, whose solution

$$u(z) = \cosh z = 1 + \frac{z^2}{2!} + \frac{z^4}{4!} + \cdots$$

we consider.

From the differential equation we obtain the recurrence relation

$$a_m^{(2)} - a_m = 0.$$

Here indices may be shifted and some judicious recombinations made using (5.4) with $s = 1$ to yield

$$a_{|m-1|} = a_{m+1} + 2m\, a_m^{(1)}.$$

It is this relation, along with (5.4) (for $s = 0$)

$$a_{|m-1|}^{(1)} = 2m\, a_m + a_{m+1}^{(1)}$$

which we used in the recurrence process, for $m = N, N-1, \cdots, 1$. Since there is no residual equation associated with $m = 0$, we need only make one pass through the above equations with starting conditions $a_N = 1$, $a_N^{(1)} = 0$, in order to generate the Chebyshev coefficients for $\cosh z$. They are displayed (for $N = 22$), along with those of $\cosh(z^{\frac{1}{2}})$ for comparison in Table 6.1. Observe how the coefficients in the case $k = 2$ clearly decay more rapidly than those in the case $k = 1$.

B. THE CASE n=1

For $n = 1$, we have $\alpha = \frac{3}{k}$, $k = 1, 2, 3$. To compute the coefficients of the auxiliary function

$$f\left(z^{\frac{1}{3}}\right) = 1 + \frac{1}{3!}z + \frac{1 \cdot 4}{6!}z^2 + \cdots$$

we employ the recurrence (obtained from the differential equation satisfied by f)

$$|a_{m-1}| = \frac{9}{2}\left[a_{m+2}^{(2)} + 3\,a_m^{(2)} + 2\,(m-1)\,a_{|m-1|}^{(1)}\right] + 6\left[a_{m+1}^{(1)} + a_{|m-1|}^{(1)}\right] - a_{m+1} \tag{6.1}$$

TABLE 6.2: Comparison of Chebyshev Coefficients for the Case $n = 1$

m	$f(z^{\frac{1}{3}})$ $(k = 1)$	m	$f(z^{\frac{2}{3}})$ $(k = 2)$	m	$f(z)$ $(k = 3)$
0	2.00556	0	2.17088	0	2.00347
1	0.16672	2	8.61475×10^{-2}	1	0.12504
2	2.77807×10^{-3}	4	7.09041×10^{-4}	2	2.60440×10^{-3}
3	1.92910×10^{-5}	6	2.44805×10^{-6}	3	4.16920×10^{-2}
4	7.30704×10^{-8}	8	4.62145×10^{-9}	4	1.04181×10^{-3}
5	1.73975×10^{-10}	10	5.49020×10^{-12}	5	1.08512×10^{-5}
6	2.84272×10^{-13}	12	4.47887×10^{-15}	6	1.73674×10^{-4}
7	3.38418×10^{-16}	14	2.66310×10^{-18}	7	2.71291×10^{-6}
8	3.06538×10^{-19}	16	1.20511×10^{-21}	8	1.88386×10^{-8}
9	2.18331×10^{-22}	18	4.28885×10^{-25}	9	3.01486×10^{-7}
10	1.25477×10^{-25}	20	1.23176×10^{-28}	10	3.42531×10^{-9}
11	5.94118×10^{-29}	22	2.91481×10^{-32}	11	1.78395×10^{-11}
12	2.35761×10^{-32}	24	5.78564×10^{-36}	12	2.85481×10^{-10}

for $m = N, N-1, \cdots, 1$, in conjunction with (5.4) when $s = 0, 1$. After one pass through the recurrence relations, the residual equation (6.1) corresponding to $m = 0$

$$\frac{9}{2}\left(a_2^{(2)} + 3\,a_0^{(2)}\right) + 3\,a_1^{(1)} - a_1 = 0$$

is satisfied to within 10^{-17} when $N = 20$. As in all of the examples presented here, we used the normalization $f(0) = 1$.

When $k = 2$, the change of variable is $x = z^{\frac{2}{3}}$. The function

$$f\left(z^{\frac{2}{3}}\right) = 1 + \frac{1}{3!}\,z^2 + \frac{4}{6!}\,z^4 + \cdots$$

is a solution of

$$9z\,u'' + 3\,u' - 4z\,u = 0.$$

From this we obtain the recurrence

$$a_{|m-1|} = \frac{9}{4}\left[a_{m+3}^{(2)} + 2\,a_{m+1}^{(2)} + a_{|m-1|}^{(2)}\right] + \frac{3}{2}\left[a_m^{(1)} + a_{m+2}^{(1)}\right] - a_{m+3} + 2\,a_{m+1}$$

for $m = N, N-1, \ldots, 1$. There is no residual since only even terms appear in our Chebyshev expansion.

In the case $k = 3$, no change of variable occurs and the differential equation is Airy's equation. Example 5.2 thoroughly discusses the computation of the Chebyshev coefficients for the auxiliary function $f(z)$ in this case.

In Table 6.2, we present the coefficients for f under the several changes of scale corresponding to $k = 1, 2, 3$. Note again that $k = 2$ gives the "best" change of variable.

C. THE CASE n=2

Here $\alpha = \frac{4}{k}$, $k = 1, 2, 3, 4$. In the case $k = 4$ (no change of variable) the differential equation of interest

$$f''(x) - x^2 f(x) = 0$$

is related to Weber's equation and has the solutions

$$\begin{aligned} y_1(x) &= 1 + \frac{2}{4!}\,x^4 + \frac{60}{8!}\,x^8 + \cdots \\ y_2(x) &= \sqrt{2}\,x\left\{1 + \frac{6}{5!}\,x^4 + \frac{28}{8!}\,x^8 + \cdots\right\} \end{aligned}$$

TABLE 6.3: Comparison of Chebyshev Coefficients for the Case $n = 2$

m	$y_1(z^{\frac{1}{4}})$ $(k = 1)$	m	$y_1(z^{\frac{1}{2}})$ $(k = 2)$	m	$y_1(z)$ $(k = 4)$
0	2.00149	0	2.08446	0	2.06332
1	8.33418×10^{-2}	2	4.24161×10^{-2}	2	4.23221×10^{-2}
2	7.44071×10^{-4}	4	1.88136×10^{-4}	4	1.07450×10^{-2}
3	2.81840×10^{-6}	6	3.55242×10^{-7}	6	9.42232×10^{-5}
4	5.87163×10^{-9}	8	3.69396×10^{-10}	8	1.19917×10^{-5}
5	7.72581×10^{-12}	10	2.42746×10^{-13}	10	6.68618×10^{-8}
6	6.99800×10^{-15}	12	1.09851×10^{-16}	12	5.67778×10^{-7}
7	4.62830×10^{-18}	14	3.63046×10^{-20}	14	2.32060×10^{-11}
8	2.33281×10^{-21}	16	9.14514×10^{-24}	16	1.47860×10^{-12}
9	9.25720×10^{-25}	18	1.81385×10^{-27}	18	4.76969×10^{-15}
10	2.96705×10^{-28}	20	2.90594×10^{-31}	20	2.43185×10^{-16}
11	7.84104×10^{-32}	22	3.83882×10^{-35}	22	6.47943×10^{-19}
12	1.73782×10^{-35}	24	4.25449×10^{-39}	24	2.76050×10^{-20}

both of which clearly have convergent Chebyshev expansions. Here, *two* passes through the recurrence relations

$$\begin{aligned} a_m^{(2)} &= a_{m+2}^{(2)} + 2(m+1)\, a_{m+1}^{(1)} \\ a_m^{(1)} &= a_{m+2}^{(1)} + 2(m+1)\, a_{m+1} \\ a_m &= 4\, a_{m+2}^{(2)} - a_{m+4} - 2\, a_{m+2} \end{aligned}$$

for $m = N-1, N-2, \ldots, 1, 0$, are needed to obtain the Chebyshev coefficients for either solution.

When $k = 1$, we have $x = z^{\frac{1}{4}}$ and the auxiliary function

$$y_1\left(z^{\frac{1}{4}}\right) = 1 + \frac{2}{4!}\, z + \frac{60}{8!}\, z^2 + \cdots$$

satisfies

$$16z\, u''(z) + 12\, u'(z) - u(z) = 0.$$

The resulting recurrence relation is

$$a_{|m-1|} = 16\, a_m^{(2)} + 4(4m-1)\, a_{|m-1|}^{(1)}, \qquad m = N, N-1, \cdots, 1. \tag{6.2}$$

When $m = 0$ in (6.2), the residual equation

$$16\, a_0^{(2)} - 4\, a_1^{(1)} - a_1 = 0$$

is satisfied to within 10^{-17} for $N = 14$.

For $k = 2$, the differential equation is

$$4z\, u''(z) + 2\, u'(z) - z\, u(z) = 0$$

whose solution

$$y_1\left(z^{\frac{1}{2}}\right) = 1 + \frac{2}{4!}\, z^2 + \frac{60}{8!}\, z^4 + \cdots$$

we consider. From the differential equation we obtain

$$a_{|m-1|} = 4\left[a_{m+1}^{(2)} + a_{|m-1|}^{(2)}\right] + 4\, a_m^{(1)} - a_{m+1}, \qquad m = N, N-1, \ldots, 1.$$

The residual equation $(m = 0)$

$$4\, a_1^{(2)} + 2\, a_0^{(1)} - a_1 = 0$$

is automatically satisfied since $y_1\left(z^{\frac{1}{2}}\right)$ is even. A comparison of the coefficients in these three cases is given in Table 6.3.

APPENDIX A: DEMONSTRATION OF THEOREM 2.2

The initial step in the proof is the determination of the fundamental matrix solution of equation (2.5), namely

$$t^{-(n+1)}\frac{dZ}{dt} = \begin{bmatrix} 0 & 1 \\ 1 & nt^{-(n+2)} \end{bmatrix} = C(t)Z \tag{A.1}$$

where

$$C(t) = \sum_{j=0}^{\infty} C_j t^{-j}.$$

Clearly, here

$$C_0 = \begin{bmatrix} 0 & 1 \\ 1 & 0 \end{bmatrix}, \; C_{n+2} = \begin{bmatrix} 0 & 0 \\ 0 & n \end{bmatrix}, \text{ and } C_j = 0 \text{ for all other } j.$$

For computational ease, we follow Sirovich (1971) and seek formal *vector* solutions of the form

$$\mathbf{z}^{(i)}(t) = \left(\sum_{j=0}^{\infty} \hat{\mathbf{z}}_j^{(i)} t^{-j}\right) e^{q^{(i)}(t)} t^{\mu_i} \tag{A.2}$$

valid for $t \in S_1$, where the open sector S_1 has central angle less than $\frac{\pi}{n+2}$. Here, the $\hat{\mathbf{z}}_j^{(i)}$ are column vectors and

$$q^{(i)}(t) = \frac{a_0^{(i)}}{n+2} t^{n+2} + a_1^{(i)} t^{n+1} + \cdots + a_{n+1}^{(i)} t \tag{A.3}$$

for $i = 1, 2$. Once the coefficients and vectors have been determined, the fundamental matrix solution to (A.1) is then constructed by setting

$$\hat{Z}(t) = \sum_{j=0}^{\infty} \hat{Z}_j t^{-j} = \left[\mathbf{z}^{(1)}(t) \quad \mathbf{z}^{(2)}(t)\right] \tag{A.4}$$

where

$$\hat{Z}_j = \left[\hat{\mathbf{z}}_j^{(1)} \quad \hat{\mathbf{z}}_j^{(2)}\right], \quad j = 0, 1, 2, \cdots$$

$$Q(t) = \begin{bmatrix} q^{(1)}(t) & 0 \\ 0 & q^{(2)}(t) \end{bmatrix}$$

$$D = \begin{bmatrix} \mu_1 & 0 \\ 0 & \mu_2 \end{bmatrix}.$$

The following relations are a consequence of substitution into the vector differential equation corresponding to the matrix equation (2.5):

$$\left(C_0 - a_0^{(i)} I\right) \hat{\mathbf{z}}_j^{(i)} = \sum_{l=1}^{j} \left(a_l^{(i)} I - C_l\right) \hat{\mathbf{z}}_{j-l}^{(i)}, \quad j = 0, 1, \cdots, n+2. \tag{A.5}$$

$$\begin{aligned} \left(C_0 - a_0^{(i)} i\right) \hat{\mathbf{z}}_j^{(i)} &= (n+2-j)\hat{\mathbf{z}}_{j-(n+2)}^{(i)} + \textstyle\sum_{l=1}^{j} \left(a_l^{(i)} I - C_l\right) \hat{\mathbf{z}}_{j-l}^{(i)} \\ j &= n+3, n+4, \ldots \qquad \text{and} \qquad i = 1, 2. \end{aligned} \tag{A.6}$$

Here we have made use of the choice $\mu_i \equiv a_{n+2}^{(i)}$.

Now, the eigenvalues of C_0 are $\lambda_1 = 1$ and $\lambda_2 = -1$. The computations for $\lambda_1 = 1$ ($i = 1$) will be done first. When $j = 0$, equation (A.5) becomes

$$\left(C_0 - a_0^{(1)} I\right) \hat{\mathbf{z}}_0^{(i)} = \begin{bmatrix} 0 \\ 0 \end{bmatrix}$$

which is satisfied if

$$a_0^{(1)} = \lambda_1 = 1 \text{ and } \hat{\mathbf{z}}_0^{(1)} = \begin{bmatrix} 1 \\ 1 \end{bmatrix}.$$

That is, $\hat{\mathbf{z}}_0^{(1)}$ is an eigenvector of C_0 corresponding to $\lambda_1 = 1$. When $j = 1$, $C_1 = 0$ and hence

$$(C_0 - I)\, \hat{\mathbf{z}}_1^{(1)} = \left(a_1^{(1)} I - C_1\right) \hat{\mathbf{z}}_0^{(1)}$$

which implies that

$$\begin{bmatrix} -1 & 1 \\ 1 & -1 \end{bmatrix} \hat{\mathbf{z}}_1^{(1)} = a_1^{(1)} \begin{bmatrix} 1 \\ 1 \end{bmatrix}. \tag{A.7}$$

From fundamental principles, this system is solvable if and only if

$$\left\langle \begin{bmatrix} 1 \\ 1 \end{bmatrix}, a_1^{(1)} \begin{bmatrix} 1 \\ 1 \end{bmatrix} \right\rangle = 0$$

that is, if and only if $a_1^{(1)} = 0$. Therefore, the only possible solution of (A.7) is

$$\hat{\mathbf{z}}_1^{(1)} = \alpha_1 \hat{\mathbf{z}}_0^{(1)}, \quad \alpha_1 \text{ constant.} \tag{A.8}$$

In like manner, when $j = 2$ we have $C_2 = 0$ (if $n > 0$) and hence,

$$(C_0 - I)\, \hat{\mathbf{z}}_2^{(1)} = a_2^{(1)} \hat{\mathbf{z}}_0^{(1)}.$$

The solvability condition now yields $a_2^{(1)} = 0$ and thus

$$\hat{\mathbf{z}}_2^{(1)} = \alpha_2 \hat{\mathbf{z}}_0^{(1)}, \quad \alpha_2 \text{ constant.}$$

Proceeding inductively, since $C_3 = \cdots = C_{n+1} = 0$, it readily follows that $a_j^{(1)} = 0$ for $j = 1, 2, \cdots, n+1$ and

$$\hat{\mathbf{z}}_j^{(1)} = \alpha_j \hat{\mathbf{z}}_0^{(1)} = \alpha_j \begin{bmatrix} 1 \\ 1 \end{bmatrix}, \quad \alpha_j \text{ constant.}$$

For $j = n+2$ the situation is slightly different. Indeed, the relation (A.5) gives

$$(C_0 - I)\,\hat{\mathbf{z}}_{n+2}^{(1)} = \left(a_{n+2}^{(1)} I - C_{n+2}\right) \hat{\mathbf{z}}_0^{(1)} \tag{A.9}$$

or

$$\begin{bmatrix} -1 & 1 \\ 1 & -1 \end{bmatrix} \hat{\mathbf{z}}_{n+2}^{(1)} = \begin{bmatrix} a_{n+2}^{(1)} \\ a_{n+2}^{(1)} - n \end{bmatrix}.$$

This equation is solvable if and only if

$$\left\langle \begin{bmatrix} 1 \\ 1 \end{bmatrix}, \begin{bmatrix} a_{n+2}^{(1)} \\ a_{n+2}^{(1)} - n \end{bmatrix} \right\rangle = 0$$

which implies that

$$\mu_1 \equiv a_{n+2}^{(1)} = \frac{n}{2}.$$

Substituting this condition into (A.9) yields

$$\begin{bmatrix} -1 & 1 \\ 1 & -1 \end{bmatrix} \hat{\mathbf{z}}_{n+2}^{(1)} = \frac{n}{2} \begin{bmatrix} 1 \\ -1 \end{bmatrix}$$

and thus

$$\hat{\mathbf{z}}_{n+2}^{(1)} = \beta \begin{bmatrix} 1 \\ 1 \end{bmatrix} + \begin{bmatrix} 0 \\ \frac{n}{2} \end{bmatrix} = \begin{bmatrix} \beta \\ \beta + \frac{n}{2} \end{bmatrix}, \quad \beta \text{ constant.}$$

When $j = n+3$, we see that $C_{n+3} = 0$ and so by (A.6)

$$(C_0 - I)\,\hat{\mathbf{z}}_{n+3}^{(1)} = -\alpha_1 \hat{\mathbf{z}}_0^{(1)} + \alpha_1 \left(\frac{n}{2} I - C_{n+2}\right) \hat{\mathbf{z}}_0^{(1)}$$

or

$$\begin{bmatrix} -1 & 1 \\ 1 & -1 \end{bmatrix} \hat{\mathbf{z}}_{n+3}^{(1)} = \alpha_1 \begin{bmatrix} \frac{n}{2} - 1 \\ -\frac{n}{2} - 1 \end{bmatrix}. \tag{A.10}$$

By the solvability condition, we have

$$\left\langle \begin{bmatrix} 1 \\ 1 \end{bmatrix}, \alpha_1 \begin{bmatrix} \frac{n}{2} - 1 \\ -\frac{n}{2} - 1 \end{bmatrix} \right\rangle = 0$$

from which it follows that $\alpha_1 = 0$. Therefore, by (A.8), $\hat{\mathbf{z}}_1^{(1)} = \vec{0}$ and by (A.10)

$$\hat{\mathbf{z}}_{n+3}^{(1)} = \alpha_{n+3} \begin{bmatrix} 1 \\ 1 \end{bmatrix}.$$

Another simple induction argument leads to

$$\begin{aligned} \alpha_2 = \alpha_3 = \cdots = \alpha_{n+1} = 0 \\ \hat{\mathbf{z}}_2^{(1)} = \hat{\mathbf{z}}_3^{(1)} = \cdots = \hat{\mathbf{z}}_{n+1}^{(1)} = \vec{0} \end{aligned}$$

and

$$\hat{\mathbf{z}}_j^{(1)} = \alpha_j \hat{\mathbf{z}}_0^{(1)}, \quad j = n+3, n+4, \ldots, 2(n+2)-1.$$

When $j = 2(n+2)$, (A.6) becomes

$$(C_0 - I)\, \hat{\mathbf{z}}_{2(n+2)}^{(1)} = -(n+2)\hat{\mathbf{z}}_{n+2}^{(1)} + \left(\frac{n}{2} I - C_{n+2}\right) \hat{\mathbf{z}}_{n+2}^{(1)} \tag{A.11}$$

or

$$\begin{bmatrix} -1 & 1 \\ 1 & -1 \end{bmatrix} \hat{\mathbf{z}}_{2(n+2)}^{(1)} = \begin{bmatrix} -\left(2 + \frac{n}{2}\right)\beta \\ -\left(2 + \frac{n}{2}\right)\left(\beta + \frac{n}{2}\right) \end{bmatrix}.$$

Applying the now familiar solvability condition, we are led to

$$\beta = \frac{-n(3n+4)}{8(n+2)}.$$

Thus, it must be that

$$\hat{\mathbf{z}}_{n+2}^{(1)} = \begin{bmatrix} \frac{-n(3n+4)}{8(n+2)} \\ \frac{n(n+4)}{16(n+2)} \end{bmatrix}.$$

With this value of β, (A.11) implies

$$\hat{\mathbf{z}}_{2(n+2)}^{(1)} = \gamma \begin{bmatrix} 1 \\ 1 \end{bmatrix} + \begin{bmatrix} 0 \\ -\beta\left(2 + \frac{n}{2}\right) \end{bmatrix} = \begin{bmatrix} \gamma \\ \gamma - \beta\left(2 + \frac{n}{2}\right) \end{bmatrix} \tag{A.12}$$

where γ is a constant.

In analogy with our previous reasoning, it is easy to show that

$$\alpha_j = 0, \quad \hat{\mathbf{z}}_j^{(1)} = \vec{0}, \quad j = n+3, \ldots, 2(n+2)-1.$$

The arbitrary constant γ can then be determined from the solvability condition for the system defined by (A.6) when $j = 3(n+2)$. We find

$$\gamma = \frac{-n(n+4)(3n+4)(5n+8)}{128(n+2)^2}$$

and so by (A.12)

$$\hat{\mathbf{z}}^{(1)}_{2(n+2)} = \begin{bmatrix} \frac{-n(n+4)(3n+4)(5n+8)}{128(n+2)^2} \\ \frac{n(n+4)(3n+4)(3n+8)}{128(n+2)^2} \end{bmatrix}.$$

At this stage, we observe that we could continue this process and obtain expressions for all $\hat{\mathbf{z}}^{(1)}_{p(n+2)}$, $p = 1, 2, 3, \ldots$. Of course, $\hat{\mathbf{z}}^{(1)}_j = \vec{0}$ for all other j.

Summarizing, we see that for $\lambda_1 = 1$:

$$a_0^{(1)} = 1,\ a_j^{(1)} = 0,\ j = 1, 2, \ldots, n+1,\ a_{n+2}^{(1)} = \frac{n}{2}$$

$$\hat{\mathbf{z}}_0^{(1)} = \begin{bmatrix} 1 \\ 1 \end{bmatrix},\ \hat{\mathbf{z}}^{(1)}_{n+2} = \begin{bmatrix} \frac{-n(3n+4)}{8(n+2)} \\ \frac{n(n+4)}{16(n+2)} \end{bmatrix},\ \hat{\mathbf{z}}^{(1)}_{2(n+2)} = \begin{bmatrix} \frac{-n(n+4)(3n+4)(5n+8)}{128(n+2)^2} \\ \frac{n(n+4)(3n+4)(3n+8)}{128(n+2)^2} \end{bmatrix}.$$

Moreover, $\mu_1 = \frac{n}{2}$ and from (A.3), $q^{(1)}(t) = \frac{t^{n+2}}{n+2}$. Hence, to three terms,

$$\begin{aligned} \mathbf{z}^{(1)}(t) &= t^{\frac{n}{2}} e^{\frac{t^{n+2}}{n+2}} \left\{ \begin{bmatrix} 1 \\ 1 \end{bmatrix} + \begin{bmatrix} \frac{-n(3n+4)}{8(n+2)} \\ \frac{n(n+4)}{16(n+2)} \end{bmatrix} t^{-(n+2)} \right. \\ &\left. + \begin{bmatrix} \frac{-n(n+4)(3n+4)(5n+8)}{128(n+2)^2} \\ \frac{n(n+4)(3n+4)(3n+8)}{128(n+2)^2} \end{bmatrix} t^{-2(n+2)} + O\left[t^{-3(n+2)}\right] \right\} \end{aligned} \tag{A.13}$$

as $t \to \infty$, $t \in S_1$.

In order to obtain $\mathbf{z}^{(2)}(t)$, we consider the relations (A.5) and (A.6) for $i = 2$ and

$$a_0^{(2)} = \lambda_2 = -1, \quad \hat{\mathbf{z}}_0^{(2)} = \begin{bmatrix} 1 \\ -1 \end{bmatrix}.$$

Proceeding as we did in the previous case gives

$$a_j^{(2)} = 0, \quad \hat{\mathbf{z}}_j^{(2)} = \vec{0},\ j \neq p(n+2),\ p = 1, 2, 3, \cdots$$

$$\mu_2 \equiv a_{n+2}^{(2)} = \frac{n}{2} = \mu_1, \quad q^{(2)}(t) = -\frac{t^{n+2}}{n+2}$$

and

$$\hat{\mathbf{z}}^{(2)}_{n+2} = \begin{bmatrix} \frac{n(3n+4)}{8(n+2)} \\ \frac{n(n+4)}{16(n+2)} \end{bmatrix},\ \hat{\mathbf{z}}^{(2)}_{2(n+2)} = \begin{bmatrix} \frac{-n(n+4)(3n+4)(5n+8)}{128(n+2)^2} \\ \frac{-n(n+4)(3n+4)(3n+8)}{128(n+2)^2} \end{bmatrix}.$$

Similar to before, when we combine everything

$$\mathbf{z}^{(2)}(t) = t^{\frac{n}{2}}e^{-\frac{t^{n+2}}{n+2}}\left\{\begin{bmatrix}1\\-1\end{bmatrix}+\begin{bmatrix}\frac{n(3n+4)}{8(n+2)}\\\frac{n(n+4)}{16(n+2)}\end{bmatrix}t^{-(n+2)}\right.$$

$$\left.+\begin{bmatrix}\frac{-n(n+4)(3n+4)(5n+8)}{128(n+2)^2}\\\frac{-n(n+4)(3n+4)(3n+8)}{128(n+2)^2}\end{bmatrix}t^{-2(n+2)}+O\left[t^{-3(n+2)}\right]\right\}$$

as $t \to \infty$, $t \in S_1$. Finally, the fundamental matrix solution of (A.1) is

$$Z(t) = t^{\frac{n}{2}}\left\{\begin{bmatrix}1 & 1\\1 & -1\end{bmatrix}+\begin{bmatrix}\beta & -\beta\\\beta+\frac{n}{2} & \beta+\frac{n}{2}\end{bmatrix}t^{-(n+2)}\right.$$

$$\left.+\begin{bmatrix}\gamma & \gamma\\\gamma-\beta(2+\frac{n}{2}) & \beta(2+\frac{n}{2})\end{bmatrix}t^{-2(n+2)}+O\left(t^{-3(n+2)}\right)\right\}\begin{bmatrix}e^{\frac{t^{n+2}}{n+2}} & 0\\0 & e^{-\frac{t^{n+2}}{n+2}}\end{bmatrix}$$

as $t \to \infty$, $t \in S_1$.

To complete our demonstration we need to replace t by $2^{\frac{1}{n+2}}x^{\frac{1}{2}}$ and recover

$$Y(x) = \begin{bmatrix}0 & 1\\1 & 0\end{bmatrix}W(x) = \begin{bmatrix}0 & 1\\1 & 0\end{bmatrix}\begin{bmatrix}1 & 0\\0 & x^{-\frac{n}{2}}\end{bmatrix}Z = \begin{bmatrix}0 & x^{-\frac{n}{2}}\\1 & 0\end{bmatrix}Z.$$

Carrying out the prescribed steps and dividing by the multiplicative constant $2^{\frac{n}{2(n+2)}}$ leads to the normalized fundamental matrix solution

$$Y(x)$$
$$\sim\begin{bmatrix}x^{-\frac{n}{4}}\exp\left(\frac{2}{n+2}x^{\frac{n+2}{2}}\right) & -x^{-\frac{n}{4}}\exp\left(-\frac{2}{n+2}x^{\frac{n+2}{2}}\right)\\x^{\frac{n}{4}}\exp\left(\frac{2}{n+2}x^{\frac{n+2}{2}}\right) & x^{\frac{n}{4}}\exp\left(-\frac{2}{n+2}x^{\frac{n+2}{2}}\right)\end{bmatrix}$$

$$+\frac{1}{2}\begin{bmatrix}(\beta+\frac{n}{2})x^{-\frac{(3n+4)}{4}}\exp\left(\frac{2}{n+2}x^{\frac{n+2}{2}}\right) & (\beta+\frac{n}{2})x^{-\frac{(3n+4)}{4}}\exp\left(-\frac{2}{n+2}x^{\frac{n+2}{2}}\right)\\\beta x^{-\frac{n+4}{4}}\exp\left(\frac{2}{n+2}x^{\frac{n+2}{2}}\right) & -\beta x^{-\frac{(n+4)}{4}}\exp\left(-\frac{2}{n+2}x^{\frac{n+2}{2}}\right)\end{bmatrix}$$

$$+\frac{1}{4}\begin{bmatrix}(\gamma-\beta(2+\frac{n}{2}))x^{-\frac{(5n+8)}{4}}\exp\left(\frac{2}{n+2}x^{\frac{n+2}{2}}\right) & (\beta(2+\frac{n}{2})-\gamma)x^{-\frac{(5n+8)}{4}}\exp\left(-\frac{2}{n+2}x^{\frac{n+2}{2}}\right)\\\gamma x^{-\frac{(3n+8)}{4}}\exp\left(\frac{2}{n+2}x^{\frac{n+2}{2}}\right) & \gamma x^{-\frac{(3n+8)}{4}}\exp\left(-\frac{2}{n+2}x^{\frac{n+2}{2}}\right)\end{bmatrix}$$
$$+\cdots$$

as $x \to \infty$, $x \in S$ (In view of the transformation from t to x, the sector S is centered just as S_1, but is twice as large). Since, by Theorem 2.1 the first column of (A.14) contains asymptotic representations for u_1 and $\frac{du_1}{dx}$, and the second column those for u_2 and $\frac{du_2}{dx}$, the proof is complete.

APPENDIX B. DEMONSTRATION OF LEMMA 4.1

LEMMA 4.1 If $u(x)$ is a solution of $u'' - x^n u = 0$ with $x = z^{\frac{1}{\alpha}}$, and

$$a_m = \frac{2}{\pi} \int_{-1}^{1} \frac{u\left(z^{\frac{1}{\alpha}}\right) T_m(z)}{\sqrt{1-z^2}}\, dz, \qquad m = 0, 1, 2, \ldots \tag{B.1}$$

is the mth Chebyshev coefficient of $u\left(z^{\frac{1}{\alpha}}\right)$, then for each fixed m, a_m is analytic in α, for $\Re\alpha > 0$.

PROOF: Since $u(x)$ is entire, we know that

$$u(x) = \sum_{j=0}^{\infty} c_j x^j$$

converges absolutely for $|x| < \infty$ and uniformly for $|x| \leq R$, for any $R > 0$. Therefore, the same may be said of the series

$$u\left(z^{\frac{1}{\alpha}}\right) = \sum_{j=0}^{\infty} c_j z^{\frac{j}{\alpha}}. \tag{B.2}$$

Substituting (B.2) into the formula (B.1) yields

$$a_m = \frac{2}{\pi} \sum_{j=0}^{\infty} c_j \left\{ \int_{-1}^{1} \frac{z^{\frac{j}{\alpha}} T_m(z)}{\sqrt{1-z^2}}\, dz \right\} = \frac{2}{\pi} \sum_{j=0}^{\infty} c_j\, F_j(\alpha; m) \tag{B.3}$$

where the uniform convergence of (B.2) justifies the interchange of summation and integration.

Now, since $|T_m(z)| \leq 1$ for all nonnegative integer m, $-1 \leq z \leq 1$,

$$\begin{aligned} \left| \int_0^1 \frac{z^{\frac{j}{\alpha}} T_m(z)}{\sqrt{1-z^2}}\, dz \right| &\leq \int_0^1 \frac{z^{\Re(\frac{j}{\alpha})}}{\sqrt{1-z^2}}\, dz \\ &\leq \int_0^1 \frac{dz}{\sqrt{1-z^2}} \\ &= \frac{\pi}{2}. \end{aligned} \tag{B.4}$$

The last inequality comes from the fact that $\Re(\frac{j}{\alpha}) \geq 0$ for $\Re\alpha > 0$.

Also,

$$\left|\int_{-1}^{0} \frac{z^{\frac{i}{\alpha}}\, T_m(z)}{\sqrt{1-z^2}}\, dz\right| = \exp\left[-j\pi \cdot \frac{\Im\alpha\, \mathrm{sgn}(\Im\alpha)}{(\Re\alpha)^2 + (\Im\alpha)^2}\right] \cdot \left|\int_{0}^{1} \frac{\tau^{\frac{i}{\alpha}}\, T_m(-\tau)}{\sqrt{1-\tau^2}}\, d\tau\right|$$
$$\leq \frac{\pi}{2} \qquad (B.5)$$

(where $z = \tau \exp\{-i\pi\, \mathrm{sgn}(\Im\alpha)\}$) by the same reasoning.

(B.4) and (B.5) together imply that

$$|F_j(\alpha;m)| = \left|\int_{-1}^{1} \frac{z^{\frac{i}{\alpha}}\, T_m(z)}{\sqrt{1-z^2}}\, dz\right| \leq \pi$$

for all α such that $\Re\alpha > 0$. Furthermore, the series

$$\sum_{j=0}^{\infty} |c_j|$$

converges since $u\left(z^{\frac{1}{\alpha}}\right)$ as given by (B.2) is entire. This, and the fact that

$$|c_j\, F_j(\alpha;m)| \leq \pi\, |c_j|, \qquad j = 0,1,2,\cdots \qquad (B.6)$$

imply by the Weierstrass M-test (see DePree and Oehring (1969), for example) that

$$a_m = \frac{2}{\pi} \sum_{j=0}^{\infty} c_j\, F_j(\alpha;m)$$

converges uniformly for all α, $\Re\alpha > 0$. The inequality (B.6) guarantees that this convergence is absolute as well. Since the series (B.3) defining a_m converges absolutely and uniformly for all α, $\Re\alpha > 0$, and each of the $F_j(\alpha;m)$ is analytic in α in this domain, we have that a_m is analytic in α for these values of α.

REFERENCES

1. M. Abramowitz and I. A. Stegun, *Handbook of Mathematical Functions*, Dover, New York, 1972.

2. British Association for the Advancement of Science, Bessel Functions - Part II, *Mathematical Tables* **X**, Cambridge University Press, 1952.

3. N.G. DeBruijn, *Asymptotic Methods in Analysis*, Dover, New York, 1981.

4. C. W. Clenshaw, The Numerical Solution of Linear Differential Equations in Chebyshev Series, *Proc. Camb. Phil. Soc.* **53** (1957), 134-149.

5. C. W. Clenshaw, *Chebyshev Series for Mathematical Functions*, Nat'l. Phys. Lab. Math. Tables, Vol. 5, H. M. Stationery Office, London, 1963.

6. C. W. Clenshaw and S. M. Picken, *Chebyshev Series for Bessel Functions of Fractional Order*, Nat'l Phys. Lab. Math. Tables, Vol. 8, H. M. Stationery Office, London, 1966.

7. J.A. Cochran, *Applied Mathematics,* Wadsworth, California, 1982.

8. J.A. Cochran and R. M. Zolnowski, Clenshaw's Method for Chebyshev Approximation with an Application to Airy Functions, Bell Telephone Laboratories MM68-4133-31, 1968.

9. P.J. Davis, *Interpolation and Approximation*, Blaisdell, New York, 1963.

10. J.D. DePree and C. Oehring, *Elements of Complex Analysis*, Addison-Wesley, Massachusetts, 1969.

11. D. Elliott, The Evaluation and Estimation of the Coefficients in the Chebyshev Series, *Math. Comp.*, **18** (1964), 274-284.

12. D. Elliott and G. Szekeres, Some Estimates of the Coefficients in the Chebyshev Series Expansion of a Function, *Math. Comp.* **19**, (1965), 25-32.

13. A. Erdélyi, *Asymptotic Expansions*, Dover, New York, 1956.

14. L. Fox and I. B. Parker, *Chebyshev Polynomials in Numerical Analysis*, Oxford University Press, London, 1968.

15. W. Gautschi, Computational Aspects of Three-Term Recurrence Relations, *SIAM Review* **9** (1) (1967), 24-82.

16. D. Gottlieb and S.A. Orszag, *Numerical Analysis of Spectral Methods: Theory and Applications*, Society for Industrial and Applied Mathematics, Philadelphia, 1977.

17. P.-F. Hsieh and Y. Sibuya, On the Asymptotic Integration of Second-Order, Linear Ordinary Differential Equations with Polynomial Coefficients, *J. Math. Anal. and Appl.* **16** (1966), 84-103.

18. E. L. Ince, *Ordinary Differential Equations*, Dover, New York, 1956.

19. C. Lanczos, *Applied Analysis*, Pitman, London, 1957.

20. G. F. Miller, On the Convergence of the Chebyshev Series for Functions Possessing a Singularity in the Range of Representation, *SIAM J. Numer. Anal.* **3** (3) (1966), 390-409.

21. F.W.J. Olver, Error Analysis of Miller's Recurrence Algorithm, *Math. Comp.* **18** (1964), 65-74.

22. F.W.J. Olver, *Asymptotics and Special Functions*, Academic Press, New York, 1974.

23. Lord Rayleigh (J. W. Strutt), The Incidence of Light Upon a Transparent Sphere of Dimensions Comparable with the Wave-length, *Proc. Roy. Soc. London, Ser. A* **84** (1910), 25-46.

24. T. J. Rivlin, *The Chebyshev Polynomials*, Wiley, New York, 1974.

25. R. E. Scraton, A Method for Improving the Convergence of Chebyshev Series, *Comp. J.*, **13** (2) (1970), 201-202.

26. L. Sirovich, *Techniques of Asymptotic Analysis*, Springer-Verlag, New York, 1971.

27. G. Szegö, *Orthogonal Polynomials*, American Mathematical Society, New York, 1939.

28. L. W. Thomé, Zur Theorie der linearen Differentialgleichungen, *Journal für die Reine und Angewandte Mathematik* **95** (1883), 44-98.

29. W. Wasow, *Asymptotic Expansions for Ordinary Differential Equations*, Dover, New York, 1965.

30. J. Wimp, *Computation with Recurrence Relations*, Pitman, Boston, 1984.

Practical Methods for the Uniform Asymptotic Evaluation of Oscillating Integrals with Several Coalescing Saddle Points

J. N. L. Connor Professor, Department of Chemistry,
The University of Manchester, Manchester, England

I. INTRODUCTION

Asymptotic techniques [1] play a crucial role in the analysis and development of short-wavelength scattering theories [2-6]. Some typical areas of application for these high frequency theories include: the propagation of electromagnetic, water and acoustic waves, collisions of atoms, molecules and heavy nuclear ions, and scattering from surfaces.

One important asymptotic technique that is extensively used to analyse short-wavelength scattering is the uniform asymptotic evaluation of oscillating integrals with several coalescing saddle points [1,7-10]. The integrals under consideration have the form

$$I(\boldsymbol{\alpha}) = \int_{-\infty}^{\infty} g(t)\exp[if(\boldsymbol{\alpha};t)/\hbar]dt, \qquad \hbar\to 0, \tag{1}$$

where g(t) and $f(\boldsymbol{\alpha};t)$ are analytic functions, with $\boldsymbol{\alpha} \equiv (\alpha_1,\alpha_2,...)$ a set of real parameters. In addition, $f(\boldsymbol{\alpha};t)$ is real for real t and g(t) may also depend on $\boldsymbol{\alpha}$, although this has not been indicated explicitly.

When the saddle point method [1,8,10] is applied to Eq. (1), it is well known that the main contribution to $I(\boldsymbol{\alpha})$ comes from the neighbourhood of the saddle points t_i of $f(\boldsymbol{\alpha};t)$, which are given by

$$\partial f(\boldsymbol{\alpha};t)/\partial t = 0 \quad \text{for } t = t_i(\boldsymbol{\alpha}), \qquad i=1,2,\ldots \, . \tag{2}$$

The positions of the saddle points depend on $\boldsymbol{\alpha}$, and can move close together or coalesce as $\boldsymbol{\alpha}$ varies. A uniform asymptotic treatment takes into account this coalescence, and expresses the result in terms of certain *canonical integrals* and their partial derivatives [11,12].

For n coalescing saddle points, with $n\geq 2$, the canonical integral is [11,12]:

$$C_{n+1}(\mathbf{a}) = \int_{-\infty}^{\infty} \exp\left[i(u^{n+1} + \sum_{k=1}^{n-1} a_k u^k)\right]du. \tag{3}$$

Equation (3) will be called the *cuspoid* canonical integral, because the exponent is the canonical form for the cuspoid catastrophes in the language of catastrophe theory [13].

In practical applications of the uniform asymptotic theory to physical problems, it is necessary to numerically evaluate $C_{n+1}(\mathbf{a})$ and its partial derivatives,

$$\partial C_{n+1}(\mathbf{a})/\partial a_1, \ \partial C_{n+1}(\mathbf{a})/\partial a_2, \ldots, \ \partial C_{n+1}(\mathbf{a})/\partial a_{n-1}, \tag{4}$$

as well as express their arguments $\{a_k\}$ in terms of the exponent $f(\boldsymbol{\alpha};t)$ of the original integral (1). This is only easy to do for the simplest case of n=2 (two coalescing saddle points) [14], for which $C_3(a_1)$ is essentially the regular Airy function (or fold canonical integral):

$$\mathrm{Ai}(x) = (2\pi)^{-1} \int_{-\infty}^{\infty} \exp[i(u^3/3+xu)]du. \tag{5}$$

In a 1975 survey on 'Unsolved problems in the asymptotic estimation of special functions', Olver [15] pointed out that the development of practical techniques to handle the case n=3 (three coalescing saddle points) was an important problem that needed solving.

The purpose of the present paper is to outline practical techniques that have been developed at the University of Manchester for the uniform asymptotic evaluation of oscillating integrals with three (n=3) and four (n=4) coalescing saddle points (the methods actually apply to any n>2 in principle) [16-23]. For three coalescing saddle points, the canonical integral is usually called Pearcey's integral [24-26] (or cusp canonical integral):

$$P(x,y) = \int_{-\infty}^{\infty} \exp[i(u^4+xu^2+yu)]du. \tag{6}$$

For four coalescing saddle points, the canonical integral is known as the swallowtail canonical integral [19,20]:

$$S(x,y,z) = \int_{-\infty}^{\infty} \exp[i(u^5+xu^3+yu^2+zu)]du. \tag{7}$$

Section II gives, to lowest order, the uniform asymptotic approximation for the integral $I(\boldsymbol{\alpha})$. Methods for the numerical evaluation of the cuspoid canonical integral, $C_n(\mathbf{a})$, and its partial derivatives are presented in Sect. III. The methods that have been developed are of two main types: the solution of differential equations, and the use of complex contour quadrature techniques. The evaluation of the arguments $\{a_k\}$ of the canonical integrals in terms of the exponent $f(\alpha;t)$ is considered in Sect. IV. Applications to problems in chemical physics and some additional discussion is given in Sect. V. A summary and conclusions appears in Sect. VI.

II. UNIFORM ASYMPTOTIC APPROXIMATION

In this section, we present the equations for the uniform asymptotic approximation to the integral $I(\alpha)$. The case of two coalescing saddle points was first analysed by Chester et al. [14] (see also Refs. [27,28]); then the theory was extended to handle an arbitrary number of coalescing saddle points [11,12,29-32]. There are many reviews of different aspects of short-wavelength scattering that emphasize the important role of uniform approximations in the theory [5,13,33-43].

Suppose that $f(\alpha;t)$ possesses n coalescing saddle points. The first step in the asymptotic analysis is the introduction of a new variable u, by locally mapping $f(\alpha;t)$ onto a polynomial [11,12]:

$$f(\alpha;t) = u^{n+1} + \sum_{j=1}^{n-1} \zeta_j u^j + A. \tag{8}$$

The new variable u is a function of α and t, i.e., $u = u(\alpha;t)$, but the new real parameters $\{\zeta_j\}$ and A depend just on α. Equation (8) defines an exact local one-to-one uniformly analytic change of variables, provided the saddle points on either side of Eq. (8) correspond

$$t = t_i(\alpha) \leftrightarrow u = u_i(\alpha), \qquad i=1,2,\ldots,n. \tag{9}$$

The original integral $I(\alpha)$ can now be written

$$I(\alpha) = \exp(iA/\hbar) \int_{-\infty}^{\infty} g(t) \frac{dt}{du} \exp\left[i(u^{n+1} + \sum_{j=1}^{n-1} \zeta_j u^j)/\hbar\right] du.$$

The terms of lowest order in the uniform asymptotic expansion are obtained by making the approximation [11,12]

$$g(t)(dt/du) \approx \sum_{k=0}^{n-1} q_k u^k. \tag{10}$$

Note that these terms are *not* the leading terms in the Taylor series expansion. The uniform asymptotic cuspoid approximation for the integral $I(\alpha)$ is then given by [11,12]

$$I(\alpha) \sim \exp(iA/\hbar) \sum_{k=0}^{n-1} \hbar^{(k+1)/(n+1)} q_k$$

$$\times\, U_k(\hbar^{-n/(n+1)}\zeta_1, \hbar^{-(n-1)/(n+1)}\zeta_2, \ldots, \hbar^{-2/(n+1)}\zeta_{n-1}), \tag{11}$$

where

$$U_k(\zeta_1, \zeta_2, \ldots, \zeta_{n-1}) \equiv \int_{-\infty}^{\infty} u^k \exp\left[i\left(u^{n+1} + \sum_{j=1}^{n-1} \zeta_j u^j\right)\right] du. \tag{12}$$

It is clear that $U_k(\zeta_1,\zeta_2,\ldots,\zeta_{n-1})$ for $k = 0,1,\ldots,n-1$ is closely related to the cuspoid canonical integral $C_{n+1}(\mathbf{a})$ and its first-order partial derivatives.

Equation (11) shows there are three problems that must be overcome before the uniform asymptotic cuspoid approximation can be considered a practical tool for applications to short-wavelength scattering. The three problems are:

(a) It is necessary to devise methods for the accurate and efficient numerical evaluation of the cuspoid canonical integral (3) and its first-order partial derivatives (4).

(b) Techniques for calculating the mapping parameters $\{\zeta_j\}$ and A in terms of the original exponent $f(\alpha;t)$ must be found.

(c) The expansion coefficients $\{q_k\}$ must be determined. The $\{q_k\}$ are found by substituting the saddle point condition (9) into Eq. (10), with the resulting jacobian factor $(dt/du)_{t_i}$ evaluated by differentiating the mapping equation (8) twice [11,12]. It is then found that the $\{q_k\}$ are obtained from a set of n *linear* equations. The solution of these linear equations is in principle straightforward, so the determination of the $\{q_k\}$ will not be considered any further.

The next two Sèctions are devoted to methods for solving problems (a) and (b) respectively.

III. NUMERICAL EVALUATION OF THE CUSPOID CANONICAL INTEGRAL

In this section, we describe two methods we have developed for the numerical evaluation of the cuspoid canonical integral

$$C_n(\mathbf{a}) = \int_{-\infty}^{\infty} \exp[i f_n(\mathbf{a};u)]du, \quad \mathbf{a} = (a_1, a_2, \ldots, a_{n-2}), \tag{13}$$

and its (n-2) first-order partial derivatives, where the exponent is

$$f_n(\mathbf{a};u) = u^n + \sum_{k=1}^{n-2} a_k u^k, \qquad n \geq 3. \tag{14}$$

In practical applications, relatively small values of n (e.g., n=3-6) are the most important. The case n=3 is the Airy integral, whose numerical evaluation is well known.

A. Contour Integral Method

The integrand of $C_n(\mathbf{a})$ exhibits infinitely rapid oscillations as $u \to \pm\infty$, making a direct numerical quadrature impossible. The contour integral method exploits the fact that the integrand of $C_n(\mathbf{a})$ is an entire function of u. We can then use Cauchy's theorem and Jordan's Lemma to choose the most convenient path in the complex u plane along which to do the numerical quadrature.

1. ***Method.*** First we write Eq. (13) as the sum of two integrals [18]

$$C_n(\mathbf{a}) = \int_{0}^{\infty} \exp[i f_n(\mathbf{a};u)]du + \int_{0}^{\infty} \exp[i f_n(\mathbf{a};-u)]du. \tag{15}$$

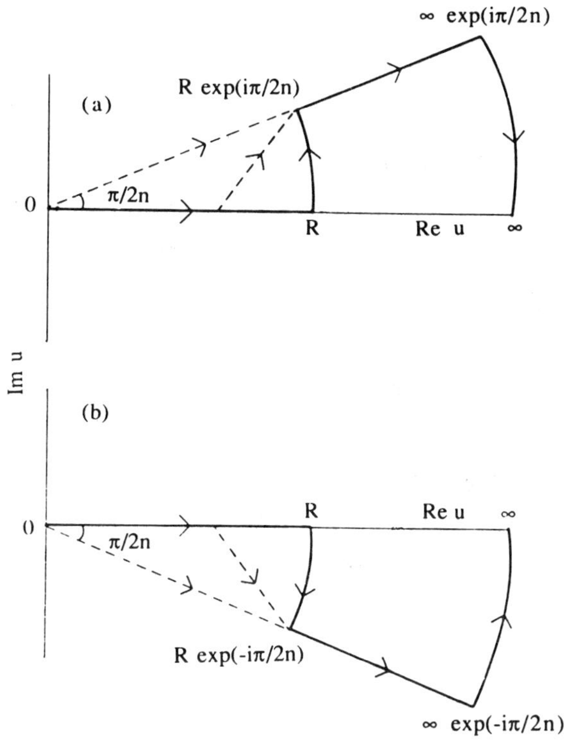

FIG. 1 *Contours in the complex u plane: (a) for n even in both integrals of Eq. (15) and for n odd in the first integral, and (b) for n odd in the second integral of Eq. (15).*

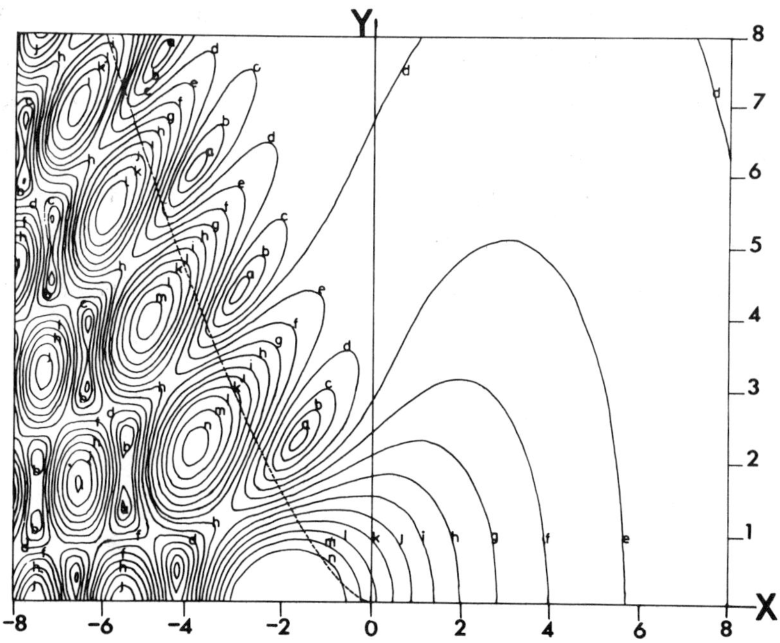

FIG. 2 *Contour plot of* $|P(x,y)|$. *Only the upper half plane is plotted because of the relation* $P(x,-y) = P(x,y)$. *The caustic* $8x^3 + 27y^2 = 0$ *is shown as a dashed line. The contours are labelled from* $a = 0.15$ *to* $n = 2.10$ *in steps of 0.15.*

Consider now the first integral in Eq. (15). We replace the contour along the real axis by the one illustrated in Fig. 1(a) as a heavy line [18]. The new contour proceeds from 0 to a point R on the real axis, then along an arc of a circle from R to R $\exp(i\pi/2n)$, and finally along the ray R $\exp(i\pi/2n)$ to $\infty \exp(i\pi/2n)$. The contribution from $\infty \exp(i\pi/2n)$ to ∞ is zero by an application of Jordan's Lemma. Along the ray $r \exp(i\pi/2n)$ with $r > R$, the modulus of the integrand $\exp[if_n(\mathbf{a};u)]$ eventually decreases rapidly, like $\exp(-u^n)$.

At first sight, it might be thought that using the direct ray from 0 to $\infty \exp(i\pi/2n)$ would be a better choice for the numerical quadrature. To investigate this point [18], consider the following form for Pearcey's integral,

$$P(x,y) = \int_0^\infty 2\cos(yu)\exp[i(u^4+xu^2)]du.$$

For positive values of x, the integrand is well behaved, and the direct ray from 0 to $\infty \exp(i\pi/8)$ is satisfactory for the numerical quadrature. For negative x, however, the term $\exp(ixu^2)$ can become very large in magnitude, before the term $\exp(iu^4)$ eventually dominates. In this situation, as x becomes more negative, violent oscillations of the integrand along the direct ray cause a quadrature to lose accuracy and eventually fail [16,18].

For the choice of contour shown in Fig. 1(a), the first integral in Eq. (15) reduces to the numerical evaluation of two finite integrals

$$\int_0^R \exp[if_n(\mathbf{a};u)]du, \tag{16}$$

and

$$\int_R^{R\exp(i\pi/2n)} \exp[if_n(\mathbf{a};u)]du, \tag{17}$$

and one infinite integral

$$\int_{R\exp(i\pi/2n)}^{\infty\exp(i\pi/2n)} \exp[i f_n(\mathbf{a};u)]du, \tag{18}$$

which can also be written in the form

$$\exp(i\pi/2n) \int_R^{\infty} \exp\left[-r^n + \sum_{k=1}^{n-2} i a_k \exp(ik\pi/2n) r^k\right] dr. \tag{19}$$

We have not yet fixed a value for R. Suppose the modulus of the integrand of Eq. (19) is required to be smaller than exp(-d) with d>1. Then we can choose R to be the largest real root of the polynomial [18]

$$r^n + \sum_{k=1}^{n-2} a_k \sin(k\pi/2n) r^k = d.$$

In many cases, we can chose R sufficiently large that the integral (19) is negligibly small. We are then left with just the two finite integrals (16) and (17) to evaluate. However we have to be careful not to choose R too large in case the oscillatory integral (16) becomes difficult to evaluate. Thus in practice, a compromise value of R may need to be used.

The contour shown in Fig. 1(a) can also be used for the second integral in Eq. (15) when n is an even integer. Otherwise the contour $0 \to R \to R \exp(-i\pi/2n) \to \infty \exp(-i\pi/2n)$ is the appropriate choice. It is illustrated in Fig. 1(b).

2. ***Discussion.*** The method outlined above has some useful properties:

(a) The method is simple and straightforward, and easily programmed on a computer. Results of high accuracy can be obtained using standard library routines (e.g. gaussian integration or adaptive integrators for oscillating non-singular integrands).

(b) The method is readily applied for different values of n. In particular, a detailed analysis of the properties of $C_n(\mathbf{a})$ for each value of n is not required.

(c) The partial derivatives of $C_n(\mathbf{a})$ can be calculated using the same contours.

(d) Other contours can be used when it is numerically convenient to do so. For example, a "corner cutting" contour that leaves the real axis at a point less than R (see Fig. 1), or the direct ray from 0 to R $\exp(\pm i\pi/2n)$ when the integrand is free of violent oscillations. Also different contours in the sector $0 < \arg u < \pi/n$ can be used to evaluate the infinite integral (18), since the integral is convergent in the same sector.

(e) The method can be generalized to evaluate numerically the two dimensional elliptic and hyperbolic umbilic canonical integrals [23].

B. Differential Equation Method

This method computes $C_n(\mathbf{a})$ and its partial derivatives by numerically integrating a set of differential equations satisfied by $C_n(\mathbf{a})$. The derivation of the differential equations for an arbitrary value of n is a fairly lengthy exercise [22], so we will just quote the formulae actually used in the computations for the cases n=3,4,5.

1. Airy Integral. To introduce the method, consider first n=3. The linear second-order differential equation satisfied by Ai(x) is well known to be

$$\frac{d^2\mathrm{Ai}(x)}{dx^2} = x\mathrm{Ai}(x). \tag{20}$$

Then Ai(x) and $\mathrm{Ai}'(x) \equiv d\mathrm{Ai}(x)/dx$ can be calculated by numerically integrating Eq. (20) from x=0 to the value of interest. To do this, the initial conditions Ai(0) and Ai′(0) must be known. They are readily obtained from the exact series representation of Ai(x). Although this method works (for $x < 0$ [19]), it is not used in practice because more efficient techniques are available.

2. Pearcey's Integral. Next consider P(x,y). The idea is to start the integration at (0,0) and for y=0 integrate to (x,0), with the initial conditions at the origin being obtained from the exact series representation of P(x,y) [16,17,19,25]. Then, in the second step, integrate from (x,0) to (x,y) keeping x constant, using as initial conditions the information calculated in the first step.

The linear differential equation in y satisfied by P(x,y) is [16,17,19]

$$4\frac{\partial^3 P}{\partial y^3} - 2x\frac{\partial P}{\partial y} - iyP = 0. \tag{21}$$

In order to integrate this equation from (x,0) to (x,y), we need to know the initial values P(x,y), $\partial P(x,y)/\partial y$ and $\partial^2 P(x,y)/\partial y^2$ at y=0.

Now P(x,0) satisfies the linear differential equation [16,17,19]

$$4\frac{\partial^2 P}{\partial x^2} + 2ix\frac{\partial P}{\partial x} + iP = 0, \qquad y = 0, \tag{22}$$

whose solution yields P(x,0) and $\partial P(x,0)/\partial x$. In practice, it is not necessary to integrate numerically Eq. (22) from (0,0) to (x,0) because it has an analytic solution, namely [16,17]

$$P(x,0) = \frac{\pi|x|^{1/2}\exp(-ix^2/8)}{4\sin(\pi/4)} \times \left[e^{i\pi/8}J_{-1/4}(x^2/8) - \text{sign}(x)e^{-i\pi/8}J_{1/4}(x^2/8)\right]. \tag{23}$$

Equation (23) is the first initial condition needed for the y equation. From the definition of P(x,y) [Eq. (6)], it is readily shown that the second initial condition is

$$\left.\frac{\partial P(x,y)}{\partial y}\right|_{y=0} = 0. \tag{24}$$

The third initial condition for the y equation can be obtained from the identity

$$\frac{\partial^2 P}{\partial y^2} = i\frac{\partial P}{\partial x}, \tag{25}$$

and by differentiating Eq. (23) for P(x,0). It is then found that [16]

$$\left.\frac{\partial^2 P(x,y)}{\partial y^2}\right|_{y=0} = \frac{i\pi|x|^{3/2}\exp(-ix^2/8)}{16\sin(\pi/4)}$$

$$\times \left\{i\left[e^{-i\pi/8}J_{1/4}(x^2/8)-\text{sign}(x)e^{i\pi/8}J_{-1/4}(x^2/8)\right]\right.$$

$$\left.-\left[\text{sign}(x)e^{i\pi/8}J_{3/4}(x^2/8) + e^{-i\pi/8}J_{-3/4}(x^2/8)\right]\right\}. \tag{26}$$

Equations (23), (24) and (26) provide the initial conditions for the integration of Eq. (21) from (x,0) to (x,y). Note that $\partial P/\partial y$ and $\partial P/\partial x$ [via the identity (25)] are obtained at the same time as P(x,y).

*3. **Swallowtail Integral.*** Finally we present the differential equations needed for the evaluation of S(x,y,z) [19]. The first step is to integrate three simultaneous ordinary linear differential equations for S(x,y,z), H(x,y,z) and G(x,y,z) where

$$H(x,y,z) \equiv \partial S(x,y,z)/\partial z, \tag{27}$$

and

$$G(x,y,z) \equiv \partial H(x,y,z)/\partial z, \tag{28}$$

from the origin (0,0,0) to (x,0,0):

$$\left.\begin{aligned} 5\frac{\partial H}{\partial x} + 3xG &= 0, \\ 5\frac{\partial G}{\partial x} - 3x\frac{\partial S}{\partial x} - S &= 0, \\ 25\frac{\partial^2 S}{\partial x^2} - 9x^2 G + 10H &= 0, \end{aligned}\right\} \quad y = z = 0. \tag{29}$$

The initial conditions for these equations are obtained from the exact series representation of S(x,y,z), and are

$$S(0,0,0) = (2/5)\Gamma(1/5)\cos(\pi/10),$$

$$H(0,0,0) = (2/5)\Gamma(2/5)\cos(7\pi/10),$$

$$G(0,0,0) = -(2/5)\Gamma(3/5)\cos(3\pi/10),$$

$$\left.\frac{\partial S(x,0,0)}{\partial x}\right|_{x=0} = -(2/5)\Gamma(4/5)\cos(\pi/10).$$

Integrating Eq. (29) from (0,0,0) to (x,0,0) produces the following information

$$S(x,0,0), \tag{30}$$

$$\frac{\partial S(x,0,0)}{\partial x} = -i\left.\frac{\partial H(x,y,0)}{\partial y}\right|_{y=0}, \tag{31}$$

$$H(x,0,0), \tag{32}$$

$$G(x,0,0) = i\left.\frac{\partial S(x,y,0)}{\partial y}\right|_{y=0}, \tag{33}$$

where the identities

$$\frac{\partial^2 S(x,y,z)}{\partial z^2} = i\frac{\partial S(x,y,z)}{\partial y}, \tag{34}$$

$$\frac{\partial^3 S(x,y,z)}{\partial z^3} = -\frac{\partial S(x,y,z)}{\partial x}, \tag{35}$$

and the definitions (27) and (28) have been used. The quantities (30)-(33) form the initial conditions for the second step, which is to integrate two simultaneous linear ordinary differential equations for S(x,y,0) and H(x,y,0) from (x,0,0) to (x,y,0) keeping x constant. The equations for this step are [19]:

$$\left.\begin{aligned} 5\frac{\partial^2 S}{\partial y^2} + 3ix\frac{\partial S}{\partial y} + 2iyH = 0, \\ 5\frac{\partial^2 H}{\partial y^2} + 3ix\frac{\partial H}{\partial y} - 2y\frac{\partial S}{\partial y} - S = 0, \end{aligned}\right\} \quad z = 0, \tag{36}$$

The solution of these equations at (x,y,0) are [again using Eqs. (34) and (35)]

$$S(x,y,0), \tag{37}$$

$$\frac{\partial S(x,y,0)}{\partial y} = -i\left.\frac{\partial^2 S(x,y,z)}{\partial z^2}\right|_{z=0}, \tag{38}$$

$$H(x,y,0), \tag{39}$$

$$\frac{\partial H(x,y,0)}{\partial y} = -i\left.\frac{\partial^3 S(x,y,z)}{\partial z^3}\right|_{z=0}, \tag{40}$$

which form the initial conditions for the next step.

In the final step, the following fourth order linear differential equation for S(x,y,z)

$$5\frac{\partial^4 S}{\partial z^4} - 3x\frac{\partial^2 S}{\partial z^2} - 2iy\frac{\partial S}{\partial z} + zS = 0, \tag{41}$$

is integrated from (x,y,0) to (x,y,z) keeping x and y constant. Notice that $\partial S/\partial x$ and $\partial S/\partial y$ are obtained along with S and $\partial S/\partial z$ from Eq. (41), because of the identities (34) and (35) [19].

The differential equations satisfied by the cuspoid canonical integral, $C_n(\mathbf{a})$, for an arbitrary value of n, have also been worked out [22]. As for the cases of Ai(x), P(x,y) and S(x,y,z), a set of linear coupled ordinary differential equations with linear coefficients are obtained for each step in the sequence $C_n(0,0,...,0) \rightarrow C_n(0,0,...,a_{n-2}) \rightarrow ... \rightarrow C_n(0,a_2,...,a_{n-2}) \rightarrow C_n(a_1,a_2,...,a_{n-2})$. The initial conditions for a given step are obtained from the solutions of the previous step [22].

4. ***Discussion.*** The differential equation method has the following advantages and disadvantages:

(a) The derivatives $\partial C_n(\mathbf{a})/\partial a_1$, $\partial C_n(\mathbf{a})/\partial a_2$,..., $\partial C_n(\mathbf{a})/\partial a_{n-2}$ are obtained at the same time as $C_n(\mathbf{a})$.

(b) If the differential equations are solved by a step-by-step numerical technique, then values of $C_n(\mathbf{a})$ and its derivatives are also obtained at intermediate points between the origin (0,0,...,0) and the final point $\mathbf{a} = (a_1,a_2,...,a_{n-2})$. The differential equation method is therefore an efficient way of generating grids of data for use in contour or perspective plots.

(c) The method is more complicated to program on a computer than the complex contour approach.

(d) The analysis to derive the differential equations and their initial conditions is non-trivial.

(e) The stability of the differential equations during the numerical integration is an important practical consideration. Calculations for n = 3 (Airy integral) [19], n = 4 (Pearcey's integral) [16,17] and n = 5 (swallowtail) [19] indicate that the method is numerically stable when *all* the saddle points of $f_n(\mathbf{a};u)$ are real, because then the independent solutions of the differential equations are of oscillatory type. In regions with saddle points that are complex valued, the independent solutions of the differential equations contain exponentially increasing solutions, which eventually limit the accuracy to which $C_n(\mathbf{a})$ and its derivatives can be computed.

C. Results for P(x,y) and S(x,y,z)

In this section, we show some plots of |P(x,y,)| and |S(x,y,z)| that have been obtained by the complex contour and differential equation methods of Sects. III.A and III.B respectively.

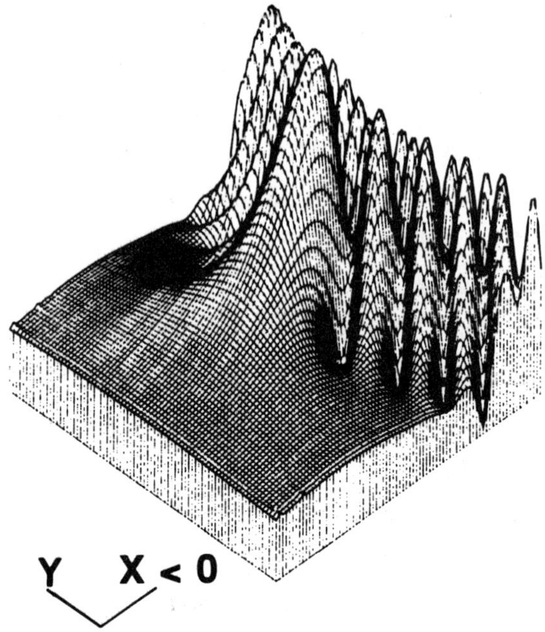

FIG. 3 *Perspective plot of* $|P(x,y)|$.

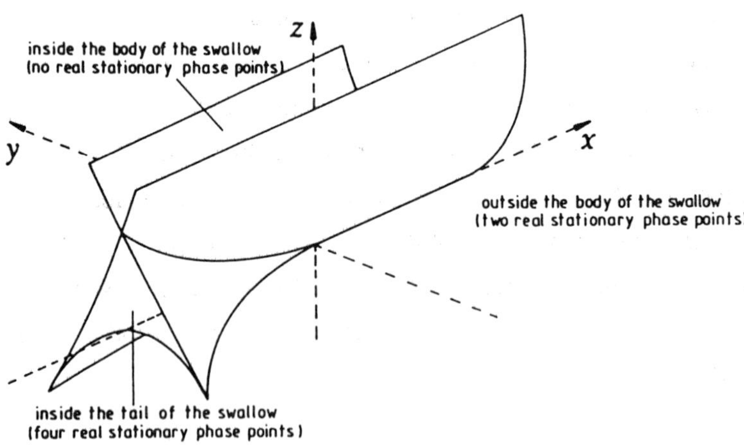

FIG. 4 *Caustic associated with the swallowtail canonical integral.*

1. Pearcey's Integral. Figure 2 shows a contour plot of $|P(x,y)|$. The caustic curve

$$8x^3 + 27y^2 = 0,$$

on which two real saddle points have coalesced, is shown as a dashed line (At x=0, y=0, all three saddles have coalesced). Note that it is only necessary to plot $|P(x,y)|$ for positive values of y because of the relation

$$P(x,y) = P(x,-y).$$

The interference structure in the region to the left of the caustic, where there are three real saddle points, is clearly visible. The interference effects rapidly dampen out on the right of the caustic. In this region, there is only one real saddle point. A three dimensional perspective plot of $|P(x,y)|$ is shown in Fig. 3.

A contour plot of arg P(x,y) can be found in Ref. [17], which also shows perspective and contour plots of $|\partial P/\partial x|$ and $|\partial P/\partial y|$. A contour plot of arg $\partial P/\partial x$ and arg $\partial P/\partial y$ is given on p. 129 of Ref. [42]. A table of values of P(x,y), $\partial P/\partial x$ and $\partial P/\partial y$ for the grid x = -8.0(2.0)8.0 and y = 0.0(2.0)8.0 is reported in Ref. [18]. The asymptotics of P(x,y) have been studied recently by Stamnes and Spjelkavik [42,44], Kaminski [45,46] and Paris [47].

2. ***Swallowtail Canonical Integral.*** The swallowtail canonical integral has four saddle points. The caustic surface in (x,y,z) space on which two or more *real* saddle points have coalesced is shown in Fig. 4. Inside the body of the swallow, there are no real saddle points (called stationary phase points in Fig. 4). Outside the body there are two real saddle points, whilst the region inside the tail corresponds to four real saddle points. Along the cuspoidal edges of the caustic, three saddle points have coalesced and at the origin (0,0,0) all four have done so. The line of self-intersection of the caustic corresponds to two pairs of coalesced real saddle points. Note also that S(x,y,z) satisfies the relation

$$S(x,y,z) = S^*(x,-y,z).$$

Perspective plots of $|S(x,y,z)|$ for fixed values of x,y and z are shown in Figs. 5-7. Alongside each plot is drawn the appropriate section of the swallowtail caustic. The numbers on the caustic sections are the number of real saddle points in the various regions (these always change by two or four on crossing a caustic line - an exception is for the z = 0 section).

The interference structure present in the plots for $|S(x,y,z)|$ can be understood in terms of the number of real saddle points present in each region. For example, inside the body of the swallow, where there are no real saddle points, $|S(x,y,z)|$ becomes exponentially small. The richest structure is observed inside the tail, where there are four real saddle points. Notice also the folds and (Pearcey) cusps in the regions of two and three coalescing saddle points respectively.

Some discussion of the zeros of S(x,y,z) can be found in Ref. [20], and perspective plots of $|\partial S/\partial x|$, $|\partial S/\partial y|$ and $|\partial S/\partial z|$, together with tables of values of S(x,y,z) and its derivatives are reported in Ref. [20]. The asymptotic behaviour of S(x,y,z) has been investigated by Kaminski [45].

IV. CALCULATION OF THE PARAMETERS OF THE CUSPOID MAP

This Section considers the second major problem that must be solved before the uniform asymptotic cuspoid approximation (11) can be applied to practical problems, namely the calculation of the parameters $\{\zeta_j\}$ and A in terms of the exponent $f(\alpha;t)$ of the original oscillatory integral (1).

From Eq. (8), the (local) cuspoid map is

$$f(\alpha;t) = u^{n+1} + \sum_{j=1}^{n-1} \zeta_j u^j + A. \tag{42}$$

The condition used to determine $\{\zeta_j\}$ and A is that the saddle points on either side of Eq. (42) must correspond

$$t = t_i(\alpha) \leftrightarrow u = u_i(\zeta), \qquad i=1,2,\ldots,n. \tag{43}$$

Note that the n saddle points $\{u_i\}$ are functions of the n-1 (unknown)

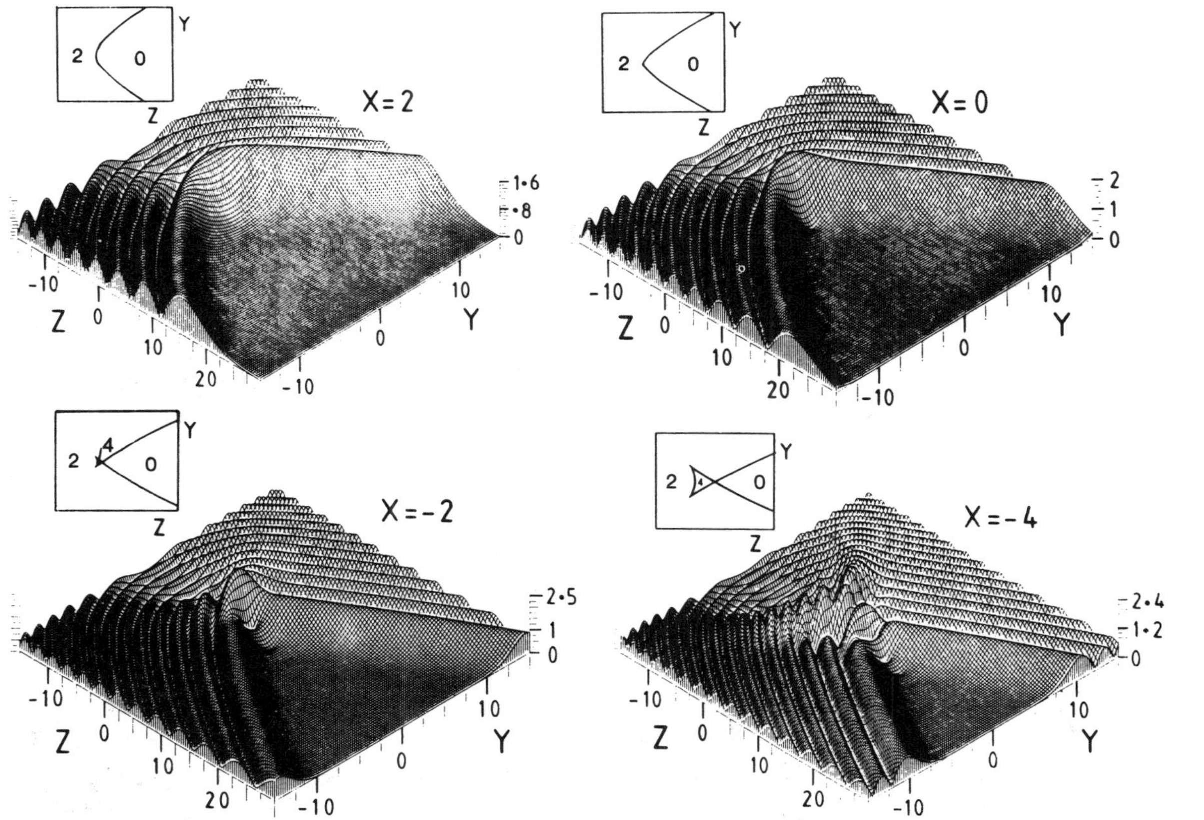

FIG. 5 *Perspective plots of* $|S(x,y,z)|$ *for* $x = 2, 0, -2, -4$, *with the corresponding caustic sections. The numbers on the caustic plots are the number of real saddle points in different regions of* (x,y,z) *space.*

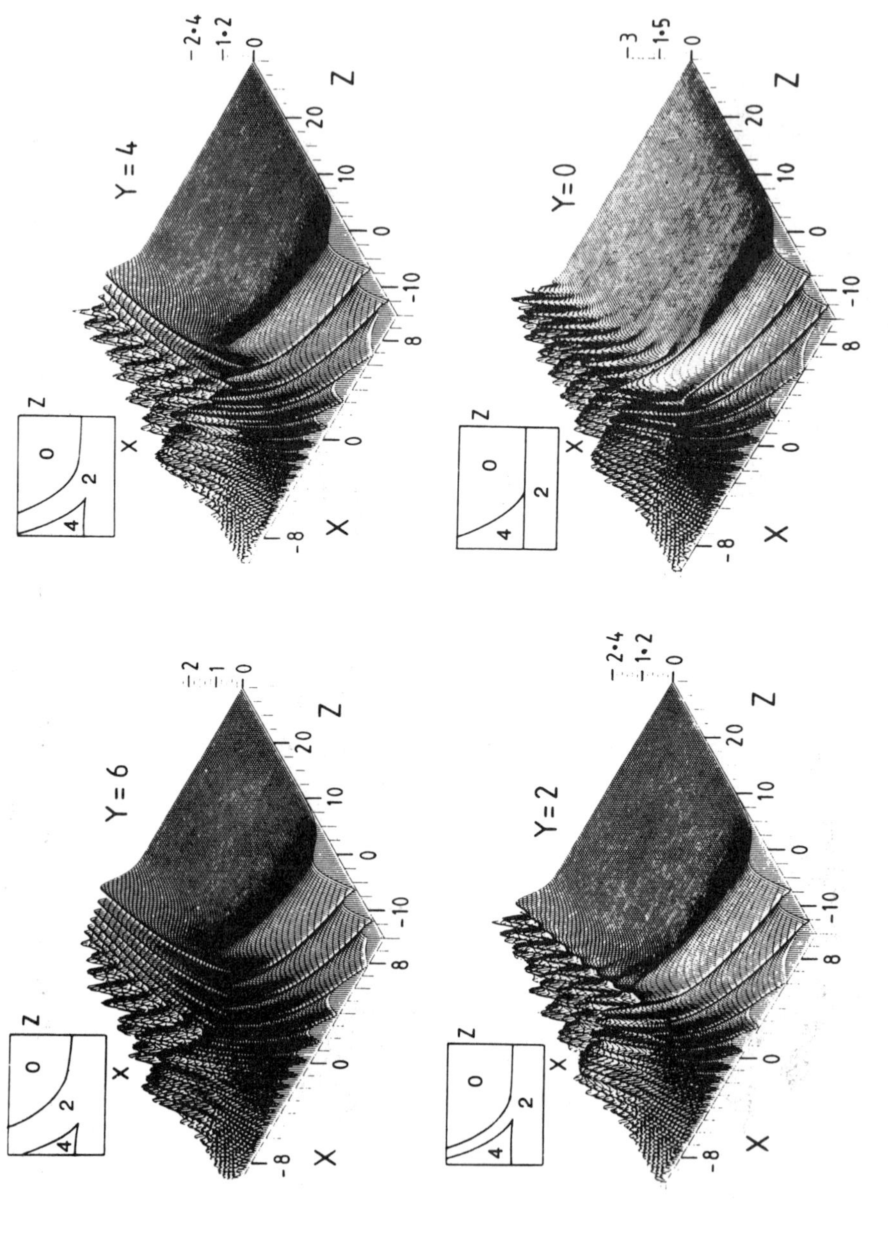

FIG. 6 *Same as Figure 5 except for the y sections with y = 6,4,2,0.*

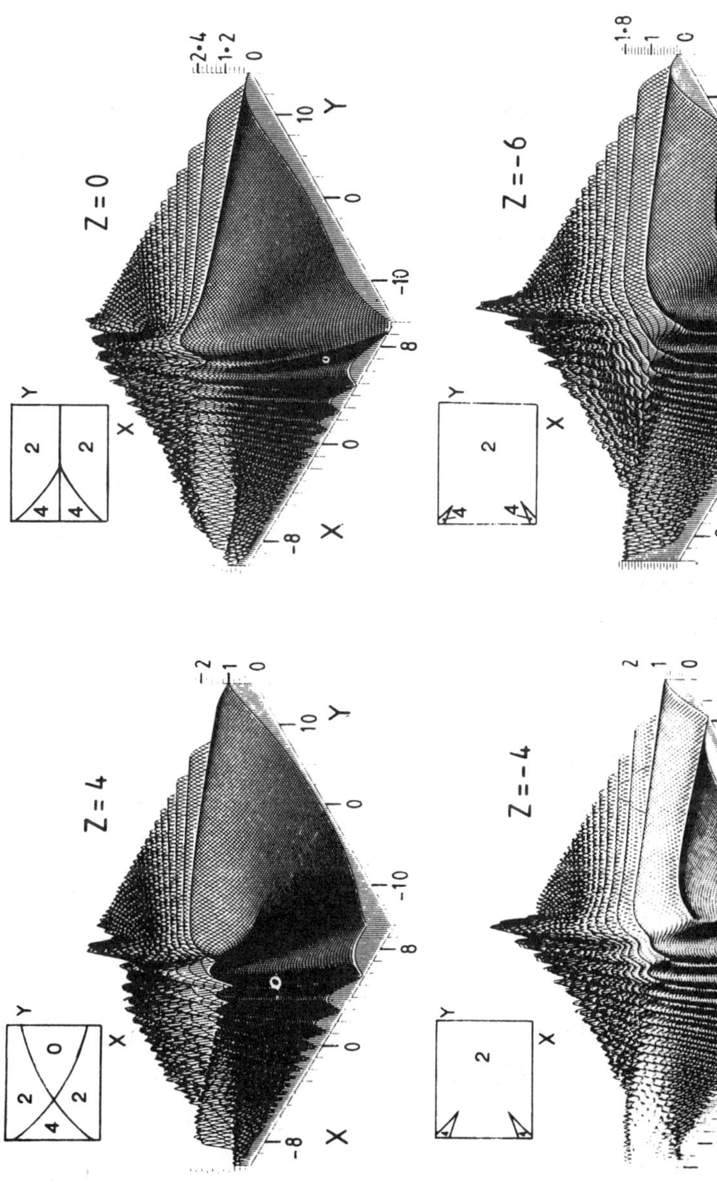

FIG. 7 Same as Figure 5 except for the z sections with z = 4,0,-4,-6.

parameters $\{\zeta_j\}$. In practical problems with small value of n=2-5, the labelling of the $\{u_i\}$ can usually be done by inspection. For example, the paths of the $\{t_i(\alpha)\}$ in the complex t plane can be examined as α varies. It is also often helpful to inspect the Taylor expansion of $f(\alpha;t)$ to (n+1)th order, because then the t and u planes become identical up to a scale transformation.

Inserting the saddle points (43) into Eq. (42) gives

$$f_i = u_i^{n+1} + \sum_{j=1}^{n-1} \zeta_j u_i^j + A, \qquad i=1,2,\ldots,n, \tag{44}$$

where the notation $f_i \equiv f(\alpha;t_i)$ has been used. The relation (44) comprises n non-linear equations, from which $\{\zeta_j\}$ and A are to be determined in terms of the $\{f_i\}$. Unfortunately, no explicit solution of these equations is known for the case of arbitrary n. Because of this, we will consider the cases n = 2,3,4 separately.

A. Two Coalescing Saddle Points

It is convenient to write the (fold) mapping in the slightly modified form [48]

$$f(\alpha;t) = u^3/3 - \zeta u + A, \tag{45}$$

with $\alpha \equiv \alpha_1$. The saddle points on the right hand side are given by $u^2 - \zeta = 0$, i.e., $u_1 = \zeta^{1/2}$ and $u_2 = -\zeta^{1/2}$, and we make the correspondence

$$t = t_1 \leftrightarrow u = u_1 \ (= \zeta^{1/2}), \tag{46}$$

and

$$t = t_2 \leftrightarrow u = u_2 \ (= -\zeta^{1/2}). \tag{47}$$

(This is appropriate when $f''(\alpha;t_1) > 0$ and $f''(\alpha;t_2) < 0$ [48]). Inserting the saddle points (46) and (47) into Eq. (45) gives two simultaneous equations, which are readily solved for ζ and A. We find

$$A = (f_1 + f_2)/2, \tag{48}$$

and

$$\zeta = [3(f_2 - f_1)/4]^{2/3}. \tag{49}$$

When the saddle points t_1 and t_2 are real, A and ζ are also both real, with the positive branch chosen for ζ. When t_1 and t_2 are complex valued (and also complex conjugates of one another), A and ζ are again real, but now the negative branch is used for ζ. Equations (48) and (49) were first obtained by Chester et al. [14].

B. Three Coalescing Saddle Points

For this case, we write the cusp map in the form [17]

$$f(\alpha,\beta;t) = u^4/4 - \xi u^2 - \eta u + A, \tag{50}$$

where $\alpha \equiv \alpha_1$ and $\beta \equiv \alpha_2$. The saddle points are the three solutions of the cubic equation

$$u_i^3 - 2\xi u_i - \eta = 0, \qquad i=1,2,3. \tag{51}$$

Before proceeding to derive explicit expressions for ξ, η and A, it is first useful to recall some facts about the solutions of Eq. (51). When $27\eta^2 < 32\xi^3$, the theory of cubic equations shows that Eq. (51) has three real roots which are all distinct. When $27\eta^2 = 32\xi^3$, two of the three real roots are equal (if $\eta = \xi = 0$, all three roots are equal), whereas if $27\eta^2 > 32\xi^3$, one root is real and the other two are complex conjugates. Evidently ξ must be positive (or zero) when Eq. (51) has three real roots, whereas ξ can be negative or positive when there is only one real root.

Inserting the saddle points into Eq. (50) gives

$$f_i = u_i^4/4 - \xi u_i^2 - \eta u_i + A, \qquad i=1,2,3, \tag{52}$$

as the set of equations to be solved for ξ, η and A. Equation (52) can be simplified by first multiplying Eq. (51) by u_i:

$$u_i^4 = 2\xi u_i^2 + \eta u_i, \qquad i=1,2,3. \qquad (53)$$

Substituting Eq. (53) into Eq. (52) then leads to the simpler form

$$f_i = -(\xi/2)u_i^2 - (3\eta/4)u_i + A, \qquad i=1,2,3. \qquad (54)$$

The procedure to obtain explicit solutions for ξ, η and A uses results from the theory of equations and the theory of symmetric polynomials [17,49,50]. First we construct the following linear, quadratic and cubic symmetric combinations of the f_i

$$\bar{f} = (f_1+f_2+f_2)/3, \qquad (55)$$

$$\Sigma = -(1/3)[f_1-\bar{f})(f_2-\bar{f}) + (f_2-\bar{f})(f_3-\bar{f}) + (f_1-\bar{f})(f_3-\bar{f})], \qquad (56)$$

$$\Pi = (f_1-\bar{f})(f_2-\bar{f})(f_3-\bar{f}). \qquad (57)$$

Notice that $\bar{f}$, Σ and Π are always real, even when f_1 and f_2 (say) are complex conjugates.

Equations (56) and (57) can be expressed in a different way using the theory of symmetric polynomials, namely [17,50]

$$\Sigma = -(1/3)[3/2)\bar{f}^2 - (f_1^2+f_2^2+f_3^2)/2], \qquad (58)$$

and

$$\Pi = (f_1^3+f_2^3+f_3^3)/3 - \bar{f}^3 - 6\bar{f}\Sigma. \qquad (59)$$

Equation (54) shows that the right hand sides of Eqs. (58) and (59) can be evaluated provided sums of powers of the roots of the type

$$s_r = u_1{}^r + u_2{}^r + u_3{}^r, \qquad r=1,2,\ldots,6,$$

can be expressed in terms of ξ and η. The s_r can be found by applying Newton's theorem to the cubic equation (51). We then obtain [17,50]

$$\bar{f} = A - 2\xi^2/3, \tag{60}$$

$$\Sigma = \xi^4/9 + 3\eta^2\xi/4, \tag{61}$$

$$\Pi = 2\xi^6/27 - 27\eta^4/64 - 5\eta^2\xi^3/4. \tag{62}$$

The parameter η can be eliminated from Eq. (61) and (62) to give a quartic equation for ξ^2

$$(\xi^2)^4 - 6\Sigma(\xi^2)^2 - 4\Pi\xi^2 - 3\Sigma^2 = 0,$$

which has the solution

$$\xi^2 = \Sigma^{1/2}\{\mathrm{sign}(\Pi)(1-\mu)^{1/2} + [2+\mu+2(1+\mu+\mu^2)^{1/2}]^{1/2}\}, \tag{63}$$

where

$$\mu = (1 - \Pi^2\Sigma^{-3}/4)^{1/3}.$$

The quantity μ can also be expressed in terms of ξ and μ using Eqs. (61) and (62)

$$\mu = \frac{(32\xi^3 - 27\eta^2)}{(8\xi^3+54\eta^2)} \left[\frac{27\eta^2}{32\xi^3}\right]^{1/3}.$$

From the discussion of the roots of Eq. (50), it follows that $\mu > 0$ when all three roots u_i are real and distinct, $\mu = 0$ when two roots are equal, and μ

< 0 when there is only one real root. In particular when $\mu \geq 0$, Eq. (63) has a unique positive solution for ξ. When there is only one real root, ξ can be positive or negative. The correct sign for ξ in this situation can be obtained by continously following the roots through the caustic $27\eta^2 = 32\xi^3$ from the bright side (three real roots) to the dark side (one real root).

Having found ξ, the second parameter η is obtained from Eq. (61)

$$\eta = \pm[(4\Sigma/3 - 4\xi^4/27)/\xi]^{1/2}, \qquad \xi \neq 0.$$

The sign of η can be arbitrarily chosen to be positive close to one of the caustic lines and negative near the other one. Finally A is calculated from Eq. (60)

$$A = \bar{f} + 2\xi^2/3.$$

C. Four Coalescing Saddle Points

We will use the following form for the swallowtail map [20]

$$f(\alpha,\beta,\gamma;t) = u^5/5 + xu^3 + yu^2 + zu + A,$$

where $\alpha \equiv \alpha_1$, $\beta \equiv \alpha_2$, $\gamma \equiv \alpha_3$. The four non-linear equations determining x,y,z and A are therefore

$$f_i = u_i^5/5 + xu_i^3 + yu_i^2 + zu_i + A, \qquad i=1,2,3,4, \tag{64}$$

where the u_i are the roots of the quartic polynomial

$$u_i^4 + 3xu_i^2 + 2yu_i + z = 0, \qquad i=1,2,3,4. \tag{65}$$

Multiplying Eq. (65) by u_i and substituting into Eq. (64) leads to the simpler set of equations

$$f_i = (2x/5)u_i^3 + (3y/5)u_i^2 + (4z/5)u_i + A, \quad i=1,2,3,4. \tag{66}$$

The algebraic method described for the cusp can be generalized to the swallowtail [20] (and formally for any value of n [22]). The procedure is to construct first the real quantities [20]

$$\bar{f} = \tau_1\{f_i\}/4, \tag{67}$$

$$\Sigma = \tau_2\{f_i-\bar{f}\}/6, \tag{68}$$

$$\Xi = \tau_3\{f_i-\bar{f}\}/4, \tag{69}$$

$$\Pi = \tau_4\{f_i-\bar{f}\}, \tag{70}$$

where the $\tau_i\{\cdot\}$ are the elementary symmetric polynomials

$$\tau_1\{r_i\} = r_1 + r_2 + r_3 + r_4,$$

$$\tau_2\{r_i\} = r_1r_2 + r_1r_3 + r_1r_4 + r_2r_3 + r_2r_4 + r_3r_4,$$

$$\tau_3\{r_i\} = r_1r_2r_3 + r_1r_2r_4 + r_1r_3r_4 + r_2r_3r_4,$$

$$\tau_4\{r_i\} = r_1r_2r_3r_4.$$

Then using the theory of symmetric polynomials and Newton's formulae for the quartic Eq. (65), the right hand sides of Eqs. (67) - (70) can be evaluated. The results are [20]

$$\bar{f} = A - (3/2)xy, \tag{71}$$

$$\Sigma = [(18/5)x^5 - 6x^3z - (23/4)x^2y^2 + (8/3)xz^2 + 3y^2z]/5, \tag{72}$$

$$\Xi = (y/25)[81x^6 - 153x^4z - 41x^3y^2 + 88x^2z^2 + 36xy^2z + (27/5)y^4 - 16z^3]. \tag{73}$$

$$\Pi = [243x^7y^2 - 513x^5y^2z + (249/16)x^4y^4 + 16x^4z^3 + 344x^3y^2z^2 + (27/2)x^2y^4z - (128/5)x^2z^4 + (162/5)xy^6 - (336/5)xy^2z^3 - (27/5)y^4z^2 + (256/25)z^5]/25. \tag{74}$$

The step just outlined involves a huge amount of algebraic manipulation, and is only feasible if a computer algebra program such as REDUCE or MACSYMA is used.

Equations (72) - (74) are three simultaneous equations for x,y,z, which contain many additional solutions for (x,y,z) in addition to the particular solution we are interested in. This causes severe practical problems when attempting a general numerical solution of these equations.

However there is a particular case where Eqs. (72) - (74) simplify considerably, namely when y=0. Equations (72) - (74) become

$$\Sigma = [(18/5)x^5 - 6x^3z + (8/3)xz^2]/5, \tag{75}$$

$$\Xi = 0,$$

$$\Pi = [16x^4z^3 - (128/5)x^2z^4 + (256/25)z^5]/25. \tag{76}$$

The theory of resultants can then be used to deduce a single polynomial equation for x, and one for z. These can be solved to yield the required value of (x,0,z) [20].

A second case where Eqs. (72) - (74) are useful is when (x,y,z) actually lies on the caustic, for then z=z(x,y) is constrained to a maximum of three possible (real) values, as illustrated in Fig. 4. It is then practical to solve Eqs. (72) - (74) numerically for (x,y,z(x,y)) [20].

The algebraic method outlined above is most useful when (x,y,z) lies on the caustic and for the case y = 0. In order to find (x,y,z) in other situations, an iterative method has been developed [20]. The procedure is to form the following differences

$$f_1 - f_2, \qquad f_3 + f_4 - f_1 - f_2, \qquad f_3 - f_4, \tag{77}$$

where the t_i are labelled so that the differences (77) are either purely real or purely imaginary. Using Eq. (66), we can then construct the matrix equation

$$\begin{bmatrix} u_1^3-u_2^3 & u_1^2-u_2^2 & u_1-u_2 \\ u_3^3+u_4^3-u_1^3-u_2^3 & u_3^2+u_4^2-u_1^2-u_2^2 & u_3+u_4-u_1-u_2 \\ u_3^3-u_4^3 & u_3^2-u_4^2 & u_3-u_4 \end{bmatrix} \begin{bmatrix} 2x \\ 3y \\ 4z \end{bmatrix} = 5 \begin{bmatrix} f_1-f_2 \\ f_3+f_4-f_1-f_2 \\ f_3-f_4 \end{bmatrix}. \tag{78}$$

The differences in the 3×3 matrix are also either purely real or purely imaginary. Equation (78) is then solved by iteration. Starting from an initial guess (x_0,y_0,z_0), the $u_i(x_0,y_0,z_0)$ are calculated from Eq. (65). Then the linear equations (78) are solved to yield new values (x_1,y_1,z_1), and so on. Note that the iterative method fails for (x,y,z) on the caustic, i.e., it is complementary to the algebraic method.

Further practical details concerning the algebraic and iterative methods can be found in Ref. [20]. The formal generalization of both methods for the case of an arbitrary number of coalescing saddle points is described in Ref. [22].

V. APPLICATIONS

This Section briefly describes some applications of the methodology of Sects. III and IV to problems in chemical physics.

A. Two Coalescing Saddle Points

There have been many applications of the uniform Airy approximation to short-wavelength scattering problems since its introduction by Chester et al. [14]. Applications in chemical physics include: the rainbow effect in the elastic scattering of atoms [48,51,52], the occurrence of rotational rainbows in atom-molecule rotationally inelastic collisions [53] and molecule-surface scattering [54], the analysis of vibrational transitions in the semiclassical theory of atom-molecule collisions (see for example, Refs. [55-58]), the theory of chemical reactions (e.g., Ref. [59]), electron detachment in negative ion collisions [60], Penning ionization [61], the pressure broadening of spectral lines [62], and the analysis of quasimolecular orbital X-ray spectra in collisions of Cl^{16+} with Ar in the energy range 2.5-20 MeV [63,64].

The uniform asymptotic theory for one dimensional integrals described in Sect. II, can also be extended to a class of multi-dimensional integrals of the form

$$I_n(\alpha) = \int_{-\infty}^{\infty} \int_{-\infty}^{\infty} \cdots \int_{-\infty}^{\infty} g(t_1, t_2, \ldots, t_n)$$

$$\times \exp[if(\alpha; t_1, t_2, \ldots, t_n)/\hbar] dt_1 dt_2 \ldots dt_n,$$

$$\equiv \int_{-\infty}^{\infty} g(\mathbf{t}) \exp[if(\alpha; \mathbf{t})/\hbar] d\mathbf{t}.$$

When $f(\alpha;\mathbf{t})$ possesses two coalescing saddle points, the uniform asymptotic approximation to $I_n(\alpha)$ is a generalization of Eq. (11) [13,65]. This approximation has been applied to the evaluation of Franck-Condon factors [66], inelastic atom-atom scattering amplitudes [66], the transfer of electronic energy in $He(^1S)+Ne^{++}(^1D)$ collisions [67] and the evaluation of resonance Raman scattering amplitudes [68].

B. Three Coalescing Saddle Points

The first application of the uniform Pearcey approximation was to the theory of cusped rainbows in the elastic scattering of He^+ by Ne [17]. In this problem, the classical deflection function possesses a maximum and a minimum. Both real and complex valued classical trajectory data were included in the semiclassical analysis, and excellent agreement with conventional partial wave results was obtained [17].

Wolf and Korsch [69] have applied the techniques developed in Ref. [17] to analyse rotational rainbow scattering for anisotropic interaction potentials that possess an attractive well (see also Ref. [70]).

C. Four Coalescing Saddle Points

The uniform swallowtail approximation has been used to calculate semiclassical reaction probabilities [21] for the collinear chemical laser reaction

$$H + F_2 \rightarrow HF + F.$$

Real and complex valued classical trajectory data were included in the computations, and good agreement was obtained with full quantum mechanical results [21].

Another (simpler) application of the uniform swallowtail approximation has been to the asymptotic evaluation of the butterfly canonical integral [20]

$$B(a,b,c,d) = \int_{-\infty}^{\infty} \exp[i(t^6/6+at^4+bt^3+ct^2+dt)]dt.$$

The butterfly integral has five saddle points. If t_1 is chosen to be sufficiently far removed from t_2, t_3, t_4, t_5, then the swallowtail map can be applied in the region of the four coalescing saddle points t_i, i=2,3,4,5. 156 sets of (a,b,c,d) butterfly data were used in the uniform asymptotic evaluation of B(a,b,c,d) in order to test the numerical procedures described in Sects. III and IV. Good agreement in all cases was obtained with the exact values of B(a,b,c,d), which were obtained by the complex contour quadrature method of Sect. III.A [20].

VI. SUMMARY AND CONCLUSIONS

Many short wavelength scattering theories require the uniform asymptotic evaluation of one dimensional oscillating integrals with several coalescing saddle points. It is then necessary to numerically evaluate the cuspoid canonical integrals and their partial derivatives, as well as to determine the arguments of the integrals from thc saddle point information.

Two methods have been developed for the numerical evaluation of the canonical integrals and their partial derivatives. The first method exploits Cauchy's theorem and Jordan's Lemma to choose the most convenient path in the complex plane along which to do the numerical quadrature. The second method solves sets of linear differential equations (with linear coefficients) satisfied by the cuspoid canonical integrals. The initial conditions for a given step are obtained from the solutions of the previous step.

The arguments of the canonical integrals can be determined analytically for the uniform Airy and Pearcey approximations. Algebraic techniques that exploit the theory of symmetric polynomials and the theory of equations are also useful for determining the arguments of the uniform swallowtail approximation when (x,y,z) lies on the caustic, and for the case y=0. In other situations, an iterative method can be used.

The uniform Airy approximation has been applied to many short-wavelength scattering problems. The uniform Pearcey and uniform swallowtail approximations can now also be regarded as practical tools for the numerical evaluation of oscillating integrals with three and four coalescing saddle points respectively.

ACKNOWLEDGMENTS

I thank my collaborators Drs. P.R. Curtis, D. Farrelly and R.A.W. Young for their help with the research described in this review. The support of the Science and Engineering Research Council (U.K.) is also gratefully acknowledged.

REFERENCES

1. F.W.J. Olver, *Asymptotics and special functions*, Academic Press, New York, 1974.

2. D. Ludwig, "Uniform asymptotic expansions for wave propagation and diffraction problems", *SIAM Review*, **12**, 325 (1970).

3. J.B. Keller, "Rays, waves and asymptotics", *Bull. Am. Math. Soc.*, **84**, 727 (1978).

4. J.B. Keller, "Progress and prospects in the theory of linear wave propagation", *SIAM Review*, **21**, 229 (1979).

5. J.N.L. Connor, "Semiclassical theory of elastic scattering", *Semiclassical methods in molecular scattering and spectroscopy*, Proceedings of the NATO Advanced Study Institute held in Cambridge, England, in

September, 1979, edited by M.S. Child, Reidel, Dordrecht, 1980, pp. 45-107.

6. N. Bleistein, *Mathematical methods for wave phenomena*, Academic Press, Orlando, 1984.

7. D.S. Jones, "Asymptotic behaviour of integrals", *SIAM Review*, **14**, 286 (1972).

8. M.V. Fedoriuk, *Metod perevala (The saddle point method)*, Nauka, Moscow, 1977.

9. R. Wong, "Error bounds for asymptotic expansions of integrals", *SIAM Review*, **22**, 401 (1980).

10. N. Bleistein and R.A. Handelsman, *Asymptotic expansions of integrals*, Dover, New York, 1986.

11. F. Ursell, "Integrals with a large parameter. Several nearly coincident saddle-points". *Proc. Cambridge Philos. Soc.*, **72**, 49 (1972).

12. J.N.L. Connor, "Semiclassical theory of molecular collisions: many nearly coincident classical trajectories", *Mol. Phys.*, **27**, 853 (1974). On p. 861, lines 25 and 26, the words "even" and "odd" should be interchanged.

13. J.N.L. Connor, "Catastrophes and molecular collisions", *Mol. Phys.*, **31**, 33 (1976).

14. C. Chester, B. Friedman and F. Ursell, "An extension of the method of steepest descents", *Proc. Cambridge Philos. Soc.*, **53**, 599 (1957).

15. F.W.J. Olver, "Unsolved problems in the asymptotic estimation of special functions", *Theory and application of special functions*, Proceedings of an Advanced Seminar sponsored by the Mathematics Research Center, The University of Wisconsin-Madison, March 31 - April 2, 1975, edited by R.A. Askey, Academic Press, New York, 1975, pp. 99-142.

16. J.N.L. Connor and D. Farrelly, "Molecular collisions and cusp catastrophes: Three methods for the calculation of Pearcey's integral and its derivatives", *Chem. Phys. Lett.*, **81**, 306 (1981).

17. J.N.L. Connor and D. Farrelly, "Theory of cusped rainbows in elastic scattering: Uniform semiclassical calculations using Pearcey's integral", *J. Chem. Phys.*, **75**, 2831 (1981). On p. 2837, 2nd column, line 27, for "Fig. 6" read "Ref. 6".

18. J.N.L. Connor and P.R. Curtis, "A method for the numerical evaluation of the oscillatory integrals associated with the cuspoid catastrophes: Application to Pearcey's integral and its derivatives", *J. Phys. A*, **15**, 1179 (1982).

19. J.N.L. Connor, P.R. Curtis and D. Farrelly, "A differential equation method for the numerical evaluation of the Airy, Pearcey and swallowtail canonical integrals and their derivatives", *Mol. Phys.*, **48**, 1305 (1983). In Eq. (3.46), for $(4/5)\Gamma(3/5)$ read $(2/5)\Gamma(3/5)$. One line below Eq. (3.50), for $H(x,y,z) = \partial S(x,y,z)/\partial x$ read $H(x,y,z) = \partial S(x,y,z)/\partial z$.

20. J.N.L. Connor, P.R. Curtis and D. Farrelly, "The uniform asymptotic swallowtail approximation: practical methods for oscillating integrals with four coalescing saddle points", *J. Phys. A*, **17**, 283 (1984). In Table 4, for c = -20.957 read c = -20.597.

21. J.N.L. Connor, P.R. Curtis and C.J. Edge and A. Laganà, "The uniform asymptotic swallowtail approximation: Application to the collinear $H+F_2$ chemical reaction", *J. Chem. Phys.*, **80**, 1362 (1984).

22. J.N.L. Connor and P.R. Curtis, "Differential equations for the cuspoid canonical integrals", *J. Math. Phys.*, **25**, 2895 (1984).

23. J.N.L. Connor, P.R. Curtis and R.A.W. Young (unpublished results).

24. L. Brillouin, "Sur une méthode de calcul approchée de certaines intégrales, dite méthode de col", *Ann. Sci. l'Ecole Norm. Sup. (Paris)*, 3rd Ser., **33**, 17 (1916).

25. T. Pearcey, "The structure of an electromagnetic field in the neighbourhood of a cusp of a caustic", *Phil. Mag.*, **37**, 311 (1946).

26. J.N.L. Connor, "Semiclassical theory of molecular collisions: three nearly coincident classical trajectories", *Mol. Phys.*, **26**, 1217 (1973).

27. B. Friedman, "Stationary phase with neighbouring critical points", *J. Soc. Ind. Appl. Math.*, **7**, 280 (1959).

28. F. Ursell, "Integrals with a large parameter. The continuation of uniformly asymptotic expansions", *Proc. Cambridge Philos. Soc.*, **61**, 113 (1965).

29. D. Ludwig, "Uniform asymptotic expansions at a caustic", *Commun. Pure Appl. Math.*, **19**, 215 (1966).

30. N. Bleistein, "Uniform asymptotic expansions of integrals with many nearby stationary points and algebraic singularities", *J. Math. Mech.*, **17**, 533 (1967).

31. S.O. Rice, "Uniform asymptotic expansions for saddle point integrals - Applications to a probability distribution occurring in noise theory", *Bell Syst. Tech. J.*, **47**, 1970 (1968).

32. J. Martin, "Integrals with a large parameter and several nearly coincident saddle points: the continuation of uniformly asymptotic expansions", *Proc. Cambridge Philos. Soc.*, **76**, 211 (1974).

33. M.V. Berry, "Waves and Thom's theorem", *Adv. Phys.*, **25**, 1 (1976).

34. M.V. Berry and C. Upstill, "Catastrophe Optics: Morphologies of caustics and their diffraction patterns", *Prog. Optics*, **18**, 257 (1980).

35. M.V. Berry, "Singularities in waves and rays", *Physics of Defects*, Les Houches, session XXXV, 28 Juillet-29 Août 1980, édité par R. Balian, M. Kléman, J.-P. Poirier, North-Holland, Amsterdam, 1981, pp. 452-543.

36. I. Stewart, "Applications of catastrophe theory to the physical sciences", *Physica D*, **2**, 245 (1981).

37. J.M. Arnold, "Oscillating integral theory for uniform representation of wave functions", *Radio Sci.*, **17**, 1181 (1982).

38. A.S. Dickinson and D. Richards, "Classical and semiclassical methods in inelastic heavy-particle collisions", *Adv. Atom. Mol. Phys.*, **18**, 165 (1982).

39. G. Dangelmayr and W. Güttinger, "Topological approach to remote sensing", *Geophys. J. Roy. Astr. Soc.*, **71**, 79 (1982).

40. Yu. A. Kravtsov and Yu. I. Orlov, "Caustics, catastrophes and wave fields", *Usp. Fiz. Nauk.*, **141**, 591 (1983) [English translation: *Sov. Phys. Usp.*, **26**, 1038 (1983)].

41. J.B. Delos, "Semiclassical calculation of quantum mechanical wavefunctions", *Adv. Chem. Phys.*, **65**, 161 (1986).

42. J.J. Stamnes, *Waves in focal regions*, Hilger, Bristol, 1986.

43. Yu. A. Kravtsov, "Rays and caustics as physical objects", *Prog. Optics*, **26**, 227 (1988).

44. J.J. Stamnes and B. Spjelkavik, "Evaluation of the field near a cusp of a caustic", *Opt. Acta*, **30**, 1331 (1983).

45. D. Kaminski, "Asymptotic expansions of some canonical diffraction integrals", Ph.D. thesis, University of Manitoba, 1986.

46. D. Kaminski, "Asymptotic expansion of the Pearcey integral near the caustic", *SIAM J. Math. Anal.* (in press).

47. R.B. Paris (private communication) 1988.

48. J.N.L. Connor and R.A. Marcus, "Theory of semiclassical transition probabilities for inelastic and reactive collisions. II. Asymptotic evaluation of the S matrix", *J. Chem. Phys.*, **55**, 5636 (1971).

49. R.L. Holford, "Modifications to ray theory near cusped caustics", Bell Telephone Laboratories, Whippany, N.J. 07981 (unpublished report) 1972.

50. D.H. Peregrine and R. Smith, "Nonlinear effects upon waves near caustics", *Philos. Trans. Roy. Soc.* A, **292**, 341 (1979).

51. M.V. Berry, "Uniform approximation for potential scattering involving a rainbow", *Proc. Phys. Soc.*, **89**, 479 (1966).

52. J.N.L. Connor, D. Farrelly and D.C. Mackay, "Complex angular momentum analysis of diffraction scattering in atomic collisions", *J. Chem. Phys.*, **74**, 3278 (1981).

53. H.J. Korsch and R. Schinke, "A uniform semiclassical sudden approximation for rotationally inelastic scattering", *J. Chem. Phys.*, **73**, 1222 (1980).

54. R. Schinke, "Rotational rainbows in diatom (solid) surface scattering", *J. Chem. Phys.*, **76**, 2352 (1982).

55. J. Stine and R.A. Marcus, "Theory of semiclassical transition probabilities for inelastic and reactive collisions. IV. Classically inaccessible transitions calculated by integration along complex-valued trajectories", *Chem. Phys. Lett.*, **15**, 536 (1972).

56. J.N.L. Connor, "Semiclassical theory of molecular collisions: real and complex-valued classical trajectories for collinear atom Morse oscillator collisions, *Mol. Phys.*, **28**, 1569 (1974).

57. J.N.L. Connor, "Meldola medal lecture. Molecular collisions and the semiclassical approximation", *Chem. Soc. Rev.*, **5**, 125 (1976).

58. J.N.L. Connor and W.J.E. Southall, "An improved Bessel uniform approximation of non-integer order for use in semiclassical collision theory: Application to vibrationally inelastic molecular collisions", *J. Phys. Chem.*, **90**, 3599 (1986).

59. J.N.L. Connor, C.J. Edge and A. Laganà, "Semiclassical and quasiclassical calculation of reaction probabilities for collinear $X+F_2 \rightarrow XF+F$ (X=Mu,H,D,T)", *Mol. Phys.*, **46**, 1231 (1982).

60. T.S. Wang and J.B. Delos, "Electron detachment in negative ion collisions. III. Model calculations", *Phys. Rev.* A, **33**, 3832 (1986).

61. R.J. Bieniek, "Semiclassical uniform approximation in Penning ionization", *J. Phys. B*, **7**, L266 (1974).

62. J. Szudy and W.E. Baylis, "Unified Franck-Condon treatment of pressure broadening of spectral lines", *J. Quant. Spectrosc. Radiat. Transfer*, **15**, 641 (1975).

63. M. Meron, B.M. Johnson, K.W. Jones, R. Schuch, H. Schmidt-Böcking and I. Tserruya, "Investigations of molecular orbitals with accel-decel beams", *Nucl. Instr. Methods B*, **10/11**, 64 (1985).

64. R. Schuch, M. Meron, B.M. Johnson, K.W. Jones, R. Hoffmann, H. Schmidt-Böcking and I. Tserruya, "Quasimolecular x-ray spectroscopy for slow Cl^{16+}-Ar collisions", *Phys. Rev.* A, **37**, 3313 (1988).

65. J.N.L. Connor, "Asymptotic evaluation of multidimensional integrals for the S matrix in the semiclassical theory of inelastic and reactive molecular collisions", *Mol. Phys.*, **25**, 181 (1973).

66. J.N.L. Connor, "Uniform semiclassical evaluation of Franck-Condon factors and inelastic atom-atom scattering amplitudes", *J. Chem. Phys.*, **74**, 1047 (1981).

67. P. Brault, O. Vallée, N. Tran Minh, J. Chapelle, "Uniform semiclassical treatment of the radial coupling term in the adiabatic basis: Application to the excitation transfer $He(^1S)+Ne^{2+}(^1D) \rightarrow He(^1S) + Ne^{2+}(^1S)$", *Phys. Rev.* A, **37**, 2318 (1988).

68. O. Atabek and R. Lefebvre, "Semiclassical estimate of resonance Raman scattering amplitudes II. Standing wave Green's function procedure", *Mol. Phys.*, **45**, 161 (1982).

69. F. Wolf and H.J. Korsch, "Fold and cusp catastrophe structures in rotational rainbow scattering", *J. Chem. Phys.*, **81**, 3127 (1984).

70. H.J. Korsch and F. Wolf, "Rainbow catastrophes in inelastic molecular collisions", *Comments Atom. Mol. Phys.*, **15**, 139 (1984).

New Inequalities for the Zeros of Confluent Hypergeometric Functions

Luigi Gatteschi Professor, University of Torino, Department of Mathematics, Torino, Italy

1. Introduction.

In two recent papers [4,5] we have obtained inequalities for the zeros of Jacobi polynomials and for the zeros of Laguerre polynomials. Here we will apply the same technique to derive bounds for the zeros of confluent hypergeometric functions $\Phi(a,c;x)$ and $\Psi(a,c;x)$. Since the Laguerre polynomials $L_n^{(\alpha)}(x), n = 0,1,2,\ldots$, $\alpha > -1$, can be expressed in terms of the confluent hypergeometric functions, that is

$$L_n^{(\alpha)}(x) = \binom{n+\alpha}{n} \Phi(-n, 1+\alpha; x) = \frac{(-1)^n}{n!}\Psi(-n, 1+\alpha; x), \tag{1.1}$$

it follows that some of the results contained in this paper can be considered as an extension of the results obtained in [5] for the zeros of Laguerre polynomials.

AMS(MOS)subject classifications. Primary 33A70, secondary 34C10,65D20.

This work was supported by the Consiglio Nazionale delle Ricerche of Italy and by the Ministero della Pubblica Istruzione of Italy.

The main tool that we need for our purposes is the well-known Sturm comparison theorem in the following form given by Szegö [10,p.19].

Theorem 1.1.(Sturm's comparison theorem). Let $f(x)$ and $F(x)$ be functions continuous in $x_0 < x < X_0$, with $f(x) \leq F(x)$. Let the functions $y(x)$ and $Y(x)$, both not identically zero, satisfy the differential equations

$$y'' + f(x)y = 0, \qquad Y'' + F(x)Y = 0, \tag{1.2}$$

respectively. Let x' and x'', $x' < x''$, be two consecutive zeros of $y(x)$. Then the function $Y(x)$ has at least one zero in the interval $x' < x < x''$ provided $f(x) \not\equiv F(x)$ in $[x', x'']$.

The statement also holds for $x' = x_0[y(x_0 + 0) = 0]$ if the additional condition

$$\lim_{x \to x_0 + 0} \{y'(x)Y(x) - y(x)Y'(x)\} = 0, \tag{1.3}$$

is satisfied (similarly for $x'' = X_0$).

It will be useful to recall another theorem which assures the existence of at least one zero in the interval $x_0 < x < x'$.

Theorem 1.2 (see Watson [14],p.518). Let $f(x)$, $F(x)$, $y(x)$ and $Y(x)$ have the same meaning as in Theorem 1.1 .Let x', $x_0 < x' < X_0$, be the first zero of $y(x)$ to the right of x_0. If we assume that $f(x) \leq F(x)$ and that

$$\lim_{x \to x_0+0} \frac{Y(x)}{y(x)} = 1 \quad , \quad \lim_{x \to x_0+0} \frac{Y'(x)}{y'(x)} = 1 \quad , \tag{1.4}$$

then $| \ y(x) \ |$ exceeds $| \ Y(x) \ |$ so long as x lies between x_0 and the first zero of $Y(x)$, so that the first zero of $Y(x)$ in the interval is on the left of the first zero x' of $y(x)$.

A similar statement holds for the interval $x'' < x < X_0$, where x'', $x_0 < x'' < X_0$, is the first zero of $y(x)$ on the left of X_0.

The differential equations that we shall assume as comparison equations are the ones used by Erdélyi and Swanson [2] in deriving uniform asymptotic approximations for the Whittaker functions

$$\begin{aligned} M_{k,m}(x) &= e^{-x/2}x^{m+1/2}\Phi(m - k + \frac{1}{2}, 2m + 1; x), \\ W_{k,m}(x) &= e^{-x/2}x^{m+1/2}\Psi(m - k + \frac{1}{2}, 2m + 1; x). \end{aligned} \tag{1.5}$$

Such equations can also be obtained by applying Olver's [8] theory to the asymptotic study of the Whittaker differential equation near the singularity $x = 0$ and near the turning point $x = 4k$.

We notice that, except at $x = 0$ and $x = \infty$, the first Whittaker function $M_{k,m}(x)$ has the same zeros as the corresponding Kummer function $\Phi(a, c; x)$and the second Whittaker function $W_{k,m}(x)$ has the same zeros as the corrisponding Tricomi function $\Psi(a, c; x)$. Therefore we shall use throughout this paper the parameters a and c together with the Whittaker parameters k and m , that is we shall set

$$k = \frac{c}{2} - a, \qquad m = \frac{c-1}{2}. \tag{1.6}$$

The distribution and the properties of the zeros of the functions $\Phi(a, c; x)$ and $\Psi(a, c; x)$ are studied in detail in the monographs of Tricomi [12,13] and Slater [9]. Here we recall only the main properties of the real zeros. Such zeros occur when a is negative and c is positive. More precisely the following theorem holds.

Theorem 1.3 (see Tricomi[13],§3.4). Let $P(a, c)$ and $P^*(a, c)$ denote the number of the positive zeros of $\Phi(a, c; x)$ and $\Psi(a, c; x)$ respectively. Then, for $c > 0$ we have

$$\begin{array}{ll} P(a,c) = 0, \quad (a \geq 0) & P(a,c) = -[a], \quad (a < 0) \\ P^*(a,c) = P(a,c), & (a < 0, c \geq 1) \\ P^*(a,c) = P(a-c+1, 2-c), (a < 0, 0 < c < 1). & \end{array}$$

It follows that the values of the difference $P(a, c) - P^*(a, c)$ can be only 0 or 1.

In the sequel a and c will be assumed to satisfy the conditions

$$a < 0 \quad , \quad c > 0. \tag{1.7}$$

Theorem 1.4 (see Tricomi [13], p.45). If $k^2 > m^2 - \frac{1}{4}$, that is if $c(a - \frac{1}{2}) < a^2$, the functions $\Phi(a, c; x)$ and $\Psi(a, c; x)$ cannot have more than one zero outside the interval

$$\left(2k - \sqrt{4k^2 - 4m^2 + 1} \quad , \quad 2k + \sqrt{4k^2 - 4m^2 + 1}\right).$$

Theorem 1.5 (see Tricomi [13], p.46).Let ϕ_r, $r = 1, 2, \ldots$, be the zeros of $\Phi(a, c; x)$ in increasing order. Then

$$\phi_r > \frac{j_{c-1,r}^2}{4k}, \tag{1.8}$$

for $r = 1, 2, \ldots$ and where $j_{c-1,r}$ is the r-th positive zero of the Bessel function $J_{c-1}(x)$. Furthermore, provided that $2k \geq j_{c-1,r}$, we have

$$\phi_r \leq \frac{j_{c-1,r}^2}{2k + \sqrt{4k^2 - j_{c-1,r}^2}} . \tag{1.9}$$

An asymptotic result is given by the formula (see Tricomi [11])

$$\phi_r = \frac{j^2_{c-1,r}}{4k}\left\{1 + \frac{2c(c-1) + j^2_{c-1,r}}{48k^2}\right\} + O(k^{-5}) \quad , \tag{1.10}$$

for $k \to \infty$ and fixed r.

A simplified form of another asymptotic approximation obtained by Tricomi [13 ,p. 47] yields the following theorem:

Theorem 1.6 Let x_r be the root of the equation

$$x - \sin x = \frac{3 - 4a - 4r}{4k}\pi. \tag{1.11}$$

Then , as $k \to \infty$,

$$\phi_r = 4k\cos^2\frac{x_r}{2} + O(k^{-1}), \tag{1.12}$$

for all the zeros in the interval $(4t'k, 4t''k)$ with t' and t'' fixed in (0,1).

Few results are known for the zeros of Tricomi's $\Psi(a, c; x)$function.

Theorem 1.7 (see Tricomi [13], p.46). Let $c \geq 3/2$ and let $a_s, s = 1, 2, \ldots$, be the zeros, in decreasing order $0 > a_1 > a_2 > \ldots$, of the Airy function $\mathrm{Ai}(x)$. Then for the zeros $\psi_1, \psi_2, \ldots$, in increasing order, of $\Psi(a, c; x)$we have

$$\psi_r < \left\{(4k)^{1/2} + 2^{-1/3}(4k)^{-1/6}a_{-[a]-r+1}\right\}^2, \tag{1.13}$$

for $r = 1, 2, \ldots, -[a]$.

2. Upper bounds for the zeros of Φ and Ψ.

The differential equation

$$\frac{d^2u}{dt^2} + \left\{4k^2\left(\frac{1}{t} - 1\right) + \frac{1 - 4m^2}{4t^2}\right\} u = 0, \tag{2.1}$$

with k and m defined by (1.6) , is satisfied by

$$\begin{aligned} u_1(t) &= e^{-2kt}(4kt)^{c/2}\Phi(a, c; 4kt) = M_{k,m}(4kt), \\ u_2(t) &= e^{-2kt}(4kt)^{c/2}\Psi(a, c; 4kt) = W_{k,m}(4kt). \end{aligned} \tag{2.2}$$

Now we observe that the cylinder function

$$z(t) = \left(\frac{f(t)}{f'(t)}\right)^{1/2} C_{2m}\big(f(t)\big) = \tag{2.3}$$
$$= \left(\frac{f(t)}{f'(t)}\right)^{1/2} \Big\{J_{2m}\big(f(t)\big)\cos(\gamma\pi) - Y_{2m}\big(f(t)\big)\sin(\gamma\pi)\Big\},$$

with $0 \le \gamma < 1$, satisfies the differential equation

$$\frac{d^2z}{dt^2} + F(t)z = 0 \quad , \tag{2.4}$$

where

$$F(t) = \frac{1}{2}\frac{f'''}{f'} - \frac{3}{4}\left(\frac{f''}{f'}\right)^2 + \left(\frac{1}{4} - 4m^2\right)\left(\frac{f'}{f}\right)^2 + f'^2. \tag{2.5}$$

The equation (2.4) can be used , by assuming

$$f(t) = 2k\{(t-t^2)^{1/2} + \arcsin t^{1/2}\} \quad , \quad 0 < t < 1 \quad , \tag{2.6}$$

as a comparison equation to derive , by means of Sturm's method , inequalities for the zeros ϕ_r and ψ_r of $\Phi(a,c;x)$ and $\Psi(a,c;x)$ respectively.

This requires the study of the function

$$G(t,2m) = F(t) - \left\{4k^2\left(\frac{1}{t} - 1\right) + \frac{1-4m^2}{4t^2}\right\} \quad , \tag{2.7}$$

for $0 < t \le 1$, or more simply of the function

$$G^*(t,2m) = \frac{tG(t,2m)}{1-t} \quad ,$$

which is analytic at $t = 0$.

The function $G(t,2m)$, with the same notation , has been considered in [5] where the following lemma was proved.

Lemma 2.1 . Let $G(t,2m)$ be the function defined by (2.7). In the interval $0 < t < 1$, $G(t,2m)$ has at least one zero if $m^2 > 1/4$ and is negative if $m^2 \le 1/4$.

The property $G(t,2m) < 0$ for $0 < t < 1$ and $m^2 \le 1/4$, that is for $0 < c \le 2$, enables us to compare the zeros of the two solutions $u_1(t)$ and $u_2(t)$ of equation (2.1) with those of two suitable solutions of (2.4).

We first consider $\Phi(a, c; x)$ and take into account that, from the results of Erdélyi and Swanson [2,(10.1)], the asymptotic representation

$$u_1(t) \cong 2^{1/2}\Gamma(c)k^{c/2}\left(\frac{f(t)}{f'(t)}\right)^{1/2} J_{c-1}(f(t)) \quad , \tag{2.8}$$

holds, as $k \to \infty$, uniformly for $0 < t \leq 1 - \epsilon$ where ϵ is a fixed positive number. Thus we set $\gamma = 0$ in (2.3) and say $z_1(t)$ the corresponding integral of (2.4). We notice that

$$\left(\frac{f(t)}{f'(t)}\right)^{1/2} = (2t)^{1/2}(1 + \frac{1}{6}t + \ldots) \quad , \quad 0 < t < 1 \quad .$$

Therefore, by means of the series representation of $J_{c-1}(x)$, we obtain

$$z_1(t) = t^{c/2}(a_0 + a_1 t + \ldots) \quad , \qquad 0 < t < 1 \quad , \tag{2.9}$$

with $a_0 \neq 0$.

Now, let $0 < c \leq 2$ and let τ_r, $r = 0, 1, 2, \ldots$, be the zeros of $z_1(t)$ in $0 \leq t < 1$. We have

$$\tau_0 = 0 \quad , \quad 2k\{(\tau_r - \tau_r^2)^{1/2} + \arcsin \tau_r^{1/2}\} = j_{c-1,r} \quad , \quad r = 1, 2, \ldots, s, \tag{2.10}$$

where

$$s = \left[\frac{1}{4} - a\right] . \tag{2.11}$$

This follows by observing that $f(t)$ is a positive increasing function in $0 \leq t \leq 1$ varying from 0 to $k\pi$ and that (see Watson [14], p.497) the number of the positive zeros of $x^{1-c}J_{c-1}(x)$ between 0 and $s\pi + (c-1)\pi/2 + \pi/4$ is exactly s. Then $j_{c-1,s}$ belongs to $(0, k\pi]$ if

$$s + \frac{c-1}{2} + \frac{1}{4} \leq k = \frac{c}{2} - a \quad ,$$

that is if the integer s satisfies the inequality

$$s \leq \frac{1}{4} - a \quad .$$

The condition (1.3), which is required when we apply Theorem 1.1 to the interval $[0, \tau_1]$, is satisfied if $c > 0$ since, from the first of (2.2) we have

$$u_1(t) = t^{c/2}(b_0 + b_1 t + \ldots),$$

and consequently, by using (2.9), we find that

$$u_1(t)z_1'(t) - z_1(t)u_1'(t) = O(t^c), \qquad t \to 0.$$

It may thus be inferred that each interval

$$\tau_{r-1} < t < \tau_r \quad , \quad r = 1, 2, \ldots, s,$$

where s is given by (2.11), contains at least one zero of $\Phi(a, c; 4kt)$. In other words: for the zeros $\phi_r, r = 1, 2, \ldots, s$, of $\Phi(a, c; x)$, if $0 < c \leq 2$, we have

$$\phi_r < 4k\tau_r \quad , \quad r = 1, 2, \ldots, s.$$

The above result can be stated in the following form.

Theorem 2.1. Let $a < 0$ and $0 < c \leq 2$. Let x_r be the root of the equation

$$x - \sin x = \pi - \frac{j_{c-1,r}}{k} \quad , \quad k = \frac{c}{2} - a, \tag{2.12}$$

where $j_{c-1,r}$ is the r -th positive zero of $J_{c-1}(x)$. Then the r -th zero ϕ_r of $\Phi(a, c; x)$ satisfies the inequality

$$\phi_r < 4k\cos^2\frac{x_r}{2} \quad , \quad r = 1, 2, \ldots, s, \tag{2.13}$$

with $s = [1/4 - a]$.

Indeed, by setting $t^{1/2} = \cos\theta$, the equation $f(t) = j_{c-1,r}$ becomes

$$2\theta - \sin 2\theta = \pi - \frac{j_{c-1,r}}{k}.$$

Thus, for the zeros $\tau_r, r = 1, 2, \ldots, s$, defined by (2.10), we have

$$\tau_r = \cos^2\frac{x_r}{2}.$$

Remark 2.1 The integer s defined by (2.11) can assume only the two values $-[a]-1$ or $-[a]$. Therefore, Theorem 2.1 furnishes inequalities for all the zeros of $\Phi(a, c; x)$ with the only exception for the largest zero which must be excluded if $s = -[a] - 1$.

Let us now consider the function $u_2(t)$ defined by the second of (2.2). The following asymptotic representation (see Erdélyi and Swanson [2], (10.2)) holds as $k \to \infty$, uniformly for $0 < t \leq 1 - \epsilon$,

$$u_2(t) \cong 2\pi^{1/2}k^{k+1/2}e^{-k}\left(\frac{f(t)}{f'(t)}\right)^{1/2} \cdot \left\{\cos(a\pi)J_{c-1}\big(f(t)\big) - \sin(a\pi)Y_{c-1}\big(f(t)\big)\right\}, \tag{2.14}$$

where $f(t)$ is given by (2.6). Thus, we set $\gamma = a - [a]$ in (2.3) and say $z_2(t)$ the corresponding integral of (2.4). Further, by virtue of the relationship (see Tricomi [13], p.21)

$$\Psi(a-c+1, 2-c; x) = x^{c-1}\Psi(a, c; x), \tag{2.15}$$

we may assume, without loss of generality, $1 \leq c \leq 2$.

The procedure for deriving inequalities satisfied by the zeros $\psi_1, \psi_2, \ldots$ of $\Psi(a, c; x)$ is perfectly analogous to that used for the zeros of $\Phi(a, c; x)$. We give only a sketch of this procedure.

From the well-known asymptotic representations for $t \to 0$

$$Y_0(t) = \frac{2}{\pi}\log t + 0(1) \quad ,$$
$$Y_\nu(t) = \frac{-\Gamma(\nu)}{\pi}\left(\frac{t}{2}\right)^{-\nu}\left(1 + O(t^\alpha)\right), \quad \alpha = \min(2, 2\nu) \quad , \quad \nu > 0 \quad ,$$

it is straightforward to verify that both $z_2(t)$ and $u_2(t)$ vanish at $t = 0$ and that the condition (1.3) required by Sturm's theorem is satisfied.

Next, we consider the zeros $\tau_r, r = 0, 1, 2, \ldots,$ of $z_2(t)$ in $0 \leq t < 1$, that is we set

$$\tau_0 = 0 \quad , \quad 2k\left\{(\tau_r - \tau_r^2)^{1/2} + \arcsin \tau_r^{1/2}\right\} = \xi_{c-1,r} \quad , \quad r = 1, 2, \ldots, \tag{2.16}$$

where $\xi_{c-1,r}$ denotes the r -th positive zero of the cylinder function

$$C_{c-1}(x) = \cos(a - [a])\pi J_{c-1}(x) - \sin(a - [a])\pi Y_{c-1}(x). \tag{2.17}$$

Then, applying Sturm's theorem, the final result can be stated in the following form.

Theorem 2.2 Let $a < 0$ and $1 \leq c \leq 2$. Let x_r be the root of the equation

$$x - \sin x = \pi - \frac{\xi_{c-1,r}}{k} \quad , \quad k = \frac{c}{2} - a \quad , \tag{2.18}$$

where $\xi_{c-1,r}$ is the r-th positive zero of the function $C_{c-1}(x)$ defined by (2.17). Then the r-th zero ψ_r of $\Psi(a, c; x)$ satisfies the inequality

$$\psi_r < 4k\cos^2\frac{x_r}{2} \quad , \quad r = 1, 2, \ldots, -[a]. \tag{2.19}$$

If $0 < c < 1$ we apply the transformation formula (2.15).

For the proof it suffices to show that the right-hand member of (2.18) is positive for $r = 1, 2, \ldots, -[a]$. Indeed, from two Schafheithin's type inequalities (see Watson [14], pp.490,492) we have

$$\xi_{c-1,r} < (r-1)\pi + \frac{7}{8}\pi + \frac{1}{4}(c-1)\pi - (a-[a])\pi \quad , \quad \text{if} \quad 1 \le c \le \frac{3}{2} \quad ,$$
$$\xi_{c-1,r} < (r-1)\pi - \frac{1}{4}\pi + \frac{1}{2}(c-1)\pi - (a-[a])\pi \quad , \quad \text{if} \quad \frac{3}{2} < c \le 2 \quad ;$$

whence for all values of r, $1 \le r \le -[a]$,

$$k\pi - \xi_{c-1,r} > \left(\frac{c}{4} - r + \frac{3}{8} - [a]\right)\pi > 0 \quad , \quad \text{if} \quad 1 \le c \le \frac{3}{2} \quad ,$$
$$k\pi - \xi_{c-1,r} > \left(-r + \frac{7}{4} - [a]\right)\pi > 0 \quad , \quad \text{if} \quad \frac{3}{2} < c \le 2 \quad .$$

3. Other bounds for the zeros of $\Phi(a, c; x)$ and $\Psi(a, c; x)$.

We shall use in this section, as comparison equation associated to (2.1), the differential equation

$$\frac{d^2v}{dt^2} + H(t)v = 0 \quad , \tag{3.1}$$

with

$$H(t) = \frac{1}{2}\frac{h'''}{h'} - \frac{3}{4}\left(\frac{h''}{h'}\right)^2 - hh'^2 \quad , \tag{3.2}$$

which is satisfied by

$$\begin{aligned} v_1(t) &= \left(h'(t)\right)^{-1/2}\mathrm{Ai}\left(h(t)\right) \quad , \\ v_2(t) &= \left(h'(t)\right)^{-1/2}\mathrm{Bi}\left(h(t)\right) \quad , \end{aligned} \tag{3.3}$$

where $\mathrm{Ai}(x)$ and $\mathrm{Bi}(x)$ are the Airy functions of first and the second kind respectively.

For the properties of these functions we refer to Miller [7] and to Abramowitz and Stegun's [1] handbook. Here we recall only that $\mathrm{Ai}(x)$, $\mathrm{Ai}'(x)$, $\mathrm{Bi}(x)$, $\mathrm{Bi}'(x)$, each have an infinity of zeros on the negative real axis and no other real zeros. Further, as $x \to \infty$,

$$\begin{aligned} &\mathrm{Ai}(x) \cong \frac{1}{2}\pi^{-1/2}x^{-1/4}\exp(-\frac{2}{3}x^{3/2}) \quad , \quad \mathrm{Ai}'(x) \cong -\frac{1}{2}\pi^{-1/2}x^{1/4}\exp(-\frac{2}{3}x^{3/2}); \\ &\mathrm{Bi}(x) \cong \pi^{-1/2}x^{-1/4}\exp(\frac{2}{3}x^{3/2}) \quad , \quad \mathrm{Bi}'(x) \cong \pi^{-1/2}x^{1/4}\exp(\frac{2}{3}x^{3/2}). \end{aligned} \tag{3.4}$$

In order to compare the equation (3.1) with the confluent equation (2.1) we assume in (3.2)

$$h(t)=\begin{cases} -(3k)^{2/3}\{\arccos t^{1/2}-(t-t^2)^{1/2}\}^{2/3} & , \quad 0\le t\le 1, \\ (3k)^{2/3}\{(t^2-t)^{1/2}-\operatorname{arccosh} t^{1/2}\}^{2/3} & , \quad t>1. \end{cases} \tag{3.5}$$

We find

$$H(t)=\frac{5}{4}\frac{t-1}{t}k^2h^{-3}(t)+\frac{3-8t}{16t^2(1-t)^2}+\frac{4k^2(1-t)}{t} \quad , \quad t>0, \tag{3.6}$$

and the following lemma holds.

Lemma 3.1 (see Gatteschi [5]). In the interval $0<t<+\infty$ the function

$$Q(t,2m)=H(t)-\left\{4k^2\left(\frac{1}{t}-1\right)+\frac{1-4m^2}{4t^2}\right\} \quad , \tag{3.7}$$

with $H(t)$ defined by (3.6) , is negative if $m^2\le 1/16$, is positive if $m^2\ge 1/9$ and has exactly one zero if $1/16<m^2<1/9$.

We now observe that the asymptotic representation of the function $u_1(t)$, defined by the first of (2.2), given by Erdélyi and Swanson [2,(10.3)] can be written in terms of the function $h(t)$,defined by (3.5) , in the form

$$u_1(t) \cong 2\Gamma(2m+1)k^{-m+1/2}\big(h'(t)\big)^{-1/2} \cdot\Big\{e^{(m-k+1/6)\pi i}\mathrm{Ai}\big(h(t)e^{2\pi i/3}\big)+e^{-(m-k+1/6)\pi i}\mathrm{Ai}\big(h(t)e^{-2\pi i/3}\big)\Big\},$$

as $k\to\infty$ uniformly for $0<\epsilon\le t<\infty$, with fixed ϵ.

Thus, by using the formula [1,p.446]

$$\mathrm{Ai}(ze^{\pm 2\pi i/3})=\frac{1}{2}e^{\pm\pi i/3}\{\mathrm{Ai}(z)\mp i\mathrm{Bi}(z)\},$$

we easily obtain

$$u_1(t) \cong 2\Gamma(2m+1)k^{-m+1/2}\big(h'(t)\big)^{-1/2} \cdot\Big\{\cos(m-k+\frac{1}{2})\pi\mathrm{Ai}\big(h(t)\big)+\sin(m-k+\frac{1}{2})\pi\mathrm{Bi}\big(h(t)\big)\Big\},$$

that is , using (1.6) ,

$$u_1(t) \cong 2\Gamma(c)k^{c/2}\big(h'(t)\big)^{-1/2}\Big\{\cos(a\pi)\mathrm{Ai}\big(h(t)\big)+\sin(a\pi)\mathrm{Bi}\big(h(t)\big)\Big\}. \tag{3.8}$$

This representation shows that a solution of (3.1) which is appropriate for the comparison with the solution $u_1(t)$ of (2.1) is given by

$$v_3(t) = \big(h'(t)\big)^{-1/2}\Big\{\cos(a\pi)\mathrm{Ai}\big(h(t)\big) + \sin(a\pi)\mathrm{Bi}\big(h(t)\big)\Big\}. \tag{3.9}$$

Therefore , being the function $h(t)$ negative and decreasing in $0 < t < \infty$, the application of Sturm's theorem will be based on the consideration of the negative zeros of the function

$$A(x) = \cos(a\pi)\mathrm{Ai}(x) + \sin(a\pi)\mathrm{Bi}(x). \tag{3.10}$$

This function has only negative zeros and the following lemma holds.

Lemma 3.2 . Let $\alpha_r, r = 1, 2, \ldots$, be the negative zeros, in decreasing order, of the function $A(x)$ defined by (3.10). Then

$$\alpha_r > -\left\{\frac{3}{2}\left(r - \frac{1}{24} - \gamma\right)\pi\right\}^{2/3} \quad , \quad \gamma = a + \frac{1}{6} - \left[a + \frac{1}{6}\right], \tag{3.11}$$

with the possible exception for the first zero.

For the proof we shall refer to the only positive zeros of $A(-x)$. By using the known representations of Airy's functions in terms of Bessel functions

$$\begin{aligned} \mathrm{Ai}(-x) &= \frac{1}{3}\sqrt{x}\,\{J_{1/3}(\xi) + J_{-1/3}(\xi)\} \quad , \\ \mathrm{Bi}(-x) &= \sqrt{\frac{x}{3}}\,\{J_{-1/3}(\xi) - J_{1/3}(\xi)\} \quad , \end{aligned} \qquad \xi = \frac{2}{3}x^{3/2}$$

and the formula

$$J_{-1/3}(\xi) = \frac{1}{2}J_{1/3}(\xi) - \frac{\sqrt{3}}{2}J_{-1/3}(\xi) \quad ,$$

we obtain from (3.10)

$$A(-x) = \sqrt{\frac{x}{3}}\left\{\cos\left(a + \frac{1}{6}\right)\pi J_{1/3}(\xi) - \sin\left(a + \frac{1}{6}\right)\pi Y_{1/3}(\xi)\right\}.$$

Hence it suffices to study the positive zeros of the cylinder function

$$C_{1/3}(\xi) = \cos(\gamma\pi)J_{1/3}(\xi) - \sin(\gamma\pi)Y_{1/3}(\xi) \quad , \quad 0 \le \gamma < 1 \quad ,$$

with γ defined as in (3.11).

A theorem of Schafheitlin's type (see Watson [14], p.490) states that the only positive zeros of $C_{1/3}(\xi)$ lie in the intervals

$$\left(s\pi + \frac{11}{12}\pi - \gamma\pi, s\pi + \frac{23}{24}\pi - \gamma\pi\right) \quad ,$$

where $s = 0, 1, 2, \ldots$, except that, if γ is sufficiently near to 1, there may be one zero in the interval

$$\left(\frac{23}{24}\pi - \gamma\pi, \pi - \gamma\pi\right) \quad .$$

It follows that the positive zeros $c_{1/3,r}$, $r = 1, 2, \ldots$, of $C_{1/3}(\xi)$ satisfy the inequality

$$c_{1/3,r} < (r-1)\pi + \frac{23}{24}\pi - \gamma\pi \quad ,$$

with the possible exception of the first zero. This proves the lemma.

Now, we can state the first result of this section.

Theorem 3.1 . Let $a < 0$ and $c > 0$. Let x_r be the root of the equation

$$x - \sin x = \frac{2}{3k}\left(-\alpha_{-[a]-r+1}\right)^{3/2} \quad , \quad k = \frac{c}{2} - a \quad , \tag{3.12}$$

where α_j is the j-th negative zero of the function $A(x)$ defined by (3.10). Then, provided that

$$\frac{2}{3k}\left(-\alpha_{-[a]-r+1}\right)^{3/2} \le \pi \quad , \tag{3.13}$$

for the r-th zero ϕ_r of $\Phi(a, c; x)$we have

$$\phi_r > 4k\cos^2\left(\frac{x_r}{2}\right), \qquad \text{if } \frac{1}{2} \le c \le \frac{3}{2}, \tag{3.14}$$

$$\phi_r < 4k\cos^2\left(\frac{x_r}{2}\right), \qquad \text{if } 0 < c \le \frac{1}{3} \quad \text{or} \quad c \ge \frac{5}{3}. \tag{3.15}$$

We first remark that the condition (3.12) is satisfied at least for $r = 2, 3, \ldots, -[a]$. Indeed , applying inequality (3.11) we have

$$\frac{2}{3k}\left(-\alpha_{-[a]-r+1}\right)^{3/2} < \frac{-[a] - r + 23/24 - \gamma}{c/2 - a}\pi < \pi,$$

for $r = 2, 3, \ldots, -[a]$, while in the case $r = 1$ the corresponding inequality

$$\frac{-[a] - 1/24 - \gamma}{c/2 - a}\pi \le \pi$$

may or may not be true .

Next we give only the essential steps of the proof of the theorem.

In the case $1/2 \leq c \leq 3/2$, i.e. $m^2 \leq 1/16$, the function $v_3(t)$ defined by (3.9) and (3.5) has at least $-[a]-1$ zeros in the interval (0,1). These zeros can be obtained by solving the equation

$$\arccos t^{1/2} - (t-t^2)^{1/2} = \frac{1}{3k}(-\alpha_j)^{3/2}, \quad k = \frac{c}{2} - a, \tag{3.16}$$

for $j = 1, 2, \ldots, -[a]-1$.

Now let $\tau_1 > \tau_2 > \ldots$ be the zeros, in decreasing order , of $v_3(t)$. According to Lemma 3.1 , $Q(t, 2m)$ is negative for $0 < t < \infty$ if $m^2 \leq 1/16$, hence we can apply Theorem 1.1 . Furthermore , using (3.4) and the derivation formula

$$\frac{d}{dx}\Phi(a,c;x) = \frac{a}{c}\Phi(a+1,c+1;x),$$

we can see , multiplying $v_3(t)$ by the constant factor $2\Gamma(c)k^{c/2}$, that the conditions (1.4) of Theorem 1.2 are satisfied at $t = \infty$. Then , we conclude that each interval

$$\tau_1 < t < \infty, \quad \tau_{i+1} < t < \tau_i, \quad i = 1, 2, \ldots,$$

provided it belongs to the interval $0 < t < \infty$ must contain at least one zero t_j of (3.16). More precisely we have

$$t_j > \tau_{-[a]-j+1}.$$

By setting $t^{1/2} = \cos(x/2)$ in (3.16) we easily derive (3.12) and (3.14). Similar arguments prove (3.15).

The derivation of inequalities for the function $\Psi(a,c;x)$ does not present difficulties. We use the Erdélyi-Swanson expansion [2,(10.4)] for $u_2(t)$, i.e.

$$u_2(t) \cong 2^{3/2}\pi^{1/2}k^{k+1/2}e^{-k}\big(h'(t)\big)^{-1/2}\mathrm{Ai}\big(h(t)\big), \tag{3.17}$$

that we compare with the function $v_1(t)$ defined by the first of (3.3). Since in this case we must consider the zeros of Airy function $\mathrm{Ai}(x)$, it is convenient to replace Lemma 3.2 with the more precise result:

Lemma 3.3 (see Gatteschi [5]). Let a_r, $r = 1, 2, \ldots$, be the zeros in decreasing order of $\mathrm{Ai}(x)$.Then

$$-\left\{\frac{3}{8}\left(4r - \frac{5}{6}\right)\pi\right\}^{2/3} < a_r < -\left\{\frac{3}{8}(4r-1)\pi\right\}^{2/3}, \tag{3.18}$$

for $r = 1, 2, \ldots$.

Then, the application of Sturm's method leads easily to the final result.

Theorem 3.2 Let $a < 0$ and $c \geq 1$.Let x_r be the root of the equation

$$x - \sin x = \frac{2}{3k}\left(-a_{-[a]-r+1}\right)^{3/2}, \quad k = \frac{c}{2} - a, \tag{3.19}$$

where a_j is the j-th negative zero of the Airy function $\mathrm{Ai}(x)$. Then for the r-th zero ψ_r of $\Psi(a, c; x)$we have

$$\psi_r > 4k\cos^2\left(\frac{x_r}{2}\right), \quad \text{if} \quad 1 \leq c \leq \frac{3}{2}, \tag{3.20}$$

$$\psi_r < 4k\cos^2\left(\frac{x_r}{2}\right), \quad \text{if} \quad c \geq \frac{5}{3}, \tag{3.21}$$

where $r = 1, 2, \ldots, -[a]$.

Here the condition $c \geq 1$, by virtue of the transformation formula (2.15) , is not restrictive.

4. Particular cases and numerical results.

A particularly interesting case of the confluent hypergeometric functions is given by the Laguerre polynomials $L_n^{(\alpha)}(x)$, which occur when $a = -n, n = 0, 1, 2, \ldots$, and $c = \alpha + 1, \alpha > -1$. This case has been considered in [5], were some lower and upper bounds for the zeros $\lambda_{n,r}^{(\alpha)}$ are obtained. Furthermore, in a more recent paper [6], we have also derived two asymptotic approximations for $\lambda_{n,r}^{(\alpha)}$ from two new uniform asymptotic expansions of Laguerre polynomials given by Frenzen and Wong [3].

Here we recall only this very simple result:

Theorem 4.1 (see Gatteschi [5]). Let $-1/2 \leq \alpha \leq 1/2$ and let $X(y)$ be the function that we obtain upon inverting

$$x - \sin x = y. \tag{4.1}$$

Then, for the zeros $\lambda_{n,r}^{(\alpha)}, r = 1, 2, \ldots, n$, of $L_n^{(\alpha)}(x)$ we have

$$\begin{aligned} \lambda_{n,r}^{(\alpha)} &> 4k\cos^2\left\{\frac{1}{2}X\left(\frac{2}{3k}(-a_{n-r+1})^{3/2}\right)\right\}, \\ \lambda_{n,r}^{(\alpha)} &< 4k\cos^2\left\{\frac{1}{2}X\left(\pi - \frac{j_{\alpha,r}}{k}\right)\right\}, \end{aligned} \tag{4.2}$$

where $k = n + (\alpha + 1)/2$, a_s is the s-th zero of $\mathrm{Ai}(x)$ and $j_{\alpha,r}$ is the r-th positive zero of $J_\alpha(x)$.

An analogous result can be obtained for the positive zeros of the parabolic cylinder function $D_\nu(x)$.

This function is defined, for $\mathrm{Re}x > 0$, by

$$\begin{aligned} D_\nu(x) &= 2^{\nu/2} \exp\left(-\frac{x^2}{4}\right) \Psi\left(-\frac{\nu}{2}, \frac{1}{2}; \frac{x^2}{2}\right) \\ &= 2^{(\nu-1)/2} \exp\left(-\frac{x^2}{4}\right) \Psi\left(\frac{1-\nu}{2}, \frac{3}{2}; \frac{x^2}{2}\right). \end{aligned} \tag{4.3}$$

According to Theorem 1.3, the number of the positive zeros of $D_\nu(x)$ is given by $-[(1-\nu)/2]$. Hence we shall assume $\nu > 1$.

We notice that, for $a = (1-\nu)/2$ and $c = 3/2$, the cylinder function defined by (2.17) reduces to

$$C_{1/2}(x) = \left(\frac{2}{\pi x}\right)^{1/2} \sin\left\{x + \left(\frac{1-\nu}{2} - \left[\frac{1-\nu}{2}\right]\right)\pi\right\}.$$

It follows that the zeros $\xi_{1/2,r}$, $r = 1, 2, \ldots,$ of $C_{1/2}(x)$ are given by

$$\xi_{1/2,r} = \frac{2r + \nu - 1}{2}\pi + \left[\frac{1-\nu}{2}\right]\pi. \tag{4.4}$$

Then, applying the Theorems 2.2 and 2.3 to the function $\Psi\left((1-\nu)/2, 3/2; x^2/2\right)$, we easily obtain the following inequalities for the zeros $\delta_{\nu,r}$ of $D_\nu(x)$.

Theorem 4.2. Let $\nu > 1$ and let $X(y)$ defined as in Theorem 4.1. Then for the zeros $\delta_{\nu,r}$, $r = 1, 2, \ldots, -[(1-\nu)/2]$, of $D_\nu(x)$ we have

$$\begin{aligned} \delta_{\nu,r} &> 2^{3/2} k^{1/2} \cos\left\{\frac{1}{2} X\left(\frac{2}{3k}\left(-a_{p-r+1}\right)^{3/2}\right)\right\}, \\ \delta_{\nu,r} &< 2^{3/2} k^{1/2} \cos\left\{\frac{1}{2} X\left(\pi - \frac{2r - 2p + \nu - 1}{2k}\pi\right)\right\}, \end{aligned} \tag{4.5}$$

where

$$p = -\left[\frac{1-\nu}{2}\right], \qquad k = \frac{2\nu + 1}{4}. \tag{4.6}$$

The very sharp numerical results that we can obtain by using the inequalities derived in this paper are shown in the following three tables. The tables show also the accuracy in the lower bound and in the upper bound, that is the number

$$\rho(r) = -\log_{10}\left|1 - \frac{B(r)}{E(r)}\right|,$$

where $B(r)$ is the bound, lower or upper, and $E(r)$ is the exact value of the corresponding zero.

Table 1. Zeros of $\Phi\ (-10.2, 1.2; x)$.

r	Lower bound	$\rho(r)$	Exact value	Upper bound	$\rho(r)$
1	0.165000	1.5	0.169799	0.169858	3.5
2	0.786024	2.2	0.791262	0.791541	3.4
3	1.881949	2.5	1.887414	1.888108	3.4
4	3.478498	2.8	3.484133	3.485499	3.4
5	5.617098	3.0	5.622872	5.625284	3.4
6	8.362000	3.2	8.367804	8.371870	3.3
7	11.81372	3.3	11.81972	11.82655	3.2
8	16.14021	3.4	16.14630	16.15829	3.1
9	21.65471	3.5	21.66088	21.68470	3.0
10	29.10059	3.7	29.10684	29.17388	2.6
11	42.32961	3.8	42.33595	....	.....

Table 2. Zeros of $\Psi\ (-10.2, 1.2; x)$.

r	Lower bound	$\rho(r)$	Exact value	Upper bound	$\rho(r)$
1	0.0005563	0.1	0.0024096	0.0024104	3.5
2	0.2520249	1.7	0.2569568	0.2570456	3.5
3	0.9665834	2.3	0.9718739	0.9722191	3.5
4	2.1601412	2.6	2.1656431	2.1664486	3.4
5	3.8611891	2.8	3.8668534	3.8683934	3.4
6	6.1149896	3.0	6.1207879	6.1234714	3.4
7	8.9915954	3.2	8.9975081	9.0020135	3.3
8	12.601914	3.3	12.607926	12.615531	3.2
9	17.133732	3.4	17.139833	17.153403	3.1
10	22.950168	3.6	22.956350	22.984472	2.9
11	30.983546	3.7	30.989809	31.081205	2.5

The Table 1 has been obtained by using Theorems 2.1 and 3.1 for the upper and lower bound respectively. Note that,since $s = 10$, the upper bound for ϕ_{11} must be excluded.

The Theorems 2.2 and 3.2 have been used for Table 2. The Table 3 has been obtained by applying Theorem 4.2.

Table 3. Positive zeros of $D_{10.5}(x)$.

r	Lower bound	$\rho(r)$	Exact value	Upper bound	$\rho(r)$
1	0.7103801	2.8	0.7114143	0.7117877	3.3
2	1.6736923	3.2	1.6746707	1.6756411	3.2
3	2.6770228	3.6	2.6777488	2.6796521	3.1
4	3.7609004	3.8	3.7615196	3.7654436	3.0
5	5.0253045	4.0	5.0258308	5.0379143	2.6

Remark 4.1. We notice that the Tricomi formula (1.12) can be written, by means of the function $X(y)$ defined in Theorem 4.1, in the form

$$\phi_r \cong 4k\cos^2\left\{\frac{1}{2}X\left(\frac{3-4a-4r}{4k}\pi\right)\right\}, \tag{4.7}$$

as $k \to \infty$ for all the zeros ϕ_r lying in the interval $(4t'k, 4t''k)$ with t' and t'' fixed constants, $0 < t' < t'' < 1$.

Now, by using the well-known asymptotic representation (see Watson[14],p.506] of the large zeros c_r of a cylinder function, it is easy to see that the bounds (2.13) and (3.14), (3.15) for the zeros ϕ_r are in fact asymptotic approximations.

More precisely, we have from Theorem 2.1

$$\phi_r \cong 4k\cos^2\left\{\frac{1}{2}X\left(\pi - \frac{j_{c-1,r}}{k}\right)\right\}, \tag{4.8}$$

as $k \to \infty$ for all the zeros lying in the interval $(0, 4t''k)$, and from Theorem 3.1

$$\phi_r \cong 4k\cos^2\left\{\frac{1}{2}X\left(\frac{2}{3k}\left(-\alpha_{-[a]-r+1}\right)^{3/2}\right)\right\}, \tag{4.9}$$

as $k \to \infty$ for all the zeros in the interval $(4t'k, 4k)$. Both approximations (4.8) and (4.9) coincide, as $k \to \infty$, with Tricomi's approximation (4.7).

Similarly, the bounds for the zeros given in Theorems 2.2 and 3.2 are asymptotic approximations which coincide with the approximation

$$\psi_r \cong 4k\cos^2\left\{\frac{1}{2}X\left(\frac{3-4[a]-4r}{4k}\pi\right)\right\}, \tag{4.10}$$

as $k \to \infty$ for all the zeros lying in the interval $(4t'k, 4t''k)$.

Acknowledgments. I would like to express my sincere thanks to Dr. Isabella Cravero and to Miss Ornella Consani for carrying out many numerical computations and for carefully reading the manuscript.

References

[1] - M.Abramowitz and I.A.Stegun,*Handbook of Mathematical Functions*, Applied Mathematics Series,55, National Bureau of Standards, Washington,DC, 1964.

[2] - A.Erdélyi and C.A.Swanson,*Asymptotic forms of Whittaker's confluent hypergeometric functions*,Mem.Amer.Math.Soc.No.25,Providence, RI,1957.

[3] - C.L.Frenzen and R.Wong,*Uniform asymptotic expansions of Laguerre polynomials*,SIAM J.Math.Anal.19(1988),1232-1248.

[4] - L.Gatteschi,*New inequalities for the zeros of Jacobi polynomials*,SIAM.J. Math. Anal.18(1987),1549-1562.

[5] - L.Gatteschi,*Some inequalities for the zeros of Laguerre polynomials*,Numerical Methods and Approximation Theory,Proc.Int.Conf. Niš/Yugoslavia 1987,Univ.of Niš (1988),23-38.

[6] - L.Gatteschi,*Uniform approximations for the zeros of Laguerre polynomials*,Numerical Mathematics, Proc.Int.Conf.Singapore 1988, Internat. Ser. Numer. Math. 86(1988), 137-148.

[7] - J.C.P.Miller,*The Airy integral*,British Assoc. Adv. Sci. Mathematical Tables, Part-vol.B,Cambridge Univ. Press,Cambridge,1946.

[8] - F.W.J.Olver,*Asymptotics and Special Functions*, Academic Press,New York, 1974.

[9] - L.J.Slater,*Confluent hypergeometric functions*,Cambridge Univ.Press,London and New York,1960.

[10] - G.Szegö,*Orthogonal Polynomials*,Colloquium Publications, Vol.23,4th ed., American Mathematical Society,Providence,RI,1975.

[11] - F.G.Tricomi,*Sugli zeri delle funzioni di cui si conosce una rappresentazione asintotica*,Ann.Mat.Pura Appl.(4)26(1947-48),283-300.

[12] - F.G.Tricomi,*Funzioni ipergeometriche confluenti*,Monografie Matematiche 1,Ediz. Cremonese,Roma,1954.

[13] - F.G.Tricomi,*Fonctions hypergéométriques confluentes*, Mémorial des Sciences Mathématiques,CXL,Gauthier-Villars,Paris,1960.

[14] - G.N.Watson,*A Treatise on the Theory of Bessel Functions*,2nd ed.,Cambridge Univ.Press,Cambridge,1966.

How (Un)stable Are Vandermonde Systems?

Walter Gautschi Professor, Department of Computer Sciences, Purdue University, West Lafayette, Indiana

ABSTRACT. Results on the condition number of Vandermonde type matrices obtained during the last 25 years are reviewed. Equal emphasis is given to real and complex nodes. Recent work dealing with nodes placed sequentially on circular and elliptic contours in the complex plane receives special attention.

I. INTRODUCTION

Many problems in applied and numerical analysis eventually boil down to solving large systems of linear algebraic equations. Since the matrices and right-hand sides of such systems are typically the result of (sometimes extensive) computations, they are subject to an unavoidable level of noise caused by the rounding errors committed during their generation. It is then a matter of practical concern trying to estimate the effect of such uncertainties upon the solution of the system.

A common answer – and one which we shall adopt in the sequel – concerning any nonsingular system

$$Ax = b, \quad \det A \neq 0, \tag{1.1}$$

is to compute (or estimate) the *condition number*

$$\operatorname{cond} A = \|A\| \cdot \|A^{-1}\| \tag{1.2}$$

of the system, where $\|\cdot\|$ denotes a suitable matrix norm. Norms, for matrices

Work supported, in part, by the National Science Foundation under grant CCR-8704404.

$A = [a_{ij}]$, that will be used here are the ∞ - *norm*,

$$||A||_\infty = \max_i \sum_j |a_{ij}|, \tag{1.3}$$

the *Euclidean* (or spectral) *norm*,

$$||A||_2 = \sqrt{\rho(AA^H)}, \tag{1.4}$$

where $\rho(\cdot)$ denotes spectral radius, and the *Frobenius norm*,

$$||A||_F = \sqrt{\mathrm{tr}(AA^H)} = \sqrt{\sum_{i,j} |a_{ij}|^2}\ . \tag{1.5}$$

If $\varepsilon_x = ||\delta x||/||x||$ is the relative error in the solution x of (1.1), caused by relative errors $\varepsilon_A = ||\delta A||/||A||$, $\varepsilon_b = ||\delta b||/||b||$ in the system, then the condition number in (1.2) indicates how much larger ε_x is compared to ε_A and ε_b, that is, roughly speaking,

$$\varepsilon_x \approx (\mathrm{cond}\ A)(\varepsilon_A + \varepsilon_b). \tag{1.6}$$

It always seemed important to us that the conditioning of matrices be investigated for many special classes of matrices. In this spirit, we began, 25 years ago, to take up the class of Vandermonde matrices. The original motivation came from unpleasant experiences with the computation of Gauss type quadrature rules from the moments of the underlying weight function. The sensitivity of the problem then indeed depends on the condition of certain (confluent) Vandermonde matrices with real nodes. Since then, we have intermittently looked at the conditioning of such matrices, considering not only real, but also complex nodes, and have enlarged the class of matrices by including Vandermonde-like matrices involving polynomial systems other than the system of powers. Here we present a brief survey of results obtained over the years, including also some original material (in Sections IV, V and VI).

To establish terminology and notation, we call a *Vandermonde matrix* a matrix of the form

$$V_n = \begin{bmatrix} 1 & 1 & \cdots & 1 \\ z_1 & z_2 & \cdots & z_n \\ \cdot & \cdot & \cdots & \cdot \\ z_1^{n-1} & z_2^{n-1} & \cdots & z_n^{n-1} \end{bmatrix}, \quad z_i \in \mathbb{C}, \quad n > 1, \tag{1.7}$$

where z_i are pairwise distinct real or complex numbers called the *nodes*. More generally, a *Vandermonde-like matrix*, with nodes z_i, is a matrix of the form

$$V_n = \begin{bmatrix} p_0(z_1) & p_0(z_2) & \cdots & p_0(z_n) \\ p_1(z_1) & p_1(z_2) & \cdots & p_1(z_n) \\ \cdot & \cdot & \cdots & \cdot \\ p_{n-1}(z_1) & p_{n-1}(z_2) & \cdots & p_{n-1}(z_n) \end{bmatrix}, \tag{1.8}$$

where $\{p_k\}$ is a system of linearly independent polynomials, often with $p_k \in \mathbf{P}_k$, the class of polynomials of degree k. Such matrices (or their transposed) are encountered, for example, when one deals with polynomial interpolation or with interpolatory approximation of linear functionals (see, e.g., [1]), the form (1.7) or (1.8) occurring depending on the choice of basis elements in polynomial spaces. Vandermonde systems with matrix (1.7) are also an important ingredient in Remes' algorithm for constructing best uniform polynomial approximations.

A brief outline of the paper is as follows. Sections II–IV are devoted to ordinary Vandermonde matrices V_n, (1.7). We begin in Section II with some basic inequalities for $\|V_n^{-1}\|_\infty$. These are then applied in Section III to obtain estimates for the condition number $\mathrm{cond}_\infty V_n$ for certain real node configurations. The bottom line here is that *V_n is ill-conditioned when all nodes are real.* Indeed, the condition number, in many cases (and perhaps always), grows exponentially with the order n of the matrix. The scenario changes drastically if one allows complex nodes. The roots of unity, for example, give rise to perfectly conditioned Vandermonde matrices. Other sequences of nodes on the unit circle are studied in Section IV. Some, like the Van der Corput sequence, perform nearly as well, and do so in a linear sequential, rather than triangular, fashion. Others, like "quasi-cyclic" sequences, do much worse. The remaining two sections discuss Vandermonde-like matrices with the polynomials p_k in (1.8) chosen to be orthogonal polynomials. In Section V we consider special real, as well as arbitrary complex nodes, in Section VI nodes placed sequentially on elliptic contours in the complex plane, and a Chebyshev system of polynomials p_k.

II. A BASIC INEQUALITY FOR INVERSES OF VANDERMONDE MATRICES

The inversion of a linear system with the Vandermonde matrix (1.7) as coefficient matrix can be easily described in terms of the elementary Lagrange interpolation polynomials

$$\ell_\lambda(z) = \prod_{\substack{\mu=1 \\ \mu\neq\lambda}}^{n} \frac{z - z_\mu}{z_\lambda - z_\mu}, \qquad \lambda = 1,2,\ldots,n, \tag{2.1}$$

associated with the nodes $z_1, z_2, \ldots, z_n$. Indeed, if we expand ℓ_λ in powers of z,

$$\ell_\lambda(z) = \sum_{\mu=1}^{n} u_{\lambda\mu} z^{\mu-1}, \tag{2.2}$$

the inversion is accomplished by multiplying the μth equation by $u_{\lambda\mu}$, $\mu = 1,2,\ldots,n$, and adding up the results. In view of $\ell_\lambda(z_\nu) = \delta_{\lambda\nu}$ (the Kronecker delta), this will express the λth unknown linearly in terms of the right-hand members of the system. Hence,

$$V_n^{-1} = [u_{\lambda\mu}]_{\substack{1 \le \lambda \le n \\ 1 \le \mu \le n}}. \tag{2.3}$$

Combining (2.1) and (2.2) yields

$$u_{\lambda 1} + u_{\lambda 2} z + \cdots + u_{\lambda n} z^{n-1} = \pi_\lambda \prod_{\substack{\mu=1 \\ \mu \ne \lambda}}^{n} (z - z_\mu), \tag{2.4}$$

where

$$\pi_\lambda = \prod_{\mu \ne \lambda} (z_\lambda - z_\mu)^{-1}. \tag{2.5}$$

Therefore, we have the alternative representation

$$u_{\lambda\mu} = (-1)^{\mu-1} \pi_\lambda \sigma_{n-\mu}^{\lambda}(z_1, \ldots, z_{\lambda-1}, z_{\lambda+1}, \ldots, z_n), \tag{2.6}$$

where σ_m^λ denotes the mth elementary symmetric function in the $n-1$ variables z_μ with z_λ removed.

Theorem 2.1. *For arbitrary $z_\nu \in \mathbb{C}$, with $z_\nu \ne z_\mu$ if $\nu \ne \mu$, there holds*

$$\max_\lambda \prod_{\mu \ne \lambda} \frac{\max(1, |z_\mu|)}{|z_\lambda - z_\mu|} < \| V_n^{-1} \|_\infty \le \max_\lambda \prod_{\mu \ne \lambda} \frac{1 + |z_\mu|}{|z_\lambda - z_\mu|}, \tag{2.7}$$

where V_n is the matrix in (1.7). The upper bound is attained if $z_\mu = |z_\mu| e^{i\theta}$, $\mu = 1,2,\ldots,n$, for some fixed $\theta \in \mathbb{R}$.

Proof (Sketch). The upper bound in (2.7), and the statement about equality, follow from (2.6) and from a simple fact (see the Lemma in [7, p.118]) about elementary symmetric functions $\sigma_m = \sigma_m(x_1, \ldots, x_n)$ in n variables, namely that $\sum_{m=0}^{n} |\sigma_m| \le \prod_{\nu=1}^{n} (1 + |x_\nu|)$, with equality precisely if all x_ν lie on the same ray through the origin.

For the lower bound we use the fact that

$$\sum_{\mu=0}^{n} |a_\mu| \ge |a_n| \prod_{\nu=1}^{n} \max(1, |\zeta_\nu|) \tag{2.8}$$

holds for any polynomial $p(z) = a_0 + a_1 z + \cdots + a_n z^n$, $a_n \ne 0$, having the zeros ζ_1,

$\zeta_2, \ldots, \zeta_n$, with equality if and only if $p(z) = a_n z^n$. This is a simple consequence of Jensen's formula (see [10, §2]). Applying (2.8) to the polynomial (of degree $n-1$) in (2.4) then easily yields the lower bound in (2.7); see again [10] for details. □

III. REAL NODES

An important case in which equality holds for the upper bound in (2.7) is when all nodes are nonnegative, $z_\nu = x_\nu \geq 0$. Then, letting

$$p_n(z) = \prod_{\nu=1}^{n} (z - x_\nu) \tag{3.1}$$

denote the node polynomial, we can write (2.7) in the form

$$\| V_n^{-1} \|_\infty = \frac{|p_n(-1)|}{\min_\nu \{(1 + x_\nu)|p_n'(x_\nu)|\}} \qquad (x_\nu \geq 0). \tag{3.2}$$

The techniques used in the first part of the proof of Theorem 2.1 can be adapted (cf. [8, Theorem 4.3]) to also deal with the case of real nodes located symmetrically with respect to the origin: $x_\nu \in \mathbf{R}$ with $x_\nu + x_{n+1-\nu} = 0$ for $\nu = 1,2,\ldots, n$. In place of (3.2), one obtains

$$\| V_n^{-1} \|_\infty = \frac{|p_n(i)|}{\min\limits_{x_\nu \geq 0} \left\{ \dfrac{1 + x_\nu^2}{1 + x_\nu} |p_n'(x_\nu)| \right\}} \qquad (x_\nu + x_{n+1-\nu} = 0,\ \ x_\nu \in \mathbf{R}). \tag{3.3}$$

Since for given nodes x_ν the norm $\| V_n \|_\infty$ is easily calculated, the results (3.2), (3.3) allow us to evaluate the condition number $\text{cond}_\infty V_n$ exactly in the respective cases. For example, if $|x_\nu| \leq 1$ for all ν, then

$$\text{cond}_\infty V_n = n \| V_n^{-1} \|_\infty \qquad (|x_\nu| \leq 1). \tag{3.4}$$

We illustrate (3.2)–(3.4) with a number of examples, ordered in decreasing severity of ill-conditioning.

Example 3.1. Harmonic nodes $x_\nu = 1/\nu$, $\nu = 1,2,\ldots, n$.

Here, an easy calculation gives $|p_n(-1)| = n+1$, and letting $\delta_\nu = (1 + x_\nu)|p_n'(x_\nu)|$, one finds

$$\delta_\nu = \frac{(\nu + 1)!(n - \nu)!}{\nu^n n!}, \quad \nu = 1,2,\ldots, n.$$

There follows

$$\min_\nu \delta_\nu \le \delta_n = \frac{n+1}{n^n}.$$

(Actually, this holds with strict inequality, for $n > 2$, as can be shown by a more detailed analysis). Consequently, by (3.4) and (3.2),

$$\text{cond}_\infty V_n > n^{n+1} \quad (x_\nu = 1/\nu). \tag{3.5}$$

Note that the condition number in (3.5) grows more rapidly than $n!$, which is far worse than the condition of the notorious Hilbert matrix, which grows "only" exponentially!

Example 3.2. Equidistant nodes on [0,1]: $x_\nu = \dfrac{\nu-1}{n-1}$, $\nu = 1,2,\ldots, n$.

Defining δ_ν as in the previous example, an elementary computation gives

$$|p_n(-1)| = \frac{(2n-2)!}{(n-1)!(n-1)^{n-1}}, \quad \delta_\nu = \frac{(n+\nu-2)(\nu-1)!(n-\nu)!}{(n-1)^n}.$$

Putting $\nu = \kappa n$, $0 < \kappa < 1$, and studying $\delta_{\kappa n}$ for $n \to \infty$, reveals that, asymptotically, $\delta_{\kappa n}$ is a minimum when $\kappa = \frac{1}{2}$, and $\delta_{\frac{1}{2}n} \sim \pi en\,(2e)^{-n}$ as $n \to \infty$. Combining this in Eq. (3.2) with the asymptotic expression for $|p_n(-1)|$, obtained by Stirling's formula, and noting that $\|V_n\|_\infty = n$, yields

$$\text{cond}_\infty V_n \sim \frac{\sqrt{2}}{4\pi} \cdot 8^n, \quad n \to \infty. \tag{3.6}$$

We are now down to exponential growth, but expect that the rate of growth can still be reduced by placing the nodes symmetrically with respect to the origin. This is confirmed in the next example.

Example 3.3. Equidistant nodes on [−1,1], $x_\nu = 1 - \dfrac{2(\nu-1)}{n-1}$, $\nu = 1,2,\ldots, n$.

Here we use (3.3). An asymptotic analysis similar to the one in the previous example, but more involved, shows that [8, Example 6.1]

$$\text{cond}_\infty V_n \sim \frac{1}{\pi} e^{-\frac{1}{4}\pi} e^{n(\frac{1}{4}\pi + \frac{1}{2}\ln 2)}, \quad n \to \infty. \tag{3.7}$$

Note that the exponential growth rate is now $\exp\left[\frac{1}{4}\pi + \frac{1}{2}\ln 2\right] = 3.1017\ldots,$

compared to 8 in the asymmetric case of Example 3.2.

Can we do better if we take the Chebyshev nodes in (–1,1)?

Example 3.4. Chebyshev nodes $x_\nu = \cos((2\nu - 1)\pi/2n)$, $\nu = 1,2,\ldots,n$. Applying (3.3), one can prove [8, Example 6.2] that

$$\operatorname{cond}_\infty V_n \sim \frac{3^{3/4}}{4}(1+\sqrt{2})^n, \quad n \to \infty. \tag{3.8}$$

Here the growth rate $1 + \sqrt{2} = 2.4142\ldots$ is indeed smaller than for equally spaced symmetric nodes, but not by a whole lot.

Seeing the condition of V_n continually improving through the series of examples above, one cannot help wondering whether there is an optimal set of real nodes $x^T = [x_1, x_2, \ldots, x_n]$ (say, with $x_1 > x_2 > \cdots > x_n$), and if so, what they are and what optimal growth rate of $\operatorname{cond}_\infty V_n$ they produce as $n \to \infty$. As far as the existence of the optimum is concerned, the answer is easily seen to be affirmative (cf. [9]). We even conjecture (but have no proof as yet) that the optimal nodes are unique, subject to the above ordering. If this were true, it would follow [9, Theorem 3.1] that the optimal node configuration is symmetric with respect to the origin. Seen in this light, the recent result [14, Theorem 3.1]

$$\operatorname{cond}_\infty V_n > 2^{n/2} \quad (n > 2,\ x_\nu \text{ symmetric}) \tag{3.9}$$

is of interest, since it shows that, accepting the above conjecture, the condition of Vandermonde matrices grows exponentially for *any* set of real nodes. Nevertheless, the growth rate indicated in (3.9) is not believed to be sharp, and the search for the optimal growth rate remains an interesting open problem. There is a result analogous to (3.9) for arbitrary positive nodes, namely [14, Theorem 2.1]

$$\operatorname{cond}_\infty V_n > 2^{n-1} \quad (n > 1,\ x_\nu \ge 0). \tag{3.10}$$

Both bounds in (3.9) and (3.10) can be slightly sharpened (cf. Theorems 3.2 and 2.2, respectively, in [14]).

IV. COMPLEX NODES

The fact that real nodes lead to ill-conditioned Vandermonde matrices is not surprising if one considers that powers constitute, as is well known, a poor basis for polynomial approximation on the real line; see, e.g., the discussion of near linear dependence in [3, pp. 119–120], or of the conditioning of the power basis in [11]. In contrast, replacing the powers by Chebyshev polynomials, and considering the corresponding

Vandermonde-like matrices (1.8), can lead to perfectly conditioned matrices if one chooses the (real) nodes appropriately; cf. Section V below.

Now it so happens that the powers are indeed "Chebyshev polynomials" on any disc in the complex plane centered at the origin, in the sense of deviating least from zero (in the uniform norm on the disc, or on the circumference of the disc) among all monic polynomials of the same degree. Therefore, one expects better conditioning of ordinary Vandermonde matrices if one allows the nodes z_ν to be complex.

If we measure the condition in the Euclidean norm (1.4), and consider for simplicity the *unit* disc, then the nth roots of unity indeed minimize $\text{cond}_2 V_n$; in fact,

$$\text{cond}_2 V_n = 1 \ \text{if} \ z_\nu = z_\nu^{(n)} = e^{i(\nu-1)2\pi/n}, \qquad \nu = 1,2,\ldots,n. \tag{4.1}$$

This is an easy consequence of the orthogonality of trigonometric functions. The roots of unity, therefore, would seem to be an ideal choice for work on the unit disc, if it weren't for the fact that they form a *triangular array* of nodes, i.e., for each n, as indicated in (4.1), there are n distinct nodes $z_\nu^{(n)}$ which change as n is increased by 1. In applications to interpolation and quadrature, this would require that the function to be interpolated, or integrated, be obtained on a two-dimensional set of points. It is an interesting question to ask how well one can do with a *linear array* of nodes on the unit circle.

One naive answer to this is to first note that the set of kth roots of unity, for k even, contains as subset the $(k/2)$th roots of unity. We may therefore generate a linear sequence of nodes by adjoining to the 2^{k-1}th roots of unity every other 2^kth root of unity, going around the circle in the positive direction, and doing this for $k = 1,2,3,\ldots$. More precisely, if

$$2^{k-1} < \nu \le 2^k \tag{4.2}$$

for some $k \ge 1$, then

$$z_1 = 1, \qquad z_\nu = e^{2\pi i(2(\nu-2^{k-1})-1)/2^k}. \tag{4.3}$$

We call this sequence, for lack of a better word, the *quasi-cyclic sequence* on the circle. (Such sequences have been used by Eiermann, Niethammer and Varga [4, p. 522] in the context of semiiterative methods for systems of linear algebraic equations.) With this choice of nodes, whenever n is a power of 2, the corresponding Vandermonde matrix V_n is perfectly conditioned, but otherwise, there is a chance, especially for large n, that the condition may deteriorate significantly. The bounds in Theorem 2.1, unfortunately, are too far apart to give much useful information. We therefore computed the condition number for V_n numerically, using, as seems natural on the circle, the Euclidean matrix norm (1.4). The results for $\text{cond}_2 V_n$ are depicted on a logarithmic scale in Figure 4.1 for $3 \le n \le 64$. As expected, the condition number shoots up to considerable heights between two successive (large) powers of 2.

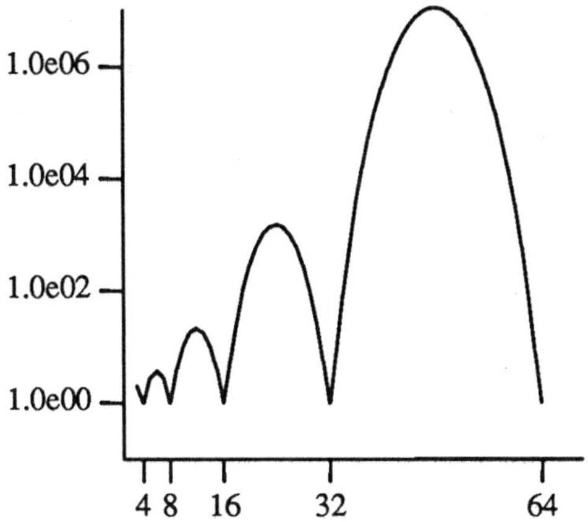

FIG. 4.1 *The condition of Vandermonde matrices* (1.7) *for* n=3(1)64 *with nodes taken from the quasi-cyclic sequence* (4.3).

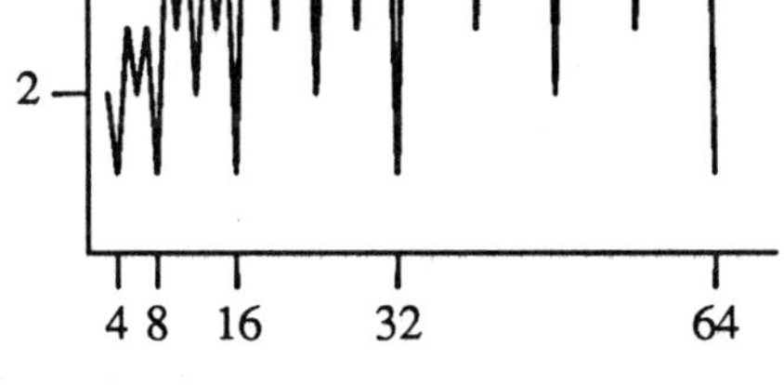

FIG. 4.2 *The condition of Vandermonde matrices* (1.7) *for* n=3(1)64 *with nodes taken from the Van der Corput sequence* (4.6).

How, then, can we avoid these large peaks? More specifically, for an integer n satisfying $2^{k-1} < n < 2^k$, in which order should we adjoin the set of 2^{k-1}th roots of unity by alternate 2^kth roots of unity such that $\max_{2^{k-1} < n < 2^k} \text{cond}_2 V_n$ is minimized? We don't know the solution to this problem, but a good candidate for an optimal (or nearly optimal) node sequence is obtained as follows.

For any integer $\nu \geq 0$, written in binary form

$$\nu = \sum_{j=0}^{\infty} c_j 2^j, \quad \nu_j \in \{0,1\}, \tag{4.4}$$

define the fraction $c_\nu \in [0,1)$ by

$$c_\nu = \sum_{j=0}^{\infty} \nu_j 2^{-j-1}. \tag{4.5}$$

The sequence $\{c_\nu\}_{\nu=0}^{\infty}$ is known as the *Van der Corput sequence.* We then take as nodes

$$z_\nu = e^{2\pi i c_{\nu-1}}, \quad \nu = 1,2,3,\ldots . \tag{4.6}$$

(Such nodes were used to good advantage by Fischer and Reichel [5, p. 228] in connection with the Richardson iteration method; see also Fischer and Reichel [6], Reichel and Opfer [17].) It is easily seen that for $n = 2^{k-1}$ the set $\{z_\nu: \nu = 1,2,\ldots, n\}$ consists of all the nth roots of unity, just like in the quasi-cyclic case. For values of n between 2^{k-1} and 2^k, however, the nodes (4.6) are picked in a zigzag manner from the 2^kth roots of unity, rather than cyclically around the circle as in (4.3). This achieves a more evenly distributed set of nodes, and one can hope that the condition number $\text{cond}_2 V_n$ remains correspondingly smaller. This is indeed confirmed by a computation for $3 \leq n \leq 64$, the results of which are summarized in Figure 4.2. (A similar picture, extended through $n = 148$, has previously been published by Reichel and Opfer [17].) Further computations [15] reveal a rather astonishing pattern for the eigenvalues and eigenspaces of the matrix $V_n V_n^H$ – a Hermitian Toeplitz matrix – which served as an inspiration for the work in [2]. There, it is proved, in particular, that all eigenvalues of $V_n V_n^H$ are powers of 2, the largest, $\lambda_{\max}$, always being equal to $\lambda_{\max} = 2^k$, and the smallest, $\lambda_{\min} = 2^\ell$, where $\ell = 0$ if n is odd, and $0 < \ell \leq k$ otherwise. There follows

$$\text{cond}_2 V_n \leq 2^{k/2} < \sqrt{2n} \quad (2^{k-1} < n \leq 2^k, \ z_\nu \text{ as in (4.6)}), \tag{4.7}$$

with equality on the left holding for every odd n. The various "levels" exhibited in Figure 4.2 thus have heights $2^{k/2}$, $k = 2,3,4,\ldots$. Comparison of Figures 4.2 and 4.1 clearly illustrates the significant improvement achieved by the Van der Corput sequence over the quasi-cyclic sequence. Similar phenomena on ellipses (and also on intervals) will be discussed in Section VI.

V. VANDERMONDE-LIKE MATRICES INVOLVING ORTHOGONAL POLYNOMIALS

As observed at the beginning of Section IV, the choice of orthogonal polynomials as bases in problems of approximation on the real line leads to Vandermonde-like matrices (1.8) which can be expected to have better condition than ordinary Vandermonde matrices. It is the purpose of this section to study the condition of such matrices in the case where

$$p_k(z) = p_k(z\,; d\sigma), \quad k = 0,1,2,\ldots, \tag{5.1}$$

are *orthonormal polynomials* with respect to some (positive) measure $d\sigma$ on the real line,

$$\int_{\mathbb{R}} p_r(x)p_s(x)d\sigma(x) = \delta_{rs} = \begin{cases} 1, & r = s, \\ 0, & r \neq s. \end{cases} \tag{5.2}$$

In most applications, the nodes z_ν are real and contained in the support of $d\sigma$,

$$z_\nu = x_\nu \in \mathbf{R}, \quad x_\nu \in \operatorname{supp} d\sigma. \tag{5.3}$$

A choice that appears particularly natural is that of the zeros of $p_n(\cdot\,; d\sigma)$,

$$x_\nu = x_\nu^{(n)}, \quad p_n(x_\nu\,; d\sigma) = 0, \quad \nu = 1,2,\ldots, n. \tag{5.4}$$

In this case the condition $\mathrm{cond}_F V_n$ of V_n (in the Frobenius norm (1.5)) can be expressed very simply in terms of *Christoffel numbers* $\gamma_\nu = \gamma_\nu^{(n)}(d\sigma)$ belonging to the measure $d\sigma$, i.e., in terms of the weights in the Gauss-Christoffel quadrature formula

$$\int_{\mathbb{R}} f(x)d\sigma(x) = \sum_{\nu=1}^{n} \gamma_\nu^{(n)} f(x_\nu^{(n)}) + R_n(f), \quad R_n(\mathbf{P}_{2n-1}) = 0. \tag{5.5}$$

(As is well known, $\gamma_\nu^{(n)} > 0$ for $\nu = 1,2,\ldots, n$.) Indeed, we have

Theorem 5.1. *The condition of* V_n *in* (1.8), *where* p_k *are the orthonormal polynomials* (5.1), (5.2) *and the nodes* x_ν *given by* (5.4), *equals*

$$\mathrm{cond}_F V_n = \left[\sum_{\nu=1}^{n} \gamma_\nu \sum_{\nu=1}^{n} \frac{1}{\gamma_\nu} \right]^{1/2}, \tag{5.6}$$

where $\gamma_\nu = \gamma_\nu^{(n)}(d\sigma)$ *are the Christoffel numbers of* $d\sigma$, *and the norm used in* (5.6) *is the Frobenius norm* $\|\cdot\|_F$ *in* (1.5).

The proof rests on the fact that $1/\gamma_\nu^{(n)}$ are the squares of the singular values of V_n, which in turn is a consequence of the discrete orthogonality property of orthogonal polynomials (cf. [12]). Note also that $\text{cond}_F V_n \geq n$ for any set of positive numbers γ_ν.

Our first example is the analogue of the example involving the roots of unity, in the sense that it achieves optimality.

Example 5.1. The Chebyshev measure $d\sigma(x) = (1-x^2)^{-1/2}dx$ on $[-1,1]$.

Here, the Christoffel numbers $\gamma_\nu^{(n)}$ are all equal to π/n. Indeed, the Chebyshev measure is the only measure for which this is true for all n. (There are other measures, however, for which equality of Christoffel numbers holds for selected values of n; see, e.g., [13, §6], [16]). It then follows from (5.6) that $\text{cond}_F V_n = n$, i.e., V_n is optimally conditioned.

Nevertheless, optimality is achieved at a price: a *triangular* array of nodes (just like earlier with roots of unity). We will show in Section VI how one can find a *linear* array of nodes that also produces well-conditioned matrices V_n, (1.8).

Not much worse than the Chebyshev measure are those of Legendre and Chebyshev of the second kind.

Example 5.2. $d\sigma(x) = dx$ and $d\sigma(x) = (1-x^2)^{1/2}dx$ on $[-1,1]$.

Here one computes from (5.6) the following condition numbers for selected values of n:

TABLE 1 *The condition of Vandermonde-like matrices for Legendre and 2nd-kind Chebyshev polynomials*
(Numbers in parentheses indicate decimal exponents.)

n	Legendre	Chebyshev 2nd kind
5	5.362(0)	5.916(0)
10	1.155(1)	1.483(1)
20	2.494(1)	3.924(1)
40	5.367(1)	1.071(2)
80	1.148(2)	2.976(2)

In stark contrast, Laguerre and Hermite polynomials give rise to extremely ill-conditioned matrices V_n, for example, $\text{cond}_F V_{40} = 1.924(30)$ and 3.699(14) in the two respective cases. This is due to the presence of very small Christoffel numbers.

If z_ν are arbitrary complex nodes, one can prove a result similar to, but weaker than, (5.6); it involves the Christoffel function, rather than Christoffel numbers. We recall that the *Christoffel function* (for some measure $d\sigma$) is defined by

$$\gamma_n(z_0; d\sigma) = \min_{\substack{p \in \mathbb{P}_{n-1} \\ p(z_0)=1}} \int_{\mathbb{R}} |p(x)|^2 d\sigma(x), \quad z_0 \in \mathbb{C}, \tag{5.7}$$

where the minimum is over all complex polynomials of degree $\le n-1$ taking on the value 1 at z_0. Alternatively,

$$[\gamma_n(z\,;d\sigma)]^{-1} = \sum_{k=0}^{n-1} |p_k(z\,;d\sigma)|^2. \tag{5.7'}$$

In place of (5.6) we then have [12]

$$\operatorname{cond}_F V_n \ge \left[\sum_{\nu=1}^{n} \gamma_n(z_\nu\,;d\sigma) \sum_{\nu=1}^{n} \frac{1}{\gamma_n(z_\nu\,;d\sigma)}\right]^{1/2}. \tag{5.8}$$

To prove (5.8), and at the same time give a version of (5.8) involving equality, one must first of all invert the matrix V_n in (1.8). This can be done similarly as in Section II for powers by expanding the fundamental Lagrange polynomial (2.1) not in powers, as in (2.2), but in the orthogonal polynomials $p_k(z) = p_k(z\,;d\sigma)$,

$$\ell_\nu(z) = \sum_{\mu=1}^{n} a_{\nu\mu} p_{\mu-1}(z), \qquad \nu = 1,2,\ldots,n. \tag{5.9}$$

Then as before, one finds

$$V_n^{-1} = A, \qquad A = [a_{\nu\mu}]. \tag{5.10}$$

Now

$$\begin{aligned}\int_{\mathbb{R}} \sum_{\nu=1}^{n} |\ell_\nu(x)|^2 d\sigma(x) &= \int_{\mathbb{R}} \sum_{\nu}\sum_{\mu} a_{\nu\mu} p_{\mu-1}(x) \sum_{\lambda} \bar{a}_{\nu\lambda} p_{\lambda-1}(x) d\sigma(x)\\ &= \sum_{\nu}\sum_{\mu,\lambda} a_{\nu\mu}\bar{a}_{\nu\lambda} \int_{\mathbb{R}} p_{\mu-1}(x) p_{\lambda-1}(x) d\sigma(x)\\ &= \sum_{\nu,\mu} |a_{\nu\mu}|^2\end{aligned}$$

on account of the orthonormality of the p_k. Consequently,

$$\|V_n^{-1}\|_F = \left[\int_{\mathbb{R}} \sum_{\nu=1}^{n} |\ell_\nu(x)|^2 d\sigma(x)\right]^{1/2}. \tag{5.11}$$

On the other hand,

$$\|V_n\|_F = \left[\sum_{\nu=1}^{n}\sum_{\mu=1}^{n} |p_{\mu-1}(z_\nu)|^2\right]^{1/2}, \tag{5.12}$$

which, on account of (5.7′), gives

$$||V_n||_F = \left[\sum_{\nu=1}^{n} \frac{1}{\gamma_n(z_\nu\,;d\sigma)}\right]^{1/2}. \tag{5.13}$$

The assertion (5.8) now follows by multiplying the two expressions in (5.11) and (5.13) and observing that

$$\int_{\mathbb{R}} |\ell_\nu(x)|^2 d\sigma(x) \geq \gamma_n(z_\nu\,;d\sigma)$$

by (5.7), since $\ell_\nu \in \mathbf{P}_{n-1}$ and $\ell_\nu(z_\nu) = 1$.

We note, however, that the product of (5.11) and (5.12) is exactly *equal* to $\text{cond}_F V_n$, and can easily be computed, at least for conventional measures $d\sigma$, the first factor by Gaussian quadrature and the other by recurrence.

Analogous results can be derived, in essentially the same way, for orthogonal polynomials $\{p_k(\cdot\,;d\sigma)\}$ that are *not* normalized (for example, for *monic* polynomials). Letting

$$d_k^2 = \int_{\mathbb{R}} p_k^2(x;d\sigma)d\sigma(x), \quad k = 0,1,2,\ldots, \tag{5.14}$$

and denoting $D = \text{diag}(d_0, d_1, \ldots, d_{n-1})$, the appropriate matrix norm to be used is then

$$||A||_{F,D} = ||D^{-1}AD||_F \tag{5.15}$$

(which clearly satisfies all the axioms of a matrix norm, including submultiplicativity, $||AB||_{F,D} \leq ||A||_{F,D}||B||_{F,D}$). In place of (5.11), one obtains

$$||V_n^{-1}||_{F,D} = \left\{\int_{\mathbb{R}} \sum_{\nu=1}^{n} \frac{1}{d_{\nu-1}^2} |\ell_\nu(x)|^2 d\sigma(x)\right\}^{1/2}, \tag{5.16}$$

and (5.12) must be modified to read

$$||V_n||_{F,D} = \left\{\sum_{\nu=1}^{n} d_{\nu-1}^2 \sum_{\mu=1}^{n} \frac{1}{d_{\mu-1}^2} |p_{\mu-1}(z_\nu)|^2\right\}^{1/2}. \tag{5.17}$$

The condition $\text{cond}_{F,D} V_n$ is then again computable as the product of (5.16) and (5.17), and can be estimated from below by

$$\mathrm{cond}_{F,D} V_n \geq \left[\sum_{\nu=1}^{n} \frac{\gamma_n(z_\nu; d\sigma)}{d_{\nu-1}^2} \sum_{\nu=1}^{n} \frac{d_{\nu-1}^2}{\gamma_n(z_\nu; d\sigma)} \right]^{1/2}. \tag{5.18}$$

For orthonormal polynomials, we have $D = I$, and the results (5.16)–(5.18) reduce to (5.11), (5.12) and (5.8).

VI. VANDERMONDE-LIKE MATRICES INVOLVING CHEBYSHEV POLYNOMIALS ON ELLIPSES

We have noted in Section IV that the powers are "Chebyshev polynomials" (i.e., monic polynomials of minimum uniform norm) on the disc. It is similarly known that the (monic) polynomials

$$p_0(z) = 1, \quad p_k(z) = 2\rho^{k/2} T_k \left[\frac{z}{2\sqrt{\rho}} \right], \quad k = 1, 2, \ldots \quad (0 < \rho \leq 1), \tag{6.1}$$

where T_k denotes the Chebyshev polynomial of the first kind, are the "Chebyshev polynomials" on the ellipse $\mathcal{E}_\rho$ with boundary given by

$$\partial \mathcal{E}_\rho = \{ z : z = e^{i\theta} + \rho e^{-i\theta}, \ 0 \leq \theta \leq 2\pi \} \tag{6.2}$$

if $0 < \rho < 1$, and on the interval $[-2,2]$ (the limit of (6.2) as $\rho \to 1$), when $\rho = 1$; cf. [17]. (The ellipse $\mathcal{E}_\rho$ is scaled so as to have capacity 1.) This suggests the study of Vandermonde-like matrices (1.8) with polynomials p_k given by (6.1) and nodes located on the elliptic contour (6.2), either in quasi-cyclic order, or in the order determined by the Van der Corput sequence. Thus, in the former case, with ν given as in (4.2), and assuming $0 < \rho < 1$,

$$z_1 = 1 + \rho, \quad z_\nu = e^{i\theta_\nu} + \rho e^{-i\theta_\nu}, \quad \theta_\nu = 2\pi(2(\nu - 2^{k-1}) - 1)/2^k, \tag{6.3}$$

and in the latter case,

$$z_\nu = e^{2\pi i c_{\nu-1}} + \rho e^{-2\pi i c_{\nu-1}}, \tag{6.4}$$

where $\{c_\nu\}_{\nu=0}^{\infty}$ is the Van der Corput sequence (4.4), (4.5). In the limit case $\rho=1$, these formulae have to be slightly modified, since we do not want to run back and forth through the interval $[-2,2]$. We then assume, in the quasi-cyclic case,

$$1 + 2^{k-1} < \nu \leq 2^k + 1 \quad (k \geq 1), \tag{6.5}$$

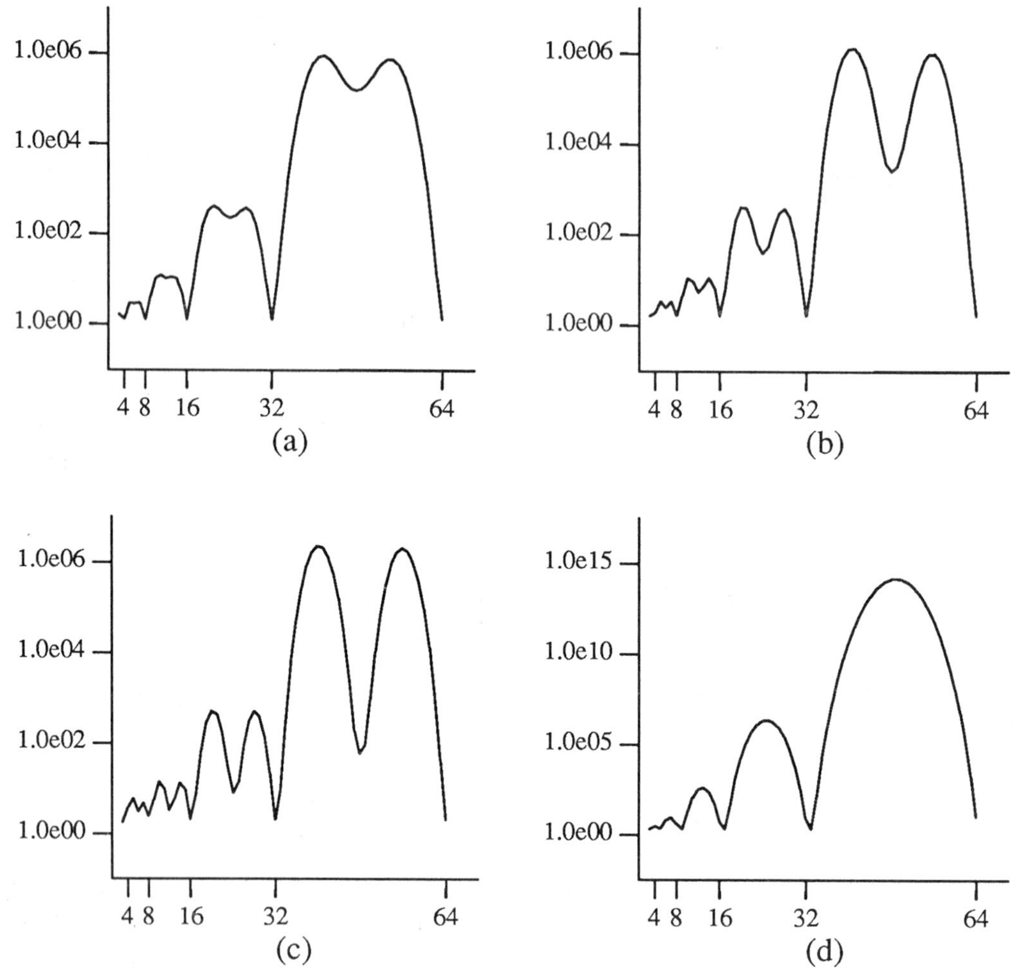

FIG. 6.1 *The condition of Vandermonde-like matrices* (1.8) *for* n=3(1)64 *involving "Chebyshev polynomials" on the ellipse* $\mathcal{E}_\rho$ *and nodes taken from the quasi-cyclic sequence* (6.3) *resp.* (6.3$_1$).
(a) $\rho = .25$ (b) $\rho = .5$ (c) $\rho = .75$ (d) $\rho = 1$.

and define

$$z_1 = -2, \quad z_2 = 2, \quad z_\nu = 2\cos \pi(2(\nu - 2^{k-1}) - 3)/2^k, \quad \nu = 3,4,\ldots ; \qquad (6.3_1)$$

in the case of the Van der Corput sequence, we let

$$z_1 = -2, \quad z_{\nu+1} = 2\cos(\pi c_{\nu-1}), \quad \nu = 1,2,3,\ldots . \qquad (6.4_1)$$

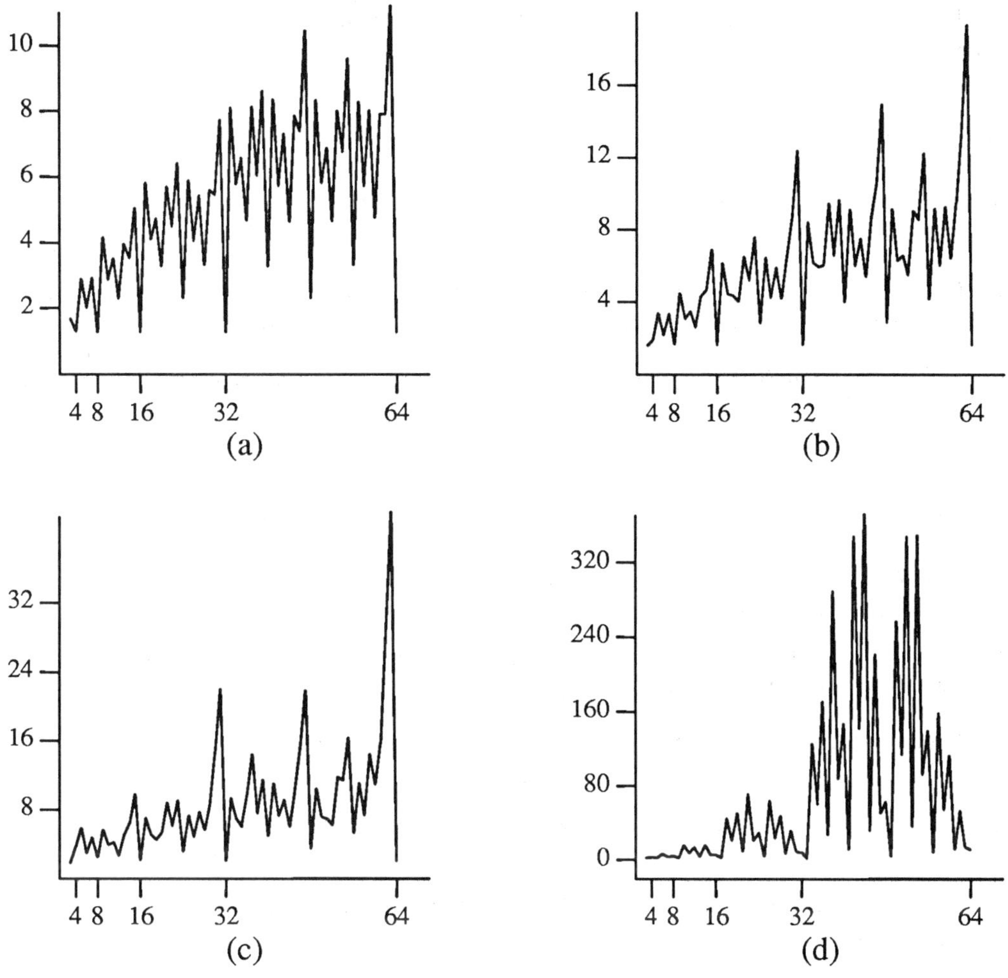

FIG. 6.2 *The condition of Vandermonde-like matrices* (1.8) *for n=3(1)64 involving "Chebyshev polynomials" on the ellipse* $\mathcal{E}_\rho$ *and nodes taken from the Van der Corput sequence* (6.4) *resp.* (6.4_1).
(a) $\rho = .25$ (b) $\rho = .5$ (c) $\rho = .75$ (d) $\rho = 1$.

We computed $\text{cond}_2 V_n$ for $3 \le n \le 64$ in the case of ellipses $\mathcal{E}_\rho$ with ρ = .25(.25).75 (for the case ρ=0, see Section IV), and for the segment [–2,2] (i.e., the case ρ=1). The results are shown graphically, on a logarithmic scale, in Figure 6.1 for nodes given by (6.3) [resp. (6.3_1)], and in Figure 6.2, on a linear scale, for the nodes (6.4) [resp. (6.4_1)]. As can be seen, Van der Corput sequences again perform significantly better than quasi-cyclic sequences. The condition, in fact, is known to grow at most polynomially in n, if $0 \le \rho < 1$, and at most like $n^{O(\log n)}$ if $\rho = 1$; cf. [17, Section 3]. In the case of quasi-cyclic sequences, when $0 < \rho < 1$, it is interesting to observe two large peaks between successive powers of 2, in contrast to the cases ρ=0 and ρ=1, which exhibit only one (a surprisingly large one when ρ=1).

ACKNOWLEDGMENT. The author is indebted to Professor Lothar Reichel for useful comments.

REFERENCES

1. Björck, A. and Elfving, T.: Algorithms for confluent Vandermonde systems, *Numer. Math.*, v. 21, 1973, pp. 130–137.

2. Córdova, A., Gautschi, W. and Ruscheweyh, S.: Vandermonde matrices on the circle: spectral properties and conditioning, submitted for publication.

3. Dahlquist, G. and Björck, Å.: *Numerical Methods,* Prentice-Hall, Englewood Cliffs, N.J., 1974.

4. Eiermann, M., Niethammer, W. and Varga, R.S.: A study of semiiterative methods for nonsymmetric systems of linear equations, *Numer. Math.,* v. 47, 1985, pp. 505–533.

5. Fischer, B. and Reichel, L.: A stable Richardson iteration method for complex linear systems, *Numer. Math.*, v. 54, 1988, pp. 225–242.

6. ________ and ________: Newton interpolation in Fejér and Chebyshev points, *Math. Comp.*, v. 53, 1989, to appear.

7. Gautschi, W.: On inverses of Vandermonde and confluent Vandermonde matrices, *Numer. Math.*, v. 4, 1962, pp. 117–123.

8. ________: Norm estimates for inverses of Vandermonde matrices, *Numer. Math.*, v. 23, 1975, pp. 337–347.

9. ________: Optimally conditioned Vandermonde matrices, *Numer. Math.*, v. 24, 1975, pp. 1–12.

10. ________: On inverses of Vandermonde and confluent Vandermonde matrices III, *Numer. Math.*, v. 29, 1978, pp. 445–450.

11. ________: The condition of polynomials in power form, *Math. Comp.*, v. 33, 1979, pp. 343–352.

12. ________: The condition of Vandermonde-like matrices involving orthogonal polynomials, *Linear Algebra Appl.*, v. 52/53, 1983, pp. 293–300.

13. ________: On some orthogonal polynomials of interest in theoretical chemistry, *BIT,* v. 24, 1984, pp. 473–483.

14. ________ and Inglese, G.: Lower bounds for the condition number of Vandermonde matrices, *Numer. Math.*, v. 52, 1988, pp. 241–250.

15. Opfer, G.: personal communication.

16. Peherstorfer, F.: On Gauss quadrature formulas with equal weights, *Numer. Math.*, v. 52, 1988, pp. 317–327.

17. Reichel, L. and Opfer G.: Chebyshev-Vandermonde systems, *Math. Comp.*, to appear.

Singular Point and Exponential Asymptotics

Floyd B. Hanson[*] Professor, Department of Mathematics, Statistics, and Computer Science, University of Illinois at Chicago, Chicago, Illinois

I. INTRODUCTION

Models of population dynamics with extinction, stable points and small stochastic effects lead to singular perturbations problems for differential equations with singular points and turning points. Both singular points and turning points give rise to resonance phenomena. Uniform methods and exponential precision asymptotics (in the sense of Meyer, 1980) have been important in the formal asymptotic analysis of these problems by Tier and Hanson (1981), and Wazwaz and Hanson (1986a,b). The foundations of this formal analysis are reviewed and given further justification. Basic properties of exponential ordering are presented. The concept of complete asymptotic expansions (in the sense of Watson, 1911), including both exponentially larger and exponentially smaller terms, is reformulated with multiple scaling techniques for the purpose of placing complete

[*]Work supported in part by the National Science Foundation Grant DMS-8806099.

expansions on a much firmer basis relative to that for algebraic asymptotic expansions (in the sense of Poincaré, 1886).

A major disadvantage, as well as advantage, of algebraic asymptotic expansions (in the sense of Poincaré, 1886) has been the neglect of exponentially small (or smaller) terms. However, neglect of exponentially smaller terms usually corresponds to the neglect of one of the independent, fundamental solutions for a second (or higher) order ODE. This makes algebraic asymptotic expansions defective, compared to the full analytic theory of ODE. Complete asymptotic expansions, by including exponentially small terms with their own expansion, are a formal method for fixing this deficiency, for both numerical and asymptotic purposes. However, uniqueness problems arise, because it is difficult to formulate a proper algorithm for separating the exponentially smaller terms from the exponentially larger terms. Other attempts to improve the concept of complete asymptotic expansions have been unsatisfactory (cf. Gregory, 1980, and Olver, 1980a). Unfortunately, these attempts, and some of the associated criticisms, present a *red herring.* In this paper, the complete or exponential asymptotic expansions are viewed from the perspective of multiple scaling techniques, such that the complete expansion is recovered as a Poincaré expansion in an extended multiple variable space. In addition, the perspective preserves the richness of ODE theory, including expansions of independent solutions.

The results of Hanson and Tier (1981), and Wazwaz and Hanson (1986b, 1988) include applications for the asymptotic development of the spectrum, the statistical moments and the distribution of exit times. Even though this work employs formal asymptotic expansions, it draws inspiration from the more rigorous error bound analysis of Olver.

The focus of this paper will be the singular perturbation problem for the second order ordinary differential equation,

$$\frac{\varepsilon}{2}xp(x;\varepsilon)u'' + x(a-x)q(x;\varepsilon)u' = su \ , \tag{1}$$

on $0 < x < b$ with $0 < \varepsilon \ll 1$ as a small parameter. In terms of the underlying applications, $u = u(x;\varepsilon)$ corresponds the Laplace transform of the exit time distribution and x corresponds to a population size. The point $x = 0$ denotes extinction of the population and the point $x = a$ denotes a stable point or carrying capacity in the interior for a deterministic, nearly logistic drift (coefficient of first order) with $0 < a < b$. In mathematical terms, the point $x = a$ is a second order turning point for the problem because $x = a$ is a zero of the first order coefficient, and $x = 0$ is a regular singular point of the equation. The coefficients $p(x;\varepsilon)$ and $q(x;\varepsilon)$ are assumed to be positive, several times differentiable in x, and to have asymptotic power series expansions in ε,

$$p(x;\varepsilon) \sim \sum_{k=0}^{\infty} p_k(x)\varepsilon^k \,, \tag{2}$$

$$q(x;\varepsilon) \sim \sum_{k=0}^{\infty} q_k(x)\varepsilon^k \,, \tag{3}$$

as $\varepsilon \to 0^+$. The parameter s corresponds to the Laplace transform variable, but can also be interpreted as an eigenvalue parameter. In this paper, the two boundary conditions, at $x = 0$ and at $x = a$, are left unspecified. However, boundary conditions are, in general, restricted at boundaries where the ODE is singular. Hanson and Wazwaz (1988) showed that de Groen (1987) made unintended major changes in the singular boundary at $x = 0$ by making a small perturbation in the drift (coefficient of u') in (1).

Special cases of Equation (1) have been derived from limits of branching processes by Ludwig (1975) and Tier and Hanson (1981). Particular singular perturbation problems for (1) have been studied by Ludwig (1975, 1976), Hanson and Tier (1981), and Tier and Hanson (1981) using matched asymptotic expansions. Wazwaz and Hanson (1986a,b) utilized a two-part uniform method to obtain more accurate results for the general problem for (1).

The following sections of this paper include discussions of the notation of exponential precision asymptotics, the multiple scaling approach to complete asymptotic expansions, and a brief review of the dual uniform method of Wazwaz and Hanson (1986b) for (1).

II. BASIC NOTATION OF EXPONENTIAL ASYMPTOTICS

In the limit of a parameter becoming small, the conventional order symbols, such as $8\varepsilon^2 + \varepsilon^5 = \mathcal{O}(\varepsilon^2)$ with *big oh* and $6\varepsilon^{3.5} = o(\varepsilon^3)$ with *little oh* as $\varepsilon \to 0^+$, indicate what dominant behavior is like in terms of the basic expressions in the small parameter, without having to be concerned with subdominant or miscellaneous constants. When exponential dependence is present, its behavior, depending on the form, may dominate algebraic or even algebraic-logarithmic terms. Thus, an exponential order notation, that de-emphasizes subdominant algebraic-logarithmic terms, would have a utility similar to that of the conventional order symbols. Olver (1980a) discusses another view of this hiding, de-emphasizing or *figleaf* effect that order symbols possess, i.e., an error bound investigation into the information hidden by the order symbol *figleaf* has important implications for justifying methods and extending them.

Olver (1964), discussing error bounds for cylinder functions, pointed out that exponentially small does not necessarily mean numerically small. O'Malley (1974), Olver (1964, 1978, 1980a), and Hanson and Wazwaz (1986b) demonstrate the need to use exponentially ordered terms for computing resonance conditions. Meyer (1980) has played a major role in emphasizing the importance of *exponential precision asymptotics*, as opposed to *algebraic asymptotics*, in a wide variety of applications: wave propagation in Meyer (1975), geophysical fluid dynamics in Lozano and Meyer (1976), semi-classical scattering in Meyer (1976a), and adiabatic invariance in Meyer (1976b).

Wazwaz and Hanson (1986b) proposed the following exponential precision asymptotics notation to clearly express the exponential dependence of the spectrum and resonance conditions for the singular point, turning point resonance problem for (1).

A. *SAME EXPONENTIAL ORDER*

Let

$$f(\varepsilon) = \mathrm{ord}_e(g(\varepsilon)) \quad \text{as} \quad \varepsilon \to 0^+ \tag{4}$$

denote that f is the *same exponential order* as g if there exists some finite number n_0 such that

$$\varepsilon^{-n}\frac{f(\varepsilon)}{g(\varepsilon)} \to \begin{cases} 0 & \text{for } n < n_0 \\ \textit{unbounded} & \text{for } n > n_0 \end{cases} \quad \text{as} \quad \varepsilon \to 0^+ \ . \tag{5}$$

The symbol ord_e is analogous to the Kaplun (1967) *same order* symbol *ord*, and has the added advantage that it can be used with less than or greater than signs. For example, $\varepsilon^m \exp(a/\varepsilon) = \text{ord}_e(\exp(a/\varepsilon))$ as $\varepsilon \to 0^+$ for some constant a, because $\varepsilon^{m-n} \exp(a/\varepsilon)/\exp(a/\varepsilon)$ approaches 0 for $n < m$, and is unbounded for $n > m$ as $\varepsilon \to 0^+$. Similarly, $F(\eta) = \text{ord}_e(G(\eta))$ as $\eta \to +\infty$, if $\eta^n F(\eta)/G(\eta)$ has the same composite limiting behavior as in (5), but as $\eta \to +\infty$.

B. *SAME OR SMALLER EXPONENTIAL ORDER*

Let

$$f(\varepsilon) \leq \text{ord}_e(g(\varepsilon)) \quad \textit{or} \quad f(\varepsilon) = \mathcal{O}_e(g(\varepsilon)) \quad \text{as} \quad \varepsilon \to 0^+ \tag{6}$$

denote that f is the *same or smaller exponential order* as g if

$$\varepsilon^{-n}\frac{f(\varepsilon)}{g(\varepsilon)} \to 0 \quad \text{for all } n < n_0 \text{ for some } n_0 \quad \text{as} \quad \varepsilon \to 0^+ \quad ; \tag{7}$$

n_0 is not necessarily finite. For example,

$$\varepsilon^m \ln^p(\varepsilon) e^{\frac{a}{\varepsilon}} \leq \text{ord}_e(e^{\frac{a}{\varepsilon}}) \quad \text{as} \quad \varepsilon \to 0^+ \quad , \tag{8}$$

where a is independent of ε, because

$$\varepsilon^{m-n} \ln^p(\varepsilon) \to 0 \quad \text{for all } n < m \quad \text{as} \quad \varepsilon \to 0^+ \ . \tag{9}$$

Similarly, $F(\eta) \leq \text{ord}_e(G(\eta))$ as $\eta \to +\infty$, if $\eta^n F(\eta)/G(\eta) \to 0$ for all $n < n_0$ for some n_0 as $\eta \to +\infty$.

C. *SMALLER EXPONENTIAL ORDER*

Let

$$f(\varepsilon) < \text{ord}_e(g(\varepsilon)) \quad \textit{or} \quad f(\varepsilon) = o_e(g(\varepsilon)) \quad \text{as} \quad \varepsilon \to 0^+ \tag{10}$$

denote that f is of *smaller exponential order* than g if

$$\varepsilon^{-n}\frac{f(\varepsilon)}{g(\varepsilon)} \to 0 \quad \text{for all } n \quad \text{as} \quad \varepsilon \to 0^+ \ . \tag{11}$$

For example,

$$\varepsilon^m \ln^p(\varepsilon) e^{\frac{a}{\varepsilon}} < \text{ord}_e(e^{\frac{b}{\varepsilon}}) \quad \text{as} \quad \varepsilon \to 0^+ \quad , \tag{12}$$

where $a < b$ are independent of ε, because

$$\varepsilon^{m-n} \ln^p(\varepsilon) e^{-\frac{(b-a)}{\varepsilon}} \to 0 \text{ for all } n \quad \text{as} \quad \varepsilon \to 0^+ \ . \tag{13}$$

Hence, $o_e(f) = \mathcal{O}_e(f)$, but $o_e(f) < \text{ord}_e(f)$ at the same time. Also, $\mathcal{O}_e(f) + o_e(f) = \mathcal{O}_e(f)$. Alternatively, let $g(\varepsilon) > \text{ord}_e(f(\varepsilon))$ if $f(\varepsilon) < \text{ord}_e(g(\varepsilon))$; let $f(\varepsilon) \neq \text{ord}_e(g(\varepsilon))$ if $f < \text{ord}_e(g)$ or $g < \text{ord}_e(f)$.

Further, o_e satisfies the same squeezing lemma as does the usual *little oh*, and this is more conveniently expressed in terms of ord_e.

Squeezing Lemma of Exponential Orders.

$$\text{ord}_e(f) < \text{ord}_e(g) \Longrightarrow \exists \ \ h(\varepsilon) \ \ \ni \text{ord}_e(f) < \text{ord}_e(h) < \text{ord}_e(g) \ . \tag{14}$$

This result follows from taking the geometric mean $h = h(\varepsilon) = \sqrt{f(\varepsilon)g(\varepsilon)}$, and noting that $f/h = h/g = \sqrt{f/g} \to 0$, because by hypothesis $f/g \to 0$ (positive functions have been assumed for simplicity). ∎

Similarly, $F(\eta) < \text{ord}_e(G(\eta))$ as $\eta \to +\infty$, if $\eta^n F(\eta)/G(\eta) \to 0$ for all n as $\eta \to +\infty$.

The function $F(\eta)$ is *exponentially independent* of $G(\eta)$ as $\eta \to +\infty$ if $F(\eta) \neq \text{ord}_e(G(\eta))$ as $\eta \to +\infty$.

Lemma: Exponentially Independent $\Longrightarrow$ *Linearly Independent.*

Let F and G be exponentially independent according to the hypothesis. By definition, suppose the exponential inequality in the direction $F(\eta) < \text{ord}_e(G(\eta))$ as $\eta \to +\infty$. Next, suppose the contrary result that F and G are linearly dependent, so there is some nonzero constant C such that $F(\eta) = CG(\eta)$. Hence, $F(\eta) = \text{ord}_e(G(\eta))$, contradicting the hypothesis and so proving the lemma. ∎

A pair $\{F_1(\eta), F_2(\eta)\}$ of functions is said to be an *exponentially ordered pair* if

$$F_2(\eta) < \text{ord}_e\left(F_1(\eta)\right) \quad \text{as} \ \eta \to +\infty \ , \tag{15}$$

or any other appropriate limit. The exponentially larger function F_1 is called the *dominant* function and the exponentially smaller function F_2 is called *recessive* or *subdominant.*

In general, exponential ordering is **not** preserved upon differentiation. Consider the counter-example: $F_1(\eta) = \cos(\exp(-\eta))$ and $F_2(\eta) = \exp(-\eta)$ as $\eta \to +\infty$. Clearly, $F_2 < \text{ord}_e(F_1)$, because $F_1 \sim 1$. However, $F_1'(\eta) = \exp(-\eta)\sin(\exp(-\eta))$ and $F_2'(\eta) = -\exp(-\eta)$, so $F_1'(\eta) < \text{ord}_e(F_2'(\eta))$, because $F_1'/F_2' \sim -\exp(-\eta) \to 0$ as $\eta \to +\infty$. In this example, the exponential ordering is, in fact, reversed.

III. EXPONENTIAL ASYMPTOTIC EXPANSIONS

The principal difficulty, in implementing exponential precision asymptotics to simultaneously expand the coefficients of exponentially small and exponentially large terms (or two terms, one of which is exponentially smaller than the other), is the problem of distinguishing the exponentially smaller term in the presence of the exponentially larger term. Gregory (1980) has analyzed the difficulties with theories of double and other multiple asymptotic expansions, where the method depends upon computing all of the exponentially larger terms to compute the exponentially smaller terms. Olver (1980a) also criticizes multiple expansions, among other unsatisfactory practices. However, singular perturbation problems, by definition, have the property that no one asymptotic series (in the sense of Poincaré, 1886, including generalizations to asymptotic sequences) is a suitable approximation over the whole domain. Uniform composite or other uniform expansions for standard singular perturbation problems combine at least two different Poincaré asymptotic expansions, such as matched inner and outer asymptotic expansions.

It is essential for many singular perturbation problems that exponentially small terms be included in the analysis to obtain correct results, as Meyer (1980) demonstrates with many physical examples. Also, when an expansion term contains an exponentially small function as a multiplying factor, the exponential order of that term can not be determined until the coefficient of that factor is determined, usually during matching (a later stage in the asymptotic analysis). This can occur because an exponentially large term can appear to be an exponentially small function in one asymptotic region and exponentially large in another region. A good example is found in the problem corresponding to (1) in this paper. Away from the singular point and turning point, Hanson and Tier (1981) found that the LG-WKB *outer* approximation (the regular outer, setting $\varepsilon = 0$, can not match turning point or singular point approximations) has a dominant solution of the form

$$u_1(x;\varepsilon) = \mathrm{ord}_e(e^{\frac{2F_0(x)}{\varepsilon}}) \text{ as } \varepsilon \to 0^+ , \tag{16}$$

where

$$F_0(x) = \int_a^x \frac{(t-a)q_0(t)}{p_0(t)}\, dt \tag{17}$$

is the squared *natural metric* of the problem. The second LG-WKB *outer* approximation has the recessive property that $u_2 = \mathrm{ord}_e(1)$ as $\varepsilon \to 0^+$ away from $x = 0$ and $x = a$. Due to the fact that both $p_0(x)$ and $q_0(x)$ are positive, $F_0(x)$ is positive, except at the turning point a where it attains its minimum of zero. Thus, $\exp(2F_0(x)/\varepsilon)$ is exponentially large, except in a small neighborhood of the turning point a. As the transition or turning point region is approached,

$$e^{\frac{2F_0(x)}{\varepsilon}} \sim e^{\frac{\zeta^2}{2}} > \mathrm{ord}_e(1) \text{ for } \mid \zeta \mid = \sqrt{\frac{2q_0(a)}{p_0(a)}}\frac{\mid x-a \mid}{\sqrt{\varepsilon}} \gg 1 , \tag{18}$$

where ζ is the stretched variable in the transition region. The result (18) shows that the exponentially large approximation in the outer region can be matched with the exponentially large term in the transition inner region for large ζ. In contrast, as the singular point region near $x = 0$ is approached, the exponentially large outer approximation, relative to its value at $x = 0$, behaves like

$$e^{\frac{2(F_0(x)-F_0(0))}{\varepsilon}} \sim e^{-\xi} < \mathrm{ord}_e(1) \text{ for } \xi = \frac{2aq_0(0)}{p_0(0)}\frac{x}{\varepsilon} \gg 1 , \tag{19}$$

where ξ is the stretched variable of the singular point inner region. This shows that the exponentially large term of the outer region will be matched with the term of the inner region containing the exponentially small function $\exp(-\xi)$ as a factor for large ξ. The exponentially large constant factor $\exp(2F_0(0)/\varepsilon)$ would be hidden in the unmatched constant coefficient of the exponentially small function for large ξ. Hence, this large constant factor has been divided out in (19) for realism. In a similar manner, the $\mathrm{ord}_e(1)$, but exponentially lesser, function u_2 of the outer region matches the $\mathrm{ord}_e(1)$, but exponentially greater, function of the inner region. In summary, it is possible to have *cross-over matching* between the exponentially larger function of one region with the exponentially smaller function of the other region. The eventual exponential order of a term will not be known in general until the unmatched constants are determined in the later stages of the matched asymptotic expansion procedure.

Upon going from a general solution composed of two linearly independent solutions to asymptotic analysis with single Poincaré expansions, the linear superposition of the two linearly independent solutions is lost if the two solutions are also exponentially independent. The situation is unsatisfactory and much more analysis is needed to place exponential asymptotics on a basis similar to that of ODE theory. The improved asymptotic theory need not be based on infinite asymptotic expansions, because only a finite number of terms are usually used in applications.

The idea for exponential ordered pair definition in (15) is related to the notion of a *fundamental or satisfactory pair of solutions* of Miller (1950). This is illustrated (see the example in Olver, 1974) by the pair $\{\exp(+z), \exp(-z)\}$ which is a numerically and asymptotically satisfactory pair for large z. However, the pair $\{\cosh(z), \sinh(z)\}$ is not satisfactory when $Re[z] \gg 1$, because $\cosh(z) = \mathrm{ord}_e(\sinh(z))$ and thus the pair is numerically indistinct. The pair $\{\exp(+z), \exp(-z)\}$ is numerically very distinct because two functions are exponentially ordered.

However, exponential ordering (or exponential independence) alone is not enough, because the exponentially larger term will not be unique due to the fact that it can contain exponentially smaller terms (see Olver, 1980b). For instance, the pair $\{\cosh(z), \exp(-z)\}$ is an exponentially ordered pair as $z \to +\infty$, but $\cosh(z)$ has an exponentially small component $\frac{1}{2}\exp(-z)$ that would need to be combined with $\exp(-z)$ in any matching of terms with $\exp(-z)$. Problems with the set $\{\cosh(z), \exp(-z)\}$ are quite transparent, because this example can easily be decomposed into an exponentially large term, without any exponentially small component, and an exponentially small term. There is usually no problem in selecting a unique exponentially small term. However, for general second order ODEs such as (1), the separation of purely exponentially larger from the exponentially smaller terms is much less clear.

A. *LG-WKB APPROXIMATION FOR MODEL COMPARISON EQUATION*

The Liouville-Green or LG-WKB approximation (see Olver, 1974) can be used to discover the appropriate exponentially ordered pair. First, $\eta = \eta(x)$ is used as a generalized independent variable, and the normal transformation $\hat{u}(\eta) = \exp(\phi(\eta)/\hat{\varepsilon})\hat{v}(\eta)$ is used to transform the appropriate model comparison or canonical ODE,

$$\hat{\varepsilon}\hat{u}'' + \hat{p}(\eta)\hat{u} + \hat{q}(\eta)\hat{u} = 0 , \tag{20}$$

say, to the normal form,

$$\hat{v}'' = \frac{\mathcal{P}(\eta;\hat{\varepsilon})}{\hat{\varepsilon}^2}\hat{v} , \tag{21}$$

with

$$\mathcal{P}(\eta;\hat{\varepsilon}) = \frac{\hat{p}^2}{4} + \hat{\varepsilon}\left(\frac{\hat{p}'}{2} - \hat{q}\right) . \tag{22}$$

(The hat (^) notation is used here for variables that are used in another form elsewhere in this paper.) Model (20) is the general form of a comparison equation that includes comparison equations for both singular point and turning point uniform methods of Wazwaz and Hanson (1986a) under quite general restrictions, and thus should be useful in many other applications. The normal form (21) is used here to facilitate the typical calculations needed for the current and similar expansions. Let $\hat{p}$ be nonzero, two times continuously differentiable, while $\hat{q}$ is continuously differentiable on the region of interest. Also, $\hat{p}$ is assumed to be integrable at $\eta = 0$, so its integral will be of one sign, as $\hat{p}$ has one sign. The region of interest is assumed to be an exponential region, so $\mathcal{P} = \mathcal{Q}^2 > 0$. Oscillatory regions are too difficult to treat here. The leading term LG-WKB approximation of two independent solutions is

$$\widehat{V}_1(\eta;\hat{\varepsilon}) = e^{+|\phi|/\hat{\varepsilon}}V_1 \quad\text{and}\quad \widehat{V}_2(\eta;\hat{\varepsilon}) = e^{-|\phi|/\hat{\varepsilon}}V_2 , \tag{23}$$

with principal exponent

$$\phi(\eta) \equiv -\tfrac{1}{2}\int_0^{\eta} \hat{p}(t)dt \ , \tag{24}$$

with exponentially order one coefficients (with expansions generated by regular perturbation methods),

$$\begin{aligned} V_1(\eta;\hat{\varepsilon}) &= |\hat{p}|^{-(1+\theta)/2}e^{+\theta\psi}[1+\hat{\varepsilon}\sigma_1] \ , \\ V_2(\eta;\hat{\varepsilon}) &= |\hat{p}|^{-(1-\theta)/2}e^{-\theta\psi}[1+\hat{\varepsilon}\sigma_2] \ , \end{aligned} \tag{25}$$

where $\theta = \mathrm{sgn}(\phi) = \mathrm{sgn}(\eta)\mathrm{sgn}(\hat{p})$, $\psi = \psi(\eta) = \int dt\, \hat{q}(t)/\hat{p}(t)$, and $\sigma_j = \sigma_j(\eta;\hat{\varepsilon}) = \mathcal{O}(1)$, for $j = 1$ or 2. According to the assumptions on $\hat{p}$, both ϕ and η have one sign on the region of interest, so $\theta^2 \equiv 1$. In (23) only the exponentially ordered part, the current focus, is displayed explicitly, and this exponential term is usually called the *geometric optics approximation* (see Bender and Orszag, 1978). The special form of the LG-WKB approximations, given in (23) to (25) here, are different from the general form given by Olver (1974), because the explicit display of the exponentially larger and smaller terms is important here. The general solution to the normal form problem will be denoted by

$$v = K_1\widehat{V}_1 + K_2\widehat{V}_2 \ , \tag{26}$$

where the K_j are constants of integration, for $j = 1$ or 2. The general solution of the original model (20) is denoted by

$$\hat{u}(\eta;\hat{\varepsilon}) = e^{\frac{\phi}{\hat{\varepsilon}}}[K_1 e^{\frac{+|\phi|}{\hat{\varepsilon}}}V_1 + K_2 e^{\frac{-|\phi|}{\hat{\varepsilon}}}V_2] \ , \tag{27}$$

with particular solutions satisfying the exponential ordering

$$e^{\frac{(1-\theta)\phi}{\hat{\varepsilon}}}V_2 = \mathrm{ord}_e(e^{\frac{(1-\theta)\phi}{\hat{\varepsilon}}}) < \mathrm{ord}_e(e^{\frac{(1+\theta)\phi}{\hat{\varepsilon}}}V_1) = \mathrm{ord}_e(e^{\frac{(1+\theta)\phi}{\hat{\varepsilon}}}) \ , \tag{28}$$

because the sign of the exponent $\pm\theta\phi \equiv \pm|\phi|$ distinguishes the dominant from the recessive solutions. Thus, the LG-WKB exponentially ordered pair is

$$\{\exp((1+\theta)\phi/\hat{\varepsilon}), \exp((1-\theta)\phi/\hat{\varepsilon})\} \ , \tag{29}$$

for the model ODE (20).

If U_1 is a *dominant solution* for equation (20) then

$$U_1(\eta;\hat{\varepsilon}) \sim K_1 e^{\frac{(1+\theta)\phi(\eta)}{\hat{\varepsilon}}} |\hat{p}|^{\frac{-(1+\theta)}{2}} e^{+\theta\psi}[1+\hat{\varepsilon}\sigma_1(\eta;0)], \tag{30}$$

(in the sense of Poincaré, 1886), exhibiting only the first two of the exponentially larger terms. If U_2 is a *recessive solution*, then

$$U_2(\eta;\hat{\varepsilon}) \sim K_2 e^{\frac{(1-\theta)\phi(\eta)}{\hat{\varepsilon}}} |\hat{p}|^{\frac{-(1-\theta)}{2}} e^{-\theta\psi}[1+\hat{\varepsilon}\sigma_2(\eta;0)], \tag{31}$$

exhibiting only the first two of the exponentially smaller terms.

B. *MULTI-SCALE EXPONENTIAL ASYMPTOTIC EXPANSIONS*

As previously demonstrated, the successful matching of terms may depend on knowing both exponentially large and small terms to exponential precision. Important for this is the notion of complete asymptotic expansions (in the sense of Watson, 1911), whereby all exponentially large and small terms are included in the expansion, although the justification is far from satisfactory (in the sense of Olver, 1964, 1974, 1980a, regarding error bounds). Dingle (1973) has elaborated on the notion of complete asymptotic expansions, requiring the complete expansion to satisfy essentially the same conditions as the expanded function.

A serious deficiency with asymptotic expansions, (in the sense of Poincaré, 1886), is the loss of independence of solutions of the original second order ODE, when one of the solutions is of smaller exponential order. Asymptotic theory desperately needs the power of linear superposition for linear ODE of order greater than one. The loss of independence, with terms of smaller exponential order, also results in numerical inaccuracies, especially near Stokes lines (i.e., *if you are too close to Poincaré, you can not be too close to Stokes*). Olver (1964, 1980a) presents a number of examples of numerical inaccuracies resulting from the use of Poincaré algebraic asymptotic expansions, but these inaccuracies can be corrected by adding the exponentially small terms and justifying the addition by error bounds. Olver (1980a, p. 195) refers to a compound expansion that is valid (in the sense of Poincaré, 1886).

The use of the concept of multiple scaling is proposed to reformulate the notion of complete asymptotic expansions, and recover the loss of independence due to Poincaré. For multiple scaling or multiple variable expansions (cf. Kevorkian and Cole, 1981), it is recognized that the exponential or geometric optics factors vary on a much faster scale than the other factors of the solution. It is these exponential or geometric optics factors that preserve the independence of the particular solutions. Meyer (1980) cites the use of multiple scaling as one approach in handling the *WKB paradox*, due to the *excessive preoccupation with algebraic asymptotic expansions*. Meyer and Painter (1984) also reveal a two-variable asymptotic structure for the oscillator modulation problem. Hence, the original problem will be extended or embedded in a larger space, in which the geometric optics factors depend on an additional variable. Let this additional independent variable, obtained from the LG-WKB approximation, be

$$\xi = \frac{\theta\phi(\eta)}{\hat{\varepsilon}} = \frac{|\phi(\eta)|}{\hat{\varepsilon}} > 0 \ , \tag{32}$$

so the dependent variable transforms to the normal dependent variable

$$v(\xi, \eta; \hat{\varepsilon}) \equiv e^{-\theta\xi}\hat{u}(\eta; \hat{\varepsilon}) = K_1 e^{+\xi} V_1 + K_2 e^{-\xi} V_2 \ , \tag{33}$$

using (27) for $\hat{u}$ in the form implied by the LG-WKB approximations. The partial derivative of the new variable in terms of the fast scale is

$$\frac{\partial v}{\partial \xi}(\xi, \eta; \hat{\varepsilon}) = K_1 e^{+\xi} V_1 - K_2 e^{-\xi} V_2 \ . \tag{34}$$

The coefficients of $\exp(+\xi)$ and $\exp(-\xi)$, respectively, can be calculated from the above formulas, yielding

$$K_1 V_1(\eta; \hat{\varepsilon}) = -\tfrac{1}{2} W[v, e^{-\xi}] \ \text{ and } \ K_2 V_2(\eta; \hat{\varepsilon}) = -\tfrac{1}{2} W[e^{+\xi}, v] \ , \tag{35}$$

where W is the Wronskian with respect to the independent variable ξ with other variables held fixed,

$$W[f(\xi), g(\xi)] \equiv f(\xi) g'(\xi) - f'(\xi) g(\xi) \ , \tag{36}$$

and $W[e^{+\xi}, e^{-\xi}] = -2 \neq 0$.

The solution $\hat{u}(\eta; \hat{\varepsilon})$ to model (20) is said to have a *complete asymptotic expansion,*

$$\hat{u}(\eta;\hat{\varepsilon}) \approx e^{\theta\xi}\left[e^{+\xi}\sum_{j=0}^{\infty}a_j(\eta)\hat{\varepsilon}^j + e^{-\xi}\sum_{j=0}^{\infty}b_j(\eta)\hat{\varepsilon}^j\right] , \tag{37}$$

as $\hat{\varepsilon} \to 0^+$, with fixed η and $\xi = |\phi(\eta)|/\hat{\varepsilon}$, if

$$K_1V_1(\eta;\hat{\varepsilon}) \sim \sum_{j=0}^{\infty}a_j(\eta)\hat{\varepsilon}^j \quad \text{and} \quad K_2V_2(\eta;\hat{\varepsilon}) \sim \sum_{j=0}^{\infty}b_j(\eta)\hat{\varepsilon}^j , \tag{38}$$

(in the sense of Poincaré, extended to coefficients depending on parameters), as $\hat{\varepsilon} \to 0^+$, with fixed η and ξ, and

$$W[1,\widetilde{V}_1] = -\frac{\partial\widetilde{V}_1}{\partial\xi} \equiv 0 , \tag{39}$$

where $\widetilde{V}_1(\xi,\eta;\hat{\varepsilon}) = V_1(\eta;\hat{\varepsilon})$, with respect to ξ.

The purpose of the condition $W[1,\widetilde{V}_1] \equiv 0$ is to exclude non-purely dominant functions of the form

$$e^{+\xi}\widetilde{V}_1 = e^{+\xi}f_1(\eta;\hat{\varepsilon}) + e^{-\xi}g_1(\eta;\hat{\varepsilon}) , \tag{40}$$

with respect to ξ, since $W[1,\widetilde{V}_1] = -2\exp(-2\xi)$ in this case. This condition, $W[1,\widetilde{V}_1] \equiv 0$, is probably the weakest, but most critical, part of the definition, because it requires the identification and eradication of that part of a dominant giving rise to exponentially smaller terms, if they exist.

Henceforth, the symbol $\approx$ will be used here to indicate a complete expansion, even when only a few of the leading terms are displayed. The symbol $\approx$ is in contrast to the symbol $\sim$ used for single Poincaré type expansions that are not valid for complete expansions having both exponentially larger and exponentially smaller component expansions. Olver (1974, 1980a) cautions that error bounds are preferred to complete expansions. Olver (1964, p.181) also remarks on the difficulty of placing the notion of complete asymptotic expansion (in the sense of Watson, 1911) on a firm mathematical foundation. However, Olver (1980a) uses complete asymptotic expansions, calling them compound asymptotic expansions, for accurate expansions of Bessel's functions. Typically, the complete expansion for a special function is derived from its integral representations by complex analysis.

The solution U_1 is called a *purely dominant solution* for equation (20), if

$$U_1(\eta;\varepsilon) \approx K_1 e^{(\theta+1)\xi} \sum_{j=0}^{\infty} a_j(\eta)\hat{\varepsilon}^j \,, \tag{41}$$

i.e., its complete expansion contains only dominant terms. Clearly, purely dominant implies dominant, but the converse is untrue. However, purely dominant implies asymptotic, i.e.,

$$U_1 \approx K_1 e^{(\theta+1)\xi} \sum_{j=0}^{\infty} a_j\hat{\varepsilon}^j \implies U_1 \sim K_1 e^{(\theta+1)\xi} \sum_{j=0}^{\infty} a_j\hat{\varepsilon}^j \,. \tag{42}$$

Recessive is the same as *purely recessive* (i.e. the complete expansion contains only recessive terms), because a recessive solution can not contain dominant terms and still be recessive. In addition, recessive is equivalent to asymptotic, i.e.,

$$U_2 \approx K_2 e^{(\theta-1)\xi} \sum_{j=0}^{\infty} b_j\hat{\varepsilon}^j \iff U_2 \sim K_2 e^{(\theta-1)\xi} \sum_{j=0}^{\infty} b_j\hat{\varepsilon}^j \,. \tag{43}$$

Actual substitution of the multiple variable extension of the solution, with the geometric optics factor in the fast (long) variable $\xi = |\phi(\eta)|/\hat{\varepsilon}$, into model ODE (20) permits the interpretation of the complete asymptotic expansion as an ordinary asymptotic expansion (in the sense of Poincaré, with extensions to parametric dependent coefficients), as $\hat{\varepsilon} \to 0^+$ with ξ and η fixed. The requirement that the dominant coefficient K_1V_1 have no ξ dependence is still critical. Let

$$\widetilde{U}(\xi,\eta;\hat{\varepsilon}) = \hat{u}(\eta;\hat{\varepsilon}) \tag{44}$$

be the extension of the solution for model (20); then $\widetilde{U}$ satisfies the partial differential equation in the multiple variables,

$$\widetilde{U}_{\xi\xi} - 2\theta\widetilde{U}_{\xi} = \hat{\varepsilon}\left[\frac{4\theta}{\hat{p}}\widetilde{U}_{\xi\eta} + \frac{2\theta\hat{p}'}{\hat{p}^2}\widetilde{U}_{\xi} - \frac{4}{\hat{p}}\widetilde{U}_{\eta} - \frac{4\hat{q}}{\hat{p}^2}\widetilde{U}\right] - \frac{4\hat{\varepsilon}^2}{\hat{p}^2}\widetilde{U}_{\eta\eta} \,. \tag{45}$$

An algebraic asymptotic expansion for $\widetilde{U}$ in powers of $\hat{\varepsilon}$, as $\hat{\varepsilon} \to 0^+$ with ξ and η

fixed, is assumed, i.e.,

$$\widetilde{U}(\xi,\eta;\hat{\varepsilon}) = \sum_{j=0}^{n} \widetilde{U}_j(\xi,\eta)\hat{\varepsilon}^j + \mathcal{O}(\hat{\varepsilon}^{n+1}) \,, \tag{46}$$

and, in fact, results in a *multi-scale exponential asymptotic expansion* in ξ, η and $\hat{\varepsilon}$. Substitution of the expansion (46) into the PDE (45) results in the leading order version of (45)

$$\widetilde{U}_{0\,\xi\xi} - 2\theta\widetilde{U}_{0\,\xi} = 0 \,, \tag{47}$$

whose general solution is

$$\widetilde{U}_0(\xi,\eta) = \widetilde{A}_0(\eta)e^{2\theta\xi} + \widetilde{B}_0(\eta) \,. \tag{48}$$

The exponentials $\exp(2\theta\xi)$ and $\exp(0) = 1$ are related to the exponentially ordered LG-WKB pair (29) for model (20) by

$$\left\{e^{(\theta+1)\xi}, e^{(\theta-1)\xi}\right\} = \begin{cases} \left\{e^{+2\xi}, 1\right\}, & \theta = +1 \\ \left\{1, e^{-2\xi}\right\}, & \theta = -1 \end{cases} \tag{49}$$

recalling that $\theta = \text{sgn}(\phi)$ and $\xi = \theta\phi/\hat{\varepsilon}$. The set $\{\exp(2\theta\xi), 1\}$ is convenient for multiple scale integration (note either $(\theta+1)$ or $(\theta-1)$ is zero in the exponents of (49)), but is not necessarily exponentially ordered in the order of dominant to recessive. The η dependence of the amplitudes $\widetilde{A}_0(\eta)$ and $\widetilde{B}_0(\eta)$ is determined by non-resonance requirements for $\widetilde{U}_1(\xi,\eta)$ in the next order step. In general, the amplitude functions $\widetilde{U}_j(\xi,\eta)$ are determined uniquely by requiring that no terms arise with *resonance* coefficients ξ or $\xi\exp(2\theta\xi)$ to each order $\hat{\varepsilon}^j$ (a multiple scaling non-resonance condition). The full leading approximation after satisfying (45) to first order in $\hat{\varepsilon}$ is

$$\widetilde{U}(\xi,\eta;\hat{\varepsilon}) \sim e^{2\theta\xi}A_0|\hat{p}(\eta)|^{-1}e^{\psi(\eta)} + B_0e^{-\psi(\eta)}, \tag{50}$$

with error $\mathcal{O}(\hat{\varepsilon})$ as $\hat{\varepsilon} \to 0^+$ with ξ and η fixed. A_0 and B_0 are genuine constants of integration. The result in (50) is equivalent to the leading order approximation of both exponential terms in the complete asymptotic expansion given in (37), ignoring the constants. The specific physical optics terms are given in (30) and (31) for the dominant and recessive solutions, respectively.

The general coefficient of the asymptotic expansion is calculated in the usual way,

$$\widetilde{U}_n(\xi,\eta) = \lim_{\substack{\hat{\varepsilon} \to 0^+ \\ \xi,\eta \text{ fixed}}} \frac{\left[\widetilde{U}(\xi,\eta;\hat{\varepsilon}) - \sum_{j=0}^{n-1} \widetilde{U}_j(\xi,\eta)(\hat{\varepsilon})^j\right]}{(\hat{\varepsilon})^n}, \tag{51}$$

with coefficients depending on parameters and with asymptotic sequence $\{(\hat{\varepsilon})^n\}$. The relationship to the coefficients of the complete asymptotic form in (37) is then

$$\widetilde{U}_n(\xi,\eta) = e^{(\theta+1)\xi} a_n(\eta) + e^{(\theta-1)\xi} b_n(\eta) , \tag{52}$$

where $a_j(\eta)$ and $b_j(\eta)$ are the dominant and recessive coefficients from the complete asymptotic form in (37).

Although the multiple scale approach here gives only the LG-WKB results, as in the leading order *multi-scale exponential asymptotic expansion* (50), these results are exactly what is needed in the matching procedure. Interpreting the complete asymptotic expansion in terms of the space of independent multiple variables $(\xi,\eta,\hat{\varepsilon})$ permits viewing the complete exponential expansion as a Poincaré algebraic expansion, with ξ and η fixed, (in the sense of formal asymptotics). It also preserves the linear independence of the original set of fundamental solutions of the ODE. The exponential coefficients are directly connected to the linear independence of the set of fundamental solutions of the ODE. However, justification of the interpretation requires the connection formulas between the standard solutions for the singular or turning points and their asymptotic expansions away from those points. In addition, rigorous justification (in the sense of Olver) requires error bounds.

C. *CONSTRUCTION OF STANDARD SOLUTIONS*

The next step is to construct a *purely exponentially ordered pair*, containing a purely dominant solution and a recessive solution, from given complete asymptotic expansions for two linearly independent solutions. An exponentially ordered pair is not essential for complete matching, but such a pair makes the task of matching easier. However, the critical part of the notion of a complete asymptotic expansion, as presented in the previous section, depends on the existence of a purely dominant solution. Starting from a fundamental set of solutions $\{U_1, U_2\}$ having a known set of complete asymptotic expansions, it is desirable to construct a set of standard functions (in the sense of Miller, 1950, 1952) such that one of the solutions is purely dominant and the other is recessive. Standard solutions have also been utilized by Olver (1977) for the multiple turning point problem.

For the second order model ODE (20), the purely exponentially ordered set is

$$\{S_1, S_2\} = \{e^{\frac{(1+\theta)\phi}{\hat{\varepsilon}}} V_1, e^{\frac{(1-\theta)\phi}{\hat{\varepsilon}}} V_2\} \; , \tag{53}$$

as deduced from the LG-WKB method. However, only the asymptotic expansions for the set $\{S_1, S_2\}$ have been deduced (see expansions for V_1 and V_2 in (38) with, $K_1 = K_2$) and the connections of the asymptotic expansions to the actual solutions with known complete asymptotic expansions are needed. The two LG-WKB asymptotic expansions are used to define the complete expansions. In general, let two candidate standard solutions, $\widehat{S}_1$ and $\widehat{S}_2$, be written in the form of the general solution as

$$\widehat{S}_1 = A_1 U_1 + A_2 U_2 \quad \text{and} \quad \widehat{S}_2 = B_1 U_1 + B_2 U_2 \; , \tag{54}$$

with constants A_j and B_j, in terms of the fundamental solutions, U_1 and U_2, having known complete expansions. Here, the goal is to make $\widehat{S}_1$ and $\widehat{S}_2$ into *standard solutions* (in the sense of Miller, 1950, 1952), so the first is exponentially larger than the second and contains no component of the second, i.e. a *purely exponentially ordered pair.* The known *complete asymptotic expansions* (in the sense of Watson, 1911) for the fundamental set are assumed to be given by

$$U_1 \approx D_1 S_1 + D_2 S_2 \quad \text{and} \quad U_2 \approx E_1 S_1 + E_2 S_2 \; , \tag{55}$$

with constants D_i and E_i, where the theoretically standard set $\{S_1, S_2\}$ is used as short-hand for both purely dominant and recessive expansions of solutions,

$$S_1 \approx e^{(1+\theta)\phi/\hat{\varepsilon}} \sum_{j=0}^{\infty} a_j(\eta)\hat{\varepsilon}^j \quad \text{and} \quad S_2 \sim e^{(1-\theta)\phi/\hat{\varepsilon}} \sum_{j=0}^{\infty} b_j(\eta)\hat{\varepsilon}^j . \tag{56}$$

Substitution of (55) into (54) leads to

$$\widehat{S}_1 \approx (A_1 D_1 + A_2 E_1)S_1 + (A_1 D_2 + A_2 E_2)S_2 \quad \text{and} \quad \widehat{S}_2 \approx (B_1 D_1 + B_2 E_1)S_1 + (B_1 D_2 + B_2 E_2)S_2 . \tag{57}$$

Equation (57) illustrates the difficulty of not working with standard solutions with pure exponential expansions as S_1 and S_2, because the complete expansions must be decomposed into their pure exponential component expansions at the time of matching, and the complexity of the coefficient products in (57) barely indicates the computational difficulty. Hence, forcing $\widehat{S}_1 \approx S_1$ and $\widehat{S}_2 \approx S_2$, respectively, produces the desired purely exponentially ordered pair of solutions, with the *tuned* constants

$$A_1 = \frac{E_2}{Det_s}, \quad A_2 = \frac{-D_2}{Det_s}, \quad B_1 = \frac{-E_1}{Det_s}, \quad B_2 = \frac{D_1}{Det_s}, \qquad Det_s = \begin{vmatrix} D_1 & E_1 \\ D_2 & E_2 \end{vmatrix} \neq 0 , \tag{58}$$

assuming the linear independence of the $\{U_1, U_2\}$ set carries over to their complete asymptotic expansions, (in the sense of Dingle, 1973). Thus, it is possible to find a set of standard solutions or purely exponentially ordered set of solutions, given the complete expansion for a linearly independent set of solutions. However, this procedure would not be satisfactory in the rigorous sense of Olver (1974), without the support of error bounds. On the other hand, Olver (1964) notes that error bounds are not available for many applications. In addition, the error bound requirement is unsatisfactory (not in the sense of Olver) for practitioners of formal asymptotic expansions, where they work closely, or in competition, with applied scientists on the same applications.

IV. DUAL UNIFORM METHOD

The two-part or dual uniform method of Wazwaz and Hanson (1986a) is discussed in this section for the singular point, turning point problem for (1) as an example of exponential asymptotic expansions. First, the motivation for this *dual uniform method* is examined. Problem (1) is transformed to normal form by eliminating the first derivative term using the transformation,

$$u(x;\varepsilon) = e^{\frac{\phi(x;\varepsilon)}{\varepsilon}} v(x;\varepsilon) \,, \tag{59}$$

where

$$\phi(x;\varepsilon) = F(x;\varepsilon) \equiv \int_a^x \frac{(t-a)q(t;\varepsilon)}{p(t;\varepsilon)}\, dt, \tag{60}$$

whose leading term is the squared *natural metric* term of Hanson and Tier (1981) given in (17). The specific coefficient of the normal form $v'' = \mathcal{P}v/\varepsilon^2$ from (21) is

$$\mathcal{P}(x;\varepsilon) = \left(\frac{(x-a)q(x;\varepsilon)}{p(x;\varepsilon)}\right)^2 + \frac{2\varepsilon s}{xp(x;\varepsilon)} - \varepsilon\left(\frac{q}{p} + (x-a)(\frac{q}{p})'\right)(x;\varepsilon) \,. \tag{61}$$

The zeros of $\mathcal{P}(x;\varepsilon)$, $x = z(\varepsilon)$ such that $\mathcal{P}(z(\varepsilon);\varepsilon) = 0$, and also the poles of $\mathcal{P}(x;\varepsilon)$, serve to characterize the solutions of (1). The LG-WKB exponential approximation is not valid in the neighborhood of the zeros and poles of $\mathcal{P}$. The local behavior near the zeros and poles determine the kind of connection formulas to the LG-WKB approximations. The singular point at $x = 0$ introduces the only pole in $\mathcal{P}(x;\varepsilon)$.

Near the turning point $x = a$, the first term in (61) for $\mathcal{P}$ behaves nearly like a quadratic, implying a scaling of ord$(\sqrt{\varepsilon})$. Near the pole $x = 0$, the second term nearly behaves like a hyperbola, implying a scaling of ord(ε). Due to the fact that the scaling is ord$(\sqrt{\varepsilon})$ near the turning point at $x = a$ and ord(ε) near the singular point at $x = 0$, there is little hope of finding a single uniform approximation for the whole domain $(0, b)$, such as found by Lynn and Keller (1970) or Olver (1977) for the multiple turning point problem.

A. *OUTLINE OF SINGULAR AND TURNING POINT UNIFORM METHODS*

In this section, an outline of the two uniform approximations of Wazwaz and Hanson (1986a) is presented to facilitate an examination of the method. In this outline, only the form of the analysis is emphasized, because the technical details are found in Wazwaz and Hanson (1986a). Hanson and Tier (1981), as well as Wazwaz and Hanson (1986a), partition the domain $(0, b)$ into four regions B, L, T, and R as indicated in Figure 1.

The composite subdomain LTR excludes a neighborhood of the singular boundary, and the uniform reduction theorem given by O'Malley (1974) applies. The uniform representation of the solution u of (1) on the region LTR is written as a linear combination of Miller's (1964) standardized parabolic cylinder functions U and V in Wazwaz and Hanson (1986a).

On the composite subdomain BL, which includes the singular boundary layer and anything else to the left of a small neighborhood of the turning point, Hanson

Singular Boundary Layer	Left Outer Region	Turning Point Layer	Right Outer Region
B	L	T	R

$\longleftarrow$ BL $\longrightarrow$

$\longleftarrow$ LTR $\longrightarrow$

$\uparrow$ 0 $\qquad\qquad$ $\uparrow$ a $\qquad\qquad$ $\uparrow$ b

$\longrightarrow$ $\qquad\qquad$ $\longleftrightarrow$

$x = \mathcal{O}(\varepsilon)$ $\qquad$ $|x - a| = \mathcal{O}(\sqrt{\varepsilon})$ $\qquad$ $x \longrightarrow$

FIG. 1 Partitioning of the domain into composite asymptotic regions BL and LTR , composed of subregions B, L, T, and R

and Wazwaz (1986a) developed a uniform representation of the solution u of equation (1) in terms of standardized linear combinations of Whittaker functions, (in the sense of Miller, 1950, 1952).

In this expository outline, the two uniform representations will be treated in parallel, to clearly exhibit the similarities and differences of the singular point uniform representation and the turning point uniform representation. The dual uniform representation from Hanson and Wazwaz (1986a) is thus

$$u(x;\varepsilon) = \begin{cases} P(z(x);\varepsilon)\widetilde{W}(Z(x;\varepsilon)) + Q(z(x);\varepsilon)\widetilde{W}'(Z(x;\varepsilon)) & \text{on } BL \\ M(y(x);\varepsilon)\tilde{w}(Y(x;\varepsilon)) + \frac{\varepsilon}{\varepsilon_T}N(y(x);\varepsilon)\tilde{w}'(Y(x;\varepsilon)) & \text{on } LTR \end{cases} \tag{62}$$

The order one uniform variable on BL is

$$z(x) = 2\left[F_0(0) - F_0(x)\right] , \tag{63}$$

where F_0 is the squared metric function of Hanson and Tier (1981) given here in (17). The corresponding BL stretched variable is $Z(x;\varepsilon) = z(x)/\varepsilon$. The order one uniform variable on LTR is

$$y(x) = \operatorname{sgn}(x-a)\sqrt{\frac{2F_0(x)}{F_0''(a)}} . \tag{64}$$

The corresponding stretched uniform variable on LTR with $\operatorname{ord}(\sqrt{\varepsilon})$ stretching is

$$Y(x;\varepsilon) = \frac{y(x)}{\varepsilon_T} \quad \text{with} \quad \varepsilon_T = \sqrt{\frac{\varepsilon}{2F_0''(a)}} . \tag{65}$$

The two *Langer* variables Z on BL and Y on LTR are related through $F_0(x)$ in (17) by the identity

$$Z(x;\varepsilon) = \frac{2F_0(0)}{\varepsilon} - \frac{Y^2(x;\varepsilon)}{2} . \tag{66}$$

The more weakly varying amplitude functions, P with Q on BL and M with N on LTR, are assumed to have Poincaré expansions as $\varepsilon \to 0^+$ and are given by

$$
\begin{array}{llllll}
P(z;\varepsilon) & \sim \sum_j P_j(z)\varepsilon^j & \text{and} & Q(z;\varepsilon) & \sim \sum_j Q_j(z)\varepsilon^j & \text{on } BL, \\
M(y;\varepsilon) & \sim \sum_j M_j(y)\varepsilon^j & \text{and} & N(y;\varepsilon) & \sim \sum_j N_j(y)\varepsilon^j & \text{on } LTR.
\end{array}
\tag{67}
$$

The model comparison equations forming the basis for the uniform approximations in (62) are given by

$$
\begin{array}{llllll}
Z\widetilde{W}''(Z) & + \; Z\widetilde{W}'(Z) & + \; k\widetilde{W}(Z) & = 0 & \text{on } BL, \\
\tilde{w}''(Y) & - \; Y\tilde{w}'(Y) & - \; c\tilde{w}(Y) & = 0 & \text{on } LTR,
\end{array}
\tag{68}
$$

where k and c are parameters. The two comparison equations in (68) capture the essential behavior of the original model around the singular point and turning point, respectively. In order to see the relationship to the general LG-WKB model used in (20), the comparison equation (68) is written in terms of the order one variables $z = \varepsilon Z$ and $y = \varepsilon_T Y$,

$$
\begin{array}{llllll}
\varepsilon z\widetilde{W}_{zz} & + \; z\widetilde{W}_z & + \; k\widetilde{W} & = 0 & \text{on } BL, \\
\varepsilon_T^2\tilde{w}_{yy} & - \; y\tilde{w}_y & - \; c\tilde{w} & = 0 & \text{on } LTR.
\end{array}
\tag{69}
$$

The correlation to the model used for the LG-WKB is presented in Table 1 below. The fact that $\theta = -1$ on BL means that the comparison equation (69a) on BL will have one LG-WKB solution that is $\text{ord}_e(1)$ and another solution that is $o_e(1)$ or exponentially small. On LTR , $\theta = +1$, so the comparison equation (69b) will have one LG-WKB solution that is exponentially large and another solution that is $\text{ord}_e(1)$.

TABLE 1 *Correlation of variables and parameters between (20) for the LG-WKB and (69).*

Region	$\hat{\varepsilon}$	η	$\hat{u}$	$\hat{p}(\eta)$	$\hat{q}(\eta)$	$\phi(\eta)$	$\psi(\eta)$	θ
BL	ε	z	$\widetilde{W}$	1	$\frac{k}{\eta}$	$-\frac{\eta}{2}$	$k\ln\lvert\eta\rvert$	-1
LTR	ε_T^2	y	$\tilde{w}$	$-\eta$	$-c$	$+\frac{\eta^2}{4}$	$c\ln\lvert\eta\rvert$	$+1$

The solutions of both scaled comparison equations in (68) can be expressed in terms of standard solutions (in the sense of Miller, 1950, 1952),

$$\begin{aligned} \widetilde{W}(Z) &= e^{-\frac{Z}{2}}[C \cdot W_1(k;Z) + D \cdot W_2(k;Z)] && \text{on } BL, \\ \tilde{w}(Y) &= e^{\frac{+Y^2}{4}}[A \cdot U(c-0.5;Y) + B \cdot V(c-0.5;Y)] && \text{on } LTR, \end{aligned} \tag{70}$$

where U and V are the standardized parabolic cylinder functions of Miller (1964) for $Y > 0$, with U being recessive and V being dominant as $Y \to +\infty$. U and V are not standard solutions for $Y < 0$. W_1 and W_2 are standardized Whittaker functions (in the sense of Miller, 1950, 1952) constructed by Wazwaz and Hanson (1986a) from the usual Whittaker functions, using the techniques discussed in an earlier section.

The four constants, A, B, C, and D, in (70) are undetermined at this step. Two constants can be eliminated in favor of others using specific boundary conditions for the second order ODE (1). The remaining constants must be determined by matching the two uniform approximations (62). Hence, a single Poincaré asymptotic expansion in each region would, in general, serve to eliminate only one of the remaining constants. Exponential asymptotic precision is thus essential for determining the two matching constants. This is not just due to the use of two uniform approximations, because Hanson and Tier (1981) needed exponential precision to match solutions for their non-uniform, four-region approximations. The uniform reduction method used by O'Malley (1974) for the pure turning point problem required only two constants, both determinable by the two-point boundary conditions.

The two parameters in (68) will have expansions such as

$$k = k(\varepsilon) \approx \sum_j k_j \varepsilon^j \quad \text{and} \quad c(\varepsilon) \approx \sum_j c_j \varepsilon^j , \tag{71}$$

depending on the application. In the resonance application of Wazwaz and Hanson (1986b), a more general expansion was required that relied on exponential asymptotics, as well as algebraic asymptotics.

First, the uniform expansions in (62) are substituted into the main equation (1) with the expansions (67) for the amplitude coefficients and (71) for the parameters. Next, the second derivatives, $\widetilde{W}''(Z)$ and $\tilde{w}''(Y)$, are eliminated using the comparison equations (68), removing a functional dependency. Terms are collected as coefficients of powers of ε, and these coefficients are further decomposed as coefficients of the independent functions $\widetilde{W}$ and $\widetilde{W}'$ on BL or $\tilde{w}$ and $\tilde{w}'$ on LTR. The ODE then has the expanded form

$$\begin{aligned} 0 &= \textstyle\sum_j[\widetilde{P}_j(z)\widetilde{W}(Z) + \widetilde{Q}(z)\widetilde{W}'(Z)]\varepsilon^j \quad \text{on } BL, \\ 0 &= \textstyle\sum_j[\widetilde{M}_j(y)\tilde{w}(Y) + \widetilde{N}(y)\tilde{w}'(Y)]\varepsilon^j \quad \text{on } LTR, \end{aligned} \tag{72}$$

where $\widetilde{P}$, $\widetilde{Q}$, $\widetilde{M}$, and $\widetilde{N}$ are the corresponding coefficients, whose particular values are given in the equations of Wazwaz and Hanson (1986a). It follows from the independence of the functions and their first derivatives in the fast variables Z on BL and Y on LTR that the coefficients of $\varepsilon^j\widetilde{W}$, $\varepsilon^j\widetilde{W}'$ on BL and $\varepsilon^j\tilde{w}$, $\varepsilon^j\tilde{w}'$ on LTR must vanish on their respective domains, i.e.,

$$\begin{aligned} \widetilde{P}_j(z) &= 0 \quad \text{and} \quad \widetilde{Q}_j(z) = 0 \quad \text{on } BL, \\ \widetilde{M}_j(y) &= 0 \quad \text{and} \quad \widetilde{N}_j(y) = 0 \quad \text{on } LTR. \end{aligned} \tag{73}$$

These conditions then imply relations for $P_j(z)$, $Q_j(z)$, $M_j(y)$, $N_j(y)$, and the parameters k_j and c_j if needed. The first few of these coefficients are given by Wazwaz and Hanson (1986a), with recurrence relations given for other values of j.

When the form of the uniform approximations is determined, there remain constants that have to be determined. Half can be obtained from the specified boundary conditions (natural or otherwise), and the other half must be determined from matching the two uniform approximations. It is the matching step where the exponential precision representation is critical, as was emphasized earlier. The two uniform approximations share validity on the overlapping sub-region L, which excludes neighborhoods of the singular point and the turning point.

Preliminary to the matching step as $\varepsilon \to 0^+$ on L, where $Y < 0$, reflection formulas of Miller (1964) must be used to relate the parabolic cylinder functions of negative arguments to those of positive arguments where the solutions are standard (in the sense of Miller, 1950). These reflection formulas can be written in the form

$$\begin{aligned} V(a,Y) &= \sin(\pi a)V(a,-Y) + \frac{\cos^2(\pi a)\Gamma(a+0.5)}{\pi}U(a,-Y), \\ U(a,Y) &= \frac{\pi}{\Gamma(a+0.5)}V(a,-Y) - \sin(\pi a)U(a,-Y), \end{aligned} \tag{74}$$

useful for expansions as $Y \to -\infty$. The first argument a of V and U is the parameter of Miller's form of the parabolic cylinder functions. The gamma functions introduced into L by these reflection formulas play a critical role in the resonance problem of Wazwaz and Hanson (1986b). The resulting standardized or complete expansions for the reflected functions as $Y \to -\infty$ are then

$$\begin{aligned} V(a,-Y) &\approx e^{+\frac{Y^2}{4}}|Y|^{+a-0.5}\sqrt{\frac{2}{\pi}}\left[1+\frac{(0.5-a)_2}{2Y^2}+\cdots\right] , \\ U(a,-Y) &\sim e^{-\frac{Y^2}{4}}|Y|^{-a-0.5}\left[1+\frac{(a+0.5)_2}{2Y^2}+\cdots\right] , \end{aligned} \tag{75}$$

according to Miller (1964), where $(a)_j \equiv a\cdot(a+1)\cdots(a+j-1)$ is the Pochhammer symbol. The corresponding complete expansions for the standardized Whittaker functions of Wazwaz and Hanson (1986b), are

$$\begin{aligned} W_2(k;Z) &\approx e^{\frac{+Z}{2}}Z^{-k}\sum_{j=0}^{\infty}\frac{(+k)_j(k+1)_j}{j!(+Z)^j} , \\ W_1(k;Z) &\sim e^{\frac{-Z}{2}}Z^{+k}\sum_{j=0}^{\infty}\frac{(-k)_j(1-k)_j}{j!(-Z)^j} , \end{aligned} \tag{76}$$

in order of purely dominant and recessive solutions, where $(k)_j$ is again the Pochhammer symbol. Equations (75) and (76) are LG-WKB exponential approximations of their respective standard solutions, with indicated order corrections.

Substituting the expansions (76) and (75) into the standard decompositions of (70), and then substituting the resulting expansions into the dual uniform representations in (62), produces two overlapping asymptotic representations $u(x;\varepsilon)$ on L in the form

$$u \approx \begin{cases} U_{BL}(Z,z,\varepsilon) &= Z^{-k_0}\left[\widetilde{D}_0\widetilde{P}_0(z)+\varepsilon\widetilde{P}_1(z)+\mathcal{O}(\varepsilon^2)\right] \\ &+\ e^{-Z}Z^{+k_0}\left[\widetilde{C}_0\widetilde{Q}_0(z)+\varepsilon\widetilde{Q}_1(z)+\mathcal{O}(\varepsilon^2)\right] \quad \text{from } BL , \\ & \\ U_{LTR}(Y,y,\varepsilon) &= e^{\frac{+Y^2}{2}}|Y|^{c_0-1}\left[\widetilde{B}_0\widetilde{N}_0(y)+\varepsilon\widetilde{N}_1(y)+\mathcal{O}(\varepsilon^2)\right] \\ &+\ |Y|^{-c_0}\left[\widetilde{A}_0\widetilde{M}_0(y)+\varepsilon\widetilde{M}_1(y)+\mathcal{O}(\varepsilon^2)\right] \quad \text{from } LTR , \end{cases} \tag{77}$$

where the functions $\widetilde{M}_j$, $\widetilde{N}_j$, $\widetilde{P}_j$ and $\widetilde{Q}_j$ are just used here to indicate linear combinations of M_i, N_i, P_i and Q_i for $i=0$ to j, in an abbreviated manner. The constants $\widetilde{A}_j$, $\widetilde{B}_j$, $\widetilde{C}_j$ and $\widetilde{D}_j$ are related to the original constants A_j, B_j, C_j and

D_j in similar combinations, but with coefficients depending on constants of lower order than j. The uniform approximations of Wazwaz and Hanson (1986a) are related to the multi-scale exponential approximations $U_{BL}(Z, z, \varepsilon)$ and $U_{LTR}(Y, y, \varepsilon)$ given in (77) by

$$\begin{aligned} u_{BL} &= U_{BL}(\frac{z(x)}{\varepsilon}, z(x), \varepsilon) \quad \text{from } BL\,, \\ u_{LTR} &= U_{LTR}(\frac{y(x)}{\varepsilon_T}, y(x), \varepsilon) \quad \text{from } LTR\,, \end{aligned} \tag{78}$$

recalling that $\varepsilon_T = \text{ord}(\sqrt{\varepsilon})$.

Wazwaz and Hanson (1986a) used *classical matching* modified for complete exponential asymptotic expansions, i.e.,

$$\begin{gathered} u_{BL} = U_{BL}(\tfrac{z}{\varepsilon}, z, \varepsilon) \approx U_{LTR}(\tfrac{y}{\varepsilon_T}, y, \varepsilon) = u_{LTR}\,, \\ \text{as } \varepsilon \to 0^+ \text{ with } z = 2F_0(0) - F_0''(a)y^2 \text{ fixed}, \end{gathered} \tag{79}$$

recalling that the symbol ($\approx$) denotes a complete exponential asymptotic expansion (in the sense of Watson, 1911), and the relationship between z and y is the order one version of (66). The corresponding *multi-scale exponential asymptotic expansions matching principle* is expressed as

$$\begin{gathered} U_{BL}(Z, z, \varepsilon) \sim U_{LTR}(Y, y, \varepsilon)\,, \\ \text{as } \varepsilon \to 0^+ \text{ with } z = 2F_0(0) - F_0''(a)y^2 \\ \text{and } Z = \frac{2F_0(0)}{\varepsilon} - \frac{Y^2}{2} \text{ fixed}, \end{gathered} \tag{80}$$

where the relationship (66) between the fast scales Z and Y is used directly. The symbol ($\sim$) in (80) for asymptotic (in the sense of Poincaré, 1886) is used here in the sense of formal asymptotic expansions and is subject to justification by error bounds analysis (in the sense of Olver, 1974).

The actual matching details to determine the matching constants are omitted here; the reader can find them in Wazwaz and Hanson (1986a).

Wazwaz and Hanson (1986b) have applied the dual uniform method to the problem of combined singular boundary resonance and turning point resonance. The eigenvalue $s = \lambda_i$ is calculated to exponential precision, with the coefficient of exponentials factors computed to first order in ε as $\varepsilon \to 0^+$. The largest or principal eigenvalue is found to be exponentially small and negative. The leading order results were found by Hanson and Tier (1981) by the four region composite asymptotic expansions. Wazwaz and Hanson (1986b) compute *near resonance* or *Ackerberg-O'Malley (1970) resonance* conditions by exponential precision asymptotics. Ackerberg-O'Malley resonance is defined to occur when the solution is not exponentially vanishing uniformly in the interior of the domain $(0, b)$, i.e.,

$$u(x;\varepsilon) \geq \text{ord}_e(1) \quad \text{as} \quad \varepsilon \to 0^+ \ , \tag{81}$$

in the exponential order notation. This resonance occurs when the parameter s is exponentially close to an eigenvalue, i. e., $s - \lambda_i < \text{ord}_e(1)$. There is a set of eigenvalues principally associated with the turning point, another set associated with the singular point and a third set under combined influence of both critical points.

Recently, Wazwaz and Hanson (1988) found the moments of the extinction time, for which $u(x;\varepsilon)$ is the moment generating function, to exponential precision. The expected extinction time is proportional to a factor that is the reciprocal of the principal eigenvalue, implying exponentially large extinction times. Due to exponential precision, the leading terms of the full distribution are calculable, and the distribution turned out to be exponential with instant initial absorption. Boundary layer corrections are given for both moments and distributions.

V. CONCLUSIONS

The exponential notation of Wazwaz and Hanson (1986b) has been expanded in this paper. Its principal use will be in formal asymptotic approximations in problems requiring exponential precision.

The more important contribution is the reformulation of complete exponential asymptotic expansion in terms of multiple scaled variables. This preserves the linear independence of solutions through the use of geometric optics factors in the LG-WKB approximation. It also permits the coefficients of the exponential asymptotic expansion to be computed in the same way as those for the Poincaré algebraic asymptotic expansion, with the fast and slow independent variable held fixed. The corresponding exponentially precise matching principal was illustrated for the dual uniform approximation of Wazwaz and Hanson (1986a). The major disadvantage to this multi-scale exponential asymptotic expansion is the need for a purely dominant solution. A method for constructing purely exponential pairs from given complete expansions was presented. This connection problem will be pursued in future developments, with error bounds in mind.

ACKNOWLEDGEMENTS

The author gratefully acknowledges prior contributions to this problem by colleagues Charles Tier and A. M. Wazwaz.

REFERENCES

1. Ackerberg, R. C., and R. E. O'Malley, Jr. (1970). *Stud. Appl. Math., 49:* 277.
2. Bender, C. M., and S. A. Orszag (1978). *Advanced Mathematical Methods for Scientists and Engineers*, McGraw-Hill, New York, p. 494.
3. de Groen, P. P. N. (1987). *Appl. Math. Letters, 1:* 49.
4. Dingle, R. B. (1973). *Asymptotic Expansions: Their Derivation and Interpretation*, Academic Press, New York.
5. Gregory, R. D. (1980). *SIAM J. Math. Anal., 11:* 115.
6. Hanson, F. B. and C. Tier (1981). *SIAM J. Appl. Math., 40:* 113.
7. Hanson, F. B. and A. M. Wazwaz (1988). *Appl. Math. Letters, 1:* 137.
8. Kaplun, S. (1967). *Fluid Mechanics and Singular Perturbations* (P. A. Lagerstrom, L. N. Howard and C. S. Liu, eds.), Academic Press, New York.
9. Kevorkian, J. and J. D. Cole (1981). *Perturbation Methods in Applied Mathematics*, Springer-Verlag, New York.
10. Lozano, C. and R. E. Meyer (1976). *Phys. Fluids, 19:* 1075.
11. Ludwig, D. (1975). *SIAM Rev., 17:* 605.
12. Ludwig, D. (1976). *SIAM-AMS Proc., 10:* 87.
13. Lynn, R. Y. S. and J. B. Keller (1970). *Comm. Pure Appl. Math., 23:* 379.
14. Meyer, R. E. (1975). *SIAM J. Appl. Math., 29:* 481.
15. Meyer, R. E. (1976a). *J. Math. Phys., 17:* 1039.
16. Meyer, R. E. (1976b). *Z. Angew. Math. Phys., 27:* 181.
17. Meyer, R. E. (1980). *SIAM Rev., 22:* 213.
18. Meyer, R. E. and J. F. Painter (1983). *Adv. Appl. Math., 4:* 145.
19. Miller, J. C. P. (1950). *Quart. J. Mech. Appl. Math., 3:* 225.

20. Miller, J. C. P. (1952). *Proc. Camb. Phil. Soc., 48:* 428.

21. Miller, J. C. P. (1964). *Handbook of Mathematical Functions*, National Bureau of Standards, Appl. Math. Ser. No. 55, M. Abramowitz and I. A. Stegun, eds., U.S. Govt. Printing Office, Washington, D. C., p. 685.

22. Olver, F. W. J. (1964). *Asymptotic Solutions of Differential Equations* (C. H. Wilcox, ed.), Wiley, New York, p. 163.

23. Olver, F. W. J. (1974). *Asymptotics and Special Functions*, Academic Press, New York.

24. Olver, F. W. J. (1977). *SIAM J. Math. Anal., 8:* 127.

25. Olver, F. W. J. (1978). *SIAM J. Math. Anal., 9:* 328.

26. Olver, F. W. J. (1980a). *SIAM Rev. 22:* 188.

27. Olver, F. W. J. (1980b). *Singular Perturbations and Asymptotics* (R. E. Meyer and S. V. Parter, eds.), Academic Press, New York, p. 317.

28. O'Malley, R. E., Jr. (1974). *Introduction to Singular Perturbations*, Academic Press, New York.

29. Poincaré, H. (1886). *Acta Math., 8:* 295.

30. Tier, C. and F. B. Hanson (1981). *Math. Biosci., 53:* 89.

31. Watson, G. N. (1911). *Phil. Trans. Roy. Soc. London, Ser. A, 211:* 279.

32. Wazwaz, A. M. and F. B. Hanson (1986a). *SIAM J. Appl. Math., 46:* 943.

33. Wazwaz, A. M. and F. B. Hanson (1986b). *SIAM J. Appl. Math., 46:* 962.

34. Wazwaz, A. M. and F. B. Hanson (1988). University of Illinois preprint.

Uniform Asymptotic Remainders

D. S. Jones Professor, Department of Mathematical Sciences, University of Dundee, Dundee, Scotland

SUMMARY

In asymptotic expansions it is common practice to discard exponentially small remainders. It is shown in this paper that benefit accrues from keeping them in a suitable form. The region of validity of the asymptotic approximation to an integral is thereby considerably extended. Also the accuracy of the approximation is improved by an appreciable factor.

I. INTRODUCTION

The last chapter of Olver's famous book on asymptotic expansions [1] is concerned with the estimation of remainder terms. The aim of the analysis is to improve the accuracy of the numerical evaluation of the terms that are normally retained in an asymptotic series. To see what is involved let us discuss an example.

Consider the integral

$$I(\lambda,z) = \int_0^1 t^\lambda e^{-zt} dt. \tag{1}$$

Integrals of this type are of frequent occurrence in asymptotic analysis and the determination of the value is desired when $|z| >> 1$. When λ is a non-negative integer the integral can be evaluated explicitly in a finite number of terms. A fortiori, the series is convergent and the question of asymptotic behaviour does not arise. Therefore we shall assume that λ is not a non-negative integer. Furthermore, we shall take λ to be real though it is probable that some of the results can be extended to complex λ. To secure convergence of the integral at the origin it is necessary to have $\lambda > -1$. For the moment we shall limit λ to negative values by requiring $-1 < \lambda < 0$; this restriction will be removed later on. The function t^λ is defined by its principal value in the complex t-plane cut along the negative axis from $-\infty$ to 0.

The integral $I(\lambda,z)$ is well known and its asymptotic evaluation can be found in Chapter 4 of Olver's book. In fact

$$I(\lambda,z) = \frac{\lambda!}{z^{\lambda+1}} - \frac{\Gamma(\lambda + 1,z)}{z^{\lambda+1}}$$

where $\Gamma(\lambda + 1,z)$ is the incomplete gamma function defined by

$$\Gamma(\lambda + 1,z) = e^{-z} z^{\lambda+1} \int_0^\infty e^{-zt}(1 + t)^\lambda dt$$

for $|\mathrm{ph}(z)| < \frac{1}{2}\pi$. From this formula Olver shows that

$$I(\lambda,z) = \frac{\lambda!}{z^{\lambda+1}} - e^{-z}\left\{ \sum_{m=0}^{M-1} \frac{\lambda!}{(\lambda - m)!z^{m+1}} + \varepsilon_M(z)\right\}$$

where $\varepsilon_M(z) = O(z^{-M-1})$ as $z \to \infty$ in the sector $|ph(z)| \leq \frac{3\pi}{2} - \delta$, $\delta > 0$. Olver derives, in his Chapter 4, various bounds for ε_M; they differ according to the range of ph(z) involved and generally become invalid as the limits of the range are approached. Our objective is twofold (a) to obtain an estimate of ε_M which is valid uniformly in ph(z) and (b) to replace the bounds by a more precise determination so that a more accurate calculation of I is feasible.

Accordingly we shall now consider the integral afresh and undertake its evaluation by a method which leads naturally to a way of tackling other types of integral.

If $ph(z) = \alpha$ the contour of integration is deformed into two straight lines joining $t = 0$ and $t = 1$ to $\infty e^{-i\alpha}$. Thus

$$I(\lambda,z) = \left(\int_0^{\infty e^{-i\alpha}} - \int_1^{1+\infty e^{-i\alpha}}\right) t^\lambda e^{-zt} dt$$

$$= \frac{\lambda!}{z^{\lambda+1}} - e^{-z}\int_0^\infty (1 + te^{-i\alpha})^\lambda e^{-|z|t} e^{-i\alpha} dt.$$

By expanding the term in parentheses in powers of t we obtain

$$I(\lambda,z) = \frac{\lambda!}{z^{\lambda+1}} - e^{-z}\sum_{m=0}^{M-1} \frac{\lambda!}{(\lambda - m)!z^{m+1}} + R_M(\lambda,z). \tag{2}$$

An appeal to Watson's lemma shows that, for fixed M, R_M is essentially of the order of $1/|z|^{M+1}$. Hence, for $|z| >> 1$, R_M is small and may be neglected in comparison with the other terms in (2), leading to the approximation

$$I(\lambda,z) \sim \frac{\lambda!}{z^{\lambda+1}} - e^{-z}\sum_{m=0}^{M-1} \frac{\lambda!}{(\lambda - m)!z^{m+1}}, \tag{3}$$

as we have already seen. It is tempting to let $M \to \infty$ in (3) but the resulting series is divergent and the expansion is only asymptotic. So M has to be chosen such that (3) commits the least error in approximating I; roughly speaking this entails $M \approx |z|$. A correction can be applied to give greater accuracy if some notion of the magnitude of the remainder R_M can be derived. When successive terms of (3) decrease in modulus until the (M+1)th is reached and thereafter do not diminish Olver calls R_M the optimal remainder. From a number of particular cases examined in Olver's last chapter it can be deduced that estimating the optimal remainder is capable of improving the accuracy of the asymptotic value for I by several significant figures.

It will be noted that when $\mathrm{Re}(z) > 0$ the first term of (3) is dominant and the series is recessive. For $\mathrm{Re}(z) < 0$ the series is dominant and the first term of (3) is recessive. Around the imaginary axis there is a transition zone where both elements play a part. In general, this switching of emphasis forces different estimates of R_M in different regions. The purpose of the following investigation is to estimate R_M in somewhat more general circumstances than in Olver's book and to supply a formula which is uniformly valid in z for $|z| \gg 1$.

2. A TRANSFORMATION

The derivation of (2) has skipped over a snag which must now be mentioned. When α increases through π the deformation of the contour into two straight lines is not immediately feasible because the lines are on the next sheet of the Riemann surface of t^λ and so one would appear to have to keep an integral along the real axis between 0 and 1 with a singularity of the integrand along the path which invalidates (2). The difficulty can be overcome by observing that augmenting the phase of z by 2π has no effect on (1) or by defining I when $\mathrm{ph}(z) > \pi$ by an equivalent integral in the next sheet of t^λ. In either case there is a different representation on either side of π and this is inconvenient for attempts to provide uniformly valid formulae. Consequently, (2) will be substantiated by an alternative route.

Since the series in (3) is of little or no importance for $|\alpha| \leq \pi/4$ we shall start by assuming that

$$\frac{1}{4}\pi + \delta < \alpha < \frac{5}{4}\pi - \delta$$

with $\delta > 0$. The region $-5\pi/4 < \alpha < -\pi/4$ is also covered by taking the complex conjugate of the expression for I.

Instead of deforming the integral in (1) into the straight lines to $\infty e^{-i\alpha}$ we select the lines going to $\infty e^{i\beta}$ where $\beta = -3\pi/4$. since $|\alpha + \beta| < \frac{1}{2}\pi$ the deformation is legitimate and

$$I(\lambda,z) = \frac{\lambda!}{z^{\lambda+1}} - e^{-z}J(\lambda,z) \tag{4}$$

where

$$J(\lambda,z) = \int_0^\infty (1 + te^{i\beta})^\lambda \exp(i\beta - zte^{i\beta})dt.$$

Because of the choice of β and the restriction on λ

$$J(\lambda,z) = \int_0^\infty \exp(i\beta - zte^{i\beta})\int_0^\infty \frac{u^{-1-\lambda}}{(-1-\lambda)!} \times$$

$$\exp\{-\tfrac{1}{2}\pi i\lambda - i(1 + te^{i\beta})u\}dudt$$

$$= \frac{e^{-\frac{1}{2}\pi i\lambda}}{(-1-\lambda)!}\int^\infty \frac{u^{-1-\lambda}e^{-iu}}{z + iu}du$$

on interchanging the order of integration, which is permissible since $\mathrm{Re}\{(z + iu)e^{i\beta}\} > 0$. Deform the contour of integration into the negative imaginary axis. In the deformation a pole is passed over when $\alpha > \pi$. Hence

$$J(\lambda,z) = \frac{2\pi i e^{z+\lambda\pi i}}{(-1-\lambda)!z^{\lambda+1}}H(\alpha - \pi) + \frac{1}{(-1-\lambda)!}\int_0^\infty \frac{t^{-1-\lambda}e^{-t}}{t + z}dt \tag{5}$$

where H(x) is the Heaviside step function. For future reference we note that, on account of the reflection formula for the factorial function,

$$\frac{2\pi i e^{z+\lambda\pi i}}{(-1-\lambda)!z^{\lambda+1}} = \frac{\lambda!}{z^{\lambda+1}}(1 - e^{2\pi i\lambda})e^{z}. \tag{6}$$

Next, consider the behaviour of

$$K(\lambda,z) = \int_0^\infty \frac{t^{-1-\lambda}e^{-t}}{t+z}\,dt. \tag{7}$$

Expansion of the denominator in powers of t gives

$$K(\lambda,z) = \sum_{m=0}^{M-1}(m-1-\lambda)!\frac{(-)^m}{z^{m+1}} + \frac{(-)^M}{z^M}\int_0^\infty \frac{t^\mu e^{-t}}{t+z}\,dt \tag{8}$$

where $\mu = M - 1 - \lambda$.

The integral in (8) represents the remainder to the asymptotic expansion of K. Successive terms in the series diminish in absolute value when $m - \lambda < |z|$. Hence an optimal remainder is obtained when M is the smallest integer such that $\mu \geq |z|$. It will be assumed from now on that M has been chosen in this way. Then an estimate of the integral is needed when μ and $|z|$ are large parameters with $\mu/|z|$ about 1.

Make the substitution

$$\mu\,\ell n(t/\mu) - t + \mu = -r^2/2\mu \tag{9}$$

on the understanding that r is positive when $t > \mu$ and negative when $t < \mu$. Then

$$\int_0^\infty \frac{t^\mu e^{-t}}{t+z}\,dt = e^{\lambda}\int_{-\infty}^\infty \frac{e^{-r^2/2\mu}}{t+z}\frac{dt}{dr}\,dr$$

where

$$\gamma = \mu\,\ell n\mu - \mu. \tag{10}$$

To allow for the pole at t = -z put

$$\frac{1}{t+z}\frac{dt}{dr} = \frac{B}{r-r_0} + C + k(r) \tag{11}$$

where k(r) is a regular function of r such that k(0) = 0 and r_0 is defined by

$$\mu \ln(-z/\mu) + z + \mu = -r_0^2/2\mu. \tag{12}$$

In (12), -z is understood to be $ze^{-i\pi}$ in the evaluation of the logarithm. To be consistent with $\mu \geq |z|$ and the rule adopted in (9) the real part of r_0 must be negative. It then follows that $\text{Im}(r_0) > 0$ when $\alpha > \pi$ and r_0 lies in the second quadrant. Likewise, r_0 is in the third quadrant when $\alpha < \pi$. Then $r \to r_0$ as $t \to -z$. Putting $r = r_0$ in (11) we have, via (9) and (12), B = 1 while substituting r = 0 provides

$$C = \frac{1}{\mu+z} + \frac{1}{r_0}.$$

Now, for $\text{Im}(r_0) < 0$,

$$\int_{-\infty}^{\infty} \frac{e^{-r^2/2\mu}}{r-r_0}dr = -i\int_{-\infty}^{\infty} e^{-r^2/2\mu}\int_0^{\infty} e^{iy(r-r_0)}dy\,dr$$

$$= -i(2\pi\mu)^{\frac{1}{2}}\int_0^{\infty} e^{-iyr_0-\mu y^2/2}dy$$

after interchanging the order of integration. The last integral can be expressed in terms of

$$\text{erfc}(w) = \frac{2}{\pi^{\frac{1}{2}}}\int_w^{\infty} e^{-y^2}dy$$

with the result that

$$\int_{-\infty}^{\infty} \frac{e^{-r^2/2\mu}}{r-r_0}dr = -\pi i e^{-r_0^2/2\mu}\text{erfc}\{ir_0/(2\mu)^{\frac{1}{2}}\}$$

when $\text{Im}(r_0) < 0$. If $\text{Im}(r_0) > 0$ only the sign of i has to be changed throughout. Furthermore

$$\text{erfc}\{-\ ir_0/(2\mu)^{\frac{1}{2}}\} = 2 - \text{erfc}\{ir_0/(2\mu)^{\frac{1}{2}}\}. \tag{13}$$

Hence

$$\int_{-\infty}^{\infty} \frac{e^{-r^2/2\mu}}{r - r_0} dr = \pi i e^{-r_0^2/2\mu}[2H(\alpha - \pi) - \text{erfc}\{ir_0/(2\mu)^{\frac{1}{2}}\}].$$

Accordingly, from (12)

$$\int_0^{\infty} \frac{t^{\mu} e^{-t}}{t + z} dt = \pi i(-z)^{\mu} e^{z}[2H(\alpha - \pi) - \text{erfc}\{ir_0/(2\mu)^{\frac{1}{2}}\}]$$

$$+ (2\pi\mu)^{\frac{1}{2}} e^{\gamma}\left(\frac{1}{\mu + z} + \frac{1}{r_0}\right) + e^{\gamma}\int_{-\infty}^{\infty} k(r)e^{-r^2/2\mu} dr. \tag{14}$$

Since the contributions of the saddle-point at $r = 0$ and of the pole at $r = r_0$ have already been accounted for in (14) we can expect that the correction due to k is of less significance. Its influence can be assessed by observing that its regularity permits the expansion

$$k(r) = \sum_{p=1} a_p r^p.$$

Thus there is the estimate

$$\int_{-\infty}^{\infty} k(r)e^{-r^2/2\mu} dr = \tfrac{1}{2}!(2\mu)^{\frac{3}{2}} a_2 + \tfrac{3}{2}!(2\mu)^{\frac{5}{2}} a_4 + \dots \quad . \tag{15}$$

The coefficients a_p can be determined by taking derivatives of (11) and putting $r = 0$. The procedure involves no difficulties of principle but the detailed calculation is tedious. However, the structure of the coefficients can be described in a simple fashion. No singularity is present in (11) as r and r_0 vary; in particular, there is none when both are zero. Therefore, the same must be true of the derivatives. But the n th derivative of the term containing B will produce something like $1/r_0^{n+1}$ when

$r = 0$. This would create a singularity when $r_0 = 0$ unless it was balanced by a corresponding singularity from the left-hand side. Consequently, the left-hand side must furnish the singular part of the expansion of $1/r_0^{n+1}$ in powers of $\mu + z$. It is thereby found that

$$a_2 = \frac{1}{r_0^3} + \frac{1}{(\mu + z)^3} - \frac{1}{\mu(\mu + z)^2} + \frac{1}{12\mu^2(\mu + z)}, \tag{16}$$

$$a_4 = \frac{1}{r_0^5} + \frac{1}{(\mu + z)^5} - \frac{5}{3\mu(\mu + z)^4} + \frac{25}{36\mu^2(\mu + z)^2}$$

$$- \frac{1}{36\mu^3(\mu + z)^2} + \frac{1}{864\mu^4(\mu + z)} . \tag{17}$$

These formulae may be verified by direct computation of the derivatives of (11).

It is evident from (16) and (17) that a_2 is of the order of $1/\mu^3$ while a_4 is of the order of $1/\mu^5$ except possibly when $\mu + z$ is small. For small $\mu + z$ it can be shown that

$$a_2 \approx \frac{-1}{540\mu^3} + \frac{\mu + z}{288\mu^4} + \frac{23(\mu + z)^2}{6048\mu^5}, \tag{18}$$

$$a_4 \approx -\frac{25}{18144\mu^5} \tag{19}$$

and the order is unaffected. In the same circumstances

$$\frac{1}{r_0} + \frac{1}{\mu + z} \approx \frac{1}{3\mu} + \frac{\mu + z}{12\mu^2} + \frac{23(\mu + z)^2}{540\mu^3} + \frac{353(\mu + z)^3}{12960\mu^4}$$

$$+ \frac{589(\mu + z)^4}{30240\mu^5} \tag{20}$$

which can be helpful in the computation of C.

The combination of (5), (8), (14) and (15) leads to

$$J(\lambda,z) \simeq \sum_{m=0}^{M-1} \frac{\lambda!}{(\lambda-m)!z^{m+1}} + \lambda!(1-e^{2\pi i\lambda})\frac{e^z}{2z^{1+\lambda}}\mathrm{erfc}\{ir_0/(2\mu)^{\frac{1}{2}}\}$$

$$+ \frac{(-)^M e^\lambda (2\pi\mu)^{\frac{1}{2}}}{(-1-\lambda)!z^M}\{\frac{1}{\mu+z} + \frac{1}{r_0} + \mu a_2 + 3\mu^2 a_4\} \qquad (21)$$

where the expansion (15) has been terminated at a_4 in view of (16)-(19) and the symbol $\simeq$ signifies that strict equality no longer holds. Naturally, extra terms can be retained if desired. $I(\lambda,z)$ now follows from (4).

3. DISCUSSION

Firstly, note that when $\lambda = 0$ all the terms in (21) except the first series disappear. The survivor is the exact quantity required for I. Therefore, although the formula developed for I has been based on the assumption that $-1 < \lambda < 0$ it is, in fact, valid for $-1 < \lambda \leq 0$.

Secondly, the asymptotic expression for erfc permits writing

$$\lambda!(1 - e^{2\pi i\lambda})\frac{e^z}{2z^{1+\lambda}}\mathrm{erfc}\{ir_0/(2\mu)^{\frac{1}{2}}\}$$

$$\sim \frac{(-)^{M-1} e^\gamma (2\pi\mu)^{\frac{1}{2}}}{(-1-\lambda)!z^M r_0}\left(1 + \frac{\mu}{r_0^2} + \frac{3\mu^2}{r_0^4} + \ldots\right) \qquad (22)$$

when $|r_0/(2\mu)^{\frac{1}{2}}| \gg 1$ and $\mathrm{Im}(r_0) < 0$. For $\mathrm{Im}(r_0) > 0$ add

$$\lambda!(1 - e^{2\pi i\lambda})e^z/z^{1+\lambda}$$

to the right-hand side of (22). This additional term appears when $\alpha > \pi$ and its effect is to correct the first term $\lambda!/z^{\lambda+1}$ of $I(\lambda,z)$ to $\lambda!e^{2\pi i\lambda}/z^{\lambda+1}$; such a change is necessary because

$I(\lambda,z)$ is unaffected by altering the phase of z by 2π. Observe that all other powers of z in our formula are integers so they do not have to be modified. It is interesting that the switch in the first term of $I(\lambda,z)$ occurs where it is negligible compared with the rest which are exponentially large. Yet, if it had been omitted, there would have been trouble as the phase of z grew. The transition takes place on a Stokes line $\mathrm{Im}(z) = 0$ in the terminology of Olver.

The series in (22) cancels precisely the powers of $1/r_0$ in $C, a_2, a_4, \ldots$. Hence, when $|r_0/(2\mu)^{\frac{1}{2}}|$ is not small i.e. when $\mathrm{ph}(z)$ is not too close to π, $J(\lambda,z)$ can be estimated from

$$J(\lambda,z) \simeq \sum_{m=0}^{M-1} \frac{\lambda!}{(\lambda - m)!z^{m+1}} + \lambda!(1 - e^{2\pi i\lambda})\frac{e^z}{z^{1+\lambda}} H(\alpha - \pi)$$

$$+ \frac{(-)^M e^{\lambda}(2\pi\mu)^{\frac{1}{2}}}{(-1-\lambda)!z^M} \{\frac{1}{\mu + z} + \mu(a_2 - \frac{1}{r_0^3}) + 3\mu^2(a_4 - \frac{1}{r_0^5})\} \quad (23)$$

The uniformly valid formula that has been derived has been demonstrated for $\pi/4 < \mathrm{ph}(z) < 5\pi/4$. However, there is no difficulty in confirming (by analytic continuation starting from a different value of β) that it holds down to $\mathrm{ph}(z) = 0$ provided that r_0, as defined by (12), changes continuously from its value near $\mathrm{ph}(z) = \pi$. In effect, this is accomplished by having $\mathrm{Im}(r_0) < 0$. Actually, in the domain $0 \leq \mathrm{ph}(z) \leq \pi/4$, r_0 does not really play a role because (23) can be used there when $|z|$ is large enough. Indeed, this observation extends the formula (in the form (23)) into $\mathrm{ph}(z) < 0$. Of course, this is not true when $\mathrm{ph}(z)$ approaches $-\pi$ when an erfc must be included. The upper limit of $5\pi/4$ can be similarly extended to 2π by keeping $\mathrm{Im}(r_0) > 0$ where relevant.

The merit of our uniform remainder can be assessed by considering what happens when $|z| = 5$, taking advantage of (23) for (say) $0 \leq \mathrm{ph}(z) \leq 3\pi/4$. If the remainder, i.e. erfc, C, a_2, a_4, is excluded leaving the normal asymptotic series the accuracy of the approximation is about 2 significant figures. Going down as far as C adds another 1-2 significant figures while each of a_2 and a_4 does likewise. Thus our formula gives a dramatic improvement in the accuracy with which $I(\lambda,z)$ is estimated for this value of $|z|$ throughout the range of ph(z), the gain being from 2 significant figures to 6-8. No doubt a like amelioration could be achieved by including the contributions of a_6 and a_8 but it is not obvious that the additional effort is justified.

4. THE ERROR INTEGRAL

The choice $\lambda = -\frac{1}{2}$ in the foregoing leads, after slight manipulation, to a formula for the error integral. It can be expressed as

$$\operatorname{erfc}(z^{\frac{1}{2}}) \simeq e^{-z}\sum_{m=0}^{M-1}\frac{1}{(-\frac{1}{2}-m)!\,z^{m+\frac{1}{2}}} + \operatorname{erfc}\{ir_0/(2\mu)^{\frac{1}{2}}\}$$

$$- \frac{(-)^M e^{\gamma-z}\mu^{\frac{1}{2}}}{(2\pi)^{\frac{1}{2}}z^{M+\frac{1}{2}}}\left\{\frac{1}{\mu+z}+\frac{1}{r_0}+\mu a_2+3\mu^2 a_4\right\}.$$

This result is interesting in specifying the remainder for a function in terms of the function itself. Of course, the arguments of the function on the two sides of the equation are not by any means the same. For instance, when $\mathrm{ph}(z^{\frac{1}{2}})$ is near $\frac{1}{2}\pi$ the value of r_0 will not be far from zero. Nevertheless, it is a little unusual to have the correction to an asymptotic expansion expressed via itself, as will occur when $r_0/\mu^{\frac{1}{2}}$ is not small.

5. SOME GENERALISATIONS

The uniform remainder has been found for a range of ph(z) subject to the restriction $-1 < \lambda \leq 0$. This constraint will now be weakened. If n is a positive integer

$$I(\lambda + n,z) = -\frac{e^{-z}}{z} + \frac{\lambda + n}{z} I(\lambda + n - 1,z) \tag{24}$$

by integration by parts. Repeated application of (24) enables $I(\lambda + n,z)$ to be expressed in terms of $I(\lambda,z)$. Thereby a uniform remainder for $I(\lambda + n,z)$ is established. The formula for $I(\lambda + n,z)$ consists of e^{-z} times a finite series of powers of $1/z$ plus $I(\lambda,z)$ multiplied by a similar finite series. Properties of the factorial function allow the normal asymptotic terms to be combined into the normal expansion for $I(\lambda + n,z)$. Unless n is large there seems little point in attempting any recombination of the terms in the uniform remainder. The multiplying factor is easy to calculate and the remainder for $I(\lambda,z)$ has to be evaluated only once for any given value of z. If n is large it may be desirable to reconsider the integral anyway.

An alternative approach is to use

$$I(\lambda + n,z) = -\frac{d}{dz} I(\lambda + n - 1,z).$$

This is a less convenient way of arriving at $I(\lambda + n,z)$ because it involves taking repeated derivatives of the remainder in $I(\lambda,z)$. On the other hand, it does offer the possibility of finding remainder terms for integrals like

$$\int^{\infty} \frac{t^{-\lambda-1} e^{-t}}{(t + z)^n} dt$$

on account of the structure of $J(\lambda,z)$.

A derivative with respect to λ introduces a logarithm into the integrand. The derivative of the remainder requires

$$\frac{\partial r_0}{\partial \lambda} = -\frac{r_0}{\mu} - \frac{\mu + z}{r_0} \tag{25}$$

deduced from (12). Then, for $-1 < \lambda \leq 0$ and $0 \leq \text{ph}(z) < 5\pi/4$

$$\int_0^1 t^\lambda \ln t\, e^{-zt} dt = \frac{\lambda!}{z^{\lambda+1}}\{\psi(\lambda)-\ln z\}-e^{-z}\sum_{m=0}^{M-1}\frac{\lambda!}{(\lambda-m)!z^{m+1}}\{\psi(\lambda)-\psi(\lambda-m)\}$$

$$-\frac{\lambda!}{2z^{1+\lambda}}[\{\psi(\lambda)-\ln z\}(1-e^{2\pi i\lambda})-2\pi i e^{2\pi i\lambda}\}\operatorname{erfc}\{ir_0/(2\mu)^{\frac{1}{2}}\}$$

$$-\frac{(-)^M(2\pi\mu)^{\frac{1}{2}}e^{\gamma-z}}{(-1-\lambda)!z^M}[\{\psi(-1-\lambda)-\ln\mu-\frac{1}{2\mu}\}\{\frac{1}{\mu+z}+\frac{1}{r_0}+\mu a_2+3\mu^2 a_4\}$$

$$+\frac{r_0}{2\mu^2}+\frac{\mu+z}{\mu r_0}+\frac{1}{(\mu+z)^2}+\frac{1}{\mu r_0}+\frac{\mu+z}{r_0^3}-a_2+\mu\frac{\partial a_2}{\partial\lambda}-6\mu a_4+3\mu^2\frac{\partial a_4}{\partial\lambda}] \qquad (26)$$

where

$$\lambda!\psi(\lambda) = \frac{d}{d\lambda}\lambda! \; .$$

The partial derivatives of a_2 and a_4 can be readily written down from (16), (17) and (26) but are rather lengthy.

Obviously, higher powers of the logarithm can be incorporated by the same process; the results become steadily more complicated. The range of λ can be extended above zero via derivatives of (24).

By means of these formulae uniform remainders can be obtained for any integrand which is the sum of a finite number of terms of the type $t^\lambda(\ln t)^n$. Thus we can cope with the sort of term that is of frequent occurrence in asymptotic analysis. In the next section is considered an integrand which is rather more general in some respects.

6. AN ALTERNATIVE APPROACH

The integral to be discussed in this section is

$$I(f,z) = \int_0^1 f(t)e^{-zt}dt \qquad (27)$$

where f(t) is analytic on the interval except possibly at the origin where it may have a singularity of the sort that has been the subject of preceding sections. It will be assumed that f can be extended analytically to $+\infty$ so that the Laplace transform

$$F(s) = \int_0^{\infty} f(t)e^{-st}dt$$

exists when the real part of s is big enough. In effect, the extension of f has to be exponentially bounded at infinity. F(s) will also denote the analytic continuation of the integral to other parts of the complex s-plane.

By virtue of the Mellin inversion formula

$$I(f,z) = \frac{1}{2\pi i}\int_{c-i\infty}^{c+i\infty} F(s)\,\frac{e^{s-z}-1}{s-z}\,ds$$

with c selected so that all the singularities of F(s) lie to the left of the contour of integration. Since c can be increased if necessary it can be arranged that s = z is also to the left of the contour. Deform the contour to the right for the part of the integrand which does not involve the factor e^{s-z}. Then

$$I(f,z) = \frac{e^{-z}}{2\pi}\int_{c-i\infty}^{c+i\infty}\frac{F(s)e^{s}}{s-z}\,ds$$

with all singularities to the left of the contour.

Now suppose that the only singularity of F(s) is a branch point at the origin such that $F(s)s^{\rho}$ is regular for some positive real ρ. Logarithmic branch points are consequently not allowed but it will be seen later that this restriction can be eased. Temporarily, the position of z will be limited so that $\pi > ph(z) \geq 0$ though this limitation of z will be relaxed shortly.

Take the branch line of F(s) along the negative real axis. Deform the contour into a loop round the branch line so as to pass over the pole s = z is the process. Then

$$I(f,z) = F(z) + \frac{e^{-z}}{2\pi i}\int_{-\infty}^{(0+)} \frac{F(s)e^{s}}{s - z}\, ds.$$

Expanding the denominator in powers of s we obtain

$$I(f,z) = F(z) - e^{-z}\sum_{m=0}^{n-1} \frac{f^{(m)}(1)}{z^{m+1}} + \frac{e^{-z}}{2\pi i z^{n}}\int_{-\infty}^{(0+)} \frac{F(s)s^{n}}{s - z} e^{s} ds \qquad (28)$$

The earlier experience with remainders indicates that n should be chosen as the smallest integer such that $n + \rho \geq |z|$. The loop can then be transformed into the negative real axis. Write $F_1(s) = F(s)s^{\rho}$ so that F_1 is regular in the domain under consideration. Then

$$\int_{-\infty}^{(0+)} \frac{F(s)s^{n}}{s - z} e^{s} ds = (-)^{n}(e^{-\pi i\rho} - e^{\pi i\rho})\int_{0}^{\infty} \frac{F_1(-t)t^{n-\rho}}{t + z} e^{-t} dt. \qquad (29)$$

Redefine μ by $\mu = n - \rho$; then the integral differs from that in (8) only by the presence of the factor $F_1(-t)$. The substitution (9) can be employed again. Instead of (11) we have

$$\frac{F_1(-t)}{t + z}\frac{dt}{dr} = \frac{F_1(z)}{r - r_0} + \frac{F_1(-\mu)}{\mu + z} + \frac{F_1(z)}{r_0} + h(r)$$

where h(r) is a regular function of r such that h(0) = 0 while r_0 and z are related by (12). Accordingly,

$$I(f,z) \simeq F(z) - e^{-z}\sum_{m=0}^{n-1} \frac{f^{(m)}(1)}{z^{m+1}} - \frac{1}{2}(1 - e^{2\pi i\rho})F(z)\operatorname{erfc}\{ir_0/(2\mu)^{\frac{1}{2}}\}$$

$$+ (-)^{n}(e^{-\pi i\rho} - e^{\pi i\rho})\frac{e^{\gamma - z}}{2\pi i z^{n}}(2\pi\mu)^{\frac{1}{2}}\left\{\frac{F_1(-\mu)}{\mu + z} + \frac{F_1(z)}{r_0} + \mu b_2 + 3\mu^2 b_4\right\} \qquad (30)$$

where $h(r) = \sum_{p=1} b_p r^{p}$.

The coefficients b_p can be calculated by the same rule as for (16) and (17), bearing in mind that it is the singularity of $F_1(z)/r_0^p$ that has to be cancelled. Hence

$$b_2 = \frac{F_1(z)}{r_0^3} + F_1(-\mu)(a_2 - \frac{1}{r_0^3}) - F_1'(-\mu)\frac{z}{\mu(\mu+z)^2} + \frac{F_1''(-\mu)}{z(\mu+z)}, \tag{31}$$

$$\begin{aligned} b_4 = {} & \frac{F_1(z)}{r_0^3} + F_1(-\mu)(a_4 - \frac{1}{r_0^5}) \\ & + F_1'(-\mu)\{\frac{1}{(\mu+z)^4} - \frac{5}{3\mu(\mu+z)^3} + \frac{25}{36\mu^2(\mu+z)^2} - \frac{1}{36\mu^3(\mu+z)}\} \\ & + \frac{1}{2}F_1''(-\mu)\{\frac{1}{(\mu+z)^3} - \frac{5}{3\mu(\mu+z)^2} + \frac{25}{36\mu^2(\mu+z)}\} \\ & + \frac{1}{6}F_1'''(-\mu)\{\frac{1}{(\mu+z)^2} - \frac{5}{3\mu(\mu+z)}\} + \frac{F_1^{(iv)}(-\mu)}{24(\mu+z)} . \end{aligned} \tag{32}$$

The remainder in (30) is uniform. Therefore the restriction on ph(z) can be dropped on the understanding that, when $ph(z) > \pi$, $F(z)$ is evaluated on the appropriate Riemann sheet.

If the branch point $F(s)$ is at $s = a$ instead of the origin its effect on (30) is the following. The factor $F(z)$ is unaltered. Everywhere else z is replaced by $z - a$, $F_1(s) = F(s + a)s^\rho$ and $f^{(m)}(1)$ is replaced by $\{e^{-at}f(t)\}^{(m)}$ evaluated at $t = 1$.

The formulae (31) and (32) may be rewritten as

$$b_2 = F_1(z)a_2 - h_2(z) + \frac{h_1(z)}{\mu} - \frac{h_0(z)}{12\mu^2}, \tag{33}$$

$$b_4 = F_1(z)a_4 - h_4(z) + \frac{5h_3(z)}{3\mu} - \frac{25h_2(z)}{36\mu^2} + \frac{h_1(z)}{36\mu^3} - \frac{h_0(z)}{864\mu^4} \tag{34}$$

where

where

$$h_n(z) = \frac{1}{(\mu + z)^{n+1}} \{F_1(z) - \sum_{p=0}^{n} \frac{(\mu + z)^p}{p!} F_1^{(p)}(-\mu)\}. \qquad (35)$$

It is evident from (18), (19), (33)-(35) that

$$b_2 \approx \frac{-F_1(-\mu)}{540\mu^3} - \frac{F_1'(-\mu)}{12\mu^2} + \frac{F_1''(-\mu)}{2\mu} - \frac{1}{6}F_1'''(-\mu), \qquad (36)$$

$$b_4 \approx \frac{-25F_1(-\mu)}{18144\mu^5} - \frac{F_1'(-\mu)}{864\mu^4} + \frac{F_1''(-\mu)}{72\mu^3} - \frac{25F_1'''(-\mu)}{216\mu^2}$$

$$+ \frac{5}{72\mu}F_1^{(iv)}(-\mu) - \frac{1}{120}F_1^{(v)}(-\mu) \qquad (37)$$

when $\mu + z$ is small.

When $F(s)$ possesses poles as well as a branch point $F_1(s)$ will not be regular everywhere. Nonetheless (30) still holds so long as $F_1(s)$ is regular in a sufficiently large region and, in particular, has no pole on the branch line. However, if z approaches a pole the computation of (30) may become awkward and it may be desirable to accord the pole special treatment. For instance,

$$f(t) = e^{5t}\int_0^{t^{\frac{1}{2}}} e^{-5y^2}\,dy$$

implies that $F(s) = \pi^{\frac{1}{2}}/2(s - 5)s^{\frac{1}{2}}$ and difficulties may arise when z is near 5.

Suppose that $F(s)$ has a simple pole at $s = s_0$. With $F(s) = G(s)/(s - s_0)$ we have

$$\frac{1}{2\pi i}\int_{c-i\infty}^{c+i\infty} \frac{F(s)e^s}{s-z}\,ds = \frac{1}{z - s_0}\{\frac{1}{2\pi i}\int_{c-i\infty}^{c+i\infty} \frac{G(s)e^s}{s-z}\,ds - f(1)\}.$$

The integral can be handled as previously and so can be derived from (30) by replacing F and F_1 by G and G_1 respectively. The finite answer when z is near s_0 may still not be easy to compute because a small quantity is being divided by a small number. Yet it may be feasible to analytically factor out the zero in suitable circumstances with substantial improvement in the computation.

Poles of higher order and the presence of several different poles may be dealt with by the same device but the formulae become steadily more cumbersome to set down.

Logarithmic branch points were excluded in the derivation of (30). Relaxation of this constraint is possible when $F(s) = F_1(s)\ln s$ with F_1 regular. Deformation of the loop can still be accomplished with $\rho = 0$ so long as the factor $e^{-\pi i\rho} - e^{\pi i\rho}$ is replaced by $2\pi i$. Therefore (30) remains valid in this case on replacing $e^{-\pi i\rho} - e^{\pi i\rho}$ by $2\pi i$ and putting $\rho = 0$.

When $r_0/\mu^{\frac{1}{2}}$ is large enough (30) can be simplified somewhat by taking advantage of the known properties of erfc. Only the result for $|ph(z)| < \pi$ will be quoted since the modification for another range of the phase is easy to write down (compare (23)). The appropriate approximation is

$$I(f,z) \simeq F(z) - e^{-z}\sum_{m=0}^{n-1}\frac{f^m(1)}{z^{m+1}}$$

$$+ (-)^n(e^{-\pi i\rho} - e^{\pi i\rho})\frac{e^{\gamma-z}}{2\pi i z^n}(2\pi\mu)^{\frac{1}{2}}\{\frac{F_1(-\mu)}{\mu + z} + \mu b_2' + 3\mu^2 b_4'\} \tag{38}$$

where b_2' and b_4' are defined as in (33) and (34) but with a_2 and a_4 replaced by $a_2 - 1/r_0^3$ and $a_4 - 1/r_0^5$ respectively.

7. AN EXAMPLE

For an application of our method of determining a uniform asymptotic remainder we consider

$$f(t) = \frac{\pi^{\frac{1}{2}}}{(\upsilon - \frac{1}{2})!2^{\upsilon}}(t^2 + 2t)^{\upsilon-\frac{1}{2}} \qquad (\upsilon > -\tfrac{1}{2}). \tag{39}$$

This selection is made for a number of reasons. One of them is that the Laplace transform is well known. Another is that it reveals that the technique is more versatile than might be construed from the preceding section. Further reasons will become apparent as the analysis proceeds.

For the Laplace transform of (39) we have

$$F(s) = e^{s}K_{\upsilon}(s)/s^{\upsilon} \tag{40}$$

where K_{υ} is the standard modified Bessel function. The analogue of (28) is

$$I(f,z) = F(z) - e^{-z}\sum_{m=0}^{n-1}\frac{f^{(m)}(1)}{z^{m+1}} + \frac{e^{-z}}{2\pi i z^{n}}\int_{-\infty}^{(0+)}\frac{F(s)s^{n}}{s - z}e^{s}ds$$

with $F(s)$ as in (40). Although the only singularity of $F(s)$ is a branch point at $s = 0$, the structure of K_{υ} prevents there being any ρ such that $F(s)s^{\rho}$ is regular. Thus our theory is not immediately applicable. Notwithstanding, the loop can still be deformed into the two sides of the branch line provided that n is sufficiently big e.g. $n - 2\upsilon > -1$. In fact, the relations between modified Bessel functions enable us to establish

$$\frac{1}{2\pi i}\int_{-\infty}^{(0+)}\frac{F(s)s^{n}}{s - z}e^{s}ds = (-)^{n+1}\cos\upsilon\pi\int_{0}^{\infty}\frac{e^{-2t}t^{n-\upsilon}}{t + z}I_{-\upsilon}(t)dt$$

where I_{υ} is the other modified Bessel function. Inspection of (29) indicates that there is a similar integral here (with a different multiplying constant) so long as the identification

$$F_{1}(-t) = e^{-t}t^{\upsilon}I_{-\upsilon}(t), \qquad \rho = 2\upsilon, \quad \mu = n - 2\upsilon \tag{41}$$

is made. F_1, so defined, has the requisite regularity. Thus progression to (30) is permissible with the consequence that

$$I(f,z) \simeq \frac{e^z}{z^\upsilon} K_\upsilon(z) - e^{-z} \sum_{m=0}^{n-1} \frac{f^{(m)}(1)}{z^{m+1}}$$

$$+ \frac{\pi i}{z^\upsilon} \cos \upsilon\pi \, e^{2\upsilon\pi i+z} I_{-\upsilon}(z) \operatorname{erfc}\{ir_0/(2\mu)^{\frac{1}{2}}\}$$

$$- (-)^n \frac{\cos \upsilon\pi}{z^n} e^{\gamma-z} (2\pi\mu)^{\frac{1}{2}} \{ \frac{e^{-\mu}\mu^\upsilon I_{-\upsilon}(\mu)}{\mu + z} + \frac{e^z z^\upsilon I_{-\upsilon}(z)}{r_0}$$

$$+ \mu b_2 + 3\mu^2 b_4\} \tag{42}$$

the parameters throughout and in b_2, b_4 being as displayed in (41). Hence the analogue of (30) has been secured despite the conditions for its direct implementation not being met.

Evidently, the determination of the remainder entails the evaluation of several transcendental functions. In other words, the calculation of the correction can be more formidable than that of the original asymptotic expansion. One is reminded of the dictum, sometimes asserted in physics, that getting an extra decimal place demands 100 times the effort expended on the previous one. Fortunately, the multiplying factor is not so huge in our case but it is perforce appreciable.

To assess the numerical performance of (42) we choose $\upsilon = 1$. This is an interesting test case because $F(s)$ has a logarithmic branch point at the origin and so might cause failure of a formula which stems from an algebraic branch point when υ is not an integer. The modulus of z will be taken as 5 which might be regarded as on the margin for the validity of asymptotics. Consistency with our earlier hypotheses then necessitates $\mu = 5$, $n = 7$.

As well as (42) we will also consider the approximation to it analogous to (38); for ease of reference it will be denoted by (38) in the following tables. The first two parts of I, i.e. the uncorrected asymptotic expansion, will be designated as 'usual'. Actually, the usual approximation will be somewhat more accurate than the customary one in which an exponentially small recessive contribution is rejected in favour of the dominant series. Table 1 displays the results for the three values 5, $5e^{i\pi/4}$, 5i of z in comparison with the approximation (38). It shows that (38) is perfectly adequate for the optimal remainder, for this range of the phase of z; it improves the accuracy of the usual expansion by about 3-4 decimal places even though the usual expansion is already in pretty good agreement with the true answer.

At the other extreme is z = -5 where (42) must be employed. As can be seen from Table 2, the usual expansion is significantly less reliable here. Yet (42) brings it into line with agreement to 5 decimal places.

The example $z = 5e^{3\pi i/4}$ is in an intermediate position (see Table 3). At first glance one might think that (38) could be deployed here. Table 3 demonstrates that one must reconsider the matter. While (38) does do better than the usual expansion it is nowhere near as satisfactory as (42). The explanation lies with r_0 which is given by

$$i\left(\frac{r_0}{2\mu}\right)^{\frac{1}{2}} = 1.6816 - 1.1676i.$$

This is insufficiently large to justify the replacement of the erfc expected by (38) in spite of the phase of z being well away from π. The inference is that, as ph(z) grows from zero, (38) may be used with safety up to $\text{ph}(z) = \frac{1}{2}\pi$ but thereafter employed with caution and certainly not for $\frac{3}{4}\pi \leq \text{ph}(z) \leq \pi$. The most prudent action is, naturally, always to check r_0 before invoking (38).

Parallel results for $|z| = 3$ are exhibited in Table 4. Here the usual approximation is still not bad for $\text{Re}(z) \geq 0$ but is

TABLE 1

	$z = 5$	$z = 5i$
True	0.1174170940	-0.4059744 +0.0687402i
Usual	0.1174179218	-0.4060622 +0.06888277i
(38)	0.1174170942	-0.4059738 +0.0687393i

	$z = 5e^{i\pi/4}$
True	0.043457065 -0.121670475i
Usual	0.04345322 -0.121670560i
(38)	0.043457066 -0.121670474i

TABLE 2

	$z = -5$
True	44.144741
Usual	44.110691 +0.103027i
(42)	44.144743

TABLE 3

	$z = 5e^{3i\pi/4}$
True	-10.519861 -2.981191i
Usual	-10.52566 -2.98851i
(38)	-10.51825 -2.97985i
(42)	-10.519864 -2.981188i

TABLE 4

	$z = 3$	$z = 3i$	$z = 3e^{i\pi/4}$
True	0.2339473	-0.23545 -0.74607i	0.16316 -0.24080449i
Usual	0.2340640	-0.23381 -0.74338i	0.16344 -0.2408394i
(38)	0.2339476	-0.23553 -0.74616i	0.16314 -0.24080446i

	$z = -3$	$z = 3e^{3\pi i/4}$
True	8.7883	-0.02780 -4.0707i
Usual	8.6913 +0.2061i	-0.0557 -4.1008i
(42)	8.7879	-0.02764 -4.0709i
(38)		-0.0031 -4.0559i

less satisfactory for $Re(z) < 0$. When the optimal remainder is added, on the other hand, the agreement with the exact answer is much better over the whole range of ph(z). The dramatic effect of the uniform remainder when $z = 3e^{3i\pi/4}$ is especially noteworthy. Since 3 cannot be regarded as a particularly large number this is an illustration of how our correction can be used to bring down an asymptotic formula to values of $|z|$ that are lower than normal.

6. CONCLUSION

A method has been devised for estimating the optimal remainder in an asymptotic approximation which is uniform with respect to variations in ph(z). Numerical evidence indicates that it provides a significant amelioration of the accuracy of the approximation. Two consequences flow therefrom. One may either use the remainder to improve accuracy at fixed z or reduce $|z|$ while retaining a fixed level of accuracy. Moreover, a single formula covers the whole range of ph(z) so that different asymptotics are no longer necessary in different regions. The existence of an explicit formula also opens the door to analytic treatment that could be precluded by the absence of the remainder.

The calculation of the remainder is quite tolerable. For the integrals of the first section, which are of such frequent occurrence in asymptotics, it can be accomplished easily by a pocket calculator, at any rate in the form (23). The more complicated (42) is well within the powers of a microcomputer so long as the appropriate Laplace transform is available.

REFERENCE

1. F.W.J. Olver, Asymptotics and Special Functions, Academic Press (1974).

Period Tripling and Subharmonic Oscillations in Marangoni Flows in a Cylindrical Liquid Bridge

Nicholas D. Kazarinoff
Joseph S. Wilkowski Department of Mathematics, Manhattan College, Bronx, New York

SUMMARY

In float-zone refining, molten silicon (Si) is held by surface tension σ in a cylindrical bridge B liquified by radiation from a ring-heater midway between two solid sections of a Si crystal rod. The **grad** σ created by the **grad** θ (θ = temperature on the free surface S of B) generates Marangoni flow in B. Period tripling of periodic oscillations and subharmonics of these oscillations have been found in this flow in numerical solutions of the time-dependent Navier-Stokes-energy system in B with "free" lateral surface S maintained by the balance between σ and pressure p and with flat ends on the CRAY-X/MP-48 of NCSA at U. Illinois.

§1. INTRODUCTION.

Si is a low Prandtl number fluid (Pr = .023) as are other liquid semiconductors and metals. The low viscoscity μ = .0088 and Pr of liquid Si cause numerical simulation of an Si float-zone stiff, while Si's high θ_M = 1685°K and chemistrymake its experimental study difficult as well. Yet, crystal growers are concerned with the poor quality of zone-refined crystals.

Ciszek [1] presents pictures which show the imperfections which occur. As a crystal freezes, layers of impurity perpendicular to its axis are created. There also exist less visible layers parallel to S. About one half of the Si wafers cut from float-zone-refined crystals are unacceptable for industrial use.

Since a free surface S and flows driven by **grad** σ are essential in float-zone refining of Si, we modeled a time-dependent liquid bridge B with S "free", axially symmetric and "pinned" at the ends of B. We did not include gravity, rotation, a magnetic field, or oxides on S in our model. Previously, only Kozhoukharova and Slavchev [8] studied a non-right-circular cylinder B and solved for the steady-state (s-s) flow for both $NaNO_3$ and Si with S the curved meniscus of B. They found that, while the θ-field is distorted, the s-s flows are close to those found previously in right-circular-cylindrical float-zones; see also Cuvelier and Drieesen [2].

In our model bridge B, as the difference $\Delta\theta = \theta_H - \theta_M$ between the heater and the melting temperatures of Si increases, asymmetric, toroidal convective rolls develop around the axis of B, one in each half. There is a critical $\Delta\theta = \Delta\theta_C$, a function of the aspect ratio A (length L to radius R), above which significant periodic oscillations of S, θ and the velocity (u, v) of the fluid arise and increase in amplitude as $\Delta\theta$ increases. The convective rolls oscillate to and fro across the mid-plane of B perpendicular to the axis like a simple pendulum. For $\Delta\theta > \Delta\theta_C$ and large enough, the oscillations become highly irregular, and there are two rolls in each half-zone. Further, S is found to be concave (bowed slightly inward) for $\Delta\theta$ above $\Delta\theta_C$. Also tiny standing waves develop in S for $\Delta\theta$ just below $\Delta\theta_C$. As $\Delta\theta$ increases their amplitudes grow, and S deforms and oscillates, accomodating to the oscillations in the fluid Si (Fig.1 for A = 1/2 and $\Delta\theta = 11^oK$.). If for a fixed $\Delta\theta$ these oscillations are followed for long times, say, 3-6 minutes, then at A =1/2 and 2/3 one observes that, as the energy from the heater is diffused and convected into the interior of the zone, for $\Delta\theta$ fixed just above a certain $\Delta\theta_{CC}$ they become irregular and then become period-tripled waves with much larger amplitudes (Fig. 2 for A =1/2 and $\Delta\theta = 15^oK$.). As $\Delta\theta$ is further increased these waves give way to subharmonics and then become chaotic (Fig. 3 for A =2/3 and $\Delta\theta = 16^oK$. and Fig. 4 for A = 1/2 and $\Delta\theta = 18.5^oK$.) .

We describe our numerics in §2 and we present 'phase plane'

portraits of the oscillations for the liquid bridge B for $\Delta\theta \geq \Delta\theta_C$ in §3. In §4 we briefly discuss our results (for a disussion and comarison with others' experimental and numerical results, see [7]) and present our conclusions. Our advice to crystal growers on the basis of our simulations is to run their zone-refining at lower temperatures and smaller aspect ratios so as to reduce the intensity and complexity of the flows in the liquified zone.

§2. NUMERICS.

Since B is assumed to be axially symmetric, cylindrical coordinates (x, r) are a natural choice. Each computational cell in the two-dimensional geometry is a small rectangle. A staggered mesh is superimposed on it with grid points located as follows: for the radial velocity v, on the top (where the top is the side away from the axis of B) and the bottom; for the axial velocity u, on the left and right sides; for the pressure p and θ, at the center. The left and right ends of B, respectively, are the melting/freezing interfaces. They are assumed to be flat with $(u, v) = (0, 0)$ (no slip) and $\theta = \theta_M$ there. On the x-axis of symmetry, we set $\theta_r = p_r = v_r = 0$. On S, $r = h(x, t)$; h is unknown. Our numerical methods are a hybrid of finite difference and finite element techniques. We use an FTCS discretization inside B, the IMSL routines LEQT1B,2B to find p, and a cubic spline to approximate h(x, t).

We use the Navier-Stokes-energy (N-S-e) system in the "conservative form" developed by Harlow & Welch [6] and Welch *et al.* [11] which includes the continuity equation directly in the N-S-e-equations, factored into both the diffusive and advective terms. This has the proven advantage of reducing truncation errors that appear from using a stagggered mesh. In the Poisson equation for p we include a term which is adjusted in each cell to preserve mass (which is preserved to within 0.1%).

A distinct difference between our model and others is that we allow S to move and change shape. There are 4 boundary conditions (b.c.'s) that occur as a result : (1) normal forces on S balance (these include p, σ, κ (κ is twice the mean-curvature of S) and the stresses generated by viscosity); (2) tangential stresses on S balance (these include stresses generated by the viscosity of the moving fluid and shear stresses caused by **grad** σ); (3) the radiation-absorbtion balance on S; (4) the kinematic b. c. $h_t + uh_x = v$, which tracks the shape of S. (4) is a hyperbolic PDE, and it is numerically unstable under the FTCS scheme we use. We add a "flux-limiting" term to it (as suggested by Nichols and Hirt [9]) to make it parabolic. For the new PDE (see (4) below stability criteria $||u||_\infty \Delta x/2 \leq \delta \leq \Delta x/(2\Delta t)$ are found in the usual manner.

Thus, the first two b.c.'s on S express the tangential and normal force balances on S :

$$-(p - p_0) + (2\mu/N^2)[v_r + (h_x)^2 u_x - h_x(v_x + u_r)] = -\sigma\kappa. \quad (1)$$

$$\mu/N^2)[2h_x v_r - 2h_x u_x + (v_x + u_r)(1 - (h_x)^2)] = -\sigma_x/N, \quad (2)$$

where $p_0 = 0$ (by assumption), $N(x,t)$ is the length of the unit tangent vector at $S(x,t)$, and $\kappa(x,t) = (1/N)[(1/h(x,t)) - h_{xx}(x,t)/N^2]$; see Cuvelier and Driessen [2]. In (2) only the component σ_x of **grad** σ along S appears (we neglect σ_r). We assume on S that $\sigma = \sigma_0 + \gamma\Delta\theta$, where $\sigma_0 = \sigma$ at θ_M, and $\gamma = -d\sigma/d\theta$ is constant (-.43 dynes/cm./°K for liquid Si). From Duranceau and Brown [3], the radiation- absorption balance on S is:

$$\theta_A = \theta_M + \Delta\theta \exp[-(x - x_o)^2/d^2] \quad (\Delta\theta = \theta_H - \theta_M).$$

(3)

$$(k_m/N)[-h_x \theta_x + \theta_r] = -\sigma^*\varepsilon(\theta^4 - \theta_A^4) + \chi(\theta - \theta_A),$$

where, k_m is B's thermal conductivity, χ is the heat-transfer coefficient, ε is the emissivity ratio, and σ^* is the Stefan-Boltzmann constant. In each simulation we chose d so that at (0, R) and (L, R) $\theta_A = \theta_M + .01\Delta\theta$. Finally, the kinematic b.c. we actually use is

$$h_t = v - h_x u + \delta h_{xx}. \quad (4)$$

Naturally, the choice of δ quantitatively influences our results.

Our basic numerical scheme is a predictor-corrector variation of the standard FTCS method. In any given time step, $h(x, t)$ is determined first. Next, auxiliary values for (u, v) are found by simply advancing by one time-step Δt in the N-S equations. With the auxiliary (u, v) as predictors, and, adjusted to conserve mass in each computational cell, p is determined by IMSL's sparse system solvers LEQT1B,2B embedded in our code becuse calling subroutines is expensive of time on the CRAY-X/MP-48. This new p is then used to correct (u, v) (to conserve mass, which is obtained to within 0.1% of vol(B)). Finally, the new (u, v) is used to integrate the energy equation forward one time-step to get a new θ. Since the viscoscity of liquid Si is small, the diffusion terms are more important than the advective terms in this equation.

Our simulation is done in real time t with physical (u, v, p, h) instead of nondimensionalizing. We have achieved speeds of 1/2 hr CPU time/min. of simulation. Our reasons for using real time t are: to save cpu-time (in nondimensional t, approximately 900 times the number of iterations are needed to complete a run equivalent to 60 "real" seconds) and to have results ready for comparison with experimental ones. The initial conditions weuse are $u = v = \theta - \theta_M = 0$ in B, and we set θ_H instantaneously to the desired value, so as to introduce no bias from previous runs. The oscillations and asymmetry we found are duepurely to the interaction of S with the fluid at this θ_H and, perhaps, to numerical approximations.

§3. RESULTS.

In Fig. 1 untripled, small-amplitude periodic oscillations are shown as they develop (the spiral in Fig. 1a spirals out to the 'limit cycle'); in Fig. 1a u(.25,h(.25, t), t) is plotted against u(.5,h(.5, t), t), both at S, for a float-zone B with A = 1/2, R = 2 cm. and $\Delta\theta = 11^0$K.; and for the same zone and heater temperature the free surface position h(.25, t) is plotted against h(.5, t). In all the Figures the quantity given first in the legends at the top is plotted on the vertical axis, the second named quantity along the horizontal axis. The plot in Fig. 1a confirms that the motion of the principal convection rolls, seen in Fig. 1c for this zone, oscillate to and fro in the axial direction x like a simple pendulum. The plot in Fig. 1b shows that the oscillations of the free surface S are more complex, having two maxima and two minima in a simple period of u. The oscillations of u (and of the convective rolls) are amost symmetric about x = 0.5cm. The period of the oscillations is amost exactly 3.26 seconds. The time frame is from t = 20 sec. to t = 180 sec. in Fig.1a,b.

For A = .5 and $\Delta\theta = 16.5^0$K. the situation has changed; see Fig. 2a-d. In Fig. 2a u(.5,h(.25, t), t) is plotted against u at x = 0.5 on S for t = 20

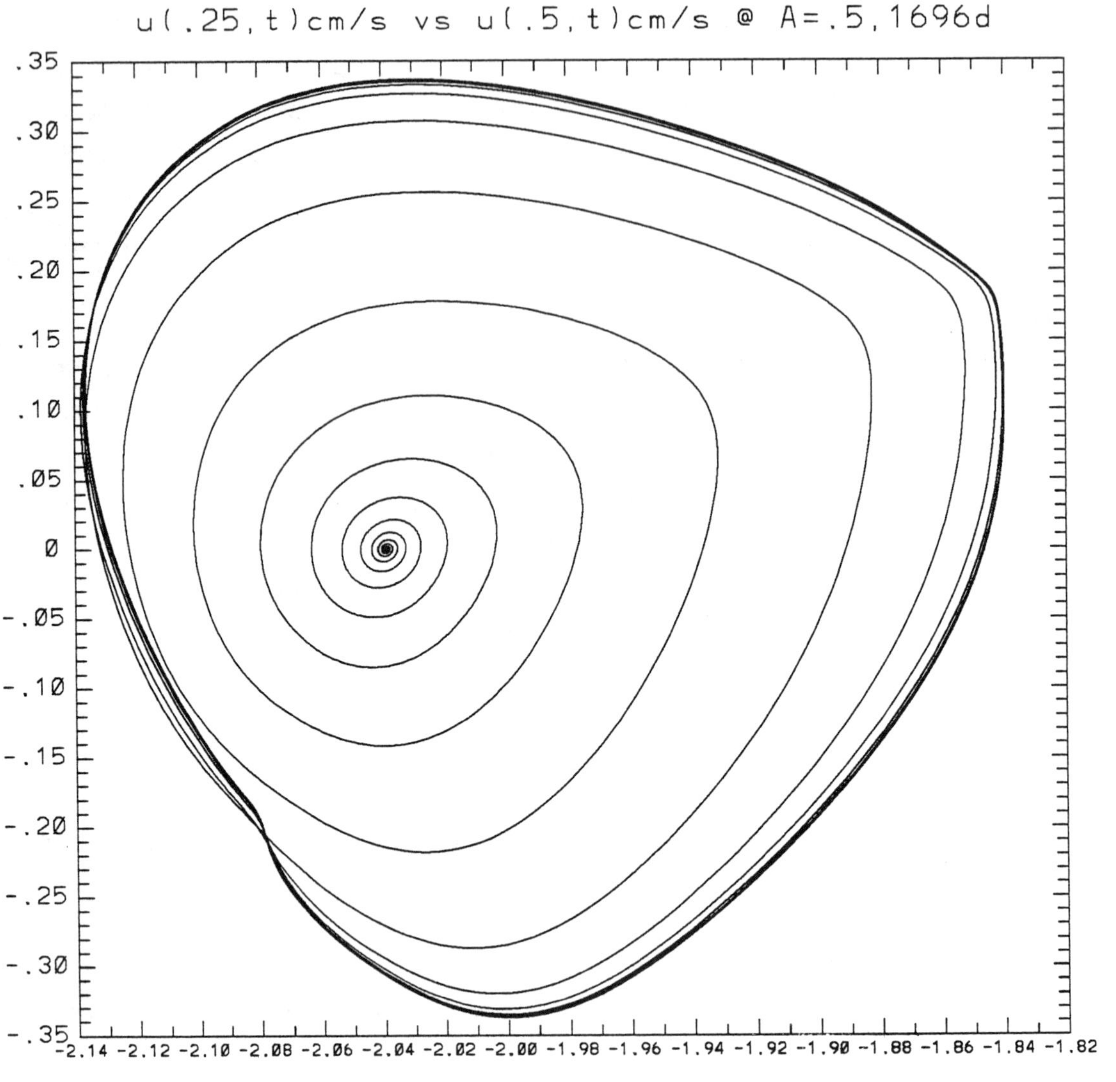
u(.25,t)cm/s vs u(.5,t)cm/s @ A=.5,1696d
.35
.3Ø
.25
.2Ø
.15
.1Ø
.Ø5
Ø
-.Ø5
-.1Ø
-.15
-.2Ø
-.25
-.3Ø
-.35
-2.14 -2.12 -2.1Ø -2.Ø8 -2.Ø6 -2.Ø4 -2.Ø2 -2.ØØ -1.98 -1.96 -1.94 -1.92 -1.9Ø -1.88 -1.86 -1.84 -1.82

Fig. 1a

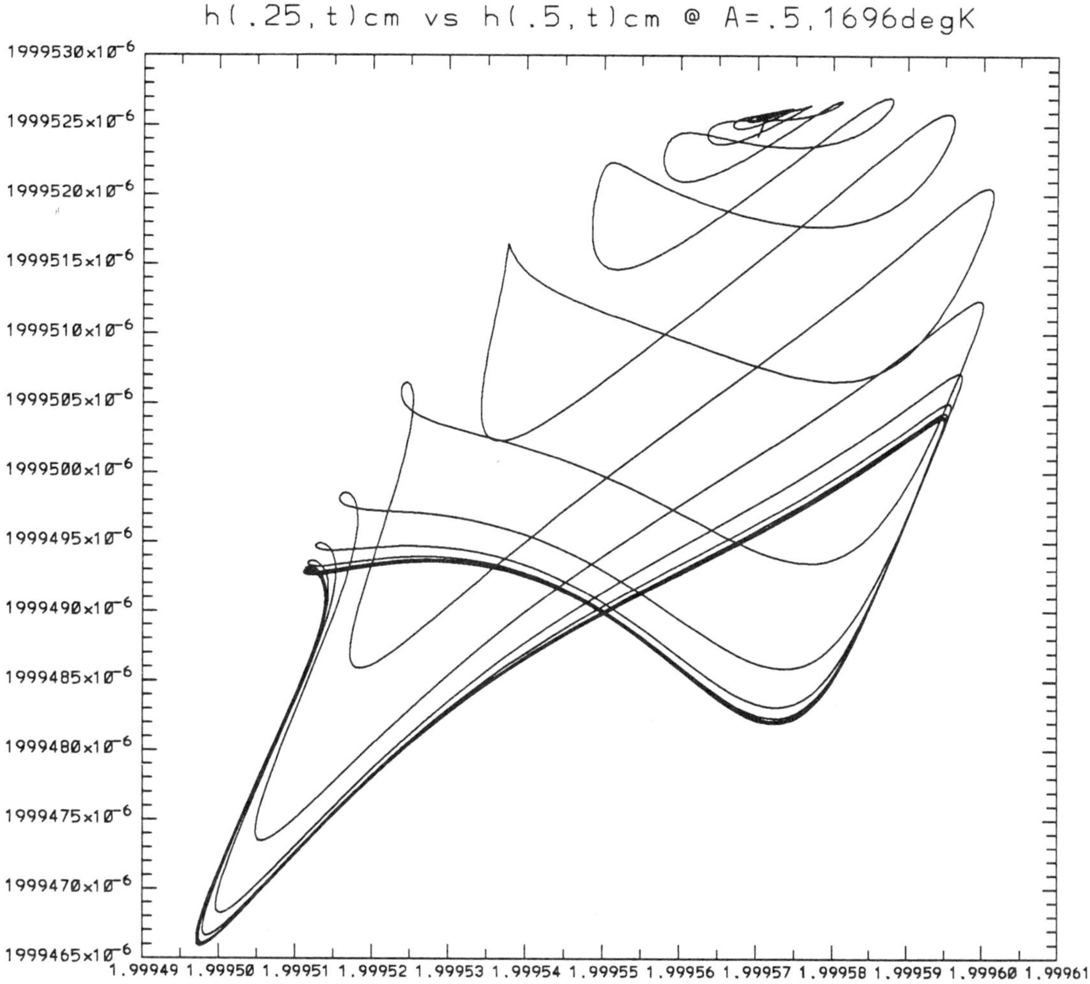
h(.25,t)cm vs h(.5,t)cm @ A=.5,1696degK
1999530x10^-6
1999525x10^-6
1999520x10^-6
1999515x10^-6
1999510x10^-6
1999505x10^-6
1999500x10^-6
1999495x10^-6
1999490x10^-6
1999485x10^-6
1999480x10^-6
1999475x10^-6
1999470x10^-6
1999465x10^-6
1.99949 1.99950 1.99951 1.99952 1.99953 1.99954 1.99955 1.99956 1.99957 1.99958 1.99959 1.99960 1.99961

Fig. 1b

Fig. 1c

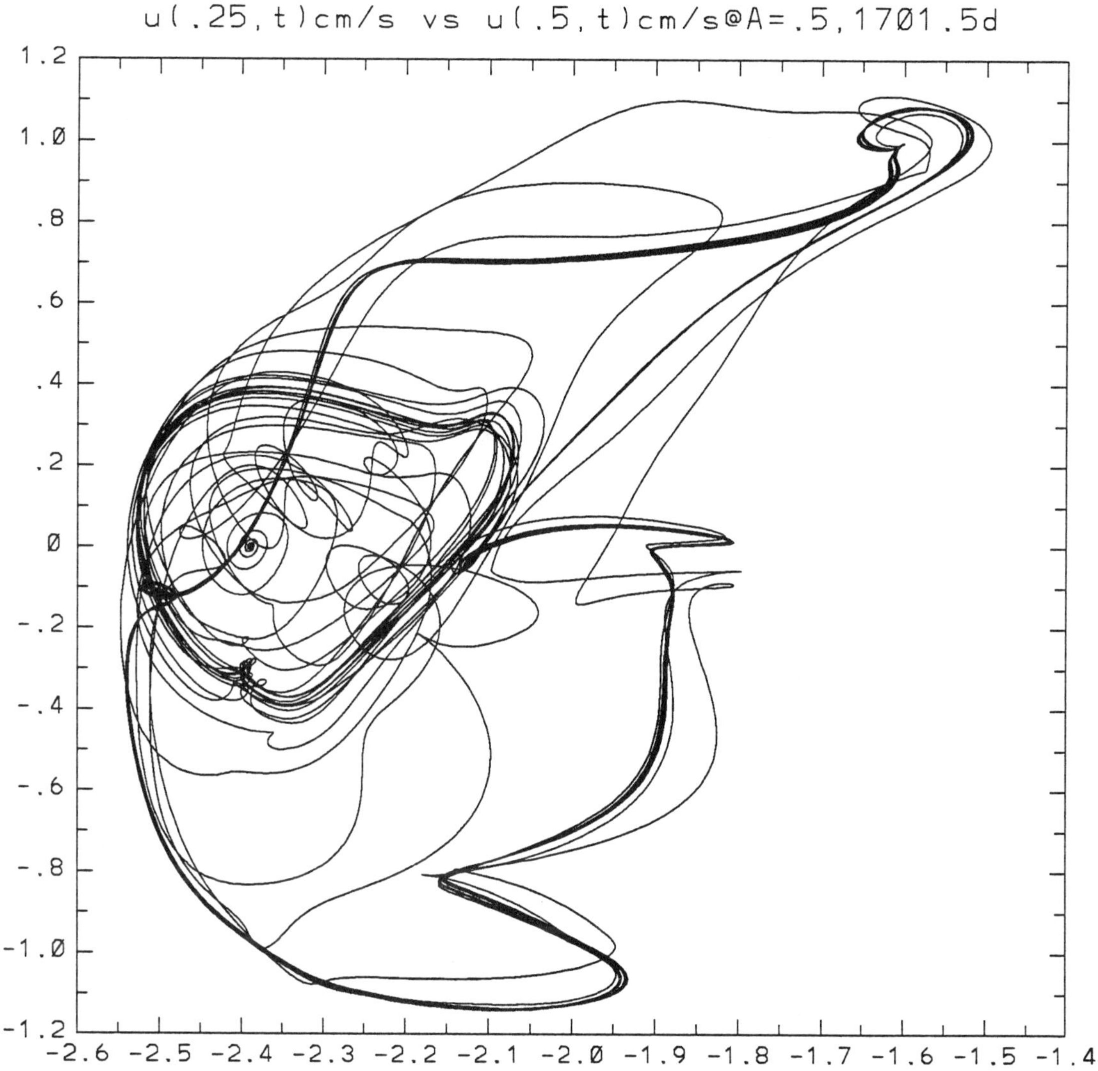
u(.25,t)cm/s vs u(.5,t)cm/s@A=.5,17Ø1.5d
1.2
1.Ø
.8
.6
.4
.2
Ø
-.2
-.4
-.6
-.8
-1.Ø
-1.2
-2.6 -2.5 -2.4 -2.3 -2.2 -2.1 -2.Ø -1.9 -1.8 -1.7 -1.6 -1.5 -1.4

Fig. 2a

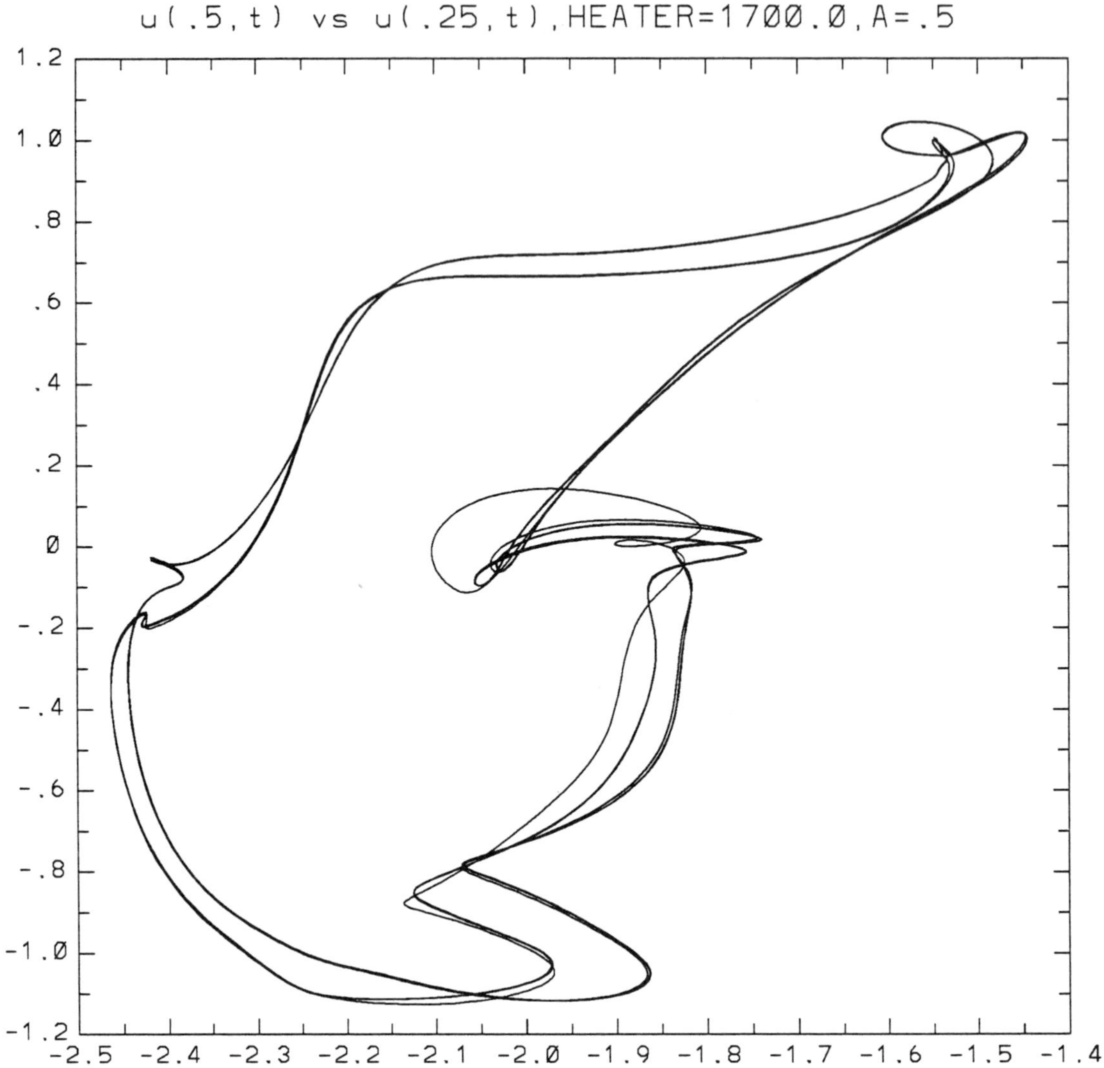
u(.5,t) vs u(.25,t),HEATER=1700.0,A=.5
1.2
1.0
.8
.6
.4
.2
0
-.2
-.4
-.6
-.8
-1.0
-1.2
-2.5 -2.4 -2.3 -2.2 -2.1 -2.0 -1.9 -1.8 -1.7 -1.6 -1.5 -1.4

Fig. 2b

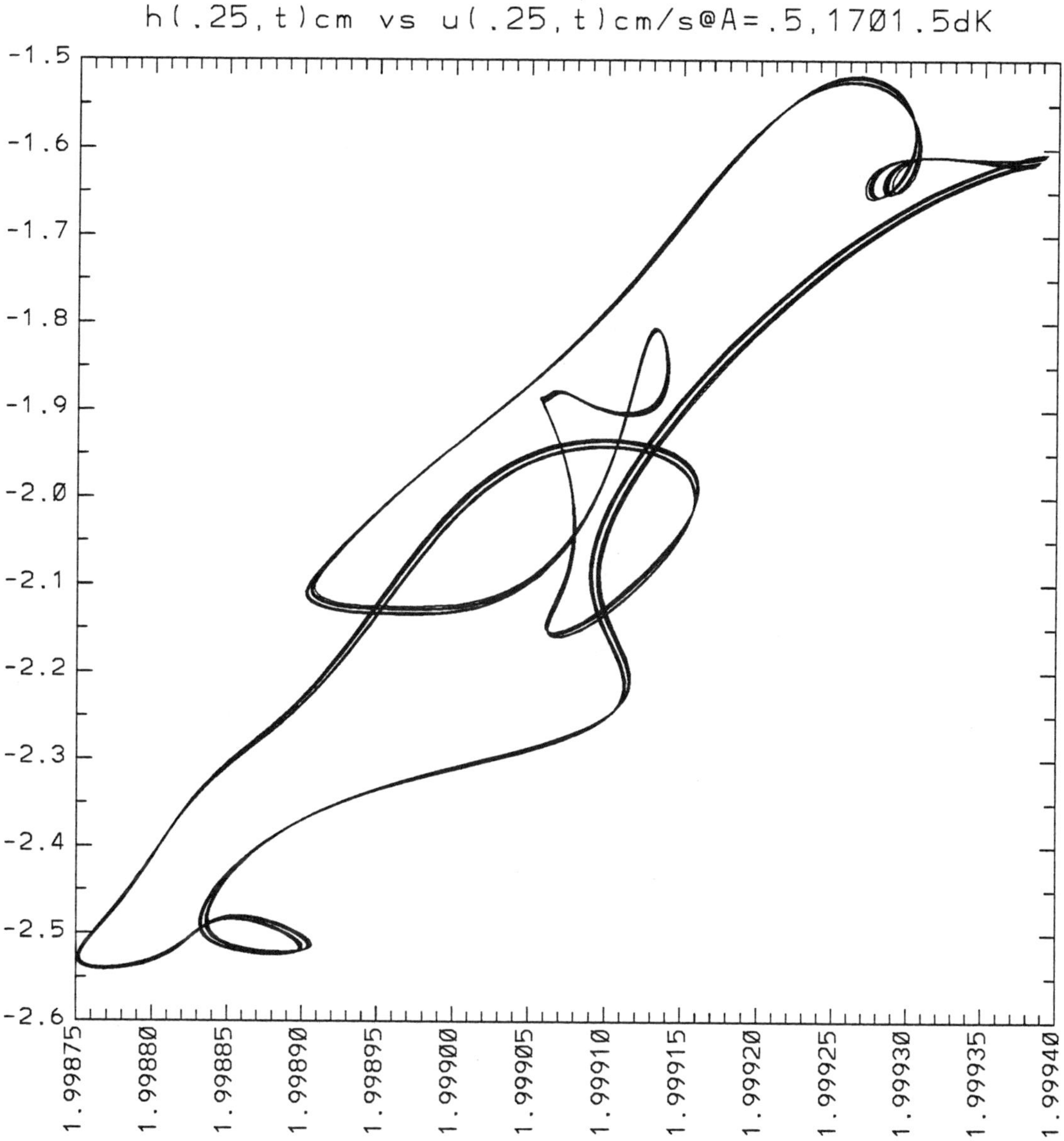
h(.25,t)cm vs u(.25,t)cm/s@A=.5,17Ø1.5dK
-1.5
-1.6
-1.7
-1.8
-1.9
-2.Ø
-2.1
-2.2
-2.3
-2.4
-2.5
-2.6
1.99875
1.9988Ø
1.99885
1.9989Ø
1.99895
1.999ØØ
1.999Ø5
1.9991Ø
1.99915
1.9992Ø
1.99925
1.9993Ø
1.99935
1.9994Ø

Fig. 2c

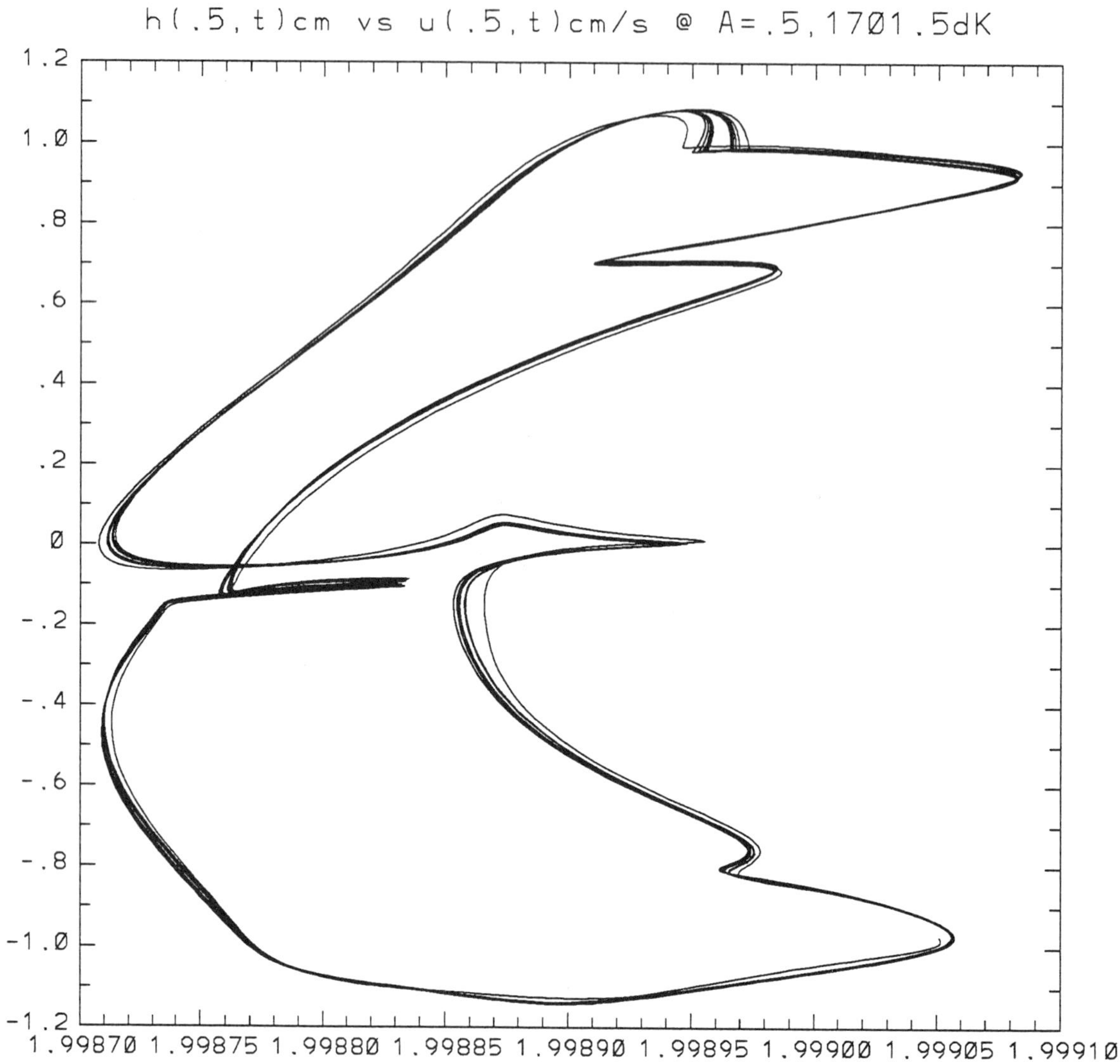
h(.5,t)cm vs u(.5,t)cm/s @ A=.5,1701.5dK
1.2
1.0
.8
.6
.4
.2
0
-.2
-.4
-.6
-.8
-1.0
-1.2
1.99870 1.99875 1.99880 1.99885 1.99890 1.99895 1.99900 1.99905 1.99910

Fig. 2d

sec. to 240 sec., and from 100 sec. to 240 sec. in Fig. 2b. Now the two and fro motion of the convective rolls is clearly not symmetric and the motion is more complex. The axial velocity u(.25,h(.25, t), t) has 8 maxima over one period of 2 x 9.85 ± 0.02 sec. (They are first subharmonics of a tripled oscillation.) The oscillations with one sixth the period occur at $\Delta\theta = 10.07^0$K. For A = 1/2, the period 3.14 sec. of the oscillations at $\Delta\theta = 12^0$K. is tripled at $\Delta\theta = 17^0$K. For these values of $\Delta\theta$ there are 4 convective rolls; two new very weak ones appear near the x-axis. In Fig. 2c u(.25, t) at S is plotted against h(.25, t) at S; in Fig. 2d the only change is x = 0.5 cm. (at the mid-circle of S) instead of x = 0.25 cm.

At A = 1/2 and $\Delta\theta = 18.5^0$K. the periodic oscillations have become rather chaotic; see Fig. 3 . The aspect ratio A strongly influences the intensity of the flows. For A = 1/2, the flow barely penetrates B away from S, and it is two-cellular. For A = 2/3, it penetrates more deeply and as $\Delta\theta$ increases the changes in the flow develop more rapidly.

Period quintupled oscillations (4th order subharmonics of period tripled waves) are shown in Fig. 4 a-c. for a B with A = 2/3 and $\Delta\theta = 16^0$K; the time t runs from 160 sec. to 240 sec. in Fig. 4. Their more compex structure is evident when one compares them to those shown in Fig. 2.

§4. CONCLUSIONS.

We believe our simulations *qualitatively* represent flows in the liquid bridges formed in float-zone refinement of Si crystals. The oscillations of (θ, u, v, p) in B's interior and on S have periods and amplitudes that vary with $\Delta\theta$ and A. Since gravity, rotation, etc. are excluded from our model, it is flexibility of S itself that enables oscillations to occur. We know of no other numerical simulations resulting in oscillations of S. Indeed, others have dismissed deformations of S as insignificant (Fu and Ostrach [4]). A sufficiently high θ_H is required for oscillations to occur. For Si, we have computed: $\Delta\theta_c = (3.7635 \pm .005)^o$K for A = 1, δ = .375, Δt = .001; while for A = 1, δ = .35, Δt = .0005, $\Delta\theta_c = (13.05 \pm .05)^o$K ; for A = 1, δ = .375, Δt = .0005, $\Delta\theta_c = (4.85 \pm .05)^o$K; and for A = 1.5, δ = 0.35, Δt = .0005, $\Delta\theta_c$ = $(4.45 \pm 0.05)^o$K. Suppressing oscillations in float-zones would produce better Si crystals. To suppress them and reduce their amplitudes if they do

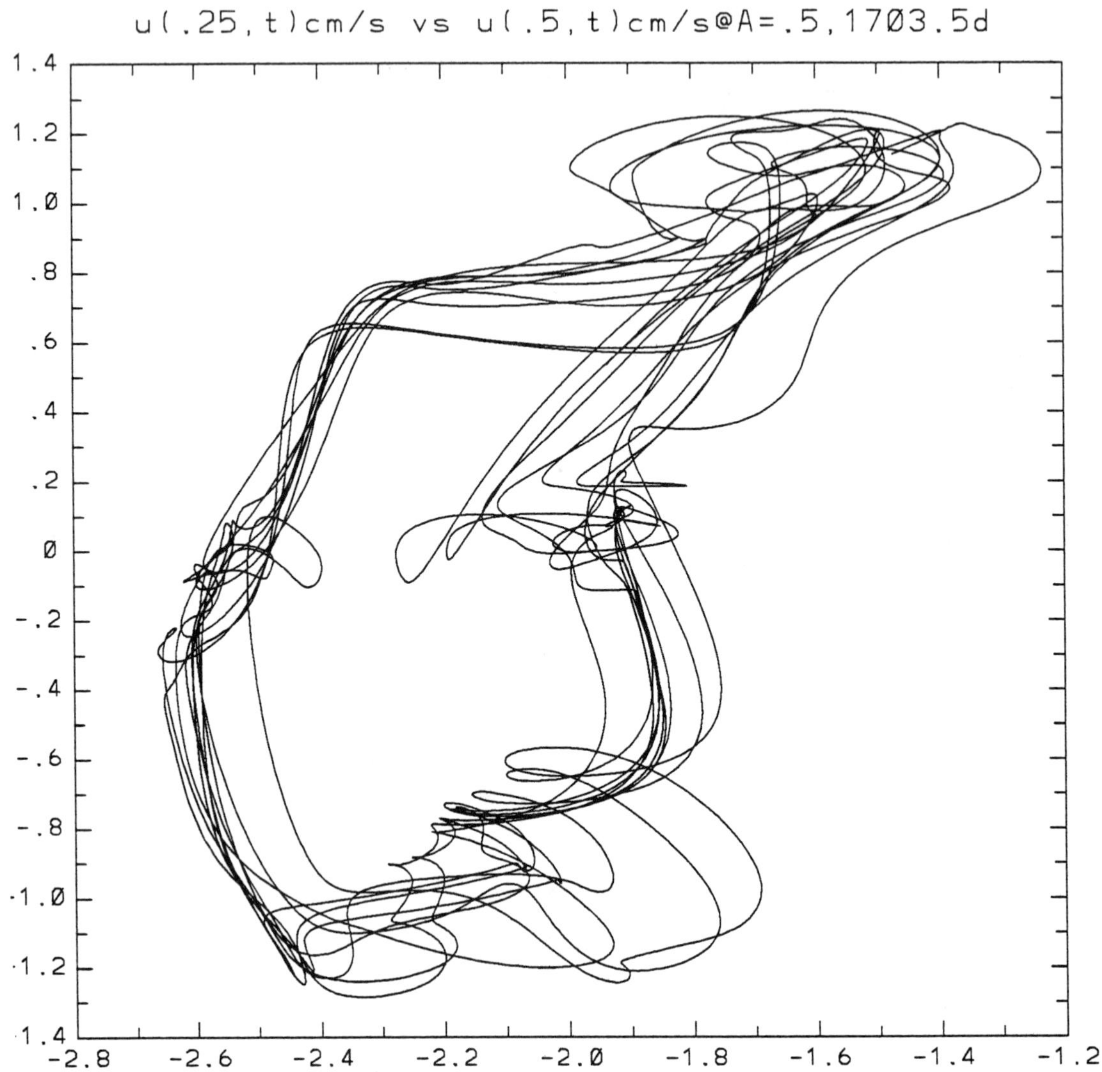
u(.25,t)cm/s vs u(.5,t)cm/s@A=.5,17Ø3.5d
1.4
1.2
1.Ø
.8
.6
.4
.2
Ø
-.2
-.4
-.6
-.8
-1.Ø
-1.2
-1.4
-2.8
-2.6
-2.4
-2.2
-2.Ø
-1.8
-1.6
-1.4
-1.2

Fig. 3a

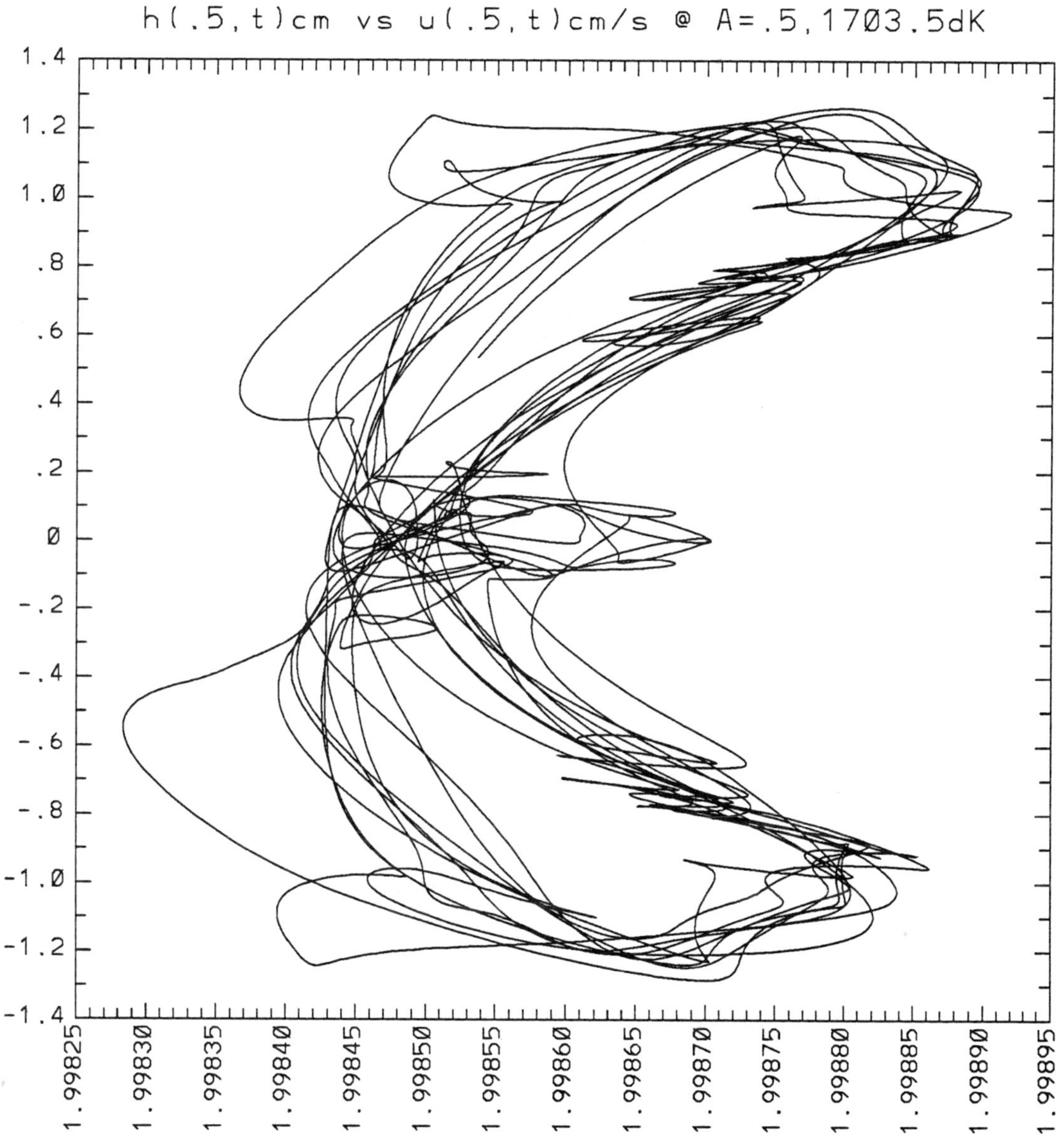
h(.5,t)cm vs u(.5,t)cm/s @ A=.5,1703.5dK
1.4
1.2
1.0
.8
.6
.4
.2
0
-.2
-.4
-.6
-.8
-1.0
-1.2
-1.4
1.99825
1.99830
1.99835
1.99840
1.99845
1.99850
1.99855
1.99860
1.99865
1.99870
1.99875
1.99880
1.99885
1.99890
1.99895

Fig. 3b

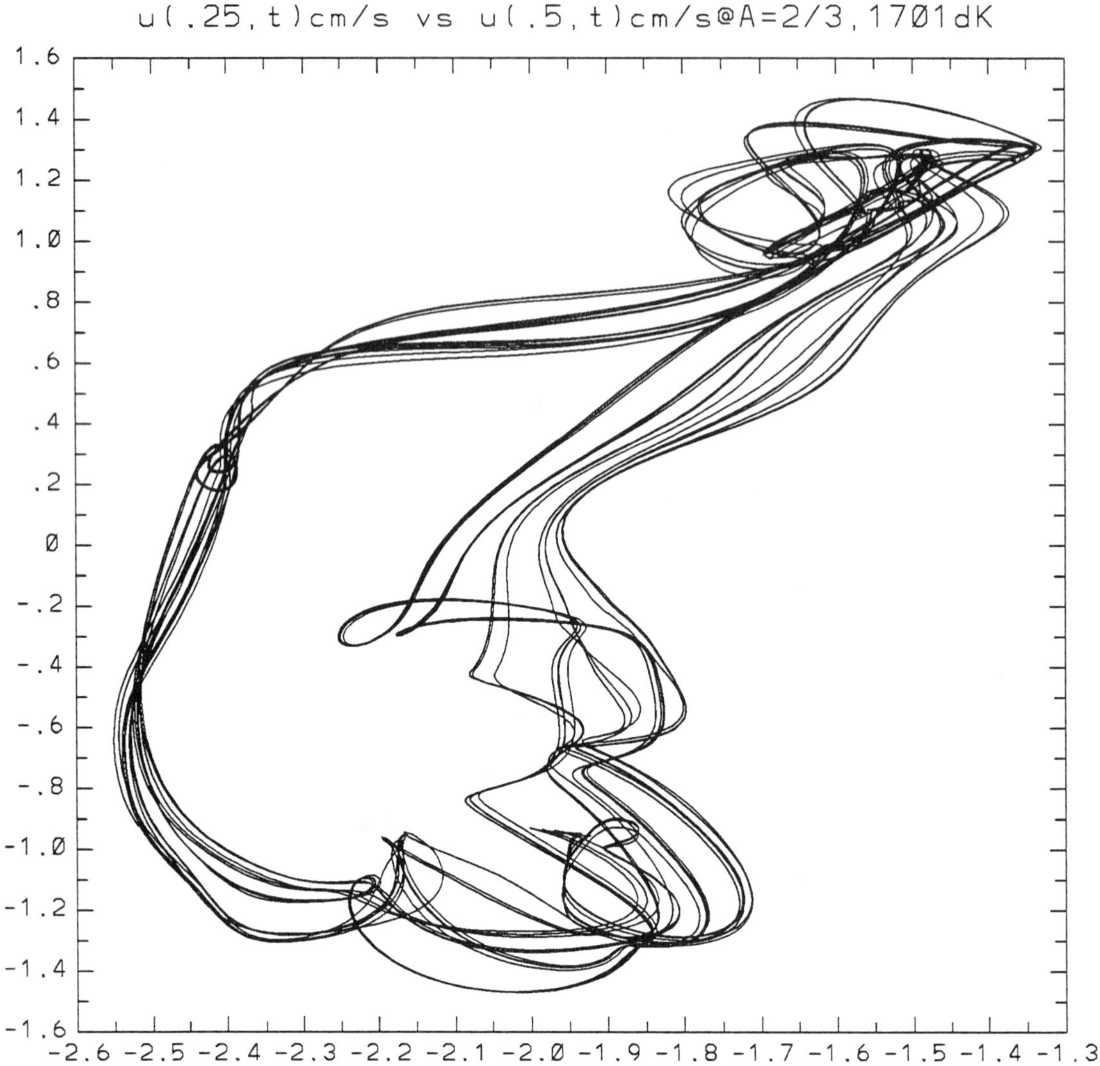
u(.25,t)cm/s vs u(.5,t)cm/s@A=2/3,17Ø1dK
1.6
1.4
1.2
1.Ø
.8
.6
.4
.2
Ø
-.2
-.4
-.6
-.8
-1.Ø
-1.2
-1.4
-1.6
-2.6 -2.5 -2.4 -2.3 -2.2 -2.1 -2.Ø -1.9 -1.8 -1.7 -1.6 -1.5 -1.4 -1.3

Fig. 4a

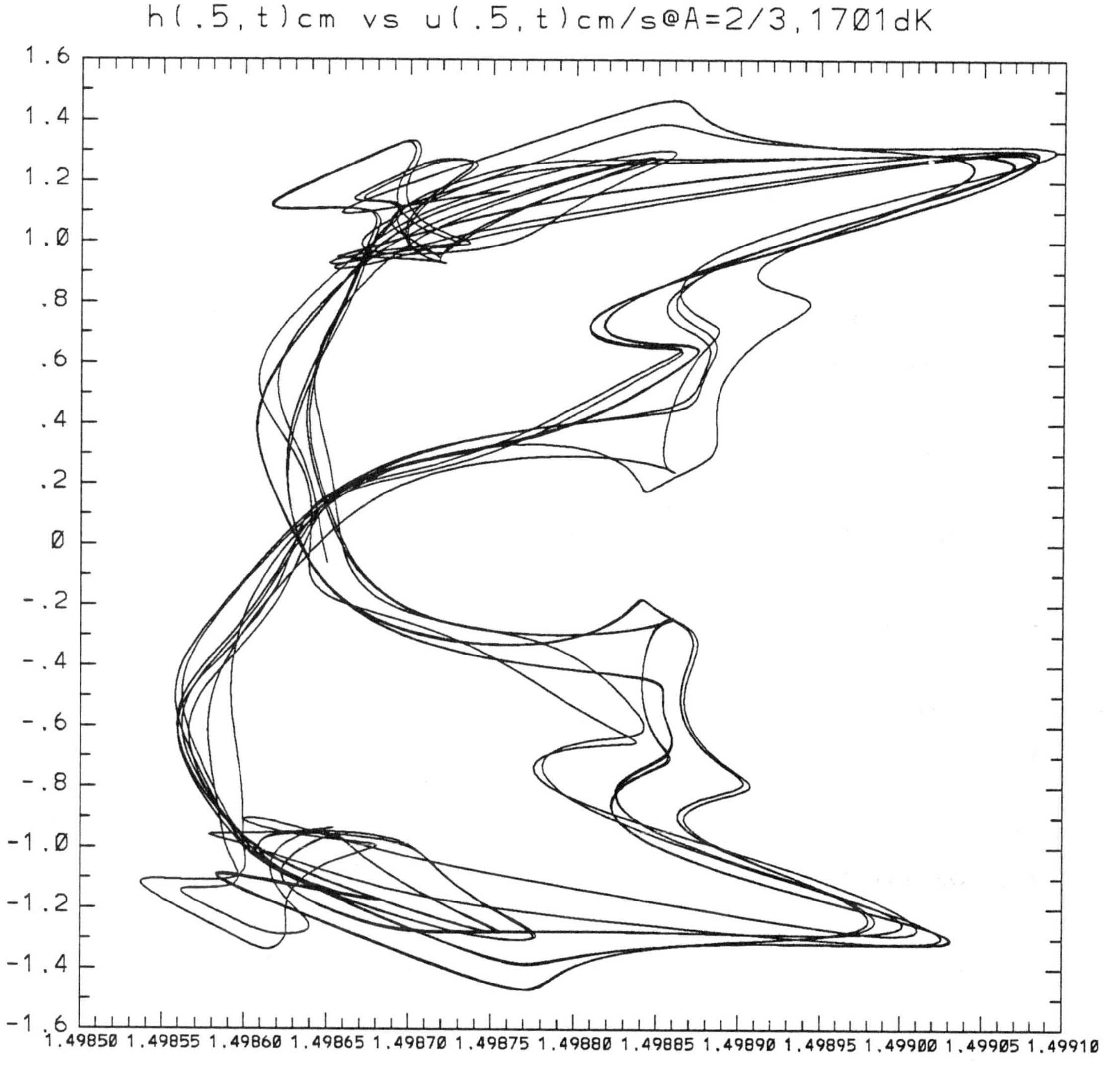
h(.5,t)cm vs u(.5,t)cm/s@A=2/3,17Ø1dK
1.6
1.4
1.2
1.Ø
.8
.6
.4
.2
Ø
-.2
-.4
-.6
-.8
-1.Ø
-1.2
-1.4
-1.6
1.4985Ø 1.49855 1.4986Ø 1.49865 1.4987Ø 1.49875 1.4988Ø 1.49885 1.4989Ø 1.49895 1.499ØØ 1.499Ø5 1.4991Ø

Fig. 4b

occur, A and θ_H should be kept as small as is feasible. The results of our numerical simulations agree qualitatively with the experimental results of Preisser, Schwabe and Scharman [10] for Marangoni flow in fluids wih Pr's = 1.0 -10..0. The only other example of period-tripling in fluid mechanics that we know of was reported recently by Hamakiotes and Berger [5] as a result of their numerical study of flows in a curved pipe (the flow of blood through the aortic arch).

We have experimented with smaller values of Δt (.001, .0005, .00025) and combinations of values for Δx and Δr, including .1, .05 and .025. There are a slight quantitative effects on the results caused by changes in these parameters; changes in Δt have the greatest effect, but the qualitative nature of our results is preserved. In particular, oscillations have occurred for whatever values we chose for these parameters.

ACKNOWLEDGEMENTS

We express warm thanks to: the National Center for Supercomputing and Applications at the U. of I., Champaign, Ill. for Grant DMS890006N on their CRAY X-MP/48; Mr R. Kufrin consultant at NCSA and Reb Carter of SUNY/B for invaluable assistance; the staff of NCSA for their generous help and hospitality; Profs. C. Doering, Clarkson U., Potsdam N. Y. , R. Krasny, U. of Mich., Ann Arbor, MI. & D. Pitman U. Minn, Minneapolis, MIN. for helpful conversations. We thank Prof. Roderick Wong for the oppurtunity to present these results at this meeting in honor of our distinguished friend of many years, Professor Frank Olver.

REFERENCES

1. Cizek, T. F., *J. Crystal Growth* **10**(1971), 263-268.
2. Cuvelier, C. & Driessen, J. M., *J. Fluid Mech.* **169**(1986), 1-26.
3. Duranceau, J. & Brown, R., *J. Crystal Growth* **75**(1986), 367-389.
4. Fu, B.-I. & Ostrach, S., Rept.7 TAS/TR 82-169, Case-Western Reserve U., 1982.
5. Hamakiotes, C. C. and Berger, S. A., *Physical Review Letters* **62**(1989), No.11, 1270-1273.

6. Harlow, F. H. & Welch, J. E., *Physics of Fluids* **8** (1965), 2182-2189.

7. Kazarinoff, N. D. and Wilkowski, J. S., *Physics of Fluids A* (Letters) **1**(1989), 625-627.

8. Kozhoukharova, Z. & Slavchev, S., *J. Crystal Growth* **74**(1986), 236-246.

9. Nichols, B. D. & Hirt, C. W., *J. Comp. Phy.* **12**(1973), 234- 246.

10. Preisser, F., Schwabe, D. and Scharman,, A. J., *J. Fluid Mech.* **126** (1983), 545-567.

11. Welch, J. E., Harlow, F. H., Shannon, J. P. and Daly, B. J., Los Alamos Sci. Lab. Rept. LA-3425, 1966.

Positive Solutions for Degenerate and Nondegenerate Elliptic Systems: Existence and Numerical Approximations

Anthony W. Leung Professor, Department of Mathematical Sciences, University of Cincinnati, Cincinnati, Ohio

Guangwei Fan Professor, Department of Mathematical Sciences, University of Cincinnati, Ohio

1. INTRODUCTION

This article is concerned with positive solutions for degenerate and nondegenerate elliptic systems of the form

$$\Delta\Psi(w_i) + f_i(x, w_1, w_2) = 0 \quad \text{in } D, \ 1 = 1,\ 2$$

with homogeneous Dirichlet conditions $w_i = 0$ in D. The function $\Psi(s)$ satisfies the conditions: $\Psi \in C^1[0,\infty)$, $\Psi(0) = 0$ and $\Psi'(s) > 0$ for $s > 0$. Problems of this nature are of interest in reaction-diffusion processes in biology and chemistry. For example, the case for $\Psi(u) = u^m$, $m > 1$ or $m \in (0,1)$ for single parabolic equations (i.e. $u_t = \Delta u^m + f(x,u)$) has been studied recently for porous medium analysis and population dynamics (cf. [2], [5], [10], [14] and [15]). As $t \to \infty$ these solutions tend to solutions of the corresponding elliptic scalar equation. Our hypothesis on f is different from those in the papers mentioned above. For example, f is not necessarily Lipschitz in u and may depend discontinuously on x. Moreover, our emphasis here are on the case of degenerate systems and numerical approximation.

In section 2, we discuss some existence and uniqueness theorems for scalar equations, which will be used in section 4 for constructing the monotone sequences. In section 3, we use the Schauder's fixed point theorem to find a positive solution in $W^{2,p}(D) \cap W_0^{1,p}(D)$ for the system between appropriate upper and lower solutions. Application to ecological prey-predator models are studied. In section 4, we construct monotone sequences of functions which will converge to upper and lower bounds for all positive solutions of the system. In case a related system has a unique positive solution, then theses sequences will converge to the solution of the system. This method is numerically adaptable to finite difference approximation equations.

2. EXISTENCE AND UNIQUENESS FOR SCALAR EQUATIONS WITH SPATIAL DISCONTINUITIES

Before our discussion for systems of equations, we first study the existence and uniqueness problems for scalar equations of the following type:

$$\Delta\Psi(w) + f(x,w) = 0 \quad \text{in } D, \tag{2.1a}$$

$$w = 0 \quad \text{on } \partial D. \tag{2.1b}$$

Here D is a bounded connected domain in $R^N (N \geq 2)$ with boundary $\partial D \in C^2$, $\Delta = \sum_{i=1}^N \partial^2/\partial x_i^2$ is the Laplace operator. The functions $\Psi: [0,\infty) \to [0,\infty)$, $f: D\times[0,\infty) \to R^1$ are assumed to satisfy the following hypotheses:

[H1] $\Psi \in C^1[0,\infty)$, $\Psi(0) = 0$, and $\Psi'(s) > 0$ for $s > 0$.

[H2] There is a bounded interval $[0,b]$ such that

i) $f \in L^\infty(D\times[0,b])$;

ii) for any fixed $x \in D$ a.e., the function $f(x,y)$ is continuous in y for all $y \in [0,b]$;

iii) There is a constant $M > 0$ such that $f(x,y_2) - f(x,y_1) \geq -M(\Psi(y_2)-\Psi(y_1))$ for $x \in D$ a.e., $0 \leq y_1 \leq$

$y_2 \le b$.

[H3] For each fixed $x \in D$ a.e., the function $f(x,y)/\Psi(y)$ is a strictly monotonic increasing or decreasing function in y for $y \in (0,b]$.

The hypotheses and results in this section will lead to insights for the study of systems. They will also be needed for the construction of monotone approximating sequences in section 4. To fix ideas, we give the following definitions.

Definition 2.1. Let u be an integrable function in D and α be any multi-index, then a locally integrable function v in D is called the αth weak derivative of u if it satisfies

$$\int_D v\phi dx = (-1)^{|\alpha|}\int_D uD^\alpha\phi dx, \qquad \text{for all } \phi \in C_0^{|\alpha|}(D).$$

We write $v = D^\alpha u$ and note that $D^\alpha u$ is uniquely determined up to sets of measure zero.

Definition 2.2. Let k be a nonnegative integer and $1 \le p < \infty$, the space $W^{k,p}(D)$ consists of all functions in the real space $L^p(D)$ whose weak derivatives of all orders $\le k$ exist and belong to $L^p(D)$. The space $W^{k,p}(D)$ is normed by

$$\|u\|_{k,p} = \{\Sigma_{|\alpha|\le k}\int_D |D^\alpha u(x)|^p dx\}^{1/p}.$$

Denote by $W_0^{k,p}(D)$ the completion in the space $W^{k,p}(D)$ of the subset $C_0^\infty(D)$. It is well known that both $W^{k,p}(D)$ and $W_0^{k,p}(D)$ are Banach space.

Definition 2.3. A function $w \in C(\bar{D})$ is called a nonnegative solution of (2.1) if $w(x) \ge 0$ in $\bar{D}$, and $\Psi(w) \in W^{2,p}(D) \cap W_0^{1,p}(D)$ $(p > N)$ satisfies (2.1), where the derivatives are taken in the weak sense. If, in addition, $w(x) > 0$ in D, then w is called a positive solution of (2.1).

We first state an existence result for nonnegative solution between weak upper and lower solution in the sense of (2.2) below.

Lemma 2.1. Suppose that [H1], [H2] are satisfied. Assume that there are functions $\underline{w}$, $\overline{w}$ in $C(\overline{D})$ with $0 \le \underline{w} \le \overline{w} \le b$ in $\overline{D}$, and $\Psi(\underline{w})$, $\Psi(\overline{w})$ are in $W^{1,p}(D)$ $(p > N)$ satisfying the inequalities:

$$-\int_D \nabla\Psi(\underline{w})\nabla\phi dx + \int_D f(x,\underline{w})\phi dx \ge 0, \quad \underline{w} = 0 \text{ on } \partial D, \tag{2.2a}$$

$$-\int_D \nabla\Psi(\overline{w})\nabla\phi dx + \int_D f(x,\overline{w})\phi dx \le 0, \tag{2.2b}$$

for all $\phi \in C_0^1(D)$, $\phi \ge 0$. Then there exists at least one nonnegative solution w of (2.1) satisfying $\underline{w} \le w \le \overline{w}$ in $\overline{D}$.

(This is a special case of Theorem 3.1 below, the proof is therefore omitted).

We now define another sense of upper and lower solutions which will be used in section 4.

Definition 2.4. A pair of functions $\underline{w}$, $\overline{w}$ in $C(\overline{D})$, with $\Psi(\underline{w})$, $\Psi(\overline{w})$ in $W^{2,p}(D)$ $(p > N)$, are respectively called strong, positive lower and upper solutions of (2.1) if they satisfy the inequalities:

$$\Delta\Psi(\underline{w}) + f(x,\underline{w}) \ge 0, \quad \text{a.e. in } D, \tag{2.3a}$$

$$\Delta\Psi(\overline{w}) + f(x,\overline{w}) \le 0, \quad \text{a.e. in } D \tag{2.3b}$$

$$0 \le \underline{w} \le \overline{w} \le b \text{ in } \overline{D}, \ 0 < \underline{w} \text{ in } D.$$

where the derivatives are taken in the weak sense.

Lemma 2.2. Suppose that [H1], [H2] are satisfied. Assume that there are strong, positive lower and upper solutions $\underline{w}$ and $\overline{w}$ of (2.1). Then there exists at least one positive solution w of (2.1) satisfying $\underline{w} \le w \le \overline{w}$ in $\overline{D}$.

(This is an immediate result of Lemma 2.1 since (2.3) implies (2.2).)

With the addition of hypothesis [H3], we have a uniqueness result.

Theorem 2.1. Assume all the hypotheses in Lemma 2.1. In addition, suppose [H3] is valid and $\underline{w} > 0$ in D. Then there exists a unique positive solution w^* of (2.1) satisfying $0 < \underline{w} \le w^* \le \overline{w}$ in D. (See [9] for proof.)

Corollary 2.1. Assume all the hypotheses in Lemma 2.2. In addition, suppose [H3] is valid. Then there exists a unique positive solution w^* of (2.1) satisfying $0 < \underline{w} \le w^* \le \overline{w}$ in D.

3. SYSTEMS AND ECOLOGICAL MODELS

In this section, we will study the existence of positive solutions for elliptic systems of the following type;

$$\begin{aligned} &\Delta\Psi(w_1) + f_1(x,w_1,w_2) = 0 \quad \text{a.e. in } D,\\ &\Delta\Psi(w_2) + f_2(x,w_1,w_2) = 0 \quad \text{a.e. in } D, \qquad (3.1)\\ &w_1 = w_2 = 0 \quad \text{on } \partial D \end{aligned}$$

where the derivatives are taken in the weak sense, and D is a bounded domain in R^N $(N \ge 2)$ with boundary $\partial D \in C^2$; and Ψ: $[0,\infty) \to [0,\infty)$, f_i: $D\times[0,\infty)\times[0,\infty) \to R^1$ are functions satisfying the following assumptions:

$[\hat{H}1]$ $\Psi \in C^1[0,\infty)$, $\Psi(0) = 0$ and $\Psi(s) > 0$ for $s > 0$.

$[\hat{H}2]$ There are two positive constants b_1 and b_2 such that

i) $f_i \in L^\infty(D\times[0,b_1]\times[0,b_2])$ for $i = 1, 2$;

ii) for fixed $x \in D$ a.e. the functions $f_i(x,y_1,y_2)$ are continuous in (y_1,y_2) for all $(y_1,y_2) \in [0,b_1]\times[0,b_2]$, $i = 1, 2$;

iii) there is a constant $M > 0$ such that

$f_1(x,\xi,y_2) - f_1(x,\eta,y_2) \ge -M(\Psi(\xi)-\Psi(\eta))$ for $x \in D$ a.e. $0 \le \eta \le \xi \le b_1$, $y_2 \in [0,b_2]$;

$f_2(x,y_1,\xi) - f_2(x,y_1,\eta) \ge -M(\Psi(\xi)-\Psi(\eta))$ for $x \in D$ a.e. $0 \le \eta \le \xi \le b_2$, $y_1 \in [0,b_1]$.

Definition 3.1. A pair of continuous functions (w_1,w_2) in $C(\overline{D})$ is called a positive solution of (3.1) if $w_i > 0$ in D, $\Psi(w_i) \in W^{2,p}(D) \cap W_0^{1,p}(D)$ $(p > N)$ and (3.1) holds.

Theorem 3.1. Assume hypotheses $[\hat{H}1]$ and $[\hat{H}2]$. Suppose that there are functions $\underline{w}_i(x)$, $\overline{w}_i(x)$ (i=1,2) in $C(\overline{D})$ with $\underline{w}_i > 0$ in D, $\underline{w}_i = 0$ on ∂D, $\Psi(\underline{w}_i)$, $\Psi(\overline{w}_i)$ in $W^{1,p}(D)$ $(p > N)$ satisfying the inequalities:

$$-\int_D \nabla\Psi(\underline{w}_1)\nabla\phi dx + \int_D f_1(x,\underline{w}_1,w_2)\phi dx \geq 0, \quad \text{for } \underline{w}_2 \leq w_2 \leq \overline{w}_2 \tag{3.2a}$$

$$-\int_D \nabla\Psi(\overline{w}_1)\nabla\phi dx + \int_D f_1(x,\overline{w}_1,w_2)\phi dx \leq 0, \quad \text{for } \underline{w}_2 \leq w_2 \leq \overline{w}_2 \tag{3.2b}$$

$$-\int_D \nabla\Psi(\underline{w}_2)\nabla\phi dx + \int_D f_2(x,w_1,\underline{w}_2)\phi dx \geq 0, \quad \text{for } \underline{w}_1 \leq w_1 \leq \overline{w}_1 \tag{3.2c}$$

$$-\int_D \nabla\Psi(\overline{w}_2)\nabla\phi dx + \int_D f_2(x,w_1,\overline{w}_2)\phi dx \leq 0, \quad \text{for } \underline{w}_1 \leq w_1 \leq \overline{w}_1 \tag{3.2d}$$

for all $\phi \in C_0^1(D)$, $\phi \geq 0$. Here $w_i = w_i(x)$ are assumed to be continuous in $\overline{D}$, and $0 \leq \underline{w}_i \leq w_i \leq \overline{w}_i \leq b_i$ in $\overline{D}$. Then there exists at least one positive solution $(w_1^* . w_2^*)$ of (3.1) satisfying $\underline{w}_i \leq w_i \leq \overline{w}_i$ in $\overline{D}$.

Proof. The complete proof is given in [9]; and we only outline the proof here. Let $\underline{u}_i = \Psi(\underline{w}_i)$, $\overline{u}_i = \Psi(\overline{w}_i)$, $X_i = \{u \in C(\overline{D}) | \underline{u}_i \leq u_i \leq \overline{u}_i$ in $\overline{D}\}$, i = 1, 2; and let M be described as in $[\hat{H}2, \text{iii}]$. The set $X_1 \times X_2$ is a bounded closed convex set in the Banach space $C(\overline{D}) \times C(\overline{D})$. We define the map T: $X_1 \times X_2 \to X_1 \times X_2$ as follows:

$$T(u_1,u_2) = (v_1,v_2) \text{ for } (u_1,u_2) \in X_1 \times X_2$$

where $v_1, v_2 \in W^{2,p}(D) \cap W_0^{1,p}(D) \subset C(\overline{D})$ $(p > N)$ and (v_1,v_2) is uniquely determined as the solution of the (decoupled) system:

$$\Delta v_i - Mv_i + f_i(x,\Psi^{-1}(u_1),\Psi^{-1}(u_2)) + Mu_i = 0 \text{ in D, } i = 1, 2. \tag{3.3}$$

(Here the derivatives are taken in the weak sense). We can show by means of $[\hat{H}1]$, $[\hat{H}2]$, (3.2) and the weak maximum principle that T is a continuous compact operator from $X_1 \times X_2$ into itself. Schauder's fixed point theorem asserts that T has a fixed point (u_1^*, u_2^*) in $X_1 \times X_2$. It follows from (3.3) that

$$\begin{aligned} &\Delta u_1^* + f_1(x, \Psi^{-1}(u_1^*), \Psi^{-1}(u_2^*)) = 0 \quad \text{a.e. in } D, \\ &\Delta u_2^* + f_2(x, \Psi^{-1}(u_2^*), \Psi^{-1}(u_2^*)) = 0 \quad \text{a.e. in } D, \\ &u_1^* = u_2^* = 0 \quad \text{on } \partial D. \end{aligned} \tag{3.4}$$

The fact that (u_1^*, u_2^*) is in $X_1 \times X_2$ implies that (u_1^*, u_2^*) is in $W^{2,p}(D) \cap W_0^{1,p}(D)$ and $\underline{u}_i \le u_i^* \le \overline{u}_i$ in $\overline{D}$ for i = 1, 2. Consequently, $(w_1^*, w_2^*) = (\Psi^{-1}(u_1^*), \Psi^{-1}(u_2^*))$ is a positive solution of (3.1) with $\underline{w}_i \le w_i^* \le \overline{w}_i$ in $\overline{D}$.

Definition 3.2. The functions $\underline{w}_i(x)$, $\overline{w}_i(x)$ (i = 1, 2) in $C(\overline{D})$ with $\underline{w}_i(x) > 0$ in D, $\underline{w}_i = 0$ on ∂D, $\Psi(\underline{w}_i)$, $\Psi(\overline{w}_i)$ (i = 1, 2) in $W^{2,p}(D)$ (p > N) are called strong, positive lower and upper solutions of (3.1) if they satisfy the inequalities:

$$\Delta\Psi(\underline{w}_1) + f_1(x, \underline{w}_1, w_2) \ge 0 \quad \text{a.e. in } D \text{ for } \underline{w}_2 \le w_2 \le \overline{w}_2 \tag{3.5a}$$

$$\Delta\Psi(\overline{w}_1) + f_1(x, \overline{w}_1, w_2) \le 0 \quad \text{a.e. in } D \text{ for } \underline{w}_2 \le w_2 \le \overline{w}_2 \tag{3.5b}$$

$$\Delta\Psi(\underline{w}_2) + f_2(x, w_1, \underline{w}_2) \ge 0 \quad \text{a.e. in } D \text{ for } \underline{w}_1 \le w_1 \le \overline{w}_1 \tag{3.5c}$$

$$\Delta\Psi(\overline{w}_2) + f_2(x, w_1, \overline{w}_2) \le 0 \quad \text{a.e. in } D \text{ for } \underline{w}_1 \le w_1 \le \overline{w}_1 \tag{3.5d}$$

where the derivatives are taken in weak sense. Here $w_i = w_i(x)$ are assumed to be continuous in $\overline{D}$, and $0 \le \underline{w}_i \le w_i \le \overline{w}_i \le b_i$ in $\overline{D}$.

The following corollary is sometimes more readily applicable than Theorem 3.1.

Corollary 3.2. Assume hypotheses $[\hat{H}1]$ and $[\hat{H}2]$. Suppose that there are strong, positive lower and upper solutions $(\underline{w}_1, \underline{w}_2)$ and $(\overline{w}_1, \overline{w}_2)$ of (3.1). Then there exists at least one positive solution (w_1^*, w_2^*) of (3.1) satisfying $\underline{w}_i \le w_i^* \le \overline{w}_i$ in $\overline{D}$, i = 1, 2.

Proof. This is an immediate result of Theorem 3.1 since (3.5) implies (3.2).

We now apply the results above to a prey-predator ecological model with degenerate density-dependent diffusion

$$\Delta u^m + u(a(x) - bu^k - cv) = 0 \quad \text{in } D \tag{3.6a}$$
$$\Delta v^m + v(e(x) + fu - gv^k) = 0 \quad \text{in } D \tag{3.6b}$$
$$u = v = 0 \quad \text{on } \partial D. \tag{3.6c}$$

Here D is a bounded connected domain in R^N $(N \geq 2)$ with boundary $\partial D \in C^2$, and m, k, b, c, f, g are positive constants with $1 + k > m > 1$. We assume that a(x), e(x) are two positive functions in $L^\infty(D)$ with

$$\underline{a} \overset{\text{def}}{=} \operatorname{ess.\,inf.}_{x\in D} a(x) > 0 \text{ and } \underline{e} \overset{\text{def}}{=} \operatorname{ess.\,inf.}_{x\in D} e(x) > 0 \tag{3.7}$$

For convenience, we denote $\bar{a} = \operatorname{ess.\,sup.}_{x\in D} a(x)$ and $\bar{e} = \operatorname{ess.\,sup.}_{x\in D} e(x)$. The following theorem gives sufficient condition for the coexistence of the two species. If we compare with the results in [7] for the nondegenerate case, we see that the conditions here are much readily satisfied. For example, there is no need for the intrinsic growth rates a(x) and e(x) to be larger than the principal eigenvalue for the domain D. Other related references can be founded in the introduction and [6], [8].

Theorem 3.3. Assume $1 + k > m > 1$, hypothesis (3.7) and

$$g(\underline{a}/c)^k > \bar{e} + f(\bar{a}/b)^{1/k}, \tag{3.8}$$

then there exists a positive solution (u,v) of (3.6) with $u, v \in C(\bar{D})$ and $u^m, v^m \in W^{2,p}(D) \cap W_0^{1,p}(D)$ $(p > N)$. Moreover, the solution satisfies $0 < u \leq (\bar{a}/b)^{1/k}$, $0 < v \leq [g^{-1}(\bar{e} + f(\bar{a}/b)^{1/k})]^{1/k}$ in D.

Proof. We will apply Corollary 3.2. Let $b_1 = (\bar{a}/b)^{1/k}$, $b_2 = [g^{-1}(\bar{e} + f(\bar{a}/b)^{1/k})]^{1/k}$. Define

$\Psi(s) = s$ for $s \geq 0$;

$f_1(x,y_1,y_2) = y_1(a(x) - by_1^k - cy_2)$,

$$f_2(x, y_1, y_2) = y_2(e(x) + fy_1 - gy_2^k) \text{ for } (x, y_1, y_2) \in D\times[0,\infty)\times[0,\infty).$$

Then one can readily verify that $[\hat{H}1]$ and $[\hat{H}2]$ are satisfied. To construct lower and upper solutions $(\underline{u},\underline{v})$, $(\bar{u},\bar{v})$, we let $\lambda_1 > 0$ be the principal eigenvalue for the problem $\Delta w + \lambda w = 0$ in D, $w = 0$ on ∂D, and $\phi(x)$ be the principal eigenfunction. Then we have $\phi(x) > 0$ in D and $\phi(x) = 0$ on ∂D. We define $\underline{u} = \underline{v} = (\delta\phi)^{1/m}$ in D for a small $\delta > 0$. Thus they satisfy $\underline{u} = \underline{v} = 0$ on ∂D, $\underline{u} = \underline{v} > 0$ in D. Also, we define $\bar{u} = b_1$. $\bar{v} = b_2$. By choosing $\delta > 0$ sufficiently small, we can verify that $(\underline{u},\underline{v})$ and $(\bar{u},\bar{v})$ are strong, positive lower and upper solutions of (3.6) by means of hypotheses (3.8). The theorem is valid by applying Corollary 3.2. For details, see [9].

4. MONOTONE SCHEME FOR APPROXIMATIONS

In this section, we will construct monotone sequences closing in to our solutions of system (3.6). The method also applies to the general case (3.1). Before our construction of monotone sequences, we need the following two lemmas concerning uniqueness and comparison. We use the same hypotheses concerning the constants m ,k and domain D as in section 3 (essentially, $1 + k > m > 1$). The method can also be adapted to finite difference approximations for systems of elliptic equations.

Lemma 4.1. Let $q(x) \in L^\infty(D)$, $\underline{q} \overset{\text{def}}{=} \inf_{x\in D} q(x) > 0$ and $\hat{p}$ be a positive constant. Suppose that $\underline{u}$ and $\bar{u}$ are strong, positive lower and upper solutions (as described in Definition 2.4) of

$$\begin{aligned} &\Delta u^m + u[q(x) - \hat{p}u^k] = 0 , \quad x \in D \\ &u = 0 \quad \text{on } \partial D \end{aligned} \tag{4.1}$$

with $0 \le \underline{u} \le \bar{u} \le b \overset{\text{def}}{=} \max_{x\in D} \bar{u}(x)$ in $\bar{D}$. Then there exists a positive solution u of (4.1) satisfying $\underline{u} \le u \le \bar{u}$ in $\bar{D}$. Moreover, if u_i (i = 1,2) are positive solutions of (4.1), then $u_1 \equiv u_2$.

One can readily verify that all the hypotheses [H1] to [H3] in section 2 are satisfied for equation (4.1). The first part of the lemma is therefore an immediate consequence of Corollary 2.1. For the part concerning uniqueness, we first establish the existence of a positive solution $\hat{u}$ of (4.1) satisfying $\hat{u} \geq u_i$ in $\bar{D}$, $i = 1, 2$. Then we note that the hypotheses [H1] to [H3] are still satisfied with the choice of b increased to be the $\max_{x \in D} \hat{u}(x)$. We can then proceed to prove that $u_i \equiv \hat{u}$ in D for $i = 1, 2$ by a procedure as in Theorem 2.1.

Lemma 4.2. Let $q_i \in L^{\infty}(D)$, $i = 1, 2$, and $q_1(x) \geq q_2(x) \geq \underline{q}_2 > 0$ a.e. in D, where $\underline{q}_2 = \inf_{x \in D} . q_2(x) > 0$. Suppose that u_i are positive solutions of

$$\Delta u^m + u[q_i(x) - \hat{p}u^k] = 0 \,, \quad x \in D, \tag{4.2}$$
$$u = 0 \quad \text{on } \partial D$$

Then $u_1(x) \geq u_2(x)$ in D. Here, $\hat{p}$ is a positive constant.

Proof. The function u_1 is a strong, positive upper solution of the problem: $\Delta u^m + u[q_2(x) - \hat{p}u^k] = 0$, $x \in D$, $u = 0$ on ∂D, because $\Delta u_1^m + u_1[q_2(x) - \hat{p}u_1^k] = u_1[-q_1(x) + \hat{p}u_1^k + q_2(x) - \hat{p}u_1^k] \leq 0$, a.e. in D. Let $\phi(x)$ be as described in the proof of Theorem 3.3. The function $v = (\delta\phi)^{1/m}$ satisfies $\Delta v^m + v[q_2(x) - \hat{p}v^k] \geq (\delta\phi)^{1/m}[-\lambda_1(\delta\phi)^{1-1/m} + \underline{q} - \hat{p}(\delta\phi)^{k/m}] \geq 0$ a.e. in D for small $\delta > 0$; it is therefore a strong, positive lower solution for the problem described above. From Lemma 4.1, we conclude that there exists a solution $\hat{u}_2$ of the problem with $0 < (\delta\phi)^{1/m} \leq \hat{u}_2 \leq u_1$ in D. From the uniqueness part of Lemma 4.1, we must have $\hat{u}_2 \equiv u_2$. Consequently, we obtain $u_2 \leq u_1$ in D.

We now assume all the hypotheses in Theorem 3.3, and proceed to construct monotone sequences converging to the positive solutions of (3.6) (as described in Theorem 3.3). Define $\underline{u} = \underline{v} = (\delta\phi)^{1/m}$ for sufficiently small $\delta > 0$, $\bar{u} = (\bar{a}/b)^{1/k}$, $\bar{v} = [g^{-1}(\bar{e} + f(\bar{a}/b)^{1/k})]^{1/k}$. First let u_1 be the unique positive solution of the problem $\Delta u^m +$

$u(a(x) - bu^k) = 0$ in D, $u = 0$ on ∂D with $\underline{u} \le u_1 \le \bar{u}$, and v_1 be the unique positive solution of the problem $\Delta v^m + v(e(x) + fu_1 - gv^k) = 0$ in D, $v = 0$ on ∂D with $\underline{v} \le v_1 \le \bar{v}$. With the aid of the following lemma, We now define inductively the positive solutions u_i, v_i, $i = 2, 3, \ldots$, as follows:

$$\Delta u_i^m + u_i(a(x) - bu_i^k - cv_{i-1}) = 0 \quad \text{in } D \tag{4.3a}$$
$$\Delta v_i^m + v_i(e(x) + fu_i - gv_i^k) = 0 \quad \text{in } D \tag{4.3b}$$
$$u_i = v_i = 0 \quad \text{on } \partial D, \tag{4.3c}$$

Lemma 4.3. The functions u_i and v_i as described in (4.3) and above are well defined for each $i = 1, 2, \ldots$. Moreover $\underline{u} \le u_i \le u_1 \le \bar{u}$, $\underline{v} \le v_i \le v_1 \le \bar{v}$.

Proof. The lemma is clearly true for $i = 1$. Assume that it is true for $i = j - 1$. We have $a(x) - cv_{j-1}(x) \ge a(x) - cv_1(x) \ge \underline{a} - c\bar{v} > 0$ by (3.8). By Lemma 4.1, u_i, v_i are uniquely defined and $\underline{u} \le u_i \le \bar{u}$, $\underline{v} \le v_i \le \bar{v}$ in $\bar{D}$. Comparing the equations satisfied by u_j and u_1 by means of Lemma 4.2, we conclude $\underline{u} \le u_j \le u_1$ in D. Again, comparing the equations satisfied by v_j and v_1 by means of Lemma 4.2, we conclude that $v_j \le v_1$ for $x \in \bar{D}$.

We next observe some order relations among the sequences u_i, v_i, $i = 1, 2, \ldots$.

Lemma 4.4. For each nonnegative integer, the following are true:

$$u_{2n+2} \le u_{2n+4} \le u_{2n+3} \le u_{2n+1}, \quad v_{2n+2} \le v_{2n+4} \le v_{2n+3} \le v_{2n+1} \tag{4.4}$$

for all $x \in \bar{D}$.

Proof. We first consider the case when $n = 0$. Using the equations which u_1, u_3 satisfy and utilizing Lemma 4.2, we conclude $u_3 \le u_1$ in $\bar{D}$. This in turn, using the same method, implies that $v_3 \le v_1$ in $\bar{D}$. Again, this implies that $u_2 \le u_4$ and then $v_2 \le v_4$ in $\bar{D}$. To establish comparison between u_3 and u_4 or v_3 and v_4, we keep applying Lemma 4.2

to first assert $u_2 \le u_1$ in $\bar{D}$, then $v_2 \le v_1$, then $u_2 \le u_3$, then $v_2 \le v_3$, then $u_4 \le u_3$ and finally $v_4 \le v_3$ (all inequalities being true in $\bar{D}$). Inequalities (4.4) are thus true for n = 0. Assume that (4.2) are true for $n = j - 1$, using the above method and $v_{2j} \le v_{2j+2}$ we deduce the following in order: $u_{2j+3} \le u_{2j+1}$, $v_{2j+3} \le v_{2j+1}$, $u_{2j+2} \le u_{2j+4}$, $v_{2j+2} \le v_{2j+4}$. Then using $u_{2j+2} \le u_{2j+1}$ we deduce the following in order: $v_{2j+2} \le v_{2j+1}$, $u_{2j+2} \le u_{2j+3}$, $v_{2j+2} \le v_{2j+3}$, $u_{2j+4} \le u_{2j+3}$, $v_{2j+4} \le v_{2j+3}$. (All inequalities are true in $\bar{D}$) This proves (4.2) for n = j.

Lemma 4.4 clearly implies that $0 \le \underline{u} \le u_2 \le u_4 \le u_6 \ldots \le u_5 \le u_3 \le u_1 \le \bar{u}$, and $0 \le \underline{v} \le v_2 \le v_4 \le v_6 \ldots \le v_5 \le v_3 \le v_1 \le \bar{v}$ for all $x \in \bar{D}$. Define $u^* = \lim_{n\to\infty} u_{2n+1}$, $u_* = \lim_{n\to\infty} u_{2n}$, $v^* = \lim_{n\to\infty} v_{2n+1}$, $v_* = \lim_{n\to\infty} v_{2n}$. The functions u_*, u^*, v_*, v^* are in $C(\bar{D})$ and ${u_*}^m$, ${u^*}^m$, ${v_*}^m$, ${v^*}^m$ are in $W^{2,p}(D) \cap W_0^{1,p}(D)$ and $(w_1, w_2, w_3, w_4) = (u_*, u^*, v_*, v^*)$ is a positive solution for the problem:

$$\begin{gathered}\Delta w_1^m + w_1(a(x) - bw_1^k - cw_4) = 0,\ \Delta w_2^m + w_2(a(x) - bw_2^k - cw_2) = 0,\\ \Delta w_3^m + w_3(e(x) + fw_1 - gw_3^k) = 0,\ \Delta w_4^m + w_4(e(x) + fw_2 - gw_4^k) = 0 \quad (4.5)\\ w_i = 0 \quad \text{on } \partial D,\ i = 1,\ldots,\end{gathered}$$

(See [9] for the assertion on the regularity of w_i). Note we have $w_i > 0$ for all $x \in D$.

We next investigate some relationship between problem (3.6) and problem (4.5).

Theorem.4.1. $u_* = u^*$ in $\bar{D}$ iff $v_* = v^*$ in $\bar{D}$.

Proof. Suppose $u_* = u^*$ in D, then v_* and v^* will be positive solutions to the same boundary value problem. By Lemma 4.1, we conclude that $v_* = v^*$ in $\bar{D}$. Similarly, the converse is true.

Theorem 4.2. Suppose that $u_* = u^*$ in $\bar{D}$ (or $v_* = v^*$ in $\bar{D}$), then the boundary value problem (3.6) has a unique positive solution (u,v) with

$u, v \in C(\bar{D})$, u^m, $v^m \in W^{2,p}(D) \cap W_0^{1,p}(D)$ $(p > N)$, $0 \le \underline{u} \le u \le \bar{u}$, $\underline{v} \le v \le \bar{v}$ in $\bar{D}$.

Proof. Note that the existence of such a solution has been established in Theorem 3.3, even without the assumption that $u_* = u^*$ in $\bar{D}$. It remain to prove that under this assumption we must have $(u,v) = (u_*,v_*)$. We can use Lemma 4.2 to conclude that $u_2 \le u \le u_1$ in $\bar{D}$; similarly $v_2 \le v \le v_1$ in $\bar{D}$. Repeated applications of Lemma 4.2 will, by induction, show that $u_{2n+2} \le u \le u_{2n+1}$, $v_{2n+2} \le v \le v_{2n+1}$, $x \in \bar{D}$, for each nonnegative integer n. Clearly, if $u_* = u^*$ (or $v_* = v^*$), then $(u,v) = (u_*,v_*)$.

The following theorem has been proved above.

Theorem 4.3. Any positive solution (u,v) of (3.6) with u, $v \in C(\bar{D})$, u^m, $v^m \in W^{2,p}(D) \cap W_0^{1,p}(D)$ $(p > N)$, $\underline{u} \le u \le \bar{u}$, $\underline{v} \le v \le \bar{v}$ in $\bar{D}$, must satisfy: $u_* \le u \le u^*$, $v_* \le v \le v^*$ in $\bar{D}$.

Theorem 4.4. Suppose that the boundary value problem (4.5) has at most one positive solution, then $u_* = u^*$, $v_* = v^*$ in $\bar{D}$. That is, the boundary value problem (3.6) has a unique positive solution.

Proof. Note that both (u_*, u^*, v_*, v^*) and (u^*, u_*, v^*, v_*) are positive solutions of (4.5). Hence the uniqueness implies that $u_* = u^*$ and $v_* = v^*$ in $\bar{D}$. The final conclusion follows from Theorem 4.2.

Reference

[1] D. G. Aronson, Density-dependent interaction-diffusion systems, dynamics and modeling of reaction systems, Publ. Math. Res. Center, Univ. Wisconsin, Academic Press, New York-London 44: 161-176 (1980).

[2] D. G Aronson, M. G. Crandall and L. A. Peleitier, Stabilization of solutions of a degenerate nonlinear diffusion problem, Nonl. Anal. T.M.A. 6: 1001-1022 (1982).

[3] P. DeMettoni, A. Schiaffino and A. Tesei, Attractive properties

of nonnegative solutions of a class of nonlinear degenerate parablic problems, Ann. Mat. Pura. Appl. 136: 35-48 (1984).

[4] D. Gilbarg and N. Trudinger, Elliptic Differential Equations of Second Order, third edition, Springer-Verlag, Berlin, 1986.

[5] M. E. Gurtin and R.C. MacCamy, On the diffusion of biological populations, Math. Biosci, 33: 35-49 (1977).

[6] T. G. Hallam, Population dynamics in a homogeneous environment, Math. Ecological, ed. T. G. Hallam and S. Levin, Biomathematics 17, Springer-Verlag, Berlin 1986.

[7] A. Leung, Monotone schemes for semilinear elliptic systems related to ecology, Math. Meth. Appl. Sci. 4: 272-285 (1982).

[8] A. Leung, Systems of Nonlinear Partial Differential Equations, Applications to Biology and Engineering, to appear, D. Reidel, Kluwer Academic Publishers.

[9] A. Leung and G. Fan, Existence of positive solutions for elliptic systems, degenerate and nondegenerate ecological models, to appear (1989).

[10] L. A. Peleiter, The porous medium equations, Applications of Nonlinear Analysis in the Physical Science, ed. H. Amann, N. Bazley, K. Kirchgassner, Pitman, London, 229-241 (1981).

[11] M. A. Pozio and A. Tesei, Degenerate parabolic problems in population dynamics, Japan J. of Applied Mathematics 2: 351-380 (1985).

[12] M. A. Pozio and A. Tesei, Support properties of solutions for a class of degenerate parabolic problems, Comm. in PDE 12: 47-75 (1987).

[13] D. Sattinger, Monotone methods in nonlinear elliptic and parabolic boundary value problems, Indiana U. Math. J. 21: 979-1000 (1972).

[14] H. Schatzman, Stationary solutions and asymptotic behavior of a quasilinear degenerate parabolic equation, Indiana Univ. J. 33: 1-30 (1984).

[15] J. Spruck, Uniqueness in a diffusion model of population biology, Comm. Partial Diff. Eqs. 8: 1605-1620 (1983).

Observable Tunneling in Several Dimensions

R. E. Meyer Professor, Department of Mathematics, University of Wisconsin–Madison, Madison, Wisconsin

I. SUMMARY

It will be outlined how exponential-precision asymptotics can be carried from ordinary differential equations to genuine partial differential equations. To keep the account brief, it will be done at the hand of spectral prediction for integrable Hamiltonian wave equations with double-well potential. The approach is to combine recent insights into the basic essence of Stokes phenomena with a revision of Keller's spectral method of EBK Quantization and with complex Geometrical Optics. The resulting spectral formulae are simple, explicit and exponentially sophisticated.

This research was supported by the National Science Foundation under Grant DMS-8521687.

II. INTRODUCTION

Exponential asymptotics was conceived (Meyer, 1973) for the purpose of tracing a simple, direct road to the prediction of physically observable quantities for which the asymptotic expansion is zero to all orders. The simplest example, perhaps, is wave propagation through a "stratified" medium in which the signal speeds depends one coordinate, $c = c(x) \in C^\infty(\mathbb{R})$. What one wants to know primarily is how much of the wave is transmitted and how much is reflected? The ratio of the length scale of $c(x)$ to the wavelength scale is normally a large parameter $1/\varepsilon$, and the solution of the wave equation has an asymptotic expansion in powers of ε, but the solution is hardly ever observable (and never, with any accuracy). What one can observe and wishes to predict are the reflection and transmission coefficients R and T. However, when the asymptotic expansion of the solution is used to calculate R, it turns out to be zero to all orders in ε. Since $|T|^2 + |R|^2 = 1$, no useful information at all has emerged. A simple way of predicting R and T was given by Meyer (1975, 1980).

A more sophisticated example are waves in a periodic potential. Their spectrum is continuous and consists of exponentially narrow bands separated by wider gaps. For ordinary differential equations these spectra were predicted to first approximation for small ε by Harrell (1978) and Weinstein and Keller (1985). This suffices in more dimensions when the variables can be separated, but as soon as such strict symmetry is perturbed, o.d.e. theory fails completely. In several dimensions, Harrell (1980) and Maslov and Fedoryuk (1981) have established an abstract theory for small ε, but it is not well suited to quantitative prediction. The objective here will be to sketch a rather simple (even if not straightforward) way of predicting the observable features of spectra for problems of such type when the wave equation is a nonseparable partial differential equation (Meyer, 1989a,b,c).

It should be emphasized that the genuine scientific objective in these problem classes is to predict certain functionals of the solutions which can be observed, rather than the solutions themselves. These functionals depend on the boundary conditions and on the structure of the wave equation, but they depend on the details of the wave equation only in some averaged sense. Instead of the exact solutions, therefore, one can work with simpler functions which differ grossly from the solutions, provided they meet the boundary conditions, respect the underlying structure, and give the right averages.

There is a sense in which this approach recalls fashion in clothes. They cover a beautiful and complicated body, but need not be leotards. Usually, we prefer them to be tailored to fit in just a few critical places, but to be otherwise loose for comfort. Hidden deeply inside, there is a skeleton carrying all of this. The body, which we forgo seeing in detail, recalls the beautiful wave functions solving the full problem exactly. The clothes correspond to the approximations we use in practice, and the structural skeleton is represented by the characteristic function of Hamilton and Jacobi. It is one of the decisive reasons for the success of the present enterprise that the approach taken deviates from the usual one by taking pains to construct the skeleton exactly (Sections II-V, VII), but avoiding nonessential difficulties by keeping the approximation clothes (Section VI) very loose indeed, except at the boundary of the wave domain.

III. HAMILTONIAN WAVE EQUATIONS

To make the account more understandable, it will be restricted to Hamiltonian operators of Schroedinger type,

$$H(-i\varepsilon\nabla, \mathbf{x}) = -\varepsilon^2\nabla^2 + V(\mathbf{x}).$$

They have an associated Hamilton-Jacobi equation (Goldstein, 1959)

$$H(\nabla S, \mathbf{x}) = |\nabla S|^2 + V(\mathbf{x}) = E \tag{1}$$

with eigenparameter E for a "characteristic function" $S(\mathbf{x})$ and canonical equations

$$d\mathbf{x}/dt = \nabla_p H, \qquad d\mathbf{p}dt = -\nabla_x H, \qquad \mathbf{p} \equiv \nabla S \tag{2}$$

for the phase-space trajectories $(\mathbf{p}, \mathbf{x})(t)$ of the Hamiltonian. On the trajectories,

$$H = const = E, \qquad dS/dt = 2|\mathbf{p}|^2, \tag{3}$$

and these together with the canonical equations are the only effective approach to the calculation of the characteristic function when the variables cannot be separated. Note that this approach does not depend on linearity of the Hamiltonian operator.

The associated wave equation $H\psi = E\psi$ has solutions

$$\psi(\mathbf{x}) = A_e(\mathbf{x}) exp[iS(\mathbf{x})/\varepsilon] \tag{4}$$

if and only if

$$i\varepsilon A_e \nabla^2 A_e = \nabla(A_e^2 \nabla S). \tag{5}$$

For present purposes, the Hamiltonian is assumed integrable, i.e., all the trajectories lie on recognizable invariant tori, which fill all of phase space. That is not the generic case, almost all the trajectories of non-integrable Hamiltonians tend to be chaotic. The Theoretical Chemists, however, have accumulated recent numerical evidence (Skodje, Bororondo and Reinhardt, 1985, Jaffe, 1988) that the chaos can be largely ignored because, for reasons not yet understood, the chaos averages out almost completely for spectral purposes.

IV. DOUBLE WELLS

To shorten the present account, it will be given at the instance of the canonical example (Harrell, 1978) of a symmetrical "double well" (Fig. 1) on a domain $\Omega \subset \mathbb{R}^2$ with a potential V even in one Cartesian coordinate x. It promotes concreteness greatly to take V also to depend on only that coordinate and to take $V(x)$ real analytic on $\mathbb{R}$. To fix the ideas further, let $V(x)$ have just one maximum (at $x = 0$) and just one minimum in $x > 0$ (Fig. 1).

It is well-known, and rather obvious from Fig. 1, that such a wave equation should have eigenfunctions odd in x (in physical partance, a Fermion pair) so that

$$\psi(\mathbf{x}) = 0 \text{ at } x = 0, \tag{6F}$$

and eigenfunctions even in x (a Boson pair) with

$$\partial\psi/\partial x = 0 \text{ at } x = 0. \tag{6B}$$

This makes it possible to confine attention to the halfspace $x > 0$.

For a potential more or less like that in Fig. 1, there is a radiation condition at $|x| = \infty$, which (4) already suggests (correctly) to be

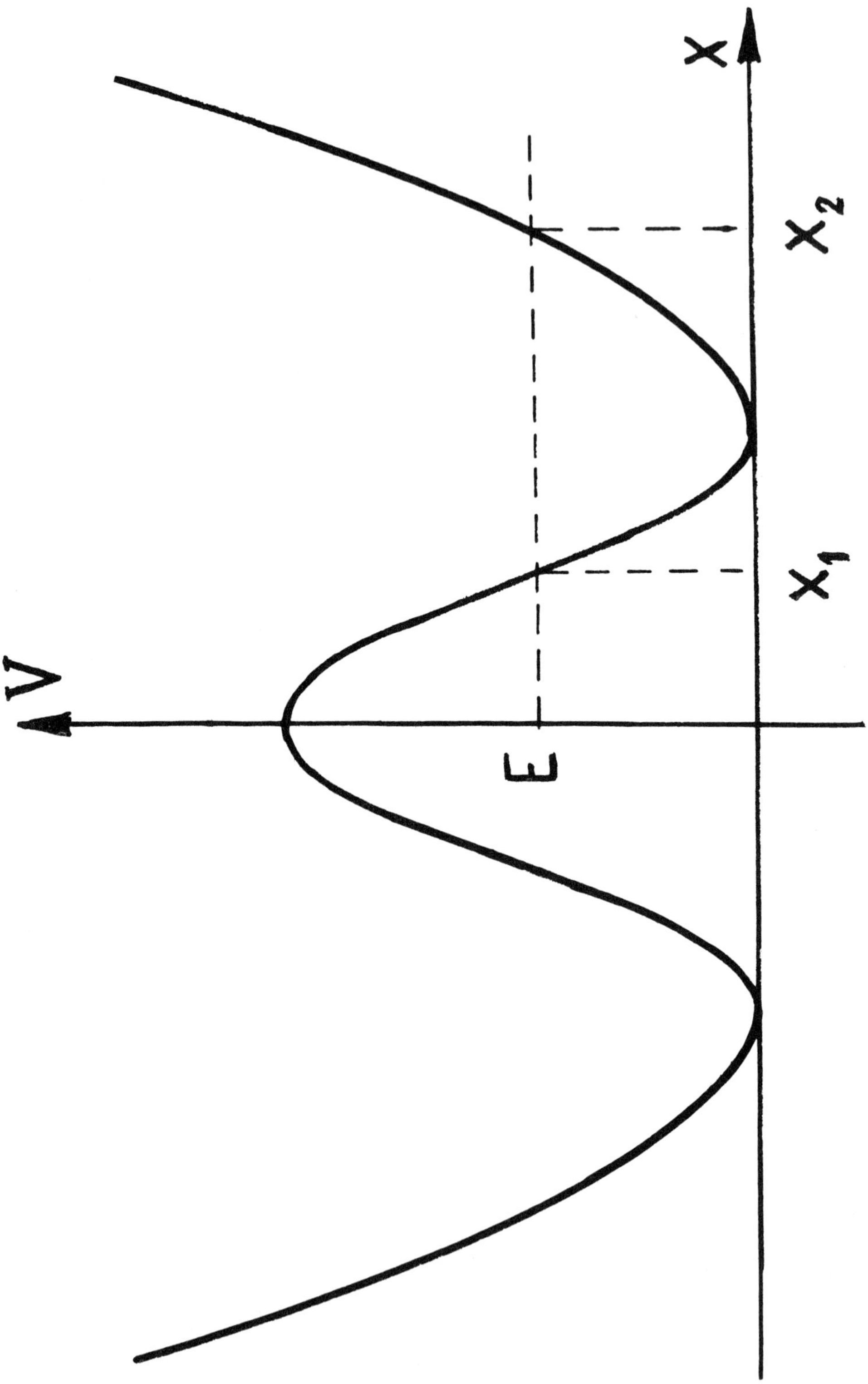

FIG. 1 *Symmetrical double-well potential with "turning points" (in $x > 0$) at a level* $E - c_t^2$.

$$Im\, S > 0 \text{ for sufficiently large } |x|. \tag{7}$$

To complete the formulation, the wave domain will be taken to be a strip bounded by sidewalls at $y = b_i(x)$ (with $b_i(x)$ even and real analytic) at which conventional boundary conditions

$$\psi(x, b_i(x)) = 0 \tag{8D}$$

or

$$\partial\psi/\partial\nu = 0 \text{ at } y = b_i(x) \tag{8N}$$

are specified, where ν denotes the direction normal to the sidewall.

This completes the formulation. It may be noted that the spectrum depends seriously on the conditions (6)-(8), but any one spectral feature depends on the potential $V(x)$ only in an average sense and hence, many details of the exact wave functions are also not germane directly to spectral predictions.

V. COMPLEX GEOMETRICAL OPTICS

The main purpose of all the simplifying assumptions is to cast the canonical equations into a concrete and explicit form from which many of the decisive qualitative facts can be read off lucidly. Along any one trajectory, (2) and (3) will be found readily by (1) to take the form

$$p_z \equiv \frac{\partial S}{\partial y} = \text{const} = c_t, \qquad p_1 \equiv \frac{\partial S}{\partial x} = [F(x;c_t)]^{1/2} \tag{9}$$

with

$$F(x;c_t) \equiv E - V(x) - c_t^2, \tag{10}$$

$$y(x) - y_\alpha = c_t \int_{x_\alpha}^{x} [F(s;c_t)]^{-1/2} ds, \tag{11}$$

$$S(x, y(x)) = S(x_\alpha, y_\alpha) + c_t[y(x) - y_\alpha] + \int_{x_\alpha}^{x} [F(s;c_t)]^{1/2} ds \tag{12}$$

in terms of the *parameter* x along the particular trajectory with "label" c_t; (x_α, y_α) is any one "initial" point on it.

It should be emphasized right away that x is a dubious parameter because it is not singlevalued on a trajectory, but having to imagine (9)-(12) on a Riemann surface is an acceptable price for their explicitness. Indeed, any root x_b of the "reduced potential" F is a branch point of p_1, y and S in (9)-(12), but otherwise, they are analytic in the parameter x. It will be noticed, on the other hand, that such roots of F depend on c_t, so different trajectories will have them at different x_b and $y_b = y(x_b)$.

At any one real value of E above the minimum and below the maximum of the potential (Fig. 1) and on any one trajectory, the real x-interval between the roots (Fig. 1) $x_1(c_t) > 0$ and $x_2(c_t) > x_1(c_t)$ of the reduced potential $F = E - c_t^2 - V(x)$ is called "the well"; the interval $(0, x_1(c_t))$ is called "the barrier" and the interval $(x_2(c_t), \infty)$, the far field. In the well, $F > 0$ and the trajectory equations (9)-(12) have real solutions. Since the Hamiltonian is taken to be integrable, these trajectories over the well lie on real invariant tori in the phase space of $\mathbf{x}$ and $\mathbf{p} = \nabla S$. The projection of a trajectory onto $\mathbf{x}$-space is called "ray" and is often helpful for the visualization and illustration of phase-space geometry and topology. In as much as the tori are bounded manifolds without boundary, the rays must have envelopes, usually called "caustics" (Keller, 1958).

Actually, the tori have gaps at the sidewalls $y = b_i(x)$, which are bridged by (8D) or (8N). These leave $|\partial S/\partial \nu|$ invariant, but reverse the sign of $\partial S/\partial \nu$; accordingly, c_t changes sign there and, when $b_i(x) \neq$ const, $c_t \equiv \partial S/\partial y$ also changes magnitude by an amount dependent on the local value of db_i/dx. Therefore, c_t is only piecewise constant on trajectories and hence also, the trajectory functions $p_1(x)$, $y(x)$ and $S(x, y(x))$ in (9)-(12) are only piecewise analytic. The rays, by contrast, have no gaps and in the well, the boundary conditions can be thought of (Keller and Rubinow, 1960) in terms of ray reflection according to the familiar Snell law of optics.

In the barrier and in the far field, on the other hand, $F < 0$ and the trajectory equations (9)-(12) have only complex-valued solutions. Those trajectory components lie on complex invariant tori in complex phase space and their projections onto position space are complex-valued rays. They are much harder to visualize and bring surprises. For instance, it is quite easy to see (Meyer, 1989a) from (9)-(12) that these complex rays arising from the continuation of the well trajectories into barrier and far field, cannot intersect the real sidewalls! Accordingly, the attractive interpretation of the boundary conditions (8) in terms of ray reflection fails in complex Geometrical Optics. It is a bonus, on the other hand, that since these complex trajectory components cannot intersect the sidewalls, the trajectory functions $p_1(x)$, $y(x)$ and $S(x, y(x))$ in (9)-(12) are analytic in the barrier and in the far field, rather than merely piecewise analytic.

It is perhaps an even bigger surprise that the ray envelope cannot coincide normally with the branch locus of the trajectories and rays! That is seen easily from (9)-(11) because (9) shows the branch locus to be that of roots of the reduced potential F in (10) and (11) shows that $dx/dy = 0$ there on the trajectories and rays (Fig. 2): the ray tangents are here parallel to the y-

FIG. 2 *Two real rays near the left end of the well; they are locally parallel to the y-direction at the respective branch points of S, but not, at the ray envelope.*

direction. But by (10), the roots of F occur at different x on trajectories with different values of c_t and therefore, the tangent to ray and envelope, where they touch, cannot be normally in the y-direction (Fig. 2). The situation is clarified further when one sketches the rays in the space of real x and complex y (Fig. 3). When the phase-space trajectories are projected onto the real x, y-plane, then there is a strip (Fig. 3) covered by the projections of both the barrier and well components of the trajectories.

The symmetries of separable wave problems preclude such a distinction between ray envelopes and branch loci, but it is a normal feature of nonseparable wave problems and presents us with an ambiguity of the caustic concept. Here it should be recalled that Geometrical Optics serves, at best, to construct a very nonuniform approximation and near the envelope and the branch locus, it has no direct significance in terms of wave functions ψ. The envelope of the ray pattern, moreover, is a singularity merely of the torus projection, not of the invariant torus itself, and can therefore have only minor relevance. The branch locus, however, carries a marked singularity of the trajectories and torus, which is a salient topological feature of the torus (Fig. 3) with its separate components connected in possibly curious ways.

Furthermore, since the wave functions ψ are singlevalued, but the characteristic function S has branch points at the roots of the reduced potential F, the amplitude function A_e in (4) must also have corresponding branch points at the roots of F. The "approximate" amplitude functions A introduced in the next Section will be chosen to have the same branch points and branch structure as the true amplitude functions A_e, even though the func-

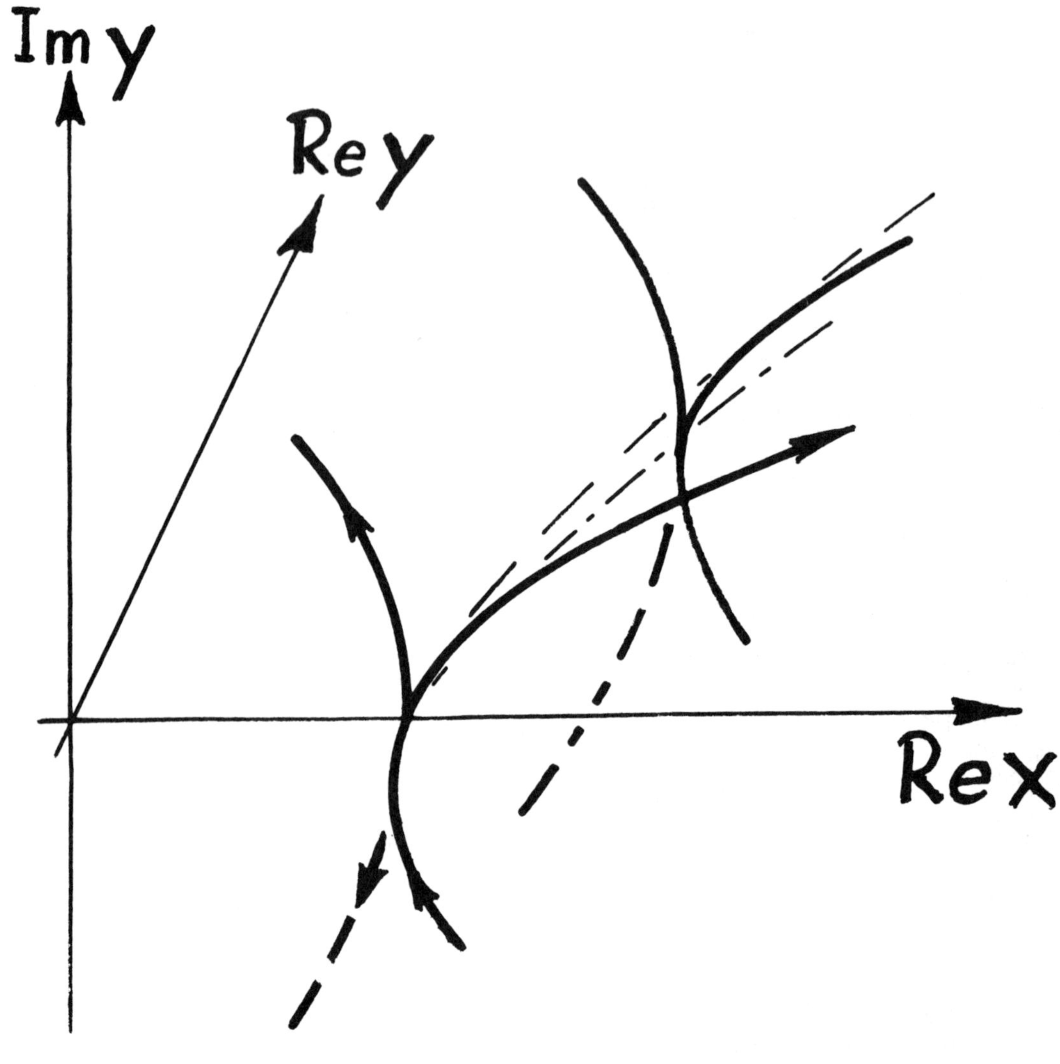

FIG. 3 *Real and complex rays near a well boundary in the space of real x and complex y.*

tions A do not approximate the A_e at all anywhere near the branch loci or envelopes. The branch loci must be where the Stokes phenomena are anchored (Meyer, 1989d), which are an important feature of the structural skeleton of wave processes hidden from view by the singlevaluedness of the wave functions. The term "caustic" will therefore be used in the following for the branch loci, rather than the ray envelopes.

VI. APPROXIMANDS

It is now proposed to approximate the solutions ψ of the wave equation in a certain sense by the help of linear combinations of very nonuniform approximands of the form

$$\phi = A(\mathbf{x}) \exp[iS(\mathbf{x})/\varepsilon] \tag{13}$$

with functions A satisfying only the truncation

$$\nabla(A^2 \nabla S) = 0 \tag{14}$$

of the true "transport equation" (5). They are therefore naive approximands of the type often associated with the terms WKB or shortwave asymptotics or Geometrical Optics. If used correctly, however, they yield a rigorous (Harrell, 1980, Maslov and Fedoryuk, 1981) approximation to ψ away from caustics and other branch loci of the characteristic function S. They have, moreover, the "double-asymptotic" (Olver, 1974, Maslov and Fedoryuk, 1981) property that the approximation improves as either ε gets smaller or the distance from the nearest branch point of S gets larger. This is of interest here because, except for the lowest and highest eigenvalues E_n between $\min V$ and $\max V$ (Fig. 1), most of the boundary of the wave domain lies at a considerable distance from the caustics. To approximate the conditions (6)-(8) well, requires then an approximation to ψ only rather far from the caustics. This is not always sufficient for dispensing with reliance on $\varepsilon << 1$ because there may well be additional singularities of S in the complex plane of the parameter x in (9)-(12) away from the real axis. In the situations of most scientific interest, however, those additional singularities tend also to lie far from where (6)-(8) need be applied and it is then possible to rely primarily on distance from the branch loci of S, so that the approximations to be constructed now are then not really shortwave or conventional WKB approximations at all: they do not refer to any small parameter.

It is well-known (Keller, 1958, Keller and Rubinow, 1960) in Geometrical Optics that (14) implies

$$A^2|\nabla S|J = \text{const} = A^{*2} \text{ along a trajectory,}$$

where J denotes a Jacobian of the transformation between Cartesian and ray coordinates. If this is combined with (9) and (13), it is found (Meyer, 1989b) that each trajectory component can be allotted an approximand pair like

$$\phi_j^{out-} = A_{j-}^* F_j^{-1/4} J^{-1/2} \exp\{\tfrac{i}{\varepsilon}[S_j^-(x, y(x)) - S_j(c_t)]\}, \tag{15}$$

$$\phi_j^{in+} = B_{j+}^* F_j^{-1/4} J^{-1/2} \exp\{-\tfrac{i}{\varepsilon}[S_j^+(x, y(x)) - S_j(c_t)]\}.$$

Here the super- and subscripts $+$, $-$ refer to the sign allotted to c_t (because (3) depends only on c_t^2); the subscript j identifies the branch of S and the associated branch of $F^{-1/4}$ and through this, also the caustic associated with that branch of S; and $S_j(c_t)$ denotes the value of S at that caustic. Finally, the superscripts "out" and "in", which correspond to the sign in front of i/ε in the braces of (15), remind us which ϕ_j approximates a wave "outgoing" from the caustic along the Stokes manifold L_j of the branch and which, approximates a wave so "incident" on the caustic. To emphasize their distinction from a single characteristic function S with its analytic continuation, the branches S_j needed for correct approximation will now be called "action functions".

VII. STOKES PHENOMENON

The next question must be: how can an association of action functions be found which resolves the Stokes phenomena correctly in several dimensions? In a way, of course, this is the crux of the whole enterprise. The answer lies in the canonical equations (2), (3) for the individual trajectories. Their solution, displayed in (9)-(12), is a quartet of analytic functions of one variable, namely the parameter x along the trajectory, in a neighborhood of any caustic point of the trajectory (except caustic points at the sidewalls). The caustic point of a trajectory is therefore a branch point of those analytic functions of the parameter in the familiar sense of classical complex analysis. Accordingly, the multivaluedness of those functions (9)-(12) can be characterized in the strictest local sense at the branch point itself. That this is a basic key to the resolution of Stokes phenomena for ordinary differential equations has been demonstrated in great generality (Meyer and Painter, 1983). Hence, if we choose an arbitrary trajectory and, for the time being, confine our attention myopically to just that trajectory, then the Stokes phenomenon for partial differential equations does not differ from that for ordinary differential equations!

In the context of the double well (Fig. 1), the caustic points are mere "simple turning points" and the rigorous answer to the "connection problem" for them has been known for 60 years (Zwaan, 1929): Three (anti-) Stokes Lines L_j issue from each such turning point (Fig. 4) and each of them has an associated "approximation domain" D_j on which the functions (15) approximate a fundamental system of the wave equation. The domain D_j contains most of the Stokes line L_j, but must be bounded away from any other Stokes line and from any turning point or singular point of S; and the further it is so bounded away, the better are the error bounds on the approximation for fixed ε. Finally, if the Stokes lines L_j issuing from the same simple turning point (Fig. 4) are numbered anticlockwise (mod 3) and if the functions (15) approximate an exact solution ψ on D_j by $\phi_j^{out-} + \phi_j^{in+}$, then the *same* solution ψ is approximated on D_{j+1} (Fig. 4) by $\phi_{j+1}^{out-} + \phi_{j+1}^{in+}$ with

$$F_{j+1}(x) = e^{-2\pi i} F_j(x), \qquad S_{j+1} = e^{-\pi i} S_j$$

if and only if

$$A_j^* = A_{j+1}^* + iB_{j+1}^*, \qquad B_j^* = iA_{j+1}^*. \tag{16}$$

In particular, if we choose $\arg F_0(x) = 0$, say, on the real x-axis in the well (x_1, x_2) (Fig. 4) then $F_1(x) = F_0(x)\exp(-2\pi i)$ is the correct choice on the domain D_1 of the barrier (Fig. 4). The corresponding approximands (15) with the action function (12) for $F_1(x)$ furnish an approximation to the same wave function ψ on a barrier component of our particular trajectory as those for $F_0(x)$ furnish for ψ on a well component of this trajectory, provided the respective amplitude constants are related by the "connection formula" (16).

This makes it possible to ask how the different trajectory components are connected (Fig. 3), which is relevant because the explicit spectral formulae ("Quantization Rules" in physical parlance) depend primarily on topo-

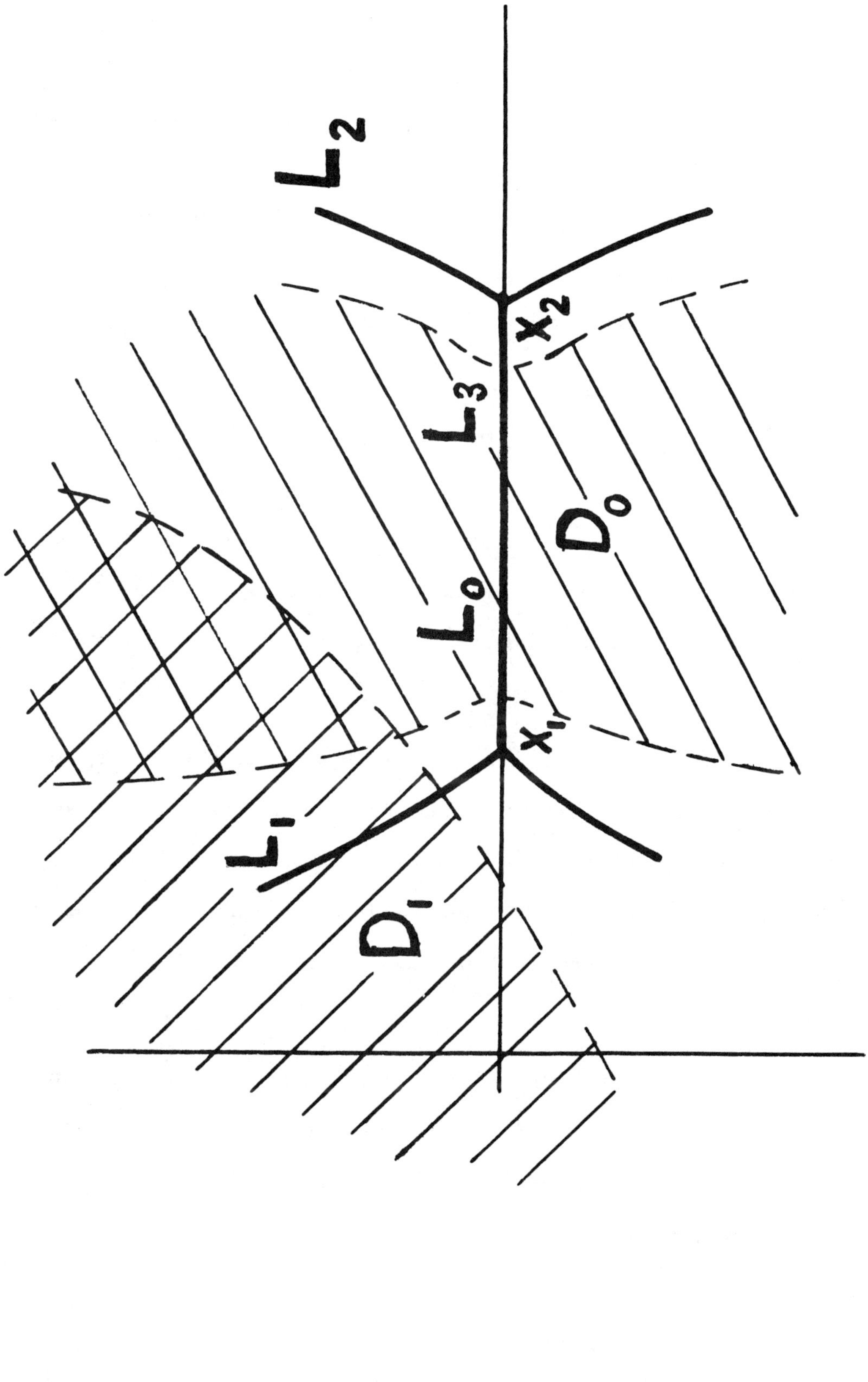

FIG. 4 *Stokes lines* L_j *and approximation domains* D_j *of a trajectory in the complex plane of its parameter* x.

logical aspects of the invariant tori (Keller, 1958, Keller and Rubinow, 1960, Keller, 1985). A wave moving rightward in the well (Figs. 1, 4) is specified by $A_0^* \neq 0$, $B_0^* = 0$ in (15) and by (16), its continuation into the barrier has $A_1^* = 0$, $B_1^* \neq 0$. The real trajectory component in the well associated with ϕ_0^{out} through S_0 therefore connects across the caustic at x_1 with the complex trajectory component in the barrier associated with ϕ_1^{in} (which is an exponentially "dominant" approximand) through $-S_1$.

By contrast, a wave moving leftward in the well (Figs. 1, 4) is specified by $A_0^* = 0$, $B_0^* \neq 0$ and by (16), its continuation into the barrier has $iA_1^* = B_1^* \neq 0$ and involves therefore *both* the dominant and recessive approximands for the barrier. The real trajectory component in the well associated with ϕ_0^{in} through $-S_0$ therefore *splits* at the caustic point x_1 (Figs. 1-4) of the trajectory to *connect with both* complex trajectory components in the barrier!

The trajectory connection is a little simpler at the caustic point x_2 between well and far field because the radiation condition (7) admits only one complex trajectory component int he far field. It may be worth remarking, however, that periodic potentials exhibit the trajectory split at both edges of each well, so that any trajectory then suffers a countably infinite number of splits.

To complete the association of action functions for a definite wave function ψ requires also a connection relation across the well, but that is exactly the same as in Keller's spectral method (Keller, 1958, Keller and Rubinow, 1960, Maslov, 1972, Keller, 1985) for a single-well potential involving no tunneling and no complex Geometrical Optics. The first approximation to that connection relation is obtained (Keller, 1958, Keller and Rubinow, 1960, Maslov, 1972, Keller, 1985) by the help of a familiar phase integral

$$\Lambda = \int_{x_1^*}^{x_2^*} [F_0(x; c_t(x))]^{1/2} dx \tag{17}$$

where x_i^* are the caustic points on the real x-axis (Fig. 1) and $c_t(x)$ denotes the distribution of the trajectory "label" c_t along the real interval (x_1^*, x_2^*) obtained by standard, real Geometrical Optics for the well.

VIII. EBK RULE

This leaves us in front of the last hurdle: how to account correctly for the tunneling and the consequent, small split between the Fermion and Boson eigenvalues by application of the symmetry conditions (6F) or (6B) at the barrier center. It is a perplexing hurdle because all the information obtained up to this point concerns approximation along the trajectories, but (9)-(11) show immediately that the complex rays over the barrier cannot intersect the symmetry line $x = 0$, $b_2(0) < y < b_1(0)$ (Fig. 3), where the symmetry conditions (6F,B) apply. Evidently, to reach the symmetry line requires an elaborate computation in complex phase space $\mathbb{C}^4$.

Without it, all we can do at this point is to apply the symmetry conditions tentatively on each trajectory component at the complex $y(0)$ where that component intersects the plane $x = 0$ (Fig. 5). But, that is also sufficient. Since the potential and sidewall shapes are even in x, the complex invariant torus component over the right half of the barrier is necessarily accompanied by a symmetrical, complex invariant torus component over the left half of the barrier (Fig. 6). Hence, if (6F) applies at $x = 0$, $y = y(0)$ to the

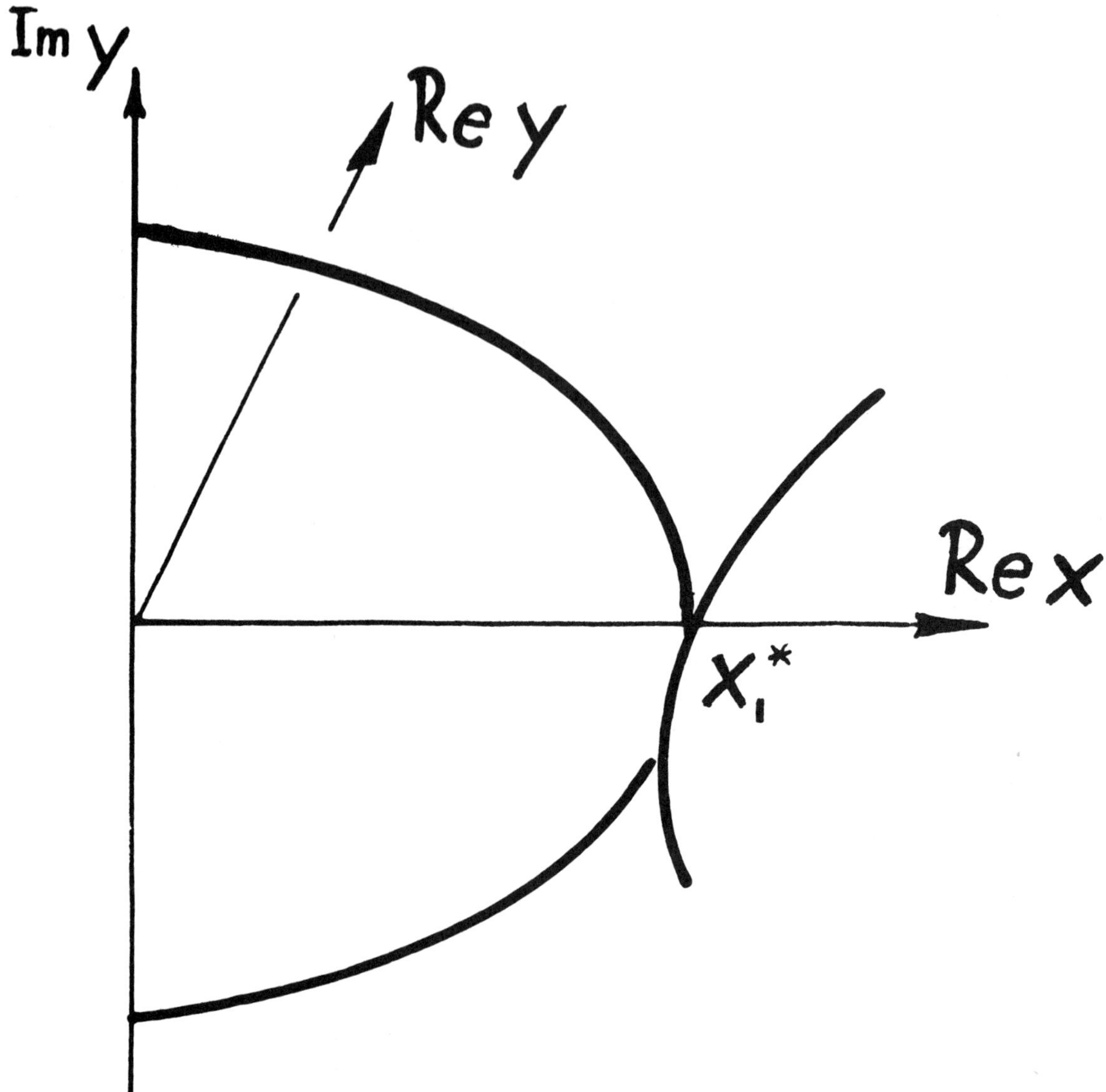

FIG. 5 *Two complex rays over the part of the barrier in $x \geq 0$, sketched in the space of real x and complex y.*

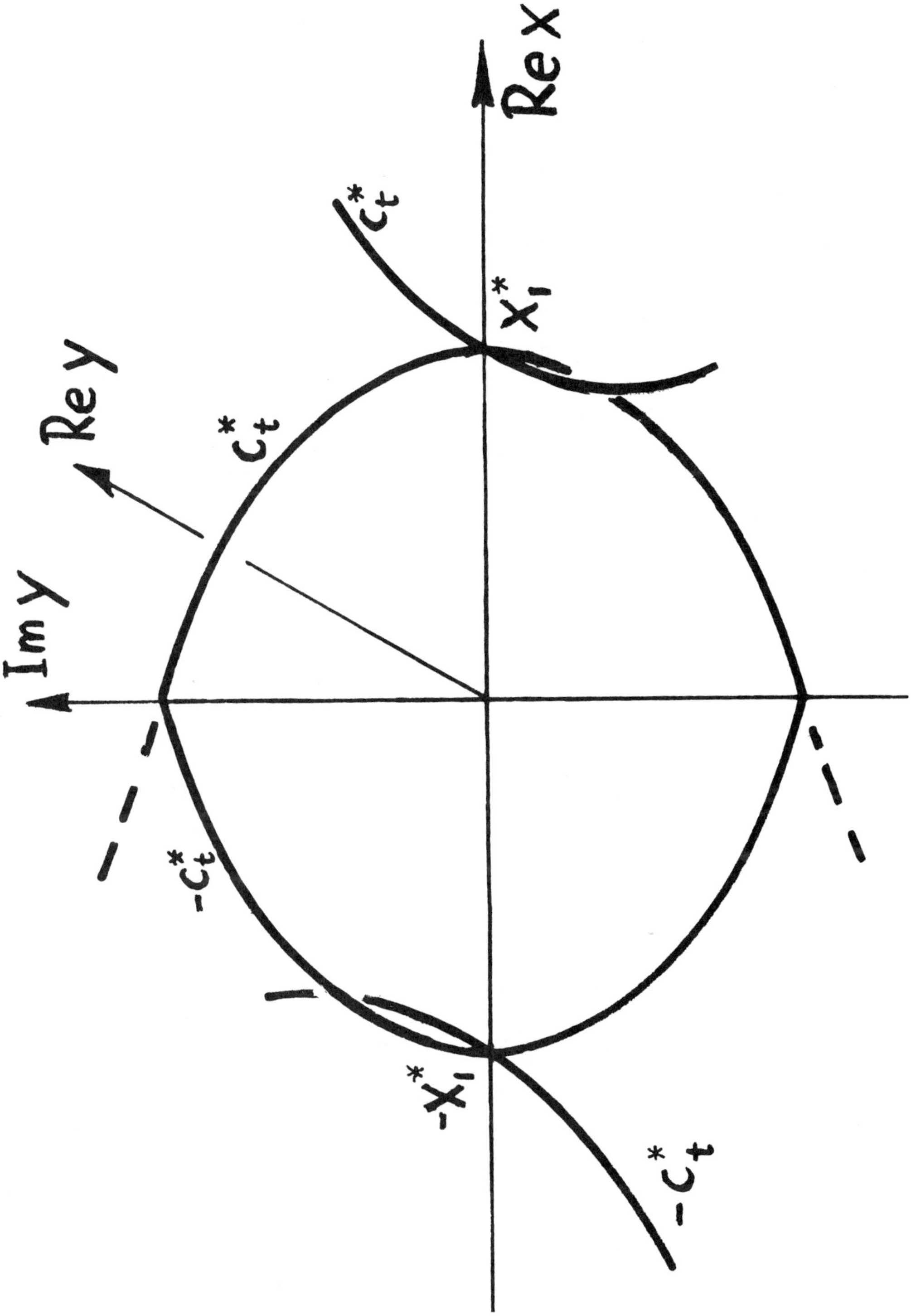

FIG. 6 *A complex ray system over the whole barrier in the space of real x and complex y with the continuation of the complex ray by real rays over the wells.*

approximation on each trajectory on the righthand half of the barrier torus, then it determines an approximation odd in x on the complete barrier torus. The lefthand half of the barrier torus carries this odd approximation to the lefthand well (Fig. 5), so that the approximation on the complete set of components of each trajectory respects the antisymmetry of the "real" Fermion pair of waves in the two wells. This, rather than (6F) directly, is the relevant symmetry concept for the double-well problem of the wave equation for Fermions; and similarly, for the Boson pair.

Now, if (6F) is applied to $\phi_1^{out-} + \phi_1^{in+}$ at the complex point $(0, y(0))$ of the trajectory branch (Fig. 3) labelled $\pm c_t$ and carrying this approximand pair, it follows immediately from (15) that

$$B_{1+}^*/A_{1-}^* = -\exp[-\Gamma(c_t)/\varepsilon], \tag{18}$$

$$\Gamma(c_t) = 2\int_0^{x_1(c_t)} |F(x; c_t)|^{1/2} dt,$$

where $x_1(c_t) > 0$ is the caustic value of x on this trajectory. When (6F) is applied to the other trajectory branch (Fig. 3) with the same label c_t, then the same formula (18) is obtained for B_{1-}^*/A_{1+}^*. The integral in (18) is a definite integral at fixed c_t along either of the two trajectory branches (Fig. 3) and Figure 6 shows that $\Gamma(c_t)$ is actually half of $\int(\nabla S)\cdot d\mathbf{s}$ around the complete, complex barrier torus on the trajectory labelled $\pm c_t$. But then, it must have the same value on each such fundamental path around the barrier torus and must equal

$$\Gamma = 2\int_0^{x_1^*} |E - c_t^{*2} - V(x)|^{1/2} dx, \tag{19}$$

where x_1^* is the caustic point on the real x-axis at the junction of well and barrier and c_t^*, is the value of c_t at x_1^*. Both x_1^* and c_t^* are already determined by the standard, real Geometrical Optics of the well. Moreover, since the integrand in (19) is analytic on the path, Γ can be evaluated at real x, so that it requires no reference to the complex extension of the physical potential specified at the very start in Section III; it adds only a computational effort of zero measure to the Geometrical Optics for a single well without tunneling. (The last sentence applies specifically to the problem class discussed here, but is a remarkable result despite this qualification.)

Application of (6B) gives similarly

$$B_{1+}^*/A_{1-}^* = B_{1-}^*/A_{1+}^* = e^{-\Gamma/\varepsilon}. \tag{20}$$

Apart from (18) or (20), the A_j^* and B_j^* must, of course, satisfy (16) and the standard connection across the well (Keller, 1958, Keller and Rubinow, 1960) involving (17) and also the connection relation analogous to (16) at the right-hand caustic x_2 (Figs. 1, 4), as well as the radiation condition (7) equivalent to $A^* = 0$ in the far field (Meyer, 1989c). All these together form a system of four linear algebraic equations for the A_j^*, B_j^* (Meyer, 1989c), which have a nontrivial solution if and only if

$$\Lambda/\varepsilon + \mu_0 + \mu_1 + \mu_2 = n_1\pi, \quad n_1 = 0, 1, 2, \ldots \tag{21}$$

$$\mu_0 = \mp \exp(-\Gamma/\varepsilon), \quad \mu_1 = \mu_2 = -\pi/4,$$

where Λ and Γ are given by (17) and (19), respectively, μ_1 and μ_2 are the Keller-Maslov indices (Keller, 1958, Keller and Rubinow, 1960, Maslov, 1972, Percival, 1977, Keller, 1985) arising directly from the caustic Stokes phenomena, and μ_0 is the novel index contributed by the "tunneling" interaction between the disjoint wells; the upper sign in μ_0 is associated with the antisymmetrical eigenfunctions (Fermions) and the lower sign, with the symmetrical ones (Bosons).

At first sight, (21) might appear inconsistent by the conventional rules of mathematical asymptotics, because Λ and Γ are only first approximations to the exact connection integrals. For fixed n_1, $V(x)$ and $b_i(x)$, in particular, Λ is but the "first term of an asymptotic expansion" in powers of ε, so that it looks as if the "exponentially small" μ_0 would be swamped by the "further terms" in Λ when ε is small enough. Olver (1974) has long shown us, however, that the conventional asymptotic rules are much too superficial. There is a simple, brief line of argument, based merely on the most obvious qualitative properties of the error structure, which proves (Meyer, 1989c) that μ_0 gives the rigorous first approximation to the Fermion-Boson split, regardless of "the order" to which a better approximation for Λ be pushed. The split can be calculated independently of the joint eigenlevel, just as it can be observed independently of the joint energy level.

However, (21) by itself does not predict discrete eigenvalues because it is independent of the sidewall boundary conditions (8N) or (8D). Those can be satisfied in the real well by Keller's (1958) original analysis of single-well spectra. It contributes a well-known further condition,

$$\frac{1}{\varepsilon}\int_{b_2(x_0)}^{b_1(x_0)} |c_t(y)|dy + \mu_+ + \mu_- = n_2\pi, \quad n_2 = 0, 1, 2, \ldots, \tag{22}$$

where the subscripts + and − refer to the sidewalls $y = b_1(x)$ and $y = b_2(x)$, respectively, and μ_+, μ_- are the Keller indices (Keller and Rubinow, 1960), which take the value zero at a boundary where (8N) applies and the value $-\pi/2$, where (8D) applies. Keller's (1958) homology argument shows why

(22) does not depend on the (real) value of x_0 chosen for the integral; and $|c_t(y)|$ in (22) is $|\partial S/\partial y|$ as function of the real y over the well traverse at x_0, which is given by standard, real Geometrical Optics.

Together, (21) and (22) are the "EBK Quantization Rule" for the double well in two dimensions with potential depending on only one coordinate. They determine a discrete spectrum dependent on the counting parameter $n = (n_1, n_2)$ and can, at the same level of approximation, be rewritten readily (Meyer, 1989c) as formulae for the mean eigenlevel $\frac{1}{2}(E_{nF} + E_{nB})$ and the energy split $E_F - E_B$, but *no more*. The mean eigenvalue, in fact, is that which each of the two identical wells (Fig. 1) would have if they did not interact. The energy split

$$E_F - E_B = 2\varepsilon e^{-\Gamma/\varepsilon}/(d\Lambda/dE)$$

is "exponentially small", but at reasonable values of ε, it need not be small quantitatively when $|n|$ is large, as it is for most of the eigenvalues.

The new EBK Rule (21), (22) differs from the known such Rules (Keller, 1958, Keller and Rubinow, 1960, Percival, 1977, Keller, 1985) in a disagreeable conceptual respect. Those earlier rules have the attractively simple interpretation that the simplevaluedness of wave functions requires the canonical action to change by an integer multiple of 2π around each path of an homology basis of the invariant torus. Such an interpretation fails when the interaction between disjoint wells has an observable effect: the new index shows that the canonical action change around the well-torus is then modified by a transcendental function of an action change around the complex barrier torus.

IX. OTHER POTENTIALS

The double well in two dimensions with potential dependent only on one co-ordinate is very helpful for a concrete explanation of the main line of quantitative analysis common to a much larger class of wave spectra. For instance, the analysis for the double well or periodic potential in a three-dimensional duct of rectangular cross-section differs very little, it merely adds to the spectral prediction a second rule of the form (22), which is well-known from standard EBK theory (Keller and Rubinow, 1960) of single wells in three dimensions. When waves in a duct of more general cross-section are considered, the basic analysis is not changed, but the numerical aspects of the standard Geometrical Optics of the well become much more laborious; by comparison, the added numerical work for the prediction of the tunneling effect remains quite minor.

In many contexts, sidewalls are not germane and potentials $V(x, y, z)$ need to be considered. This causes no substantial change in the basic analysis, as long as $V(x, y, z)$ is analytic in the three variables, is strictly even in one, and grows sufficiently at large $|x|$ to assure clear radiation conditions. Indeed, the analysis covers quadruple and sextuple wells, for which an index of the new type arises in two, or in all three, EBK rules to predict the more complicated eigenvlaue split.

The analyticity of the potential, moreover, is not of primary importance, but has been assumed mainly for the ease with which it affords clarity of structure. It has been used decisively only in the neighborhood of the caustic branch manifolds of the characteristic function to document very clearly the still somewhat unfamiliar, strictly local approach to the essence and resolution of the Stokes phenomena (Meyer and Painter, 1983, Meyer, 1989d). This simple, direct approach to them appears almost indispensable for "exponential precision" in more than one dimension. It can also deal with a very general class (Meyer and Painter, 1983) of branch structures including those arising from singularities of the potential, rather than from mere "turning points". Certainly, potentials in C^N with finite N can be treated by the line of analysis here outlined, but they are unrealistic in science. Nonanalytic potentials in C^∞, however, cause serious additional difficulties and must be expected to generate a more sophisticated, transcendental form of the new index (Stengle, 1985).

Nonlinear wave equations open up so large a field that concrete, quantitative results can be anticipated only for specific sub-classes. Serious additional difficulties arise even then, but the analysis here outlined has been designed especially to cope with some of the central obstacles then encountered. First, the characteristic function of Hamilton and Jacobi, and its exact Geometrical Optics, are not restricted in any way to linear wave operators. Secondly, the need to focus the analysis on a single, arbitrarily chosen trajectory becomes even more imperative for nonlinear operators (Meyer, 1976). Thirdly, the ability to resolve Stokes phenomena on a strictly local level on individual trajectories is needed critically for nonlinear wave equations; indeed, connection depends only on the Frechet derivative of H at the particular turning point of the trajectory.

Even for linear operators, of course, it is easy to conceive of classes of potentials of a generality for which the labor of computation prohibits more concrete results than Harrell's (1980) and Maslov's (1983). Fortunately, such pathological potentials are rarely the object of quantitative observation in science.

REFERENCES

Goldstein, H. (1959). *Classical Mechanics*, Addison-Wesley, Reading, MA.

Harrell, E. M. (1978). *Commun. Math. Phys.*, **60**: 73.

Harrell, E. M. (1980). *Commun. Math. Phys.*, **75**: 239.

Jaffe, C. (1988). *J. Chem. Phys.*, **88**: 7603.

Keller, J. B. (1958). *Ann. Phys.* (NY), **4**: 180.

Keller, J. B. (1985). *SIAM Rev.*, **27**: 485.

Keller, J. B. and Rubinow, S. I. (1960). *Ann. Phys.* (NY), **9**: 24.

Maslov, V. P. (1972). *Theorie des Perturbations*, Dunod, Paris.

Maslov, V. P. (1983). *Russ. Math. Surv.*, **38:6**: 1.

Maslov, V. P. and Fedoryuk, M. V. (1981). *Semiclassical Approximation in Quantum Mechanics*, Reidel, Dordrecht.

Meyer, R. E. (1973). *J. Appl. Math. Phys.*, **24**: 293.

Meyer, R. E. (1975). *SIAM J. Appl. Math.*, **29**: 481.

Meyer, R. E. (1976). *J. Appl. Math. Phys.*, **27**: 181.

Meyer, R. E. (1980). *SIAM Rev.*, **22**: 213.

Meyer, R. E. (1989a). On exponential asymptotics for nonseparable wave equations in several dimensions, Part I, to appear.

Meyer, R. E. (1989b). On exponential asymptotics for nonseparable wave equations in several dimensions, Part II, to appear.

Meyer, R. E. (1989c). On exponential asymptotics for nonseparable wave equations in several dimensions, Part III, to appear.

Meyer, R. E. (1989d). A simple explanation of the Stokes phenomenon, to appear in *SIAM Rev.*

Meyer, R. E. and Painter, J. F. (1983). *SIAM J. Math. Anal.*, **14**: 450.

Olver, F. W. J. (1974). *Asymptotics and Special Functions*, Academic Press, New York.

Percival, I. C. (1977). *Adv. Chem. Phys.*, **36**: 1.

Skodje, R. T., Borondo, F., and Reinhardt, W. P. (1985). *J. Chem. Phys.*, **82**: 461. Skodje, R. T. and Borondo, F. (1986). *J. Chem. Phys.*, **84**: 1533.

Stengle, G. (1985), *SIAM J. Math. Anal.*, **16**: 371.

Weinstein, M. I. and Keller, J. B. (1985). *SIAM J. Appl. Math.*, **45**: 45 and ibid. (1987), **47**: 941.

Zwaan, A. (1929). *Arch. Neerlandaises Sci. Exactes Natur. 3A*, **12**: 1.

On Stokes' Phenomenon and Converging Factors

Frank W. J. Olver Professor, Institute for Physical Science and Technology, University of Maryland, College Park, Maryland

Dedicated to my friends and fellow workers in asymptotic analysis, numerical analysis and special functions.

ABSTRACT

Recently M. V. Berry has provided a new interpretation of Stokes' phenomenon in which the change in form of a compound asymptotic expansion occurs smoothly, albeit very rapidly, as a Stokes line is crossed. Berry's analysis is based on R. B. Dingle's theory of terminants, and is therefore quite formal. In this paper new analysis is developed to place the theories of both Dingle and Berry on rigorous mathematical foundations. The analysis is illustrated by application to Macdonald's modified Bessel function, the generalized exponential integral and one of the confluent hypergeometric functions, but the method, like the methods of Dingle and Berry, is quite general.

This research has been supported by the U.S. National Science Foundation under Grant DMS 87-23039.

1. INTRODUCTION AND SUMMARY

Discontinuities in compound asymptotic expansions in the neighbourhood of an irregular singular point were first explored and described in a series of papers by Sir George Stokes, beginning with [12]. The need for discontinuities in the representation of analytic and *a fortiori* smooth functions was hard to accept,[1] especially because there is no single line or curve in the complex plane that can be identified as the exact place where the discontinuities occur. In consequence, there has always been an element of mystery associated with this "phenomenon" of Stokes, whether one approaches the subject from the standpoint of rigorous mathematical analysis or formal physical applications.

With a view to arriving at a fuller understanding of the phenomenon, I applied a theory of error bounds in a paper presented at the 1978 SIAM Spring Meeting, held at the University of Wisconsin [11]. The example chosen was Macdonald's modified Bessel function $K_\nu(z)$ with fixed real or complex values of the order ν, and large absolute values of the complex argument z. Let us use the symbols $\sum_1$ and $\sum_2$ to denote the formal series

$$\sum_1 = \left(\frac{\pi}{2z}\right)^{\frac{1}{2}} e^{-z} \sum_{s=0}^{\infty} \frac{A_s(\nu)}{z^s}, \qquad \sum_2 = \left(\frac{\pi}{2z}\right)^{\frac{1}{2}} e^{z} \sum_{s=0}^{\infty} (-)^s \frac{A_s(\nu)}{z^s}, \tag{1.1}$$

where

$$A_0(\nu) = 1, \quad A_s(\nu) = \frac{(4\nu^2-1^2)(4\nu^2-3^2)\cdots\{4\nu^2-(2s-1)^2\}}{s!\; 8^s}, \quad s \geq 1. \tag{1.2}$$

Also, let δ denote an arbitrarily small positive constant. Then in the sense of Poincaré

$$K_\nu(z) \sim \sum_1 \tag{1.3}$$

[1] Not least by Stokes himself; compare a letter to his fiancée, dated March 19, 1857 and reprinted in [13].

as $|z| \to \infty$, uniformly with respect to $\text{ph } z$ in the sector $|\text{ph } z| \le \frac{3}{2}\pi - \delta$. On the other hand, as $z \to \infty$ in the sector $\frac{1}{2}\pi + \delta \le \text{ph } z \le \frac{5}{2}\pi - \delta$ we have the compound expansion

$$K_\nu(z) \sim \sum_1 + 2i \cos(\nu\pi) \sum_2, \tag{1.4}$$

again uniformly valid (in Poincaré's sense) with respect to $\text{ph } z$. In this example the discontinuity occurs in the coefficient of $\sum_2$. In the common sector of validity, given by $\frac{1}{2}\pi + \delta \le \text{ph } z \le \frac{3}{2}\pi - \delta$, either (1.3) or (1.4) may be used on the grounds that the contribution of $\sum_2$ is uniformly exponentially small compared with that of $\sum_1$. However, in the sector $|\text{ph } z| \le \frac{1}{2}\pi - \delta$, $\sum_2$ dominates $\sum_1$; hence the omission of $\sum_2$ from the asymptotic expansion of $K_\nu(z)$ is mandatory. $\sum_2$ is also dominant in the sector $\frac{3}{2}\pi + \delta \le \text{ph } z \le \frac{5}{2}\pi - \delta$; consequently it is mandatory here to have the coefficient $2i \cos(\nu\pi)$ associated with $\sum_2$.

In which part of the common sector $\frac{1}{2}\pi + \delta \le \text{ph } z \le \frac{3}{2}\pi - \delta$ should we prefer the expansion (1.3), and in which part the expansion (1.4)? This was the question addressed in [11]. Except for a region **P** that is bounded by the imaginary axis and a parabola having the real axis as axis, I showed that for all sufficiently large $|z|$ the optimum form of (1.3) will yield greater or lesser accuracy than the optimum form of (1.4), according as $\text{ph } z \lessgtr \pi$; see Fig. 1. As a corollary, if it is decided that the change from (1.3) to (1.4) is to take place along a ray emanating from the origin, then this ray has to be the "Stokes line" $\text{ph } z = \pi$, since this is the only ray that is completely contained in **P**.

Combined with the error bounds given in [11], the result just stated is fully satisfactory for the purpose of constructing a computing algorithm for generating $K_\nu(z)$ for large values of $|z|$. However, it leaves unresolved the actual situation inside the region **P**. We know that just before we enter **P** from the second quadrant the coefficient of the series $\sum_2$ should be zero, whereas when we emerge from **P** into the third quadrant this coefficient needs to be $2i \cos(\nu\pi)$.

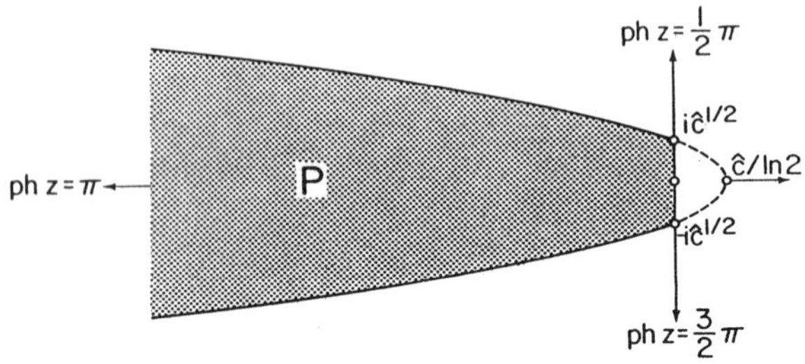

FIG. 1 *Region of uncertainty in the z-plane.*
$\hat{c} = (\frac{1}{2}\ln 2)^2 + \frac{1}{16} + \frac{1}{2}(\pi+1)|\nu^2 - \frac{1}{4}| + \frac{1}{2}|\nu^2 + \frac{1}{12}|.$
Reproduced with permission from Reference [11]. (Copyright 1980 by the Society for Industrial and Applied Mathematics. All rights reserved.)

In a recent paper [2], M. V. Berry has adopted the view that the coefficient of $\sum_2$ should be regarded as continuous function of ph z, rather than a discontinuous constant. This is achieved by terminating the series in equations (1.1): $\sum_1$ is truncated at its optimum number of terms, given by $n = \text{int}[2|z|] + 1$,[2] and $\sum_2$ is truncated at its first term. The required coefficient S_n, say, is defined *exactly* by the equation

$$K_\nu(z) = \left(\frac{\pi}{2z}\right)^{\frac{1}{2}} e^{-z} \sum_{s=0}^{n-1} \frac{A_s(\nu)}{z^s} + S_n \times 2i \cos(\nu\pi)\left(\frac{\pi}{2z}\right)^{\frac{1}{2}} e^z. \qquad (1.5)$$

Berry calls S_n the *Stokes multiplier*. Clearly S_n is a piecewise

---

[2] "int" means "integer part of".

continuous function of $|z|$, owing to the relation between n and $|z|$, and a continuous function of ph z. Berry claims that if $|z|$ is large and fixed and ph z increases from values somewhat below π to values somewhat above π, then S_n changes rapidly, *and of course continuously*, from 0 to 1. This interesting result is obviously consistent with the results of [11], as described above.

Section 2 of the present paper describes in detail Berry's method, as applied to the function $K_\nu(z)$. The required approximation to the Stokes multiplier S_n for large $|z|$ is derived by application of R. B. Dingle's theory of terminants. This theory is well-known to physicists, but is quite formal. In §§3 and 4 I indicate how to put Berry's method, and therefore incidentally Dingle's theory, on a rigorous analytical foundation. At the same time I show how to improve Berry's approximation for S_n, and also indicate briefly how to link the new results with the rigorous theory of converging factors that was presented in [10], Chapter 14. Some conclusions are drawn in the final section, §5.

The full proofs of the results of §§3 and 4 are quite long. In consequence, only the main steps are sketched; further details will be furnished elsewhere.

2. BERRY'S THEORY

Hankel's integral representation for Macdonald's function can be expressed in the form

$$K_\nu(z) = \frac{\sqrt{\pi}\left(\frac{1}{2}z\right)^\nu}{\Gamma\left(\nu+\frac{1}{2}\right)} \int_1^\infty e^{-zt}(t^2-1)^{\nu-\frac{1}{2}}\,dt, \tag{2.1}$$

valid when

$$\operatorname{Re}\nu > -\frac{1}{2}, \qquad |\text{ph } z| < \frac{1}{2}\pi;$$

see, for example, [10], p. 254. On replacing t by $1 + 2t$ we have

$$K_\nu(z) = \sqrt{\pi}\,\frac{(2z)^\nu e^{-z}}{\Gamma\left(\nu+\frac{1}{2}\right)} \int_0^\infty e^{-2zt}t^{\nu-\frac{1}{2}}(1+t)^{\nu-\frac{1}{2}}dt. \tag{2.2}$$

Then rotating the integration path through an angle $-\pi$ and replacing t by $te^{-\pi i}$ we obtain the analytic continuation of $K_\nu(z)$ to the sector $\frac{1}{2}\pi < \mathrm{ph}\, z < \frac{3}{2}\pi$, given by

$$K_\nu(z) = -i\sqrt{\pi}\, e^{-\nu\pi i} \frac{(2z)^\nu e^{-z}}{\Gamma\left(\nu+\frac{1}{2}\right)} \int_{\mathscr{C}} e^{2zt} t^{\nu-\frac{1}{2}} (1-t)^{\nu-\frac{1}{2}} dt. \tag{2.3}$$

The integration path $\mathscr{C}$ is from $t = 0$ to $t = +\infty$, but unless $\nu - \frac{1}{2}$ is a nonnegative integer $\mathscr{C}$ has to pass above the singularity at $t = 1$; see Fig. 2. The branches of $t^{\nu-\frac{1}{2}}$ and $(1-t)^{\nu-\frac{1}{2}}$ are continuous along $\mathscr{C}$ and take their principal values in the neighbourhood of $t = 0$.

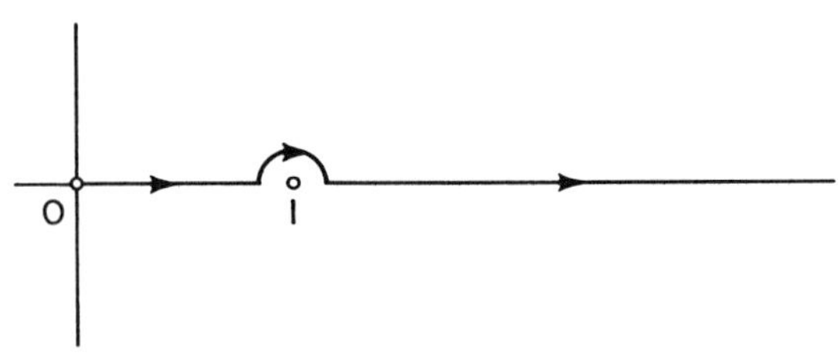

FIG. 2 *t-plane. Contour* $\mathscr{C}$.

The Maclaurin expansion of $(1-t)^{\nu-\frac{1}{2}}$ is given by

$$(1-t)^{\nu-\frac{1}{2}} = \sum_{s=0}^{\infty} (-)^s \binom{\nu - \frac{1}{2}}{s} t^s, \qquad |t| < 1.$$

Substituting this series in (2.3) and integrating term by term, as in Laplace's method, we derive

$$K_\nu(z) \sim -i\sqrt{\pi}\, e^{-\nu\pi i} \frac{(2z)^\nu e^{-z}}{\Gamma\left(\nu+\frac{1}{2}\right)} \sum_{s=0}^{\infty} (-)^s \binom{\nu - \frac{1}{2}}{s} \frac{\Gamma\left(s+\nu+\frac{1}{2}\right)}{(2ze^{-\pi i})^{s+\nu+\frac{1}{2}}} \tag{2.4}$$

as $z \to \infty$, valid when $\frac{1}{2}\pi + \delta \le \text{ph } z \le \frac{3}{2}\pi - \delta$.[3] This expansion is, of course, simply a rewritten form of (1.3) for this range of ph z.

Now for large s

$$(-)^s \binom{\nu - \frac{1}{2}}{s} \Gamma\left(s+\nu+\frac{1}{2}\right) = \frac{\Gamma(s)}{\Gamma\left(\frac{1}{2} - \nu\right)} \left\{1+O\left(\frac{1}{s}\right)\right\}. \tag{2.5}$$

Hence if $s \ge n$,

$$(-)^s \binom{\nu - \frac{1}{2}}{s} \Gamma\left(s+\nu+\frac{1}{2}\right) = \frac{\Gamma(s)}{\Gamma\left(\frac{1}{2} - \nu\right)} \left\{1+O\left(\frac{1}{n}\right)\right\} \tag{2.6}$$

as $n \to \infty$, uniformly with respect to s. Following Dingle [4], [5] and [6], Chapter 21, Berry terminates the divergent series (2.4) at its n th term, replaces the coefficients in the remainder by the right-hand side of (2.6) and then interprets the remainder by Borel summation. This yields (1.5), with

$$S_n = \left\{1 + O\left(\frac{1}{n}\right)\right\}\hat{S}_n \tag{2.7}$$

and

$$\hat{S}_n = \frac{\sec(\nu\pi)e^{-2z}}{2i\Gamma\left(\nu+\frac{1}{2}\right)\Gamma\left(\frac{1}{2} - \nu\right)} \int_{\mathscr{C}} e^{2zt} \frac{t^{n-1}}{1-t}\, dt = \frac{e^{-2z}}{2\pi i} \int_{\mathscr{C}} e^{2zt} \frac{t^{n-1}}{1-t}\, dt. \tag{2.8}$$

Again, these equations are valid when $\frac{1}{2}\pi + \delta \le \text{ph } z \le \frac{3}{2}\pi - \delta$.

---

[3]Compare [10], Chapter 4, Theorem 6.1. For each choice of δ the indentation of $\mathscr{C}$ at $t = 1$ has to be sufficiently small. Observe that we cannot use Watson's lemma here because $\mathscr{C}$ does not coincide with the positive real axis.

Another way of viewing the result just obtained is to apply Laplace's method to the last integral and insert the result in (1.5) by means of (2.7). This reproduces the asymptotic expansion (2.4), except that in the terms beyond the n th the coefficients $(-)^s\binom{\nu-\frac{1}{2}}{s}\Gamma\left(s+\nu+\frac{1}{2}\right)$ are replaced by the right-hand side of (2.6). This is exactly what we expect, since Borel summation is formally equivalent to the inversion of Watson's lemma, or in the present case Laplace's method.

The next step is to approximate S_n, or equivalently, $\hat{S}_n$. As in the theory of converging factors it is crucial to choose n to be either at, or close to, its optimum value. The optimum situation occurs when the infinite series $\sum_1$ is truncated just before its numerically smallest term: from (1.1), (1.2) and (1.5) we see that for large $|z|$ this corresponds to $n = \text{int}[2|z|] + 1$. Berry sets

$$2z = -A - iB, \qquad n - 1 = A + \mu, \tag{2.9}$$

where A, B and μ are real, A being large and positive and $|\mu|$ being bounded. Since

$$\mu = \text{int}\left[\sqrt{A^2+B^2}\right] - A = \text{int}\left[A + \frac{B^2}{2A} - \frac{B^4}{8A^3} + \cdots\right] - A,$$

this assumption on μ implies $B^2/(2A)$ is bounded. Thus we are working in a region of the z-plane similar to that indicated in Fig. 1. On substituting in (2.8) by means of (2.9) we have

$$\hat{S}_n = \frac{e^{A+iB}}{2\pi i}\int_{\mathscr{C}} e^{-(A+iB)t}\,\frac{t^{A+\mu}}{1-t}\,dt.$$

To obtain an approximation for large A we first shift the singularity of the integrand to the origin by the change of variable $t = 1 + \tau$. This gives

$$\hat{S}_n = -\frac{1}{2\pi i}\int_{-1}^{\infty} e^{-A\{\tau-\ell n(1+\tau)\}}(1+\tau)^{\mu}e^{-iB\tau}\,\frac{d\tau}{\tau},$$

where the path passes above the origin. Again, following Laplace's method we expect that as $A \to \infty$ the major contribution to this integral derives from the neighbourhood of $\tau = 0$. We therefore expand locally in powers of τ, and extend the lower limit of integration to $-\infty$. This leads to the approximation

$$\hat{S}_n \sim -\frac{1}{2\pi i}\int_{-\infty}^{\infty} e^{-\frac{1}{2}A\tau^2}\left(1+\mu\tau+\frac{1}{3}A\tau^3\right)e^{-iB\tau}\,\frac{d\tau}{\tau}. \tag{2.10}$$

By taking the indentation of the path above the origin to be a semicircle and then letting the radius tend to zero, we see that

$$-\frac{1}{2\pi i}\int_{-\infty}^{\infty} e^{-\frac{1}{2}A\tau^2}\cos(B\tau)\frac{d\tau}{\tau} = \frac{1}{2}.$$

In the remaining contributions the singularity at the origin is removable, and again on discarding the odd parts of the integrand we see that (2.10) is exactly equivalent to

$$\hat{S}_n \sim \frac{1}{2} + \frac{1}{2\pi}\int_{-\infty}^{\infty} e^{-\frac{1}{2}A\tau^2}\,\frac{\sin(B\tau)}{\tau}\,d\tau + \frac{i}{2\pi}\int_{-\infty}^{\infty} e^{-\frac{1}{2}A\tau^2}\left(\mu+\frac{1}{3}A\tau^2\right)\cos(B\tau)d\tau,$$

where the integration paths are now along the real axis. Evaluating the last two integrals, we arrive at the desired approximation in the form

$$\hat{S}_n \sim \frac{1}{2} + \frac{1}{2}\,\mathrm{erf}\left(\frac{B}{\sqrt{2A}}\right) + i\,\frac{e^{-B^2/(2A)}}{\sqrt{2\pi A}}\left(\mu + \frac{1}{3} - \frac{B^2}{3A}\right), \tag{2.11}$$

where, as usual, erf denotes the error function:

$$\mathrm{erf}(\zeta) = \frac{2}{\sqrt{\pi}}\int_0^{\zeta} e^{-t^2}dt. \tag{2.12}$$

This is, essentially, Berry's approximation to the Stokes multiplier for the present problem. He goes on to observe that owing to the factor $(2\pi A)^{-1/2}$ the first two terms dominate, and hence, in consequence of (2.7),

$$S_n \sim \frac{1}{2} + \frac{1}{2}\,\mathrm{erf}\left(\frac{B}{\sqrt{2A}}\right). \tag{2.13}$$

The extreme values of the error function are given by

$$\mathrm{erf}(-\infty) = -1, \qquad \mathrm{erf}(+\infty) = 1,$$

and as the following table shows these extremes are approached rapidly:

ζ	$-\infty$	-2	-1	0	1	2	∞
$\mathrm{erf}(\zeta)$	-1	-0.995	-0.843	0	0.843	0.995	1

Hence even though we have assumed $|B|\sqrt{2A}$ to be bounded, the approximation (2.13) indicates that S_n increases smoothly from the value 0 when $B/\sqrt{2A}$ is moderately large and negative, to $\frac{1}{2}$ when $B/\sqrt{2A} = 0$, and finally to 1 when $B/\sqrt{2A}$ is moderately large and positive. This is exactly what we expect, because the corresponding change in $\mathrm{ph}\, z$ is from values just less than π to values just greater than π; compare the first of equations (2.9). The importance of Berry's result is that it quantifies the change in S_n as $\mathrm{ph}\, z$ varies with $|z|$ fixed. Firstly, the change is continuous. Secondly, it is very rapid because $|\arctan(B/A)|$ is vanishingly small when $A \to \infty$ with $|B|/\sqrt{2A}$ bounded. Berry calls the quantity $B/\sqrt{2A}$ the *Stokes variable* because of its key role in the approximation of S_n.

Although we have shown how to treat Stokes' phenomenon only in the vicinity of $\mathrm{ph}\, z = \pi$ for Macdonald's function, the method is quite general. For example, Berry applies it to the asymptotic expansion of Dawson's integral $e^{-z^2}\int_0^z e^{t^2}dt$ and the Airy function of the second kind $\mathrm{Bi}(z)$, near $\mathrm{ph}\, z = 0$ in both cases. Furthermore, in both cases he computes S_n directly from its defining equation (compare (1.5)) for various values of z, and demonstrates that the numerical results agree well with the corresponding values derived from his approximation (compare (2.13)). The essential requirement for the method to succeed for a given asymptotic expansion is that for large s the s th term can be approximated by an expression of the form $M\Gamma(s-\beta+1)F^{\beta-1-s}$, where M, F and β are independent of s; compare (2.4) and (2.5). When this condition is fulfilled, Dingle's theory is applicable and yields an integral for the Stokes multiplier similar to (2.8), from which approximations of the forms (2.11) and (2.13) again follow.

From the mathematical standpoint there are two major problems in putting this valuable formal analysis on a rigorous basis. Firstly, there is the justification of Dingle's theory used in the construction of an appropriate integral representation of the Stokes multiplier. Secondly, there is the approximation to this multiplier in terms of the error function. Had we considered the generalized exponential integral in place of Macdonald's function, the first problem would not have arisen. Because of this, and also because the results for the generalized exponential integral play a fundamental role in the other cases, we begin with the generalized exponential integral, deferring the more difficult problem of justifying Dingle's theory until §4.

3. GENERALIZED EXPONENTIAL INTEGRAL

For real or complex values of p and z, other than $z = 0$, the function $E_p(z)$ is defined by

$$E_p(z) = z^{p-1} \int_z^\infty \frac{e^{-t}}{t^p}\, dt, \tag{3.1}$$

with appropriate choices of integration path and branches of the multivalued functions. In the sector $|\mathrm{ph}\, z| < \frac{1}{2}\pi$, this definition is equivalent to

$$E_p(z) = e^{-z} \int_0^{\infty} \frac{e^{-zt}}{(1+t)^p} \, dt, \tag{3.2}$$

where the integration path is now the positive real axis, and $(1+t)^p$ assumes its principal value. Our concern in this section is the asymptotic behaviour of $E_p(z)$ for large $|z|$ and fixed p. By applying Watson's lemma to the last integral and extending the region of validity of the result by analytic continuation,[4] we obtain

$$E_p(z) \sim \frac{e^{-z}}{z} \sum_{s=0}^{\infty} (-)^s \frac{(p)_s}{z^s}, \qquad |\text{ph } z| \le \frac{3}{2}\pi - \delta, \tag{3.3}$$

where $(p)_s$ is Pochhammer's notation for $p(p+1)\cdots(p+s-1)$ and δ is again an arbitrarily small positive constant.

To derive asymptotic expansions of $E_p(z)$ for other ranges of ph z, we may use the continuation formula[5]

$$E_p(z) = \frac{2\pi i e^{-p\pi i}}{\Gamma(p)} z^{p-1} + E_p(ze^{-2\pi i}). \tag{3.4}$$

Thus for the sector $\frac{1}{2}\pi + \delta \le \text{ph } z \le \frac{7}{2}\pi - \delta$ we have

$$E_p(z) \sim \frac{2\pi i e^{-p\pi i}}{\Gamma(p)} z^{p-1} + \frac{e^{-z}}{z} \sum_{s=0}^{\infty} (-)^s \frac{(p)_s}{z^s}. \tag{3.5}$$

Comparison of (3.3) and (3.5) shows that Stokes' phenomenon occurs in the vicinity of ph $z = \pi$. We now investigate this region in detail.[6]

We shall need an alternative integral representation, given by

---

[4]See, for example, [10], Chapter 4, §3.

[5]Provable by use of (3.6) below.

[6]Stokes' phenomenon also occurs in the vicinities of the other Stokes lines ph $z = -\pi$, $\pm 3\pi$, $\pm 5\pi, \ldots$, and similar methods can be used in these cases.

$$E_p(z) = \frac{z^{p-1}e^{-z}}{\Gamma(p)} \int_0^\infty \frac{e^{-zt}t^{p-1}}{1+t}\,dt, \qquad |\text{ph } z| < \tfrac{1}{2}\pi. \tag{3.6}$$

When p is a positive integer this formula can be obtained from the integral (3.2) by repeated integration by parts; see, for example, [9], §3.[7] However, as long as z^{p-1} and t^{p-1} have their principal values it follows by application of the convolution theorem for the Laplace transform that formula (3.6) is valid in the half-plane $\text{Re } p > 0$. Next, if n denotes an arbitrary nonnegative integer, we have identically

$$\frac{1}{1+t} = 1 - t + t^2 - \cdots + (-)^{n-1}t^{n-1} + \frac{(-)^n t^n}{1+t}, \qquad t \neq -1.$$

The result of substituting in (3.6) by means of this expansion and then integrating term by term can be expressed in the form

$$E_p(z) = \frac{e^{-z}}{z} \sum_{s=0}^{n-1} (-)^s \frac{(p)_s}{z^s} + \frac{2\pi i e^{-p\pi i}}{\Gamma(p)} z^{p-1} T_{n+p}(z), \tag{3.7}$$

where

$$T_q(z) = \frac{e^{q\pi i}\Gamma(q)}{2\pi i} \frac{E_q(z)}{z^{q-1}}. \tag{3.8}$$

Although we have assumed that $\text{Re } p > 0$ and $|\text{ph } z| < \frac{1}{2}\pi$ in deriving (3.7), analytic continuation with respect to p and z removes both of these restrictions, other than $z \neq 0$. Accordingly, we now have an explicit formula for the n th remainder term in the asymptotic expansion (3.3) expressed as a multiple $T_{n+p}(z)$ of the "extra term" $2\pi i e^{-p\pi i} z^{p-1}/\Gamma(p)$ in the expansion (3.5). In Berry's

---

[7] The integral (3.6), with p a positive integer, plays a fundamental role in Dingle's theory [4] and [6], Chapter 21: his converging factor $\Lambda_s(z)$ is defined by

$$\Lambda_s(z) = \frac{1}{s!} \int_0^\infty \frac{\varepsilon^s e^{-\varepsilon}}{1+(\varepsilon/z)}\,d\varepsilon, \quad |\text{ph } z| < \pi, \quad s = 0,1,2,\ldots .$$

In our present notation $\Lambda_s(z) = z e^z E_{s+1}(z)$.

terminology, $T_{n+p}(z)$ is the Stokes multiplier.

From (3.8) and (3.6) it is seen that

$$T_q(z) = \frac{e^{q\pi i}}{2\pi i} e^{-z} \int_0^\infty \frac{e^{-zt} t^{q-1}}{1+t} dt, \tag{3.9}$$

valid when $|\mathrm{ph}\, z| < \frac{1}{2}\pi$ and $\mathrm{Re}\, q > 0$. As in the case of the integral (2.2) we rotate the path of integration in (3.9) through an angle $-\pi$, and then replace t by $te^{-\pi i}$. This yields

$$T_q(z) = \frac{e^{-z}}{2\pi i} \int_{\mathscr{C}} \frac{e^{zt} t^{q-1}}{1-t} dt, \tag{3.10}$$

where the contour $\mathscr{C}$ is again as indicated in Fig. 2. This formula is valid when $\mathrm{Re}\, q > 0$ and z lies in the sector of interest, given by $\frac{1}{2}\pi < \mathrm{ph}\, z < \frac{3}{2}\pi$.

As in §§1 and 2, we consider the multiplier $T_{n+p}(z)$ when the infinite series in (3.7) is truncated just before its numerically smallest term. The corresponding value of n is given by $n = \mathrm{int}[\hat{n}] + 1$, where $|p + \hat{n}| = |z|$. Accordingly, we set

$$z = \rho e^{i(\theta+\pi)}, \qquad n = \rho - p + \alpha, \tag{3.11}$$

where ρ is a large positive parameter and $|\theta|$ and $|\alpha|$ are bounded, θ being real. This gives

$$T_{n+p}(z) = T_{\rho+\alpha}(\rho e^{i(\theta+\pi)}) = \frac{\exp(\rho e^{i\theta})}{2\pi i} \int_{\mathscr{C}} \exp(-\rho e^{i\theta} t) \frac{t^{\rho+\alpha-1}}{1-t} dt. \tag{3.12}$$

For large ρ, the integrand in (3.12) has a saddle-point at $t = e^{-i\theta}$ that coalesces with the simple pole at $t = 1$ when $\theta = 0$. Problems of this type have been considered by several writers, especially van der Waerden [14], Bleistein [3] and Jones [8]. To construct an asymptotic expansion for $T_{n+p}(z)$ when ρ is large that is uniform with respect to θ in a closed interval containing $\theta = 0$, we follow these authors and employ a quadratic change of integration variable of the form

$$e^{i\theta}t - \ell n\ t = \frac{1}{2}w^2 + icw + d, \tag{3.13}$$

where c and d are chosen in such a way that the singularity $t = 1$ corresponds to $w = 0$, and the saddle-point $t = e^{-i\theta}$ corresponds to $w = -ic$. Accordingly, we have

$$d = e^{i\theta}, \quad \frac{1}{2}c^2 = 1 + i\theta - e^{i\theta}. \tag{3.14}$$

In solving the last equation for c as a function of θ we choose the continuous branch with the Maclaurin expansion that begins

$$c = \theta + \frac{1}{6}i\theta^2 - \frac{1}{36}\theta^3 + \cdots . \tag{3.15}$$

Values of $\frac{1}{2}c^2$ and c for $\theta \in [-\frac{1}{2}\pi, \frac{1}{2}\pi]$ are indicated on Figs. 3 and 4.

By carrying out the conformal mapping from the t-plane to the w-plane, and making appropriate deformations of integration path, we find that

$$T_{n+p}(z) = \frac{\exp(\rho e^{i\theta})}{2\pi i} \int_{-ic-\infty}^{-ic+\infty} e^{-\rho(\frac{1}{2}w^2+icw+d)} \frac{f(\theta,\alpha,w)}{w}\, dw, \tag{3.16}$$

where

$$f(\theta,\alpha,w) = t^{\alpha-1} \frac{w}{1-t} \frac{dt}{dw}, \tag{3.17}$$

and the integration path is parallel to the real axis, indented if necessary with a loop to pass above the origin; see Fig. 5. This result is valid in a θ-interval that includes $\left[-\frac{1}{2}\pi, \frac{1}{2}\pi\right]$.

The function $f(\theta,\alpha,w)$ is free from singularity at $w = 0$, and we may therefore set

$$f(\theta,\alpha,w) = f(\theta,\alpha,0) + wg(\theta,\alpha,w), \tag{3.18}$$

where $g(\theta,\alpha,w)$ is analytic at $w = 0$. On substituting in (3.16) by means of this equation and observing that $\exp(\rho e^{i\theta}) = e^{\rho d}$, we obtain

$$T_{n+p}(z) = \frac{f(\theta,\alpha,0)}{2\pi i} \int_{-ic-\infty}^{-ic+\infty} e^{-\rho(\frac{1}{2}w^2+icw)} \frac{dw}{w} +$$

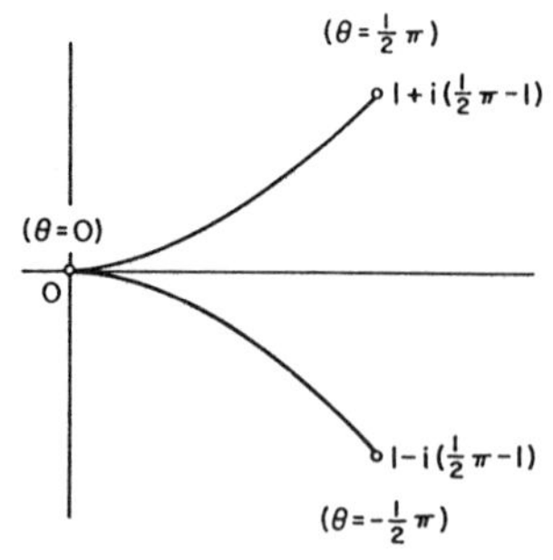

FIG. 3 $\frac{1}{2}c^2$-*plane.*

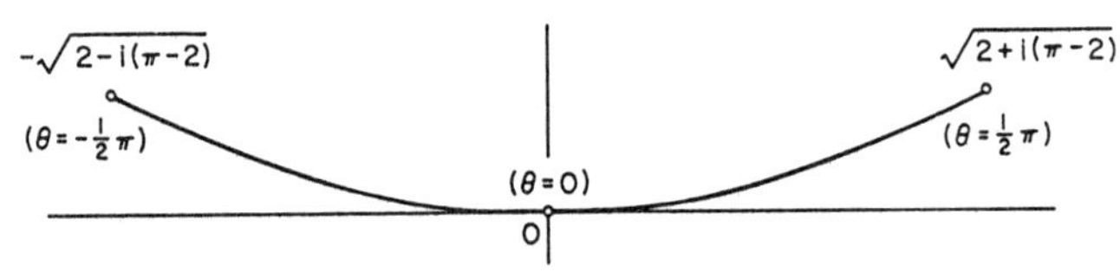

FIG. 4 *c-plane.*

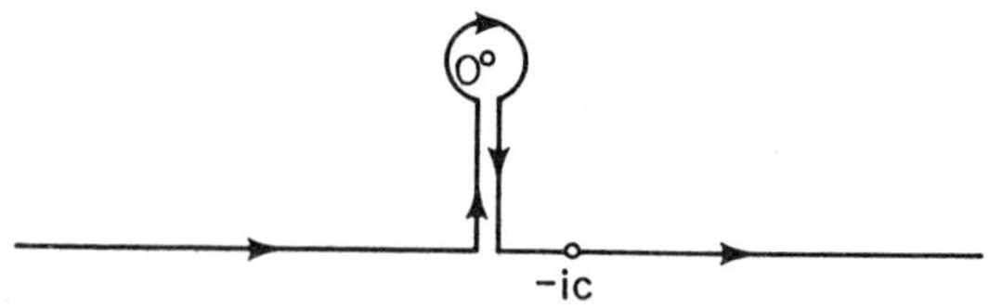

FIG. 5 *w-plane. Path for the integrals in equations (3.16) and (3.19).*

$$+ \frac{1}{2\pi i} \int_{-ic-\infty}^{-ic+\infty} e^{-\rho(\frac{1}{2}w^2+icw)} g(\theta,\alpha,w)dw. \tag{3.19}$$

In the first term in the right member of (3.19) the value of $f(\theta,\alpha,0)$ is found by letting $w \to 0$ in (3.17) and using L'Hospital's rule:

$$f(\theta,\alpha,0) = -1. \tag{3.20}$$

Furthermore, the contour integral can be evaluated in terms of the error function; thus

$$\int_{-ic-\infty}^{-ic+\infty} e^{-\rho(\frac{1}{2}w^2+icw)} \frac{dw}{w} = -\pi i\left\{1 + \operatorname{erf}\left(c\sqrt{\tfrac{1}{2}\rho}\,\right)\right\}. \tag{3.21}$$

In the remaining contribution to $T_{n+p}(z)$ in (3.19), the indentation of the integration path is no longer necessary, because the integrand is analytic at $w = 0$. The Taylor-series expansion of $g(\theta,\alpha,w)$ at the new saddle-point $w = -ic$ has the form

$$g(\theta,\alpha,w) = \sum_{s=0}^{\infty} g_s(\theta,\alpha)(w+ic)^s, \tag{3.22}$$

in which the coefficients $g_s(\theta,\alpha)$ are continuous functions of θ and α. Multiplying by $\exp\{-\rho(\frac{1}{2}w^2+icw)\}$ and integrating term by term, we find that

$$\int_{-ic-\infty}^{-ic+\infty} e^{-\rho(\frac{1}{2}w^2+icw)} g(\theta,\alpha,w)dw \sim e^{-\frac{1}{2}\rho c^2} \sum_{s=0}^{\infty} \Gamma\left(s+\frac{1}{2}\right) g_{2s}(\theta,\alpha)\left(\frac{2}{\rho}\right)^{s+\frac{1}{2}}, \tag{3.23}$$

as $\rho \to \infty$ uniformly with respect to $\theta \in \left[-\frac{1}{2}\pi, \frac{1}{2}\pi\right]$ and bounded values of $|\alpha|$.

Combination of (3.19), (3.20), (3.21) and (3.23) yields

$$T_{n+p}(z) \sim \frac{1}{2} + \frac{1}{2}\,\mathrm{erf}\left(c\sqrt{\tfrac{1}{2}\rho}\right) - i\,\frac{e^{-\frac{1}{2}\rho c^2}}{\sqrt{2\pi\rho}} \sum_{s=0}^{\infty} \left(\frac{1}{2}\right)_s g_{2s}(\theta,\alpha)\left(\frac{2}{\rho}\right)^s . \tag{3.24}$$

In combination with (3.7), this is the desired result.

We observe that when $\theta = 0$ we have $c = 0$. Accordingly (3.24) reduces to

$$T_{n+p}(z) \sim \frac{1}{2} - \frac{i}{\sqrt{2\pi\rho}} \sum_{s=0}^{\infty} \left(\frac{1}{2}\right)_s g_{2s}(0,\alpha)\left(\frac{2}{\rho}\right)^s . \tag{3.25}$$

Thus the error in the commonly-used mean value $\frac{1}{2}$ for the multiplier $T_{n+p}(z)$ in these circumstances is $O\left(\rho^{-\frac{1}{2}}\right)$, that is, $O\left(|z|^{-\frac{1}{2}}\right)$.

Next, from Figs. 3 and 4 we see that if ρ is large and fixed and θ increases from 0 to $\frac{1}{2}\pi$, then the real parts of both c and $\frac{1}{2}c^2$ are positive and increasing. In consequence, $e^{-\frac{1}{2}\rho c^2}$ becomes very small, $\mathrm{erf}\left(c\sqrt{\frac{1}{2}\rho}\right) \to 1 + O\left(e^{-\frac{1}{2}\rho c^2}\Big/\sqrt{\frac{1}{2}\pi\rho c^2}\right)$, and the multiplier $T_{n+p}(z)$ tends towards 1 rapidly. Similarly as θ decreases from 0 to $-\frac{1}{2}\pi$, $T_{n+p}(z)$ tends towards zero rapidly. This is exactly the kind of behaviour described by Berry; indeed, in the present circumstances his formal approximation to the Stokes multiplier is given by

$$T_{n+p}(z) \sim \frac{1}{2} + \frac{1}{2}\,\mathrm{erf}\left(\sin\theta\sqrt{\tfrac{1}{2}\rho\sec\theta}\right);$$

compare (2.13) with $A + iB = \rho e^{i\theta}$. Since

$$c = \theta + \frac{1}{6}i\theta^2 - \frac{1}{36}\,\theta^3 + \cdots ,$$

$$\sin\theta\sqrt{\sec\theta} = \theta + \frac{1}{12}\,\theta^3 + \cdots ,$$

our "Stokes variable" $c\sqrt{\frac{1}{2}\rho}$ is approximated by $\sin\theta\sqrt{\frac{1}{2}\rho\sec\theta}$ when θ is small. An advantage of our result, however, is that it furnishes a uniform asymptotic expansion for $T_{n+p}(z)$ in a

θ-interval that includes $\left[-\frac{1}{2}\pi, \frac{1}{2}\pi\right]$. *Inter alia* we no longer need assume that $|B|/\sqrt{2A}$ is bounded.

The coefficients $g_{2s}(\theta,\alpha)$ in the infinite series in (3.24) can be calculated from formulae (3.13), (3.17), (3.18), (3.20) and (3.22) in the usual manner; see, for example, [10], Chapter 9, §§10.1 and 12.2. The first one is found to be

$$g_0(\theta,\alpha) = \frac{i}{c} + \frac{e^{-i\alpha\theta}}{1-e^{-i\theta}}, \tag{3.26}$$

this expression being replaced by its limiting value

$$g_0(0,\alpha) = \frac{2}{3} - \alpha \tag{3.27}$$

at $\theta = 0$.

In the case $p = 1$ the last result can be confirmed by means of the asymptotic expansion of the converging factor for $E_1(xe^{\pi i})$, with x large and positive, that was derived in [10], Chapter 14, §4. In the notation of this reference we have

$$-e^{-x}E_1(xe^{\pi i}) = i\pi e^{-x} + \sum_{s=0}^{n-1} \frac{s!}{x^{s+1}} + C_n(x)\frac{n!}{x^{n+1}}, \tag{3.28}$$

where

$$C_n(x) = x - n - \frac{1}{3} + O(n^{-1}) \tag{3.29}$$

as $n \to \infty$, uniformly with respect to bounded values of $|x - n|$. Comparing this result with (3.7) and setting $z = xe^{\pi i}$, $p = 1$, we see that

$$T_{n+1}(xe^{\pi i}) = \frac{1}{2} + \frac{n!}{2\pi i}\frac{e^x}{x^{n+1}}C_n(x).$$

Then on substituting $x = \rho$ and $n = \rho - 1 + \alpha$ (compare (3.11)), and using (3.29) and Stirling's formula, we obtain

$$T_{n+1}(z) = \frac{1}{2} + \frac{i(\alpha-\frac{2}{3})}{(2\pi\rho)^{\frac{1}{2}}} + O\left(\frac{1}{\rho^{\frac{3}{2}}}\right),$$

in agreement with (3.25) and (3.27).

4. CONFLUENT HYPERGEOMETRIC FUNCTION

Instead of proceeding now to Macdonald's function, we shall consider one of the confluent hypergeometric functions:

$$U(a,a-b+1,z) = \frac{1}{\Gamma(a)} \int_0^\infty e^{-zt} t^{a-1} (1+t)^{-b} dt, \tag{4.1}$$

where the integration path is along the real axis and t^{a-1} and $(1+t)^{-b}$ assume their principal values. This integral converges when $\operatorname{Re} a > 0$ and $|\operatorname{ph} z| < \frac{1}{2}\pi$; elsewhere $U(a,a-b+1,z)$ is defined by analytic continuation. Since

$$K_\nu(z) = \sqrt{\pi}(2z)^\nu e^{-z} U(\nu+\tfrac{1}{2}, 2\nu+1, 2z),$$

results for Macdonald's function will follow as a special case; see below.[8]

By application of Watson's lemma we have

$$U(a,a-b+1,z) \sim z^{-a} \sum_{s=0}^{\infty} (-)^s \frac{(a)_s (b)_s}{s! z^s} \tag{4.2}$$

as $z \to \infty$ in the sector $|\operatorname{ph} z| \le \frac{3}{2}\pi - \delta$ $\left(< \frac{3}{2}\pi\right)$, for any fixed values of a and b.[9] The relevant continuation formula is supplied by

$$U(a,a-b+1,z) = e^{-2a\pi i} U(a,a-b+1,ze^{-2\pi i})$$
$$+ \frac{2\pi i e^{-a\pi i}}{\Gamma(a)\Gamma(b)} e^z U(1-b,a-b+1,ze^{-\pi i}). \tag{4.3}$$

Hence we derive

---

[8] Also included as special cases are Dawson's integral and the Airy function of the second kind. As noted in §2, these functions were used as examples in Berry's paper [2]. Dawson's integral is also included as a special case of the generalized exponential integral treated in §3.

[9] The restriction $\operatorname{Re} a > 0$ is removed in the usual way; compare, for example, [10], Chapter 4, §§3, 5.

$$U(a,a-b+1,z) \sim z^{-a} \sum_{s=0}^{\infty} (-)^s \frac{(a)_s (b)_s}{s! z^s}$$

$$-2\pi i \frac{e^{-(a+b)\pi i}}{\Gamma(a)\Gamma(b)} z^{b-1} e^z \sum_{s=0}^{\infty} \frac{(1-a)_s (1-b)_s}{s! z^s} \tag{4.4}$$

as $z \to \infty$ in the sector $\frac{1}{2}\pi + \delta \le \text{ph } z \le \frac{5}{2}\pi - \delta$.

Stokes' phenomenon occurs in the vicinity of $\text{ph } z = \pi$. We first seek a suitable representation for the remainder term in the asymptotic expansion (4.2). In [7] and [10], Chapter 14, this representation was obtained by use of the form of Taylor's theorem that is obtained by repeated integration by parts. This approach does not succeed for the present problem and the first step is to use instead Cauchy's integral formula.

On rotating the path of integration in (4.1) through an angle $-(\pi-\gamma)$, where γ is an arbitrary constant that satisfies $0 < \gamma \le \frac{1}{2}\pi$, and replacing t by $te^{-\pi i}$, we obtain

$$U(a,a-b+1,z) = \frac{e^{-a\pi i}}{\Gamma(a)} \int_0^{\infty e^{i\gamma}} e^{zt} t^{a-1} (1-t)^{-b} dt, \tag{4.5}$$

valid when $\frac{1}{2}\pi - \gamma < \text{ph } z < \frac{3}{2}\pi - \gamma$. We apply Taylor's theorem in the form

$$(1-t)^{-b} = \sum_{s=0}^{n-1} \frac{(b)_s}{s!} t^s + \chi_n(t), \tag{4.6}$$

in which n is an arbitrary nonnegative integer and

$$\chi_n(t) = \frac{t^n}{2\pi i} \int_{\mathcal{L}} \frac{(1-w)^{-b}}{w^n (w-t)} dw.$$

Here $\mathscr{L}$ is a simple closed contour that encloses $w = 0$ and $w = t$, but not $w = 1$; see, for example [1], Chapter 4, §3.1. Now suppose that Re $b > 0$. Then the integral of $(1-w)^{-b}w^{-n}(w-t)^{-1}$ around any large circle vanishes as the radius of the circle tends to infinity. Accordingly, $\mathscr{L}$ may be deformed into the loop integral depicted in Fig. 6, giving

$$\chi_n(t) = \frac{t^n}{2\pi i}\int_{\infty}^{(1-)} \frac{(1-w)^{-b}}{w^n(w-t)}\,dw. \tag{4.7}$$

On the upper side of the loop $(1-w)^{-b} = e^{b\pi i}(w-1)^{-b}$, where $(w-1)^{-b}$ has its principal value. Similarly, on the lower side of the loop $(1-w)^{-b} = e^{-b\pi i}(w-1)^{-b}$. Hence if, also, Re $b < 1$, we may collapse the loop on to the interval $[1,\infty)$. This yields

$$\chi_n(t) = \frac{\sin(b\pi)}{\pi}\, t^n \int_1^{\infty} \frac{(w-1)^{-b}}{w^n(w-t)}\,dw. \tag{4.8}$$

We now substitute in (4.5) by means of (4.6) and (4.8), and then integrate term by term to obtain

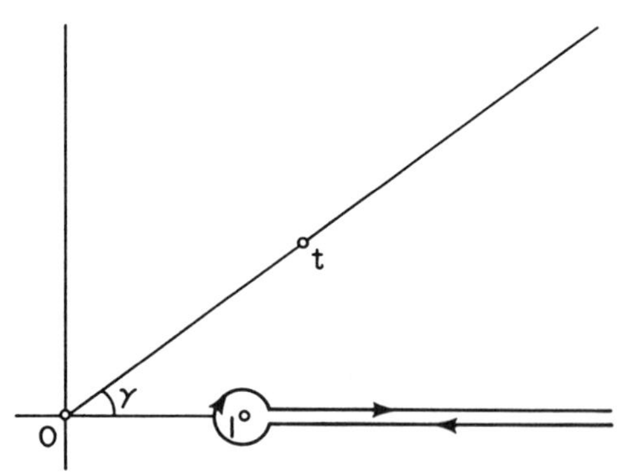

FIG. 6 *w-plane. Path for the integral (4.7).*

$$U(a,a-b+1,z) = z^{-a} \sum_{s=0}^{n-1} (-)^s \frac{(a)_s(b)_s}{s!z^s} + R_n(z), \tag{4.9}$$

where

$$R_n(z) = \frac{e^{-a\pi i}\sin(b\pi)}{\pi\Gamma(a)} \int_0^{\infty e^{i\gamma}} e^{zt}t^{n+a-1}dt \int_1^\infty \frac{(w-1)^{-b}}{w^n(w-t)}\,dw, \tag{4.10}$$

valid when $\frac{1}{2}\pi - \gamma < \text{ph } z < \frac{3}{2}\pi - \gamma$, $\text{Re } a > 0$ and $0 < \text{Re } b < 1$. This is the desired explicit representation of the n th remainder term in the expansion (4.2). Our objective now is to determine the asymptotic behaviour of this remainder term when the asymptotic expansion (4.2) is truncated at or near its numerically smallest term. Corresponding to (3.11) we set

$$z = \rho e^{i(\theta+\pi)}, \qquad n = \rho - a - b + 1 + \alpha, \tag{4.11}$$

where θ is real, and seek the asymptotic behaviour of $R_n(z)$ as $\rho \to \infty$ for fixed values of a and b, and bounded values of $|\theta|$ and $|\alpha|$.

Provided that $\text{Re } a < \text{Re } b$ we may convert the repeated integral (4.10) into a double integral, and then take new variables $\tau = t/w$, $v = w - 1$. This leads to

$$R_n(z) = \frac{e^{-a\pi i}\sin(b\pi)}{\pi\Gamma(a)} \int_0^{\infty e^{i\gamma}} \frac{\tau^{n+a-1}}{1-\tau}\, e^{z\tau}d\tau \int_0^\infty v^{-b}(1+v)^{a-1}e^{z\tau v}dv.$$

The last major step is to substitute for $(1+v)^{a-1}$ by means of the binomial expansion

$$(1+v)^{a-1} = \sum_{s=0}^{\infty} (-)^s \frac{(1-a)_s}{s!} v^s,$$

and then integrate term by term. This yields

$$R_n(z) \sim \frac{e^{-a\pi i}\sin(b\pi)}{\pi\Gamma(a)} \sum_{s=0}^{\infty} (-)^s \frac{(1-a)_s\Gamma(s+1-b)}{s!}$$

$$\times \int_0^{\infty e^{i\gamma}} \frac{\tau^{n+a-1}e^{z\tau}}{(1-\tau)(z\tau e^{-\pi i})^{s-b+1}}\, d\tau.$$

The integrals in this expansion may be identified in terms of the function $T_q(z)$ introduced in §3; compare equation (3.10). Using also the recurrence and reflection formulae for the Gamma function, we arrive at

$$R_n(z) \sim -2\pi i \frac{e^{-(a+b)\pi i}}{\Gamma(a)\Gamma(b)} z^{b-1}e^z \sum_{s=0}^{\infty} \frac{(1-a)_s(1-b)_s}{s!} \frac{T_{n-s+a+b-1}(z)}{z^s} \tag{4.12}$$

as $\rho \to \infty$, where z and n are related to ρ by (4.11). This is a uniform asymptotic expansion in the sense that if a and b are fixed and the infinite sum is truncated at its m th term, then the error in the resulting approximation to $R_n(z)$ is $z^{b-1}e^zO(\rho^{-m})$, that is, $z^{b-m-1}e^zO(1)$, uniformly with respect to $\theta \in \left[-\frac{1}{2}\pi, \frac{1}{2}\pi\right]$ and bounded values of $|\alpha|$. All of the restrictions on a and b that were introduced during the analysis, given by $0 < \text{Re } a < \text{Re } b < 1$ are removable by analytic continuation. This elegant expansion is the desired result; compare (4.4) and (4.9). In particular, we note that except for the presence of the functions $T_{n-s+a+b-1}(z)$, the right-hand side of (4.12) is identical with the second line of (4.4).

To compare our result with that of Berry in the same circumstances, we truncate (4.12) at its first term, given by

$$R_n(z) = -2\pi i \frac{e^{-(a+b)\pi i}}{\Gamma(a)\Gamma(b)} z^{b-1}e^z\left\{T_{n+a+b-1}(z) + O\left(\frac{1}{z}\right)\right\}.$$

If we now set $p = a + b - 1$, then equations (4.11) become identical with (3.11); hence we may replace $T_{n+a+b-1}(z)$ by the uniform asymptotic expansion (3.24). Again, on taking only the dominant terms we arrive at

$$R_n(z) = -2\pi i \frac{e^{-(a+b)\pi i}}{\Gamma(a)\Gamma(b)} z^{b-1} e^z \left\{ \frac{1}{2} + \frac{1}{2}\operatorname{erf}\left(c\sqrt{\tfrac{1}{2}\rho} \right) + e^{-\frac{1}{2}\rho c^2} O\left(\rho^{-\frac{1}{2}}\right) \right\},$$

where c is defined as in §3; see equations (3.14) and (3.15). Berry's approximation has the same form, except that c is replaced by $\sin\theta\sqrt{\sec\theta}$. Similar comments to those made towards the end of §3 again apply.

Lastly, on specializing the main results of this section to the case of Macdonald's function we find that the Stokes multiplier S_n defined by (1.5) has the uniform expansion

$$S_n \sim \sum_{s=0}^{\infty} (-)^s \frac{A_s(\nu)}{z^s} T_{n-s}(2z),$$

as $z \to \infty$ in the sector $\frac{1}{2}\pi \le \operatorname{ph} z \le \frac{3}{2}\pi$. Here n and z are related by $n = 2|z| + \alpha$, where $|\alpha|$ is bounded. This is a rigorous, and more precise, version of Berry's result (2.13).

5. CONCLUSIONS

The purpose of this paper has been to provide a rigorous mathematical basis both for R. B. Dingle's well-established theory of terminants and a recent new interpretation of Stokes' phenomenon by M.V. Berry. The theories of these authors were outlined in §2 by means of an application to Macdonald's modified Bessel function $K_\nu(z)$ of large argument.

In §3 we investigated the asymptotic expansion of the generalized exponential integral $E_p(z)$ for large $|z|$ in the vicinity of $\operatorname{ph} z = \pi$ by constructing an integral representation for a so-called Stokes multiplier. The integrand has a coalescing saddle-point and simple pole, and by employing a quadratic change of integration variable we were able to construct a uniform asymptotic expansion for the Stokes multiplier in terms of the error function. This analysis corresponds to the steps in Berry's method that led from the integral representation (2.8) to the approximations (2.11) and (2.13). We also noted briefly the connection of the new results to the theory of Airey's converging factors.

In §4 we considered the asymptotic expansion of the confluent hypergeometric function $U(a,a-b+1,z)$ for fixed a and b and large $|z|$ in the vicinity of $\text{ph}\, z = \pi$. By employing Cauchy's integral formula for the remainder term in Taylor's theorem, we constructed a double integral for the n th remainder term $R_n(z)$ in the asymptotic expansion of $U(a,a-b+1,z)$. Then by a change of integration variables we were able to derive a uniform asymptotic expansion for $R_n(z)$ in a series of generalized exponential integrals. This analysis corresponds to the steps in the Dingle-Berry theory that led from the expansion (2.4) to the representations (2.7) and (2.8). Subsequently, we observed that if we restrict ourselves to the dominant term and substitute the approximations that were found in §3 for the generalized exponential integral, then we obtain an approximation for $R_n(z)$ in terms of the error function similar to that of Berry.

Although we used the confluent hypergeometric function in §4 to illustrate our analysis, the method is quite general and indeed is applicable in the same circumstances that the methods of Dingle and Berry can be used.

Full details of the proofs outlined in §§3, 4 are too lengthy to include in the present paper, and will appear elsewhere.

ACKNOWLEDGMENT

I am indebted to Dr. M. V. Berry for permission to make use of his results before they were published.

REFERENCES

1. L. V. Ahlfors, *Complex analysis*, Second edition, McGraw-Hill, New York, 1966.

2. M. V. Berry, "Uniform asymptotic smoothing of Stokes's discontinuities", *Proc. Roy. Soc. London Ser. A* 422: 7-21 (1989).

3. N. Bleistein, "Uniform asymptotic expansions of integrals with stationary point near algebraic singularity", *Comm. Pure Appl. Math.* 19: 353-370 (1966).

4. R. B. Dingle, "Asymptotic expansions and converging factors I. General theory and basic converging factors", *Proc. Roy. Soc. London Ser. A* 244: 456-475 (1958).

5. R. B. Dingle, "Asymptotic expansion and converging factors IV. Confluent hypergeometric, parabolic cylinder, modified Bessel, and ordinary Bessel functions", *Proc. Roy. Soc. London Ser. A* 249: 270-283 (1958).

6. R. B. Dingle, *Asymptotic expansions: their derivation and interpretation*, Academic Press, London, 1973.

7. H. Jeffreys, "The remainder in Watson's lemma", *Proc. Roy. Soc. London Ser. A* 248: 88-92 (1958).

8. D. S. Jones, "Asymptotic behavior of integrals", *SIAM Rev.* 14: 286-317 (1972).

9. G. F. Miller, *Tables of generalized exponential integrals*, N.P.L. Math. Tables, Vol. 3, H. M. Stationery Office, London, 1960.

10. F. W. J. Olver, *Asymptotics and special functions*, Academic Press, New York, 1974.

11. F. W. J. Olver, "Asymptotic approximations and error bounds", *SIAM Review* 22: 188-203 (1980).

12. G. G. Stokes, "On the discontinuity of arbitrary constants which appear in divergent developments", *Trans. Cambridge Philos. Soc.* 10: 105-128 (1857). Reprinted in *Mathematical and Physical Papers*, Vol. 4, 77-109, Cambridge University Press, London, 1904.

13. G. G. Stokes, *Memoir and scientific correspondence*, selected and arranged by J. Larmor, Vol. I, 62 (1907), Cambridge University Press, London.

14. B. L. van der Waerden, "On the method of saddle points", *Appl. Sci. Res. Ser. B* 2: 33-45 (1951).

Singularly Perturbed Boundary Value Problems Viewed in the Liénard Plane

J. D. Allen†
Robert E. O'Malley, Jr.* Department of Mathematical Sciences, Rensselaer Polytechnic Institute, Troy, New York

1. INTRODUCTION

Consider the autonomous scalar equation

$$\epsilon\ddot{u} + f(u)\dot{u} + g(u) = 0$$

on the interval $0 \leq t \leq T$ with the boundary values

$$u(0) = A \qquad \text{and} \qquad u(T) = B$$

† On leave of absence from Northwestern College, Orange City, Iowa 51041, USA

* This work was supported in part by the U. S. National Science Foundation and the U. S. Army Research Office under grants DMS-85-04034 and DAAL 03-86-K-0112.

being prescribed. We shall seek the asymptotic behavior of solutions $u(t, \epsilon)$ as $\epsilon \to 0^+$ for various values of A, B, and T, assuming smoothness of f and g. Such boundary value problems and their generalizations in terms of partial differential equations are important in many applications, as well as a challenge to numerical analysts due to the likelihood of having narrow boundary and interior layer regions where solutions change rapidly.

We shall adopt a geometric approach to finding the limiting solutions by introducing the Liénard variable

$$z = \epsilon \dot{u} + F(u)$$

where

$$F(u) = \int_0^u f(s)ds,$$

and examining trajectories in the $u - z$ plane. This change of variables, presumably original to Liénard (1928), is commonly used to study nonlinear oscillations for corresponding initial value problems, especially to obtain periodic solutions. Our second-order equation is, then, equivalent to the fast-slow system

$$\begin{cases} \epsilon \dot{u} = z - F(u) \\ \dot{z} = -g(u). \end{cases}$$

This system will have rest points on the "characteristic curve"

$$\Gamma : \qquad z = F(u)$$

when $g(u) = 0$, and corresponding trajectories $z(u)$ will move between the vertical lines $u = A$ and $u = B$ satisfying the differential equation

$$\frac{dz}{du} = \frac{\epsilon g(u)}{F(u) - z}$$

taking "time"

$$T = \epsilon \int_A^B \frac{du}{z - F(u)}$$

since $dt = -\frac{dz}{g(u)} = \frac{\epsilon du}{F(u) - z}$. A typical trajectory is pictured as

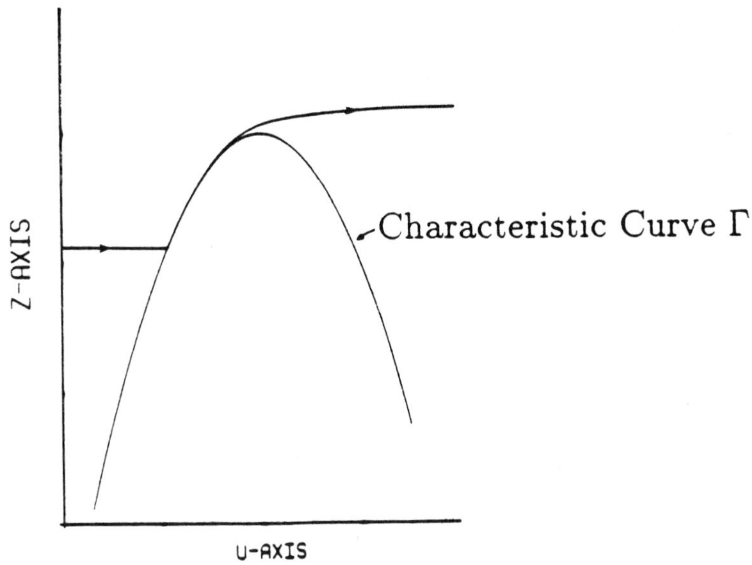

The corresponding solution graph might be given for small ϵ by

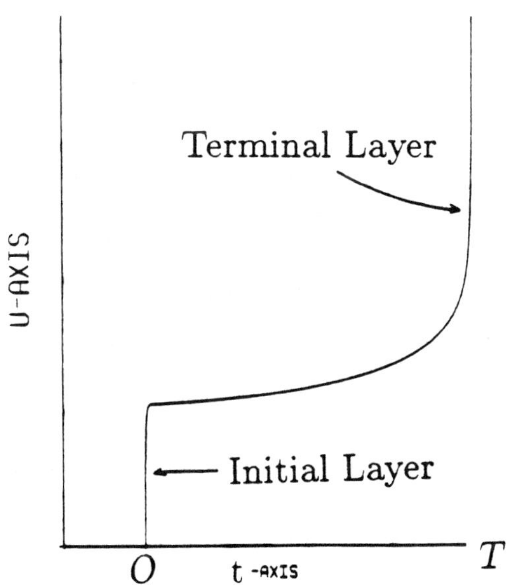

Motion is rapid, and nearly horizontal, except within an $O(\epsilon)$ neighborhood of the characteristic curve Γ. This is so because $\dot{u} = O\left(\frac{1}{\epsilon}\right)$ is then large and $\frac{dz}{du} = O(\epsilon)$ is correspondingly small. Further, u will increase above Γ and decrease below Γ, while z will increase for $g < 0$ and decrease for $g > 0$. In order to use up the allotted time T, bounded trajectories must generally spend most time near arcs of Γ, indeed sometimes remaining near rest points (i.e., zeroes of g) for quite a while.

We note that other phase planes could also be used. The traditional $u - \dot{u}$ plane would, for example, yield the system

$$\begin{cases} \epsilon\dot{v} = -f(u)v - g(u) \\ \\ \dot{u} = v. \end{cases}$$

Diener (1981) found it convenient to rescale the fast-variable v by introducing $V = \epsilon v$. This corresponds to solving the differential equation

$$\frac{dV}{du} = -f(u) + \frac{\epsilon g(u)}{V}$$

in the $u - V$ "plane of observability." Lutz and Goze (1981) and Wallet (1985) also used the Liénard plane to study related problems. We, indeed, owe a substantial debt to the Strasbourg group of non-standard analysts for their significant realization that much could be easily learned about the limiting solutions to singularly perturbed problems by critically examining such phase-plane trajectories (cf. Zvonkin and Shubin (1984) and Diener and Diener (1987)).

2. RELATIONS TO SOME CLASSICAL RESULTS

(a). The well-known theory of Coddington and Levinson (1952) shows that the two-point problem for $u(t, \epsilon)$ will have a unique limiting solution

$U_0(t)$ for $t > 0$ which satisfies the reduced problem

$$f(U_0)\dot{U}_0 + g(U_0) = 0, \qquad U_0(T) = B,$$

provided $f(U_0(t))$ remains positive for $T \geq t \geq 0$. Moreover, there will then be an $O(\epsilon)$ thick initial t interval, or "initial layer", where u varies from $u(0) = A$ to nearly $U_0(0)$ according to the limiting inner problem

$$\begin{cases} \dfrac{d^2u}{d\tau^2} + f(u)\dfrac{du}{d\tau} = 0 \\ \\ u(0) = A, \quad u(\infty) = U_0(0) \end{cases}$$

on $\tau \geq 0$ obtained by introducing the stretched time $\tau = t/\epsilon$ and invoking a mild boundary layer stability assumption, such as requiring the positivity of $f(u)$ between $u = A$ and $U_0(0)$.

Because of the sign of $\dot{u}$ off Γ, we naturally call that portion Γ^+ of Γ where $F(u)$ is increasing **attractive** and that portion Γ^- where $F(u)$ is decreasing (at least, potentially) **repulsive**. Such conclusions are classically obtained by noting that any rest point of the equation $\epsilon\frac{du}{dt} = z - F(u)$ is stable to perturbations when $F'(u) = f(u)$ is positive. If f remains positive throughout the u interval of interest, the limiting solution of the two-point problem will move instantaneously and horizontally in the Liénard plane toward Γ, which it will then follow for T units of time. Along $\Gamma = \Gamma^+$, we will have

$$Z_0 = F(U_0) \qquad \text{and} \qquad U_0(T) = B.$$

Differentiation with respect to t then implies that U_0 will satisfy the terminal value problem

$$f(U_0)\dot{U}_0 + g(U_0) = 0, \qquad U_0(T) = B$$

on $0 < t < T$. In particular, note that the value $U_0(0)$ could be well-approximated using standard numerical integration routines. In the case $A > U_0(0)$, the graphs of the solution (for small values of ϵ) in the Liénard and $u - t$ planes are as follows:

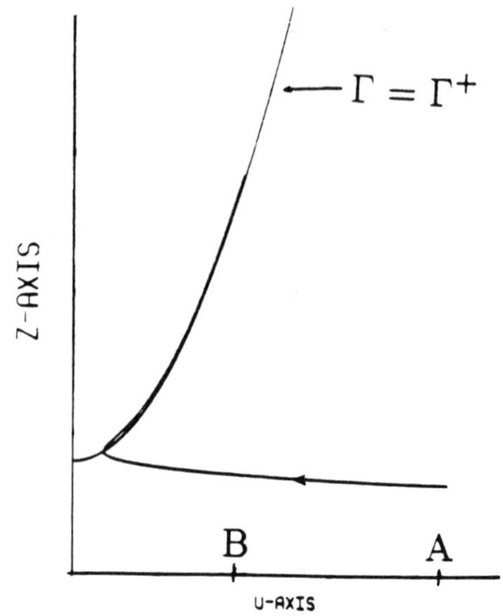
$\Gamma = \Gamma^+$
Z-AXIS
B
A
U-AXIS

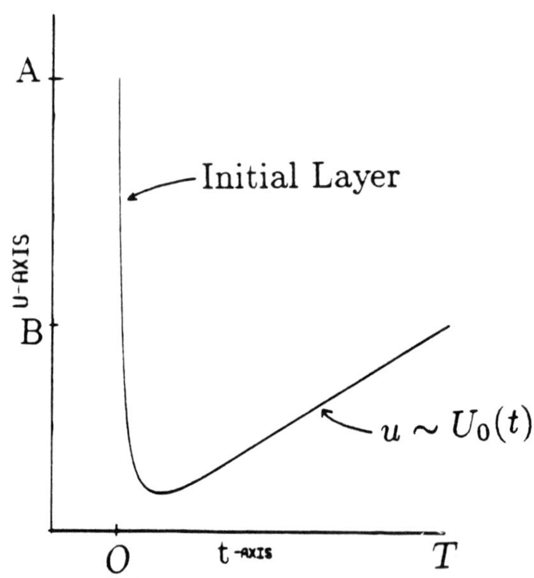
A
Initial Layer
U-AXIS
B
$u \sim U_0(t)$
O
t-AXIS
T

If $A < Y_0(0)$, the solution $u(t, \epsilon)$ would, instead, increase monotonically for ϵ sufficiently small. Likewise, the solution would, instead, feature a terminal boundary layer if $f(u)$ was everywhere negative.

(b). Lorenz (1982) used inverse-monotonicity arguments to show (among other results for nonautonomous equations) that our two-point problem has a unique solution when

$$g'(u) \leq 0 \qquad \text{and} \qquad g(0) = 0.$$

Moreover, he showed that the limiting solution could be obtained numerically by using "upwinding," i.e. by using an appropriate one-sided difference approximation for the derivative $\dot{u}$, selected according to the sign of $f(u)$. Related work is reported in Osher (1981) and Lorenz (1984). These results are remarkable in that they do not restrict the number or type of zeroes of f. The limiting solution obtained can be discontinuous at such turning points. The permitted jumps are, indeed, restricted, and elsewhere, the limiting solution U_0 will satisfy the reduced equation $f(U_0)\dot{U}_0 + g(U_0) = 0$.

In the Liénard plane, rest points $(\bar{u}, \bar{z})$ occur when

$$g(\bar{u}) = 0 \qquad \text{and} \qquad \bar{z} = F(\bar{u}).$$

If $ug(u) < 0$ for $u \neq 0$, the origin will be the only rest point. The eigenvalues λ of the local linearized system will satisfy

$$\epsilon\lambda^2 + f(0)\lambda + g'(0) = 0,$$

so they will have opposite signs when $g'(0) < 0$, whether or not $f(0) = 0$, and the origin will be a saddle point. When $g(0) = 0$ and $g'(0) > 0$, however, the eigenvalues will have the sign opposite to $f(0)$ if that is nonzero and they will be purely imaginary if $f(0) = 0$. Thus, the rest point will then be a node, a center, or a spiral.

We can anticipate that the study of all possible solution paths in the Liénard plane should allow us to obtain results in a wide variety of situations.

3. SOME EXAMPLES

(a). $g(u) \equiv 0$.

Here

$$\begin{cases} \epsilon \dot{u} = z - F(u) \\ \dot{z} = 0, \end{cases}$$

so all points of $\Gamma : z = F(u)$ are rest points. Further, trajectories must lie on a line

$$z(t) \equiv k.$$

To reach $u(T) = B$, we will need k to be greater then the maximum F_M of $F(u)$ between $u = A$ and B if $B > A$. Then, $u(t)$ will increase monotonically throughout $0 \le t \le T$. (We would analogously find that z would have to lie below Γ if $B < A$.) Because the solution can be implicitly expressed as

$$t = \epsilon \int_A^u \frac{du}{k - F(u)}$$

and t must reach T when $u = B$, k must be within an $O(\epsilon)$ neighborhood of F_M. In particular, if $F(u) < F_M$ for $A \le u < B$, u must feature a rapid initial jump from A to $U_0(t) = B$ for $t > 0$. Otherwise, $U_0(t)$ will be a monotonic function which jumps between values e where $F(e) = F_M$, dividing up the time T according to the relative strengths of the poles of $\frac{1}{F_M - F(u)}$. If, for example, there were two such maxima e_1 and e_2 with

$F(u) \sim F_M - c_1(u - e_1)^2$ near e_1 and $F(u) \sim F_M - c_2(u - e_2)^2$ near e_2, we would obtain a limiting solution

$$U_0(t) = \begin{cases} e_1, & 0 < t < \tilde{t} \\ e_2, & \tilde{t} < t < T, \end{cases}$$

where the determination of $\tilde{t}$ would depend upon the detailed asymptotic evaluation of $u(t, \epsilon)$ and the resulting relations between c_1, c_2, and $\tilde{t}$. In the Liénard plane, our phase portrait would be given by

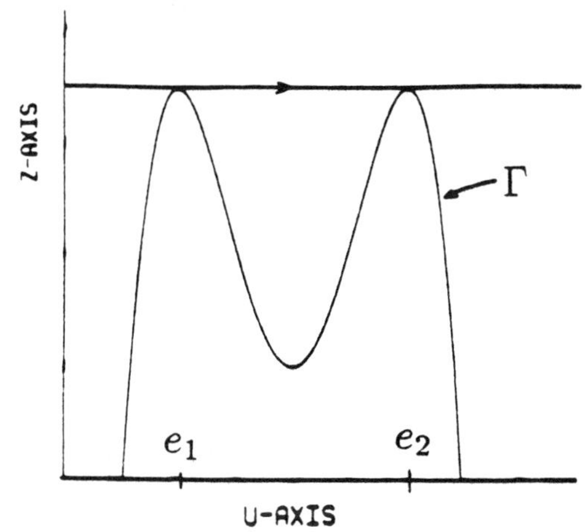

In the $y - t$ plane, the graph of our unique solution (for ϵ sufficiently small) would look like

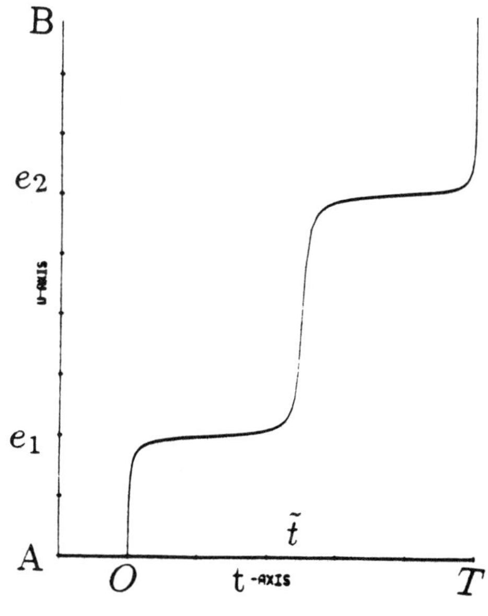

(b). $f(u) = -g(u) = u$

Solutions of this classical Lagerstrom-Cole model

$$\epsilon\ddot{u} + u\dot{u} - u = 0$$

have been studied in Cochran (1962), Cole (1968), Howes (1978), Kevorkian and Cole (1981), Lagerstrom (1988), and elsewhere. Representative trajectories in the Liénard plane could be readily obtained on a graphics terminal by integrating forward and backwards in time from appropriately chosen initial points. Thus, we would obtain the phase portrait

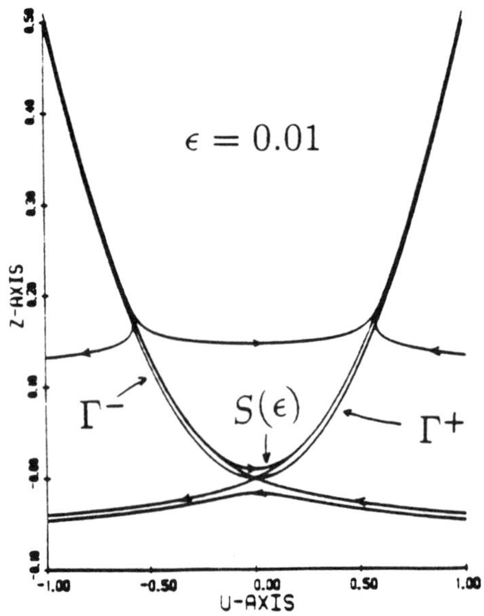

Here, the characteristic curve $\Gamma : z = \frac{1}{2}u^2$ is a parabola and the sole rest point is a saddle at the origin . Trajectories cross Γ vertically, and those (with $u < 0$) rapidly become horizontal as they move away from Γ^-. For $u > 0$, such paths then move along Γ^+. Note the existence of the special trajectory

$$S(\epsilon): \quad z = \frac{1}{2}u^2 + \epsilon,$$

slightly above Γ, along which $\dot{u} = 1$, and of separatrices between Γ and $S(\epsilon)$ which enter or leave the rest point.

To go from $u(0) = A < 0$ to $u(T) = B > 0$, for example, a trajectory must remain above Γ, so $u(t)$ will increase monotonically. Note that the travel time from A to B on $S(\epsilon)$ will be $B - A$. Above (below) $S(\epsilon)$, it would be shorter (longer). For faster transit, a solution path must nearly follow $S(\epsilon) \sim \Gamma^-$ from $u = A$ until it jumps rapidly and horizontally to $S(\epsilon) \sim \Gamma^+$, after which it follows $S(\epsilon)$ to $u = B$. The limiting trajectory would be determined by $\dot{U}_0 = 1, U_0(0) = A, 0 \leq t < \tilde{t}$, and $\dot{U}_0 = 1, U_0(T) = B$ for $\tilde{t} \leq t \leq T$. Thus, we would have

$$U_0(t) = \begin{cases} t + A & \text{for } 0 \leq t < \tilde{t} \\ t + B - T & \text{for } \tilde{t} < t \leq T. \end{cases}$$

Note that these limits respectively satisfy the left and right reduced problems, defined via the limiting equation and the appropriate end condition. Since the jump at $\tilde{t}$ is horizontal and Γ is symmetric with respect to the z-axis, we would need $U_0(\tilde{t}^-) = -U_0(\tilde{t}^+)$, so

$$\tilde{t} = \frac{1}{2}(T - B - A).$$

To keep $\tilde{t}$ within $(0, T)$ and $U_0(\tilde{t}^-) < 0$, we need to restrict A and B so that

$$-T < A + B < T$$

and

$$T < B - A.$$

The resulting solution has the shock layer form pictured.

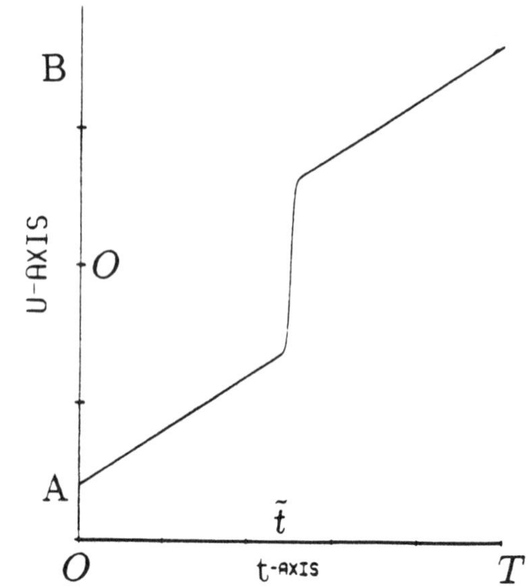

For slower motions (than along $S(\epsilon)$) between any $A < 0$ and $B > 0$, the trajectory would have to pass between $S(\epsilon)$ and the nearby separatrix. The lower the z-intercept attained, the more time is used. To move from $u(0) = A$ to the origin along Γ^- takes (asymptotically) $-A$ units of time and to move from the origin to $u(T) = B$ along Γ^+ takes B time units, so the corresponding limiting solution is given by

$$U_0(t) = \begin{cases} t + A, & 0 \le t \le -A \\ 0, & -A \le t \le T - B \\ t + B - T, & T - B \le t \le T \end{cases}$$

provided $T > B - A$. This time-consuming solution, then, utilizes the trivial singular solution $U_0^s(t) \equiv 0$ of the limiting equation, in addition to the other solutions along which $\dot{U}_0 = 1$. The solution involves two corner layers where u is continuous, but $\dot{u}$ is discontinuous.

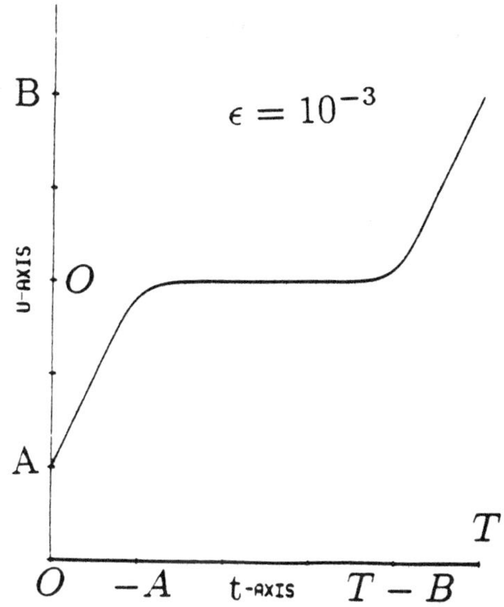

Proceeding in a straightforward fashion, it is easy to find a unique trajectory corresponding to any set of A, B, and T values. If A and B are positive, with $B > T$, the Coddington and Levinson theory guarantees that the limiting solution will be given by $U_0(t) = t + B - T$. If, however, $0 < B < T$, the limit $U_0(t) = t + B - T$ must be preceded by the trivial one, i.e. there will be a thin initial layer where nonuniform convergence from A to $U_0^s(t) \equiv 0$ occurs, followed by a corner layer joining U_0^s and $U_0(t)$.

(c). $f(u) = g(u) = u$

The equation

$$\epsilon \ddot{u} + u\dot{u} + u = 0$$

is a steady-state version of Burgers' equation and is therefore of some physical interest (cf. Murray (1973)). For it, the characteristic curve Γ is again

$z = \frac{1}{2}u^2$, but the rest point at the origin is now a center. The Liénard plane portrait is symmetric with respect to the z axis and there is now the special trajectory

$$S(\epsilon): \qquad z = \frac{1}{2}u^2 - \epsilon,$$

slightly below Γ, on which $\dot{u} = -1$. Pictorially, we have

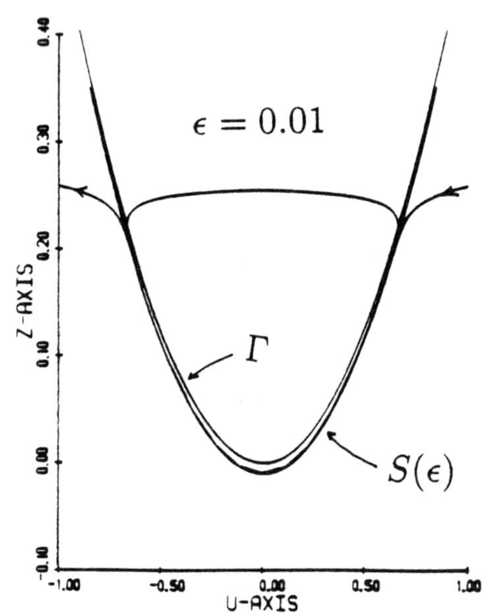

Note that above Γ, trajectories form closed paths which correspond to periodic solutions. For a solution following $S(\epsilon)$, we would have $B = A - T$. If $B < 0 < A$ and $B < A - T$, we would need a trajectory on which $\dot{u} < -1$. The phase portrait suggests that a unique solution (with endpoint layers) will exist, corresponding to a trajectory below $S(\epsilon)$. If, however, $B > A - T$, we would need $\dot{u} > -1$, which corresponds to paths which are (parts of) periodic orbits. Since the vertical lines $u = A$ and B cross any such trajectory twice, we may have four-fold ways of satisfying our boundary conditions. Likewise, since each trajectory could encircle the origin more than one time, we must then expect many solutions of small amplitude. As Lutz and Goze (1981) show, there is a solution (sometimes many) for each set of A, B, and T values. Along them, we will have $\dot{U}_0(t) = -1$, except possibly for regularly spaced jumps. Further, the amplitudes of these oscillations can be determined. We note that the trivial singular solution of the reduced problem is largely irrelevant compared to the $\dot{U}_0 = -1$ solutions. Moreover, although the Lorenz sign condition on g' does not imply uniqueness, the two-point problem does have a unique solution for many A, B, and T values.

We note that somewhat similar phase plane analyses are effective for studying conservative systems (where $f(y) \equiv 0$) (cf. Carrier and Pearson (1968) and O'Malley (1976)), often leading to multiple solutions with regularly-spaced "spikes".

(d). $f(u) = u(1 - u^2)$ and $g(u) = -u$.

Lorenz (1982) introduced the equation

$$\epsilon\ddot{u} + u(1 - u^2)\dot{u} - u = 0,$$

which is equivalent to the system

$$\begin{cases} \epsilon\dot{u} = z - F(u) \\ \dot{z} = u \end{cases}$$

for $F(u) = \frac{1}{4}u^2(2 - u^2)$. All information is contained in the phase portrait:

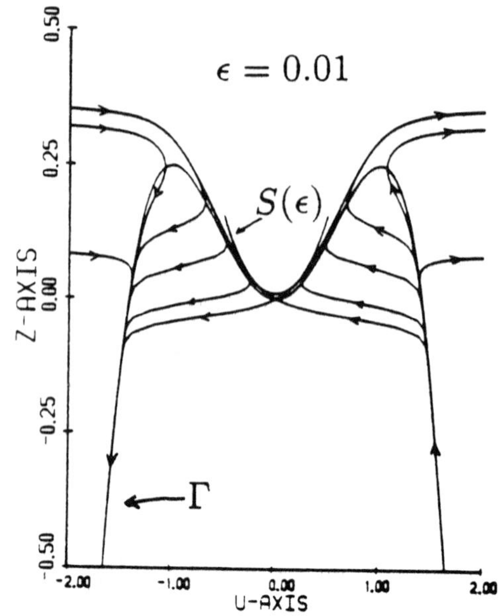

We note that the two-point problem has a unique solution, according to the Lorenz condition, for any prescribed boundary values. As for the Cole-Lagerstrom equation, the origin is a saddle point, with separatrices as indicated. It is natural to again look for a slow manifold near Γ, expressed as

$$S(\epsilon): \qquad z = F(u) + \epsilon h(u, \epsilon),$$

along which $\dot{u}$ is bounded. Since $\dot{z} = (f(u) + \epsilon h'(u, \epsilon))\dot{u} = u$ there, we will need

$$\dot{u} = \frac{z - F(u)}{\epsilon} = h(u, \epsilon) = \frac{u}{f(u) + \epsilon h'(u, \epsilon)}.$$

We can expand $h(u, \epsilon)$ as a power series in ϵ and use this relation to obtain

$$h(u, \epsilon) = \frac{1}{1 - u^2} - \frac{2\epsilon}{(1 - u^2)^4} + \frac{20\epsilon^2}{(1 - u^2)^7} - \cdots,$$

corresponding to

$$\dot{u} = \left(1 - u^2 + \frac{2\epsilon}{(1 - u^2)^2} - \frac{16\epsilon^2}{(1 - u^2)^5} + \cdots\right)^{-1}.$$

The manifold is well-defined for $|u| < 1$. It ceases to be slow, however, as the turning points $u = \pm 1$ are approached. This should be anticipated, since the Liénard portrait shows that these maxima of F are where fast horizontal motion (for $|u| > 1$) changes to slow motion along Γ (for $|u| < 1$) or vice versa. We note that introduction of such slow manifolds has been very effective in related control theory (cf. Spong et al (1987) and Sharkey and O'Reilly (1988)).

Along Γ, the reduced equation $U_0(1 - U_0^2)\dot{U}_0 - U_0 = 0$ will be satisfied, so if $U_0(t) \neq 0$, integration implies that we must have

$$U_0 - \frac{1}{3}U_0^3 = t + C$$

for some constant C. Detailed study indeed shows that the limiting solution will always jump among the trivial singular solution $U_0^s(t) \equiv 0$ of the reduced equation and the four implicit solutions of the cubic equation determined by the boundary conditions $U_0(0) = A, U_0(0) = -1, U_0(T) = B$, and $U_0(T) = 1$. The reason the boundary values $U_0(0) = -1$ and $U_0(T) = 1$ are used is that these values correspond to maxima of F where rapid motion changes to slow or slow motion to fast.

To illustrate some possibilities, consider the two-point problem with $A > \sqrt{2}$ and $B > 1$. We will obtain the solution for increasing values of T. The first possibility (corresponding to Coddington and Levinson's result with f negative) involves motion along the right-most arc of Γ, followed by rapid, horizontal motion to B (providing a terminal boundary layer.) Thus, the limiting solution for $0 \leq t < T$ will be given implicitly as the solution $U_0^L(t)$ to $U_0 - \frac{1}{3}U_0^3 = t + A - \frac{1}{3}A^3$ with $U_0(0) = A$. The Liénard portrait, however, shows that such a trajectory is only possible for

$$U_0^L(t) \geq 1.$$

Thus, a monotonically decreasing limiting solution is attained provided

$0 \leq T \leq T_1 \equiv \int_A^1 (1-u^2)du$ (since $dt = \frac{dz}{u} = \frac{f(u)}{u} du$ on Γ). For somewhat larger values of T, one would again follow $U_0^L(t)$ on some interval $0 \leq t < \tilde{t}$ (where u varies from A to some $U_0^L(\tilde{t})$ between 1 and $\sqrt{2}$), then would jump rapidly from the outer right arc of Γ to its inner right arc, then would move upward along Γ to $u = 1$, and, finally, would move rapidly and horizontally right to B. Note that the second limit $\bar{U}_0^R(t)$ would satisfy $(1-U_0^2)\dot{U}_0 = 1$ and $U_0(T) = 1$, so $\bar{U}_0^R - \frac{1}{3}(\bar{U}_0^R)^3 = t + \frac{2}{3}$ for $\tilde{t} < t < 1$. Moreover, since the jump is horizontal at the shock location $\tilde{t}$, we will need

$$F(U_0^L(\tilde{t})) = F(\bar{U}_0^R(\tilde{t})).$$

This Rankine-Hugoniot type condition (cf. Smoller (1983)) uniquely determines $\tilde{t}$ provided $1 < U_0^L(\tilde{t}) < \sqrt{2}$. The limiting solution with an interior shock layer and a terminal boundary layer is pictured as:

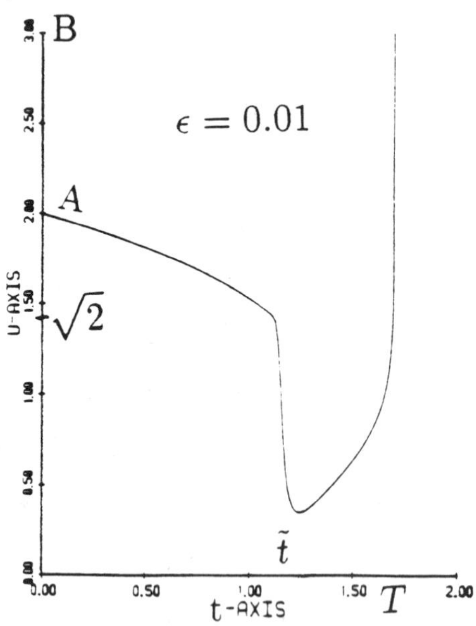

The time required,

$$T = \int_{A}^{U_0^L(\tilde{t})} (1 - u^2)du + \int_{\bar{U}_0^R(\tilde{t})}^{1} (1 - u^2)du,$$

increases monotonically as the z level of the horizontal jump decreases from $F(1) = \frac{1}{4}$ to $F(0) = F(\sqrt{2}) = 0$. Thus, this type of solution is appropriate for these $A - B$ values as long as $T_1 < T < T_2 \equiv \int_A^{\sqrt{2}} (1 - u^2)du + \int_0^1 (1 - u^2)du$. For $T > T_2$, the solution follows $U_0^L(t)$ until $U_0^L(\hat{t}) = \sqrt{2}$, jumps rapidly left to $U_0^S(t) \equiv 0$, remains there for $T - T_2$ units of time, then follows $\bar{U}_0^R(t)$ until $\bar{U}_0^R(T) = 1$, and then jumps rapidly right to $u(T) = B$. Thus, we have

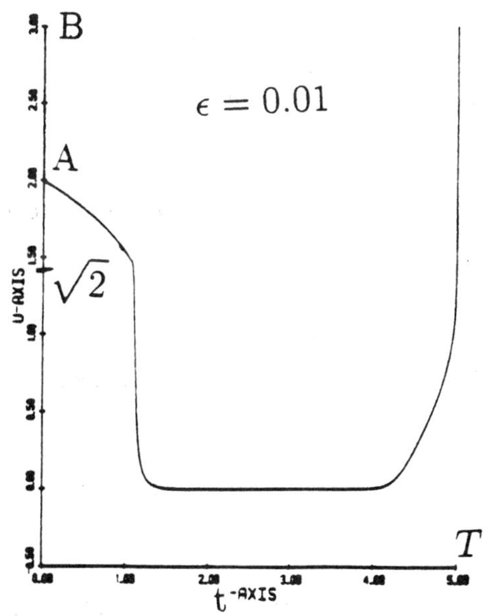

We note that Lorenz calculated the solution to such a problem using upwinding.

(e). $f(u) = u$ and $g(u) = -u^2$.

For this example, the system

$$\begin{cases} \epsilon \dot{u} = z - \dfrac{1}{2}u^2 \\ \dot{z} = u^2 \end{cases}$$

has a degenerate equilibrium point at the origin. Lorenz' condition suggests that solutions remaining in the right-half plane are unique. A look at possible trajectories, however, suggests the possibility of three solutions for large enough values of A and B. The problem is now being analyzed numerically and analytically.

REFERENCES

G. F. Carrier and C. E. Pearson (1968), *Ordinary Differential Equations*, Blaisdell, Waltham, Mass.

J. A. Cochran (1962), "Problems in Singular Perturbation Theory", Ph.D. thesis, Stanford University.

E. A. Coddington and N. Levinson (1952), "A boundary value problem for a nonlinear differential equation with a small parameter," Proc. Amer. Math. Soc. **3**, 73-81.

J. D. Cole (1968), *Perturbation Methods in Applied Mathematics*, Ginn-Blaisdell, Boston.

F. Diener (1981), "Méthode du plan d'observabilité," these , Institute de Recherche Mathématique Avancée, Strasbourg.

F. Diener and M. Diener (1987), *Fleuvres*, C.N.R.S. - Unité Associée No. 212, Paris.

F. A. Howes (1978), "Boundary-interior layer interactions in nonlinear singular perturbation theory," Memoirs, Amer. Math. Soc. 203.

J. Kevorkian and J. D. Cole (1981), *Perturbation Methods in Applied Mathematics*", Springer-Verlag, New York.

P. A. Lagerstrom (1988), *Matched Asymptotic Expansions*, Springer-Verlag, New York.

A. Liénard (1928), Etude des oscillations entretenues, Rev. Gén. d' Élect. 23, 901-946.

J. Lorenz (1982), *Nonlinear boundary value problems with turning points and properties of difference schemes*, Lecture Notes in Math. 942, Springer-Verlag, New York, 150-169.

J. Lorenz (1984), "Analysis of difference schemes for a stationary shock problem," SIAM J. Numer. Anal. **21**, 1038-1053.

R. Lutz and M. Goze (1981), *Nonstandard Analysis, Lecture Notes in Math. 881, Springer-Verlag, New York.*

J. D. Murray (1973), "On Burgers' model equation for turbulence," J. Fluid Mech. 59(2), 263-279.

R. E. O'Malley, Jr. (1976), "Phase plane solutions to some singular perturbation problems", J. Math. Anal. Appl. **54**, *449-466.*

S. Osher (1981), "Nonlinear singular perturbation problems and one-sided difference schemes", SIAM J. Numer. Anal. **21**, *263-270.*

P. M. Sharkey and J. O'Reilly (1988) "Composite control of nonlinear singularly perturbed systems: A geometric approach," Int. J. Control 48, 2491-2506.

J. Smoller (1983), Shock Waves and Reaction-Diffusion Equations, Springer-Verlag, New York.

M. W. Spong, K. Khorasani, and P. Kokotovic (1987), "An integral manifold approach to the feedback control of flexible joint robots," IEEE J. Robotics and Automation 3, 291-300.

G. Wallet (1985) "Dérive lente du champ de Liénard," *Mathematiques Finitaires et Analyse Non-Standard, Journées S.M.F., Luminy, 51-64.*

A. K. Zvonkin and M. A. Shubin (1984), "Nonlinear analysis and singular perturbations of ordinary differential equations," *Russian Math. Surveys 39(2), 69-131.*

On the Asymptotic Theory of the Orr-Sommerfeld Problem

W. H. Reid Professor, Department of Mathematics, University of Chicago, Chicago, Illinois

I. INTRODUCTION

The stability of two-dimensional parallel shear flows with respect to infinitesimal disturbances is governed by the well-known Orr-Sommerfeld equation

$$(\mathrm{i}\,\alpha R)^{-1}(\mathrm{D}^2-\alpha^2)^2\phi = (U-c)(\mathrm{D}^2-\alpha^2)\phi - U''\phi, \tag{1}$$

where $\phi(z)e^{\mathrm{i}\,\alpha(x-ct)}$ is the stream function of the disturbance in the usual normal mode analysis, $U(z)$ is the basic velocity distribution, R is the Reynolds number, and $\mathrm{D} = \mathrm{d}/\mathrm{d}z$. When viscous effects are neglected, Eq. (1) reduces to Rayleigh's equation

$$(U-c)(\mathrm{D}^2-\alpha^2)\phi - U''\phi = 0 \tag{2}$$

from which Rayleigh showed that a necessary (but not sufficient) condition for instability is that the basic velocity distribution should have an inflection point. Flows without an inflection point are therefore stable in the absence of viscosity.

One normally thinks of viscosity as having a stabilizing influence and that is certainly true in general but, as Reynolds first suggested, for flows without an inflection point it may actually be the *cause* of instability. In such circumstances the instability is said to be "weak" in the sense that it occurs only for large values of R or, more precisely, for large values of αR and it is for this reason that asymptotic methods of approximation are of importance in dealing with the Orr-Sommerfeld equation and the related eigenvalue problem.

If we now let z_c be a point at which $U - c = 0$ and if $U_c' \equiv U'(z_c) \neq 0$ then z_c is a simple turning point of Eq. (1); it is also a regular singular point of Eq. (2). Our goal then is to obtain asymptotic approximations to the solutions of Eq. (1) which are uniformly valid in a full (complex) neighborhood of z_c. From an asymptotic point of view, the solutions of Eq. (1) are of three types. There are two solutions of dominant-recessive type which behave somewhat like Airy functions; there is one solution of balanced type which behaves somewhat like a Scorer function; and there is one solution of well-balanced type, $\phi_1(z)$ say, which does not exhibit the Stokes phenomenon and which can be approximated by a convergent expansion in powers of $1/\alpha R$.

In this article, therefore, I wish to discuss a number of aspects of this problem. In the first instance it is necessary to determine what the general structure of the asymptotic expansions should be. Next there is a somewhat technical matter as to how the slowly-varying coefficients in the expansions can be determined. And, finally, there is the important notion of first approximations as applied not only to the solutions of the Orr-Sommerfeld equation but also to the related eigenvalue relation.

The general structure of the expansions depends, of course, on the form of the basic velocity distribution. To illustrate this effect we will consider three basic flows: plane Couette flow for which $U(z) = z$ $(-1 \leq z \leq 1)$, plane

Poiseuille flow for which $U(z) = 1 - z^2$ $(-1 \le z \le 1)$, and the asymptotic suction boundary-layer profile for which $U(z) = 1 - \mathrm{e}^{-z}$ $(0 \le z < \infty)$.

II. THE HEURISTIC THEORY

This theory was developed principally by Heisenberg, Tollmien and Lin during the period from 1924 to 1947 and it has been summarized by Lin (1955). For the present purposes it is sufficient to consider only those approximations which, in the language of the method of matched asymptotic expansions, are of outer type. Thus, in the first instance we let

$$\phi(z) \sim \phi^{(0)}(z) + \epsilon^3 \phi^{(1)}(z) + \cdots, \tag{3}$$

where $\phi^{(0)}(z)$ satisfies Eq. (2), i.e. the "reduced" or "inviscid" form of the Orr-Sommerfeld equation. Here we have let $\epsilon = (\mathrm{i}\,\alpha R U_c')^{-1/3}$.

One solution of Eq. (2), $\phi_1^{(0)}(z)$ say, is analytic at z_c and it may therefore be expected to provide a uniform approximation to $\phi_1(z)$ with an error of the order of ϵ^3. The other solution, $\phi_2^{(0)}(z)$ say, has a logarithmic branch point at z_c and is commonly called the singular inviscid solution. These then are the outer approximations of so-called "inviscid" type.

The outer approximations of "viscous" type are obtained by applying the WKBJ method to Eq. (1) and they are given by

$$\begin{aligned}\phi_{3,4}(x) \sim \frac{1}{2}\pi^{-1/2}\epsilon^{5/4}(U-\mathrm{c})^{-5/4}\exp(\mp\frac{2}{3}\epsilon^{-3/2}\eta^{3/2})\times \\ \times\{G_0(z) \mp \epsilon^{3/2}G_1(z) + \cdots\},\end{aligned} \tag{4}$$

where $G_0(z) =$ constant $= 1$ (say) and

$$\eta(z) = \Big[\frac{3}{2}\int_{z_\mathrm{c}}^{z}\Big(\frac{U-\mathrm{c}}{U_\mathrm{c}'}\Big)^{1/2}\mathrm{d}z\Big]^{2/3}. \tag{5}$$

In Eq. (4), the term involving $G_1(z)$ was not considered in the original version of the heuristic theory but it is in fact needed for the determination of one of the slowly-varying coefficients in the fully uniform approximations.

III. A PRELIMINARY TRANSFORMATION

As in the case of second-order equations, we first make a preliminary transformation of Liouville type. This transformation can, of course, be arranged so that the resulting equation is in normal form, i.e. it does not contain a third derivative term. Although the requirement that the transformed equation be in normal form is attractive theoretically it does unfortunately lead to some unnecessary complications. Instead we simply let

$$\phi(z) = \chi(\eta), \tag{6}$$

where $\eta(z)$ is defined by Eq. (5). Equation (1) then becomes

$$\epsilon^3(\chi^{\text{iv}} + f_0\chi''') - (\eta + \epsilon^3 f_1)\chi'' - (g_0 + \epsilon^3 g_1)\chi' - (h_0 + \epsilon^3 h_1)\chi = 0, \tag{7}$$

where $f_0, f_1, \dots, h_1$ are holomorphic in a bounded domain of the η-plane containing the origin. The general structure of the theory depends crucially on the value of $g_0(0)$ and to a lesser extent on the value of $h_0(0)$. In the case of the Orr-Sommerfeld equation, to which we shall restrict our attention, we have

$$g_0(0) = 0 \qquad \text{and} \qquad h_0(0) = -U_c''/U_c'. \tag{8}$$

In attempting to develop an asymptotic theory for Eq. (7), it would be natural to suppose that the relevant comparison equation would necessarily be of fourth order. This was the approach first used by Wasow (1953b) and later in greater generality by Lin and Rabenstein (1960). The latter authors showed that the comparison equation must be of the form

$$\epsilon^3 u^{\text{iv}} - (\eta u'' + \alpha u' + \beta u) = 0. \tag{9}$$

The solutions of this equation have integral representations of Laplace type from which, as Rabenstein (1958) has shown, it is possible to obtain the inner and outer expansions of the solutions. What are needed, of course, are approximations to the solutions which are uniformly valid in a bounded domain of the (complex) η-plane containing the origin and these are extremely difficult to derive from the integral representations.

Another difficulty with this approach is that α and β in Eq. (9) must be allowed to have asymptotic expansions of the form

$$\alpha \equiv \alpha(\epsilon) \sim \sum_{s=0}^{\infty} \alpha_s \epsilon^{3s}. \tag{10}$$

Furthermore, the coefficients α_s and β_s are not known *a priori* but must be determined as part of the solution by certain regularity conditions.

In what follows, therefore, I shall attempt to describe an alternate approach, based on the use of certain generalized Airy functions, which avoids most of the technical difficulties associated with the comparison equation approach.

IV. PLANE COUETTE FLOW

When $U(z) = z$, Eq. (1) can be written in the form

$$\{\epsilon^3(\mathrm{D}^2 - \alpha^2) - \eta\}(\mathrm{D}^2 - \alpha^2)\phi = 0, \tag{11}$$

where $\eta = z - c$ and D now stands for $\mathrm{d}/\mathrm{d}\eta$. This equation has two solutions of well-balanced type but none of balanced type and it thus fails to exhibit one of the essential features of the general case. Nevertheless, it does provide valuable insight into the structure of the asymptotic expansions of the solutions of dominant-recessive type.

As Wasow (1953a) first showed, the solutions of Eq. (11) have integral representations of Laplace type. In the case of the solutions of dominant-recessive type, their integral representations can be reduced to a form to which one can then apply the method of Chester, Friedman and Ursell, and to that extent the asymptotic theory of Eq. (11) may be regarded as complete (see Drazin and Reid 1981, pp. 261–263). For more general velocity distributions, however, some other method must be used to infer the general structure of the uniform expansions.

To exploit the symmetries of Eq. (11), let us consider three solutions of dominant-recessive type, $V_k(\eta)$ $(k = 1, 2, 3)$, each of which is recessive in a non-overlapping sector of angle $2\pi/3$. From the integral representations of these solutions it is easy to show that their inner expansions can be expressed to all orders in terms of the generalized Airy functions

$$A_k(\zeta, p) = \frac{1}{2\pi i} \int_{L_k} t^{-p} \operatorname{ext}(\zeta t - \frac{1}{3} t^3)\, dt, \tag{12}$$

where $\zeta = \eta/\epsilon$, $p = 0, \pm 1, \pm 2, \ldots$ and L_k are the usual Airy contours. A crucial step in the present argument is based on the observation that these Airy functions satisfy the recursion formula

$$A_k(\zeta, p-3) - \zeta A_k(\zeta, p-1) + (p-1) A_k(\zeta, p) = 0. \tag{13}$$

If we now let p take on the values 2, 1 and 0 (say), then Eq. (13) shows that to any finite order the inner expansion can be rewritten in terms of these three Airy functions together with polynomial coefficients. And this in turn suggests that the uniform expansions must be of the form

$$V_k(\eta) = \mathcal{A}(\eta, \epsilon) A_k(\zeta, 2) + \epsilon^2 \mathcal{B}(\eta, \epsilon) A_k(\zeta, 1) + \epsilon \mathcal{C}(\eta, \epsilon) A_k(\zeta, 0), \tag{14}$$

where $\mathcal{A}$, $\mathcal{B}$ and $\mathcal{C}$ have expansions of the form

$$\mathcal{A}(\eta, \epsilon) = \sum_{s=0}^{\infty} A_s(\eta) \epsilon^{3s}. \tag{15}$$

The structure of the expansions (14) suggest that in a "first approximation" we have simply

$$V_k(\eta) \sim \mathcal{A}_0(\eta) A_k(\zeta, 2) + \epsilon^2 \mathcal{B}_0(\eta) A_k(\zeta, 1) + \epsilon \mathcal{C}_0(\eta) A_k(\zeta, 0) \tag{16}$$

in which the relative error associated with each term is of order ϵ^3. Thus, in this approximation, it is necesary to determine three slowly-varying coefficients. One way of doing this is to derive and then solve the differential equations which they satisfy. That is particularly simple in the case of $\mathcal{A}_0$ but much less so in the case of $\mathcal{B}_0$ and $\mathcal{C}_0$. One can show, almost "by inspection", that $(\mathrm{D}^2 - \alpha^2)(\eta \mathcal{A}_0) = 0$ and the requirement that $\mathcal{A}_0$ be analytic at

$\eta = 0$ then shows that $\eta\mathcal{A}_0$ must be a multiple of $\sinh\alpha\eta$. Thus, if we fix the normalization of $\mathcal{A}_0$ by requiring that $\mathcal{A}_0(0) = 1$ then we have

$$\mathcal{A}_0(\eta) = \frac{\sinh\alpha\eta}{\alpha\eta}. \tag{17}$$

In the case of $\mathcal{B}_0$ and $\mathcal{C}_0$, a simpler and more illuminating approach is to use the matching technique first suggested by Lin (1957). In this approach one expands the Airy functions in Eq. (16) for $k = 1$ (say) and then matches the result to the WKBJ approximation to $\phi_3(z)$ as given by Eq. (4). To first order we have simply $\mathcal{A}_0 + \eta\mathcal{C}_0 = G_0$. The requirement that $\mathcal{C}_0$ also be analytic at $\eta = 0$ implies that we must set $G_0 = \mathcal{A}_0(0) = 1$ and hence

$$\mathcal{C}_0(\eta) = \frac{1}{\eta}\Big(1 - \frac{\sinh\alpha\eta}{\alpha\eta}\Big). \tag{18}$$

Continuing the matching to second order, we find that $\mathcal{B}_0$ is essentially a regularized form of $G_1(z)$. It is automatically analytic at $\eta = 0$ and is given by

$$\mathcal{B}_0(\eta) = \frac{1}{\eta^2}\Big(1 - \frac{\sinh\alpha\eta}{\alpha\eta}\Big) + \alpha^2. \tag{19}$$

Thus, the asymptotic theory of the Orr-Sommerfeld equation for plane Couette flow is both reasonably simple and reasonably complete. And we now wish to consider how these ideas can be extended to more general velocity profiles.

V. PLANE POISEUILLE FLOW

Let us consider first the solutions of dominant-recessive type for general velocity profiles. To infer the general structure of the uniform expansions for these solutions, it is necessary to first consider their inner expansions. To three orders it is found (Reid 1972) that the inner expansions can be expressed in terms of the generalized Airy functions $A_k(\zeta, p)$. To proceed to higher orders it is necessary to consider a further generalization of Airy functions defined by

$$A_k(\zeta, p, q) = \frac{1}{2\pi i}\int_{L_k} t^{-p}(\log t)^q \exp(\zeta t - \frac{1}{3}t^3)\,dt, \tag{20}$$

where $q = 1, 2, \ldots$.

At the fourth order it is found that the inner expansions of the solutions of dominant-recessive type contain a term of the form

$$-\beta_1 A_k(\zeta, 2, 1), \quad \text{where} \quad \beta_1 = \frac{U_c^{\mathrm{iv}}}{U_c'} - \frac{U_c'''U_c''}{U_c'^2}. \tag{21}$$

From these results I was led to conjecture (Reid 1974) that the uniform expansions must be of the form

$$\begin{aligned} V_k(\eta) = {} & \mathcal{A}(\eta,\epsilon)A_k(\zeta,2,0) + \epsilon^2\mathcal{B}(\eta,\epsilon)A_k(\zeta,1,0) + \epsilon\mathcal{C}(\eta,\epsilon)A_k(\zeta,0,0) \\ & + \epsilon^3\{\mathcal{D}(\eta,\epsilon)A_k(\zeta,2,1) + \epsilon^2\mathcal{E}(\eta,\epsilon)A_k(\zeta,1,1) \\ & + \epsilon\mathcal{F}(\eta,\epsilon)A_k(\zeta,0,1)\} + \cdots, \end{aligned} \tag{22}$$

and Baldwin (1981) has confirmed this in greater detail.

Two important points emerge from this discussion. First, for general velocity profiles the "first approximations" are of the form given by Eq. (16). Second, for plane Poiseuille flow $\beta_1 = 0$ and this suggests that the uniform expansions, to at least a "second approximation," are entirely in terms of $A_k(\zeta, p)$. To still higher orders it is not known whether or not the uniform expansions involve the generalized Airy functions $A_k(\zeta, p, q)$ with $q > 0$.

To deal with the solutions of balanced type it is necessary to introduce the functions

$$B_k(\zeta, p, q) = \frac{1}{2\pi i}\int_{\infty_k}^{0+} t^{-p}(\log t)^q \exp(\zeta t - \frac{1}{3}t^3)\,dt, \tag{23}$$

where $\infty_k = \infty \exp[2(k-1)\pi i/3]$. These functions are essentially similar to Scorer functions. Again it is convenient to consider three solutions of balanced type which will be denoted by $U_k(\eta)$. A consideration of the inner expansions of these solutions suggests that the corresponding "first approximations" to the solutions of balanced type are of the form

$$\begin{aligned} U_k(\eta) \sim {} & \mathcal{G}_0(\eta) - \epsilon(U_c''/U_c')\{\mathcal{A}_0(\eta)B_k(\zeta,2,1) \\ & + \epsilon^2\mathcal{B}_0(\eta)B_k(\zeta,1,1) + \epsilon\mathcal{C}_0(\eta)B_k(\zeta,0,1)\}, \end{aligned} \tag{24}$$

where $\mathcal{A}_0, \mathcal{B}_0$ and $\mathcal{C}_0$ are the same as in Eq. (16).

Thus, to obtain first approximations to the solutions of balanced and dominant-recessive type, it is necessary to determine just four slowly-varying coefficients. This can be done in a manner similar to that described in Sec. IV for plane Couette flow. Specifically, $\mathcal{A}_0(\eta)$ can be expressed in terms of the regular inviscid solution $\phi_1^{(0)}(z)$, $\mathcal{G}_0(\eta)$ can be expressed in terms of the *regular part* of the singular inviscid solution $\phi_2^{(0)}(z)$, and $\mathcal{B}_0(\eta)$ and $\mathcal{C}_0(\eta)$ can then be obtaind by using the matching technique described in Sec. IV.

VI. THE ASYMPTOTIC SUCTION PROFILE

The approximations which have been discussed so far are uniformly valid only in a bounded domain of the z-plane. For flows of the boundary-layer type, however, we require approximations which remain uniformly valid on the semi-infinite interval $0 \leq z < \infty$. This can be achieved by a simple redefinition of the large parameter along the lines discussed by Olver (1954) for second-order equations.

To identify the relevant large parameter, observe that as $z \to \infty$ the Orr-Sommerfeld equation becomes

$$(\mathrm{D}^2 - \alpha^2)(\mathrm{D}^2 - \beta^2)\phi = 0, \tag{25}$$

where

$$\beta = \{\mathrm{i}\,\alpha R(1-c) + \alpha^2\}^{1/2} \quad \text{with } \operatorname{Re}(\beta) > 0. \tag{26}$$

Thus we choose β rather than $\epsilon^{-3/2}$ as the large parameter of the problem and then rewrite the Orr-Sommerfeld equation in the form

$$(\mathrm{D}^2 - \alpha^2)^2\phi - (\beta^2 - \alpha^2)\Big\{\frac{U-c}{1-c}(\mathrm{D}^2 - \alpha^2)\phi - \frac{U''}{1-c}\phi\Big\} = 0. \tag{27}$$

For this equation the relevant Langer variable is

$$\eta(z) = \Big\{\frac{3}{2}\int_{z_c}^{z}\Big(\frac{U-c}{1-c}\Big)^{1/2}\mathrm{d}\,z\Big\} \tag{28}$$

and the asymptotic theory of Eq. (27) can then be developed along the lines already described. For further details see Lakin and Reid (1982).

There is, however, one further important point to note. In the case of the asymptotic suction profile, for which $U(z) = 1 - e^{-z}$, the factor β_1 also vanishes (cf. Eq. (21)). More generally, Baldwin (1970, 1985) has shown that in this case the solutions of the Orr-Sommerfeld equation have integral representations of Barnes type and in the latter paper he further showed that the uniform expansions of dominant-recessive type are of the form given by Eq. (14), i.e. they involve only $A_k(\zeta, p)$. Similarly, the uniform expansions of balanced type involve only $B_k(\zeta, p, 1)$.

From this discussion it is evident, irrespective of the form of the basic velocity profile, that the solutions of dominant-recessive type have "first approximations" of the form given by Eq. (16) and that the corresponding solutions of balanced type are given by Eq. (24). It is also evident that the form of the higher approximations depends in a sensitive way on the form of the basic velocity profile and there seems to be no way, other than studying the inner expansions of the solutions, to infer the general structure of such higher approximations.

VII. THE EIGENVALUE RELATION

Having obtained uniform first approximations to the solutions of the Orr-Sommerfeld equation, it might be thought that the derivation of the corresponding approximation to the eigenvalue relation would be perfectly straightforward. That indeed is the conventional wisdom, but there is one important difficulty associated with what might be called the "truncation problem." This is largely a consequence of the different asymptotic behavior of the A- and B-type Airy functions.

To illustrate the main point involved, consider plane Poiseuille flow. For this flow the odd modes are known to be stable and for the even modes we can take the boundary conditions in the form

$$\phi(-1) = \phi'(-1) = 0 \quad \text{and} \quad \phi'(0) = \phi'''(0) = 0. \tag{29}$$

After some preliminary simplification, it is found that the eigenvalue relation involves only the generalized Airy functions $A_1(\zeta, p)$ and $B_3(\zeta, p, 1)$, and it is at this stage that the truncation problem arises.

This problem can be resolved, however, by using certain identities satisfied by the A- and B-type Airy functions. For example, it is found that $B_3(\zeta, 1, 1)$ appears only in the form of a second-order Wronskian with $A_1(\zeta, 1)$ and it can then be eliminated by using the identity

$$\mathcal{W}[A_1(\zeta,1), B_3(\zeta,1,1)] = 2A_1(\zeta,0,1) + (\gamma - 4\pi i)A_1(\zeta,0), \tag{30}$$

where γ is Euler's constant. Thus, the rapidly-varying terms in the eigenvalue relation can all be expressed in terms of the A-type Airy functions. In a "first approximation" to the eigenvalue relation we then have

$$\begin{aligned} A_1(\zeta,1) + \epsilon\{C_1(\eta)A_1(\zeta,-1) + C_2(\eta)A_1(\zeta,-1,1)\} \\ + \epsilon^2\{C_3(\eta)A_1(\zeta,0) + C_4(\eta)A_1(\zeta,0,1)\} = 0, \end{aligned} \tag{31}$$

where η and ζ are to be evaluated at $z = -1$. All of the slowly-varying coefficients in this equation are expressible in terms of $\mathcal{A}_0(\eta)$, $\mathcal{B}_0(\eta)$, $\mathcal{C}_0(\eta)$ and $\mathcal{G}_0(\eta)$ which, in turn, can be expressed in terms of the heuristic approximations.

The main point of this discussion is that although the first approximations to the solutions of dominant-recessive type involve only $A_k(\zeta, p)$. The corresponding approximation to the eigenvalue relation also involves $A_1(\zeta, p, 1)$. In particular, it does not involve the B-type Airy functions and in this way the truncation problem is resolved.

VIII. CONCLUDING REMARKS

Most of the theory discussed thus far has been purely formal. The claim has been made that the proposed approximations are uniformly valid in a full complex neighborhood of the turning point, and one may well ask to what extent that claim is justified.

In the first instance, the claim is based simply on the general structure of the approximations: everything seems to fit together in a very natural way! Next, there is some numerical evidence. Lakin, Ng and Reid (1978) made a calculation of the curve of marginal stability for plane Poiseuille flow, using

Eq. (31), and obtained results which are virtualy indistinguishable from the best numerical results available at that time. For example, they obtained a value of the minimum critical Reynolds number of 5769.7 as compared to the "exact" numerical value of 5772.22. Later, Lakin and Reid (1982) made similar calculations for the asymptotic suction profile and they also found good agreement with the results of direct numerical calculations. As to the individual solutions, Ng and Reid (1984) considered what are essentially "second approximations" for both plane Couette flow and plane Poiseuille flow. A comparison of the results with the first approximations described in this paper provides some additional evidence that they are indeed uniformly valid.

As to rigorous results, Baldwin (1981) has succeeded in obtaining error bounds for the uniform approximations to the solutions of dominant-recessive type along the lines which Olver (1974) has done so successfully for second-order equations. The bounds which Baldwin obtained would certainly seem to be of the correct form but, unfortunately, they are rather complicated; even numerical evaluation appears difficult and it is therefore not possible as yet to assess how realistic they are.

To obtain rigorous error bounds for the approximations of balanced type is much more difficult and for the eigenvalue relation all but impossible. Thus, although the formal theory is reasonably complete, much still remains to be done to provide a rigorous justification for the proposed approximations.

REFERENCES

Baldwin, P. (1970). *Mathematatika* **17**: 206.

Baldwin, P. (1981). *Mathematatika* **28**: 116

Baldwin, P. (1985). *Proc. Roy. Soc. (Lond.)* **A399**: 321

Drazin, P. G. and Reid, W. H. (1981). *Hydrodynamic Stability*, Cambridge University Press.

Lakin, W. D., Ng, B. S. and Reid, W. H. (1978). *Phil. Trans. Roy. Soc. (Lond.)* **A289**: 347.

Lakin, W.D. and Reid, W. H. (1982). *Quart. J. Mech. Appl. Math.* **35**: 69.

Lin, C.C. (1955). *The Theory of Hydrodynamic Stability*, Cambridge University Press.

Lin, C.C. (1957). "On the Stability of the Laminar Boundary Layer," Proceedings of the Symposium on Naval Hydrodynamics, National Research Council Publication 515, Washington, D.C., pp. 353–371.

Lin, C. C. and Rabenstein, A. L. (1960). *Trans. Amer. Math. Soc.* **94**: 24.

Ng, B. S. and Reid, W. H. (1984). *Quart. J. Mech. Appl. Math.* **37**: 373.

Olver, F. W. J. (1954). *Phil. Trans. Roy. Soc. (Lond.)* **A247**: 307.

Olver, F. W. J. (1974). *Asymptotics and Special Functions*, Academic Press, New York.

Rabenstein, A. L. (1958). *Arch. Rat. Mech. Anal.* **1**: 418.

Reid, W. H. (1972). *Studies in Appl. Math.* **51**: 341.

Reid, W. H. (1974). *Studies in Appl. Math.* **53**: 217.

Wasow, W. (1953a). *J. Res. Nat. Bur. Stand.* **51**: 195.

Wasow, W. (1953b). *Annals Math.* **58**: 222.

Gevrey Property of Formal Solutions in a Parameter

Yasutaka Sibuya Professor, Department of Mathematics, University of Minnesota, Minneapolis, Minnesota

1. PRELIMINARIES

Let s be a nonnegative real number. A formal power series $\Sigma_{m\geq 0} a_m(x)\, \varepsilon^m$ is said to be of *Gevrey order* s *uniformly on* D if there exist two nonnegative numbers C_0 and C_1 such that

$$| a_m(x) | \leq C_0 (m!)^s C_1{}^m \tag{1.1}$$

for $x \in D$ and $m = 0, 1, \ldots$.

Set

$$V = \{ \varepsilon ; \ \alpha_0 < \arg \varepsilon < \beta_0, \ 0 < | \varepsilon | < \delta_0 \}$$

and

$$W = \{ \varepsilon ; \ \alpha \le \arg \varepsilon \le \beta, \ 0 < |\varepsilon| \le \delta \} .$$

A function $f(x,\varepsilon)$ is said to admit an asymptotic expansion

$$\Sigma_{m\ge 0} a_m(x) \varepsilon^m$$

of *Gevrey order* s *as* $\varepsilon \to 0$ *in* V *uniformly on* D if
(i) $\Sigma_{m\ge 0} a_m(x) \varepsilon^m$ is of Gevrey order s uniformly on D,
(ii) for each W such that $\alpha_0 < \alpha < \beta < \beta_0$ and $0 < \delta < \delta_0$,
there exist two nonnegative numbers K_W and L_W such that

$$| f(x,\varepsilon) - \Sigma_{N\ge m\ge 0} a_m(x) \varepsilon^m | \le K_W((N+1)!)^s L_W^{N+1} |\varepsilon|^{N+1} \tag{1.2}$$

for $x \in D$, $\varepsilon \in W$ and $N = 0, 1, \dots$.

A function $f(x,\varepsilon)$ is said to be flat of Gevrey order s as $\varepsilon \to 0$ in V uniformly on D if f admits an asymptotic expansion of Gevrey order s as $\varepsilon \to 0$ in V uniformly on D and the expansion of f is 0, i.e.

$$| f(x,\varepsilon) | \le K_W((N+1)!)^s L_W^{N+1} |\varepsilon|^{N+1} \tag{1.3}$$

for $x \in D$, $\varepsilon \in W$ and $N = 0, 1, \dots$.

Proposition 1.1: *A function* $f(x,\varepsilon)$ *is flat of Gevrey order* s *as* $\varepsilon \to 0$ *in* V *uniformly on* D *if and only if, for each* W *such that* $\alpha_0 < \alpha < \beta < \beta_0$ *and* $0 < \delta < \delta_0$*, there exist a nonnegative number* γ_W *and a positive number* λ_W *such that*

$$| f(x,\varepsilon) | \le \gamma_W \exp(-\lambda_W | \varepsilon |^{-k}) \qquad \textit{for} \quad (x,\varepsilon) \in D \mathrm{x} W \tag{1.4}$$

where

$$k = 1/s \quad . \tag{1.5}$$

It is evident that (1.4) implies (1.3). To derive (1.4) from (1.3) it suffices to choose a suitable N for each ε.

Remark 1.2: A function $f(x,\varepsilon)$ is identically zero on $D \times V$ if
(i) f is holomorphic in $\varepsilon \in V$ for each fixed x in D,
(ii) f is flat of Gevrey order s as $\varepsilon \to 0$ in V uniformly on D,
(iii) $\beta_0 - \alpha_0 > \pi s$.

2. MAIN THEOREM

We consider a system of ordinary differential equations:

$$\varepsilon^\sigma dy/dx = f(x,\varepsilon) + A(x,\varepsilon)y + \Sigma_{|\mathbf{p}|\ge 2} f_{\mathbf{p}}(x,\varepsilon)\, y^{\mathbf{p}} , \tag{2.1}$$ +

where
(I) σ is a positive integer; y, f and $f_{\mathbf{p}}$ are n-dimensional vectors; A is an nxn matrix; $\mathbf{p} = (p_1, ..., p_n)$ and the p_j are nonnegative integers; $|\mathbf{p}| = p_1 + ... + p_n$; $y^{\mathbf{p}} = y_1^{p_1} y_2^{p_2} .. y_n^{p_n}$, and y_j is the j-th entry of the column vector y ,
(II) the entries of f, $f_{\mathbf{p}}$ and A are holomorphic in (x,ε) for

$$| x | < r_0, \qquad | \varepsilon | < r_1 \ , \tag{2.2}$$

where r_0 and r_1 are positve numbers,
(III) the series $\Sigma_{|\mathbf{p}|\ge 2} f_{\mathbf{p}}(x,\varepsilon)\, y^{\mathbf{p}}$ is convergent uniformly on each compact subset of a domain:

$$| x | < r_0, \quad | \varepsilon | < r_1 \ , \quad | y | = \max_{n \geq j \geq 1} | y_j | < r_2 \ , \tag{2.3}$$

where r_2 is a positive number,
(IV) $f(0,0) = 0$,

(V) $A(0,0)$ is invertible.

In this paper, we shall prove the following result:

Theorem 2.1: *System (2.1) admits a unique formal solution:*

$$y = \sum_{m \geq 0} a_m(x)\, \varepsilon^m \quad , \tag{2.4}$$

where the coefficients $a_m(x)$ *are holomorphic and bounded for*

$$| x | < r_3 \tag{2.5}$$

for some r_3 *such that* $0 < r_3 < r_0$. *Furthermore this formal series is of Gevrey order* $s = 1/\sigma$ *uniformly on domain* (2.5) *if we choose* r_3 *sufficiently small.*

3. LEMMAS

We shall reduce theorem 2.1 to the following lemma. Fix an open disc $| x | < \rho_0$ $(\rho_0 > 0)$, and consider a covering of $\{ \varepsilon \,;\, 0 < | \varepsilon | < \rho_0 \}$ by a finite number of open sectors

$$\mathcal{S}_j = \{ \varepsilon \,; \quad \alpha_j < \arg \varepsilon < \beta_j, \quad 0 < | \varepsilon | < \rho_0 \},$$

$$(j = 1, 2, \ldots, p), \tag{3.1}$$

where $\alpha_j < \alpha_{j+1}$ and $\alpha_p < \alpha_1 + 2\pi$. We assume that

$$\mathcal{S}_j \cap \mathcal{S}_{j+1} \neq \emptyset \qquad (\mathcal{S}_{p+1} = \mathcal{S}_1),$$
$$\mathcal{S}_j \cap \mathcal{S}_h = \emptyset \quad \text{otherwise for } j \neq h. \tag{3.2}$$

Set

$$U_j = \mathcal{S}_j \cap \mathcal{S}_{j+1} \qquad (j = 1, \ldots, p). \tag{3.3}$$

Lemma 3.1: *Assume that* p *funtions* $\phi_1(x,\varepsilon), \ldots, \phi_p(x,\varepsilon)$ *satisfy the following conditions:*
(i) $\phi_j(x,\varepsilon)$ *is continuous and bounded on* $D \times \mathcal{S}_j$, *and holomorphic in* $\varepsilon \in \mathcal{S}_j$ *for each fixed* $x \in D$,
(ii) $|\ \phi_j(x,\varepsilon) - \phi_{j+1}(x,\varepsilon)\ | \leq \gamma \exp(-\lambda|\varepsilon|^{-k})$ *on* $D \times U_j$,
where $\gamma \geq 0$, $\lambda > 0$ *and* $k > 0$ *are suitable numbers.*

Then, for each j , *the function* $\phi_j(x,\varepsilon)$ *admits an asymptotic expansion of Gevrey order*

$$s = 1/k \tag{3.4}$$

as $\varepsilon \to 0$ *in* $\mathcal{S}_j$ *uniformly on* D.

To prove this lemma we shall use the following lemma:

Lemma 3.2: *Assume that* p *funtions* $\delta_1(x,\varepsilon), \ldots, \delta_p(x,\varepsilon)$ *satisfy the following conditions:*
(i) $\delta_j(x,\varepsilon)$ *is continuous and bounded on* $D \times U_j$, *and holomorphic in* $\varepsilon \in U_j$ *for each fixed* $x \in D$,
(ii) $|\ \delta_j(x,\varepsilon)\ | \leq \gamma \exp(-\lambda|\varepsilon|^{-k})$ *on* $D \times U_j$,
where $\gamma \geq 0$, $\lambda > 0$ *and* $k > 0$ *are suitable numbers.*

Then, there exist p *funtions* $\psi_1(x,\varepsilon), \ldots, \psi_p(x,\varepsilon)$ *such that*
(a) $\psi_j(x,\varepsilon)$ *is continuous on* $D \times \mathcal{S}_j$, *and holomorphic in* $\varepsilon \in \mathcal{S}_j$ *for each fixed* $x \in D$,

(b) *for each* j, *the function* $\psi_j(x,\varepsilon)$ *admits an asymptotic expansion of Gevrey order* $s = 1/k$ *as* $\varepsilon \to 0$ *in* $\mathscr{S}_j$ *uniformly on* D,

(c) $\delta_j(x,\varepsilon) = \psi_j(x,\varepsilon) - \psi_{j+1}(x,\varepsilon)$ *on* $D \times U_j$.

Remark 3.3: The asymptotic expansion of ψ_j is independent of j.

If we apply lemma 3.2 to

$$\delta_j(x,\varepsilon) = \phi_j(x,\varepsilon) - \phi_{j+1}(x,\varepsilon) \quad (j = 1, \ldots, p), \tag{3.5}$$

we get

$$\phi_j(x,\varepsilon) - \phi_{j+1}(x,\varepsilon) = \psi_j(x,\varepsilon) - \psi_{j+1}(x,\varepsilon) \quad \text{on } D \times U_j \tag{3.6}$$

and hence

$$\phi_j(x,\varepsilon) - \psi_j(x,\varepsilon) = \phi_{j+1}(x,\varepsilon) - \psi_{j+1}(x,\varepsilon) \quad \text{on } D \times U_j . \tag{3.7}$$

Define a function ϕ by

$$\phi(x,e) = \phi_j(x,\varepsilon) - \psi_j(x,\varepsilon) \quad \text{on } D \times \mathscr{S}_j \quad (j = 1, \ldots, p). \tag{3.8}$$

Then ϕ is continuous on $D \times \{\varepsilon ; |\varepsilon| < \rho_0\}$ and holomorprhic in $\varepsilon \in \{\varepsilon ; |\varepsilon| < \rho_0\}$ for each fixed $x \in D$. Hence ϕ is a sum of a convergent power series in ε with coefficients continuous in $x \in D$. Since $\phi_j(x,\varepsilon) = \psi_j(x,\varepsilon) + \phi(x,\varepsilon)$, lemma 3.1 follows immediately. For a proof of lemma 3.2 see Ramis (1978) and Sibuya (1981).

4. PROOF OF THEOREM 2.1

The existence and uniqueness of formal solution (2.4) can be verified easily.

For every direction θ in the ε-plane, there exist a sector

$$\mathcal{S}(\theta) = \{ \varepsilon \; ; \; |\arg(\varepsilon) - \theta| < \alpha(\theta), \; 0 < |\varepsilon| < \omega(\theta) \}$$

and a disc

$$\Delta(\theta) = \{ x \; ; \; |x| < r(\theta) \}$$

such that system (2.1) has a solution $\phi(x,\varepsilon;\theta)$ admitting series (2.4) as an asymptotic expansion in the conventional sense as $\varepsilon \to 0$ in $\mathcal{S}(\theta)$ uniformly on $\Delta(\theta)$ (cf. Y.Sibuya (1958)).

We can choose a finite number of $\theta_1, \ldots, \theta_p$ in such a way that p sectors

$$\mathcal{S}_j = \{ \varepsilon \; ; \; |\arg(\varepsilon) - \theta_j| < \alpha(\theta_j), \; 0 < |\varepsilon| < \rho_0 \} \qquad (j=1, \ldots, p) \tag{4.1}$$

satisfy conditions (3.2) and (3.3), and that $\{ \mathcal{S}_j ; j = 1, \ldots, p\}$ is a covering of $\{ \varepsilon ; \; 0 < |\varepsilon| < \rho_0 \}$. Let us fix a ρ_0 by

$$0 < \rho_0 \leq \min_{p \geq j \geq 0} \omega(\theta_j) . \tag{4.2}$$

Set

$$\phi_j(x,\varepsilon) = \phi(x,\varepsilon;\theta_j) \qquad (j = 1, \ldots, p) \tag{4.3}$$

and

$$D_0 = \{ x \; ; |x| < r \leq \min_{p \geq j \geq 1} r(\theta_j) \} . \tag{4.4}$$

To apply lemma 3.1 to functions (4.3), it suffices to show that

$$|\ \phi_j(x,\varepsilon) - \phi_{j+1}(x,\varepsilon)\ | \leq \gamma \exp(-\lambda|\varepsilon|^{-\sigma}) \quad \text{on } D \times U_j , \tag{4.5}$$

for sufficiently small ρ_0 and r, where $\gamma \geq 0$ and $\lambda > 0$ are suitable numbers and

$$U_j = \mathcal{S}_j \cap \mathcal{S}_{j+1} . \tag{4.6}$$

If estimate (4.5) is proved, then lemma 3.1 implies that ϕ_j admits an asymptotic expansion of Gevrey order $s = 1/\sigma$. Since formal solution (2.4) is the expansion of ϕ_j, we can complete the proof of theorem 2.1.

Hereafter, we shall derive estimate (4.5). To do this, set

$$F(x,\varepsilon,y) = \Sigma_{|\mathbf{p}| \geq 2}\, f_{\mathbf{p}}(x,\varepsilon)\, y^{\mathbf{p}} \tag{4.7}$$

and

$$\delta_j(x,\varepsilon) = \phi_j(x,\varepsilon) - \phi_{j+1}(x,\varepsilon) \qquad (j = 1, \ldots, p). \tag{4.8}$$

Then

$$\begin{aligned} \varepsilon^{\sigma}\, d\delta_j(x,\varepsilon)/dx = {} & A(x,\varepsilon)\, \delta_j(x,\varepsilon) \\ & + F(x,\varepsilon, \phi_{j+1}(x,\varepsilon) + \delta_j(x,\varepsilon)) \\ & - F(x,\varepsilon, \phi_{j+1}(x,\varepsilon)) . \end{aligned} \tag{4.9}$$

Note that δ_j is flat in U_j uniformly on D_0, and that

$$[\int_0^1 (\partial F/\partial y)\,(x,\varepsilon,\ \phi_{j+1}(x,\varepsilon) + t\delta_j(x,\varepsilon))dt]\ \delta_j(x,\varepsilon)$$

$$= F(x,\varepsilon,\ \phi_{j+1}(x,\varepsilon) + \delta_j(x,\varepsilon)) - F(x,\varepsilon,\ \phi_{j+1}(x,\varepsilon))\ . \qquad (4.10)$$

Therefore $v = \delta_j(x,\varepsilon)$ is a flat solution to the linear system

$$\varepsilon^{\sigma}\, dv/dx = [A(x,\varepsilon) + B(x,\varepsilon)]v \qquad \text{on } D_0 \times U_j\ , \qquad (4.11)$$

where

$$B(x,\varepsilon) = \int_0^1 (\partial F/\partial y)\,(x,\varepsilon,\ \phi_{j+1}(x,\varepsilon) + t\delta_j(x,\varepsilon))dt. \qquad (4.12)$$

Since $A(0,0)$ is invertible and $B(0,0) = 0$, we can derive (4.5) by utilizing a conventional method such as "block-diagonalization theorem" (cf. Y.Sibuya (1958)).

This work was supported in part by grants from the National Science Foundation and was done at Department of Mathematics, University of Southern California, Los Angeles, during the author's sabbatical leave in 1988/89.

REFERENCES

1. J.-P.Ramis (1978), Asterisque, 59-60, pp.173-204.

2. Y.Sibuya (1958), J. Fac. Sci., Univ. Tokyo, 7, pp. 527-540.

3. Y.Sibuya (1981), SIAM J. Math. Anal., 12, pp. 653-668.

A Riccati Approach to the Airy Equation

Donald R. Smith Professor, Department of Mathematics, University of California, San Diego, La Jolla, California

ABSTRACT

A Riccati transformation is used, first to obtain a fundamental solution for the Airy equation $f''(z) = zf(z)$ on the region $-\pi/3 \leq \mathrm{ph}(z) \leq \pi/3$, and then to obtain convenient representations for lateral connection matrices for the study of the equation in other regions of the complex plane.

1. INTRODUCTION

The integral

$$\int_0^\infty \cos\left(\tfrac{1}{2}\pi(t^3 - xt)\right)dt \equiv 2\left(\frac{\pi^2}{12}\right)^{1/3} Ai(-(\pi^2/12)^{1/3}x) \tag{1.1}$$

was studied by Airy (1838, 1849) in an investigation of the intensity of light in a neighborhood of a caustic. The resulting function $f = Ai(z)$ is a solution of the equation (cf. Olver (1974, Section 2.8))

$$\frac{d^2 f}{dz^2} = zf. \tag{1.2}$$

See Olver (1974) and Wasow (1965, 1976, 1985) for a general analysis and history of the "*Airy equation*" (1.2).

It is sometimes customary to rewrite (1.2) as the first-order system

$$\frac{d}{dz}\begin{pmatrix} f \\ g \end{pmatrix} = \begin{pmatrix} 0 & 1 \\ z & 0 \end{pmatrix}\begin{pmatrix} f \\ g \end{pmatrix} \tag{1.3}$$

with $g = f'(z)$, and then the eigenvector transformation

$$\begin{pmatrix} f \\ g \end{pmatrix} = \frac{1}{2}\begin{pmatrix} 1 & 1 \\ -\sqrt{z} & \sqrt{z} \end{pmatrix}\begin{pmatrix} u \\ v \end{pmatrix} \tag{1.4}$$

transforms (1.3) into

$$\frac{d}{dz}\begin{pmatrix} u \\ v \end{pmatrix} = \left[\begin{pmatrix} -\sqrt{z} & 0 \\ 0 & \sqrt{z} \end{pmatrix} + \frac{1}{4z}\begin{pmatrix} -1 & 1 \\ 1 & -1 \end{pmatrix}\right]\begin{pmatrix} u \\ v \end{pmatrix} \tag{1.5}$$

where the principal square root will be used here,

$$\sqrt{z} = \sqrt{|z|}e^{i\,\mathrm{ph}\,(z)/2} \quad \text{with} \quad |z| \geq 0, \ -\pi < \mathrm{ph}\,(z) \leq \pi. \tag{1.6}$$

The change of independent variable

$$\xi = (2/3)z^{3/2} \tag{1.7}$$

reduces (1.5) to

$$\frac{d}{d\xi}\begin{pmatrix} U \\ V \end{pmatrix} = \left[\begin{pmatrix} -1 & 0 \\ 0 & 1 \end{pmatrix} + \frac{1}{6\xi}\begin{pmatrix} -1 & 1 \\ 1 & -1 \end{pmatrix}\right]\begin{pmatrix} U \\ V \end{pmatrix} \tag{1.8}$$

with

$$\begin{pmatrix} U \\ V \end{pmatrix}(\xi) := \begin{pmatrix} u \\ v \end{pmatrix}(z). \tag{1.9}$$

It is shown in Smith (1985, 1986, 1987) that a Riccati transformation can be used to provide Liouville/Green approximations for solutions $f = f(x)$ of systems of equations including (1.2) and (1.5) for real z. The Riccati transformation also provides a convenient tool for a study of such equations for complex z. In the present paper this Riccati approach is illustrated for the Airy equation (1.2), (1.5). The Riccati transformation is given in Section 2 and is then used in Section 3 to obtain a fundamental solution that provides a recessive solution for (1.2) in the region

$$-\pi/3 \leq \mathrm{ph}(z) \leq \pi/3. \tag{1.10}$$

Then in Section 4 the Riccati transformation leads to convenient representations for lateral connection matrices used in the study of the Airy equation in other regions of the complex plane. The use of a Riccati transformation as illustrated here is convenient for certain aspects of the study of equations such as (1.2), but the approach has limitations which are not shared by the broader methods and analyses of Olver illustrated in the representative selection of references Olver (1961, 1974, 1977a, 1977b, 1978).

2. THE RICCATI TRANSFORMATION

The Riccati transformation

$$\begin{pmatrix} U \\ V \end{pmatrix} = \begin{pmatrix} 1 & 0 \\ -R & 1 \end{pmatrix} \begin{pmatrix} 1 & -S \\ 0 & 1 \end{pmatrix} \begin{pmatrix} \hat{U} \\ \hat{V} \end{pmatrix} \tag{2.1}$$

reduces (1.8) to the diagonal form

$$\frac{d}{d\xi}\begin{pmatrix} \hat{U} \\ \hat{V} \end{pmatrix} = \left[\begin{pmatrix} -1 & 0 \\ 0 & 1 \end{pmatrix} - \frac{1}{6\xi}\begin{pmatrix} 1 + R(\xi) & 0 \\ 0 & 1 - R(\xi) \end{pmatrix} \right] \begin{pmatrix} \hat{U} \\ \hat{V} \end{pmatrix} \tag{2.2}$$

if R and S satisfy the equations (cf. Smith (1985, Exercise 9.2.5; 1986))

$$\frac{dR}{d\xi} - 2R = -\frac{1}{6\xi}\left(1 - R^2\right) \tag{2.3}$$

and

$$\frac{dS}{d\xi} + \left(2 + \frac{R(\xi)}{3\xi}\right) S = -\frac{1}{6\xi}. \tag{2.4}$$

A recessive solution of (2.3) is obtained in the region

$$|\xi| > 0, \ |\text{ph}\ \xi| \leq \tfrac{\pi}{2} \tag{2.5}$$

as the solution of the integral equation

$$R(\xi) = R_0(\xi) - \int_\xi^\infty \frac{e^{-2(\sigma-\xi)}}{6\sigma} R(\sigma)^2 d\sigma \tag{2.6}$$

with

$$R_0(\xi) := \int_\xi^\infty \frac{e^{-2(\sigma-\xi)}}{6\sigma} d\sigma. \tag{2.7}$$

The integral defining R_0 is convergent and independent of path in the region (2.5), and indeed $R_0(\xi)$ is holomorphic in the right half-plane. Repeated integrations by parts yield

$$R_0(\xi) = \frac{1}{6}\sum_{k=0}^{n}(-1)^k \frac{k!}{(2\xi)^{k+1}} + E_n^{(0)}(\xi) \tag{2.8$_n$}$$

with remainder $E_n^{(0)}$ given by

$$E_n^{(0)}(\xi) = \frac{(-1)^{n+1}(n+1)!}{6 \cdot 2^{n+1}} \int_\xi^\infty \sigma^{-(n+2)} e^{-2(\sigma-\xi)} d\sigma. \tag{2.9$_n$}$$

A direct estimation of the integral in $(2.9)_n$ yields

$$|E_n^{(0)}(\xi)| \leq n!/\left(6 \cdot |2\xi|^{n+1}\right) \tag{2.10$_n$}$$

on the region (2.5). From $(2.9)_n$ and $(2.9)_{n+1}$ there follows

$$E_n^{(0)}(\xi) = \frac{(-1)^{n+1}(n+1)!}{6 \cdot (2\xi)^{n+1}} + E_{n+1}^{(0)}(\xi),$$

which with $(2.10)_n$ and $(2.10)_{n+1}$ implies

$$|E_n^{(0)}(\xi)| \leq \frac{(n+1)!}{3 \cdot |2\xi|^{n+2}} \tag{$(2.11)_n$}$$

uniformly on the region (2.5), so that the order of magnitude of the truncation error in $(2.8)_n$ is at most twice the order of the next term in the expansion $(2.8)_{n+1}$. The following slightly different estimate can also be obtained directly from $(2.9)_n$,

$$|E_n^{(0)}(\xi)| \leq \frac{(n+1)!}{6 \cdot cos\,(\text{ph}\ \xi) \cdot |2\xi|^{n+2}} \tag{$(2.12)_n$}$$

for $|\text{ph}\ \xi| < \pi/2$, $|\xi| > 0$. This last estimate is better than $(2.11)_n$ for $|\text{ph}\ \xi| < \pi/3$. In particular $(2.12)_n$ shows that the error is bounded by the size of the next term in the expansion if ξ is real, $\text{ph}(\xi) = 0$.

Lemma 2.1: *The Riccati differential equation (2.3) and the integral equation (2.6)-(2.7) have a common holomorphic solution R on the region*

$$\mathcal{D} := \left\{ \xi \;\middle|\; |\text{ph}(\xi)| \leq \pi/2,\ |\xi| \geq \lambda^* := 2/(\sqrt{145} - 1) \doteq 0.1811332 \right\} \tag{2.13}$$

satisfying

$$|R(\xi)| \leq \frac{1}{12|\xi|}\left(1 + \frac{1.23}{|\xi|}\right) \tag{2.14}$$

everywhere on $\mathcal{D}$. The solution is unique subject to the condition (2.14).

Proof. A routine argument using $(2.8)_0, (2.11)_0$ shows (cf. Smith (1986)) that the right-side of (2.6)—considered as an operator on the space of continuous functions R equipped with the supremum norm—maps the collection of such functions satisfying

$$|R(\xi)| \leq \frac{\rho}{12|\xi|} \quad \text{for} \quad |\xi| \geq \frac{1}{\rho - 1}, |\text{ph}\,\xi| \leq \frac{\pi}{2} \tag{2.15}$$

contractively into itself, for any fixed ρ satisfying

$$2 \cdot \frac{144}{143} \leq \rho < \rho^* := \frac{1 + \sqrt{145}}{2}. \tag{2.16}$$

(The operator maps the given collection of functions into itself if $\rho \geq 2 \cdot 144/143$ and it is a contraction with contractivity $= \rho(\rho-1)/36$ which is positive and less than 1 if $1 < \rho < \rho^*$.) It follows directly by the Banach/Picard fixed-point theorem that (2.6) has a unique solution $R(\xi)$ satisfying (2.15), for any ρ satisfying (2.16). We then let $\rho \to \rho^*$ and find that the solution satisfies (2.15) with ρ replaced by ρ^*,

$$|R(\xi)| \leq \frac{\rho^*}{12|\xi|} \quad \text{for} \quad |\xi| \geq \lambda^*. \tag{2.17}$$

Iterating the estimate for R several times in the right-side of (2.6), we obtain the improved estimate (2.14), and this leads directly to the assertion of the lemma. (The constant 1.23 appearing in (2.14) is not best possible.)

The solution R of (2.3), (2.14) given by the lemma satisfies the integral equation

$$R(\xi) = e^{2(\xi-\xi_0)}R(\xi_0) + \int_\xi^{\xi_0} \frac{e^{-2(\sigma-\xi)}}{6\sigma} d\sigma - \int_\xi^{\xi_0} \frac{e^{-2(\sigma-\xi)}}{6\sigma} R(\sigma)^2 d\sigma \tag{2.18}$$

for any fixed ξ_0 in the region $\mathcal{D}$ of (2.13). The earlier equation (2.6) corresponds to the choice $\xi_0 = \infty$ in (2.18). The method of proof of **Lemma 2.1** can be applied to (2.18) with a different choice of $\xi_0 \in \mathcal{D}$ located near $|\xi| = \lambda^*$, and in this way the solution R can be continued everywhere on the region (2.5).

The solution R has an asymptotic expansion

$$R(\xi) \sim \frac{1}{6}\sum_{k=0}^{\infty} \frac{a_k}{(2\xi)^{k+1}} \tag{2.19}$$

valid for large ξ in the region (2.5), where the coefficients a_k are obtained by inserting (2.19) into (2.6), yielding

$$a_0 = 1, \quad a_1 = -1, \quad \text{and}$$

$$a_k = (-1)^k k! \left[1 - \frac{1}{6}\sum_{\ell=0}^{k-2} \frac{(-1)^\ell}{(\ell+2)!} \sum_{m=0}^{\ell} a_m a_{\ell-m}\right] \quad \text{for} \quad k = 2, 3, \cdots. \tag{2.20}$$

The remainder $E_n(\xi)$ in the following truncated expansion (cf. $(2.8)_n$)

$$R(\xi) = \frac{1}{6}\sum_{k=0}^{n} \frac{a_k}{(2\xi)^{k+1}} + E_n(\xi) \tag{2.21}$$

can be estimated by applying the methods of the proof of **Lemma 2.1** to the integral equation for E_n obtained by inserting (2.21) into (2.6).

The function $S(\xi) \equiv S(\xi, \xi_0)$ defined for ξ, ξ_0 in the region (2.5) by the formula

$$S(\xi, \xi_0) := -\int_{\xi_0}^{\xi} \frac{e^{-2(\xi-\sigma)}}{6\sigma}\, exp\left[-\int_{\sigma}^{\xi} \frac{R(s)}{3s} ds\right] d\sigma, \tag{2.22}$$

satisfies the differential equation (2.4) as function of ξ, for any fixed ξ_0, where R is the solution of (2.3) given by **Lemma 2.1** and where $S(\xi, \xi_0)$ has been taken to vanish at $\xi = \xi_0$.

The following conjugation property (Schwarz reflection property) for R will be useful in Section 4.

Lemma 2.2: *The solution R of the Riccati differential equation (2.3) given by* **Lemma 2.1** *satisfies*

$$R(\overline{\xi}) = \overline{R(\xi)} \tag{2.23}$$

on the region (2.5), so that in particular $R(\xi)$ is real for real ξ.

Proof: It follows from (2.7) that R_0 satisfies (2.23),

$$R_0(\overline{\xi}) = \overline{R_0(\xi)},$$

so that the conjugate of equation (2.6) can be written as

$$\overline{R(\xi)} = R_0(\overline{\xi}) - \int_{\overline{\xi}}^{\infty} \frac{e^{-2(\sigma-\overline{\xi})}}{6\sigma} \left(\overline{R(\xi)}\right)^2 d\sigma. \tag{2.24}$$

The stated result (2.23) follows directly from (2.24) and the uniqueness result of **Lemma 2.1**. The result follows alternatively from the expansion (2.19)-(2.20).

If ξ_0 is real, then the function $S(\xi, \xi_0)$ of (2.22) also satisfies this conjugation property as a function of ξ.

3. A FUNDAMENTAL SOLUTION IN THE SECTOR $\mathcal{S}_0$

A fundamental solution (matrix) of the diagonal system (2.2) is given as

$$\begin{pmatrix} \hat{U}_1 & \hat{U}_2 \\ \hat{V}_1 & \hat{V}_2 \end{pmatrix} = \tag{3.1}$$
$$(2/3)^{1/6} \begin{pmatrix} \xi^{-1/6} e^{-\xi}\, exp\left(\int_\xi^\infty \frac{R(\sigma)}{6\sigma} d\sigma\right) & 0 \\ 0 & \xi^{-1/6} e^{+\xi}\, exp\left(-\int_\xi^\infty \frac{R(\sigma)}{6\sigma} d\sigma\right) \end{pmatrix}$$

and then a corresponding fundamental solution of (1.8) can be given by (2.1), (3.1) as

$$\begin{pmatrix} U_1(\xi) & U_2(\xi,\xi_0) \\ V_1(\xi) & V_2(\xi,\xi_0) \end{pmatrix} = \begin{pmatrix} \hat{U}_1(\xi) & -S(\xi,\xi_0)\hat{V}_2(\xi) \\ -R(\xi)\hat{U}_1(\xi) & [1 + R(\xi)S(\xi,\xi_0)]\, \hat{V}_2(\xi) \end{pmatrix}. \tag{3.2}$$

The resulting fundamental solution matrix for (1.3) is given with (1.4), (1.9), (3.2) as

$$\begin{aligned} \begin{pmatrix} f_1(z) & f_2(z,\xi_0) \\ g_1(z) & g_2(z,\xi_0) \end{pmatrix} &= \frac{1}{2} \begin{pmatrix} 1 & 1 \\ -\sqrt{z} & \sqrt{z} \end{pmatrix} \begin{pmatrix} U_1(\xi) & U_2(\xi,\xi_0) \\ V_1(\xi) & V_2(\xi,\xi_0) \end{pmatrix} \\ &= \frac{1}{2} \begin{pmatrix} (1 - R(\xi))\hat{U}_1(\xi) & [1 - S(\xi,\xi_0)\cdot(1 - R(\xi))]\hat{V}_2(\xi) \\ -\sqrt{z}\,(1 + R(\xi))\,\hat{U}_1(\xi) & \sqrt{z}\,[1 + S(\xi,\xi_0)\cdot(1 + R(\xi))]\,\hat{V}_2(\xi) \end{pmatrix} \end{aligned} \tag{3.3}$$

with $\xi = \xi(z) := \frac{2}{3}\, z^{3/2}$ here and below.

The first column of (3.3) provides a recessive solution of (1.3),

$$\begin{pmatrix} f_1(z) \\ g_1(z) \end{pmatrix} = \begin{pmatrix} (1 - R(\xi)) \\ -\sqrt{z}(1 + R(\xi)) \end{pmatrix} \frac{e^{-(2/3)z^{3/2}}}{2\, z^{1/4}} exp\left[\int_\xi^\infty \frac{R(\sigma)}{6\sigma} d\sigma\right] \tag{3.4}$$

for $z \in \mathcal{S}_0$,

$$\mathcal{S}_0 := \{ z \,|\, |z| > 0,\ -\pi/3 \le \mathrm{ph}(z) \le \pi/3 \}\,. \tag{3.5}$$

It follows directly from (3.4) that the resulting recessive solution $f_1(z)$ of the Airy equation (1.2) and its derivative $f_1'(z) = g_1(z)$ have asymptotic expansions of the form

$$\begin{aligned} f_1(z) &\sim \frac{1}{2\,z^{1/4}} e^{-\xi} \sum_{k=0}^{\infty} (-1)^k \frac{u_k}{\xi^k} \\ u_0 &= 1, \quad u_1 = \frac{5}{72}, \quad \cdots \end{aligned} \tag{3.6}$$

and

$$\begin{aligned} f_1'(z) &\sim -\frac{z^{1/4}}{2} e^{-\xi} \sum_{k=0}^{\infty} (-1)^k \frac{v_k}{\xi^k} \\ v_0 &= 1, \quad v_1 = -\frac{7}{72}, \quad \cdots \end{aligned} \tag{3.7}$$

for $z \in \mathcal{S}_0$, where the values of the coefficients u_k, v_k are obtained directly from (2.19) and (2.20). (The notation here and elsewhere follows that of Olver (1974).) Alternatively, once the forms of the expansions (3.6) and (3.7) are known from (3.4), the coefficients in the expansions can be determined by inserting (3.6) directly into (1.2). Readily available error estimates for R (see Section 2) can be used to provide corresponding error estimates for (3.6) and (3.7). Up to an inessential constant multiple, $f_1(z)$ agrees with the Airy function $Ai(z)$ of (1.1), with (see (3.6) and Olver (1974, Section 4.4))

$$f_1(z) = \sqrt{\pi} Ai(z). \tag{3.8}$$

Let $\mathcal{S}_1$ be the angular (rotational) translate of $\mathcal{S}_0$ through $2\pi/3$ radians,

$$\mathcal{S}_1 \equiv e^{2\pi i/3} \mathcal{S}_0 := \left\{ z \,\middle|\, |z| > 0,\ \pi/3 \le \mathrm{ph}(z) \le \pi \right\}. \tag{3.9}$$

By the basic continuation theorem for ordinary differential equations it follows that (3.3) can be extended so as to provide a fundamental solution for (1.3) not only on the region $\mathcal{S}_0$ of (3.5) but also on $\mathcal{S}_1$. The determinant of (3.3) is obtained with (3.1) as

$$det \begin{pmatrix} f_1(z) & f_2(z,\xi_0) \\ g_1(z) & g_2(z,\xi_0) \end{pmatrix} = \frac{1}{2} \tag{3.10}$$

for $z \in \mathcal{S}_0$. However, the determinant of any given fundamental solution of (1.3) is constant, independent of z (Abel's theorem), and so in particular (3.10) holds both on $\mathcal{S}_0$ and on $\mathcal{S}_1$. In the following we consider the continuation of (3.3) from $\mathcal{S}_0$ into $\mathcal{S}_1$; a similar discussion applies to the other sector $\mathcal{S}_{-1} = e^{-2\pi i/3}\mathcal{S}_0$.

4. A LATERAL CONNECTION MATRIX

Let $F_0 = F_0(z, \xi_0)$ denote the fundamental solution (3.3) on $\mathcal{S}_0$ and its continuation on $\mathcal{S}_1$,

$$F_0(z, \xi_0) := \begin{pmatrix} f_1(z) & f_2(z, \xi_0) \\ g_1(z) & g_2(z, \xi_0) \end{pmatrix}, \tag{4.1}$$

so that $F = F_0$ satisfies the equation

$$\frac{d}{dz}F = \begin{pmatrix} 0 & 1 \\ z & 0 \end{pmatrix} F \tag{4.2}$$

for $z \in \mathcal{S}_0 \cup \mathcal{S}_1$ and for any fixed $\xi_0 = (2/3)z_0^{3/2}$, $z_0 \in \mathcal{S}_0$. Note that the first column of (4.1) depends only on z and is independent of ξ_0.

The function

$$F_1 = F_1(z, \xi_1) := F_0(e^{-2\pi i/3}z, \xi_1) \tag{4.3}$$

is also a fundamental solution of (4.2) for any fixed $\xi_1 = (2/3)z_1^{3/2}$, $z_1 \in \mathcal{S}_0$ (cf. Olver (1974, p. 55)). Also, for any fundamental solution of (4.2) there holds

$$\frac{d}{dz}F^{-1} = -F^{-1}\begin{pmatrix} 0 & 1 \\ z & 0 \end{pmatrix}. \tag{4.4}$$

For $z \in \mathcal{S}_0 \cup \mathcal{S}_1$ define the fundamental solution $\hat{F}$ as

$$\hat{F}(z) \equiv \hat{F}(z, \xi_0, \xi_1) := \begin{cases} F_0(z, \xi_0) & \text{if } z \in \mathcal{S}_0, \\ F_1(z, \xi_1)C(z, \xi_0, \xi_1) & \text{if } z \in \mathcal{S}_1, \end{cases} \tag{4.5}$$

where the connection matrix C is defined as

$$C(z, \xi_0, \xi_1) := F_1^{-1}(z, \xi_1)F_0(z, \xi_0). \tag{4.6}$$

Using (4.2) for F_0 and (4.4) for F_1^{-1} there follows upon differentiation of (4.6) the result

$$\frac{d}{dz}C(z, \xi_0, \xi_1) \equiv 0, \tag{4.7}$$

so that:

$$\text{the connection matrix } C(z, \xi_0, \xi_1) \text{ is independent of } z, \quad C(z, \xi_0, \xi_1) \equiv C(\xi_0, \xi_1). \tag{4.8}$$

Hence (4.5)-(4.6) provides a fundamental solution $\hat{F}$ for (4.2). The resulting solution is smooth on $\mathcal{S}_0 \cup \mathcal{S}_1$, so that (4.3), (4.5), and (4.8) provide a convenient representation for the continuation of the fundamental solution (3.3) from $\mathcal{S}_0$ to $\mathcal{S}_1$.

We evaluate $C(\xi_0, \xi_1)$ of (4.8) by putting

$$z = \hat{z} := \hat{r}e^{i\pi/3} \tag{4.9}$$

in (4.6), and we find with (4.3) and (4.8)

$$C(\xi_0, \xi_1) = F_0^{-1}(\hat{r}e^{-i\pi/3}, \xi_1)F_0(\hat{r}e^{+i\pi/3}, \xi_0), \tag{4.10}$$

where F_0 can be evaluated here with (3.1), (3.3) and (4.1) because the arguments $\hat{r}e^{-i\pi/3}$ and $\hat{r}e^{+i\pi/3}$ are both in $\mathcal{S}_0$. In this way, with suitable choices for the parameters $\hat{r}, \xi_0$, and ξ_1 in (4.10), the Riccati results (3.1)-(3.3) on the sector $\mathcal{S}_0$ can be used to study the Airy equation in $\mathcal{S}_1$.

To illustrate the procedure we only discuss briefly how to obtain the appropriate asymptotic expansion for the Airy function $f_1(z) = \sqrt{\pi}Ai(z)$ of (3.8) in $\mathcal{S}_1$, analogous to the expansion (3.6) in $\mathcal{S}_0$. For $z \in \mathcal{S}_1$, $f_1(z)$ is given by the appropriate component $\hat{F}_{11}$ of (4.5),

$$f_1(z) = f_1(e^{-2\pi i/3}z)C_{11}(\xi_0, \xi_1) + f_2(e^{-2\pi i/3}z, \xi_1)C_{21}(\xi_0, \xi_1), \tag{4.11}$$

where the left-side and hence also the right-side here must be independent of the parameters ξ_0, ξ_1, and $\hat{r}$.

In fact one sees directly from (4.10) that the first-column components of the connection matrix $C(\xi_0, \xi_1)$ are independent of ξ_0 while the components in the second row are independent of ξ_1,

$$C(\xi_0, \xi_1) = \begin{pmatrix} C_{11}(\xi_1) & C_{12}(\xi_0, \xi_1) \\ C_{21} & C_{22}(\xi_0) \end{pmatrix}. \tag{4.12}$$

In particular the component C_{21} must be constant, independent of all parameters. From (4.10) there follows

$$C_{21} = -2\left[f_1(\hat{r}e^{i\pi/3})g_1(\hat{r}e^{-i\pi/3}) - f_1(\hat{r}e^{-i\pi/3})g_1(\hat{r}e^{i\pi/3})\right], \tag{4.13}$$

and it then follows with (3.4), Lemma 2.1 and Lemma 2.2, upon letting $\hat{r} \to \infty$ in (4.13), that

$$C_{21} = \tfrac{1}{2}e^{-i\pi/2} = -\frac{i}{2}. \tag{4.14}$$

An analogous argument yields a related result for $C_{11}(\xi_1)$, and these results can be inserted into (4.11) along with the appropriate results for f_1 and f_2 from (3.3). The expansion (2.19)-(2.20) for R then leads directly to the usual expansion for $f_1(z) = \sqrt{\pi}Ai(z)$ in $\mathcal{S}_1$; see Olver (1974, p. 117) for the resulting expansion.

REFERENCES

1. G.B. Airy (1838), On the intensity of light in the neighbourhood of a caustic, *Trans. Cambridge Philos. Soc.*, **6**: 379-402.

2. G.B. Airy (1849), Supplement to a paper On the intensity of light in the neighbourhood of a caustic, *Trans. Cambridge Philos. Soc.*, **8**: 595–599.

3. F.W.J. Olver (1961), Error bounds for the Liouville-Green (or WKB) approximation, *Proc. Cambridge Philos. Soc.*, **57**: 790–810.

4. F.W.J. Olver (1974), *Asymptotics and Special Functions*, Academic Press, New York.

5. F.W.J. Olver (1977a), Connection formulas for second-order differential equations with multiple turning points, *SIAM J. Math. Anal.*, **8**: 127–154.

6. F.W.J. Olver (1977b), Connection formulas for second-order differential equations having an arbitrary number of turning points of arbitrary multiplicities, *SIAM J. Math. Anal.*, **8**: 673–700.

7. F.W.J. Olver (1978), General connection formulæ for Liouville-Green approximations in the complex plane, *Philos. Trans. Roy. Soc. London*, **A/289**: 501–548.

8. D.R. Smith (1985), *Singular Perturbation Theory*, Cambridge Univ. Press, Cambridge.

9. D.R. Smith (1986), Liouville/Green approximations via the Riccati transformation, *J. Math. Analysis and Applications*, **116**: 147–165.

10. D.R. Smith (1987), Decoupling of recessive and nonrecessive solutions for a second-order system, *J. Diff. Eqn.*, **68**: 383–399.

11. W. Wasow (1965), *Asymptotic Expansions for Ordinary Differential Equations*, Interscience Pub. (John Wiley), New York.

12. W. Wasow (1976), *Asymptotic Expansions for Ordinary Differential Equations* (Reprint Edition), Robert E. Krieger Pub.

13. W. Wasow (1985), *Linear Turning Point Theory*, Springer-Verlag, New York.

A System of Polynomials Associated with the Chester, Friedman, and Ursell Technique

K. Soni
R. P. Soni
Department of Mathematics, University of Tennessee, Knoxville, Tennessee

1. Introduction.

Let

$$I(N) = \int h(z)\, e^{Nk(z,\alpha)}\, dz \tag{1.1}$$

where N is a large parameter and $h(z)$, $k(z,\alpha)$ are analytic functions of their arguments in the domains of interest. Let us assume further that for fixed α, the asymptotic expansion of $I(N)$ can be obtained by the method of saddle points. In general, the location of the saddle points would vary with α and it is possible that as α approaches a certain value α_0, some saddle points may coincide or even that some saddle points may coincide with an end point of the contour of integration. In such cases, the asymptotic behavior of $I(N)$ may change drastically as $\alpha \to \alpha_0$. There are many special functions whose integral representations exhibit this behavior. The question is, can we find an asymptotic expansion of $I(N)$ which holds for all α in some neighborhood of α_0 even though such an expansion is not continuous at that point. In 1956 such an expansion was given by Chester, Friedman and Ursell

[2]. They considered the case of two simple saddle points which coalesce as $\alpha \to 0$. Their technique is known as the "Method of Chester, Friedman and Ursell". In 1966, Bleistein tackled the same problem when a simple saddle point coalesces with an end point of the contour of integration and $h(z)$ has an algebraic singularity there. His technique is known as the "Bleisteins's method". Both of these techniques have two major components. First a change of variables is introduced which reduces the integral to the simplest form without obscuring its essential features. The simplified forms are

$$I(N) = \int \exp\left[N\left(\frac{1}{3}\,w^3 - b^2 w\right)\right] g(w)\,dw \tag{1.3}$$

and

$$I(N) = \int w^\gamma \exp\left[N\left(\frac{1}{2}\,w^2 - bw\right)\right] g(w)\,dw \tag{1.3}$$

respectively. We will refer to these as the standard forms of the integrals. The parameter b depends on α. In general, $b = b(\alpha)$ and $w = w(z, \alpha)$ are defined only implicitly. We see that in each case as $b \to 0$, certain critical points merge. After the standard form has been obtained, $g(w)$ is expanded in such a way that the resulting asymptotic expansion is valid for all b in some neighborhood of $b = 0$ and furthermore, the expansion is uniform in b in a certain sense. Chester, Friedman and Ursell use the expansion

$$g(w) = \Sigma\, p_n(b)\,(w^2 - b^2)^n + \Sigma\, q_n(b)\,w(w^2 - b^2)^n \tag{1.4}$$

in (1.2) and prove that the resulting expansion of $I(N)$ in terms of the Airy function and its derivative is indeed a uniform asymptotic expansion. They apply this technique to find a uniform asymptotic expansion of the Bessel function $J_\nu(\nu a)$ when ν is large and a is close to one, and they compare their results with those given by Olver ([**5**], [**6**, p. 423]). They find Olver's results, derived by using the techniques of asymptotic expansions for ordinary differential equations, more general and decidedly better but they add that since a function defined by (1.1) need not be the solution of an ordinary differential equation, both methods are needed. Copson [**3**] also discusses the uniform asymptotic expansion of $J_\nu(\nu a)$ by the Chester, Friedman and Ursell method. His treatment is quite detailed. It makes one more aware of the problems associated with the computation of successive coefficients as well as the error bounds in the asymptotic expansion of $I(N)$. When we analyze the process, we find that even though the expansion (1.4) for $g(w)$ is the most natural, there is a possibility that some alternative form of expansion may make the task somewhat simpler. At the very least, such an approach should be explored. It is with this objective that we introduce in this paper a sequence of polynomials $\{P_n(t)\}$ defined as follows:

$$\begin{cases} P_0(t) &= 1 \, , \\ P_1(t) &= t \, , \\ P_2(t) &= \frac{1}{2}\, t^2 \, , \\ P_n'(t) &= (t^2 - b^2)\, P_{n-3}(t), \;\; P_n(0) = 0, \;\; n \geq 3 \, . \end{cases} \tag{1.5}$$

Using integration by parts successively and assuming that the integrated terms vanish at each end of the contour, we find that

$$\int_C \exp\left[N\left(\frac{1}{3}w^3 - b^2 w\right)\right] (w^2 - b^2)\, P_{3n+j}(w)\, dw \tag{1.6}$$

$$= \begin{cases} 0\,, & j=0\,, \\ (-1/N)^{n+1} \int_C \exp\left[N\left(\frac{1}{3}w^3 - b^2 w\right)\right] dw, & j=1 \\ (-1/N)^{n+1} \int_C \exp\left[N\left(\frac{1}{3}w^3 - b^2 w\right)\right] w\, dw, & j=2\,. \end{cases}$$

Thus if in (1.2),

$$g(w) = a_0 + a_1 w + (w^2 - b^2) \sum_{n=0}^{\infty} c_n\, P_n(w)\,, \tag{1.7}$$

then formally,

$$I(N) \sim \left(a_0 + \sum_{n=0}^{\infty} (-1/N)^{n+1}\, c_{3n+1}\right) \mathcal{A}_0(N,b) + \left(a_1 + \sum_{n=0}^{\infty} (-1/N)^{n+1}\, c_{3n+2}\right) \mathcal{A}_1(N,b)\,,\ N \to \infty\,, \tag{1.8}$$

where

$$\mathcal{A}_j(N,b) = \int_C \exp\left[N\left(\frac{1}{3}w^3 - b^2 w\right)\right] w^j\, dw,\ j = 0,1\,. \tag{1.9}$$

Note that for $b = 0$, the polynomials $P_n(t)$ reduce to constant multiples of t^n and for an appropriate contour C, (1.8) is simply the expansion obtained by Watson's Lemma. In Watson's Lemma successive terms in the expansion of a function at the origin provide successive terms in the asymptotic expansion of the transform. As (1.8) indicates, the expansion (1.7) also enjoys the same property.

We refer to the polynomials introduced here as polynomials associated with the Chester, Friedman and Ursell technique or as polynomials associated with the Airy type integrals. Their properties are similar to the properties of the polynomials associated with the Bleistein technique where integrals of the type (1.3) are involved. These polynomials were introduced by Soni and Sleeman [**8**] and later discussed by Soni and Temme [**9**]. In the present paper, we will omit some of the proofs as they are similar to those given in [**8**] and [**9**]. In the next two sections, we discuss some of the properties of the polynomials defined by (1.5) and the associated rational functions defined in section 3. These properties are used to obtain Taylor series type expansion of analytic functions in terms of these polynomials. This is done in sections 4 and 5. In section 6 we discuss the asymptotic expansion of $I(N)$ and the error term. In section 7 we discuss the asymptotic expansion of $J_\nu(\nu a)$ near $a = 1$.

The polynomials associated with the integrals of the type

$$I(N) = \int_0^\infty t^\gamma \exp\left[-N\left\{\frac{1}{3}\, t^3 - \frac{1}{2}\, (\alpha+\beta)t^2 + \alpha\beta t\right\}\right]\, h(t)\, dt \tag{1.10}$$

where the saddle points $t = \alpha$ and $t = \beta$ approach zero independently, generalize those considered here and many of their properties can be developed in a similar manner. Here we restrict ourselves to $\gamma = 0$ and $\beta = -\alpha$.

2. Some Properties of $P_n(t)$.

REPRESENTATION. By (1.5) for $n = 0, 1, 2, \ldots,$

$$P_{3n}(t) = [\Gamma(n+1)]^{-1} \quad \left(\frac{1}{3}\, t^3 - b^2 t\right)^n \tag{2.1}$$

and for $j = 1, 2,;\; n = 0, 1, \ldots,$

$$P_{3n+j}(t) = [\Gamma(n+1)]^{-1} \int_0^t \left\{\frac{1}{3}\,(t^3 - v^3) - b^2(t - v)\right\}^n v^{j-1}\, dv\ . \tag{2.2}$$

Explicit representations for $P_{3n+1}(t)$ and $P_{3n+2}(t)$ can be obtained by using (2.2).

BOUNDS FOR $P_n(t)$. By (2.1),

$$|P_{3n}(t)| \le \frac{(|t| + |b|)^{3n}}{n!3^n}\ . \tag{2.3}$$

By (2.2), after a change of variable,

$$\begin{aligned} |P_{3n+j}(t)| &\le \frac{|t|^j}{n!3^n} \left| \int_0^1 \{t^3(1-u^3) - 3b^2 t(1-u)\}^n u^{j-1}\, du \right| \\ &\le \frac{|t|^j\{|t|^3 + 3|b|^2|t|\}^n}{n!\,3^n} \int_0^1 (1-u^3)^n u^{j-1}\, du\ . \end{aligned} \tag{2.4}$$

Since

$$\int_0^1 (1-u^3)^n u^{j-1}\, du = \frac{\Gamma(n+1)\Gamma(j/3)}{3\Gamma(n+1+j/3)}\ , \tag{2.5}$$

it follows that

$$|P_{3n+1}(t)| \le \frac{(|t| + |b|)^{3n+1}\ \Gamma(4/3)}{3^n\ \Gamma(n + \frac{4}{3})}\ , \tag{2.6}$$

$$|P_{3n+2}(t)| \le \frac{1}{2}\ \frac{(|t| + |b|)^{3n+2}\ \Gamma(5/3)}{3^n\ \Gamma(n + 5/3)}\ . \tag{2.7}$$

3. Associated Rational Functions.

Let us assume for the present that

$$\frac{1}{w-t} = \sum_{n=0}^{\infty} P_n(t)\,\phi_n(w)\,. \tag{3.1}$$

Differentiating formally with respect to t and w respectively and using (1.5), we also have

$$\frac{1}{(w-t)^2} = \phi_1(w) + t\phi_2(w) + (t^2-b^2)\sum_{n=3}^{\infty} P_{n-3}(t)\phi_n(w) \tag{3.2}$$

and

$$-\frac{1}{(w-t)^2} = \sum_{n=0}^{\infty} P_n(t)\,\phi_n'(w)\,. \tag{3.3}$$

Let $t = 0$ in (3.1) and $t = \pm b$ in (3.2). Since $P_n(0) = 0$ for $n \geq 1$, we obtain

$$\begin{cases} \phi_0(w) = \dfrac{1}{w} \\[2ex] \phi_1(w) = \dfrac{1}{2(w-b)^2} + \dfrac{1}{2(w+b)^2} \\[2ex] \phi_2(w) = \dfrac{2w}{(w^2-b^2)^2} = -\dfrac{d}{dw}\left(\dfrac{1}{w^2-b^2}\right)\,. \end{cases} \tag{3.4}$$

By using (3.4) we can show that

$$\begin{aligned} &\frac{1}{(w-t)^2} - \phi_1(w) - t\phi_2(w) \\ &= (t^2-b^2)\left\{\frac{1}{(w-t)^2(w^2-b^2)} + \frac{2w}{(w-t)(w^2-b^2)^2}\right\}\,. \end{aligned} \tag{3.5}$$

Now substitute this in (3.2) and use (3.1) and (3.3) for $(w-t)^{-1}$ and $(w-t)^{-2}$ respectively. We obtain

$$\begin{aligned} \sum_{n=0}^{\infty} P_n(t)\phi_{n+3}(w) = &-\frac{1}{(w^2-b^2)}\sum_{n=0}^{\infty} P_n(t)\phi_n'(w) \\ &+ \frac{2w}{(w^2-b^2)^2}\sum_{n=0}^{\infty} P_n(t)\phi_n(w)\,. \end{aligned} \tag{3.6}$$

If we assume the expansion in $P_n(t)$ to be unique, we must have for $n \geq 0$,

$$\begin{aligned}\phi_{n+3}(w) &= -\frac{1}{w^2-b^2}\,\phi'_n(w) + \frac{2w}{(w^2-b^2)^2}\,\phi_n(w) \\ &= -\frac{d}{dw}\left(\frac{1}{w^2-b^2}\,\phi_n(w)\right) .\end{aligned} \tag{3.7}$$

The expressions for $\phi_n(w)$ obtained above is the reason for the following definition.

DEFINITION: The sequence of rational functions $\{\phi_n(w)\}$ is defined by (3.4) and (3.7).

We want to know when the expansion (3.1) is valid and for that we need to know more about ϕ_n.

REPRESENTATION FOR $\phi_n(w)$. By induction for $n = 0, 1, 2, \ldots,$

$$\phi_{3n}(w) = \left(-\frac{d}{dw}\,\frac{1}{w^2-b^2}\right)^n \left\{\frac{1}{w}\right\} , \tag{3.8}$$

$$\phi_{3n+1}(w) = \left(-\frac{d}{dw}\,\frac{1}{w^2-b^2}\right)^n \left\{\frac{1}{2(w+b)^2} + \frac{1}{2(w-b)^2}\right\} , \tag{3.9}$$

$$\phi_{3n+2}(w) = \left(-\frac{d}{dw}\,\frac{1}{w^2-b^2}\right)^n \left\{-\frac{d}{dw}\,\frac{1}{w^2-b^2}\right\} . \tag{3.10}$$

This does not provide much information about the magnitude of ϕ_n. However, we note that

$$\begin{aligned}&\left(-\frac{d}{dw}\,\frac{1}{w^2-b^2}\right)\left(\frac{1}{w^\alpha(w-b)^\beta(w+b)^\gamma}\right) \\ &= \frac{\alpha}{w^{\alpha+1}(w-b)^{\beta+1}(w+b)^{\gamma+1}} + \frac{\beta+1}{w^\alpha(w-b)^{\beta+2}(w+b)^{\gamma+1}} \\ &+ \frac{\gamma+1}{w^\alpha(w-b)^{\beta+1}(w+b)^{\gamma+2}} .\end{aligned} \tag{3.11}$$

When α, β, γ are nonnegative, the coefficient of each term on the right is also nonnegative. Therefore, by induction again,

$$\phi_{3n}(w) = \frac{1}{(w-b)^n(w+b)^n} \sum_{\substack{\ell,m\geq 0\\ \ell+m\leq n}} \frac{a_{\ell mn}}{(w-b)^\ell(w+b)^m w^{n-\ell-m+1}}, \tag{3.12}$$

$$\phi_{3n+1}(w) = \frac{1}{(w-b)^n(w+b)^n} \left\{ \sum_{\ell=0}^{n} \frac{b_\ell}{(w-b)^\ell(w+b)^{n-\ell+2}} + \sum_{\ell=0}^{n} \frac{c_\ell}{(w-b)^{\ell+2}(w+b)^{n-\ell}} \right\}, \tag{3.13}$$

$$\phi_{3n+2}(w) = \frac{1}{(w-b)^{n+1}(w+b)^{n+1}} \sum_{\ell=0}^{n+1} \frac{d_\ell}{(w-b)^\ell(w+b)^{n+1-\ell}}, \tag{3.14}$$

$a_{\ell mn}, b_\ell, c_\ell$ and d_ℓ are all positive constants. We will use these to find bounds for ϕ_{2n+j}, $j = 0, 1, 2$.

BOUNDS FOR $\phi_n(w)$. When $b = 0$, by (3.8)-(3.10),

$$\begin{cases} \phi_{3n}(w) = \dfrac{3^n\Gamma(n+1)}{w^{3n+1}}, \quad \phi_{3n+1}(w) = \dfrac{3^n\Gamma(n+4/3)}{\Gamma(4/3)w^{3n+2}}, \\ \phi_{3n+2}(w) = \dfrac{2\cdot 3^n\Gamma(n+5/3)}{\Gamma(5/3)w^{3n+3}}. \end{cases} \tag{3.15}$$

Therefore, by (3.12) and $|w| > |b|$,

$$\begin{aligned} |\, \phi_{3n}(w) \,| &\leq \frac{1}{(|w|-|b|)^{3n+1}} \,\Sigma\, a_{\ell mn} \\ &\leq \frac{3^n\Gamma(n+1)}{(|w|-|b|)^{3n+1}}. \end{aligned} \tag{3.16}$$

Similarly,

$$|\, \phi_{3n+1}(w) \,| \leq \frac{3^n\Gamma(n+4/3)}{\Gamma(4/3)(|w|-|b|)^{3n+2}}, \tag{3.17}$$

$$|\, \phi_{3n+2}(w) \,| \leq \frac{2\cdot 3^n\Gamma(n+5/3)}{\Gamma(5/3)(|w|-|b|)^{3n+3}}. \tag{3.18}$$

By using the above bounds for $\phi_n(w)$ with the bounds for $P_n(t)$ given in (2.3), (2.6) and (2.7), we find that

$$|P_n(t)\phi_n(w)| \leq \frac{(|t|+|b|)^n}{(|w|-|b|)^{n+1}} \,, \; |w| > |b| \,. \tag{3.19}$$

Therefore

$$\overline{\lim_{n\to\infty}} \; | \, P_n(t)\phi_n(w) \, |^{\frac{1}{n}} \leq \frac{|t|+|b|}{|w|-|b|}, \; |w| > |b| \,. \tag{3.20}$$

Thus we have the following theorem (see also Thm. 2, [**9**]).

THEOREM 1. *The series* $\sum_{n=0}^{\infty} P_n(t) \; \phi_n(w)$ *converges absolutely when* $0 < (|t|+|b|)/(|w|-|b|) < 1$ *and uniformly in the regions* Ω_t, Ω_w *which satisfy*

$$\sup_{t\in\,\Omega_t, w\in\,\Omega_w} \frac{|t|+|b|}{|w|-|b|} \leq 1-\delta < 1, \; |w| > |b| \,. \tag{3.21}$$

Note that for $b = 0$, the above theorem reduces to the well known result that $\Sigma t^n/w^{n+1}$ converges absolutely in $|t/w| < 1$ and uniformly in $|t/w| \leq 1-\delta < 1$.

4. Expansion of $(w-t)^{-1}$.

In section 3 we only proved that the series $\Sigma \; P_n(t) \; \phi_n(w)$ converges in certain t and w-domains. We have not yet proved that the series converges to $(w-t)^{-1}$. That requires some additional work.

The following notation will prove to be quite useful (see [**9**, (2.2) and (2.8)]).

NOTATION. Let

$$\Theta_w\ f(w) = -\ \frac{d}{dw}\ \frac{1}{w^2 - b^2}\ f(w),\ \Theta_w^n = \Theta_w\ \theta_w^{n-1},\ n = 2, 3, \ldots, \tag{4.1}$$

and

$$I_t\ f(t) = \int_0^t (u^2 - b^2)\ f(u) du,\ I_t^n = I_t\ I_t^{n-1}, n = 2, 3, \ldots\ . \tag{4.2}$$

By induction, we have

$$I_t^n\ f(t) = \frac{1}{\Gamma(n)} \int_0^t \left\{ \frac{1}{3}\left(t^3 - u^3\right) - b^2(t - u) \right\}^{n-1} (u^2 - b^2)\ f(u) du\ . \tag{4.3}$$

To prove (3.1) we proceed as follows. By (3.5),

$$\frac{\partial}{\partial t} \left\{ \frac{1}{w - t} - \sum_{k=0}^{2} P_k(t)\ \phi_k(w) \right\} = (t^2 - b^2) \left\{ - \frac{\partial}{\partial w}\ \frac{1}{(w - t)(w^2 - b^2)} \right\} . \tag{4.4}$$

Therefore, for $|w| > |t|$,

$$\frac{1}{w - t} - \sum_{k=0}^{2} P_k(t)\ \phi_k(w) = -\ \frac{\partial}{\partial w}\ \frac{1}{w^2 - b^2} \int_0^t (u^2 - b^2)\ \frac{1}{w - u}\ du\ . \tag{4.5}$$

The constant of integration is zero because equality holds for $t = 0$. Thus, by (4.1) and (4.2),

$$\frac{1}{w - t} - \sum_{k=0}^{2} P_k(t)\ \phi_k(w) = \Theta_w\ I_t\ \frac{1}{w - t}\ . \tag{4.6}$$

Since $I_t\ P_k(t) = P_{k+3}(t)$ and $\Theta_w\ \phi_k(w) = \phi_{k+3}(w)$, by induction it follows that

$$\frac{1}{w - t} - \sum_{k=0}^{3n-1} P_k(t)\ \phi_k(w) = \Theta_w^n\ I_t^n\ \frac{1}{w - t}\ . \tag{4.7}$$

But $\Theta_w \; I_t = I_t \; \Theta_w$. Furthermore, by (3.8) and (3.12),

$$\Theta_w^n \, \frac{1}{w-t} = \sum_{\substack{\ell,n\geq 0 \\ \ell+m\leq n}} \frac{a_{\ell mn}}{(w-b)^{\ell+n}(w+b)^{m+n}(w-t)^{n-\ell-m+1}} , \tag{4.8}$$

and by (4.3), $|w| > |t|$,

$$I_t^n \, \frac{1}{(w-t)^k} = \frac{1}{\Gamma(n)3^{n-1}} \int_0^1 \left\{t^3(1-u^3) - 3b^2t(1-u)\right\}^{n-1} \frac{(t^3u^2 - tb^2)}{(w-tu)^k} \, du \, . \tag{4.9}$$

From the last equation for $k = n - \ell - m + 1$ we obtain,

$$\left| \, I_t^n \, \frac{1}{(w-t)^k} \, \right| \leq \frac{\{|t|^3 + 3|b|^2|t|\}^n}{3^n \; n!(|w| - |t|)^k} \, . \tag{4.10}$$

Therefore, by (4.8) and (3.16),

$$\begin{aligned} \left| \, I_t^n \; \Theta_w^n \, \frac{1}{w-t} \, \right| &\leq \frac{\{|t|^3 + 3|b|^2|t|\}^n}{3^n \; n!(|w| - \rho)^{3n+1}} \sum a_{\ell mn} \\ &\leq \frac{(|t| + |b|)^{3n}}{(|w| - \rho)^{3n+1}} , \end{aligned} \tag{4.11}$$

where $\rho = \max\{|b|, |t|\}$. Now by (4.7) it follows that $(w-t)^{-1} = \Sigma \, P_k(t) \; \phi_k(w)$ at least for all w, $|w| > |t| + 2\rho$. The regions in which the equality holds can be extended by analytic continuation. By using Theorem 1, we can state the following.

THEOREM 2. *Let $|b| \leq \beta$. For all w and t such that $|t| + 2\beta < |w|$,*

$$\frac{1}{w-t} = \sum_{n=0}^{\infty} P_n(t) \; \phi_n(w) \, .$$

5. Expansion Of Analytic Functions.

Let $f(t)$ be analytic in a domain D which includes a disk of radius r, $r > 2|b|$. For $|t| < r - 2|b|$,

$$\begin{aligned} f(t) &= \frac{1}{2\pi i} \int_{C_r} \frac{f(w)}{w - t}\, dw , \\ &= \frac{1}{2\pi i} \int_{C_r} f(w) \left(\sum_{n=0}^{\infty} P_n(t)\, \phi_n(w) \right) dw , \\ &= \sum_{n=0}^{\infty} c_n\, P_n(t) , \end{aligned} \tag{5.1}$$

where C_r is the contour $|w| = r$ and

$$c_n = \frac{1}{2\pi i} \int_{C_r} f(w)\, \phi_n(w)\, dw . \tag{5.2}$$

The interchange of the order of summation and integration in (5.1) is justified by the absolute convergence of the series. Thus we have a Taylor series type expansion for $f(t)$. Next we find the magnitude of the remainder. Let

$$R_n(t) = f(t) - \sum_{k=0}^{n-1} c_k\, P_k(t) . \tag{5.3}$$

By (5.1) and (3.19), $r > |t| + 2|b|$,

$$\begin{aligned} |R_n(t)| &\le \int_{C_r} |f(w)| \sum_{k=n}^{\infty} |P_n(t)\, \phi_n(w)|\, |dw|/(2\pi) \\ &\le Mr \frac{(|t| + |b|)^n}{(r - |b|)^n (r - |t| - 2|b|)} \end{aligned} \tag{5.4}$$

where $M = \max\ |f(w)|$ for $w \epsilon\ C_r$.

Another bound for $R_n(t)$ which depends on the magnitude of the nth derivative of $f(w)$ rather than on the magnitude of $f(w)$, can also be given.

In this case it is more convenient to consider $R_{3n}(t)$. We will also need a known result (see [**9**, Appendix]) which we state below.

LEMMA: If $f(w)$ is regular within and on a contour C which has the points $w = a, b, c$ in its interior, then

$$\left| \frac{1}{2\pi i} \int_C \frac{f(w)}{(w-a)^{\ell+1}(w-b)^{m+1}(w-\ell)^{n+1}} \, dw \right| \le \frac{\mathcal{M}}{(\ell+m+n+2)!} \tag{5.5}$$

where

$$\mathcal{M} = \max_{w \epsilon \, \Delta} |f^{(\ell+m+n+2)}(w)| \,, \tag{5.6}$$

Δ is the closed triangular region with vertices at $w = a, b, c$ respectively, Δ inside C.

Now we find a bound for $R_{3n}(t)$. Let $f(w)$ be analytic within and on a countour C and let $t, b, -b$ be in the interior of C. By (5.2) and (4.7),

$$\begin{aligned} R_{3n}(t) &= \frac{1}{2\pi i} \int_C f(w) \left\{ \frac{1}{w-t} - \sum_{k=0}^{3n-1} P_k(t) \, \phi_k(w) \right\} dw \,, \\ &= \frac{1}{2\pi i} \int_C f(w) \, I_t^n \, \Theta_w^n \, \frac{1}{w-t} \, dw \,, \\ &= I_t^n \left\{ \frac{1}{2\pi i} \int_C f(w) \, \Theta_w^n \, \frac{1}{w-t} \, dw \right\} . \end{aligned} \tag{5.7}$$

By (4.8) and (5.5),

$$\begin{aligned} &\left| \frac{1}{2\pi i} \int_C f(w) \, \Theta_w^n \, \frac{1}{w-t} \, dw \right| \\ &\le \sum a_{\ell mn} \left| \frac{1}{2\pi i} \int_C \frac{f(w)}{(w-b)^{\ell+n}(w+b)^{m+n}(w-t)^{n-\ell-m+1}} \, dw \right| \\ &\le \frac{\mathcal{M} 3^n \, n!}{(3n)!} \end{aligned} \tag{5.8}$$

where $\mathcal{M} = \max |f^{(3n)}(w)|$ for $w \epsilon \Delta$, the closed triangular region with vertices at $w = b, -b$ and t. In the last inequality above we have also used the fact that $\Sigma\, a_{\ell mn} = 3^n\, n!$. Recall that I_t^n is given by (4.3). By using a change of variable as in (4.9) and the bound in (5.8), we obtain,

$$|R_{3n}(t)| \leq \frac{(|t|^3 + 3|b^2 t|)^n}{(3n)!}\, \mathcal{M} \tag{5.9}$$

where $\mathcal{M} = \max |f^{(3n)}(w)|$, $w \epsilon \Delta$.

We state two other results and omit the proofs which are straightforward. These results are not needed for asymptotic expansions but theoretically they are interesting.

THEOREM 3. *Let C be a simple closed curve, positively oriented with $w = 0, b, -b$ inside C. Then $(\{P_m\}, \{\phi_n\})$ is biorthogonal on C and*

$$\frac{1}{2\pi i} \int_C P_m(w)\, \phi_m(w)\, dw = 1\ .$$

THEOREM 4. *Let $f(t)$ be regular in the annulus $r \leq |t| \leq R$, $R - r > 4\beta$, $|b| \leq \beta$. Then for $r + 2\beta < |t| < R - 2\beta$,*

$$f(t) = \sum_{k=0}^{\infty} c_k\, P_k(t) + \sum_{k=0}^{\infty} d_k\, \phi_k(t)$$

where

$$c_k = \frac{1}{2\pi i} \int_C f(w)\, \phi_k(w)\, dw\ ,$$
$$d_k = \frac{1}{2\pi i} \int_C f(w)\, P_k(w)\, dw\ ,$$

C is the circle $|w| = r'$, *positively oriented,* $\max\{r, \beta\} < r' < R$.

These two theorems are the same as those given in [**9**] where the polynomials are those associated with the Bleistein technique. We give them here to emphasize that the system $(\{P_n\}, \{\phi_n\})$ discussed here is also a biorthogonal system and Laurent type expansions can also be given.

6. Asymptotic Expansions.

Assume that $f(w) = \Sigma\, c_n\, P_n(w)$ at least in some domain which contains the points $w = 0, b, -b$ and let R_n denote the remainder defined by (5.3). If we further assume that the contour is such that (1.6) holds, then

$$\begin{aligned} I_1(N) &= \int_C e^{N(\frac{1}{3}\, w^3 - b^2 w)}\, (w^2 - b^2)\, f(w)\, dw \qquad (6.1) \\ &= \left\{ \sum_{k=0}^{n-1} (-1/N)^{k+1}\, c_{3k+1} \right\} \mathcal{A}_0(N, b) \\ &+ \left\{ \sum_{k=0}^{n-1} (-1/N)^{k+1}\, c_{3k+2} \right\} \mathcal{A}_1(N, b) \\ &+ E_{3n}(N, b)\ , \end{aligned}$$

where $\mathcal{A}_0$ and $\mathcal{A}_1$ are defined by (1.9) and E_{3n} is the error term

$$E_{3n}(N, b) = \int_C e^{N(\frac{1}{3}\, w^3 - b^2 w)} (w^2 - b^2)\, R_{3n}(w)\, dw\ . \qquad (6.2)$$

Before we can claim that (6.1) provides a uniform asymptotic expansion as $N \to \infty$, we need to examine E_{3n}. We will assume that the contour C is either any one of the Airy contours which start and end at ∞ with w^3 negative as $|w| \to \infty$ or that it starts at the origin but ends like an Airy contour. We will also assume that $f(w)$ is analytic in an appropriate sector and that $f^{(n)}(w)$, $n = 0, 1, \ldots$, do not grow faster than a function of finite exponential order in that sector. (This condition however, can be weakened in specific cases). For the contours which start at the origin we need to know the behavior of R_{3n} at that point. By (5.7), $R_{3n}(0) = 0$ for $n \geq 1$. Furthermore, for $0 \leq k < n$,

$$\left(\frac{1}{t^2 - b^2}\frac{d}{dt}\right)^k R_{3n}(t) = I_t^{n-k}\left\{\frac{1}{2\pi i}\int_C f(w)\,\Theta_w^n\,\frac{1}{w-t}\,dw\right\} . \tag{6.3}$$

Therefore,

$$\left(\frac{1}{t^2 - b^2}\frac{d}{dt}\right)^k R_{3n}(t)\,\Big|_{t=0} = 0 . \tag{6.4}$$

Using integration by parts n times in (6.2) we obtain

$$E_{3n}(N, b) = (-\frac{1}{N})^n \int_C e^{N(\frac{1}{3}w^3 - b^2 w)}(w^2 - b^2)(\frac{1}{w^2 - b^2}\frac{d}{dw})^n R_{3n}(w)dw . \tag{6.5}$$

For the type of contours that we have mentioned above, the integrated term each time vanishes at both ends. By (6.3) and (5.8),

$$\left|\left(\frac{1}{w^2 - b^2}\frac{d}{dw}\right)^n R_{3n}(w)\right| \leq \mathcal{M}\,\frac{3^n n!}{(3n)!} \tag{6.6}$$

where $\mathcal{M} = \max\ |f^{(3n)}(w)|$ for w in the closed triangular region bounded by $w = t, b, -b$. A bound for $E_{3n}(N, b)$ can now be obtained. In particular if $\mathcal{M}$ is bounded, we obtain a uniform asymptotic expansion for $I_1(N)$.

An alternative to (6.6) is to estimate a bound directly from the contour integral in (5.8). This will depend on $\max|f(w)|$. Other alternatives are to use a bound for R_{3n} given in section 5. In fact one may need more than one type of estimate in a problem, one when the parameter is close to the contour and the other when it is away from it.

7. An Application.

Chester, Friedman and Ursell describe briefly in [**2**] how the cubic transformation can be used to obtain a uniform asymptotic expansion for the Bessel function $J_\nu(\nu a)$ from its integral representation when ν is large and $0 < a \leq 1$. As mentioned earlier, Copson discusses this problem in detail [**3**, pp. 108-117]. He proves that if $a = \text{sech}\ \alpha,\ \alpha > 0$, then

$$J_\nu(\nu \text{ sech } \alpha) = \sqrt{\frac{2b}{\sinh \alpha}}\ \frac{Ai(N^{2/3}b^2)}{N^{1/3}}\ \{1 + O(\tfrac{1}{N})\} \tag{7.1}$$

where

$$N = \nu \text{ sech } \alpha,\ b^3 = \frac{3}{2}\ (\alpha \cosh \alpha - \sinh \alpha)\ . \tag{7.2}$$

The implied constant in (7.1) can be taken as bK where K is independent of b. We will now compute the second term in the asymptotic expansion of (7.1), give a representation for the expansion coefficients and by considering the error term $E_{3n}(N, b)$ show that the asymptotic expansion holds uniformly in α, $0 \leq \alpha \leq \alpha_1 < \infty$ where α_1 is an arbitrary positive number. We use Copson's work [**3**, Chapter 10] and some of the results obtained in earlier sections.

$$J_\nu(\nu \text{ sech } \alpha) = \frac{1}{2\pi i} \int_{\infty - \pi i}^{\infty + \pi i} e^{N(\sinh z - z \cosh \alpha)} dz\ , \tag{7.3}$$

$N = \nu$ sech α. By the transformation

$$\frac{1}{3}\, w^3 - b^2 w = \sinh z - z \cosh \alpha \,, \tag{7.4}$$

$$J_\nu(\nu \operatorname{sech} \alpha) = \frac{1}{2\pi i} \int_{C_1} e^{N(\frac{1}{3}\, w^3 - b^2 w)} \frac{dz}{dw}\, dw \,. \tag{7.5}$$

The parameter b is determined by (7.2) and is positive when α is positive. The curve C_1 is the curve of steepest descent through b which begins at $\infty\, e^{-\frac{1}{3}\pi i}$ and ends at $\infty\, e^{\frac{1}{3}\pi i}$. The function $w = w(z)$ is defined uniquely by (7.4) and the requirement that the points $z = \pm\alpha$ in the z-plane map onto the points $\pm b$ respectively in the w-plane. Furthermore, $w(z)$ is an analytic function of $z = x + iy$ and maps the region $D = \{z : |y| \leq 3\pi/2\}$ conformally onto the region D' in the w-plane. The region D' is shown in figure 1. (Copson considers the domain $|y| \leq \pi$ in the z-plane but we need a somewhat larger region in the w-plane than the one he consideres). Let

$$g(w) = \frac{dz}{dw} = \frac{w^2 - b^2}{\cosh z - \cosh \alpha} \,, \tag{7.6}$$

$g(w)$ is an analytic function of w^2 and α^2. Near $w^2 = b^2$,

$$g(w) = \sum_{n=0}^{\infty} k_n (w^2 - b^2)^n \,, \tag{7.7}$$

$$k_0 = \left(\frac{2b}{\sinh \alpha}\right)^{1/2} \,, \tag{7.8}$$

$$k_1 = (3b \sinh \alpha)^{-1} \left\{ \sqrt{\frac{\sinh \alpha}{2b}} - \frac{b \cosh \alpha}{\sinh \alpha} \right\} \,, \tag{7.9}$$

$$k_2 = (8b \sinh \alpha)^{-1/2} \left\{ \frac{1}{6 \sinh \alpha} + \frac{5}{12}\, \frac{1}{\sinh^3 \alpha} - \frac{5}{24 b^3} \right\} \,. \tag{7.10}$$

The coefficients k_0 and k_1 are given by Copson, k_2 can be computed by using his technique of successive differentiation [**3**, p. 114]. Let

$$g(w) = k_0 + (w^2 - b^2)\ f(w)\ . \tag{7.11}$$

Then $f(w)$ is an analytic function of w^2, $w \epsilon D'$ and near $w^2 = b^2$,

$$f(w) = \sum_{n=1}^{\infty} k_n\ (w^2 - b^2)^{n-1}\ . \tag{7.12}$$

Also by (5.1), at least for $|b|$ sufficiently small,

$$f(w) = \sum_{n=0}^{\infty} c_n\ P_n(w)\ , \tag{7.13}$$

where by (5.2),

$$\begin{aligned} c_n &= \frac{1}{2\pi i} \int_{C_r} f(w)\ \phi_n(w)\ dw\ , \\ &= \frac{1}{2\pi i} \int_{C_r} \{g(w) - k_0\}\ (w^2 - b^2)^{-1}\ \phi_n(w)\ dw\ , \\ &= \frac{1}{2\pi i} \int_{C_r} \frac{dz}{dw}\ (w^2 - b^2)^{-1}\ \phi_n(w)\ dw\ . \end{aligned} \tag{7.14}$$

C_r is a simple, closed, positively oriented contour in the interior of the region D' of the w-plane and has the points $w = 0, \pm b$ inside it. We may assume that it encloses the line segment $|Re\ w| \leq b_1,\ b_1 = w(\alpha_1),\ 0 \leq \alpha \leq \alpha_1$. Since $f(w)$ is an analytic function of w and b within and on C_r and the only possible singularities of $\phi_n(w)$ are poles at $w = 0,\ \pm b$, it follows that $c_n = c_n(b)$ are (real) analytic functions of b, $0 \leq b \leq b_1$.

Let R_n denote the remainder after n terms in the expansion of $f(w)$, (see (5.3)). In (7.5) write $\frac{dz}{dw}$ in terms of $f(w)$ by using (7.6) and (7.11). Then by (6.1),

$$
\begin{aligned}
J_\nu(\nu \operatorname{sech} \alpha) = & \left\{ k_0 + \sum_{m=0}^{n-1} (-1/N)^{m+1} c_{3m+1} \right\} N^{-1/3} Ai(N^{2/3}b^2) \\
& - \left\{ \sum_{m=0}^{n-1} (-1/N)^{m+1} c_{3m+2} \right\} N^{-2/3} Ai'(N^{2/3}b^2) \\
& + \frac{1}{2\pi i} E_{3n}(N, b) ,
\end{aligned} \tag{7.15}
$$

where we have used the fact that when the contour of integration in (1.9) is C_1,

$$
(2\pi i)^{-1} A_0(N, b) = N^{-1/3} Ai(N^{2/3}b^2) \tag{7.16}
$$

and

$$
(2\pi i)^{-1} A_1(N, b) = -N^{-2/3} Ai'(N^{2/3}b^2) . \tag{7.17}
$$

The error term $E_{3n}(N, b)$ is given by (6.5). Assuming for the present that (7.15) does provide a uniform asymptotic expansion of $J_\nu(\nu \operatorname{sech} \alpha)$, we see that the expansion coefficients are given by (7.14). The expansion (7.7) or (7.12) is useful when we want to obtain the residue at $w = \pm b$. At $w = 0$, we can use (7.6). Thus,

$$
\begin{cases}
c_0 = \dfrac{k_0}{b^2} - \dfrac{1}{\cosh \alpha - 1} , \\
c_1 = 0 , \\
c_2 = 2k_2 ,
\end{cases} \tag{7.18}
$$

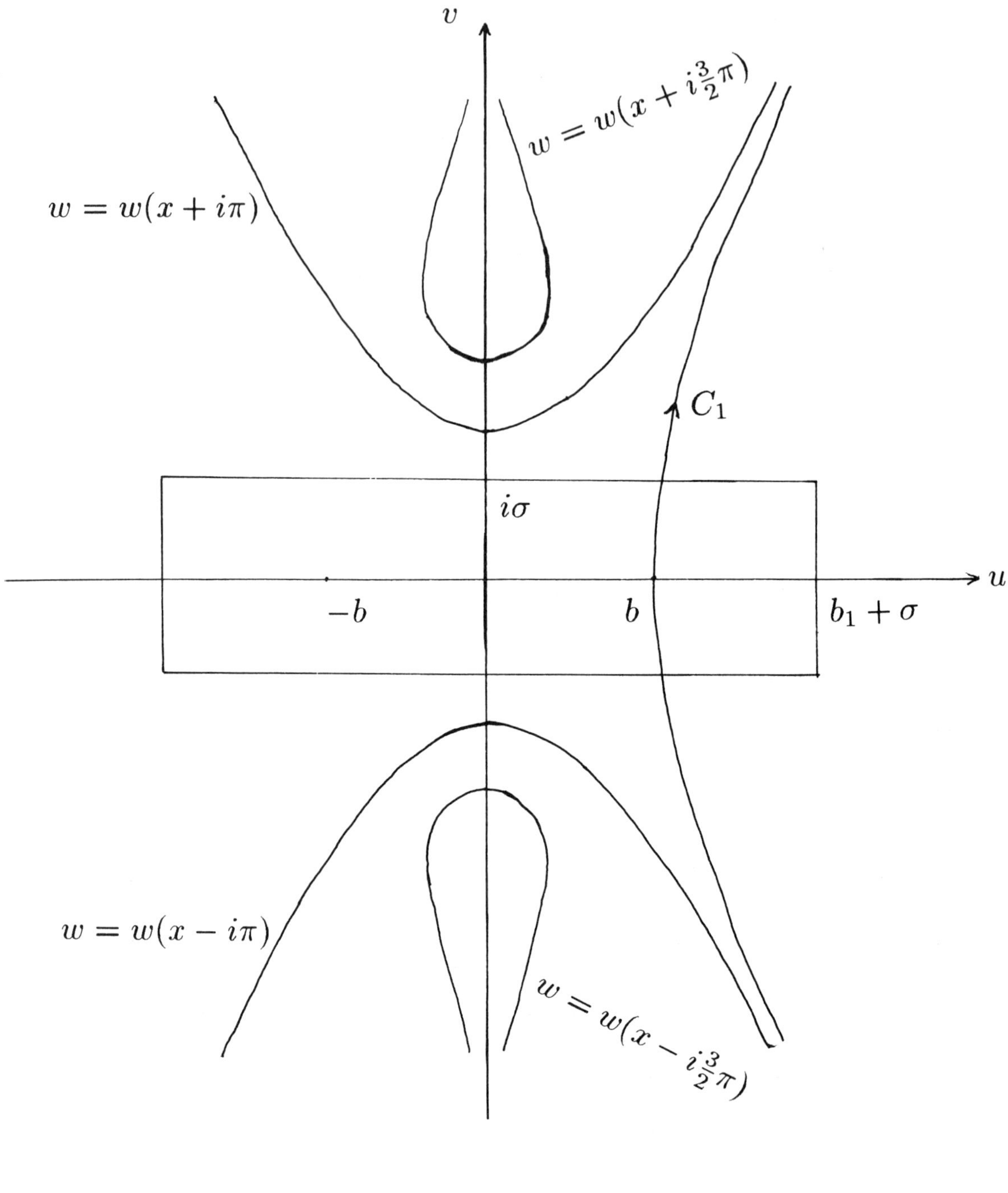

Figure 1. The w-plane

where k_0 and k_2 are given in (7.8) and (7.10).

By (7.15), for $n = 1$,

$$J_\nu(\text{ sech } \alpha) = N^{-1/3} k_0 \; Ai(N^{2/3} b^2) + N^{-5/3} 2k_2 \; Ai'(N^{2/3} b^2) + \frac{1}{2\pi i} \; E_3(N, b) \; . \tag{7.19}$$

Now we discuss E_{3n}. By (6.5),

$$E_{3n} = \left(-\frac{1}{N}\right)^n \int_{C_1} e^{N(\frac{1}{3} \; w^3 - b^2 w)} (w^2 - b^2) \; \Phi_n(w) \; dw \tag{7.20}$$

where

$$\Phi_n(w) = \left(\frac{1}{(w^2 - b^2)} \frac{d}{dw}\right)^n \; R_{3n}(w) \; . \tag{7.21}$$

We can show that $\Phi_n(w)$ is uniformly bounded on C_1. The domain D' in the w-plane is shown in Figure 1. It contains the real axis and has the curves $w = w(x \pm i\frac{3}{2}\pi), -\infty < x < \infty$ as its boundary.

The region D' depends on b. As $w \to \infty$ in D', $g(w) \sim 1/(3w)$, $0 \leq b \leq b_1$. For $z = x + iy$, $x \to \infty$, $|y| \leq \pi$, see [**3**, p. 115]. However, the same argument can be used when $|y| \leq 3\pi/2$, $|x| \to \infty$. By Ritt's theorem [**6**, p. 9] it can be shown that if $w = t$ is any point on C_1 then for $|t|$ sufficiently large, $g(t)$ and therefore $\Phi_n(t)$ is uniformly bounded. Since $w = w(b, z)$, let

$$\sigma = \min_{0 \leq b \leq b_1} \; w(b, i\pi) \; . \tag{7.22}$$

If t lies within the rectangle $|u| \leq b_1 + \sigma$, $|v| \leq \sigma$, $w = u + iv$, then we can use the bound (6.6) to conclude that $\Phi_n(t)$ is bounded uniformly in that rectangle. Hence it follows that $\Phi_n(t)$ is bounded uniformly on C_1, $0 \le b \le b_1$. Now a bound for E_{3n} can be obtained by following Copson [**3**, pp. 116-117]

$$|E_{3n}| \le K\ N^{-n-5/6}\ Ai(N^{2/3}b^2)$$

where K is independent of b, $0 \le b(\alpha) \le b_1$. We omit the details.

References.

1. N. Bleistein, *Uniform asymptotic expansion of integrals with stationary point near algebraic singularity*, Comm. Pure Appl. Math **19** (1966), 353-370.
2. C. Chester, B. Friedman and F. Ursell, *An extension of the method of steepest descents*, Proc. Cambridge Philos. Soc. **53** (1957), 599-611.
3. E.T. Copson, "Asymptotic Expansions," Cambridge Tracts in Math. and Math. Physics, No. 55, Cambridge Univ. Press, London, 1965.
4. D.S. Jones, *Asymptotic behavior of integrals*, SIAM Rev. **14** (1972), 268-317.
5. F.W.J. Olver, *The asymptotic expansion of Bessel functions of large order*, Philos. Trans. Roy. Soc. London, Ser. A, 247 (1954), 328-368.
6. F.W.J. Olver, "Asymptotics and Special Functions," Academic Press, New York, 1974.
7. L.A. Skinner, *Uniformly valid expansions for Laplace integrals*, SIAM J. Math. Anal. **11** (1980), 1058-1067.
8. K. Soni and B.D. Sleeman, *Uniform asymptotic expansions and associated polynomials*, J. Math. Anal. Appl. **123** (1987), 197-220.
9. K. Soni and N.M. Temme, *On a biorthogonal system associated with uniform asymptotic expansions*, IMAJ. (To Appear).
10. F. Ursell, *Integrals with a large parameter: Paths of descent and conformal mapping*, Proc. Cambridge Philos. Soc. **67** (1970), 371-381.
11. R. Wong, *Error bounds for asymptotic expansions of integrals*, SIAM Rev. **22** (1980), 401-435.

Rational Approximation of the Step, Filter, and Impulse Functions

Yasuhiko Ikebe [1]
Marek Kowalski [2]
Frank Stenger [3]
Department of Mathematics, University of Utah, Salt Lake City, Utah

1 Introduction

Modern design of electronic filters, such as low–pass, high–pass and band–pass filters includes the construction of a rational function which satisfies the desired specifications for cut–off frequencies, pass–band gain, transition band–width, and stop–band attenuation [1, Chapter 13]. Thus rational function approximations to the χ^*-function ($\chi^*(w) = 1$ if $-1 < w < 1$, 0 if $|w| > 1$) and the sgn function ($\operatorname{sgn} x = 1$ if $w > 1, 0$ if $w = 0, -1$ if $w < 0$) are needed. Elementary analytical methods lead to rational functions, such as

$$\frac{1}{1+\epsilon^2 T_n^2(w)} \quad \text{or} \quad \frac{1}{1+w^{2n}}$$

where $T_n(w)$ is the nth degree Chebyshev polynomial [1]. More advanced methods employ elliptic functions to derive a rational function approximation

---

[1] On leave to the University of Utah, from the University of Tsuhuba

[2] On leave to the University of Utah, from the University of Warsaw

[3] Department of Mathematics, University of Utah, Salt Lake City, UT, 84112. This work was done while this author was supported by U.S. Army Research Contract No. DAAL–03–87–K–0008

which has equiripple characteristic in both the pass and stop bands, but still requires evaluations of elliptic functions [1, p. 334, 2, 3, 6].

The purpose of this paper is to derive new rational function approximations, for the Heaviside, the filter and the impulse functions. Our main motivation of these derivations are the simplicity of the results; the formulas depend on a single parameter N which determines the degree and accuracy. Another advantage of the formulas are their simple relationships with one another; letting $H_N(w)$ denote the approximation of the Heaviside function, we have $\chi_N(w) = H_N[(1+w)/(1-w)]$ is the approximation to the χ^* function, and $H'_N(w)$, the derivative of $H_N(w)$ is the approximation to the impulse (the delta) function.

A different derivation of χ_N was obtained previously in [5], using elliptic functions. Although we could easily have started with the results in [5], we have chosen, instead, to bypass the use of elliptic functions via the direct use of Poisson's summation formula.

Finally, while the approximations of this paper do not yield the optimal ones, as described in [3,6], they are close to optimal, and they offer an advantage of simplicity of functional expression.

2 Approximation of the Heaviside Function

Let $\mathbb{R}$ denote the real line, i.e., $\mathbb{R} = (-\infty, \infty)$, let h be a positive number and for $a \in \mathbb{R}$, set

$$\operatorname{sgn} a = \begin{cases} 1 & if \quad a > 0 \\ 0 & if \quad a = 0 \\ -1 & if \quad a < 0 \end{cases} \tag{2.1}$$

Let us define $\tilde{H}(w)$, $H(w)$ and $H_N(w)$ for $w \in \mathbb{R}$ by

$$\tilde{H}(w) = \frac{1}{2} + \frac{1}{2} \operatorname{sgn} w \tag{2.2}$$

$$H(w) = \frac{1}{2} + \frac{hw}{\pi} \sum_{k=-\infty}^{\infty} \frac{e^{kh}}{w^2 + e^{2kh}} \tag{2.3}$$

and

$$H_N(w) = \frac{1}{2} + \frac{hw}{\pi} \sum_{k=-N}^{N} \frac{e^{kh}}{w^2 + e^{2kh}}, \tag{2.4}$$

where in (2.4), N is a positive integer.

Let us note at the outset, that each of the three functions $\tilde{H}, H$ and H_N satisfiy the functional relationships

$$G(\frac{1}{w}) = G(w) \tag{2.5}$$

and

$$G(-w) + G(w) = 1. \tag{2.6}$$

The first main result of our paper is

Theorem 2.1 *If $h > 0$ and $w \in \mathbb{R}$, then*

$$|\tilde{H}(w) - H(w)| \leqq \sum_{n=1}^{\infty} \frac{1}{\cosh\left(\frac{n\pi^2}{h}\right)} < \frac{2e^{-\pi^2/h}}{1 - e^{-\pi^2/h}} \tag{2.7}$$

Proof: We apply Poisson's summation formula, which states that if f is continuous on $\mathbb{R}$, and if the limit

$$F(x) = \lim_{M\to\infty} \int_{-M}^{M} e^{ixt} f(t)dt \tag{2.8}$$

exists for all $x \in \mathbb{R}$, then for all $h > 0$, we have

$$\lim_{M\to\infty} h \sum_{m=-M}^{M} f(mh) = \lim_{M\to\infty} \sum_{n=-M}^{M} F\left(\frac{2n\pi}{h}\right). \tag{2.9}$$

By taking

$$f(t) = \frac{we^t}{\pi[w^2 + e^{2t}]} \tag{2.10}$$

we have, as shown in Appendix A, that

$$F(x) = \int_{\mathbb{R}} e^{ixt} f(t)dt = \frac{e^{ix\log|w|}}{2\cosh(\pi x/2)} \operatorname{sgn}(w), \tag{2.11}$$

and so

$$\frac{hw}{\pi} \sum_{m=-\infty}^{\infty} \frac{e^{mh}}{w^2 + e^{2mh}} = \frac{1}{2} \operatorname{sgn} w \sum_{n=-\infty}^{\infty} \frac{e^{2i(n\pi/h)\log|w|}}{\cosh(\pi^2 n/h)} \tag{2.12}$$

Hence, by (2.2), (2.3) and (2.12), we have

$$|H(w) - \tilde{H}(w)| \leqq \sum_{n=1}^{\infty} \frac{1}{\cosh\left(\frac{\pi^2 n}{h}\right)} \leqq \frac{2e^{-\pi^2/h}}{1 - e^{-\pi^2/h}} \cong 2e^{-\pi^2/h} \tag{2.13}$$

We now estimate the rate of convergence of the series defining $H(w)$. To this end, we set $\delta_N(w) = H(w) - H_N(w)$, where $H_N(w)$ is defined as in (2.4), so that

$$\delta_N(w) = \frac{hw}{\pi} \sum_{|n|>N} \frac{e^{nh}}{w^2 + e^{2nh}}. \tag{2.14}$$

It then follows that

$$\delta_N(1/w) = \delta_N(w) \tag{2.15}$$

and

$$\delta_N(-w) = -\delta_N(w). \tag{2.16}$$

Theorem 2.2 *Let $\tilde{H}(w)$, $H(w)$ and $H_N(w)$ be defined as in (2.2), (2.3), and (2.4). Then for $w \in \mathbb{R}$, we have*

$$|H_N(w) - H(w)| < \frac{1}{\pi}(|w| + \frac{1}{|w|})e^{-(N+1)h} \tag{2.17}$$

$$|\tilde{H}(w) - H_N(w)| < \frac{1}{\pi}\left(|w| + \frac{1}{|w|}\right) e^{-(N+1)h} + 2e^{-\pi^2/h}. \tag{2.18}$$

Moreover, by taking

$$h = \frac{\pi}{(N+1)^{1/2}} \tag{2.19}$$

we have

$$|H_N(w) - \tilde{H}(w)| < \frac{1}{\pi}\left[|w| + 2\pi + \frac{1}{|w|}\right] e^{-\pi(N+1)^{1/2}}. \tag{2.20}$$

Proof:

By the relations (2.15) and (2.16), if suffices to consider the case of $w \geq 1$. In this case we have

$$H(w) = H_N(w) + \frac{hw}{\pi} \sum_{m=N+1}^{\infty} \frac{e^{mh}}{w^2 + e^{2mh}} + \frac{hw}{\pi} \sum_{m=-\infty}^{-N-1} \frac{e^{mh}}{w^2 + e^{2mh}}$$

$$< H_N(w) + \frac{hw}{\pi} \sum_{m=N+1}^{\infty} e^{-mh} + \frac{hw}{\pi} \sum_{m=-\infty}^{-N-1} \frac{e^{mh}}{w^2}$$

$$= H_N(w) + \frac{hw}{\pi} \frac{e^{-(N+1)h}}{1 - e^{-h}} \left[1 + \frac{1}{w^2}\right] \tag{2.21}$$

$$< H_N(w) + \frac{1}{\pi}(w + 1/w)e^{-(N+1)h}$$

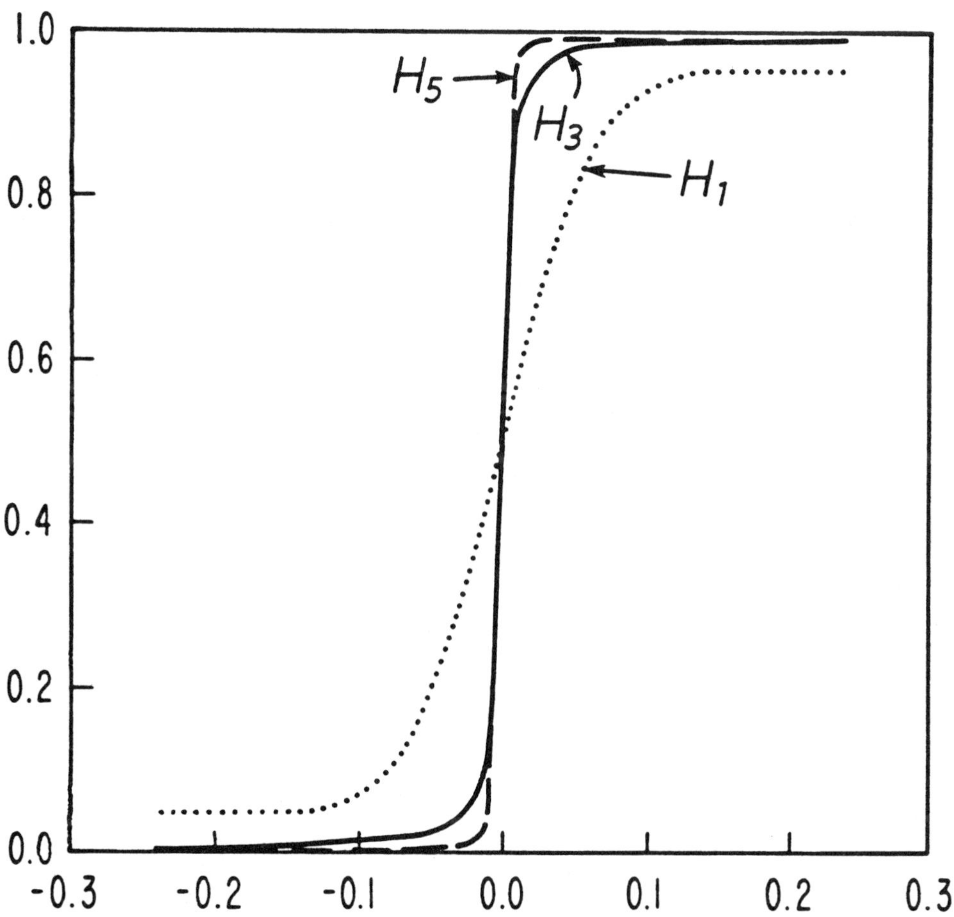

Figure 2.1: Section of the Step Function H_N, N = 1,3,5

This proves (2.17). The result (2.18) comes from combining (2.13) and (2.17). Finally (2.20) results from substituting (2.19) into (2.18).

Corresponding to $a > 1$, let us set

$$\epsilon = \frac{1}{a + \sqrt{a^2 - 1}}. \tag{2.22}$$

Hence if $-\frac{1}{\epsilon} < w < -\epsilon$ or else if $\epsilon < w < \frac{1}{\epsilon}$, then $|w| + \frac{1}{|w|} < 2a$.

We hence have proved

Corollary 2.3 *If $\tilde{H}$ and H_N are defined as in (2.2) and (2.4), and if h is selected by the formula (2.19), then for $-\frac{1}{\epsilon} < w < -\epsilon$ or for $\epsilon < w < \frac{1}{\epsilon}$, where ϵ is defined by (2.22), with $a > 1$, we have*

$$|H_N(w) - \tilde{H}(w)| < \left(\frac{2a + 2\pi}{\pi}\right) e^{-\pi(N+1)^{1/2}}. \tag{2.23}$$

3 Approximation of the Filter Function

Let us replace w in (2.2), (2.3) and (2.4) by $(1+w)/(1-w)$, and let us denote the new functions by $\tilde{\chi}(w)$, $\chi(w)$ and $\chi_N(w)$, i.e. we set

$$\tilde{\chi}(w) = \begin{cases} 0 & if \quad w < -1 \\ \frac{1}{2} & if \quad w = -1 \\ 1 & if \quad -1 < w < 1 \\ \frac{1}{2} & if \quad w = 1 \\ 0 & if \quad w > 1, \end{cases} \tag{3.1}$$

$$\chi(w) = \frac{1}{2} + \frac{h}{\pi}\left(1 - w^2\right) \sum_{k=-\infty}^{\infty} \frac{e^{kh}}{(1+w)^2 + (1-w)^2 e^{2kh}} \tag{3.2}$$

and

$$\chi_N(w) = \frac{1}{2} + \frac{h}{\pi}\left(1 - w^2\right) \sum_{k=-N}^{N} \frac{e^{kh}}{(1+w)^2 + (1-w)^2 e^{2kh}}. \tag{3.3}$$

Then, paraphrasing theorems (2.1) and (2.2), we have

Theorem 3.1 *Let $\tilde{\chi}$, χ and χ_N be defined as in (3.1), (3.2) and (3.3) above. Then for $h > 0$, and N a positive integer, we have for all $w \in \mathbb{R}$*

$$|\tilde{\chi}(w) - \chi(w)| \leqq \sum_{n=1}^{\infty} \frac{1}{\cosh(\frac{n\pi^2}{h})} < \frac{2}{1 - e^{-\pi^2/h}} \cong 2e^{-\pi^2/h}; \tag{3.4}$$

$$|\chi(w) - \chi_N(w)| \leqq \frac{2}{\pi} \left| \frac{1 + w^2}{1 - w^2} \right| e^{-(N+1)h}; \tag{3.5}$$

and

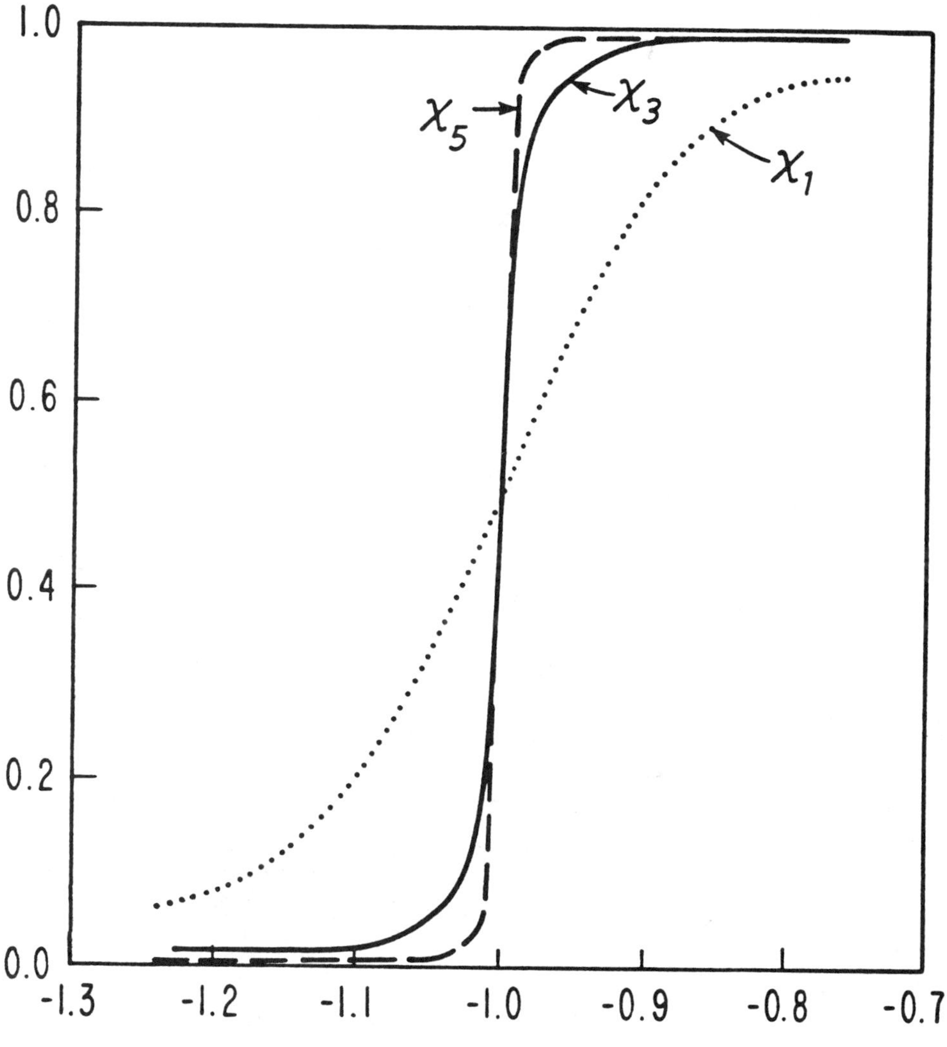

Figure 3.1: Section of the Filter Function χ_N, N = 1,3,5

$$(3.6) \qquad |\tilde{\chi}(w) - \chi_N(w)| \leqq \frac{2}{\pi}\left|\frac{1+w^2}{1-w^2}\right| e^{-(N+1)h} + 2e^{-\pi^2/h}.$$

Moreover, by taking h as in (2.19), we have

$$(3.7) \qquad |\tilde{\chi}(w) - \chi_N(w)| \leqq \frac{2}{\pi}\left|\frac{(1+w^2)}{|1-w^2|} + \pi\right| e^{-\pi(N+1)^{1/2}}.$$

Note, that whereas the bound on the differences (2.17), (2.18) and (2.20) is unbounded for $w = 0$ and $w = \infty$, the corresponding differences (3.5), (3.6) and (3.7) are unbounded at $w = \pm 1$. By Corollary 2.3 we also have

Corollary 3.2 *Let $\tilde{\chi}$ and χ_N be defined as in (3.1) and (3.3) respectively, and let h be relected by (2.19). Corresponding to $a > 1$, let δ be determined by the expression*

$$(3.8) \qquad \delta = \left(\frac{a-1}{a+1}\right)^{1/2}.$$

If either $-\delta \le w < \delta$, or else if $w \in \mathbb{R}$ and $1/\delta < |w| < \infty$, then

$$(3.9) \qquad |\tilde{\chi}(w) - \chi_N(w)| \leqq \frac{2}{\pi}[a+\pi]e^{-\pi(N+1)^{1/2}}.$$

4 Approximation of the Delta Function

The derivative of $\tilde{H}$ defined as in (2.2) is the Dirac delta function, i.e., $\tilde{H}'(w) = \delta(w)$, which satisfies the relation

$$(4.1) \qquad \int_{\mathbb{R}} f(w)\tilde{H}'(w)dw = f(0)$$

for all functions f that are continuous on $\mathbb{R}$. We may expect, then, that $H_N'(w)$ is also a close approximation to $\tilde{H}'(w)$, in the sense of (4.1). We shall see below that this is in fact the case, provided that f is not only continuous, but also bounded and absolutely integrable over $\mathbb{R}$. Indeed, if f is not absolutely integrable over $\mathbb{R}$ (a situation which we do not expect to encounter in real life), we can get some rather disturbing results, e.g., by taking $f(w) \equiv 1$ for all $w \in \mathbb{R}$, we get

$$(4.2) \qquad \int_{\mathbb{R}} f(w)H_N'(w)dw = \int_{\mathbb{R}} H_N'(w)dw = 0.$$

Although we shall study here only the approximtion of the distrubtion $\tilde{H}'$ via H_N', we mention that for $k > 1$, $H_N^{(k)}$ also provides an accurate approximatin of the distrubution $\tilde{H}^{(k)}$. Approximations of $\tilde{H}^{(k)}$ were previously obtained via orthognal series expansion, by Zemanian [7].

We shall see below, that $H'_N(0)$ is very large, and $H_N(w)$ is very small when $w \in D_N = \{w \in \mathbb{R} : R_N^{-1} < |w| < R_N\}$, where $R_N > 1$ and moreover $H'_N(0)$ and R_N increase very rapidly as N increases. On the other hand, $H'_N(w)$ becomes negative in the region $|w| > R_N$ and as $|w|$ increases in this region, $H'_N(w)$ first increases in magnitude, and then it approaches zero very slowly. One may thus ask about the behavior of $H'_N(w)$, e.g., whether there exists a sequence of numbers $\{\alpha_n\}$ such that $\alpha_n H'_N(\alpha_n w)$ approaches a fixed function as $N \to \infty$.

Previous statements of the above observations, and answers to the above questions are given in the following

Theorem 4.1 *Let H_N be defined as in (2.4). Then H'_N, the derivative of H_N, is given by*

$$H'_N(w) = \frac{h}{\pi} \sum_{m=-N}^{N} e^{mh} \frac{e^{2mh} - w^2}{(e^{2mh} + w^2)^2}. \tag{4.3}$$

Setting

$$\alpha_n = e^{-\pi(N+1)^{1/2}}, \beta_N = e^{\pi(N+1)^{1/2}} \tag{4.4}$$

and letting h be selected by (2.19), we have, as $N \to \infty$

$$\alpha_n H'_N(\alpha_n w) = \frac{1}{\pi(1 + w^2)} + \mathcal{O}(1/N^{1/2}) \tag{4.5}$$

uniformly, on $\in \mathbb{R}$, and also,

$$\beta_N H'_N(\beta_n w) = -\frac{1}{\pi(1 + w^2)} + \mathcal{O}(1/N^{1/2}) \tag{4.6}$$

uniformly on $\mathbb{R}\backslash[-\epsilon, \epsilon]$, where ϵ is an aribtrary positive number. Furthermore, if f is continous and bounded on $\mathbb{R}$, and if $\int_{\mathbb{R}} |f(w)| dw < \infty$, then

$$\int_{\mathbb{R}} H_N(w) f(w) dw = f(0) + \mathcal{O}\left(\omega(f; e^{\frac{-\pi}{2}(N+1)^{1/2}}\right) \\ + \mathcal{O}(N^{-1/2}), N \to \infty \tag{4.7}$$

where $\omega(f; , \delta)$ denotes the modulus of continuity of f:

$$\omega(f; \delta) = \sup_{x \in \mathbb{R}, h \in \mathbb{R}, |h| < \delta} |f(x + h) - f(x)|. \tag{4.8}$$

Proof of Theorem 4.1. If α_n is defined by (4.4), then we have

$$\alpha_N H'_N(\alpha_n w) = \frac{1}{\sqrt{N+1}} \sum_{k=0}^{2N} f\left\{\frac{k}{\sqrt{N+1}}; w\right\}, \tag{4.9}$$

where

$$f(x;w) = e^{\pi x}\frac{e^{2\pi x} - w^2}{(e^{2\pi x} + w^2)^2}. \tag{4.10}$$

Thus (4.9) is a Riemann approximation to the integral

$$\int_0^{(2N+1)/\sqrt{N+1}} f(x;w)dx = \frac{1}{\pi}\left(\frac{1}{1+w^2} - \frac{e^{\pi(2N+1)/\sqrt{N+1}}}{e^{2\pi(2N+1)/\sqrt{N+1}} + w^2}\right). \tag{4.11}$$

After some computation, we get

$$\left|\frac{d}{dx}f(x;w)\right| \leqq 7\pi e^{-\pi x}. \tag{4.12}$$

Hence

$$\left|\frac{1}{\sqrt{N+1}}\sum_{k=0}^{2N} f\left(\frac{k}{\sqrt{N+1}};w\right) - \int_0^{(2N+1)/\sqrt{N+1}} f(x;w)dx\right|$$

$$= \frac{1}{\sqrt{N+1}}\left|\sum_{k=0}^{2N}\left[f(\frac{k}{\sqrt{N+1}};w) - \sqrt{N+1}\int_{k/\sqrt{N+1}}^{(k+1)/\sqrt{N+1}} f(x;w)dx\right]\right.$$

$$\leqq \frac{1}{\sqrt{N+1}}\sum_{k=0}^{2N}\frac{1}{\sqrt{N+1}}\max_{k/\sqrt{N+1}\leqq x\leqq (k+1)/\sqrt{N+1}}\left|\frac{d}{dx}f(x;w)\right| \tag{4.13}$$

$$\leqq \frac{1}{N+1}\sum_{h=0}^{2N} 7\pi e^{-\pi k/\sqrt{N+1}} \leqq \frac{7\pi}{(N+1)\left[1 - e^{-\pi/\sqrt{N+1}}\right]} \leqq \frac{7e^{\pi/\sqrt{N+1}}}{\sqrt{N+1}}.$$

By (4.9), (4.10), (4.11) and (4.12),

$$\left|\alpha_N H_N'(\alpha_N w) - \frac{1}{\pi(1+w^2)}\right| \leqq \frac{e^{\pi(2N+1)/\sqrt{N+1}}}{e^{2\pi(2N+1)/\sqrt{N+1}} + w^2} \tag{4.14}$$

$$+\frac{7e^{\pi/\sqrt{N+1}}}{\sqrt{N+1}} = \mathcal{O}\left(\frac{1}{N^{1/2}}\right)$$

as $N \to \infty$, and where the bound on the extreme right–hand side of (4.14) is clearly valid uniformly for $w \in \mathbb{R}$.

We next consider $\beta_N H'(\beta_n w)$. In this case, we have

(4.15) $$\beta_N H'_N(\beta_N w) = \frac{1}{\sqrt{N+1}} g\left(\frac{k}{\sqrt{N+1}}; w\right)$$

where

(4.16) $$g(x; w) = -e^{-\pi x} \frac{w^2 - e^{-2\pi x}}{(w^2 + e^{-2\pi x})^2}$$

Thus, again $\beta_N H'_N(\beta_n w)$ is a Riemann approximation to the integral

(4.17) $$\int_0^{(2N+1)/\sqrt{N+1}} g(x; w)dx = -\frac{1}{\pi}\left[\frac{1}{1+w^2} - \frac{e^{-\pi(2N+1)/\sqrt{N+1}}}{e^{-2\pi(2N+1)/\sqrt{N+1}} + w^2}\right]$$

and

(4.18) $$\left|\frac{d}{dx} g(x; w)\right| \leqq \frac{(7+w^2)(1+w^2)}{w^6} e^{-\pi x} \quad \text{if} \quad w \neq 0.$$

Now, proceeding as in the case of $\alpha_N H'_N(\alpha_n w)$, we get

(4.19) $$\left|\frac{1}{\sqrt{N+1}} \sum_{h=0}^{2N} g\left(\frac{k}{\sqrt{N+1}}; w\right) - \int_0^{(2N+1)/\sqrt{N+1}} g(x; w)dx\right|$$
$$\leqq \frac{(7+w^2)(1+w^2)}{\pi w^6} \frac{e^{\pi/\sqrt{N+1}}}{\sqrt{N+1}}$$

This proves (4.6).

It only remains to prove (4.7). Now taking α_N as in (4.4), we have

$$\int_{\mathbb{R}} H'_N(w) f(w)dw = \int_{\mathbb{R}} \alpha_N H'_N(\alpha_N w) f(\alpha_N w)dw$$
$$= \int_{\mathbb{R}} \left[\alpha_N H'_N(\alpha_n w) - \frac{1}{\pi(1+w^2)}\right] f(\alpha_N w)dw$$

(4.20)

$$+ \int_{\mathbb{R}} \frac{\alpha_N}{\pi} \frac{f(v)}{\alpha_n^2 + v^2} dv$$
$$= \mathcal{O}\left(\frac{1}{\sqrt{N}}\right) + \int_{\mathbb{R}} \frac{\alpha_N}{\pi} \frac{f(v)}{(\alpha_N^2 + v^2)} dv$$

On the other hand,

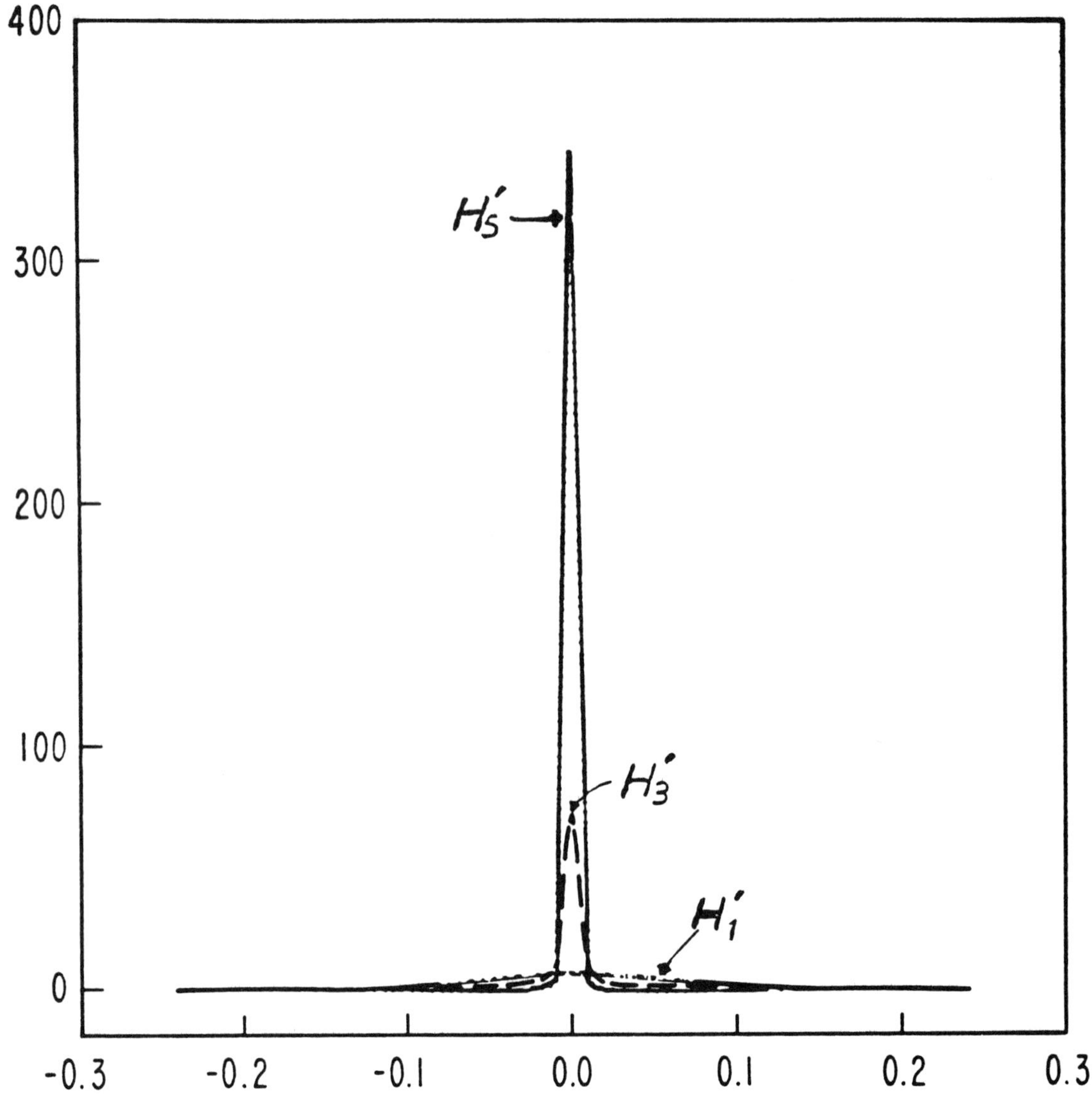

Figure 4.1: Section of the Impulse Function H_N, N = 1,3,5

$$\left|\int_{\mathbb{R}} \frac{\alpha_N f(v)dv}{\pi\,[\alpha_N^2+v^2]} - f(0)\right| = \left|\int_{\mathbb{R}} \frac{\alpha_N[f(v)-f(0)]}{\pi[\alpha_N^2+v^2]}dv\right|$$

(4.21)

$$= \left|\int_{-\sqrt{\alpha_N}}^{\sqrt{\alpha_N}} \frac{\alpha_N[f(v)-f(0)]}{\pi[\alpha_N^2+v^2]}dv\right| + 2M\int_{\sqrt{\alpha_N}}^{\infty} \frac{\alpha_N dv}{\pi[\alpha_N^2+v^2]}$$

where $M = \sup_{x\in\mathbb{R}} |f(x)|$.

But

$$\sup_{|v|\leqq\sqrt{\alpha_N}} |f(v)-f(0)| \leqq \omega(f;\sqrt{\alpha_N}) \tag{4.22}$$

and

$$\int_{\sqrt{\alpha_N}}^{\infty} \frac{\alpha_n dv}{\pi\,[\alpha_N^2+v^2]} = \tan^{-1}(\sqrt{\alpha_N}) \tag{4.23}$$

Hence we have

$$\left|\int_{\mathbb{R}} \frac{\alpha_n\ f(v)d\nu)}{\pi\,[\alpha_N^2+v^2]} - f(0)\right|$$

(4.24)

$$\leqq \omega(f;\sqrt{\alpha_N}) + \frac{2M}{\pi}\tan^{-1}(\sqrt{\alpha_N})$$

$$\leqq \omega(f;\sqrt{\alpha_N}) + \frac{2M}{\pi}\sqrt{\alpha_N}.$$

We thus get (4.7), using the definition (4.4) of α_N.

APPENDIX

A Deriviation of (2.11)

It is enough to consider the case of $w > 0$. Then

$$F(x) = \int_{\mathbb{R}} e^{ixt}\frac{we^t}{\pi[w^2+e^{2t}]}dt = \frac{1}{\pi}e^{ix\log w}\int_{\mathbb{R}} \frac{e^{i\xi x}}{\cosh\xi}d\xi, \tag{A.1}$$

where we have set $\xi = t - \log\ w$. Next we have

$$
\begin{aligned}
\int_{\mathbb{R}} \frac{e^{i\xi x}}{\cosh \xi} d\xi &= 4Re \int_{\mathbb{R}}^{\infty} \frac{e^{(ix-1)\xi}}{1+e^{-2\xi}} d\xi \\
&= 4Re \int_0^{\infty} e^{(ix-1)\xi} \left(1 - e^{-2\xi} + e^{-4\xi} - \cdots\right) d\xi \qquad \text{(A.2)} \\
&= 4\left\{\frac{1}{x^2+1} - \frac{3}{x^2+3^2} + \frac{5}{x^2+5^2} - \cdots\right\} \\
&= \frac{\pi}{\cosh(\pi x/2)},
\end{aligned}
$$

where the last identity was obtained from [4, Eq. 1.445.4].

Clearly, if $w < 0$, then we have

$$
F(x) = -\int_{\mathbb{R}} \frac{e^{ixt}|w|e^t}{\pi[w^2 + e^{2t}]} dt. \qquad \text{(A.3)}
$$

Hence (2.11) follows from (A.1), (A.2) and (A.3).

References

[1] V. K. Aatre, "Network Theory and Filter Design", John Wiley (1981).

[2] A. Andreas, "Digital Filters, Analysis and Design", McGraw–Hill (1979).

[3] W. Cauer, "Benerkung über eine Extremalaufgabe von E. Zolotareff", A. Angew. Math. Mech. 20, (1940) 358.

[4] I. S. Gradshteyn and I. M. Ryzhik, "Tables of Integrals, Series and Products", 4th Edition, Acad. Press (1965).

[5] F. Stenger, "An Analytic Function which is an Approximate Characteristic Function", SIAM J. Numr. Anal. 12 91975) 239-254.

[6] J. Todd, "Inequalities of Chebyshev, Zolotareff, Cauer and W. B. Jordan, in "Inequalities", edited by O. Shisha, Academic Press, 1967, pp. 321-329.

[7] A. H. Zemanian, "Orthonormal Series Expansion of Certain Distributions and Distributional Trunstorm Calculus", J. Math. Anal. Appl. 14 (1966) 263-276.

Polynomial Asymptotic Estimates of Gegenbauer, Laguerre, and Jacobi Polynomials

Nico M. Temme Professor, Center for Mathematics and Computer Science, Amsterdam, The Netherlands

1. Introduction

We discuss asymptotic forms of $C_n^\gamma(x)$, $L_n^\alpha(x)$, $P_n^{(\alpha,\beta)}(x)$, the Gegenbauer, Laguerre and Jacobi polynomials. The asymptotic behavior of these classical orthogonal polynomials has been the subject of several investigations. The research usually concentrates on the case that the degree n of the polynomial is the large parameter, and for all classical orthogonal polynomials the asymptotic behavior is well established now. Inside the domain of the zeros of the polynomial the behavior can be described in terms of elementary functions, such as trigonometric functions. In the domain where the transition from oscillatory to monotonic behavior occurs, familiar higher transcendental functions can be used as estimates. For example, the "first" zeros of the Jacobi polynomial can be approximated in terms of the zeros of the J-Bessel function. In Szegö (1958) several classical results can be found. For Jacobi polynomials he gives an estimate of

"Hilb's type", which is an approximation in terms of the $J-$Bessel function. In earlier papers, however, Szegö already gave more extensive expansions for Legendre and Jacobi polynomials, again in terms of the $J-$Bessel function. See ASKEY (1982).

More recent publications give uniform asymptotic expansions in larger domains for the argument, and also more terms in the expansion. ERDÉLYI (1960) derived leading terms of the expansion for Laguerre polynomials in terms of Bessel and Airy functions, and quite recently FRENZEN & WONG (1988) constructed complete expansions for this class of polynomials. ELLIOTT (1971) treated the Jacobi polynomials essentially outside the domain of orthogonallity, giving a full expansion in terms of modified Bessel functions; in FRENZEN & WONG (1985) a uniform expansion for the Jacobi polynomials is given in terms of $J-$Bessel functions, complete with error bounds for the remainder.

As mentioned earlier, the degree n of the polynomials is usually considered as the large parameter in the asymptotic expansions. In many existing cases, however, the asymptotic variable includes additional parameters as well. For instance, in ERDÉLYI (1960) and FRENZEN & WONG (1988) a natural choice for the large parameter of the Laguerre polynomial $L_n^\alpha(x)$ is $\kappa = n + (\alpha+1)/2$, although the parameter α (and hence the order of the comparison function $J_\alpha(z)$) is kept fixed in their analysis. In TEMME (1986) three different cases are considered for large values of κ. Depending on the value of x and the size of α/n, a Bessel function, a Hermite polynomial, or an Airy function is used for the Laguerre polynomials.

In the case of Jacobi polynomials $P_n^{(\alpha,\beta)}(x)$, a convenient choice is $\kappa = n+(\alpha+\beta+1)/2$, see FRENZEN & WONG (1985). See also ELLIOTT (1971) and NESTOR (1984). In this paper we consider the same asymptotic variable κ, now with the possibility that α and/or β dominate n, and with x inside the domain of orthogonallity. We use the Gegenbauer polynomial and the Laguerre polynomial as new estimates for the Jacobi polynomial.

Some of the results on orthogonal polynomials in this paper follow from earlier, sometimes more general, results in the literature. The Laguerre polynomial is in fact a special case of the Whittaker, or confluent hypergeometric function. For this class of functions many recent results are available, for instance, OLVER(1980), and very recently DUNSTER(1989). On the other hand, the Jacobi polynomial and the Gegenbauer polynomial are special cases of the Gauss hypergeometric function, for which class rather few interesting uniform expansions have been investigated.

Recently, other investigations on orthogonal polynomials with several large parameters have been published, but the approach is quite different from ours. LI-CHEN

CHEN & ISMAIL(1989) investigated the asymptotics of $P_n^{(\alpha+an,\beta+bn)}(x)$ as $n \to \infty$, and a, b, α, β, x remain fixed. The approximants are in terms of elementary functions, and the approach is based on Darboux's method applied on a generating function. They also considered the Laguerre polynomial $L_n^{\alpha+an}(b+xn)$ for large values of n. Asymptotics of Jacobi and Laguerre polynomials with similar arrangement of parameters occur in work on incomplete polynomials of SAFF & VARGA (1979), MOAK, SAFF & VARGA (1979), and MHASKAR & SAFF (1984). In the second paper the distribution of zeros of that type of Jacobi polynomial is considered.

In the present approach, the estimates of the orthogonal polynomials are obtained by using differential equations, and Liouville Green transformations thereof. This technique is extensively developed by Frank Olver. Some of the results follow from his earlier work on, for instance, Whittaker functions. The proofs of the new estimates of Jacobi polynomials will be given elsewhere. Another interesting problem is how to obtain similar estimates by using integral representations of the polynomials. We intend to consider also this point in future publications.

1.1. The Liouville-Green transformation

The techniques used in this paper are based on the Liouville-Green (LG) transformation. Starting point is a differential equation of the form

$$W'' = [\kappa^2 p(x) + q(x)]W, \tag{1.1}$$

defining the orthogonal polynomial, where κ is a large parameter, a simple function of the degree and other parameters, but independent of x. Formally we proceed as follows. A transformation of both the dependent and independent variable is introduced by writing

$$W(x) = \sqrt{\dot{x}}\, w(\xi), \tag{1.2}$$

where $\xi = \xi(x)$, and the dot indicates differentiation with respect to ξ. Then (1.1) becomes

$$\ddot{w} = [\kappa^2 \dot{x}^2 p(x) + \psi(\xi)]w, \tag{1.3}$$

where

$$\psi(\xi) = \dot{x}^2 q(x) + \sqrt{\dot{x}}\, \frac{d^2}{d\xi^2} \frac{1}{\sqrt{\dot{x}}}. \tag{1.4}$$

Depending on the nature of the function p and on the asymptotic phenomena to be investigated, the function $\xi(x)$ is defined by replacing $\dot{x}^2 p(x)$ by a simple rational function $r(\xi)$. That is, we write

$$\dot{x}^2 p(x) = r(\xi) \tag{1.5}$$

and the new differential equation becomes

$$\ddot{w} = [\kappa^2 r(\xi) + \psi(\xi)]w. \tag{1.6}$$

The function $r(\xi)$ is chosen such that it mimics features of $p(x)$ that are essential for the asymptotic behavior of $W(x)$. As a rule, when we take $\psi = 0$, (1.6) reduces to the differential equation of a known special function that plays the part of a basic approximant in the asymptotic estimate.

For example, for the orthogonal polynomials on the interval $[-1,1]$, p has two simple zeros x_1, x_2 in this interval, and p is negative in (x_1, x_2). Then the zeros of the orthogonal polynomial occur in the same interval. Under the influence of some parameters we may have the situation that $x_1 \to x_2$. When no other zeros or poles of p are nearby the interval (x_1, x_2), one may consider for that particular case the transformation (1.5) in the form

$$\dot{x}^2 p(x) = \xi^2 - \rho^2,$$

where ρ depends on x_1, x_2. When $r(\xi) = \xi^2 - \rho^2$ and $\psi = 0$, equation (1.6) reduces to the differential equation for the parabolic cylinder function. Under further special circumstances, this function reduces to a Hermite polynomial.

In this paper we transform the differential equations defining the classical orthogonal polynomials $P_n^{(\alpha,\beta)}(x)$, $L_n^\alpha(x)$, $C_n^\gamma(x)$, the Jacobi, Laguerre and Gegenbauer polynomials, to a form in which another orthogonal polynomial acts as a basic approximant, for instance a Hermite polynomial. In these cases, the x–domain is the complete domain of orthogonallity. We also mention estimates in which other comparison functions occur, and which describe the transition of the oscillatory to the monotonic regions. In all cases we concentrate on real values of the parameters.

2. Gegenbauer polynomials

The Gegenbauer polynomials, or ultraspherical polynomials, can be defined by the generating function

$$(1 - 2xz + z^2)^{-\gamma} = \sum_{n=0}^{\infty} C_n^\gamma(x) z^n, \quad -1 < x < 1, \quad |z| < 1. \tag{2.1}$$

An explicit representation is

$$C_n^\gamma(x) = \frac{1}{\Gamma(\gamma)} \sum_{k=0}^{\lfloor n/2 \rfloor} \frac{(-1)^k \Gamma(\gamma + n - k)}{k!(n-2k)!} (2x)^{n-2k}, \tag{2.2}$$

where $\lfloor m \rfloor$ is the integer number satisfying $\lfloor m \rfloor \le m < \lfloor m \rfloor + 1$, with $m \in \mathbb{R}$. These polynomials satisfy the differential equation

$$(1 - x^2)y'' - (2\gamma + 1)xy' + n(n + 2\gamma)y = 0. \tag{2.3}$$

A simple transformation

$$W(x) = (1 - x^2)^{(2\gamma+1)/4} C_n^\gamma(x) \tag{2.4}$$

gives for W the equation

$$(1 - x^2)W'' + \left[(n + \gamma)^2 + \frac{2 + 4\gamma - 4\gamma^2 + x^2}{4(1 - x^2)}\right] W = 0.$$

This can be written in the form

$$W'' = \kappa^2 \frac{x^2 - x_0^2}{(1 - x^2)^2} W - \frac{x^2 + 3}{4(1 - x^2)^2} W, \tag{2.5}$$

with

$$\kappa = n + \gamma, \quad x_0^2 = \frac{\kappa^2 - (\gamma - \frac{1}{2})^2}{\kappa^2}. \tag{2.6}$$

We assume that $\gamma > 1/2$, and $n = 0, 1, 2, \ldots$; x_0 is the positive number defined by the above equation. We have $x_0 \in [0, 1]$. For the asymptotic problem we assume that κ is large.

Especially interesting is the behavior of the polynomials when x crosses the values $\pm x_0$ (turning points of (2.5)) and ± 1 (singular points). When $\gamma >> n$ the turning points (and all zeros of the Gegenbauer polynomial) tend to zero and coalesce. When $n >> \gamma$ the turning points coalesce with the singularities at ± 1. In both cases, the asymptotic behavior of the polynomials cannot be described in terms of elementary functions.

The Gegenbauer polynomial is a function of three variables. In the asymptotic problem we try to find approximating functions of at most two variables. This gives a proper reduction of variables, as pointed out by OLVER (1975-B). It is not possible to construct one approximation of two variables with which we can describe the behavior of the polynomials as $\kappa \to \infty$, uniformly with respect to x on the entire real axis. For the transition near $x = \pm 1$ we need a J–Bessel function as approximant.

2.1. Estimation in terms of a Hermite polynomial

Consider as a simpler form of (2.5), especially if κ is large and x is bounded away from ± 1, the equation

$$u'' = \kappa^2(x^2 - x_0^2)u, \tag{2.7}$$

with solution

$$u(x) = D_\nu(x\sqrt{2\kappa}\,), \quad 2\nu + 1 = \kappa\, x_0^2,$$

where $D_\nu(x)$ is one of the Weber parabolic cylinder functions. This function is related to the Hermite polynomial. We have for integer values of n

$$H_n(z) = 2^{n/2} e^{z^2/2} D_n(z\sqrt{2}\,), \tag{2.8}$$

where the Hermite polynomial is given by

$$H_n(x) = n! \sum_{k=0}^{\lfloor n/2 \rfloor} \frac{(-1)^k}{k!(n-2k)!}(2x)^{n-2k}. \tag{2.9}$$

Observe that when $n << \gamma$ the quantity ν behaves like $\nu = n + \mathcal{O}(\gamma^{-1})$. So it appears that the "first approximation" (2.7) does not exactly yield a Hermite polynomial.

In this section we construct a first approximation to (2.5) based on a LG transformation, which exactly reduces to the equation for the Hermite polynomial. Then, by using the theorems of OLVER (1975-A), the asymptotic estimation of the Gegenbauer polynomials can be described in more detail. This gives an asymptotic interpretation of the well-known limit relation

$$\lim_{\gamma\to\infty} \gamma^{-n/2} C_n^\gamma(x/\sqrt{\gamma}\,) = \frac{1}{n!} H_n(x), \tag{2.10}$$

which can be verified by considering (2.2) and (2.9), and the limit of each coefficient of the Gegenbauer polynomial.

The points $\pm x_0$ are two turning points of (2.5). By hypothesis x_0 is bounded away from the singularities ± 1. In this subsection the conditions on x, x_0 read

$$x \in (-1,1), \quad x_0 \in [0,b], \quad \text{with} \quad b \text{ fixed in } (0,1).$$

The condition $x_0 \le b$ implies

$$\gamma \ge \frac{\frac{1}{2} + n\sqrt{1-b^2}}{1-\sqrt{1-b^2}}.$$

Usually, n is considered as the large parameter; from the above condition and $\kappa \to \infty$ it follows that $\gamma \to \infty$, and that n may be large, but not larger than $\mathcal{O}(\gamma)$.

We now transform (2.5) into an equation with as dominant part an equation of the form (2.7). Considering the basic steps outlined in (1.1) - (1.6), we perceive that the appropriate LG transformation reads

$$(\eta^2-\rho^2)\left(\frac{d\eta}{dx}\right)^2=\frac{x^2-x_0^2}{(1-x^2)^2},\quad W(x)=\sqrt{\dot{x}}\,w(\eta),\tag{2.11}$$

where the dot indicates differentiation with respect to η, and the non-negative number ρ is defined by

$$\int_{-\rho}^{\rho}\sqrt{\rho^2-\eta^2}\,d\eta=\int_{-x_0}^{x_0}\frac{\sqrt{x_0^2-x^2}}{1-x^2}\,dx.$$

Evaluation of both integrals yields

$$1/2\pi\rho^2=\pi\left(1-\sqrt{1-x_0^2}\right).$$

It is convenient to introduce parameters ν, μ by writing

$$x_0=\sin\mu,\quad \nu=\cos\mu,\quad \mu\in[0,1/2\pi].\tag{2.12}$$

Then we have, using (2.6),

$$\rho=\sqrt{\frac{2n+1}{\kappa}}=2\sin(1/2\mu).\tag{2.13}$$

The relation between η and x is one-to-one, with the points $-x_0$ and x_0 corresponding respectively to $-\rho$ and ρ. On taking the positive square root in the first of (2.11) and integrating, we obtain

$$\int_{-\infty}^{\eta}\sqrt{\bar{\eta}^2-\rho^2}\,d\bar{\eta}=\int_{-1}^{x}\frac{\sqrt{\bar{x}^2-x_0^2}}{1-\bar{x}^2}\,d\bar{x},\quad -1<x\le -x_0$$

$$\int_{-\rho}^{\eta}\sqrt{\rho^2-\bar{\eta}^2}\,d\bar{\eta}=\int_{-x_0}^{x}\frac{\sqrt{x_0^2-\bar{x}^2}}{1-\bar{x}^2}\,d\bar{x},\quad -x_0\le x\le x_0$$

$$\int_{\rho}^{\eta}\sqrt{\bar{\eta}^2-\rho^2}\,d\bar{\eta}=\int_{x_0}^{x}\frac{\sqrt{\bar{x}^2-x_0^2}}{1-\bar{x}^2}\,d\bar{x},\quad x_0<x<1.$$

From these relations it follows that η is an increasing function of x with $\eta=-\infty$ and $+\infty$ corresponding to $x=-1$ and $+1$, respectively.

If $x_0\ne 0$, the relation between x en η, defined in (2.11) is given by

(*i*) $-\rho\le\eta\le\rho,\quad -x_0\le x\le x_0$:

$$1/2\left[\eta\sqrt{\rho^2-\eta^2}+\rho^2\arcsin\frac{\eta}{\rho}\right]=\arcsin\frac{x}{x_0}-\nu\arctan\frac{\nu x}{\sqrt{x_0^2-x^2}};\tag{2.14}$$

(*ii*) $\eta \geq \rho, \quad x_0 \leq x < 1$:

$$\tfrac{1}{2}\left[\eta\sqrt{\eta^2-\rho^2} - \rho^2 \operatorname{arccosh}\frac{\eta}{\rho}\right] = -\operatorname{arccosh}\frac{x}{x_0} + \nu \operatorname{arctanh}\frac{\sqrt{x^2-x_0^2}}{\nu x}; \tag{2.15}$$

(*iii*) $\eta \leq -\rho, \quad -1 < x \leq -x_0$:

$$-\tfrac{1}{2}\left[\eta\sqrt{\eta^2-\rho^2} + \rho^2 \operatorname{arccosh}\frac{-\eta}{\rho}\right] = -\operatorname{arccosh}\frac{-x}{x_0} - \nu \operatorname{arctanh}\frac{\sqrt{x^2-x_0^2}}{\nu x}. \tag{2.16}$$

Alternatively, when $x_0 = 0$, so that $\rho = 0$, we find

$$\tfrac{1}{2}\eta^2 = -\tfrac{1}{2}\ln(1-x^2), \quad \text{or} \quad \eta = \sqrt{-\ln(1-x^2)},$$

with $\operatorname{sign}(\eta) = \operatorname{sign}(x)$. Whether or not $x_0 = 0$, the differential equation (2.5) transforms into the form (1.6):

$$\ddot{w} = \kappa^2(\eta^2-\rho^2)w + \psi(\eta)w, \tag{2.17}$$

where

$$\psi(\eta) = -\dot{x}^2 \frac{x^2+3}{4(1-x^2)^2} + \sqrt{\dot{x}}\,\frac{d^2}{d\eta^2}\frac{1}{\sqrt{\dot{x}}}.$$

By carrying out the differentiations by means of the first of (2.11), we arrive at the formula

$$\psi(\eta) = \frac{3\eta^2+2\rho^2}{4(\eta^2-\rho^2)^2} + \frac{(\eta^2-\rho^2)(1-x^2)}{4(x^2-x_0^2)^3}\left(4x^2x_0^2 - 3x^2 - 2x_0^2 + x_0^4\right).$$

The right-hand side is to be replaced by its limiting form when $\eta = \pm\rho$. It follows from Olver's theory that ψ is a continuous function of $\eta \in \mathbb{R}$, but for this special case we can prove that η is an analytic function of $x \in (-1,1)$ and $x_0 \in [0,1)$, and that ψ is an analytic function of η. For instance, at $\eta = 0$ and as $x_0 \to 0$, we have $\psi(0) = -\tfrac{1}{32}\, x_0^2 + \mathcal{O}(x_0^4)$.

We need to know the asymptotic behavior of $x(\eta)$ as $\eta \to \pm\infty$. From (2.15) we derive that

$$\tfrac{1}{2}\eta^2 - \tfrac{1}{4}\rho^2 - \tfrac{1}{2}\rho^2\ln\left(\frac{2\eta}{\rho}\right) + \mathcal{O}(\eta^{-2}) = \tfrac{1}{2}\nu\ln(2\cot^2\mu)$$
$$- \operatorname{arccosh}\left(\frac{1}{x_0}\right) - \tfrac{1}{2}\nu\ln(1-x) + \frac{1+x_0^2}{4\nu}(1-x) + \mathcal{O}\left[(1-x)^2\right]$$

as $\eta \to +\infty$, with a similar relation with $\eta \to -\infty$. It follows that

$$1-|x| = c(x_0)|\eta|^{\rho^2/\nu}e^{-\eta^2/\nu}[1+o(1)], \quad \text{as} \quad \eta \to \pm\infty, \tag{2.18}$$

where $c(x_0)$ does not depend on η, and is given by

$$c(x_0) = \exp\frac{\rho^2/2 + \nu\ln(2\cot^2\mu) - 2\,\mathrm{arccosh}(1/x_0) + \rho^2\ln(2/\rho)}{\nu}. \tag{2.19}$$

Further,

$$\psi(\eta) = \mathcal{O}\Big(\frac{1}{1+\eta^2}\Big), \quad \eta \in \mathbb{R},$$

uniformly with respect to $x_0 \in [0, b]$.

2.2. Identification of solutions

As remarked above, the solutions of

$$\ddot{w}_0 = \kappa^2(\eta^2 - \rho^2)w_0 \tag{2.20}$$

are parabolic cylinder functions. By considering the behavior of the Gegenbauer polynomial and that of the two solutions of the differential equation for parabolic cylinder functions, it follows that only the function $D_\nu(z)$ should be used as approximant. As in (2.7) we have, using (2.13)

$$w_0(\eta) = D_\nu(\eta\sqrt{2\kappa}\,), \quad 2\nu + 1 = \kappa\rho^2 = 2n + 1.$$

It follows that the parabolic cylinder function reduces to a Hermite polynomial. Using (2.8), we obtain as a solution of (2.20)

$$w_0(\eta) = 2^{-n/2}e^{-\kappa\eta^2/2}H_n(\eta\sqrt{\kappa}\,).$$

We obtain for the complete equation (2.17)

$$w(\eta) = 2^{-n/2}e^{-\kappa\eta^2/2}H_n(\eta\sqrt{\kappa}\,) + \varepsilon(\eta).$$

Since our function ψ of (2.17) satisfies the conditions of Theorem I of Olver (1975-A, §6), we can infer that $|\varepsilon(\eta)|$ is small compared with $|w_0(\eta)|$, except in the neighborhoods of the zeros of this function, where $|\varepsilon(\eta)|$ is small compared with the envelope of $|w_0(\eta)|$. This conclusion holds uniformly with respect to $\eta \in \mathbb{R}$, $x_0 \in [0, b]$. It is also possible to construct bounds for $|\varepsilon(\eta)|$. In terms of the Gegenbauer polynomial we obtain

$$C_n^\gamma(x) = A_n^\gamma(1 - x^2)^{-(2\gamma+1)/4}\sqrt{\dot{x}}\,e^{-\kappa\eta^2/2}\big[H_n(\eta\sqrt{\kappa}\,) + \bar{\varepsilon}\big], \tag{2.21}$$

where A_n^γ does not depend on x. Again, $|\bar{\varepsilon}|$ is small (in the sense described above) when $\kappa \to \infty$, uniformly with respect to $x \in (-1, 1)$. By using

$$C_n^\gamma(1) = \frac{(2\gamma)_n}{n!}, \quad \lim_{z\to\infty} z^{-n}H_n(z) = 2^n \tag{2.22}$$

and the result in (2.18), we obtain

$$A_n^\gamma = \sqrt{\nu}\,[2c(x_0)]^{\gamma/2-1/4}2^{-n}\kappa^{-n/2}(2\gamma)_n/n!.$$

Evaluating this expression by using (2.19), we can write

$$A_n^\gamma = \kappa^{-1/4}2^{-n}e^{n/2+1/4}(\gamma-{}^1\!/\!_2)^\gamma(\gamma+n/2-1/4)^{-\gamma-n/2+1/4}(2\gamma)_n/n!,$$

with $\kappa = n+\gamma$.

It is not difficult to verify that (2.21) contains (2.10) as a special case; from Olver's theory it follows that $\bar{\varepsilon}$ tends to zero in the limit considered.

2.3. *Other asymptotic forms for large γ and/or n*

When in (2.5) x runs in a domain that contains both the points x_0 and 1, and x_0 may coalesce with the point 1, the equation resembles a particular form of the Bessel equation. For this case we can use an LG transformation of the form

$$\frac{\nu^2-\zeta}{4\zeta^2}\left(\frac{d\zeta}{dx}\right)^2 = \frac{x^2-x_0^2}{(1-x^2)^2},\quad W(x)=\sqrt{-\dot{x}}\,w(\zeta), \tag{2.23}$$

where ν is defined in (2.12), that is

$$\nu=\sqrt{1-x_0^2}=\frac{\gamma-1/2}{\kappa},\qquad \kappa=n+\gamma,$$

and the dot indicates differentiation with respect to ζ. The point $x = x_0$ is transformed to $\zeta=\nu^2$, and the point $x=1$ to $\zeta=0$. The derivative $\dot{x}$ is negative in the domain of interest. Again, κ is large, and $\gamma \geq {}^1\!/\!_2$.

The relation between x and ζ can be obtained by integrating the first of (2.23).

With these transformations the Gegenbauer equation (2.5) is transformed to

$$\ddot{w} = \left[\kappa^2\left(\frac{\nu^2}{4\zeta^2}-\frac{1}{4\zeta}\right)-\frac{1}{4\zeta^2}\right]w+\frac{\phi(\zeta)}{\zeta}\,w, \tag{2.24}$$

where

$$\frac{\phi(\zeta)}{\zeta} = -\dot{x}^2\frac{x^2+3}{4(1-x^2)^2}+\sqrt{-\dot{x}}\,\frac{d^2}{d\zeta^2}\frac{1}{\sqrt{-\dot{x}}}+\frac{1}{4\zeta^2}.$$

The function ϕ is analytic in a domain containing both the turning point $\zeta=\nu^2$ and the singular point $\zeta=0$.It can be verified that

$$\phi(\zeta)=\mathcal{O}(\zeta^{-2}),\quad \text{as}\quad \zeta\to-\infty.$$

The solution of (2.24) can be put in the form $w(\zeta) = \sqrt{\zeta}\, J_{\gamma-1/2}(\kappa\sqrt{\zeta}\,) + \varepsilon(\zeta)$. For the Gegenbauer polynomials we obtain

$$C_n^\gamma(x) = B_n^\gamma\,(1-x^2)^{-(2\gamma+1)/4}\sqrt{-\dot{x}}\,[\sqrt{\zeta}\,J_{\gamma-1/2}(\kappa\sqrt{\zeta}\,) + \varepsilon(\zeta)], \tag{2.25}$$

where B_n^γ does not depend on x. This number follows from the behavior of the $J-$Bessel function at $\zeta = 0$ and $C_n^\gamma(1) = (2\gamma)_n/n!$. We also need the number c_1 in $x = 1 + c_1\zeta + \mathcal{O}(\zeta^2)$ as $\zeta \to 0$, that is, $c_1 = -1/(1+x_0^2)$. So we obtain

$$B_n^\gamma = \frac{(2\gamma)_n}{n!}2^{\gamma+1/2}(-c_1)^{\gamma/2-1/4}\kappa^{1/2-\gamma}\Gamma(\gamma+{}^1\!/\!_2).$$

From Boyd & Dunster (1986) it follows that (2.25) provides an asymptotic estimate of the Gegenbauer polynomial with $\kappa \to \infty$, $\gamma = \mathcal{O}(n)$. The estimate holds uniformly with respect to $x \in [-x_0 + \delta, \infty)$ (and also in larger domains of the complex plane). Upper bounds for $|\varepsilon(\zeta)|$ follow from the cited reference, also for an estimate involving more terms in the expansion. Since the Gegenbauer polynomial is symmetric with respect to x, a similar expansion hold in the $x-$domain containing the points $-x_0$ and -1.

A third expansion is needed to describe the behavior for $x \geq 1$ and $n = \mathcal{O}(\gamma)$. As explained in Olver((1974), for this case an expansion can be given in terms of elementary functions.

3. Laguerre polynomials

The Laguerre polynomials have the explicit representation

$$L_n^\alpha(x) = \sum_{k=0}^{n}\binom{n+\alpha}{n-k}\frac{(-x)^k}{k!}.$$

We assume that the parameters are real with $n \geq 0$, $\alpha \geq 0$, $x \in \mathbb{R}$. For large values of

$$\kappa = n + {}^1\!/\!_2(\alpha+1) \tag{3.1}$$

the asymptotic behavior can be described in terms of the Hermite polynomial and the $J-$Bessel function, just as in the previous section. In this section we summarize the recent results obtained in Temme (1986). The results also follow from earlier results in the literature for Whittaker functions. The relation between Laguerre polynomials and Whittaker functions is

$$L_n^\alpha(z) = \frac{(-1)^n}{n!}z^{-(\alpha+1)/2}e^{z/2}W_{\kappa,\mu}(z) = \frac{\Gamma(\alpha+n+1)}{n!\Gamma(\alpha+1)}z^{-(\alpha+1)/2}e^{z/2}M_{\kappa,\mu}(z), \tag{3.2}$$

with κ as in (3.1) and $\mu = \alpha/2$. The functions $M_{\kappa,\mu}(z)$, $W_{\kappa,\mu}(z)$ are solutions of Whittaker's equation

$$\frac{d^2y}{dz^2} = \left[\frac{1}{4} - \frac{\kappa}{z} + \frac{\mu^2 - 1/4}{z^2}\right] y. \tag{3.3}$$

A first transformation $z = 4\kappa x$ yields the differential equation

$$W'' = \kappa^2 \left[\frac{4(x-x_1)(x-x_2)}{x^2} - \frac{1}{4x^2}\right] W, \tag{3.4}$$

where

$$x_1 = \frac{1-\sqrt{1-\tau^2}}{2}, \quad x_2 = \frac{1+\sqrt{1-\tau^2}}{2}, \quad \tau = \frac{\alpha}{2\kappa}, \tag{3.5}$$

with solution

$$W(x) = x^{(\alpha+1)/2} e^{-2\kappa x} L_n^\alpha(4\kappa x).$$

The x-zeros of the polynomial $L_n^\alpha(4\kappa x)$ occur within the interval $[x_1, x_2]$. When $\alpha >> n$ the parameter τ tends to unity, and the turning points x_1, x_2 are close together. This case is described by a parabolic cylinder function, which reduces for this particular case of the Laguerre polynomials to a Hermite polynomial. When $\tau \to 0$ the zeros of $L_n^\alpha(4\kappa x)$ are spread over the x-interval $(0, 1)$, and a J-Bessel function is used in this case.

3.1. Estimation in terms of a Hermite polynomial

The LG transform is defined by

$$(\eta^2 - \rho^2)\left(\frac{d\eta}{dx}\right)^2 = \frac{4(x-x_1)(x-x_2)}{x^2}, \quad W(x) = \sqrt{\dot{x}}\, w(\eta), \tag{3.6}$$

where the dot indicates differentiation with respect to η, and ρ is given by

$$\rho = \sqrt{2(1-\tau)}.$$

The relation between η and x is one-to-one, with

$$\eta(0) = -\infty, \quad \eta(x_1) = -\rho, \quad \eta(x_2) = \rho, \quad \eta(+\infty) = \infty.$$

The differential equation (3.4) transforms into

$$\ddot{w} = \kappa^2(\eta^2 - \rho^2)w + \psi(\eta)w, \tag{3.7}$$

where

$$\psi(\eta) = \frac{3\eta^2 + 2\rho^2}{4(\eta^2 - \rho^2)^2} - \frac{x(\eta^2 - \rho^2)}{64[(x-x_1)(x-x_2)]^3}(4x^3 + (1 - 4\tau^2)x + \tau^2).$$

We have

$$\psi(\eta) = \mathcal{O}(\frac{1}{\eta^2+1}), \quad \eta \in \mathbb{R},$$

uniformly with respect to $\tau \in [\tau_0, 1]$, where τ_0 is a fixed number in $(0,1)$.

By identifying solutions, and considering the asymptotic behavior of the Laguerre polynomials and the Hermite polynomials, we arrive at the estimate

$$\begin{aligned} L_n^\alpha(4\kappa x) =& \frac{(-1)^n}{n!} 2^{-\alpha-n/2-3/4}\kappa^{-\alpha/2-1/4}x^{-\alpha/2}(\frac{n+\alpha+1/2}{e})^{n+\alpha+1/2} \\ & e^{2\kappa x-\kappa\eta^2/2}\sqrt{\dot{x}}\left[H_n(\eta\sqrt{\kappa})+\bar{\varepsilon}\right] \end{aligned} \quad . \tag{3.8}$$

This representation has an asymptotic interpretation as $\kappa \to \infty$; in that case $|\bar{\varepsilon}|$ is small (in the same sense as in (2.21)), uniformly with respect to $x \in (0,\infty)$ and $t \in [\tau_0, 1]$. The latter gives for α the condition

$$\alpha \geq \frac{\tau_0}{1-\tau_0}(2n+1)$$

with τ_0 any fixed number in $(0,1)$.

3.2. Other asymptotic forms for large α and/or n.

In TEMME (1986) details are given for two other asymptotic forms of the Laguerre polynomials, one in terms of the J–Bessel function, and one in terms of an Airy function. The Bessel function is used to describe the asymptotics of $L_n^\alpha(4\kappa x)$ for $\kappa \to \infty$, uniformly with respect to $\alpha \in [0, n_1 n]$ and $x \in (-\infty, x_0]$, $x_0 = \xi_0 x_1 + (1-\xi_0)x_2$, where $x_{1,2}$ are given in (3.5), n_1 is a fixed positive number, and ξ_0 is a fixed number in $(0,1)$. The Airy function is used to describe the asymptotics uniformly with respect to $\alpha \in [0, n_2 n]$ and $x \in [x_3, \infty)$, $x_3 = \xi_3 x_1 + (1-\xi_3)x_2$, with n_2 a fixed positive number, and ξ_3 a fixed number in $(0,1)$.

Both forms follow from DUNSTER (1989), where the results are obtained in the setting of Whittaker functions. A fourth form can be given in terms of elementary functions; it is uniformly valid with respect to $x \leq 0$ and $\alpha \in [n_3 n, \infty)$, where n_3 is a fixed positive number.

In this way we have four asymptotic estimates in overlapping domains of the three real parameter domains $x \in \mathbb{R}$, $n \geq 0$, $\alpha \geq 0$. It is not difficult to verify that α may assume negative values, as long as $\kappa \to \infty$.

4. Jacobi polynomials

The Jacobi polynomials $P_n^{(\alpha,\beta)}(x)$ are solutions of the differential equation

$$(1-x^2)y'' + [(\beta-\alpha) - (\alpha+\beta+2)x]y' + n(n+\alpha+\beta+1)y = 0. \tag{4.1}$$

In fact this is a differential equation of hypergeometric type, the relation with hypergeometric functions being

$$P_n^{(\alpha,\beta)}(x) = \binom{n+\alpha}{n} F(-n, \alpha+\beta+n+1; \alpha+1; \frac{1-x}{2}), \tag{4.2}$$

which gives the explicit representation

$$P_n^{(\alpha,\beta)}(x) = \frac{\Gamma(\alpha+n+1)}{n!\Gamma(\alpha+\beta+n+1)} \sum_{k=0}^{n} \binom{n}{k} \frac{\Gamma(\alpha+\beta+n+k+1)}{\Gamma(\alpha+k+1)} \left(\frac{x-1}{2}\right)^k. \tag{4.3}$$

The following normalization and symmetry relations hold

$$P_n^{(\alpha,\beta)}(1) = \binom{n+\alpha}{n}, \quad P_n^{(\alpha,\beta)}(-x) = (-1)^n P_n^{(\beta,\alpha)}(x). \tag{4.4}$$

The Gegenbauer polynomial is a special case of the Jacobi polynomial:

$$C_n^{\gamma}(x) = \frac{(2\gamma)_n}{(\gamma+1/2)_n} P_n^{(\gamma-1/2,\gamma-1/2)}(x). \tag{4.5}$$

The function

$$W(x) = (1-x)^{(\alpha+1)/2}(1+x)^{(\beta+1)/2} P_n^{(\alpha,\beta)}(x) \tag{4.6}$$

satisfies the equation

$$W'' = \frac{Q(x,\alpha,\beta,n)}{(1-x^2)^2} W, \tag{4.7}$$

with

$$Q(x,\alpha,\beta,n) = \kappa^2(x^2-1) + {}^1\!/\!{}_2 x(\alpha^2-\beta^2) + {}^1\!/\!{}_2(\alpha^2+\beta^2) - {}^1\!/\!{}_4(3+x^2),$$

$$\kappa = n + (\alpha+\beta+1)/2. \tag{4.8}$$

We factorize the function Q in the following way

$$Q(x,\alpha,\beta,n) = \kappa^2(x-x_1)(x-x_2) - {}^1\!/\!{}_4(3+x^2), \tag{4.9}$$

where

$$x_{1,2} = \frac{\beta^2-\alpha^2 \pm \sqrt{[4\kappa^2-(\alpha+\beta)^2][4\kappa^2-(\alpha-\beta)^2]}}{4\kappa^2}, \tag{4.10}$$

where x_2 has the + sign. It is not difficult to verify that this factorization yields the same equations for the Gegenbauer polynomials in the case that $\alpha = \beta = \gamma - 1/2$.

We consider the case where the parameters α, β are positive, and the independent variable x is real. The parameter κ of (4.8) is large.

The numbers $x_{1,2}$ satisfy $-1 < x_1 < x_2 < 1$. Several limiting cases of these numbers are possible. To describe some of the phenomena, we introduce

$$\sigma = \frac{\alpha + \beta}{2\kappa}, \qquad \delta = \frac{\alpha - \beta}{2\kappa}.$$

Then we can write

$$x_{1,2} = -\sigma\delta \pm \sqrt{(1-\sigma^2)(1-\delta^2)},$$

and we have the following cases:

- when $n >> \alpha + \beta$, that is, $\sigma \to 0$, then $x_1 \to -1$, $x_2 \to +1$;
- when $\alpha + \beta >> n$, that is, $\sigma \to 1$, then x_1 and x_2 coalesce at $-\delta$;
- when $\alpha >> \beta$, that is, $\delta \to 1$, then x_1 and x_2 coalesce at $-\sigma$;
- when $\beta >> \alpha$, that is, $\delta \to -1$, then x_1 and x_2 coalesce at σ.

Note that the collisions may happen at +1 when $\beta >> \alpha >> n$, and at -1 when $\alpha >> \beta >> n$.

In the following subsections we give the setup of two LG transformations for obtaining asymptotic estimates of the Jacobi polynomials in terms of the Gegenbauer polynomials and of the Laguerre polynomials. Both forms seem to be new, but the proofs of the asymptotic nature of the approximations will be elaborated elsewhere. We demonstrate the usefulness of the estimates by comparing the zeros of the Jacobi polynomials with transformed zeros of the LG approximants for these cases: the Gegenbauer polynomials and the Laguerre polynomials.

4.1. Estimation in terms of a Gegenbauer polynomial

We define a Liouville-Green transformation for (4.7) by writing

$$\frac{\eta^2 - \rho^2}{(1-\eta^2)^2}\left(\frac{d\eta}{dx}\right)^2 = \frac{(x-x_2)(x-x_1)}{(1-x^2)^2}, \quad W(x) = \sqrt{\dot{x}}\, w(\eta), \tag{4.11}$$

where the dot indicates differentiation with respect to η, and ρ is the non-negative number defined by

$$\int_{-\rho}^{\rho} \frac{\sqrt{\rho^2 - \eta^2}}{1-\eta^2}\, d\eta = \int_{x_1}^{x_2} \frac{\sqrt{(x_2 - x)(x - x_1)}}{1 - x^2}\, dx.$$

Evaluation of both integrals yields

$$\pi(1-\sqrt{1-\rho^2}\,) = \pi\left[1 - 1/2\sqrt{(1-x_1)(1-x_2)} - 1/2\sqrt{(1+x_1)(1+x_2)}\,\right].$$

The square roots in the right-hand side can easily be evaluated; using (4.10) we obtain

$$\sqrt{(1-x_1)(1-x_2)} = \frac{\alpha}{\kappa}, \quad \sqrt{(1+x_1)(1+x_2)} = \frac{\beta}{\kappa}. \tag{4.12}$$

It follows that

$$\sqrt{1-\rho^2} = \frac{\alpha+\beta}{2\kappa}. \tag{4.13}$$

The relation between η and x is one-to-one, with the points x_1 and x_2 corresponding respectively to $-\rho$ and ρ. For the integration of the first of (4.11) we introduce the following notations

$$x_0 = 1/2(x_2 - x_1), \quad \bar{x} = 1/2(x_1+x_2), \quad \rho = \sin\mu, \quad \nu = \cos\mu,$$

$$T = \begin{cases} \arccos[(x-\bar{x})/x_0], & \text{if } x_1 < x < x_2; \\ \operatorname{arccosh}[(x-\bar{x})/x_0], & \text{if } x_2 < x < 1; \\ \operatorname{arccosh}[(\bar{x}-x)/x_0], & \text{if } -1 < x < x_1. \end{cases} \tag{4.14}$$

We also use

$$R = \begin{cases} \sqrt{(x_2-x)/(x-x_1)}, & \text{if } x_1 < x < x_2; \\ \sqrt{(x-x_2)/(x-x_1)}, & \text{if } x_2 < x < 1; \\ \sqrt{(x_1-x)/(x_2-x)}, & \text{if } -1 < x < x_1. \end{cases}$$

Integrating the first of (4.11) we obtain when $x_1 \neq x_2$, that is, $x_0 \neq 0$

(*i*) $-\rho \le \eta \le \rho, \quad x_1 \le x \le x_2$:

$$\begin{aligned} &\arcsin\frac{\eta}{\rho} - \nu \arctan\frac{\nu\eta}{\sqrt{\rho^2-\eta^2}} - 1/2\pi(1+\nu) = \\ &-T - \frac{\beta}{\kappa}\arctan\frac{\kappa(1+x_2)}{\beta R} - \frac{\alpha}{\kappa}\arctan\frac{\kappa(1-x_2)}{\alpha R}; \end{aligned} \tag{4.15}$$

(*ii*) $\rho \le \eta < 1, \quad x_2 \le x < 1$:

$$\begin{aligned} &-\operatorname{arccosh}\frac{\eta}{\rho} + \nu\,\operatorname{arctanh}\frac{\sqrt{\eta^2-\rho^2}}{\nu\,\eta} = \\ &-T + \frac{1}{2\kappa}\left[\beta\ln\frac{\kappa(1+x_2)+\beta R}{\kappa(1+x_2)-\beta R} + \alpha\ln\frac{\kappa(1-x_2)+\alpha R}{\kappa(1-x_2)-\alpha R}\right]; \end{aligned} \tag{4.16}$$

(*iii*) $-1 < \eta \leq -\rho, \quad -1 < x \leq x_1$:

$$
\begin{aligned}
&-\operatorname{arccosh}\frac{\eta}{\rho} + \nu \operatorname{arctanh}\frac{\nu\eta}{\sqrt{\eta^2-\rho^2}} = \\
&-T - \frac{1}{2\kappa}\Big[\beta \ln\frac{\kappa(1+x_1)+\beta R}{\kappa(1+x_1)-\beta R} - \alpha \ln\frac{\kappa(1-x_1)+\alpha R}{\kappa(1-x_1)-\alpha R}\Big].
\end{aligned} \tag{4.17}
$$

Alternatively, when $x_1 = x_2 = \bar{x}$, that is, $x_0 = 0$, so that $\rho = 0$, we find for $|\eta| < 1$, $|x| < 1$

$$
\ln(1-\eta^2) = (1-\bar{x})\ln\frac{1-x}{1-\bar{x}} + (1+\bar{x})\ln\frac{1+x}{1+\bar{x}}
$$

with $\operatorname{sign}(\eta) = \operatorname{sign}(x-\bar{x})$. Whether or not $x_0 = 0$, the differential equation (4.7) transforms into

$$
\ddot{w} = \Big[\kappa^2\frac{(\eta^2-\rho^2)}{1-\eta^2)^2} - \frac{\eta^2+3}{4(1-\eta^2)^2}\Big]w + \psi(\eta)w, \tag{4.18}
$$

where

$$
\psi(\eta) = -\dot{x}^2\frac{x^2+3}{4(1-x^2)^2} + \sqrt{\dot{x}}\,\frac{d^2}{d\eta^2}\frac{1}{\sqrt{\dot{x}}} + \frac{\eta^2+3}{4(1-\eta^2)^2}. \tag{4.19}
$$

For identifying the solutions we need to know the asymptotic behavior of $x(\eta)$ as $\eta \to 1$. It is straightforward to verify from (4.16) that

$$
1-x = (1-\eta)^{(\alpha+\beta)/(2\alpha)}C[1+o(1)], \quad \text{as} \quad \eta \uparrow 1,
$$

where C is not depending on η , and is given by (t is given in (4.14) with $x = 1$)

$$
\begin{aligned}
C = \exp\Big[&-\frac{2\kappa T}{\alpha} + \frac{\beta}{\alpha}\ln\frac{\sqrt{(1+x_2)(1-x_1)} + \sqrt{(1+x_1)(1-x_2)}}{\sqrt{(1+x_2)(1-x_1)} - \sqrt{(1+x_1)(1-x_2)}} + \\
&\ln\Big\{\frac{2(1-x_1)(1-x_2)}{x_2-x_1}\big(\sqrt{1-x_1}+\sqrt{1-x_2}\big)\Big\}\Big].
\end{aligned} \tag{4.20}
$$

When $\psi = 0$ the solution of (4.18) is a Gegenbauer polynomial (see (2.5)); that is,

$$
w_0(\eta) = (1-\eta^2)^{(2\gamma+1)/4}C_n^\gamma(\eta),
$$

where $\gamma = (\alpha+\beta+1)/2$. We write for the complete solution of (4.18)

$$
w(\eta) = (1-\eta^2)^{(2\gamma+1)/4}C_n^\gamma(\eta) + \varepsilon(\eta),
$$

with the expectation that $\varepsilon(\eta)$ is a small contribution as $\kappa \to \infty$. Using (2.4), (2.5) we obtain the asymptotic estimate

$$
P_n^{(\alpha,\beta)}(x) = A_n^{\alpha,\beta}\sqrt{\dot{x}}\,(1-x)^{-(\alpha+1)/2}(1+x)^{-(1+\beta)/2}(1-\eta^2)^{(2\gamma+1)/4}[C_n^\gamma(\eta)+\bar{\varepsilon}(\eta)], \tag{4.21}
$$

where $A_n^{\alpha,\beta}$ can be derived by computing the behavior of the polynomials at $x = 1$, $\eta = 1$ from (2.22) and (4.4). The result is

$$A_n^{\alpha,\beta} = 2^{(\beta-\alpha)/4} C^{\alpha/2} \sqrt{\frac{2\alpha}{\alpha+\beta} \frac{\Gamma(2\gamma)\Gamma(n+\alpha+1)}{\Gamma(\alpha+1)\Gamma(n+2\gamma)}}.$$

When $\alpha = \beta$, the functions $\psi, \varepsilon, \bar{\varepsilon}$ vanish identically; the mappings described in (4.15), (4.16), and (4.17) reduce to the identity mapping $x = \eta$, and the relation between the polynomials is then as in (4.5).

4.2. Estimation in terms of a Laguerre polynomial

A known limit is

$$\lim_{\alpha\to\infty} P_n^{(\alpha,\beta)}(2x/\alpha - 1) = (-1)^n L_n^\beta(x).$$

We give a new asymptotic estimate that includes this relation. We concentrate on an estimate that can describe the behavior of $P_n^{(\alpha,\beta)}(x)$ for x–values in a domain that contains the singular point $x = -1$ (a double pole) of (4.7) and both turning points $x_{1,2}$ defined in (4.10) and located in $(-1,1)$. Recall that both turning points coalesce with the pole when $\alpha >> \beta >> n$. Especially this situation can be described by the Laguerre polynomial.

We define an LG transformation by writing $W(x) = \sqrt{x}\, w(\eta)$ and

$$\frac{4(\eta-\rho_1)(\eta-\rho_2)}{\eta^2}\left(\frac{d\eta}{dx}\right)^2 = \frac{(x-x_2)(x-x_1)}{(1-x^2)^2}, \tag{4.22}$$

where the positive numbers $\rho_{1,2}$ are to be determined. A first condition reads

$$\int_{\rho_1}^{\rho_2} \frac{\sqrt{(\rho_2-\eta)(\eta-\rho_1)}}{\eta}\, d\eta = \int_{x_1}^{x_2} \frac{\sqrt{(x_2-x)(x-x_1)}}{1-x^2}\, dx.$$

Evaluation of both integrals gives

$$\rho_1 + \rho_2 - 2\sqrt{\rho_1\rho_2} = \frac{2n+1}{2\kappa}. \tag{4.23}$$

It is required that the relation between η and x is one-to-one for $x \in (-\infty, 1)$, with the points x_1 and x_2 corresponding respectively to ρ_1 and ρ_2. We also request that the mapping is regular at the corresponding points $x = -1$, $\eta = 0$. A necessary condition follows from $\eta/(1+x) = \mathcal{O}(1)$ as $x \to -1$. Assuming that

$$\frac{d\eta}{dx} \sim \frac{\eta}{1+x} \to L, \quad \text{as} \quad x \to -1,$$

we obtain from (4.22) the second condition

$$4\sqrt{\rho_1\rho_2} = \frac{\beta}{\kappa}, \tag{4.24}$$

where we used the second of (4.12). From those two conditions $\rho_{1,2}$ can be determined. We have

$$2\sqrt{\rho_{1,2}} = \sqrt{(2n+1+2\beta)/2\kappa} \pm \sqrt{(2n+1)/2\kappa},$$

where ρ_2 has the + sign.

The integrated form of (4.22) for the location of the parameters $\rho_1 \le \eta \le \rho_2$, $x_1 \le x \le x_2$ reads

$$\begin{aligned} &2S - 2\sqrt{\rho_1\rho_2}\arcsin\frac{(\rho_1+\rho_2)\eta - 2\rho_1\rho_2}{\eta(\rho_2-\rho_1)} - (\rho_1+\rho_2)\arcsin\frac{\rho_1+\rho_2-2\eta}{\rho_2-\rho_1} = \\ &-T - \frac{\beta}{\kappa}\arctan\frac{\kappa(1+x_2)}{\beta R} - \frac{\alpha}{\kappa}\arctan\frac{\kappa(1-x_2)}{\alpha R} - \left(\frac{2n+1}{4\kappa} - 1\right)\pi, \end{aligned} \tag{4.25}$$

where $S = \sqrt{(\rho_2-\eta)(\eta-\rho_1)}$ and the parameters of the right-hand side are as in (4.15).

The LG transformation (4.22) transforms (4.7) into the differential equation

$$\ddot{w} = \left[\kappa^2\frac{4(\eta-\rho_1)(\eta-\rho_2)}{\eta^2} - \frac{1}{4\eta^2}\right]w + \psi(\eta)w, \tag{4.26}$$

where

$$\psi(\eta) = -\dot{x}^2\frac{x^2+3}{4(1-x^2)^2} + \sqrt{\dot{x}}\,\frac{d^2}{d\eta^2}\frac{1}{\sqrt{\dot{x}}} + \frac{1}{4\eta^2}.$$

When $\psi = 0$ we denote the solution by w_0. This function can be written in terms of the Whittaker functions, just as the solution of the differential equation in (3.4), but now with parameters $\rho_{1,2}$ satisfying (4.23) and (4.24). To obtain the proper form of the solutions, we transform the above differential equation (with $\psi = 0$) into (3.3), with $4\kappa\eta$ replaced with z. Then we obtain

$$\frac{d^2y}{dz^2} = \left[\frac{1}{4} - \frac{k}{z} + \frac{m^2 - 1/4}{z^2}\right]y,$$

with solutions the Whittaker functions $W_{k,m}(z), M_{k,m}(z)$, with

$$k = \kappa(\rho_1+\rho_2) = n + \frac{\beta+1}{2}, \quad m = 2\kappa\sqrt{\rho_1\rho_2} = {}^1\!/_2\,\beta.$$

It follows that when $\psi = 0$ a solution of (4.26) is

$$w_0(\eta) = \eta^{\beta/2}e^{-2\kappa\eta}L_n^{\beta}(4\kappa\eta). \tag{4.27}$$

We write for the complete solution

$$w(\eta) = \eta^{\beta/2} e^{-2\kappa\eta} L_n^\beta(4\kappa\eta) + \varepsilon(\eta),$$

with the expectation that $\varepsilon(\eta)$ is a small contribution as $\kappa \to \infty$. We obtain an asymptotic estimate of the Jacobi polynomial in terms of the Laguerre polynomial

$$P_n^{(\alpha,\beta)}(x) = B_n^{\alpha,\beta} \sqrt{\dot{x}}\,(1-x)^{-(\alpha+1)/2}(1+x)^{-(1+\beta)/2}\eta^{\beta/2} e^{-2\kappa\eta}[L_n^\beta(4\kappa\eta) + \bar{\varepsilon}(\eta)], \quad (4.28)$$

where $B_n^{\alpha,\beta}$ can be derived by computing the behavior of the polynomials at $x = -\infty$, $\eta = -\infty$.

We expect that the above approximation holds for $\kappa \to \infty$, uniformly with respect to $x \in [-\infty, 1-\delta]$, and $x_1, x_2 \in [-1, 1-\delta]$, with δ a small fixed positive number.

5. Numerical verification by computation of zeros

For given values of n, α, β we have verified the asymptotic estimate (4.28), where the relation between η and x is given in (4.25). We computed the zeros z_m, $m = 1 \ldots n$ of the Laguerre polynomial $L_n^\beta(z)$, giving $\eta_m = z_m/(4\kappa)$, $m = 1 \ldots n$. By inverting (4.25) we obtained $x_m = x(\eta_m)$, and we compared these values with "exact" zeros $x_m^{(e)}$ of the Jacobi polynomial $P_n^{(\alpha,\beta)}(x)$. In Table 5.1 we show for $n = 10$, $\beta = 5.5$ and several values of α the corresponding number of correct decimal digits in the approximation of the zeros. That is, we show

$$\log_{10} |x_m^{(e)} - x_m|, \quad m = 1, 2, \ldots 10.$$

Table 5.1. Correct decimal digits in the approximations of zeros of $P_{10}^{(\alpha,5.5)}(x)$

m \ α	1.0	2.5	5.0	10	25	50	75	100
1	4.3	4.5	4.7	5.0	5.6	6.3	6.7	7.0
2	4.1	4.2	4.4	4.7	5.3	6.0	6.4	6.7
3	3.9	4.0	4.2	4.5	5.2	5.8	6.2	6.5
4	3.8	3.9	4.1	4.4	5.0	5.6	6.1	6.4
5	3.7	3.8	4.0	4.3	4.9	5.5	5.9	6.3
6	3.6	3.7	3.9	4.2	4.8	5.4	5.8	6.1
7	3.6	3.7	3.9	4.2	4.7	5.3	5.7	6.0
8	3.6	3.7	3.9	4.1	4.7	5.3	5.6	6.0
9	3.6	3.7	3.9	4.1	4.6	5.2	5.6	5.9
10	3.7	3.8	4.0	4.2	4.6	5.1	5.5	5.8

The values of z_m and $x_m^{(e)}$ are obtained from a numerical procedure that generates the zeros by computing the eigenvalues of a tri-diagonal matrix that corresponds with the recursion formulas of the polynomials.

From the table it follows that the zeros near $x = -1$ are slightly better approximated, but that in general the approximations are uniform with respect to m. A similar table for the asymptotic estimate (3.8) is given in [18], showing the same sharp results. We also used the numerical verification for the asymptotic estimates (2.21) and (4.21), again with satisfactory results.

Acknowledgment The author wants to thank Drs. Adri Olde Daalhuis (CWI) for his comments on earlier versions of the paper.

References

[1] R. Askey (1982), The collected works of Gabor Szegö, Vol. 2, Birkhauser.

[2] W.G.C. Boyd & T.M. Dunster (1986), Uniform asymptotic solution of a class of second order linear differential equations having a turning point and a regular singularity, with an application to Legendre functions, SIAM J. Math. Anal. **17**, 422-450.

[3] T.M. Dunster (1989), Uniform asymptotic expansions for Whittaker's confluent hypergeometric functions, SIAM J. Math. Anal. **20**, 744-760.

[4] D. Elliott (1971), Uniform asymptotic expansions of the Jacobi polynomials and an associated function, Math. Comp. **25**, 309-315.

[5] A. Erdélyi, (1960), Asymptotic forms for Laguerre polynomials, J. Indian Math. Soc. **24**, 235-250.

[6] C.L. Frenzen & R. Wong (1985), A uniform asymptotic expansions of the Jacobi polynomials with error bounds, Can. J. Math. **37**, 979-1007.

[7] C.L. Frenzen & R. Wong (1988), Uniform asymptotic expansions of Laguerre polynomials, SIAM J. Math. Anal. **19**, 1232-1248.

[8] Li-chen Chen & M.E.H. Ismail (1989), On asymptotics of Jacobi polynomials, Report ICM 89-004, Department of Mathematics, University of south Florida, Tampa.

[9] H.N. Mhaskar & E.B. Saff (1984), Polynomials with Laguerre weights in L^p, 511-523 in *Rational approximation and interpolation*, P.R. Graves-Morris & E.B. Saff (eds.), LNM 1105, Springer-Verlag.

[10] D.S. Moak, E.B. Saff & R.S. Varga (1979), On the zeros of Jacobi polynomials $P_n^{(\alpha_n,\beta_n)}(x)$, Trans. Amer. Math. Soc. **249** 159-163.

[11] J.J. Nestor III (1984), Uniform asymptotic approximations of solutions of second-order linear differential equations, with a coalescing simple turning point and simple pole. Ph.D. thesis, Applied Mathematics, University of Maryland.

[12] F.W.J. Olver (1974), Asymptotics and Special Functions, Academic Press, New York.

[13] F.W.J. Olver (1975-a), Second-order linear differential equations with two turning points, Phil. Trans. Roy. Soc. London, Ser. A **278**, 175-185.

[14] F.W.J. Olver (1975-b), Unsolved problems in the asymptotic estimation of special functions, 99-142 in *Theory and application of special functions*, R. Askey (ed.), Academic Press.

[15] F.W.J. Olver (1980), Whittaker functions with both parameters large: uniform approximations in terms of parabolic cylinder functions, Proc. Royal Soc. Edinb. **84A**, 213-234.

[16] E.B. Saff & R.S. Varga (1979), The sharpness of Lorentz's theorem on incomplete polynomials, Trans. Amer. Math. Soc. **249** 163-186.

[17] G. Szegö (1958), Orthogonal polynomials, Amer. Math. Soc. Colloq. Publ., Vol. 23, New York.

[18] N.M. Temme (1986), Laguerre polynomials: asymptotics for large degree, CWI Report AM-R8610 (submitted).

Integrals with a Large Parameter and the Maximum-Modulus Principle

F. Ursell Professor, Department of Mathematics, University of Manchester, Manchester, England

A comparatively simple method is described for finding error bounds for integrals involving one or more small parameters as well as a large positive parameter N. This method makes use of the maximum-modulus theorem of complex-variable theory and has been applied in at least three previous instances, but in each instance so much supporting analytical detail was required that the simple idea underlying the argument was perhaps not readily perceived. Here the method is applied to two simple examples, the Legendre functions $P_N(\cosh z)$ and $Q_N(\cosh z)$ near the singular point $z=0$, and the contour integral $\int g(z,\alpha)\exp\{Nf(z,\alpha)\}dz$ near the critical value $\alpha=0$ at which two saddle points coincide. In each case uniformly asymptotic expansions are obtained, involving Bessel functions and Airy functions respectively. Bounds for the error terms are found indirectly. It is observed that the coefficient functions near the critical values are analytic and slowly varying, unlike the original integrals. Uniform bounds for the coefficient functions (and hence for the error terms) can therefore be found from simpler non-uniform bounds by applying the maximum-modulus theorem. The region of validity is a finite region not shrinking to 0 when $N \to \infty$, and it is shown how this region can be extended.

1. INTRODUCTION.

The higher transcendental functions of mathematical physics either satisfy second-order differential equations with analytic coefficients, or they have integral representations, or both. Starting from the differential equation or from an integral we may use well-known techniques (described e.g. by Olver 1974) to obtain asymptotic expansions, valid when some of the variables or parameters are large. The validity of an expansion is established when a suitably small bound is established for the error term (i.e. the difference between the function and the first p terms of its asymptotic expansion). At present, error bounds are derived most frequently from a differential equation, and it has been authoritatively stated (Olver 1974, ch.10, §7.1) that these are on the whole more efficient than bounds obtained from integral representations.

In the present work a comparatively simple method will be described for finding error bounds for integrals involving a large parameter. This method makes use of the maximum-modulus theorem of complex-variable theory, stated near the end of this section, and has been applied in at least three previous instances (Ursell 1972, 1980, 1984, also Martin 1974) but in each instance so much supporting analytical detail was required that the simple idea underlying the argument was perhaps not readily perceived. It may therefore be useful to illustrate the method by means of two examples for which a simple description is possible.
These examples are
(1) the Legendre functions $P_N(\cosh z)$ and $Q_N(\cosh z)$ of large degree N in a neighbourhood of the singular point $z = 0$;
(2) the contour integral

$$\int g(z,\alpha)\, \exp\{Nf(z,\alpha)\}dz, \qquad N \to \infty,$$

where $g(z,\alpha)$ and $f(z,\alpha)$ are analytic functions of their arguments; it is supposed that two saddle points are nearly coincident when α is near the critical value $\alpha=0$. We now state the maximum-modulus principle:

Lemma 1.

Suppose that the function $F(z)$ is analytic in the closed domain D, and suppose that $|F(z)| \leqslant M$ on the boundary of D. Then $|F(z)| \leqslant M$ in the interior of D.

Our method is based on the following idea.

Suppose (as in our first example) that we require an error bound near $z = 0$. We cannot readily find this but we can find a bound when z is bounded away from 0. Then the same bound is valid near $z = 0$, provided that the argument involves only analytic functions. (In our first example the function $Q_N(\cosh z)$ is not analytic at $z = 0$, and an indirect approach is needed.)

2. THE LEGENDRE FUNCTIONS $P_N(\cosh z)$, $Q_N(\cosh z)$ when $N \to \infty$, and z is near 0.

We have

$$P_N(\cosh z) = \frac{1}{\pi} \int_{-z}^{z} e^{-ut} \, (2\cosh z - 2\cosh t)^{-\frac{1}{2}} dt, \tag{2.1}$$

$$Q_N(\cosh z) = \int_{z}^{\infty} e^{-ut} \, (2\cosh t - 2\cosh z)^{-\frac{1}{2}} dt, \tag{2.2}$$

where $u = (N+\frac{1}{2})$. These may be compared with the Bessel functions

$$I_0(uz) = \frac{1}{\pi} \int_{-z}^{z} e^{-ut} (z^2-t^2)^{-\frac{1}{2}} dt, \tag{2.3}$$

$$\frac{z}{u} I_1(uz) = \frac{1}{\pi} \int_{-z}^{z} e^{-ut} (z^2-t^2)^{\frac{1}{2}} dt, \tag{2.4}$$

$$K_0(uz) = \int_{z}^{\infty} e^{-ut} (t^2-z^2)^{-\frac{1}{2}} dt, \tag{2.5}$$

$$\frac{z}{u} K_1(uz) = \int_{z}^{\infty} e^{-ut} (t^2-z^2)^{\frac{1}{2}} dt. \tag{2.6}$$

Write $G_0(t^2,z^2) = \left[\frac{2\cosh z - 2\cosh t}{z^2-t^2}\right]^{-\frac{1}{2}}$, an analytic function of the two variables t^2 and z^2, then we can write

$$G_0(t^2,z^2) = a_0(z^2) - (t^2-z^2)\, b_0(z^2) + (t^2-z^2)^2\, G_0^{**}(t^2,z^2).$$

Since $P_N(\cosh z) = \frac{1}{\pi}\int_{-z}^{z} G_0(t^2,z^2)e^{-ut}(z^2-t^2)^{-\frac{1}{2}}dt$, we have

$$P_N(\cosh z) - a_0(z^2)\, I_0(uz) - b_0(z^2)\,\frac{z}{u}\, I_1(uz)$$

$$= \frac{1}{\pi}\int_{-z}^{z} G_0^{**}(t^2,z^2)e^{-ut}(z^2-t^2)^{3/2}dt$$

$$= \frac{1}{\pi u^2}\int_{-z}^{z} G_1(t^2,z^2)e^{-ut}(z^2-t^2)^{-\frac{1}{2}}dt, \quad \text{by integration by parts,}$$

where $G_1(t^2,z^2) = (z^2-t^2)^{\frac{1}{2}}. \dfrac{\partial^2}{\partial t^2}\left\{G_0^{**}(t^2,z^2).(z^2-t^2)^{3/2}\right\}$

is again an analytic function of t^2 and z^2. Proceeding in this way we find that

$$P_N(\cosh z) = I_0(uz)\left[a_0(z^2) + \frac{a_1(z^2)}{u^2} +\ldots+ \frac{a_{p-1}(z^2)}{u^{2p-2}}\right]$$

$$+ \frac{z}{u}\, I_1(uz)\left[b_0(z^2) + \frac{b_1(z^2)}{u^2} +\ldots+ \frac{b_{p-1}(z^2)}{u^{2p-2}}\right]$$

$$+ \frac{1}{\pi u^{2p}}\int_{-z}^{z} G_p(t^2,z^2)e^{-ut}(z^2-t^2)^{-\frac{1}{2}}dt, \tag{2.7}$$

and similarly that

$$Q_N(\cosh z) = K_o(uz)\left[a_0(z^2) + \frac{a_1(z^2)}{u^2} +\ldots+ \frac{a_{p-1}(z^2)}{u^{2p-2}}\right]$$

$$- \frac{z}{u}\, K_1(uz)\left[b_0(z^2) + \frac{b_1(z^2)}{u^2} +\ldots+ \frac{b_{p-1}(z^2)}{u^{2p-2}}\right]$$

$$- \frac{1}{u^{2p}}\int_{z}^{\infty} G_p(t^2,z^2)e^{-ut}(t^2-z^2)^{-\frac{1}{2}}dt. \tag{2.8}$$

Write $$S_p(z,u) = \frac{1}{\pi}\int_{-z}^{z} G_p(t^2,z^2)e^{-ut}(z^2-t^2)^{-\frac{1}{2}}dt, \tag{2.9}$$

$$T_p(z,u) = \int_{z}^{\infty} G_p(t^2,z^2)e^{-ut}(t^2-z^2)^{-\frac{1}{2}}dt, \tag{2.10}$$

these are the error terms except for a factor u^{-2p}. To establish the validity of the asymptotic expansions

$$P_N(\cosh z) \sim I_0(uz) \sum_0^\infty \frac{a_s(z^2)}{u^{2s}} + \frac{z}{u} I_1(uz) \sum_1^\infty \frac{b_s(z^2)}{u^{2s}}, \tag{2.11}$$

$$Q_N(\cosh z) \sim K_0(uz) \sum_0^\infty \frac{a_s(z^2)}{u^{2s}} - \frac{z}{u} K_1(uz) \sum_0^\infty \frac{b_s(z^2)}{u^{2s}}, \tag{2.12}$$

we shall show that there are constants $M_p(A)$, $M_p(B)$, independent of z and u, such that

$$|S_p(z,u)| < M_p(A)|I_0(uz)| + M_p(B) \frac{1}{u} |zI_1(uz)|, \tag{2.13}$$

$$|T_p(z,u)| < M_p(A)|K_0(uz)| + M_p(B) \frac{1}{u} |zK_1(uz)|. \tag{2.14}$$

It is evident that the inequalities (2.13) and (2.14) are in a sense best possible since they show that the error terms are of the same order as the first neglected terms. A direct approach would involve estimates of $S_p(z,u)$ and $T_p(z,u)$ for large, moderate, and small values of $|uz|$ and a comparison with the Bessel functions $I_0(uz)$, $I_1(uz)$, $K_0(uz)$ and $K_1(uz)$, and all these functions exhibit considerable variation. An indirect approach will therefore be used.

We shall show how the inequalities (2.13) and (2.14) can be established in a simple manner by use of the maximum-modulus principle. We observe that the same two asymptotic series $\sum a_s(z^2)u^{-2s}$ and $\sum b_s(z^2)u^{-2s}$ occur in both the asymptotic expansions (2.11) and (2.12), and that each term in each series is analytic near $z = 0$ although $T_p(z,u)$ and $K_0(uz)$ are not analytic at $z = 0$, which is a logarithmic branch point. These asymptotic series are not convergent and do not define functions.

Let us define functions $A_p(z,u)$, $B_p(z,u)$ by means of the equations

$$S_p(z,u) = I_0(uz)\, A_p(z,u) + \frac{z}{u} I_1(uz)\, B_p(z,u) \tag{2.15}$$

$$T_p(z,u) = K_0(uz)\, A_p(z,u) - \frac{z}{u} K_1(uz)\, B_p(z,u) \tag{2.16}$$

Then $A_p(z,u) = uz\{S_p(z,u)\, K_1(uz) + T_p(z,u)\, I_1(uz)\}$, (2.17)

$B_p(z,u) = u^2\{S_p(z,u)\, K_0(uz) - T_p(z,u)\, I_0(uz)\}$, (2.18) where the relation $uz\{I_0(uz)\, K_1(uz) + I_1(uz)\, K_0(uz)\} = 1$ has been used. The functions $T_p(z,u)$, $K_0(uz)$, $K_1(uz)$ are not single-valued near $z=0$, but we can show, by continuing analytically round $z=0$, that $A_p(z,u)$ and $B_p(z,u)$ are single-valued and can therefore be expanded in a Laurent series. It is easily seen that $A_p(z,u)$ and $B_p(z,u)$ are at most logarithmically large near $z = 0$ and must therefore be bounded and therefore analytic. (The bounds may possibly depend on u.)

Let us consider $B_p(z,u)$ which (as we have just seen) is analytic near $z = 0$. Let z vary along a circle $|z| = \rho > 0$, then we can estimate each term in (2.18) from its expression as an integral by Watson's Lemma. When $\operatorname{Re} z > 0$ on the circle it is easy to see that $B_p(z,u)$ is not exponentially large because in each product in (2.18) one term has the exponentially small factor e^{-uz} and the other the exponentially large factor e^{uz}. When $\operatorname{Re} z < 0$ the argument is slightly more complicated: it is not difficult to show that there is a complex number c such that $T_p(z,u) + cS_p(z,u)$, $K_0 + cI_0$ and $K_1 - cI_1$ are exponentially small, and it follows as before that

$$B_p(z,u) = u^2\{S_p(z,u)(K_0(uz)+cI_0(uz)) - I_0(uz)(T_p(z,u) + cS_p(z,u))\}$$

is not exponentially large. A simple calculation using Watson's Lemma now shows that in fact

$$|B_p(z,u)| \leqslant M_p(B), \text{ independent of } u, \text{ when } |z| = \rho > 0. \tag{2.19}$$

Since $B_p(z,u)$ is analytic when $|z| \leqslant \rho$ it now follows from the maximum-modulus theorem that

$$|B_p(z,u)| \leqslant M_p(B) \quad \text{when} \quad |z| \leqslant \rho,$$

and similarly, that (2.20)

$$|A_p(z,u)| \leqslant M_p(A) \quad \text{when} \quad |z| \leqslant \rho.$$

It then follows from (2.15) and (2.16) that

$$\left|S_p(z,u)\right| \leqslant M_p(A)\left|I_0(uz)\right| + M_p(B)\,\frac{1}{u}\left|zI_1(uz)\right| \tag{2.21}$$

and

$$\left|T_p(z,u)\right| \leqslant M_p(A)\left|K_0(uz)\right| + M_p(B)\,\frac{1}{u}\left|zK_1(uz)\right|, \tag{2.22}$$

which are the required results. We have thus proved that the asymptotic expansions (2.11) and (2.12) are valid when $|z| \leqslant \rho$ and $N \to \infty$. The corresponding theory for $P_N^{-m}(\cosh z)$ and $Q_N^{-m}(\cosh z)$ is given in detail by Ursell (1984).

3. INTEGRALS WITH TWO NEARLY COINCIDENT SADDLE POINTS.

The method described in this section is applicable to any number of nearly coincident saddle points, see Ursell (1972), and will be illustrated by the simplest case of two nearly coincident saddle points and a single parameter α. Thus we consider the integral

$$\int g(z,\alpha)\exp\left\{Nf(z,\alpha)\right\}dz \tag{3.1}$$

For any fixed value of α this can be estimated by the method of steepest descents, which shows that the dominant contributions come from the saddle points, which are the zeros of $\frac{\partial}{\partial z} f(z,\alpha)$. It will be assumed that there are two saddle points $z_1(\alpha)$, $z_2(\alpha)$ which coincide at $z = 0$ when $\alpha = 0$. It is found that the expansions obtained by the method of steepest descents are non-uniform when $\alpha \to 0$, here we shall obtain uniform expansions involving Airy functions.

Since Airy functions involve a cubic polynomial in the exponent it is natural to change to a new variable of integration $u = u(z,\alpha)$ such that

$$f(z,\alpha) = \frac{1}{3} u^3 - u\zeta(\alpha) + A(\alpha) \tag{3.2}$$

We want one of the three solutions, $u(z,\alpha)$ say, of this equation to be a univalent function of z in a circle $|z| = \text{const}$ which does not shrink to 0 when $\alpha \to 0$. Suppose that u satisfies this condition, then

$$\frac{\partial f}{\partial z}(z,\alpha)\,\frac{dz}{du} = (u^2-\zeta(\alpha)). \tag{3.3}$$

The left-hand side vanishes when $z = z_1(\alpha)$ and $z = z_2(\alpha)$, the right-hand side vanishes when $u = \pm\zeta^{\frac{1}{2}}(\alpha)$, these points must therefore correspond, and from (3.2) we obtain the conditions

$$f(z_1(\alpha),\alpha) = -\frac{2}{3}\,\zeta^{3/2}(\alpha) + A(\alpha), \tag{3.4}$$

$$f(z_2(\alpha),\alpha) = \frac{2}{3}\zeta^{3/2}(\alpha) + A(\alpha), \tag{3.5}$$

determining $\zeta(\alpha)$ and $A(\alpha)$. Unless $\zeta(\alpha)$ and $A(\alpha)$ take these values there can be no uniformly univalent solution of the cubic equation (3.2), thus the conditions (3.4) and (3.5) are necessary; it was shown by Chester, Friedman and Ursell (1957) that they are also sufficient. The proof of the corresponding result for m nearly coincident saddle points is due to Levinson (1961). It is therefore sufficient to consider the integral

$$I(N,\alpha) = \int G_0(z,\alpha)\exp\left\{N(\tfrac{1}{3}z^3-\alpha z)\right\}dz \tag{3.6}$$

where it will be assumed for simplicity that $G_0(z,\alpha)$ is an entire function of z. (This condition can be relaxed, see Ursell (1972).) More precisely, we consider the integrals $I_1(N,\alpha)$ $I_2(N,\alpha)$, $I_3(N,\alpha)$, in which the contours of integration go from $\infty\exp(-\frac{1}{3}\pi i)$ to $\infty\exp(\frac{1}{3}\pi i)$, from $\infty\exp(\frac{1}{3}\pi i)$ to $\infty\exp(\pi i)$, and from $\infty\exp(\pi i)$ to $\infty\exp(-\frac{1}{3}\pi i)$ respectively. (These contours will be denoted by $C(1)$, $C(2)$, $C(3)$ respectively.) Thus

$$I_1(N,\alpha) + I_2(N,\alpha) + I_3(N,\alpha) = 0. \tag{3.7}$$

We now write

$$G_0(z,\alpha) = A_0(\alpha) + zB_0(\alpha) + (z^2-\alpha)H_0(z,\alpha), \tag{3.8}$$

then

$$\begin{aligned}
&\int G_0(z,\alpha)\exp\left\{N(\tfrac{1}{3}z^3-\alpha z)\right\}dz \\
&-A_0(\alpha)\int\exp\left\{N(\tfrac{1}{3}z^3-\alpha z)\right\}dz - B_0(\alpha)\int z.\exp\left\{N(\tfrac{1}{3}z^3-\alpha z)\right\}dz \\
&= \frac{1}{N}\int H_0(z,\alpha)\frac{d}{dz}\exp\left\{N(\tfrac{1}{3}z^3-\alpha z)\right\}dz \\
&= -\frac{1}{N}\int\exp\left\{N(\tfrac{1}{3}z^3-\alpha z)\right\}\frac{d}{dz}H_0(z,\alpha)dz, \quad \text{by integration by parts,} \\
&= \frac{1}{N}\int G_1(z,\alpha)\exp\left\{N(\tfrac{1}{3}z^3-\alpha z)\right\}dz,
\end{aligned} \tag{3.9}$$

where $$G_1(z,\alpha) = -\frac{d}{dz}H_0(z,\alpha). \tag{3.10}$$

We can repeat this argument, obtaining the result

$$\int G_0(z,\alpha)\exp\left\{N(\tfrac{1}{3}z^3-\alpha z)\right\}dz$$

$$-\left[A_0(\alpha)+\frac{A_1(\alpha)}{N}+\ldots+\frac{A_{p-1}(\alpha)}{N^{p-1}}\right]\int\exp\left\{N(\tfrac{1}{3}z^3-\alpha z)\right\}dz$$

$$-\left[B_0(\alpha)+\frac{B_1(\alpha)}{N}+\ldots+\frac{B_{p-1}(\alpha)}{N^{p-1}}\right]\int z.\exp\left\{N(\tfrac{1}{3}z^3-\alpha z)\right\}dz$$

$$=\frac{1}{N^p}\int G_p(z,\alpha)\exp\left\{N(\tfrac{1}{3}z^3-\alpha z)\right\}dz, \tag{3.11}$$

where $G_p(z,\alpha)$ is defined recursively and is evidently an entire function. We note that $\int\exp\left\{N(\frac{1}{3}z^3-\alpha z)\right\}dz$ is an Airy function and that $\int z.\exp\left\{N(\frac{1}{3}z^3-\alpha z)\right\}dz$ is an Airy function derivative. To show that (3.11) is an asymptotic expansion uniformly valid when $|\alpha| \leqslant \rho$ and $N \to \infty$, it is sufficient to show that there are constants $M_p(A)$, $M_p(B)$ independent of α, such that for any contour C beginning and ending at ∞

$$\left|\int_C G_p(z,\alpha)\exp\left\{N(\tfrac{1}{3}z^3-\alpha z)\right\}dz\right|$$

$$\leqslant M_p(A)\left|\int_C \exp\left\{N(\tfrac{1}{3}z^3-\alpha z)\right\}dz\right|$$

$$\leqslant M_p(B)\left|\int_C z.\exp\left\{N(\tfrac{1}{3}z^3-\alpha z)\right\}dz\right| \tag{3.12}$$

when $|\alpha| \leqslant \rho$ and $N > N_0$.

To prove the inequality (3.12) we proceed indirectly. We observe that we have not so far specified the contour in (3.11) and (3.12), and that we obtain the same coefficients $A_s(\alpha)$, $B_s(\alpha)$ whichever of the three contours C(1), C(2), C(3) is chosen. (The suffix p will now be omitted). Define $A(N,\alpha)$, $B(N,\alpha)$ by the pair of equations

$$\int_{C(1)} G(z,\alpha)\exp\left\{N(\tfrac{1}{3}z^3-\alpha z)\right\}dz = A(N,\alpha)\int_{C(1)}\exp\left\{N(\tfrac{1}{3}z^3-\alpha z)\right\}dz$$

$$+B(N,\alpha)\int_{C(1)} z.\exp\left\{N(\tfrac{1}{3}z^3-\alpha z)\right\}dz, \tag{3.13}$$

$$\int_{C(2)} G(z,\alpha)\exp\left\{N(\tfrac{1}{3}z^3-\alpha z)\right\}dz = A(N,\alpha)\int_{C(2)}\exp\left\{N(\tfrac{1}{3}z^3-\alpha z)\right\}dz$$

$$+B(N,\alpha)\int_{C(2)} z.\exp\left\{N(\tfrac{1}{3}z^3-\alpha z)\right\}dz, \qquad (3.14)$$

Then (3.7) shows that the corresponding equation also holds for C(3). To prove (3.12) it is sufficient to show that $|A(N,\alpha| \leqslant M(A)$ and $|B(N,\alpha)| \leqslant M(B)$. We see that (in an obvious abbreviated notation)

$$A(N,\alpha).\Delta(N,\alpha) = \int_{C(1)} G.\exp\int_{C(2)} z.\exp - \int_{C(2)} G.\exp\int_{C(1)} z.\exp \qquad (3.15)$$

where $\Delta(N,\alpha) = \int_{C(1)} \exp\int_{C(2)} z.\exp - \int_{C(2)} \exp\int_{C(1)} z.\exp$ (3.16)

is the Wronskian determinant of the Airy equation and can therefore be found explicitly. In fact,

$$\frac{\partial}{\partial\alpha}\Delta(N,\alpha) = -\int_{C(1)} Nz.\exp\int_{C(2)} z.\exp + \int_{C(2)} Nz.\exp\int_{C(1)} z.\exp$$

$$= -\int_{C(1)} \exp\int_{C(2)} Nz^2.\exp + \int_{C(2)} \exp\int_{C(1)} Nz^2.\exp$$

$$= 0,$$

since $\int(z^2-\alpha)\exp\left\{N(\tfrac{1}{3}z^3-\alpha z)\right\}dz = \left[\tfrac{1}{N}\exp\left\{N(\tfrac{1}{3}z^3-\alpha z)\right\}\right] = 0.$

Thus $\Delta(N,\alpha) = \Delta(N,0), = \text{const}.N^{-1} \neq 0$, as is easily shown. To find a bound for $A(N,\alpha)$, suppose first that $|\alpha| = \rho$ is bounded away from 0, then the integrals $\int G.\exp$, $\int\exp$, $\int z.\exp$ can all be evaluated by the method of steepest descents since the saddle points $z_1(\alpha) = +\alpha^{\frac{1}{2}}$ and $z_2(\alpha) = -\alpha^{\frac{1}{2}}$ are well separated. Suppose first that $\frac{4}{3}\pi < ph(\alpha) < 2\pi$, then $\frac{2}{3}\pi < ph(z_1(\alpha)) < \pi$, and $-\frac{1}{3}\pi < ph(z_2(\alpha)) < 0$; it is now readily shown that the two paths of steepest descent from $z_1(\alpha)$ form a contour equivalent to C(2) and that the two paths of steepest descent from $z_2(\alpha)$ form a contour equivalent to C(1). The exponential factor associated with $z_1(\alpha)$ is $\exp(-\frac{2}{3}N\alpha^{3/2})$, while the exponential factor associated with $z_2(\alpha)$ is $\exp(\frac{2}{3}N\alpha^{3/2})$. We can now find a bound for each term on the right-hand side of (3.15) when $\frac{4}{3}\pi < ph(\alpha) < 2\pi$; since each term involves only one relevant saddle point we find that

$|A(N,\alpha)| \leqslant M$ when $\frac{4}{3}\pi < \mathrm{ph}(\alpha) < 2\pi$, where M depends on $G(z,\alpha)$. Similarly when $0 < \mathrm{ph}(\alpha) < 0\,\frac{2}{3}\,\pi$ the appropriate contours are $C(1)$ and $C(3)$, and when $\frac{2}{3}\pi < \mathrm{ph}(\alpha) < \frac{4}{3}\pi$ the appropriate contours are $C(2)$ and $C(3)$. In each case we find that $|A(N,\alpha)| \leqslant M(A)$ and similarly that $|B(N,\alpha)| \leqslant M(B)$. These bounds were obtained for $|\alpha| = \rho > 0$, but $A(N,\alpha)$ and $B(N,\alpha)$ are analytic when $|\alpha| \leqslant \rho$, and so by the maximum-modulus principle we have the bounds

$$|A(N,\alpha)| \leqslant M(A) \quad \text{and} \quad |B(N,\alpha)| \leqslant M(B) \quad \text{when} \quad |\alpha| \leqslant \rho. \tag{3.17}$$

It now follows from (3.130 that, for $C = C(1)$ or $C(2)$ or $C(3)$,

$$\left|\int_C G(z,\alpha)\exp\left\{N(\tfrac{1}{3}z^3-\alpha z)\right\}dz\right| \leqslant M(A)\left|\int_C \exp\left\{N(\tfrac{1}{3}z^3-\alpha z)\right\}dz\right|$$
$$+\; M(B)\left|\int_C z.\exp\left\{N(\tfrac{1}{3}z^3-\alpha z)\right\}dz\right|.$$

This is the required inequality (3.12). For further details see Ursell (1972) where the corresponding theory for a quartic exponent and three nearly coincident saddle points is presented, and where it is shown that the theory is in fact applicable to any number of nearly coincident saddle points.

4. EXTENSION OF THE DOMAIN OF VALIDITY.

We have obtained two examples of uniformly asymptotic expansions, and we now show how their validity can be extended to wider regions. Our expansions for Legendre functions were found to be valid in a circular region $|z| \leqslant \rho$, which is the most interesting region, for elsewhere expansions involving exponentials can be found by the ordinary method of steepest descents. The same expansions (2.11) and (2.12) had previously been obtained by Olver (1974) who started from Legendre's equation not from integral representations, and who was able to show that these expansions are in fact valid in unbounded regions. We now observe that the validity of our expansions may also be extended by using integrals. For we have shown above (eqns 2.15 and 2.16 with $p = 0$) that

$$P_N(\cosh z) = I_0(uz)\, A_0(z,u) + \frac{z}{u}\, I_1(uz)\, B_0(z,u), \tag{4.1}$$

where

$$A_0(z,u) = uz\{P_N(\cosh z)\, K_1(uz) + Q_N(\cosh z)\, I_1(uz)\}, \tag{4.2}$$

$$B_0(z,u) = u^2\{P_N(\cosh z)\, K_0(uz) - Q_N(\cosh z)\, I_0(uz)\}; \tag{4.3}$$

here each term on the right-hand side is given in the form of an integral. When $|z| > \frac{1}{2}\rho$ each of these integrals can be expanded by Watson's Lemma. When $\operatorname{Re} z > 0$ we find that each product of a Legendre function and a Bessel function involves a large exponential factor e^{uz} and a small exponential factor e^{-uz}, (cf §2 above), thus when $\operatorname{Re} z > 0$ we find that

$$A_0(z,u) \sim \sum_{s=0}^{\infty} \frac{a_s(z^2)}{u^{2s}}, \quad B_0(z,u) \sim \sum_{s=0}^{\infty} \frac{b_s(z^2)}{u^{2s}}, \tag{4.4}$$

and by uniqueness of Poincare asymptotic series the coefficients must be the same as in (2.11). Similarly when $\operatorname{Re} z < 0$ we may use (2.19) to obtain the same result. The behaviour of the expansions near $z = \infty$ requires further study, but will probably not involve any serious difficulties, cf. Ursell 1965, §4. Alternatively we may expand $S_p(z,u)$ and $T_p(z,u)$ directly by Watson's Lemma when $|z| > \frac{1}{2}\rho$.

Our second example was the problem of several nearly coincident saddle points; here also the validity of the expansion can be extended but the analytic continuation of the coefficients requires much more elaborate treatment, see Ursell (1965), Martin (1974).

5. DISCUSSION.

We have now applied the maximum-modulus principle to two uniformly asymptotic expansions of integrals, to show that the error after p terms is of the same order as the first neglected term. Our argument was based on the observation that the coefficient functions in these expansions are analytic, single-valued, and slowly varying, although the integrals themselves vary rapidly. It remains to be seen whether this principle can be usefully applied in numerical as well as in theoretical work.

In our first example we treated the Legendre functions $P_N(\cosh z)$ and $Q_n(\cosh z)$ for large N in a complete neighbourhood of $z = 0$; expansions involving Bessel functions were obtained. The same argument is applicable to the Legendre functions $P_N^{-m}(\cosh z)$ and $Q_N^{-m}(\cosh z)$ for fixed m, details are given by Ursell (1984).

In our second example we treated an integral with 2 nearly coincident saddle points, and the expansion involved Airy functions. The same argument is applicable to any number of nearly coincident saddle points, the integral then involves generalization of the Airy function; details are given by Ursell (1972). The same idea can also be applied to certain double integrals involving functions of 2 complex variables. An example is given by Ursell (1980).

References

Chester, C., Friedman, B and Ursell, F. (1957) An extension of the method of steepest descents. *Proc.Camb.Phil.Soc.* 53, 599-611.

Levinson, N. (1961) Transformation of an analytic function of several variables to a canonical form. *Duke Math.J.* 28, 345-353.

Martin, J. (1974) Integrals with a large parameter and several nearly coincident saddle points; the continuation of uniformly asymptotic expansions. *Proc.Camb.Phil.Soc.* 76, 211-231.

Olver, F.W.J. (1974) *Asymptotics and Special Functions* (*Academic Press*)

Ursell, F. (1965) Integrals with a large parameter. The continuation of uniformly asymptotic expansions. *Proc.Camb.Phil.Soc.* 61, 113-128.

Ursell, F. 1972 Integrals with a large parameter. Several nearly coincident saddle points. *Proc.Camb.Phil.Soc.* 72, 49-65.

Ursell, F. (1980) Integrals with a large parameter: a double complex integral with four nearly coincident saddle points. *Math.Proc.Camb.Phil.Soc.* 87, 249-273.

Ursell, F. (1984) Integrals with a large parameter: Legendre functions of large degree and fixed order. *Math.Proc.Camb.Phil.Soc.* 95, 367-380.

Finite Axial Extension and Torsion of Elastic Helicoidal Shells

Frederic Y. M. Wan Professor, Department of Applied Mathematics, University of Washington, Seattle, Washington

ABSTRACT

A finite deformation and infinitesimal strain formulation of the problem of axial extension and torsion of helicoidal shells is deduced from a general nonlinear shell theory for infinitesimal strain problems. A consistent application of the infinitesimal strain assumption further simplifies the relevant boundary value problem effectively to a linear problem; only the unknown stretch and twist parameters appear nonlinearly. The effects of finite deformations are clearly shown by examining the perturbation solutions for slightly pretwisted strips and (shallow) shells with a small pitch. In the presence of the core of the helicoid, the Poincaré-Lighthill perturbation technique is needed to avoid the spurious singularity associated with the regular perturbation solution.

The research was supported by National Science Foundation Grant DMs-8743445.

I. INTRODUCTION

Many widely used solid structures, such as turbine blades and coil springs with thin cross-sections, are special cases of helicoidal shells. A study of their structural integrity in the elastic range may be formulated as boundary value problems in shell theory. Among the problems fundamental to the understanding of the elastostatics of these structures is the problem of a helicoidal shell acted upon by equal and opposite axial forces and torques (Figure 1). This paper presents an analysis of this fundamental problem within the framework of an infinitesimal strain-finite deflection shell theory. The Saint Venant solution for this problem obtained by our analysis reduce to the known result for shells with a small pitch [3].

In circular cylindrical coordinates (r,θ,z), the middle surface of a helicoidal shell is described by $z = a\theta$ where the constant $2\pi a$ is the pitch of the right helicoid. The special case of axial extension and torsion of pretwisted elastic strips has been investigated previously by way of a linear shell theory in [2, 4, 6, 7]. The principal qualitative conclusions obtained were:

(1) While the stress and strain distributions in the shell are rotationally symmetric, the displacement field is multi-valued in the polar angle θ of the form

$$u_r = u(r) \quad , \quad u_\theta = \psi r a\theta \quad , \quad u_z = ka\theta \tag{1.1}$$

where u_r, u_θ and u_z are the radial, circumferential and axial displacement component of the middle surface of the shell, respectively, and k and ψ are two parameters corresponding to the axial strain and the angle of twist for a flat strip.

(2) The elastostatic problem may be reduced to a two point boundary value problem involving a second order (linear) ODE for the only unknown function $u(r)$ with k and ψ appearing as *forcing terms.*

(3) The membrane and inextensional bending actions of the shell are uncoupled in a well-defined way.

(4) There is no edge effect due to edge bending along the helical $(r = \text{constant})$ edges of the shell.

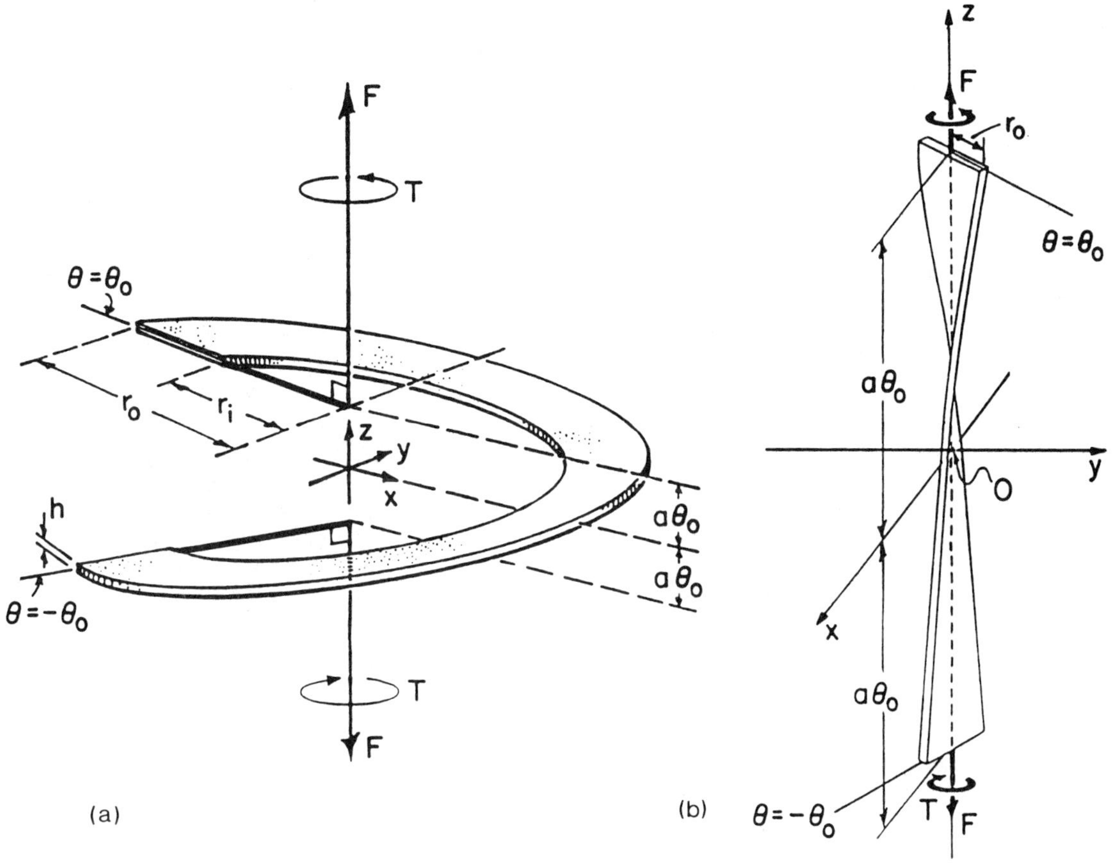

1. Helicoidal shells under axial forces and torques: (a) a ring shell sector; (b) a pretwisted strip.

The analysis for pretwisted strips was extended to a larger class of problems which includes the axial extension and torsion of helicoidal springs with wide rectangular cross-sections [6]. For a shell with small pitch-to-width ratio, a/r_0 (where r_0 is the radial distance from the axis of the shell to its outer helical edge), the latter problem was investigated earlier by a nonlinear shallow shell theory in [3].

The present paper considers the more general problem of rotationally symmetric stress state of helicoidal shells within the framework of a finite deflection-infinitesimal strain theory for shells with arbitrary pitch-to-width ratios. We will adopt also a semi-inverse procedure, used successfully in the earlier work, in the sense that the loading is assumed to be such that the stress and strain distributions in the shell will be independent of θ. Among the several nonlinear shell theories available in the literatures, we chose to follow the approach of [5] to make apparent the meaning of the parameters which appear in our semi-inverse procedure.

The requirement of rotationally symmetric strain distributions restricts the possible displacement fields and gives rise to a set of nonlinear strain displacement relations allowing for the rotationally symmetric stretching, twisting, bending and shearing actions of the shell. The subsequent analysis will be concerned exclusively with the stretching and twisting actions and the strain displacement relations are simplified by the restriction of small strain. For this class of problems, we show that the strain displacement relations must be linear in the radial displacement component u. The resulting boundary value problem has a governing differential equation which is linear in the only unknown function u; nonlinearity appears only in the *forcing terms*. In that case, the analysis of the relevant boundary value problem can be carried out exactly as that for a linear theory. Upon casting the present finite deformation problem in an appropriate form, the solution for the radial displacement, stress resultants and stress couples, can be obtained directly from the linear analysis of [6] without another set of independent calculations. The leading term perturbation solutions for the extreme cases of small and large pitch-to-width ratio are identical to the results of [3] and to the corresponding results for a shallow hyperbolic paraboloidal shell based on Marguerre's nonlinear shallow shell theory, respectively. While the principal qualitative conclusions for a linear theory carry over to our *quasilinear theory*, the latter gives additional information concerning certain critical values of the applied forces and torques when the deformation is no longer infinitesimal.

II. ANALYSIS OF STRAINS

Let ξ_1 and ξ_2 be general orthogonal surface coordinates, and $\vec{r}(\xi_1,\xi_2)$ and $\vec{n}(\xi_1,\xi_2)$ be the position vector and the unit normal of the midsurface of the shell before deformation. After deformation, $\vec{r}$ changes to $\vec{\rho}(\xi_1,\xi_2)$ with a new unit normal $\vec{v}$. In terms of $\vec{r}$ and $\vec{\rho}$ the expressions for midsurface strain and curvature change measures of an infinitesimal strain, large deformation shell theory proposed in [5] are

$$\varepsilon_1 = \frac{1}{2}\left(\frac{\vec{\rho}\,'\cdot\vec{\rho}\,'}{\alpha^2} - 1\right), \qquad \varepsilon_2 = \frac{1}{2}\left(\frac{\dot{\vec{\rho}}\cdot\dot{\vec{\rho}}}{\alpha_2^2} - 1\right), \qquad \gamma = \frac{\vec{\rho}\,'\cdot\dot{\vec{\rho}}}{\alpha_1\alpha_2} \tag{2.1}$$

$$\kappa_1 = \frac{\vec{\rho}\,'\cdot\dot{\vec{v}}}{\alpha_1^2} - \frac{1}{R_{11}}, \qquad \kappa_2 = \frac{\dot{\vec{\rho}}\cdot\vec{v}\,'}{\alpha_2^2} - \frac{1}{R_{22}},$$

$$\tau = \frac{\vec{\rho}\,'\cdot\dot{\vec{v}} + \dot{\vec{\rho}}\cdot\vec{v}\,'}{\alpha_1\alpha_2} - \frac{2}{R_{12}} \tag{2.2}$$

where primes and dots indicate differentiation with respect to ξ_1 and ξ_2, respectively, $\alpha_1 = |\vec{r}\,'|$, $\alpha_2 = |\dot{\vec{r}}|$ and R_{ij} are radii of curvature of the undeformed midsurface.

To specialize these general expressions to the case of a helicoidal shell, we take $\xi_1 = r$ and $\varepsilon_2 = \theta$. With $z = a\theta$, we have $\alpha_1 = 1$, $\alpha_2 = (a^2 + r^2)^{1/2} \equiv \alpha$, $1/R_{11} = 1/R_{22} = 0$ and $1/R_{12} = a/\alpha^2 \equiv 1/R$. The position vector $\vec{r}$ of the undeformed middle surface of a helicoidal shell is given by

$$\vec{r} = r\vec{i}_r + a\theta\vec{i}_z, \qquad \vec{i}_r = \cos\theta\,\vec{i}_x + \sin\theta\,\vec{i}_y \tag{2.3}$$

where $\vec{i}_r$, $\vec{i}_x$, $\vec{i}_y$ and $\vec{i}_z$ are unit vectors in directions indicated by the subscripts. Analogous to these, we write the position vector $\vec{\rho}$ of the deformed midsurface as

$$\vec{\rho} = U_\rho(r,\theta)\vec{i}_\rho + U_z(r,\theta)\vec{i}_z, \qquad \vec{i}_\rho = \cos\eta\,\vec{i}_x + \sin\eta\,\vec{i}_y \tag{2.4}$$

where $\eta = \eta(r,\theta)$. Observe that $d(\vec{i}_\rho)/d\eta = -\sin\eta\,\vec{i}_x + \cos\eta\,\vec{i}_y \equiv \vec{i}_\eta$.

We are interested here in the class of deformation given by

$$U_\rho = U(r) \quad , \qquad U_z = W(r) + c_z\theta \; , \qquad \eta = \phi(r) + \phi_0\theta \tag{2.5}$$

where c_z and ϕ_0 are two unknown constants. The associated strain and curvature change measures are

$$\varepsilon_r = \frac{1}{2}\left[(U')^2 + (\phi')^2U^2 + (W')^2 - 1\right] \; , \quad \varepsilon_\theta = \frac{1}{2}\left[\frac{\phi_0^2U^2 + c_z^2}{\alpha^2} - 1\right]$$

$$\gamma = \frac{\phi_0\phi' U^2 + c_z W'}{\alpha} \tag{2.6}$$

$$\kappa_r = \frac{c_zU'}{\alpha}\left[\phi'' U + 2\phi' U'\right] - \frac{U}{\alpha}\left[(U'' - (\phi')^2U)(c_z\phi' - \phi_0W') + \phi_0U'W''\right]$$

$$\kappa_\theta = -\frac{\phi_0^2U^2}{\alpha^3}(\phi_0W' - c_z\phi') \tag{2.7}$$

$$\frac{1}{2}\tau = \frac{\phi_0\phi' U^2}{\alpha^2}(c_z\phi' - \phi_0W') + \frac{c_z\phi_0(U')^2}{\alpha^2} - \frac{a}{\alpha^2} \; .$$

Note that they are all independent of θ.

For stretching and twisting shell actions, we take $W \equiv \phi \equiv 0$. In that case we have $\gamma = \kappa_r = \kappa_\theta = 0$. Upon setting

$$U = r + u(r) \quad , \quad c_z = a(1 + k) \quad , \quad \phi_0 = 1 + \psi a \; , \tag{2.8}$$

the remaining strain measures become

$$\varepsilon_\theta = \frac{r^2}{\alpha^2}\left\{\frac{u}{r}\left(1 + \frac{u}{2r}\right) + \left(\psi a + \frac{1}{2}\psi^2a^2\right)\left(1 + \frac{u}{r}\right)^2 + \frac{ka^2}{r^2}\left(1 + \frac{1}{2}k\right)\right\}$$

$$\varepsilon_r = u' \; , \qquad \tau = \frac{2a}{\alpha^2}\{k + \psi a + k\psi a\} \tag{2.9}$$

where we have omitted $u' \equiv \varepsilon_r$ compared to unity in the expression for τ to be consistent with the assumption of infinitesimal strains. Note that τ does not depend on u .

III. A QUASILINEAR THEORY FOR FINITE STRETCHING AND TWISTING

The expression for ε_θ in (2.9) is quadratic in (u/r) . We solve this equation for u/r *to get*

$$\frac{u}{r} = -1 \pm \left[\frac{1 + 2\alpha^2\left\{\varepsilon_\theta - ka^2(1 + k/2)/\alpha^2\right\}/r^2}{(1 + \psi a)^2}\right]^{1/2} \tag{3.1}$$

Given that $ka^2(1 + k/2)/\alpha^2$ *and* $\psi ar^2(1 + \psi a/2)/\alpha^2$ are necessarily of the order of the hoop strain ε_θ, only the positive square root is acceptable in the above expression for u/r. (This is so even for $\psi a + 1 << 1$.) Otherwise, we have $\varepsilon_r = u$ not small compared to unity which violates the assumption of infinitesimal strains. It follows that we have $u/r = 0(\varepsilon_\theta)$ and $1 + u/r \simeq 1$. In that case, the strain displacement relations given by (2.9) can be further simplified to

$$\varepsilon_r = u' \ , \ \varepsilon_\theta = \frac{r^2}{\alpha^2}\left[\frac{u}{r} + \psi a\left(1 + \frac{1}{2}\psi a\right) + \frac{ka^2}{r^2}\left(1 + \frac{1}{2}k\right)\right] \tag{3.2a,b}$$

$$\tau = \frac{2a}{\alpha^2}[k + \psi a + k\psi a] \ . \tag{3.2c}$$

The strain measures are related to the stress resultants and couples of the shell by a system of stress strain relations. We will take them int he form

$$\varepsilon_r = A(N_r - \nu N_\theta) \quad , \quad \varepsilon_\theta = A(N_\theta - \nu N_r) \tag{3.3}$$

$$M_{r\theta} = M_{\theta r} = \frac{1}{2}D(1 - \nu)\tau \tag{3.4}$$

where in terms of Young's modulus E , Poisson's ratio ν, and the shell thickness h , we have $A = 1/Eh$ and $D = Eh^3/12(1 - \nu)^2$. Equations (3.4) and the expression for τ in (3.2) can be combined to give

$$M_{r\theta} = M_{\theta r} = \frac{D(1-\nu)}{R}(k + \psi a + k\psi a) \tag{3.5}$$

where the unknown function u does not appear explicitly. Without the nonlinear term, the expression (3.5) is exactly the same as what we found for a linear theory [4,6].

The stress resultants and couples, together with the external radial surface load intensity p_r , and edge loads are subject to the conditions of force and moment equilibrium. Within the framework of a small strain theory we have, as a consequence of equilibrium and strain displacement relations, the following virtual work equation for the nonvanishing stress, strain and displacement measures:

$$\int_{r_i}^{r_0} (N_r \delta\varepsilon_r + N_\theta \delta\varepsilon_\theta + M_{r\theta}\delta\tau)\,\alpha dr$$

$$= \int_{r_i}^{r_0} p_r \delta u \;\alpha dr + P\delta[ka] + T\delta[\psi a] + N_0\delta[u(r_0)] - N_i\delta[u(r_i)] \quad . \tag{3.6}$$

In (3.6) P and T are the resultant force and torque at a *radial* (θ = constant) edge, and N_0 and N_i are the applied radial edge resultant at the *helical* edges, $r = r_0$ and $r = r_i$, respectively. Note that the constraint of no transverse shear deformation normally introduces fictitious corner forces. Their effects should be included in the virtual work expression (3.6). However, the contribution of these effects to the final results is of the same order as that from terms neglected in the approximations already made. Therefore, it is consistent not to include the effects of corner forces in (3.6).

From (3.2), (3.5) and (3.6), we get as conditions of local and overall equilibrium, one differential equation

$$(\alpha N_r)' - \frac{r}{\alpha} N_\theta + \alpha p_r = 0 \tag{3.7}$$

two boundary conditions at the two helical edges

$$N_r(r_0) = N_0 \quad , \quad N_r(r_i) = N_i \tag{3.8}$$

and two integrated conditions at the radial edges

$$\int_{r_i}^{r_0} \left\{ \frac{a}{\alpha}(1+k)\, N_\theta + D(1-\nu)\, \frac{2a}{\alpha^3}(1+\psi a)\,(k+\psi a+k\psi a) \right\} dr = P \tag{3.9}$$

$$\int_{r_i}^{r_0} \left\{ \frac{r^2}{\alpha}(1+\psi a)\, N_\theta + D(1-\nu)\, \frac{2a^2}{\alpha^3}(1+k)\,(k+\psi a+k\psi a) \right\} dr = T \ . \tag{3.10}$$

Other conditions at the helical edges can also be prescribed. For example, we may prescribe the displacement component u instead.

The expressions for ε_r and ε_θ in equations (3.2) may be combined with (3.3) to give N_r and N_θ in terms of the unknown function u :

$$A N_r = u' + \frac{\nu r^2}{\alpha^2}\left[\frac{u}{r} + \psi a\left(1+\frac{1}{2}\psi a\right) + \frac{k a^2}{r^2}\left(1+\frac{1}{2}k\right)\right] \tag{3.11}$$

$$A N_\theta = \nu u' + \frac{r^2}{\alpha^2}\left[\frac{u}{r} + \psi a\left(1+\frac{1}{2}\psi a\right) + \frac{k a^2}{r^2}\left(1+\frac{1}{2}k\right)\right] \ . \tag{3.12}$$

Upon substituting these expressions into (3.7), we get a second order differential equation for u :

$$u'' + \frac{r}{a^2+r^2}u' + \frac{\nu a^2 - r^2}{(a^2+r^2)^2}u = \frac{(1+\nu)a^2 r k\left(1+\frac{1}{2}k\right) + \psi a r\left(1+\frac{1}{2}\psi a\right)\left[(1+\nu)r^2 - 2\nu a^2\right]}{\left(a^2+r^2\right)^2} - A p_r \ . \tag{3.13}$$

Equation (3.13) and (3.8), expressed in terms of u by (3.11), define a two-point boundary value problem for u . It determines u up to two unknown parameters k

and ψ. These two parameters are then related to P and T by the two integral relations (3.9) and (3.10).

Equation (3.13) is the only differential equation to be solved for the problem of finite stretching and twisting given that u/r is of the order of the midsurface strains; it is *linear* in the unknown u. The effect of finite deformation appears only through the nonlinear terms involving the parameters k and ψ. With these nonlinear terms omitted from the right-hand side of the equation, we recover the governing differential equation for small deformation as obtained in [2] and elsewhere. It is also significant that the parameters k and ψ do not appear in the coefficients on the left side of the differential equation.

IV. A STRESS FUNCTION FORMULATION

We consider here a shell free of surface loads so that $p_r = 0$ and the two helical edges are free of edge tractions so that $N_i = N_0 = 0$. The two radial edges of the shell $\theta = \pm\,\theta_0$ are subject to equal and opposite resultant axial forces and torques. With the condition at the two helical edges prescribed in terms of N_r, the corresponding two-point boundary value problem for an infinitesimal deflection theory has been reformulated in [6] for the stress variable N_r. Given the form of the boundary value problem for the quasilinear theory, we will also use a similar stress function formulation to take advantage of the results already obtained in [6].

We begin by using the equilibrium equation (3.7) with $p_r \equiv 0$ to write

$$N_\theta = \frac{\alpha}{r}(\alpha N_r)' . \tag{4.1}$$

The strain measures given in terms of a single unknown function u satisfy the compatibility equation

$$\left(\frac{\alpha^2}{r}\varepsilon_\theta\right)' - \varepsilon_r = \psi a\left(1 + \frac{1}{2}\psi a\right) - \frac{a^2}{r^2}k\left(1 + \frac{1}{2}k\right) . \tag{4.2}$$

Upon expressing ε_r and ε_θ in terms of N_r by way of the stress strain relations, we get

$$\varepsilon_r = A\left[N_r - \nu\frac{\alpha}{r}(\alpha N_r)'\right] , \quad \varepsilon_\theta = A\left[\frac{\alpha}{r}(\alpha N_r)' - \nu N_r\right] . \tag{4.3}$$

Introduction of these expressions into the compatibility equation (4.2) gives us a second order differential equation for N_r :

$$N_r^{\cdot\cdot} - \frac{2a^2 - 3r^2}{r(a^2 + r^2)} N_r^{\cdot} - \frac{(1 - \nu)a^2}{(a^2 + r^2)^2} N_r = \frac{\psi a\left(1 + \frac{1}{2}\psi a\right)r^2 - a^2 k\left(1 + \frac{1}{2}k\right)}{A\,(a^2 + r^2)^2} \ . \tag{4.4}$$

The boundary conditions at the helical edges are now simply (3.8) with $N_0 = N_i = 0$:

$$N_r(r_i) = N_r(r_0) = 0 \ . \tag{4.5}$$

Having determined N_r by (4.4) and (4.5), we get N_θ from (4.1), while $M_{r\theta}$ is known from (3.5). We then use the second equation of (3.2) to determine u . Finally, the parameters k and ψ are related to P and T by way of (3.9) and (3.10).

To obtain the solution of the boundary value problem for N_r and the corresponding overall load-deformation relations, we introduce the following dimensionless quantities:

$$s = r/r_0 \ , \ s_i = r_i/r_0 \ , \ \lambda = r_0/a \ , \ \mu = h/r_0$$
$$(n_r, n_\theta) = A(N_r, N_\theta) \ , \ m = \frac{r_0 M_{r\theta}}{D(1 - \nu)} \ , \ \bar{u} = \frac{u}{r_0} \ . \tag{4.6}$$

In terms of these quantities, the differential equation (4.4) becomes

$$n_r^{\cdot\cdot} - \frac{2 - 3\lambda^2 s^2}{s(1 + \lambda^2 s^2)} n_r^{\cdot} - \frac{(1 - \nu)\lambda^2}{(1 + \lambda^2 s^2)^2} n_r = \frac{\psi r_0 \lambda^2 s^2\left(\lambda + \frac{1}{2}\psi r_0\right) - k\lambda^2\left(1 + \frac{1}{2}k\right)}{(1 + \lambda^2 s^2)^2} \tag{4.7}$$

where dots now indicate differentiation with respect to s , and the boundary conditions (4.5) become

$$n_r(1) = n_r(s_i) = 0 \ . \tag{4.8}$$

The auxiliary equations (4.1), (3.5) and (3.2b) may be written as

$$n_\theta = \frac{1+\lambda^2 s^2}{\lambda^2 s} n_r^{\bullet} + n_r \ , \quad m = \frac{k\lambda + \psi r_0(1+k)}{1+\lambda^2 s^2} \tag{4.9a,b}$$

$$\bar{u} = \frac{(1+\lambda^2 s^2)^2}{\lambda^4 s^2}\left[n_r^{\bullet} + \frac{(1+\nu)\lambda^2 s}{1+\lambda^2 s^2} n_r - \frac{k\left(1+\frac{1}{2}k\right)\lambda^2 s + \psi r_0\left(\lambda+\frac{1}{2}\psi r_0\right)\lambda^2 s^3}{(1+\lambda^2 s^2)^2}\right] \tag{4.10}$$

and the two integral relations (3.10) and (3.11) may be written as

$$\frac{PA}{r_0} = (1+k)\int_{s_i}^{1} \frac{\sqrt{1+\lambda^2 s^2}}{\lambda^2 s^2} n_r ds + \frac{\mu^2(\lambda+\psi r_0)\,[k\lambda+\psi r_0(1+k)]}{6(1+\nu)}\int_{s_i}^{1} \frac{ds}{(1+\lambda^2 s^2)^{3/2}} \tag{4.11}$$

$$\frac{TA}{r_0^2} = -\frac{\lambda+\psi r_0}{\lambda^2}\int_{\rho_i}^{1} \sqrt{1+\lambda^2 s^2}\, n_r ds + \frac{\mu^2(1+k)[k\lambda+\psi r_0(1+k)]}{6(1+\nu)}\int_{s_i}^{1} \frac{ds}{(1+\lambda^2 s^2)^{3/2}} \ . \tag{4.12}$$

The boundary value problem defined by (4.7) and (4.8) is *linear* and differs from the corresponding boundary value problem for infinitesimal deformation in [6] only in the appearance of nonlinear terms in the parameters k and ψ. We may therefore write the solution of the finite deformation problem as

$$n_r = k\left(1+\frac{1}{2}k\right)n_{rk} + \psi r_0\left(\lambda+\frac{1}{2}\psi r_0\right)n_{r\psi} \tag{4.13}$$

where n_{rk} and $n_{r\psi}$ are the same as those for the linear theory. Since these quantities have already been obtained in [6], we do not have to solve the present finite deformation problem anew. In particular, the stiffness relations (4.11) and (4.12) can be written as

$$\frac{PA}{2r_0} = (1+k)\left[k\left(1+\frac{1}{2}k\right)C^M_{Fk} + \psi r_0\left(\lambda + \frac{1}{2}\psi r_0\right)C^M_{F\psi}\right]$$

$$+\frac{\mu^2}{12(1+\nu)}(\lambda + \psi r_0)\left[k\lambda + \psi r_0(1+k)\right]\left[\frac{s}{\sqrt{1+\lambda^2 s^2}}\right]_{s_i}^{1} \tag{4.14a}$$

$$\frac{TA}{2r_0^2} = (\lambda + \psi r_0)\left[k\left(1+\frac{1}{2}k\right)C^M_{Tk} + \psi r_0\left(\lambda + \frac{1}{2}\psi r_0\right)C^M_{T\psi}\right]$$

$$+\frac{\mu^2}{12(1+\nu)}(1+k)\left[k\lambda + \psi r_0(1+k)\right]\left[\frac{s}{\sqrt{1+\lambda^2 s^2}}\right]_{s_i}^{1} \tag{4.14b}$$

where

$$C^M_{Fk} = \frac{1}{2}\int_{s_i}^{1}\frac{n_{\theta k}}{\sqrt{1+\lambda^2 s^2}}\,ds \quad , \quad C^M_{T\psi} = \frac{1}{2}\int_{s_i}^{1}\frac{\lambda s^2 n_{\theta\psi}}{\sqrt{1+\lambda^2 s^2}}\,ds$$

$$C^M_{F\psi} = \frac{1}{2}\int_{s_i}^{1}\frac{n_{\theta\psi}}{\sqrt{1+\lambda^2 s^2}}\,ds = \frac{1}{2}\int_{s_i}^{1}\frac{\lambda s^2 n_{\theta k}}{1+\lambda^2 s^2}\,ds = C^M_{Tk} \tag{4.14c}$$

(with $n_\theta + k\left(1+\frac{1}{2}k\right)n_{\theta k} + \psi r_0\left(\lambda + \frac{1}{2}\psi r_0\right)n_{\theta\psi}$) are the stiffness coefficients associated with the linear membrane shell action obtained in [6].

For a shell with a large pitch-to-width ratio so that $\lambda^2 \ll 1$, a perturbation solution in powers of λ^2 for the quasilinear theory of finite stretching and twisting is appropriate. We consider here only the special case of a pretwisted strip with $s_i = -1$. (Here, $r_i = -r_0$ corresponds to the image point of r_0 with respect to the z-axis.) Upon setting

$$\{n_r(s;\lambda), n_\theta(s;\lambda), \bar{u}(s;\lambda)\} \sim \sum_{i=0}^{\infty}\{n_{ri}(s), n_{\theta i}(s), \bar{u}_i(s)\}\lambda^{2i} \;, \tag{5.1}$$

we get from the results of [6] for $\lambda << 1$ the following leading term solution of the membrane resultants and radial displacement:

$$n_r \sim \frac{1}{2} k\left(1 + \frac{1}{2} k\right)\underline{\underline{\lambda^2}}\{-1 + s^2\} + \frac{1}{4} \psi r_0\left(\lambda + \frac{1}{2} \psi r_0\right)\underline{\underline{\lambda^2}}\{-1 + s^4\}$$

$$n_\theta \sim k\left(1 + \frac{1}{2} k\right) + \psi r_0\left(\lambda + \frac{1}{2} \psi r_0\right)s^2 \tag{5.2}$$

$$\bar{u} \sim k\left(1 + \frac{1}{2} k\right)\{-\nu s\} + \psi r_0\left(\lambda + \frac{1}{2} \psi r_0\right)\left\{-\frac{1}{3} \nu s^3\right\}$$

and from the expression for m in (4.9b) for the twisting stress couple

$$m \sim k\underline{\underline{\lambda}} + \psi r_0(1 + \underline{k}) \ . \tag{5.3}$$

For $\lambda << 1$, k is effectively the axial strain of the strip. Consistent with the infinitesimal strain approximation already made in our formulation, we should set $1 + ck = 1$ *for any* $0(1)$ quantity c. Also, we have

$$n_{rk} = 0(\lambda^2 n_{\theta k}) \quad , \quad n_{r\psi} = 0(\lambda^2 n_{\theta\psi}) \ ;$$

therefore, n_r is of order λ^2 compared to n_θ. Finally, the twisting stress σ_t induced by m_k is negligibly small relative to the direct hoop stress σ_d induced by $n_{\theta k}$, i.e.,

$$\frac{\sigma_{tk}}{\sigma_{dk}} = \frac{h}{2(1 + \nu)r_0} \frac{m_k}{n_{\theta k}} = 0(\mu\lambda) \ ,$$

so that to leading order, we should consistently omit the $k\lambda$ term in m. Altogether, these observations allow us to simplify (5.2) and (5.3) to get the following leading term solution for the stress resultants, twisting couple and radial displacement:

$$n_r \sim 0 \ , \quad n_\theta \sim k + \psi r_0\left(\lambda + \frac{1}{2} \psi_0 r\right)s^2$$

$$m \sim \psi r_0 \ , \quad \bar{u} \sim -\nu k s - \frac{1}{3} \nu \psi r_0\left(\lambda + \frac{1}{2} \psi r_0\right)s^3 \ . \tag{5.4}$$

Correspondingly the leading term overall load-deformation relations for $\lambda^2 << 1$ consistent with the simplified leading term resultants and couples in (5.4) are

$$\frac{PA}{2r_0} \sim k + \frac{1}{3}\psi r_0\left(\lambda + \frac{1}{2}\psi r_0\right) \tag{5.5a}$$

$$\frac{TA}{2r_0^2} \sim \frac{1}{3}k(\lambda + \psi r_0) + \frac{1}{5}\psi r_0(\lambda + \psi r_0)\left(\lambda + \frac{1}{2}\psi r_0\right) + \frac{\mu^2}{6(1+\nu)}\psi r_0 \,. \tag{5.5b}$$

We see from (5.5) that the shell stiffness comes mainly from its membrane action; bending is important only in the torque-twist relation and only when $|\lambda + \psi r_0| << \mu^2$ and $\left|\lambda + \frac{1}{2}\psi r_0\right| << \mu^2$.

The leading term solution (5.4) and (5.5) for $\lambda^2 << 1$ is identical to the exact solution of Marguerre's shallow shell equations for axial extension and torsion of pretwisted rectangular plates (with midsurface equation $z = \beta xy$). This exact solution is given by $w = \psi xy$ and $F = \left[12ky^2 + \psi(2\beta + \psi)y^4\right]/24A$ where w is the transverse midplane displacement of the plate, F the dual stress function and $\beta r_0 = \lambda$. Observe that there is a reciprocal relation inherent in the overall load-deformation relations in the sense that the expressions for P and T may be written as

$$P = \frac{\partial \Pi}{\partial k} \;, \quad T = \frac{\partial \Pi}{\partial \psi} \tag{5.6a,b}$$

where

$$\Pi = \frac{r_0}{A}\left\{k^2 + \frac{1}{3}k\psi r_0(2\lambda + \psi r_0) + \frac{1}{5}\left[(\psi r_0)^2\lambda^2 + (\psi r_0)^3\lambda + \frac{1}{4}(\psi r_0)^4\right] + \frac{\mu^2(\psi r_0)^2}{6(1+\nu)}\right\} \tag{5.6c}$$

with $\partial^2\Pi/\partial k\partial\psi = \partial^2\Pi/\partial\psi\partial k$. The following conclusions may be drawn from the overall load-deformation relations (5.5):

(1) For the limiting case of a flat strip with $\lambda = 0$, we have

$$n_r = 0 \quad , \quad n_\theta = k + \frac{1}{2}(\psi r_0)^2 s^2 \ ,$$

$$m = \psi r_0 \quad , \quad \overline{u} = -\nu k s - \frac{1}{6}\nu(\psi r_0)^2 s^3 \tag{5.7}$$

and

$$\frac{PA}{2r_0} = k + \frac{1}{6}(\psi r_0)^2 \ , \qquad \frac{TA}{2r_0^2} = \left[\frac{1}{3}k + \frac{1}{10}(\psi r_0)^2 + \frac{\mu^2}{6(1+\nu)}\right](\psi r_0) \ . \tag{5.8}$$

Unlike the linear case, the twisting and stretching actions of the shell remain coupled for a flat strip undergoing finite deformations.

(2) If the deformation due to the applied forces and torques is such that $\psi r_0 = -\lambda$, then the overall load-deformation relations become

$$\frac{PA}{2r_0} = k + \frac{\lambda^2}{6} \quad , \quad \frac{TA}{2r_0^2} = -\frac{\mu^2 \lambda}{6(1+\nu)} \ . \tag{5.9}$$

The expression for T in this case implies the somewhat unexpected results that the shell cannot be flattened by axial forces alone!

(3) If on the other hand, the deformation is such that $\psi r_0 = -2\lambda$, then (5.8) becomes

$$\frac{PA}{2r_0} = k \quad , \quad \frac{TA}{2r_0^2} = -\frac{\lambda}{3}\left[k + \frac{\mu^2}{1+\nu}\right] \ . \tag{5.10}$$

We have then another somewhat unexpected result. If the shell is deformed into its image shape by axial torques alone, then there would be no axial displacement associated with the deformation. In fact, there would be no membrane shell action at all in the sense that $n_\theta = n_r = \overline{u} = 0$.

(4) If $P = 0$, then we have $k = -\psi r_0\left(\lambda + \frac{1}{2}\psi r_0\right)/3$ and therewith

$$\frac{TA}{2r_0^2} \sim \psi r_0\left[\frac{2}{45}(2\lambda + \psi r_0)(\lambda + \psi r_0) + \frac{\mu^2}{6(1+\nu)}\right] \tag{5.11}$$

$$= \lambda^3(\psi a)\left[\frac{4}{45}(1 + \psi a)\left(1 + \frac{1}{2}\psi a\right) + \frac{(\mu/\lambda)^2}{6(1 + \nu)}\right] .$$

For an initially flat strip $(\lambda = 0)$, we have

$$\frac{TA}{2r_0^2} \sim \psi r_0\left[\frac{2}{45}(\psi r_0)^2 + \frac{\mu^2}{6(1 + \nu)}\right] \tag{5.12}$$

so that T is a monotone increasing function of ψ. For strips with an initial pretwist, a plot of T vs ψ is given in Figure (2) for different values of the thickness parameter $\mu = h/r_0$. For a fixed value of μ, and $\lambda \le \mu$, we see that T continues to be a monotone increasing function of ψ for all ψ. In fact, a straightforward calculation shows that T has no stationary value for $(\mu/\lambda)^2 > 4(1 + \nu)/15$ and two stationary values for smaller values of (μ/λ) attained at

$$\psi a = -1 \pm \left[\frac{1}{3} - \frac{5}{4}\frac{(\mu/\lambda)^2}{1 + \nu}\right]^{1/2} < 0 . \tag{5.13}$$

There are in principle three different possible (negative) values of ψ corresponding to the same value of the end torque T. The corresponding graph for the strain energy of the shell (Figure 3) with two stationary points at

$$\psi a = \frac{1}{2}\left\{-3 \pm \left[1 - \frac{15}{1 + \nu}\left(\frac{\mu}{\lambda}\right)^2\right]^{1/2}\right\} \tag{5.14}$$

for $(\mu/\lambda)^2 < (1 + \nu)/15$ suggests the possibility of a snap-through at some critical value of $T\,(< 0)$ for a given value of $(\mu/\lambda)^2 << 1$.

(5) If $T = 0$, we can in principle solve (5.5b) to get $\psi = \psi(k)$ and use it to eliminate ψ from (5.5a) to get $P = P(k)$. For $\mu^2/|\lambda + \psi r_0| << 1$, we have the linear stiffness relation $PA/2r_0 \sim 4k/9$ in the absence of axial torques.

VI. HELICOIDAL RING SHELLS WITH A SMALL PITCH

A helicoidal ring shell with $0 < s_i < 1$ is said to be shallow if its pitch is small so that $(\lambda s_i)^2 = (r_i/a)^2 >> 1$. For a shallow helicoidal ring shell, a

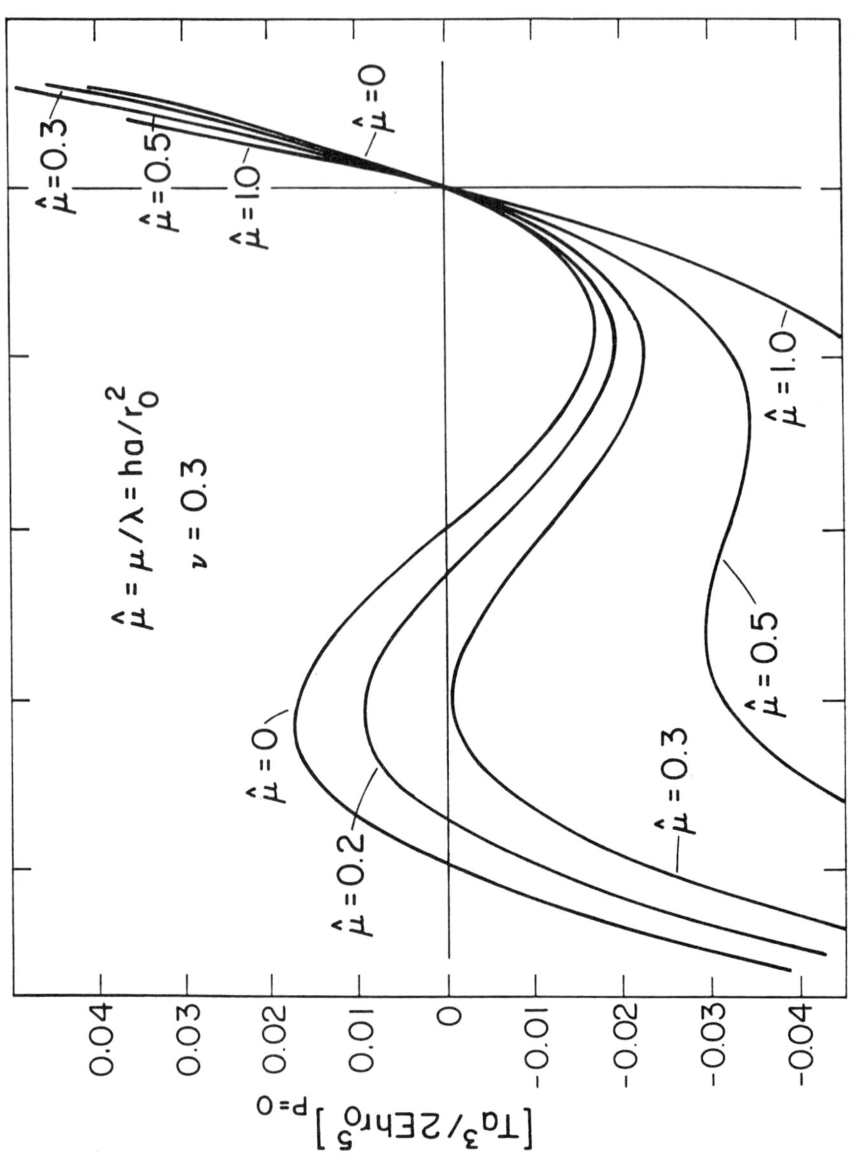

2. Torque-twist relations in the absence of axial forces.

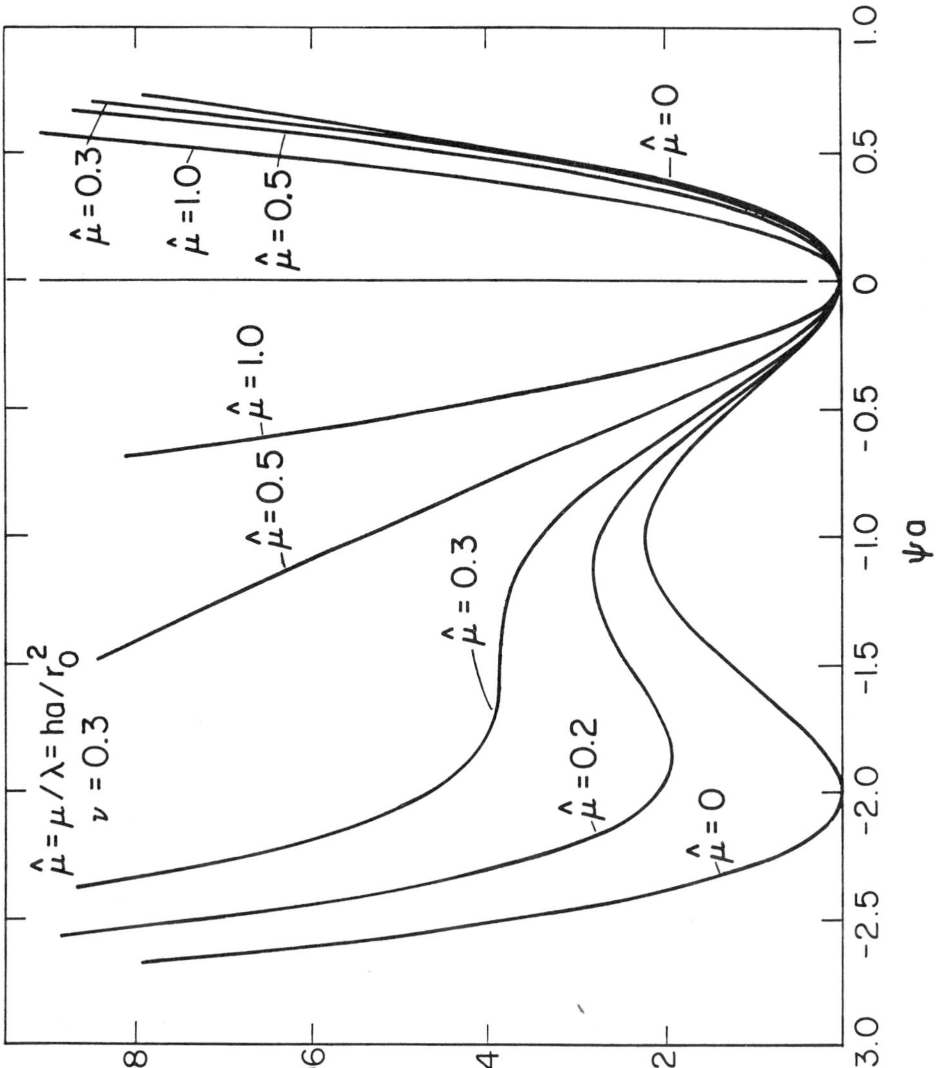

3. Energy-twist relations in the absence of axial forces.

perturbation solution in powers of $\varepsilon^2 \equiv \lambda^{-2}$ for the finite stretching and twisting problem is appropriate. Upon setting

$$\{\bar{u}(s;\lambda), n_r(s;\lambda), n_\theta(s;\lambda)\} \sim \sum_{j=0}^{\infty} \{\hat{u}_j(s), \hat{n}_{rj}(s), \hat{n}_{\theta j}(s)\}\varepsilon^{2j} , \tag{6.1}$$

we get from the results of [6] for $(r_i/a)^2 >> 1$ the following leading term solution

$$\begin{aligned} n_r \sim\ & k_\varepsilon\left(\varepsilon + \frac{1}{2} k_\varepsilon\right)\left[A_{k0} + B_{k0}s^{-2} + \frac{1}{4}s^{-2}(1 + 2\ell n s)\right] \\ & + \psi_\varepsilon r_0\left(1 + \frac{1}{2}\underline{\psi_\varepsilon r_0}\right)\left[A_{\psi 0} + B_{k0}s^{-2} + \frac{1}{2}\ell n s\right] \\ n_\theta \sim\ & k_\varepsilon\left(\varepsilon + \frac{1}{2} k_\varepsilon\right)\left[A_{k0} - B_{k0}s^{-2} + \frac{1}{4}s^{-2}(1 - 2\ell n s)\right] \\ & + \psi_\varepsilon r_0\left(1 + \frac{1}{2}\underline{\psi_\varepsilon r_0}\right)\left[A_{\psi 0} - B_{\psi 0}s^{-2} + \frac{1}{2}(1 + \ell n s)\right] \\ \bar{u}_r \sim\ & k_\varepsilon\left(\varepsilon + \frac{1}{2} k_\varepsilon\right)\left[A_{k0}(1-\nu)s - B_{k0}(1+\nu)s^{-1} + \frac{1}{4}s^{-1}\{(3+\nu) + 2(1+\nu)\ell n s\}\right] \\ & + \psi_\varepsilon r_0\left(1 + \frac{1}{2}\underline{\psi_\varepsilon r_0}\right)\left[A_{\psi 0}(1-\nu)s - B_{\psi 0}(1+\nu)s^{-1} + \frac{1}{2}s\{(1-\nu)\ell n s - 1\}\right] \end{aligned} \tag{6.2}$$

where

$$\begin{aligned} A_{k0} &= \frac{\ell n s_i}{2(1-s_i^2)} \qquad B_{k0} = -\frac{1}{4}\left[1 + \frac{2\ell n s_i}{1-s_i^2}\right] \\ A_{\psi 0} &= -B_{\psi 0} = \frac{s_i^2 \ell n s_i}{2(1-s_i^2)} \end{aligned} \tag{6.3}$$

and where $k_\varepsilon \equiv k\varepsilon$ and $\psi_\varepsilon \equiv \psi\varepsilon$ remain finite as $\varepsilon \to 0$ in order for the expressions in (2.8) to be meaningful. Also, from the expression for m in (4.9b), we have the following leading term solution for the twisting couple

$$\mathrm{m} \sim \mathrm{s}^{-2}\left[\mathrm{k}_\varepsilon\left(1 + \underline{\psi_\varepsilon \mathrm{r}_0}\right) + \psi_\varepsilon \mathrm{r}_0 \varepsilon\right] . \tag{6.4}$$

With $u_\theta = \left[\psi a + ur^{-1}(1+\psi a)\right]r\theta = \left[\psi_\varepsilon r_0 + ur^{-1}(1+\psi_\varepsilon r_0)\right]r\theta$ giving the circumferential displacement component and u/r of the order of the circumferential strain, we have for $\varepsilon << 1$, $u_\theta/r = 0(\psi a\theta) = 0(\psi_\varepsilon r_0\theta)$ also of the order of magnitude of the circumferential strain so that $1 + c(\psi_\varepsilon r_0) \simeq 1$ for any $0(1)$ quantity c. The underlined terms in (6.2) and (6.4) should therefore be omitted in order to be consistent with the approximation inherent in the present formulation of the problem. Also, we have

$$n_{\theta\psi} = 0(\psi_\varepsilon r_0) \quad , \quad m_\psi = 0(\psi_\varepsilon r_0 \varepsilon s^{-2}) \tag{6.5}$$

so that

$$\frac{\sigma_{t\psi}}{\sigma_{d\psi}} = \frac{h}{2(1+\nu)r_0}\frac{m_\psi}{n_{\theta\psi}} = 0(\mu\varepsilon) \ . \tag{6.6}$$

To the leading order, we should consistently omit the $\psi_\varepsilon r_0\varepsilon$ term in m also.

Altogether, the above two observations allow us to simplify (6.2) by omitting all underlined terms (which are nonlinear in ψ_ε) in the expressions for n_r, n_θ and $\bar{u}$ (with the resulting expression labelled (6.2')). They also allow us to simplify (6.4) to

$$m \sim k_\varepsilon s^{-2} \ . \tag{6.4'}$$

Correspondingly, the leading term overall load-deformation relations for $\varepsilon^2 << 1$, consistent with these simplified leading resultants and couple, are

$$\frac{PA}{2r_0} \sim \left\{ k_\varepsilon \frac{1-s_i^2}{s_i^2}\left[\frac{\mu^2}{24(1+\nu)}\right] + \frac{1}{16}k_\varepsilon(\varepsilon + k_\varepsilon)\left(\varepsilon + \frac{1}{2}k_\varepsilon\right)\left[1 - \left(\frac{2s_i \ell n s_i}{1-s_i^2}\right)^2\right] \right.$$

$$\left. + \frac{1}{4}\psi_\varepsilon r_0(\varepsilon + k_\varepsilon)\left(1 + \frac{1}{2}\underline{\psi_\varepsilon r_0}\right)\ell n s_i\left[1 + \frac{1+s_i^2}{1-s_i^2}\ell n s_i\right] \right\} \tag{6.7a}$$

$$\frac{TA}{2r_0^2} \sim \frac{1}{4}k_\varepsilon\left(\varepsilon + \frac{1}{2}k_\varepsilon\right)\left(1 + \underline{\psi_\varepsilon r_0}\right)\ell n s_i\left[1 + \frac{1+s_i^2}{1-s_i^2}\ell n s_i\right]$$

$$+ \frac{1}{16}(\psi_\varepsilon r_0)\left(1 - s_1^2\right)\left[1 - \left(\frac{2s_i \ell n s_i^2}{1-s_i^2}\right)^2\right] \tag{6.7b}$$

where the underlined $\psi_\varepsilon r_0$ terms are of order of the hoop strain and are to be deleted. Again, the shell stiffness comes almost exclusively from membrane shell action; but in contrast to $\lambda^2 << 1$, bending is now only important in the force-extension relation and only when $|\varepsilon + k_\varepsilon|\left|\varepsilon + \frac{1}{2}k_\varepsilon\right| = 0(\mu^2)$.

The leading term solution (6.2'), (6.4') and (6.7) for $(\varepsilon/s_i)^2 << 1$ is identical to the exact solution of Marguerre's shallow shell equations for the problem of bending and twisting of helicoidal ring shells (see [3]). It is not difficult to see that there is a reciprocity relation inherent in (6.7) in the sense that the expressions for P and T in (4.11) may be written as

$$P = \frac{\partial \widehat{\Pi}}{\partial k_\varepsilon} \quad , \quad T = \frac{\partial \widehat{\Pi}}{\partial \psi_\varepsilon} \tag{6.8}$$

where

$$\begin{aligned}\frac{A}{r_0}\widehat{\Pi} &= \frac{k_\varepsilon^2}{16}\left[\left(\varepsilon^2 + k_\varepsilon\varepsilon + \frac{1}{4}k_\varepsilon^2\right) + \frac{2\mu^2}{3(1+\nu)}\right] \\ &+ \frac{1}{2}k_\varepsilon\left(\varepsilon + \frac{1}{2}k_\varepsilon\right)\psi_\varepsilon r_0 \ell n s_i\left[1 + \frac{1+s_i^2}{1-s_i^2}\ell n s_i\right] \\ &+ \frac{1}{16}(\psi_\varepsilon r_0)^2(1-s_i^2)\left[1 - \left(\frac{2s_i \ell n s_i}{1-s_i^2}\right)^2\right]\end{aligned} \tag{6.9}$$

with $\partial^2\widehat{\Pi}/\partial k_\varepsilon \partial \psi_\varepsilon = \partial^2\widehat{\Pi}/\partial \psi_\varepsilon \partial k_\varepsilon$. The following conclusions may now be drawn from (6.7)

(1) For the limiting case $\varepsilon = 0$, the overall load-deformation relations (6.7) reduce to

$$\begin{aligned}\frac{PA}{2r_0} = k_\varepsilon \Bigg\{ &\frac{1-s_i^2}{s_i^2}\,\frac{\mu^2}{24(1+\nu)} + \frac{1}{32}k_\varepsilon^2\left[1 - \left(\frac{2s_i \ell n s_i}{1-s_i^2}\right)^2\right] \\ &+ \frac{1}{4}\psi_\varepsilon r_0 \ell n s_i\left[1 + \frac{1+s_i^2}{1-s_i^2}\ell n s_i\right]\Bigg\}\end{aligned} \tag{6.10a}$$

$$\frac{TA}{2r_0^2} \sim \frac{1}{8}k_\varepsilon^2 \ell n s_i \left[1 + \frac{1+s_i^2}{1-s_i^2}\ell n s_i\right]$$
$$+ \frac{1}{16}(\psi_\varepsilon r_0)\left(1 - s_1^2\right)\left[1 - \left(\frac{2s_i \ell n s_i}{1-s_i^2}\right)^2\right] . \tag{6.10b}$$

Unlike the linear case analyzed in [6], the twisting and stretching actions of the shell remain coupled for ring plate sector undergoing finite deformations.

(2) If the deformation due to the applied forces and torques is such that $k_\varepsilon = -\varepsilon$, then (6.7) becomes

$$\frac{PA}{2r_0} \sim k_\varepsilon \frac{1-s_i^2}{s_i^2}\frac{\mu^2}{24(1+\nu)} \tag{6.11a}$$

$$\frac{TA}{2r_0^2} \sim \frac{1}{8}k_\varepsilon \varepsilon \ell n s_i \left[1 + \frac{1+s_i^2}{1-s_i^2}\ell n s_i\right]$$
$$+ \frac{1}{16}(\psi_\varepsilon r_0)\left(1 - s_1^2\right)\left[1 - \left(\frac{2s_i \ell n s_i^2}{1-s_i^2}\right)^2\right] . \tag{6.11b}$$

The expression for P in this case implies that the shell cannot be flattened by axial torques alone.

(3) If, on the other hand, the deformation is such that $k_\varepsilon = -2\varepsilon$, then we have

$$\frac{PA}{2r_0} \sim -\varepsilon\left\{\frac{1-s_i^2}{s_i^2}\frac{\mu^2}{24(1+\nu)} + \frac{1}{4}(\psi_\varepsilon r_0)\ell n s_i\left[1 + \frac{1+s_i^2}{1-s_i^2}\ell n s_i\right]\right\} \tag{6.12a}$$

$$\frac{TA}{2r_0^2} \sim \frac{1}{16}(\psi_\varepsilon r_0)\left(1 - s_i^2\right)\left[1 - \left(\frac{2s_i \ell n s_i}{1-s_i^2}\right)^2\right] . \tag{6.12b}$$

If the shell is deformed into its mirror image shape by axial forces alone, then there would be no circumferential displacement associated with the deformation. In fact, there would be no membrane action at all given that we have $n_\theta \equiv n_r \equiv \bar{u} \equiv 0$ for this case.

(4) If $T = 0$, then we have

$$\frac{1}{4}\psi_\varepsilon r_0 \sim -\frac{k_\varepsilon\left(\varepsilon + \frac{1}{2}k_\varepsilon\right)\ell n s_i}{1 - s_i^2}\left[\frac{1 + \dfrac{1 + s_i^2}{1 - s_i^2}\ell n s_i}{1 - \left(\dfrac{2s_i \ell n s_i}{1 - s^2}\right)^2}\right] \tag{6.13}$$

$$\equiv -k_\varepsilon\left(\varepsilon + \frac{1}{2}k_\varepsilon\right)g(s_i)$$

and therewith

$$\frac{PA}{2r_0} \sim k_\varepsilon\left\{\frac{1 - s_i^2}{s_i^2}\frac{\mu^2}{24(1 + \nu)} + (\varepsilon + k_\varepsilon)\left(\varepsilon + \frac{1}{2}k_\varepsilon\right)f(s_i)\right\} \tag{6.14a}$$

where

$$f(s_i) = \frac{1}{16}\left[1 - \left(\frac{2s_i \ell n s_i}{1 - s_i^2}\right)^2\right] - g(s_i)\ell n s_i\left[1 + \frac{1 + s_i^2}{1 - s_i^2}\ell n s_i\right] \tag{6.14b}$$

with $g(s_i)$ defined in (6.13). This result was previously obtained and discussed in [3]. We merely note an additional observation on this stiffness relation as ε increases from zero. For $\varepsilon = 0$, P is a monotone increasing function of k_ε as $f(s_i)$ was found to be nonnegative in [3]. As ε increases from zero but remains less than μ, this monotone increasing property persists. However, for $\varepsilon > \mu$ and $k_\varepsilon < 0$, there are in principle three possible values of k_ε, which requires the same axial force P to maintain. However, the strain energy of the shell for the three different equilibrium configurations suggests the possibility of a snap-through at some critical value of $P > 0$.

(5) If $P = 0$, we can solve the first equation (6.7) for k_ε in terms of ψ_ε.

We can then eliminate k_ε from (6.7b) to get $T = T(\psi_\varepsilon)$. For $\mu^2/|\varepsilon + k_\varepsilon| << \left|\varepsilon + \frac{1}{2}k_\varepsilon\right|$, we have to obtain a good first approximation for the following linear stiffness relation:

$$\frac{TA}{2r_0^2} \sim \psi_\varepsilon r_0 \left\{ \frac{1}{16}\left(1 - s_i^2\right)\left[1 - \left(\frac{2s_i^2 \ell n s_i}{1 - s_i^2}\right)^2\right] - \left[\ell n s_i + \frac{1 + s_i^2}{1 - s_i^2}(\ell n s_i)^2\right]^2 \right\}$$

in the absence of axial forces.

7. OTHER PERTURBATION SOLUTIONS

Instead of a parametric series solution in the width-to-pitch ratio λ, we could have sought a perturbation solution in terms of the parameter

$$\eta = \frac{\lambda}{\sqrt{1 + \lambda^2}} = \frac{r_0}{\sqrt{a^2 + r_0^2}} . \tag{7.1}$$

Note that we have $\eta < 1$ for all $a > 0$ whether or not we have $a >> r_0$. Perturbation series of the form (5.1) with λ replaced by η (first suggested in [4]) are applicable for all (nonnegative) λ.

For a highly pretwisted strip (with $a/r_0 << 1$) which spans the range $-r_0 \leq r \leq r_0$ radially, a perturbation solution in $\varepsilon = 1/\lambda = a/r_0$ similar to that of section (6) is no longer appropriate. A leading term solution for that type of parametric series would correspond to neglecting a^2 terms in (4.4) leaving us with

$$n_0'' + \frac{3}{s} n_0' = \frac{1}{s^2} \psi_\varepsilon r_0\left(1 + \frac{1}{2} \psi_\varepsilon r_0\right) - \frac{1}{s^4} k_\varepsilon\left(1 + \frac{1}{2} k_\varepsilon\right) . \tag{7.2}$$

The leading term approximation for the dimensionless radial resultant AN_r has a singularity at $s = 0$, the center of the pretwisted strip. Yet we do not expect any singular behavior throughout the strip for $a > 0$. The singularity of the ODE (7.2) is in fact spuriously introduced by a perturbation solution process which is inappropriate for this type of problem. The appropriate perturbation scheme [1] requires that we seek the primary unknown $n \equiv AN_r$ and the auxiliary stress and displacement variables as functions of a new independent variable x and the parameter ε. The solution scheme requires that we take n and the independent variable s as functions of x and the parameter $\varepsilon = a/r_0$ and expand both in powers of ε:

$$n(s;\varepsilon) = \sum_{i=0}^{\infty} \tilde{n}_i(x)\varepsilon^{2i} \quad , \quad s = \sum_{i=0}^{\infty} s_i(x)\varepsilon^{2i} \ . \tag{7.3}$$

The details of this Poincaré-Lighthill type perturbation solution have already been worked out for the linear theory in [4] and will not be repeated here. We note only that for a shell-theoretic solution to be useful, we need $h/r_0 << a/r_0$; otherwise the stress distribution near the central axis of the helicoidal shell should be calculated by the three-dimensional theory of elasticity.

REFERENCES

1. J. Kevorkian and J.D. Cole, *Springer-Verlag, New York-Heidelberg-Berlin* (1981).
2. J.K. Knowles and E. Reissner, *Quart. Appl. Math.* 17 409-422 (1959)
3. E. Reissner, *Quart. Appl. Math.* 11, 473-483 (1954).
4. E. Reissner, *IUTAM Symp. on Shell Theory*, 434-466 (1959), Amsterdam (1960).
5. E. Reissner, *Develop. in Mech.* 3 *(Proc. 9th Midwestern Conf. on Mech.)*, 55-58 (1967), Madison, Wisconsin (1965).
6. E. Reissner and F.Y.M. Wan, *J. Math. & Phys.* 4, 1-31 (1968).
7. R.G. Sinclair, *Ph.D. Thesis, M.I.T.* (Sept. 1960).

Some Properties of Convolution Sequences and Asymptotics for the Taylor Coefficients for Products of Bessel Functions

Jet Wimp Professor, Department of Mathematics and Computer Sciences, Drexel University, Philadelphia, Pennsylvania

1. INTRODUCTION AND PRELIMINARIES

Let $\{a_n\}$, $\{b_n\}$ be sequences of real numbers. We define the convolution of $\{a_n\}$ and $\{b_n\}$ to be the sequence $\{c_n\}$ defined by

This research was supported by the National Science Foundation under Research Contract #DMS 8802381

The author thanks Doron Zeilberger, who posed a problem that led to the present investigation.

$$\{a_n * b_n\}_n := c_n = \sum_{k=0}^{n} a_k b_{n-k}\,, \qquad n=0,1,2\ldots.. \tag{1}$$

If the a_n are the coefficients of a formal power series f(t) and b_n the coefficients of another formal power series g(t), then, obviously, the convolution sequence c_n are the coefficients of the formal series f(t)· g(t).

One question is, if a_n and b_n have an asymptotic representations as, say, generalized Birkhoff-Poincaré asymptotic series (Birkhoff called such series *subnormal forms*, see Wimp(1984), Appendix B), does $\{a_n*b_n\}_n$ have a similar representation? Even if so, as the following example makes clear, the convolution may not be represented by a single such series even though the original sequences are represented by series of the same type.

Theorem 1

Let $s_1, s_2 \neq 1, 2, 3, \ldots\ldots, v_1, v_2 \neq -1, -2, -3, \ldots.$ be real numbers. Then

$$\left\{(v_1+n)^{-s_1} * (v_2+n)^{-s_2}\right\}_n \sim \frac{\Gamma(1-s_1)\,\Gamma(1-s_2)}{\Gamma(2-s_1-s_2)}\, n^{1-s_1-s_2}\left[1 + a_1/n+\ldots\right] + \zeta(s_2,v_2)n^{-s_1}\left[1 + b_1/n +\ldots\right]$$
$$+ \zeta(s_1,v_1)n^{-s_2}\left[1 + c_1/n +\ldots\right]\,. \tag{2}$$

Proof: For the special functions involved, see Erdélyi et al (1953). The proof depends on the theory of the function

$$\Phi(t,s,v) := \sum_{n=0}^{\infty} (v+n)^{-s}\, t^n\,, \qquad |t| < 1. \tag{3}$$

Using formula of the above reference, v. 1, p.29(8), shows that

$$\Phi = \Gamma(1-s)(1-t)^{s-1}[1+O(1-t)] + \zeta(s,v) + O(1-t), \quad z\text{-->}1.\ \mathrm{Re}\, t <1\ . \tag{4}$$

Forming the product of two such functions gives a function analytic in the unit disk whose coefficients are the convolution sequence, and the behavior of this function near t = 1 is easily determined from the above formula. Applying Darboux's method gives the lead terms in the theorem; it is easily seen the terms are the lead terms of complete asymptotic expansions in powers of n^{-1}.

In this case a_n and b_n have similar asymptotic representations as simple Poincaré series, e.g.,

$$a_n \sim n^{-s_1}\{1 + c_1/n + c_2/n^2 + \dots\}. \tag{5}$$

In this paper we shall discuss the case when $a_n = b_n$, i.e., the convolution of a sequence with itself, and later, we shall deal with an important case of a repeated convolution, $\{a_n*a_n*a_n* \dots *a_n\}$. It turns out that if a_n satisfies a linear homogeneous difference equation with polynomial coefficients, so does the convolution sequence. Thus, when a_n has a representation as a generalized asymptotic series, so does the convolution sequence, since the solution of such an equation may asymptotically be represented by a linear combination of such series.To show that the convolution sequence satisfies a difference equation when the original sequence does, we first we establish a lemma. We assume the differential equations or difference equations in question are non-trivial, i.e., if the order of the equation is stated to be s, then the coefficient of the sth derivative of y, or of y(n+s) in the case of a difference equation, is non-zero.

Lemma

Let $z = z(t)$ satisfy a linear homogeneous differential of order s with rational coefficients. Then $w = z^2$ satisfies a linear homogeneous differential equation with rational coefficients of order s(s+1)/2 or less.

Proof: Differentiating repeatedly we can write

$$w^{(r)} = \sum_{j=0}^{\leq r/2} \alpha_{rj}\, z^{(j)}\, z^{(r-j)}, \quad r = 0,1,2,3\dots\dots \tag{6}$$

We also have

$$z^{(m)} = \sum_{p=0}^{s-1} \beta_{m\,p}\, z^{(p)}, \quad m = 0,1,2\dots, \ \beta_{m\,p} = \delta_{m\,p}, \ m \leq s-1, \ \beta_{m\,p} \text{ rational.} \tag{7}$$

Substituting this into the sum in (6) gives

$$w^{(r)} = \sum_{j=0}^{\leq r/2} \alpha_{rj} \sum_{p,q=0}^{s-1} \beta_{j\,p}\beta_{r-j\,q}\, z^{(p)}\, z^{(q)}, \quad r = 0,1,2,3\dots \tag{8}$$

If r goes from zero to R, we can consider this a system of R+1 linear equations in the unknowns $z^{(m)}z^{(n)}$, $0 \leq m \leq n \leq s-1$, of which are s(s+1)/2. Thus taking R =

s(s+1)/2 we have a system of R+1 equations in R unknowns. The first two rows of the augmented matrix of the system are [1 0 0 ..... w], [0 2 0 0...w']. Performing Gauss-Jordan reduction will yield a matrix with, say, m rows at the bottom, each having R leading zeros, in which case setting to zero the $R+1^{st}$ entry of the $R+2\text{-}m^{th}$ row will produce a differential equation for w of order s(s+1)/2 +1-m. The worst case scenario as far as the order of the equation will be when m = 1. Note because of the non-zero leading entries in the first two rows, there will always be a legitimate differential equation. As an example, for z''-z=0, the procedure yields the differential equation w'' - 4w = 0, yet s = 3.

Theorem 2

Let the sequence a_n satisfy a linear homogeneous difference equation with polynomial coefficients. Then $\{a_n * a_n\}_n$ satisfies a linear homogeneous difference equation with polynomials coefficients.

Proof: We can write the equation as

$$\sum_{r=0}^{\sigma} p_r(n+r)\, a_{n+r} = 0 \ , \qquad n = 0,1,2..... \ . \tag{9}$$

where p_r is a polynomial and, say, $\sigma>0$, $p_\sigma \neq 0$. We define the series

$$f(t) := \sum_{n=0}^{\infty} a_n\, t^n \ , \tag{10}$$

where the sum is to be interpreted as a formal series taken from the ring of formal power series with complex coefficients. (For a treatment of how these series can be used in analysis, in particular, how formal differential equations can be solved by such series, see Wimp and Boyer (1989).) Multiplying the equation by t^n and summing gives

$$\sum_{r=0}^{\sigma} t^{-r}\, p_r(\delta)\, a_{n+r}\, t^{n+r} = 0, \qquad \delta = tD, \ D = d/dt \ , \tag{11}$$

or

$$\sum_{r=0}^{\sigma} t^{-r}\, p_r(\delta)\, \{f(t) - a_0 - a_1 t - ..... - a_{r-1} t^{r-1}\} = 0 \ . \tag{12}$$

Putting all terms which do not involve f on the right hand side gives

$$\Theta(f(t)) := \sum_{r=0}^{\sigma} t^{\sigma-r}\, p_r(\delta)\, f(t) = V(t) \ . \tag{13}$$

where V is a polynomial. There are two cases to distinguish. When V = 0, we have a differential equation for f, and the lemma tells us we then can find a differential equation for f^2. If $V \neq 0$, we form the operator $V\Theta' - V'\Theta$ which annihilates the right hand side, and we again can find a differential equation for f^2. In either case, the differential equation has polynomial coefficients, and its coefficients are the convolution of a_n with itself. We know f^2 has a formal series expansion, and when we seek to obtain it from its differential equation in the usual way, the result is a linear homogeneous difference equation for the convolution sequence.

2. THE DIFFERENTIAL EQUATION

In this section we will develop the theory of a differential equation which we will later apply to study the recurrence satisfied by the power series coefficients of $t^{-\nu r}J_\nu{}^r$, which, of course, are just the r-fold convolution of the coefficients for $t^{-\nu}J_\nu$ with themselves.

Theorem 3

Let z=z(t) satisfy the second order linear homogeneous differential equation

$$z'' + qz = 0, \qquad q = q(t), \tag{14}$$

where q is a rational function. Then

$$w = z^r \ , \tag{15}$$

satisfies the differential equation of order r+1,

$$\Theta(w) = w^{(r+1)} + a_0(t)w^{(r-1)} + a_1(t)w^{(r-2)} + \ldots + a_{r-1}(t)w = 0 \ , \tag{16}$$

where the a_j are rational functions of t.

Remark: Actually the result holds whether or not q is rational, but the theorem is not then very descriptive, since any suitably differentiable function satisfies a differential equation of arbitrary order.

Proof: We have

$$\begin{aligned}
&w=z^r,\\
&w'=rz^{r-1}z',\\
&w''=rz^{r-1}z''+r(r-1)z^{r-2}z'^2=-qrw+r(r-1)z^{r-2}z'^2,\\
&w'''=-q(6r-8)w''-q'(4r-2)w'-rw(q''+3(r-2)q^2)+r(r-1)(r-2)z^{r-3},
\end{aligned} \tag{17}$$

etc.. By induction, we see we may write

$$w^{(k)}=g_k^r(w)+r(r-1)(r-2)\ldots(r-k+1)z^{r-k}z'^k, \quad k=1,2,3\ldots \quad . \tag{18}$$

A recurrence relation for $g_k{}^r$ can be found by differentiating the above equation. We find

$$\begin{aligned}
w^{(k+1)}&=g_k^{r\,\prime}-kqr(r-1)(r-2)\ldots\ldots(r-k+1)z^{r+1-k}z'^{r-1}\\
&=g_k^{r\,\prime}-kq(r-k+1)[w^{(k-1)}-g_{k-1}^r] ,
\end{aligned} \tag{19}$$

which means that

$$g_{k+1}^r=g_k^{r\,\prime}-qk(r-k+1)[w^{(k-1)}-g_{k-1}^r], \quad k=0,1,2,\ldots\ldots, \; g_o^r=g_1^r=0. \tag{20}$$

By putting $k = r+1$ in (18) we see that w satisfies the differential equation

$$w^{(r+1)}=g_{r+1}^r \quad . \tag{21}$$

This is a differential equation of the form in the Theorem, and the formula (20) provides an algorithm for constructing the equation. Thus, for example, $w = z^2$ satisfies

$$w'''+4qw'+2q'w=0; \tag{22}$$

$w = z^3$ satisfies

$$w^{IV}+10qw''+10q'w'+(3q''+9q^2)w=0; \tag{23}$$

$w=z^4$ satisfies

$$w^{V}+20qw'''+30q'w''+(18q''+64q^2)w'+(64qq'+4q''')w=0 . \tag{24}$$

For our applications to Bessel functions we need

Theorem 4.

Let $\Omega(w)$ denote that part of Θ which is independent of q', q'', q''',... . Then for r even,

$$\Omega = DQ(D^2) := D(D^2 + qr^2)(D^2 + q(r-2)^2).....(D^2 + 4q) \ , \tag{25}$$

and for r odd

$$\Omega = P(D^2) := (D^2 + qr^2)(D^2 + q(r-2)^2).....(D^2 + q) \quad . \tag{26}$$

Proof: To obtain the portion of Θ which is independent of q' and its derivatives, we may take q to be constant in the equation

$$z'' + qz = 0 \quad , \tag{27}$$

which means

$$z = a_1 \cos(\sqrt{q}\, rt) + a_2 \sin(\sqrt{q}\, rt) \ . \tag{28}$$

Then Ω will be just the differential operator which annihilates

$$w = (\cos(\sqrt{q}\, t))^r \ . \tag{29}$$

We may write

$$w = c_1 \cos(\sqrt{q}\, rt) + c_2 \cos(\sqrt{q}\, (r-2)t) + c_3 \cos(\sqrt{q}\, (r-4)t) + ... \tag{30}$$

Forming the product of the differential operators which annihilate the successive terms of the above expression gives the theorem.

3. BESSEL FUNCTIONS

The differential equation satisfied by

$$z = \sqrt{t}\, J_v(t) \ , \tag{31}$$

is

$$w'' + qw = 0 \; , \quad q = 1 + \frac{(\frac{1}{4} - v^2)}{t^2} \; . \tag{32}$$

see Watson (1962). We assume Rev > -1. We write

$$w := z^r = \left\{\sqrt{t}\, J_v(t)\right\}^r := t^{\frac{r}{2} + vr} \sum_{n=0}^{\infty} A_n^r t^{2n} \; . \tag{33}$$

To obtain the recurrence for A_n^r, we put the above expression into the differential equation of the previous section satisfied by w, with the appropriate choice of q. We then equate like powers of t to zero, and this gives us a recurrence satisfied by A_n^r. To obtain some insight into the nature of the recurrence, we carry this process out for the case r = 4. We have

$$\begin{aligned}
w &= \sum_{n=0}^{\infty} A_n^r\, t^{2n+4v+2} \; , \\
w' &= \sum_{n=0}^{\infty} A_n^r\, (2n+4v+2)\, t^{2n+4v+1} \; , \\
w'' &= \sum_{n=0}^{\infty} A_n^r\, (2n+4v+2)(2n+4v+1)\, t^{2n+4v} \; , \\
w''' &= \sum_{n=0}^{\infty} A_n^r (2n+4v+2)(2n+4v+1)(2n+4v)\, t^{2n+4v-1} \; , \\
w^{IV} &= \sum_{n=0}^{\infty} A_n^r\, (2n+4v+2)\ldots(2n+4v-1)\, t^{2n+4v-2} \; , \\
w^{V} &= \sum_{n=0}^{\infty} A_n^r\, (2n+4v+2)\ldots(2n+4v-2)\, t^{2n+4v-3} \; .
\end{aligned} \tag{34}$$

We put these expressions into the differential equation, which in this case is

$$w^V + \left(20 + \frac{5}{t^2}\right) w''' - \frac{15}{t^3} w'' + \left(64 + \frac{32}{t^2} + \frac{31}{t^4}\right) w' - \left(\frac{32}{t^3} + \frac{32}{t^5}\right) w = 0 \; . \tag{35}$$

Equating coefficients of $t^{2n+4v-3}$ gives the following second order recurrence:

$$\begin{aligned}
&[(2n+4v+2)(2n+4v+1)(2n+4v)(2n+4v-1)(2n+4v-2) + 5(2n+4v+2)(2n+4v+1)(2n+4v) \\
&+31(2n+4v+2) - 32]A_n^4 + [40(n+2v)(2n+4v-1)(2n+4v-2) + 64(n+2v) - 32]A_{n-1}^4 \\
&+128(n+v-1)A_{n-2}^4 = 0
\end{aligned} \tag{36}$$

A certain amount of asymptotic information may be recovered from this recurrence by using the standard asymptotic theory of difference equations with polynomial coefficients, see Wimp(1984), Appendix B. To do this, we estimate the coefficients in the equation in the following way:

$$\left[32n^5 + O(n^3)\right]A_n^4 + 160\left[n^3 + n^2(6v\text{-}3/2) + O(n)\right]A_{n-1}^4 + 128(n+v-1)A_{n-2}^4 = 0 \ .$$

These terms turn out to be sufficient to determine some--- but not all--- of the lead term asymptotics. A full set (2) of asymptotically linearly independent solutions can be obtained by making the substitution

$$A_n^r \sim \frac{\lambda^n n^\theta}{n!^2}\left[1 + \frac{c_1}{n} + \frac{c_2}{n} + \ldots\right] \tag{38}$$

which means that

$$A_{n+j}^r = A_n^r \lambda^j n^{\theta-j}\left[1 + \frac{j(\theta\text{-}j\text{-}1)}{n} + O(n^{-2})\right] \ , \ j = 0,\pm 1,\pm 2 \ldots\ldots \tag{39}$$

λ and θ are determined by substituting this expression into the recurrence, dividing by n^θ and equating coefficients of n^5 and n^4. There are two possible values of λ, $\lambda = -4$ and $\lambda = -1$. To choose among them requires an additional investigation. There is a single value of θ, namely, $\theta = -4v-3/2$. The same procedure can be carried out for general r, but note that λ and the constant multiplier term cannot be determined without some further information. There is another way of doing this, but first we establish

Theorem 5

i) A_n^r satisfies a recurrence relation of order $[(r+1)/2]$;

ii) the lead term asymptotics for A_n^r may be determined from the related differential equation $\Omega(w) = 0$ if λ and the constant multiplier are known;

iii)

$$A_n^r \sim M_r(v)\frac{\left(\frac{-r^2}{4}\right)^n}{n!^2} n^{1/2-(v+1/2)r} \{1 + c_1/n + c_2/n^2 + \ldots.\}, \ n \text{---}{>}\infty \ , \tag{40}$$

where

$$M_r(\nu) = \frac{\left(\frac{r}{2}\right)^{r(\nu+1/2)}}{2^{(r-2)/2}\pi^{(r-1)/2}}, \quad r=0,1,2.... . \tag{41}$$

Proof: The first two statements are obvious from the form of the differential equation (21) and the result of substituting the series (33) into it. It is interesting to see how the majority of the asymptotics may be obtained from the difference equation because the technique to be used has fairly wide application. The finer asymptotics we will save for Section 4. We assume r is even, the computations for r odd being similar. Let r --> 2r and define

$$\Omega = DP(D^2) := \sum_{m=0}^{r} D^{2m+1} S_m^r \quad . \tag{42}$$

Substituting the series into the differential equation $\Omega(w) = 0$ and equating to zero coefficients of like powers of t yields

$$\sum_{m=0}^{r} S_m^r A_{n+m-r}^{2r} (2n-r+2\nu r+1)_{2m+1} = 0 \ . \tag{43}$$

We now make the substitution (38) for A_n^{2r}. We get

$$\sum_{m=0}^{r} (4\lambda)^m S_m^r \left\{ 1 + \frac{[m(\theta+r+2\nu r-1/2)+r(-\theta+\nu-r+1/2)]}{n} + O(n^{-2}) \right\} = 0 \ . \tag{44}$$

Equating to zero the constant term gives

$$\sum_{m=0}^{r} (4\lambda)^m S_m^r = 0, \quad \text{or } P(4\lambda) = 0, \quad \text{or } \lambda = -r^2/4,\ -(r-2)^2/4,\ -(r-4)^2/4,\ .... . \tag{45}$$

As we shall prove shortly, λ has the first value above. Note that these distinct values of λ account for all formal series solutions of the difference equation for A_n^{2r}. However, we have no idea what the remainder of the formal series correspond to. In the case where q = const., they correspond to the functions

$$\cos(\sqrt{q}\,(r-2)t), \quad \cos(\sqrt{q}\,(r-4)t), \ ..... . \tag{46}$$

In the Bessel function case, the meaning of the remaining series is obscure. At any rate, equating to zero the coefficient of 1/n gives as equation which may be written

$$(4\lambda)P'(4\lambda)(\theta+r+2\nu r-1/2) + rP(4\lambda)(-\theta+\nu-r+1/2) = 0 \ . \tag{47}$$

But since the second term is zero and the roots of $P(4\lambda)$ are simple, this means

$$\theta = -r(2\nu+1)+1/2 \ . \tag{48}$$

4. SOME ASYMPTOTICS AND TAYLOR SERIES FOR DETERMINANTS OF BESSEL FUNCTIONS

The following result will be very useful.

Theorem 6

Let $\lambda, \sigma > 0$,

$$\begin{aligned} a_n &\sim \frac{(n+1)^{-\alpha}\lambda^n}{n!^2}\left\{1+\frac{c_1}{(n+1)}+\frac{c_2}{(n+1)^2}+\dots\right\}, \quad n\text{--->}\infty , \\ b_n &\sim \frac{(n+1)^{-\beta}\sigma^n}{n!^2}\left\{1+\frac{d_1}{(n+1)}+\frac{d_2}{(n+1)^2}+\dots\right\}, \quad n\text{--->}\infty . \end{aligned} \tag{49}$$

Then

$$\begin{aligned} \{a_n * b_n\}_n &\sim \frac{(n+1)^{-\alpha-\beta-1/2}}{2\sqrt{\pi}\ n!^2}\,\sigma^{-\beta/2-1/4}\,\lambda^{-\alpha/2-1/4}\,(\sqrt{\sigma}+\sqrt{\lambda})^{\alpha+\beta+1+2n} \\ &\times\left\{1+\frac{e_1}{(n+1)}+\frac{e_2}{(n+1)^2}+\dots\right\} . \end{aligned} \tag{50}$$

Proof: We assume that $\lambda > \sigma$, the computations being much the same for $\lambda < \sigma$ and trivial for $\lambda = \sigma$. A direct computation, based on the formula Erdélyi et al v. 2, p. 170 (16), shows

$$\left\{\frac{\lambda^n}{n!\ \Gamma(n+\alpha+1)} * \frac{\sigma^n}{n!\ \Gamma(n+\beta+1)}\right\}_n = \frac{(\lambda-\sigma)^n}{\Gamma(\beta+n+1)\Gamma(\alpha+n+1)}\,P_n^{(\beta,\alpha)}\left(\frac{\lambda+\sigma}{\lambda-\sigma}\right) \tag{51}$$

The argument of the Jacobi polynomial is off the cut $[-1,1]$, so the asymptotic formula in Szegö (1959), p. 194, (8.21.29), may be used to estimate it. Next, we write

$$(n+1)^{-\alpha} = n! \left\{ \frac{1}{\Gamma(n+1+\alpha)} + \frac{f_1}{\Gamma(n+2+\alpha)} + \frac{f_2}{\Gamma(n+3+\alpha)} + \cdots \right\} , \tag{52}$$

and similarly for $(n+1)^{-\beta}$. If one incorporates remainder terms, an elementary argument shows the result just established may be extended to complete asymptotic expansions.

Remark: It is clear the result holds when λ, $\sigma < 0$ provided absolute values of these quantities are taken wherever they appear and the right hand side of (50) is multiplied by $(-1)^n$.

The proof of Theorem 5 may now be finished off by applying this result to the expression on the right hand side of (40) and using induction on r.

Determinants of the form

$$D_{\nu,r} = \left| a_{\nu+i-j} J_{\nu+i-j}(t) \right| , \quad i,j = 1,2,3,\ldots\ldots,r , \tag{53}$$

occur frequently, e.g., in combinatorics, see Zeilberger and Wimp (1989). Provided none of the determinants

$$\left| \frac{a_{\nu+i-j}}{\Gamma(\nu+1+i-j)} \right| , \quad r = 1,2,3\ldots\ldots, \tag{54}$$

is zero, it is clear we may write

$$D_{\nu,r} = t^{\nu r} \sum_{n=0}^{\infty} A_n^{\nu,r} t^{2n} , \quad A_0^{\nu,r} \neq 0 , \tag{55}$$

We wish to obtain asymptotic estimates for $A_n^{\nu,r}$, $n \to \infty$. We will assume that none of the determinants (54) vanishes and that, temporarily, ν is not an integer.

To begin, we notice that

$$D_{\nu,1} = a_\nu J_\nu(t) , \tag{56}$$

so

$$A_n^{\nu,1} = \frac{a_\nu (-1/4)^n}{2^\nu n! \Gamma(n+\nu+1)} = \frac{a_\nu (-1/4)^n n^{-\nu}}{2^\nu n!^2} \left[1 + O(\tfrac{1}{n}) \right] , \tag{57}$$

and

$$D_{\nu,2} = \frac{a_\nu^2}{4^\nu \Gamma(\nu+1)^2} \, {}_1F_2\left(\begin{matrix} 1/2+\nu \\ 1+\nu, \quad 1+2\nu \end{matrix} \middle| -t^2\right) - \frac{a_{\nu-1}a_{\nu+1}}{4^\nu \Gamma(\nu)\Gamma(\nu+2)}$$

$$\times {}_2F_3\left(\begin{matrix} 1/2+\nu, \quad 1+\nu \\ \nu, \ \nu+2, \quad 1+2\nu \end{matrix} \middle| -t^2\right), \tag{58}$$

so

$$A_n^{\nu,2} = \frac{(1/2+\nu)_n \ (-1)^n}{4^\nu \ n! \ (1+2\nu)_n \ \Gamma(\nu+1)\Gamma(\nu+n+1)} \left[a_\nu^2(n+\nu+1) - a_{\nu-1}a_{\nu+1}(n+\nu)\right]$$

$$= \frac{(-1)^n \ n^{-2\nu-1/2}}{n!^2 \sqrt{\pi}} \left[(a_\nu^2 - a_{\nu-1}a_{\nu+1}) + O(\tfrac{1}{n})\right] \quad . \tag{59}$$

We now define the determinants $d_{\nu,r}$, $r = 1,2,3....$, by $d_{\nu,1} = a_\nu 2^{-\nu}$,

$$d_{\nu,2} = \frac{(a_\nu^2 - a_{\nu-1}a_{\nu+1})}{\sqrt{\pi}},$$

$$d_{\nu,r} = \frac{\left[d_{\nu,r-1}^2 - d_{\nu-1,r-1}d_{\nu+1,r-1}\right]}{d_{\nu,r-2}}, \quad r = 3,4..... \tag{60}$$

We have

Theorem 7

Let none of the determinants (60) vanish. Then

$$A_n^{\nu,r} = \frac{d_{\nu,r} \ n^{-\nu r - (r-1)/2}}{n!} \left(\frac{-r^2}{4}\right)^n \left(1 + O(n^{-1})\right) \quad n \to \infty \tag{61}$$

Proof: Using Sylvester's identity , Wimp (1984) p. 247, on (53) shows

$$D_{v,r}\,D_{v,r-2} = D^2_{v,r-1} - D_{v-1,r-1}\,D_{v+1,r-1}\,, \qquad r = 2,3,4\,\ldots.. \tag{62}$$

This means

$$A_n^{v,r} * A_n^{v,r-2} = A_n^{v,r-1} * A_n^{v,r-1} - A_n^{v+1,r-1} * A_n^{v+1,r-1}\,, \quad r = 2,3,4\,\ldots... \tag{63}$$

Induction, which is accomplished by using Theorem 6 and the initial values of $A_n{}^{v,1}$ and $A_n{}^{v,2}$ in (57) and (59), then yields the theorem.

The case we have left unexamined, namely, when some (or all!) of the determinants $d_{v,r}$ vanish is important. Zeilberger and Wimp (1989) have treated a case when $a_v = 1$ and $v = 0$, for $r = 1,2,3$. The situation is complicated by the fact that there is a cancellation of leading terms when Theorem 6 is applied to the formula (63). Considerable work remains to be done on this case. Some of our tentative efforts are discussed in Section 5. However, it is clear that the above analysis still holds water when v is an integer, positive or negative.

5. A SINGULAR CASE

The cases where the determinants (54) vanish are important. For instance, in Zeilberger and Wimp (1989) it is shown that the nth Taylor's series coefficient for the function $|J_{i-j}(t)|$, $i,j = 1,2,3$, are essentially the number of ways of returning to the origin in 2n steps in the three dimensional constrained walk $x \ge y \ge z$. We discuss that case here, since the asymptotic description of the coefficient involves an interesting theorem.

From Luke (1969) v.1, p.241, (24) we have

$${}_2F_1\left(\begin{matrix}-n, & -n-\lambda \\ & c\end{matrix}\,;\, s\right) = \frac{\Gamma(c)}{\Gamma(a)\Gamma(c-a)}\left\{\begin{matrix}\Gamma(a)\,(n\,s)^{-a}\left[1+O(n^{-1})\right] + \Gamma(c-a)\,(n\,s e^{-\pi i})^{a-c} \\ \times\,(1-s)^{c-a+n}\left[1+O(n^{-1})\right]\end{matrix}\right\} \tag{64}$$

Now consider

$$\omega_n(t) = {}_2F_1\left(\begin{matrix}-n, & -n-\lambda \\ & \lambda+1\end{matrix}\,;\,\omega t\right) = \left[1+\sqrt{\omega t}\right]^{2n}\,{}_2F_1\left(\begin{matrix}-n, \ \lambda+1/2 \\ 2\lambda+1\end{matrix}\,;\,\frac{4\sqrt{\omega t}}{(1+\sqrt{\omega t}\,)^2}\right), \tag{65}$$

$$\omega > 0,$$

the transformation of $_2F_1$ following by Erdélyi et al (1953), v.1, p.113 (36). Let a = λ + 1/2, c = 2 λ + 1.

We find that

$$\omega_n(t) = \frac{\Gamma(\lambda+1)\, 2^{2\lambda}}{\sqrt{\pi}} \left[1+\sqrt{\omega t}\right]^{2n} \left\{ n^{-\lambda-1/2} s(t)^{-\lambda-1/2} \left[1+O(n^{-1})\right] + \right.$$

$$\left. n^{-\lambda-1/2}\, s(t)^{-\lambda-1/2}\, e^{(\lambda+1/2)\pi i} \left[1-s(t)\right]^{n+\lambda+1/2} \left[1+O(n^{-1})\right] \right\},$$

$$s(t) = \frac{4\sqrt{\omega t}}{(1+\sqrt{\omega t}\,)^2} \quad . \tag{66}$$

Note that s(1) <1. Now let

$$I_n = \int_0^1 t^{\alpha}\, (1-t)^{\beta}\, \omega_n(t)\, dt \quad , \tag{67}$$

and use Erdélyi et al (1953), v. 1, p. 67 (3). Only the first term in (66) contributes to the integral, the other being $O(\lambda^n)$, $0 < \lambda < 1$. An appeal to Laplace's method is straightforward. Further, a direct computation gives

$$I_n = \frac{\Gamma(\alpha+1)\Gamma(\beta+1)}{\Gamma(\alpha+\beta+2)}\; {}_3F_2\left(\begin{matrix} -n, & -n-\lambda, & \alpha+1 \\ & \lambda+1, & \alpha+\beta+2 \end{matrix} ; \omega \right) . \tag{68}$$

Combining these computations yields

Theorem 8

For α, β, λ > -1 and ω > 0,

$${}_3F_2\left(\begin{matrix} -n, & -n-\lambda, & \alpha+1 \\ \lambda+1, & \alpha+\beta+2 \end{matrix} ; \omega \right) \sim \frac{\Gamma(\alpha+\beta+2)\Gamma(\lambda+1)}{\Gamma(\alpha+1)2\sqrt{\pi}}\, \omega^{-\lambda/2-\beta/2-3/4}$$

$$\times (1+\sqrt{\omega}\,)^{2n+2\lambda+\beta+2}\, n^{-\lambda-\beta-3/2} \quad , \quad n\to\infty \; . \tag{69}$$

We now return to

$$D(t) = \left| J_{i-j}(t) \right|_{i,j=1,2,3}, \tag{70}$$

$$= \sum_{n=0}^{\infty} A_{2n} t^{2n} .$$

Using well-known recursion properties of Bessel functions we find that

$$D(t) = \frac{4}{t} J_1^3(t) + \frac{4}{t} J_0^2(t) J_1(t) - \frac{4}{t^2} J_0(t) J_1^2(t) . \tag{71}$$

Next, the formulas for products of Bessel functions Erdélyi et al (1953), v. 2, p.11, (49) give

$$\frac{1}{t} J_1^2(t) - \frac{1}{t} J_0^2(t) - \frac{1}{t^2} J_0(t) J_1(t) = \frac{1}{2} \sum_{k=0}^{\infty} \frac{(-1)^k (1/2)_k}{k!\, (2)_k^2} t^{2k-1} . \tag{72}$$

Forming the Cauchy product of this and the series for $J_1(t)$ gives

$$A_n = \frac{(-1)^n}{4^n\, n!\, (n+1)!}\; {}_3F_2\left(\begin{matrix} -n, \ -n-1, \ 1/2 \\ 2, \ 2 \end{matrix} ; 4 \right) . \tag{73}$$

A straightforward identification of parameters with those of the theorem ($\alpha = -1/2$, $\lambda = 1$, $\omega = 4$, $\beta = 1/2$) provides the asymptotic estimate,

$$A_n \sim \frac{\sqrt{3}\, (\frac{9}{4})^{n+2}\, (-1)^n}{\pi\, n^4\, n!^2} \left[1 + O(n^{-1}) \right], \quad n \to \infty . \tag{74}$$

However, this result depends very much on the special form of the determinant for D(t). We have not yet succeded in extending it to more general a_ν nor to higher order determinants. Our work on this problem continues.

REFERENCES

Erdélyi, A. et al (1953) *Higher transcendental functions*,3 v., McGraw-Hill, NY

Luke, Y. L. (1969) *The special functions and their approximations*, 2 v., Academic Press, NY

Szegö, G. (1959) *Orthogonal polynomials,* Amer. Math. Soc., Providence, RI

Watson, G.N. (1962) *A treatise on the theory of Bessel functions,* Cambridge U. Press, Cambridge

Wimp, J. (1984) *Computation with recurrence relations,* Longman Press, Boston

Wimp, J., and Boyer, R. (1989) "Formal series and an algorithm for computing some special determinants with elements in a ring," Appl. Num. Math. (to appear)

Zeilberger, D., and Wimp, J. (1989) "The drunkard's walk in x≥y≥z" (preprint)

A Generalization of Olver's Algorithm for Linear Difference Systems

R. V. M. Zahar Professor, Professeur Titulaire, University of Montreal, Department of Information and Operations Research, Montreal, Quebec, Canada

1. INTRODUCTION

In 1967, F.W.J. Olver [1] developed a method for computing a particular solution $\{f(n)\}$ of the difference equation

$$a_2(n)y(n+2) + a_1(n)y(n+1) + a_0(n)y(n) = b(n), \quad n=0,1,2, \ldots \qquad (1.1)$$

which is effective when $\{f(n)\}$ is asymptotically intermediate, in the sense that there exist two complementary solutions $\{e(n)\}$ and $\{g(n)\}$ of (1.1) for which

$$\lim_{n\to\infty} \frac{e(n)}{f(n)} = 0, \qquad \lim_{n\to\infty} \frac{f(n)}{g(n)} = 0. \tag{1.2}$$

Olver's algorithm is based on the solution by Gaussian elimination of the tridiagonal system of equations

$$\begin{gathered} y(0) = \alpha, \\ a_2(n)y(n+2) + a_1(n)y(n+1) + a_0(n)y(n) = b(n), \quad n=0,1,2, \ldots N-1 \\ y(N+1) = 0. \end{gathered} \tag{1.3}$$

where N is sufficiently large and where it is assumed that one initial condition $f(0) = \alpha$ is known. Incorporated in the algorithm is a technique for estimating the truncation error in the approximation $y(n)$ of $f(n)$, and of automatically predicting during the forward elimination of (1.3), the value of N which is necessary for the $y(n)$ to achieve a given precision.

Since the publication of Olver's algorithm, it has been the subject of several analyses and generalizations to scalar equations of arbitrary order [2,3,4,5]. In this article, we show that the algorithm can be directly generalized to linear difference systems (homogeneous or inhomogeneous) of the form

$$\mathbf{y}(n) = \mathbf{A}(n)\,\mathbf{y}(n-1) + \mathbf{b}(n), \qquad n \in \mathbf{Z}\,, \tag{1.4}$$

where $\mathbf{Z}$ is the set of positive integers, $\mathbf{A}(n)$ is a given nonsingular matrix of order p for each n, and where $\mathbf{y}(n)$, $\mathbf{b}(n)$ are p dimensional column vectors. We begin our analysis by examining, in the context of (1.4), the counterpart to the boundary value system (1.3), and by presenting necessary and sufficient conditions for the convergence of its solution. We then show that a method similar to Olver's, and which includes the prediction of N, can be developed for the difference system.

2. CONVERGENCE OF THE BOUNDARY VALUE SYSTEM

To facilitate the discussion, we use the notations $\mathbf{y}_1$ and $\mathbf{y}_2$ to correspond respectively to the first k and the last p-k components of a vector $\mathbf{y}$. For square matrices $\mathbf{A}$ we introduce a partition consistent with the vector partition, namely

$$\mathbf{A} = \begin{bmatrix} \mathbf{A}_{11} & \mathbf{A}_{12} \\ \mathbf{A}_{21} & \mathbf{A}_{22} \end{bmatrix},$$

where $\mathbf{A}_{11}$ is of dimension k x k, $\mathbf{A}_{12}$ is k x p-k, $\mathbf{A}_{21}$ is p-k x k and $\mathbf{A}_{22}$ is p-k x p-k. With this terminology, we can introduce our first direct generalization of system (1.3) to equations of the form (1.4) as follows.

We suppose that $\{\mathbf{f}(n)\}$ is the desired solution of (1.4) and that $\mathbf{f}_1(0) = \boldsymbol{\alpha}$ is known. The boundary value system therefore can be expressed as

$$\mathbf{y}_1(0) = \boldsymbol{\alpha}, \tag{2.1a}$$

$$\mathbf{y}(n) = \mathbf{A}(n)\,\mathbf{y}(n-1) + \mathbf{b}(n), \qquad n = 1,2,\ \ldots,\ N, \tag{2.1b}$$

$$\mathbf{y}_2(N) = \mathbf{0}, \tag{2.1c}$$

where, as a straightforward generalization of (1.3), we arbitrarily take the last p-k components of $\mathbf{y}(N)$ to be 0. We commence our analysis of this system with a brief discussion of the existence and uniqueness of its solution which, when it exists, we denote by $\mathbf{y}^{[N]}(n)$, n = 0,1, ..., N.

It follows from the nonsingularity of $\mathbf{A}(n)$ that the homogeneous counterpart of (1.4) possesses a fundamental matrix $\mathbf{H}(n)$. Let $\mathbf{G}(n) = \mathbf{A}(n)\mathbf{A}(n-1)\ \ldots\ \mathbf{A}(1) = \mathbf{H}(n)\mathbf{H}^{-1}(0)$ be the corresponding Green's function. Because $\mathbf{f}(n)$ is a particular solution of (2.1b), the solution $\mathbf{y}^{[N]}(n)$ of (2.1) exists if and only if it satisfies

$$\mathbf{y}^{[N]}(n) = \mathbf{G}(n)\left[\mathbf{y}^{[N]}(0) - \mathbf{f}(0)\right] + \mathbf{f}(n) \tag{2.2}$$

for all n = 0,1, ..., N, as well as (2.1a) and (2.1c). Employing the equality $\mathbf{y}_1^{[N]}(0) = \alpha = \mathbf{f}_1(0)$ and substituting (2.2) into (2.1c), we obtain

$$0 = \mathbf{G}_{22}(N)\left[\mathbf{y}_2^{[N]}(0) - \mathbf{f}_2(0)\right] + \mathbf{f}_2(N). \tag{2.3}$$

Because the quantities $\mathbf{G}_{22}(N)$, $\mathbf{f}_2(0)$ and $\mathbf{f}_2(N)$ are all given data, equation (2.3) immediately yields the following result.

Lemma 2.1. *For a fixed* N *the solution* $\mathbf{y}^{[N]}(n)$, n = 0,1,...,N *of boundary value system* (2.1) *exists and is unique if and only if* $\mathbf{G}_{22}^{-1}(N)$ *exists.*

Since we have taken $\mathbf{y}_2(N) = \mathbf{0}$ arbitrarily, the question of whether or not $\mathbf{y}^{[N]}(n)$ is an estimate of the desired solution $\mathbf{f}(n)$ becomes relevant. More precisely, we define convergence as follows.

Definition 2.1. The solution $\mathbf{y}^{[N]}(n)$ of (2.1) is said to *converge* to $\mathbf{f}(n)$ if for all N sufficiently large, the vectors $\mathbf{y}^{[N]}(n)$, n = 0,1, ... , N of (2.1) exist and are unique, and if for fixed n,

$$\lim_{N\to\infty} \mathbf{y}^{[N]}(n) = \mathbf{f}(n). \tag{2.4}$$

It follows from (2.3) that when $\mathbf{G}_{22}^{-1}(N)$ exists,

$$\mathbf{y}_2^{[N]}(0) = \mathbf{f}_2(0) - \mathbf{G}_{22}^{-1}(N)\,\mathbf{f}_2(N), \tag{2.5}$$

so that $\mathbf{y}_2^{[N]}(0) \to \mathbf{f}_2(0)$ as $N \to \infty$ if and only if

$$\lim_{N\to\infty} \mathbf{G}_{22}^{-1}(N)\mathbf{f}_2(N) = \mathbf{0}. \tag{2.6}$$

However, if $\mathbf{y}_2^{[N]}(0) \to \mathbf{f}_2(0)$ as $N \to \infty$, then from (2.2) and the initial condition $\mathbf{y}_1^{[N]}(0) = \mathbf{f}_1(0)$, it follows that (2.4) holds for fixed n, and convergence is achieved. The simple expression (2.6) is therefore the key condition for convergence. It is preferable, however, to re-express (2.6) in terms of a fundamental matrix, because asymptotic information about a fundamental system of solutions is often known in practice.

Theorem 2.1. *The solution* $\mathbf{y}^{[N]}(n)$ *of* (2.1) *converges to the desired solution* $\mathbf{f}(n)$ *if and only if* (1.4) *possesses a fundamental matrix* $\mathbf{H}(n)$ *such that*

(a) $\mathbf{H}_{12}(0) = \mathbf{0}$, *and* $\mathbf{H}_{22}^{-1}(N)$ *exists for all* N *sufficiently large, and*

(b) $$\lim_{N\to\infty} \mathbf{H}_{22}^{-1}(N)\ \mathbf{f}_2(N) = \mathbf{0}. \tag{2.7}$$

Proof. If $\mathbf{y}^{[N]}(n)$ converges, then by Lemma 2.1, $\mathbf{G}_{22}^{-1}(N)$ must exist for all large N and (2.6) must hold. But since $\mathbf{G}(n)$ is independent of the fundamental system chosen for $\mathbf{H}(n)$, we can take $\mathbf{H}(0) = \mathbf{I}$, the identity. In that case, $\mathbf{H}_{12}(0) = \mathbf{0}$ and $\mathbf{H}_{22}^{-1}(N) = \mathbf{G}_{22}^{-1}(N)$ so (a) and (b) follow.

Conversely, suppose that a fundamental matrix $\mathbf{H}(n)$ satisfying (a) and (b) exists. Let $\mathbf{F}(n)$ be the fundamental matrix defined by

$$\mathbf{F}(0) = \begin{bmatrix} \mathbf{I} & \mathbf{0} \\ \mathbf{0} & \mathbf{H}_{22}(0) \end{bmatrix},$$

which is nonsingular. Then by expanding the equality $\mathbf{G}(N)=\mathbf{F}(N)\mathbf{F}^{-1}(0)$, it follows that $\mathbf{G}_{22}(N) = \mathbf{H}_{22}(N)\ \mathbf{H}_{22}^{-1}(0)$ and therefore that $\mathbf{G}_{22}^{-1}(N)$ exists for all N sufficiently large. Consequently (2.5) holds, and taking limits

$$\lim_{N\to\infty} \mathbf{y}_2^{[N]}(0) = \mathbf{f}_2(0) - \lim_{N\to\infty} \mathbf{H}_{22}^{-1}(N)\mathbf{f}_2(N), \tag{2.8}$$

so that convergence follows from (b), completing the proof.

In order to demonstrate the above theorem, we now present a convergence result for Olver's system (1.3), which corresponds to the case p=2 and k=1. For a second order scalar equation, each fundamental matrix $\mathbf{F}(n)$ can be written as:

$$\mathbf{F}(n) = \begin{bmatrix} e(n) & g(n) \\ e(n+1) & g(n+1) \end{bmatrix},$$

where e(n) and g(n) are linearly independent complementary solutions of (1.1). If $e(0) \neq 0$, a fundamental matrix $\mathbf{H}(n)$ satisfying $\mathbf{H}_{12}(0) = \mathbf{0}$ can be constructed by taking

$$\mathbf{H}(n) = \begin{bmatrix} e(n) & g(n) - \dfrac{g(0)}{e(0)}\, e(n) \\ e(n+1) & g(n+1) - \dfrac{g(0)}{e(0)} e(n+1) \end{bmatrix} . \tag{2.9}$$

If $g(N+1) \neq 0$, then expression (2.7) can be written as

$$\lim_{N\to\infty} \frac{f(N+1)/g(N+1)}{1 - \dfrac{g(0)}{e(0)} \dfrac{e(N+1)}{g(N+1)}} = 0.$$

The following result is an immediate consequence of the last expression and Theorem 2.1.

Corollary 2.1 (*cf*. Olver [1], Theorem 1). *The solution* $y^{[N]}(n)$ *of* (1.3) *converges to* f(n) *if the scalar difference equation* (1.1) *possesses two complementary solutions* e(n) *and* g(n) *such that* $e(0) \neq 0$ *and*

$$\lim_{n\to\infty} \frac{e(n)}{g(n)} = 0, \qquad \lim_{n\to\infty} \frac{f(n)}{g(n)} = 0. \tag{2.10}$$

We note that the first expression of (2.10) requires that e(n) be subdominant with respect to g(n) rather than to f(n), as in (1.2). A condition stronger than that of (2.10), forcing f(n) to be at least as dominant as e(n), is usually required in applications of Olver's algorithm so that numerical stability can be achieved [1].

For most scalar equations, the conditions of Theorem 2.1 (or Corollary 2.1) are easy to verify when the desired solution is asymptotically intermediate. As a trivial example, consider a second order scalar equation written in system form as

$$\mathbf{y}(n) = \begin{bmatrix} 0 & 1 \\ -10 & 11 \end{bmatrix} \mathbf{y}(n-1) + \begin{bmatrix} 0 \\ -4(5^n) \end{bmatrix},$$

which has a fundamental matrix and a particular solution given by

$$\mathbf{F}(n) = \begin{bmatrix} 1 & 10^n \\ 1 & 10^{n+1} \end{bmatrix}, \qquad \mathbf{f}(n) = \begin{bmatrix} 5^n \\ 5^n \end{bmatrix}.$$

Let $\mathbf{f}(n)$ be the desired solution. By the construction in (2.9), a matrix $\mathbf{H}(n)$ satisfying the conditions of Theorem 2.1 (with k=1) can be easily obtained, and convergence follows.

In the case of difference systems, however, an obvious type of degeneracy can occur, as illustrated by the following example. Consider

$$\mathbf{y}(n) = \begin{bmatrix} 10 & 0 \\ 0 & 1 \end{bmatrix} \mathbf{y}(n-1) + \begin{bmatrix} -5^n \\ -4(5^{n-1}) \end{bmatrix},$$

for which

$$\mathbf{F}(n) = \begin{bmatrix} 0 & 10^n \\ 1 & 0 \end{bmatrix}, \qquad \mathbf{f}(n) = \begin{bmatrix} 5^n \\ 5^n \end{bmatrix}. \tag{2.11}$$

Denoting the complementary solutions as $\mathbf{e}(n) = [0\ 1]^T$ and $\mathbf{g}(n) = [10^n\ 0]^T$, it follows that $\mathbf{f}(n)$ is intermediate in the vector sense,

$$\lim_{n\to\infty} \frac{\|\mathbf{e}(n)\|}{\|\mathbf{f}(n)\|} = 0, \qquad \lim_{n\to\infty} \frac{\|\mathbf{f}(n)\|}{\|\mathbf{g}(n)\|} = 0.$$

(where $\|\cdot\|$ is the L_∞ norm) and so we would expect the solution of (2.1) to converge when k=1. It is clear from the expression of $\mathbf{F}(n)$ in (2.11), however, that no fundamental matrix $\mathbf{H}(n)$ exists for which both $\mathbf{H}_{12}(0)=\mathbf{0}$ and (2.7) holds. This difficulty has resulted from the simplistic notion of allowing only the first k components of $\mathbf{f}(n)$ to be specified in boundary value system (2.1). Indeed, if we specify the second component $\mathbf{f}_2(0)$ and take $\mathbf{f}_1(N) = 0$ in this example, then convergence will result. Although the correct formulation of the boundary value system is transparent for (2.11), it is difficult to discover for general difference systems, especially if the dimension p is large. The only apparent way of avoiding this difficulty is to generalize the form of the boundary conditions in (2.1), as follows.

Let $\mathbf{T}$ be a matrix of dimension k x p and rank k, $\boldsymbol{\beta}$ be a given vector of dimension k, and let $\mathbf{S}$ be p-k x p of rank p-k. Consider the modified boundary value system given by

$$\begin{aligned} \mathbf{Ty}(0) &= \boldsymbol{\beta} \\ \mathbf{y}(n) &= \mathbf{A}(n)\,\mathbf{y}(n-1) + \mathbf{b}(n), \qquad n = 1,2,\ldots,\mathbf{N}, \\ \mathbf{Sy}(N) &= \mathbf{0}. \end{aligned} \tag{2.12}$$

Although the conditions for convergence of this system can be given directly [6], we choose instead to reduce (2.12) to a system of the form (2.1).

Let $\mathbf{Q}(0)$ and $\mathbf{Q}(N)$ be permutation matrices such that $\bar{\mathbf{T}} = \mathbf{T}\mathbf{Q}^{-1}(0)$ and $\bar{\mathbf{S}} = \mathbf{S}\mathbf{Q}^{-1}(N)$ have submatrices $\bar{\mathbf{T}}_{11}$ and $\bar{\mathbf{S}}_{22}$ which are nonsingular, and let $\mathbf{z}(0) = \mathbf{V}\mathbf{Q}(0)\ \mathbf{y}(0)$ and $\mathbf{z}(N) = \mathbf{W}\mathbf{Q}(N)\ \mathbf{y}(N)$, where

$$V = \begin{bmatrix} I & \bar{\mathbf{T}}_{11}^{-1}\bar{\mathbf{T}}_{12} \\ 0 & I \end{bmatrix} \qquad \mathbf{W} = \begin{bmatrix} I & 0 \\ \bar{\mathbf{S}}_{22}^{-1}\bar{\mathbf{S}}_{21} & I \end{bmatrix}. \tag{2.13}$$

Otherwise, let $\mathbf{z}(n) = \mathbf{y}(n)$ for $n=1,2, \ldots,N-1$. Then $\mathbf{z}(n)$ satisfies the *reduced system*

$$\begin{aligned} \mathbf{z}_1(0) &= \boldsymbol{\alpha} \\ \mathbf{z}(n) &= \mathbf{B}(n)\ \mathbf{z}(n-1) + \mathbf{c}(n), \qquad N = 1,2, \ldots ,\mathbf{N}, \\ \mathbf{z}_2(N) &= \mathbf{0} \end{aligned} \tag{2.14}$$

where

$$\begin{aligned} \mathbf{B}(1) &= \mathbf{A}(1)\mathbf{Q}^{-1}(0)\mathbf{V}^{-1} & \boldsymbol{\alpha} &= \bar{\mathbf{T}}_{11}^{-1}\boldsymbol{\beta} \\ \mathbf{B}(n) &= \mathbf{A}(n), \quad n = 2, \ldots,N-1 & \mathbf{c}(n) &= \mathbf{b}(n), \quad n = 1,\ldots,N-1 \\ \mathbf{B}(N) &= \mathbf{W}\mathbf{Q}(N)\mathbf{A}(N) \end{aligned}$$

In other words, only the first two and last two equations of system (2.14) are different than those of system (2.12).

Since system (2.14) is of exactly the same form as (2.1), Theorem 2.1 can be applied to (2.14). In order that convergence can be achieved for the reduced system, however, the matrix $\mathbf{T}$ must be properly dispositioned. For instance, even if there exists a permutation matrix $\mathbf{Q}(0)$ which selects the appropriate components of $\mathbf{y}(0)$, the corresponding sub-matrix $\bar{\mathbf{T}}_{11}$ of $\bar{\mathbf{T}} = \mathbf{T}\mathbf{Q}^{-1}(0)$ must be nonsingular. As we noted, an appropriate permutation matrix for (2.11) is

$$\mathbf{Q}(0) = \begin{bmatrix} 0 & 1 \\ 1 & 0 \end{bmatrix},$$

and thus $\bar{\mathbf{T}}_{11} = t_{12}$ must be nonzero in that simple example. A general method for avoiding badly structured matrices $\mathbf{T}$ is, when possible, to *choose* $\mathbf{T}$ *such that all k-order sub-matrices are nonsingular*. A more detailed description of the proper choice of $\mathbf{T}$ is given in [7], where linear differential systems similar in form to (2.1) are considered. We discuss the choice of $\mathbf{S}$ after presenting the algorithm for solving (2.12).

3. THE GENERALIZED ALGORITHM

Because systems (2.1), (2.12) and (2.14) are of *staircase* (or almost block diagonal) form, each can be solved by Gaussian elimination, with a pivoting technique which preserves all the $\mathbf{0}$ blocks [8]. For analytical purposes, it is convenient to present the algorithm in two stages: a block decomposition method employing a limited type of block pivoting; then a complete algorithm using full pivoting.

To introduce the algorithm, we consider system (2.1), expressed in the form (3.1):

$$
\begin{bmatrix}
\mathbf{I} & & & & & & & & \\
-\mathbf{A}_{21}(1) & -\mathbf{A}_{22}(1) & \mathbf{0} & \mathbf{I} & & & & & \\
-\mathbf{A}_{11}(1) & -\mathbf{A}_{12}(1) & \mathbf{I} & \mathbf{0} & & & & & \\
 & & -\mathbf{A}_{21}(2) & -\mathbf{A}_{22}(2) & \mathbf{0} & \mathbf{I} & & & \\
 & & -\mathbf{A}_{11}(2) & -\mathbf{A}_{12}(2) & \mathbf{I} & \mathbf{0} & & & \\
 & & & & & \ddots & & & \\
 & & & & -\mathbf{A}_{21}(N) & -\mathbf{A}_{22}(N) & \mathbf{0} & \mathbf{I} \\
 & & & & -\mathbf{A}_{11}(N) & -\mathbf{A}_{12}(N) & \mathbf{I} & \mathbf{0} \\
 & & & & & & & \mathbf{I}
\end{bmatrix}
\begin{bmatrix}
\mathbf{y}_1(0) \\ \mathbf{y}_2(0) \\ \mathbf{y}_1(1) \\ \mathbf{y}_2(1) \\ \mathbf{y}_1(2) \\ \vdots \\ \mathbf{y}_2(N-1) \\ \mathbf{y}_1(N) \\ \mathbf{y}_2(N)
\end{bmatrix}
=
\begin{bmatrix}
\boldsymbol{\alpha} \\ \mathbf{b}_2(1) \\ \mathbf{b}_1(1) \\ \mathbf{b}_2(2) \\ \mathbf{b}_1(2) \\ \vdots \\ \mathbf{b}_2(N) \\ \mathbf{b}_1(N) \\ \mathbf{0}
\end{bmatrix}
$$

where the blocks of each equation (2.16) have been interchanged in order that the square matrix $\mathbf{A}_{22}(n)$ appears on the diagonal. As discussed in [9], a direct **LU** decomposition can be given for the total matrix $\mathbf{A}$ of (3.1), where

$$\mathbf{L} = \begin{bmatrix} \mathbf{I} & & & & & & & & \\ -\mathbf{A}_{21}(1) & \mathbf{I} & & & & & & & \\ -\mathbf{A}_{11}(1) & \mathbf{V}(1) & \mathbf{I} & & & & & & \\ & & -\mathbf{A}_{21}(2) & \mathbf{I} & & & & & \\ & & -\mathbf{A}_{11}(2) & \mathbf{V}(2) & \mathbf{I} & & & & \\ & & & & & \ddots & & & \\ & & & & & -\mathbf{A}_{21}(N) & \mathbf{I} & & \\ & & & & & -\mathbf{A}_{11}(N) & \mathbf{V}(N) & \mathbf{I} & \\ & & & & & & & & \mathbf{I} \end{bmatrix} \tag{3.2a}$$

$$\mathbf{U} = \begin{bmatrix} \mathbf{I} & & & & & & & & \\ & -\mathbf{R}(1) & \mathbf{0} & \mathbf{I} & & & & & \\ & & \mathbf{I} & -\mathbf{V}(1) & & & & & \\ & & & -\mathbf{R}(2) & \mathbf{0} & \mathbf{I} & & & \\ & & & & \mathbf{I} & -\mathbf{V}(2) & & & \\ & & & & & \ddots & & & \\ & & & & & -\mathbf{R}(N) & \mathbf{0} & \mathbf{I} & \\ & & & & & & \mathbf{I} & -\mathbf{V}(N) & \\ & & & & & & & \mathbf{I} & \end{bmatrix} . \tag{3.2b}$$

Here the matrices $\mathbf{V}(n)$ of dimension k x p-k, and $\mathbf{R}(n)$ of dimension p-k x p-k, satisfy $\mathbf{V}(0) = 0$ and

$$\mathbf{R}(n) = \mathbf{A}_{21}(n)\mathbf{V}(n-1) + \mathbf{A}_{22}(n) \tag{3.3a}$$

$$\mathbf{V}(n) = [\mathbf{A}_{11}(n)\ \mathbf{V}(n-1) + \mathbf{A}_{12}(n)]\ \mathbf{R}^{-1}(n) \tag{3.3b}$$

for n = 1,2, ...,N. Thus, each block of $\mathbf{L}$ and $\mathbf{U}$ can be obtained by forward recurrence of (3.3), provided that $\mathbf{R}(n)$ is nonsingular for each n.

The matrices $\mathbf{V}(n)$ and $\mathbf{R}(n)$ can also be expressed in terms of the fundamental matrix $\mathbf{H}(n)$ in Theorem 2.1, as follows. Since $\mathbf{H}(n)$ satisfies $\mathbf{H}(n) = \mathbf{A}(n)\mathbf{H}(n-1)$, we have in partitioned form,

$$\begin{aligned} \mathbf{H}_{22}(n) &= \mathbf{A}_{21}(n)\mathbf{H}_{12}(n-1) + \mathbf{A}_{22}(n)\mathbf{H}_{22}(n-1) \\ \mathbf{H}_{12}(n) &= \mathbf{A}_{11}(n)\mathbf{H}_{12}(n-1) + \mathbf{A}_{12}(n)\mathbf{H}_{22}(n-1) \end{aligned} \tag{3.4}$$

for $n \in \mathbf{Z}$. Thus, if $\mathbf{H}_{22}^{-1}(n)$ exists for all $n = 0,1, \ldots,N$, we can multiply (3.4) by $\mathbf{H}_{22}^{-1}(n-1)$, obtaining

$$\begin{aligned} \mathbf{H}_{22}(n)\mathbf{H}_{22}^{-1}(n-1) &= \mathbf{A}_{21}(n)[\mathbf{H}_{12}(n-1)\mathbf{H}_{22}^{-1}(n-1)] + \mathbf{A}_{22}(n) \\ \mathbf{H}_{12}(n)\mathbf{H}_{22}^{-1}(n-1) &= \mathbf{A}_{11}(n)[\mathbf{H}_{12}(n-1)\mathbf{H}_{22}^{-1}(n-1)] + \mathbf{A}_{12}(n) \end{aligned} \tag{3.5}$$

which is of the exact same form as (3.3). Consequently, if we note that $\mathbf{H}_{12}(0) = 0$ and $\mathbf{V}(0) = 0$, then by induction we have:

Lemma 3.1. *Suppose that a fundamental matrix* $\mathbf{H}(n)$ *satisfying the conditions of Theorem* 2.1 *exists. If* $\mathbf{H}_{22}(n)$ *is nonsingular for all* $n = 0,1, \ldots,N$, *then the matrices* $\mathbf{R}(n)$ *and* $\mathbf{V}(n)$ *of* (3.2) *satisfy the equivalences*

$$\mathbf{R}(n) = \mathbf{H}_{22}(n)\mathbf{H}_{22}^{-1}(n-1) \qquad \mathbf{V}(n) = \mathbf{H}_{12}(n)\mathbf{H}_{22}^{-1}(n). \tag{3.6}$$

Even if the matrix $\mathbf{H}(n)$ of theorem 2.1 exists, however, it may well happen that $\mathbf{H}_{22}(n)$, or equivalently $\mathbf{R}(n)$, is singular for some n, in which case equations (3.3) do not hold (as in example (2.11)). In order to ensure that an algorithm based on a decomposition of the form (3.2) is successful, it is necessary to introduce full pivoting.

We now consider the more general system (2.12) and we suppose as in (3.1) that the blocks of $\mathbf{y}(n) = \mathbf{A}(n)\mathbf{y}(n-1) + \mathbf{b}(n)$ have been interchanged. As in [6], we begin with k steps of column pivoting in the matrix $\mathbf{T}$, followed by p-k steps of row pivoting, then k steps of column pivoting, and so on until the last block of order p-k is encountered, where row pivoting is employed. Elimination by rows is performed throughout. If we let $\mathbf{Q}(n)$, $n = 0,1, \ldots,N$ be the permutation matrices which result from the column pivoting steps and which transform $\mathbf{y}(n)$ to $\bar{\mathbf{y}}(n) = \mathbf{Q}(n)\mathbf{y}(n)$, then as a result of pivoting alone, (2.12) is transformed into the equivalent system

$$\bar{\mathbf{T}}\,\bar{\mathbf{y}}(0) = \boldsymbol{\beta} \tag{3.7a}$$

$$\bar{\mathbf{y}}(n) = \bar{\mathbf{A}}(n)\bar{\mathbf{y}}(n-1) + \bar{\mathbf{b}}(n), \qquad n = 1,2, \ldots,N \tag{3.7b}$$

$$\bar{\mathbf{S}}\,\bar{\mathbf{y}}(N) = 0 \tag{3.7c}$$

where $\bar{\mathbf{T}} = \mathbf{T}\mathbf{Q}^T(0)$, $\bar{\mathbf{S}} = \mathbf{S}\mathbf{Q}^T(N)$ and for $n = 1,2, \ldots,N$,

$$\bar{\mathbf{A}}(n) = \mathbf{Q}(n)\mathbf{A}(n)\mathbf{Q}^T(n-1) \qquad \bar{\mathbf{b}}(n) = \mathbf{Q}(n)\mathbf{b}(n).$$

Convergence Theorem 2.1 therefore applies to the *reduced pivoted system* obtained from (3.7) in precisely the same manner that (2.14) is obtained from (2.12). If the conditions of the theorem are satisfied, then for N sufficiently large, the matrix of the total system (3.7) is nonsingular. Consequently, the matrices $\bar{\mathbf{R}}(n)$ corresponding to the decomposition of system (3.7) are all nonsingular, and

$$\bar{\mathbf{R}}(n) = \bar{\mathbf{H}}_{22}(n)\,\bar{\mathbf{H}}_{22}^{-1}(n-1), \tag{3.8}$$

where $\bar{\mathbf{H}}(n) = \mathbf{Q}(n)\mathbf{H}(n)$ is the fundamental matrix for the reduced pivoted system.

Equation (3.8) suggests how the matrix $\bar{\mathbf{S}}$ of (3.7c) should be chosen. Because of condition (2.7), convergence should be rapid when the submatrix $\bar{\mathbf{H}}_{22}(N)$ of the reduced pivoted system has a large determinant. But the type of pivoting we have employed tends to maximize the determinants of the matrices $\bar{\mathbf{R}}(n)$ appearing on the diagonal of the upper triangular matrix $\bar{\mathbf{U}}$. In the context of (3.8) therefore, pivoting can be interpreted as an attempt to choose, from all the p-k order submatrices of $\mathbf{H}(n)$, the $\bar{\mathbf{H}}_{22}(n)$ with a determinant that grows at the maximum rate. Consequently $\bar{\mathbf{S}}$ should be chosen so that, for the pivoted system, equation (2.7) becomes

$$\lim_{N\to\infty} \bar{\mathbf{H}}_{22}(N)\bar{\mathbf{f}}_2(N) = 0,$$

which is achieved simply by taking $\bar{\mathbf{S}} = [\mathbf{0}\ \mathbf{I}]$. In other words, we suggest that the final equation be chosen as $\bar{\mathbf{y}}_2(N) = 0$, *after* pivoting has been performed on the entire system.

The algorithm for (2.12) therefore amounts to a decomposition $\bar{\mathbf{A}} = \bar{\mathbf{L}}\,\bar{\mathbf{U}}$ of the matrix $\bar{\mathbf{A}}$ of (3.7) and the solution of $\bar{\mathbf{L}}\,\bar{\mathbf{x}} = \bar{\mathbf{b}}$ and $\bar{\mathbf{U}}\,\bar{\mathbf{y}} = \bar{\mathbf{x}}$ where $\bar{\mathbf{y}} = [\bar{\mathbf{y}}_1^T(0), \bar{\mathbf{y}}_2^T(0), \ldots, \bar{\mathbf{y}}_1^T(N), \bar{\mathbf{y}}_2^T(N)]$, and where $\bar{\mathbf{x}} = [\bar{\mathbf{x}}_1^T(0), \bar{\mathbf{x}}_2^T(0), \ldots, \bar{\mathbf{x}}_1^T(N), \bar{\mathbf{x}}_2^T(N)]$ is the vector obtained after the forward elimination process. Thus, in the terminology of (3.2), with bars to denote that pivoting is performed, $\bar{\mathbf{x}}$ satisfies

$$\begin{aligned} \bar{\mathbf{x}}_2(n-1) &= \bar{\mathbf{A}}_{21}(n)\ \bar{\mathbf{x}}_1(n-1) + \bar{\mathbf{b}}_2(n) \\ \bar{\mathbf{x}}_1(n) &= \bar{\mathbf{A}}_{11}(n)\ \bar{\mathbf{x}}_1(n-1) - \bar{\mathbf{S}}(n)\bar{\mathbf{x}}_2(n-1) + \bar{\mathbf{b}}_1(n) \end{aligned} \qquad (3.9)$$

for $n \geq 2$. We now discuss how N can be chosen before any back substitution is performed.

4. ESTIMATION OF THE TRUNCATION ERRORS

If we denote the solution of (3.7) for a given N by $\bar{\mathbf{y}}^{[N]}(n)$, then by (3.2b), the solution of $\bar{\mathbf{U}}\,\bar{\mathbf{y}} = \bar{\mathbf{x}}$ corresponds to $\bar{\mathbf{y}}_2(N) = \mathbf{0}$ and the recurrences

$$\bar{\mathbf{y}}_1^{[N]}(n) = \bar{\mathbf{V}}(n)\bar{\mathbf{y}}_2^{[N]}(n) + \bar{\mathbf{x}}_1(n) \qquad (4.1a)$$

$$\bar{\mathbf{y}}_2^{[N]}(n-1) = \bar{\mathbf{R}}^{-1}(n)\ [\bar{\mathbf{y}}_2^{[N]}(n) - \bar{\mathbf{x}}_2(n-1)] \qquad (4.1b)$$

for n = N, N-1, ..., 1. Thus, by repeated application of (4.1b) from n to N-1, we have

$$\bar{\mathbf{y}}_2^{[N]}(n) = -\sum_{s=n}^{n=1} \Big[\prod_{t=n}^{s} \bar{\mathbf{R}}^{-1}(t+1) \Big]\ \bar{\mathbf{x}}_2(s). \qquad (4.2)$$

When the conditions for convergence hold, however, the desired solution $\bar{\mathbf{f}}_2(n)$ satisfies the equation obtained from (4.2) with N replaced by ∞, or

$$\bar{\mathbf{f}}_2(n) = -\sum_{s=n}^{\infty} [\prod_{t=n}^{s} \bar{\mathbf{R}}^{-1}(t+1)]\, \bar{\mathbf{x}}_2(s). \tag{4.3}$$

Consequently, the absolute error $\mathbf{e}_2^{[N]}(n)$ in $\bar{\mathbf{y}}_2^{[N]}(n)$ satisfies

$$\mathbf{e}_2^{[N]}(n) = \bar{\mathbf{y}}_2^{[N]}(n) - \bar{\mathbf{f}}_2(n) = \sum_{s=N}^{\infty} [\prod_{t=N}^{s} \bar{\mathbf{R}}^{-1}(t+1)]\, \bar{\mathbf{x}}_2(s). \tag{4.4}$$

It follows that for a given n either the absolute error norm $\|\mathbf{e}_2^{[N]}(n)\|$ or the relative error $\|\mathbf{e}_2^{[N]}(n)\|/\|\bar{\mathbf{f}}_2(n)\|$ can be estimated as precisely as desired by taking as many terms as necessary in (4.3) and (4.4). But once $\mathbf{e}_2^{[N]}(n)$ is estimated, $\mathbf{e}_1^{[N]}(n) = \bar{\mathbf{y}}_1^{[N]}(n) - \bar{\mathbf{f}}_1(n)$ can be easily computed as follows. By (4.1a), we see that $\bar{\mathbf{y}}_1^{[N]}$ satisfies

$$\bar{\mathbf{y}}_1^{[N]}(n) = \bar{\mathbf{V}}(n)\, \bar{\mathbf{y}}_2^{[N]}(n) + \bar{\mathbf{x}}_1(n)$$

where $\bar{\mathbf{V}}(n)$ and $\bar{\mathbf{x}}_1(n)$ do not depend on N. Under the conditions of convergence, we also have

$$\bar{\mathbf{f}}_1(n) = \bar{\mathbf{V}}(n)\, \bar{\mathbf{f}}_2(n) + \bar{\mathbf{x}}_1(n),$$

and thus

$$\mathbf{e}_1^{[N]}(n) = \bar{\mathbf{V}}(n)\, \mathbf{e}_2^{[N]}(n), \tag{4.5}$$

which demonstrates that the total error in $\bar{\mathbf{y}}^{[N]}(n)$ can be estimated during the forward elimination process.

Suppose, therefore, that we wish to approximate $\mathbf{f}(n)$, $n = 0,1,\ldots, E$ to d significant decimal figures. A suggested procedure can be summarized as follows:

(i) Continue the forward recurrence (3.9) until an $N>E$ is reached for which

$$\frac{\|\mathbf{e}^{[N]}(E)\|}{\|\bar{\mathbf{f}}^{[N]}(E)\|} \le \tfrac{1}{2} \cdot 10^{-d}$$

where $\bar{\mathbf{f}}_2^{[N]}(E)$ and $\mathbf{e}_2^{[N]}(E)$ are estimated by taking only the first terms in the sums (4.3) and (4.4), and $\mathbf{e}_1^{[N]}(E)$ is calculated by (4.5).

(ii) Compute $\bar{\mathbf{y}}^{[N]}(n)$ backwards by (4.1) for $n = N, \ldots, 1$.

(iii) Verify (approximately) that

$$\max_{0 \le n \le E} \frac{\|\mathbf{e}^{[N]}(n)\|}{\|\bar{\mathbf{f}}\ (n)\|} \le \tfrac{1}{2} \cdot 10^{-d}$$

as follows. Estimate $\bar{\mathbf{f}}(n)$ by the computed values $\bar{\mathbf{y}}^{[N]}(n)$, and calculate

$$\begin{aligned} \mathbf{e}_2^{[N]}(n-1) &= \bar{\mathbf{R}}^{-1}(n)\ \mathbf{e}_2^{[N]}(n) \\ \mathbf{e}_1^{[N]}(n-1) &= \bar{\mathbf{V}}(n)\ \ \ \mathbf{e}_2^{[N]}(n) \end{aligned}$$

for $n = E, E-1, \ldots, 1$. If the error test fails for some $n \in \{0,1, \ldots, E\}$, replace E by the point E' where the maximum error occurs, and return to step (i).

The estimates can be made more precise by including several terms in (4.3) and (4.4) at step (i).

5. CONCLUDING REMARKS

We have endeavoured to present a broad outline of the recurrence theory and computational elements involved in the solution of linear difference systems of the type (1.4). A more general description, including an extensive development of the algorithms as well as several applications, will be found in [10].

Although stability of the algorithm has been mentioned only briefly in the present article, a rigorous analysis can also be given (see, for example, [9]). In general terms, the algorithm is stable if the desired solution $\bar{\mathbf{f}}(n)$ is asymptotically subdominant with respect to p-k complementary solutions, and is at least as dominant as another k complementary solutions.

An investigation of the computational details of the methods described in sections 4 and 5 indicate that the algorithm can be programmed in an efficient manner. For instance, with the pivoting we have suggested, the decomposition of each $\bar{\mathbf{R}}(n)$ as a product of $\bar{\mathbf{L}}(n)$ and $\bar{\mathbf{U}}(n)$ is automatically performed during the elimination process, and the inversion of each $\bar{\mathbf{R}}(n)$ can be effectively avoided.

The algorithm has received extensive application to the solution of stiff differential systems by the method of Taylor Series [11]. In the latter context, convergence was attained and the algorithm was uniformly successful for every one of the several thousand difference systems considered.

REFERENCES

1. F.W.J. Olver, *J. Res. Nat. Bur. Standards 71B*: 111 (1967).
2. R.V.M. Zahar, *Computational algorithms for linear difference equations*, Ph.D. dissertation (Purdue, 1968).
3. J. Oliver, *Numer. Math. 11*: 349 (1968).
4. J.R. Cash, *Math. Comp. 32*: 497 (1978).
5. D.W. Lozier, *Report NBSIR 80, Math. Analysis Division, NBS* (1980).
6. R.V.M. Zahar, *Cong. Numer. 61*: 233 (1988).
7. R.V.M. Zahar, *Cong. Numer. 61*: 221 (1988).
8. D.C.L. Lam, *Implementation of the box scheme and model analysis of diffusion-correction equations*, Ph.D. dissertation (Waterloo, 1974).
9. R.V.M. Zahar, *Dépt. d'IRO 434, Univ. de Montréal* (1982).
10. J.R. Cash and R.V.M. Zahar, *Computational aspects of linear difference systems*, In press.
11. F. Jalbert and R.V.M. Zahar, *Cong. Numer. 46*: 347 (1985).

Part II

Contributed Papers

Continental Shelf Wave Scattering: Partial Removal of the Rigid Lid

Anthony M. J. Davis Professor, Department of Mathematics, The University of Alabama, Tuscaloosa, Alabama

Continental shelf waves are coastally trapped waves that arise from the interaction of Coriolis forces with depth changes. They propagate with the coastline on the right or left in the Northern or Southern Hemispheres respectively but their group velocity may be forward (longer waves) or backward (shorter waves). They are conveniently modelled by taking the normal component of the Earth's rotation to be constant (the β-plane approximation) and regarding the ocean as a large body of rotating shallow water.

When referred to horizontal rectangular coordinates x,y, the equations of motion in the linearized shallow-water wave theory are

$$\partial u/\partial t - fv = -g\,\partial\eta/\partial x \tag{1a}$$
$$\partial v/\partial t + fu = -g\,\partial\eta/\partial y \tag{1b}$$

and the equation of continuity is

$$\partial(hu)/\partial x + \partial(hv)/\partial y + \partial\eta/\partial t = 0. \tag{2}$$

In these equations, η is the elevation of the free surface above the equilibrium level, h is the variable depth, u and v are the components of the velocity in the x,y-directions, g is the acceleration due to gravity and f is the Coriolis parameter, given by $f = 2\Omega\sin\theta$, where Ω is the Earth's angular velocity and θ is the latitude.

With f assumed constant in (1), consider the simplification of (2) by using either constant depth or the rigid lid approximation. For $h \equiv H$, the equations have solutions in which u,v and η are related multiples of $\exp[i(\kappa x - \omega t) \pm (\kappa^2 + c^2)^{1/2}y]$, where

$$c^2 = (f^2 - \omega^2)/gH. \tag{3}$$

If the third term of Eq. (2) is neglected, the rigid lid approximation and $h \equiv h(y) = h(0)e^{2by}$, the corresponding solutions are $\exp[i(\kappa x - \omega t) - by]$ times combinations of $\cos(\Gamma y)$ and $\sin(\Gamma y)$, where

$$\Gamma(\kappa) = (2b\kappa f/\omega - \kappa^2 - b^2)^{1/2}. \tag{4}$$

Now for an exponential depth profile defined by

$$h(y) = \begin{cases} He^{-2b(\ell - y)} & (0 \le y \le \ell) \\ H & (y \ge \ell), \end{cases} \tag{5}$$

the usual shelf wave formation, first described by Buchwald and

Adams (1968), adopts the rigid lid approximation everywhere and hence has available a mass transport stream function $\psi(x,y)$ defined by

$$u = \frac{1}{h}\frac{\partial\psi}{\partial y}e^{-i\omega t}, \quad v = -\frac{1}{h}\frac{\partial\psi}{\partial x}e^{-i\omega t}. \tag{6}$$

By requiring continuity at the shelf-ocean boundary $y = \ell$ and no flow across the coastline $y = 0$, the eigenvalue equation is

$$\Gamma \cot \Gamma\ell + b + |\kappa| = 0. \tag{7}$$

When $f/\omega > 1$, a finite number of real solutions exists in the range $0 < \kappa < 2bf/\omega$; none of these has Γ pure imaginary. There are also pairs of complex conjugate eigenvalues in the strip $0 < \mathrm{Re}\,\kappa < bf/\omega$ (Davis, 1987). For fixed κ, Sturm-Liouville theory is applicable and modes corresponding to different frequencies are orthogonal but a more complicated situation occurs when the frequency ω is prescribed (Huthnance, 1975).

The continental shelf is in reality punctuated by topographic and shoreline irregularities so there is a variety of scattering problems to be studied. Eigenfunction expansions are inappropriate but Green's function and Wiener-Hopf techniques have been successfully applied to equations satisfied by ψ, defined by Eq. (6) when $\partial\eta/\partial t$ is neglected everywhere.

However, certain difficulties arise in the neighborhood of $\kappa = 0$ and need to be resolved. Obviously $|\kappa|$ is not analytic but, of more obvious physical significance, the Kelvin wave in the ocean and its counterpart on the shelf, given by

$$\begin{bmatrix} u \\ v \end{bmatrix} = \begin{bmatrix} 1 \\ 0 \end{bmatrix} \exp[-i\omega t + (i\omega x - fy)/\sqrt{gH}\,], \quad g\eta = u\sqrt{gH} \quad (y > \ell) \tag{8a}$$

$$\begin{bmatrix} u \\ v \end{bmatrix} = \begin{bmatrix} 1 \\ 0 \end{bmatrix} e^{-i\omega t}, \quad g\eta = (i\omega x - fy)e^{-i\omega t} \quad (0 < y < \ell), \tag{8b}$$

are suppressed by the analysis because u must be continuous. In a recent note concerning the absence of the Kelvin wave, Buchwald and Hsieh (1988) suggested that (8b) be regarded as an inner expansion, based on the small parameter $f\ell/\sqrt{gH}$, of the outer solution (8a) near the tip of a long thin barrier parallel to the coast. An alternative idea, proposed in a recent paper (Davis, 1989) is to relax the rigid lid approximation in the ocean.

The net effect of neglecting $\partial\eta/\partial t$ in the ocean is evidently to replace the parameter c in Eq. (3) by zero i.e. to also remove the Coriolis parameter from the governing differential and thus make the Rossby radius infinite. For ℓ = 50 km, H = 2 km, $c\ell$ = .04 but ℓ may not be an appropriate length scale in the ocean which sees the shelf as a narrow strip. When the problem requires the use of Fourier transform methods, c is not dominated by all wave numbers and its retention eliminates the difficulties caused by the non-analytic behavior of $|\kappa|$ at the origin. Thus the rigid lid approximation will not be used in the ocean and the stream function, now restricted to the shelf, will not be introduced.

With the depth h given by Eq. (5) and $\partial\eta/\partial t$ neglected on the shelf, Eq. (2) becomes

$$\partial u/\partial x + \partial v/\partial y + 2bv = 0 \quad (0 \leq y \leq \ell) \tag{9a}$$

$$\partial u/\partial x + \partial v/\partial y + H^{-1}\,\partial\eta/\partial t = 0 \quad (y > \ell). \tag{9b}$$

The boundary conditions

$$u,v \text{ continuous at } y = \ell \tag{10}$$

$$v = 0 \quad \text{at } y = 0 \tag{11}$$

are satisfied, subject to the eigenvalue equation

$$\Gamma \cot \Gamma\ell + b + \kappa\,\frac{\lambda f - \kappa\omega}{\kappa f - \lambda\omega} = 0\,, \tag{12}$$

by the travelling wave solutions

$$\begin{bmatrix} u \\ v \end{bmatrix} = e^{i(\kappa x - \omega t)} \frac{e^{b(\ell - y)}}{\sin \Gamma \ell} \begin{bmatrix} -b \sin \Gamma y - \Gamma \cos \Gamma y \\ i\kappa \sin \Gamma y \end{bmatrix} (0 \le y \le \ell)$$
$$\begin{bmatrix} u \\ v \end{bmatrix} = e^{i(\kappa x - \omega t)} e^{-\lambda(y - \ell)} \begin{bmatrix} \kappa(\lambda f - \kappa\omega)/(\kappa f - \lambda\omega) \\ i\kappa \end{bmatrix}, \; (y \ge \ell) \tag{13}$$

where $\lambda = (\kappa^2 + c^2)^{1/2}$ and c and $\Gamma(\kappa)$ are given by Eqs. (3) and (4) respectively. When $|\kappa| >> c$, the last term in Eq. (12) is $0(c^2)$ different from $|\kappa|$ and hence only minor changes are made to the roots of Eq. (7). However, an important extra root is found by observing that as κ increases from 0 to $c\omega(f^2 - \omega^2)^{-1/2} = \omega/\sqrt{gH}$, the last term in Eq. (12) decreases from 0 to $-\infty$ and hence there must be exactly one solution $\kappa = k_0^+$ in this range, given approximately by

$$k_0^+ \sim \frac{\omega}{\sqrt{gH}} - \frac{\omega f}{2bgH}, \quad \Gamma(k_0^+) \sim i(b - \frac{f}{\sqrt{gH}}).$$

Here Γ is pure imaginary and the + denotes a forward travelling surface wave (positive group velocity), described now as a modified Kelvin wave because it can be viewed as a uniformly valid solution ($\kappa = \kappa_0^+$ in Eq. (13)) that approximates Eqs. (8a,b) in the respective regions.

For problems involving finite, usually small, irregularities e.g. Buchwald (1977), Chao et al (1979), Green's functions can be constructed by direct use of Fourier transforms and subsequent application of appropriate radiation conditions, determined by requiring that the solution for $|x| >> \ell$ be a superposition of forward or backward travelling shelf waves. Additionally, the direction of energy propagation on the shelf has been shown (Davis, 1981) to be exactly opposite to the phase velocity of $\psi \exp[-b(y + ixf/\omega)]$. The symmetry imposed by no flow across $y = 0$ ensures that net energy transport is always parallel to the coastline. In these calculations, the modified Kelvin wave is very weakly excited.

The Wiener-Hopf technique has been applied to problems with a semi infinite barrier or coastline e.g. Hsieh and Buchwald (1984), Davis (1987,1989). Then the neighborhood of $\kappa = 0$ is significant and the inclusion of the modified Kelvin wave is important. Consider, for example, a problem similar to that discussed by Hsieh and Buchwald (1985). The incident shelf wave $[u_I, v_I]$ defined by setting $\kappa = k_I^+$ (an eigenvalue of Eq. (12) with $I > 0$ in Eq. (13) and multiplying by $e^{b(L-\ell)}\sin[\Gamma(k_I^+)\ell]/\sin[\Gamma(k_I^+)L]$, is scattered by the semi-infinite barrier $y = L < \ell$, $x > 0$. The total field is

$$\begin{bmatrix} u_I - u \\ v_I - v \end{bmatrix} = \begin{bmatrix} u_I \\ v_I \end{bmatrix} - \frac{1}{2\pi}\int_{-\infty}^{\infty} \begin{bmatrix} U(\kappa,y) \\ V(\kappa,y) \end{bmatrix} e^{i(kx-\omega t)}\,d\kappa,$$

where Eqs. (1,9-11) and the continuity of v at $y = L$ imply that

$$\begin{bmatrix} U \\ V \end{bmatrix} = \frac{We^{b(L-y)}}{\kappa \sin \Gamma L}\begin{bmatrix} -b\sin\Gamma y - \Gamma\cos\Gamma y \\ i\kappa\sin\Gamma y \end{bmatrix} \qquad (0 \le y \le L)$$

$$\begin{bmatrix} U \\ V \end{bmatrix} = \frac{We^{-b(y-L)}}{\kappa \sin(\Gamma L+\phi)}\begin{bmatrix} -b\sin(\Gamma y+\phi) - \Gamma\cos(\Gamma y+\phi) \\ i\kappa\sin(\Gamma y+\phi) \end{bmatrix} \qquad (L \le y \le \ell)$$

$$\begin{bmatrix} U \\ Y \end{bmatrix} = W\exp[-\lambda(y-\ell)-b(\ell-L)]\frac{\sin(\Gamma\ell+\phi)}{\sin(\Gamma L+\phi)}\begin{bmatrix} (\lambda f-\kappa\omega)/(\kappa f-\lambda\omega) \\ i \end{bmatrix}, \qquad (y \ge \ell)$$

where $W \equiv W(\kappa)$, Γ is given by (4), and

$$\Gamma\cot(\Gamma\ell+\phi) + b + \kappa\left(\frac{\lambda f-\kappa\omega}{\kappa f-\lambda\omega}\right) = 0.$$

From this last equation, which defines $\cot\phi$ in terms of κ, it follows that

$$\frac{\sin(\Gamma L+\phi)}{\sin\phi} = \frac{E(\kappa,L)}{E(\kappa,0)}$$

where

$$E(\kappa,L) = \Gamma\cos[\Gamma(\ell - L)] + [b+\kappa(\frac{\lambda f - \kappa\omega}{\kappa f - \lambda\omega})]\sin\Gamma(\ell - L).$$

Thus, from Eq. (12), the zeros of $E(\kappa,L)$ determine the wave numbers of shelf waves on a shelf of width $(\ell - L)$.

The Wiener Hopf equation is obtained by applying the mixed conditions $v = v_I$ $(x > 0)$ and u continuous $(x < 0)$ at $y = L$. Thus

$$iW = V(\kappa,L) = \int_{-\infty}^{0} v\, e^{-i(\kappa x - \omega t)}dx + \int_0^\infty v_I\, e^{-i(\kappa x-\omega t)}dx$$

$$= V_+ + \frac{k_I^+}{\kappa - k_I^+} \qquad (\mathrm{Im}(\kappa - k_I^+) < 0]$$

$$[U]_{L-}^{L+} = \frac{W\Gamma\sin\phi}{\kappa\sin\Gamma L\sin(\Gamma L + \phi)} = \int_0^\infty [u]_{L-}^{L+} e^{-i(\kappa x-\omega t)}dx.$$

Since flux into the channel is possible, $u(x,L-) \to B_0 e^{-i\omega t}$ as $x \to \infty$, for some constant B_0, and hence $[U]_{L-}^{L+}$ has a term $-B_0/i\kappa$ for $\mathrm{Im}\,\kappa < 0$. To avoid the pole at the origin, define $T_- = i\kappa L[U]_{L-}^{L+}$. Then, when a small positive imaginary part is assigned to ω/f, the shelf wave numbers are displaced above or below the real axis according as the group velocity is positive or negative and there is a strip $|\mathrm{Im}\,\kappa| < d$ in which

$$V_+ + \frac{k_I^+}{\kappa - k_I^+} = T_- = (\frac{\sin\Gamma L}{\Gamma L})\frac{E(\kappa,L)}{E(\kappa,0)} \qquad (= iW).$$

Here the suffices +,- denote functions regular for $\mathrm{Im}\,\kappa > -d$ or $< d$ and, after the factorizations

$$\frac{\sin\Gamma L}{\Gamma L} = II_+(k)II_-(\kappa), \quad E(\kappa,L) = E_+(\kappa,L)E_-(\kappa,L) \text{ etc,}$$

the usual Wiener-Hopf calculations yield

$$T_-(\kappa) = (\frac{k_I^+}{\kappa - k_I^+})\frac{E_+(k_I^+,0)}{II_+(k_I^+)E_+(k_I^+,L)} \cdot \frac{E_-(\kappa,0)}{E_-(\kappa,L)II_-(\kappa)}.$$

The amplitude of the periodic flux in the channel is $B_0L = -T_-(0)$ and the inverse Fourier transforms for u and v are evaluated by deformation of the path into the upper or lower half plane according as $x >$ or < 0. There are branch cuts from ic to $i\infty$ and $-ic$ to $-i\infty$ due to the definition of λ but none arise from the definition of Γ in Eq. (4). The branch cut integrals decay algebraically as $|x| \to \infty$ and, for $x < 0$, the residues at the zeros of $E_+(\kappa,0)$ yield back scattered waves. For $x > 0$, the residue at $\kappa = k_I^+$ cancels the incident shelf wave on either side of the barrier at $y = L$. In the channel the residues at the zeros of $\Pi_-(\kappa)$ yield the admissible eigenfunctions corresponding to the zeros of $\sin \Gamma L$. For $y > L$, the residues at the zeros of $E_-(\kappa,L)$ yield forward travelling shelf waves, including the modified Kelvin wave. When u is calculated, the path is deformed below the pole at the origin in order to confine the corresponding residue contribution to $x > 0$. As expected, this is $B_0 e^{-i\omega t}$ in the channel but zero for $y > L$. Thus the scattered field is readily obtained, with no difficulties at $\kappa = 0$ and a modified Kelvin wave allowed by the analysis as a very long forward propagating shelf wave.

REFERENCES

1. V. T. Buchwald (1977) Lect. Notes Phys., 64: 188.
2. V. T. Buchwald and J. K. Adams (1968) Proc. Roy. Soc. (Lond.) A305: 235.
3. V. T. Buchwald and W. W. Hsieh (1988) J. Phys. Oceanogr. 18: 394.
4. S. Y. Chao, L. J. Pietrafesa and G. S. Janowitz (1979) J. Phys. Oceanogr. 9: 687.
5. A. M. J. Davis (1981) Philos. Trans R. Soc. (Lond.) A303: 383.
6. A. M. J. Davis (1987) Geophys. Astrophys. Fluid Dyn. 39: 25.
7. A. M. J. Davis (1989) Geophys. Astrophys. Fluid Dyn. in press.

8. W. W. Hsieh and V. T. Buchwald (1984) Geophys. Astrophys. Fluid Dyn. 28: 257.

9. W. W. Hsieh and V. T. Buchwald (1985) J. Phys. Oceanogr. 15: 524.

10. J. M. Huthnance (1975) J. Fluid Mech. 69: 689.

Some Aspects of Invariant Subspaces Computation

Luca Dieci[1] Professor, School of Mathematics, Georgia Institute of Technology, Atlanta, Georgia

Robert D. Russell[2] Professor, Department of Mathematics and Statistics, Simon Fraser University, Burnaby, British Columbia, Canada

SUMMARY

Algorithms for computing invariant subspaces are interpreted using a general framework based upon the fact that the invariant subspace problem is equivalent to solving a system of nonlinear decoupling equations. The popular eigenvalue and invariant subspace algorithms are interpreted as techniques for approximating their finite Taylor expansion; this gives an easy way to show the basic convergence properties of these algorithms. For one type of decoupling, the invariant subspace computation reduces to solving an algebraic Riccati equation. Exploiting the simple quadratic structure of its Taylor expansion, the known schemes for solving the Riccati equation, and some new ones, are interpreted, and, in particular, a modified secant method having certain computational advantages over these methods is considered.

[1]Supported in part under NSF Grant #DMS8802762.

[2]This work was supported in part under NSERC (Canada) Grant A8781.

I. INTRODUCTION

The problem of computing invariant subspaces of matrices has been extensively studied and is generally well understood. Computing (a basis for) an invariant subspace relative to a cluster of eigenvalues is often desirable because the individual eigenvalue problems can be ill-conditioned if the eigenvalues are close together and/or the eigenvectors are almost linearly dependent. Furthermore, in many applications one is only interested in achieving a blocking of eigenvalues with the same order of magnitude and/or sign of the real parts, that is, in computing their invariant subspaces (e.g., in certain optimal control problems (Kokotovic, Khalil, and O'Reilly, 1986) and in singular perturbation problems (Kreiss, Nichols and Brown, 1986)).

Much of our motivation and understanding comes from studying the ODE case (Dieci and Russell, 1987), where development of methods largely parallels that for numerical linear algebra, but for brevity of presentation we shall not emphasize that here. The corresponding problem there is to find a change of variables which produces a so-called *decoupling* of the system. This idea provides the underlying framework to completely characterize the invariant subspace problem which consists of a system of nonlinear equations involving the decoupling equation for the invariant subspace matrix of variables T_1 and its left inverse (normalization) S^*. Two important and related general approaches are choosing $S = T_1$ (a unitary transformation — orthogonal in the real case) and fixing S (basically, a Riccati type of transformation).

The *nonlinearity* of the general process is at the core of our interpretation of the invariant subspace problem (as is the case for many other processes in linear algebra). The problem of computing an invariant subspace is, in fact, the same as finding roots for a system of nonlinear equations. Taking this viewpoint, it is natural to devise iterative methods for solving such a system based upon its truncated Taylor expansion (or an opportune modification of such an expansion). In our case, the system of equations is of polynomial type, hence the expansion is finite. Approximating it, the popular computational methods and some new ones can be derived. (Peters and Wilkinson's (1979) interpretation of Newton's method to solve the eigenvalue problem is a one dimensional analogue of this approach.)

We briefly interpret the familiar orthogonal iteration and treppeniteration types of algorithms this way in §2. The advantage of the approach is not so much to prove new results about these methods as to give a unifying framework, e.g., one for which their order of convergence is generally transparent.

Several methods for refining invariant subspace estimates have previously been developed which, as remarked by Demmel (1987), essentially attempt to solve the Riccati equation, and we consider the problem of solving this equation in §3. Once more, looking at truncated Taylor expansions of this equation enables us to derive some known (and a few new) linearly convergent schemes as well as Newton's method. We also propose a modified secant iteration and compare some of its computational aspects to Newton's method, over which it can provide certain advantages. Finally, a brief interpretation of several algorithms for the eigenvector problem is given in §4.

Many of the considerations extend to the general linear operator setting. Finally, the extension of most of the ideas to the regular generalized eigenvalue problem is possible, as will be investigated elsewhere.

II. DECOUPLING FRAMEWORK

Consider the problem of computing an invariant subspace for a matrix $A = \begin{bmatrix} A_{11} & A_{12} \\ A_{21} & A_{22} \end{bmatrix} \in \mathbb{C}^{n\times n}$, where $A_{11} \in \mathbb{C}^{q\times q}$, $A_{22} \in \mathbb{C}^{p\times p}$, and $p + q = n$. We assume that A has a spectral decomposition $\Lambda = \Lambda_I + \Lambda_{II}$ and seek a matrix change of variables $T \in \mathbb{C}^{n\times n}$ such that

$$T^{-1}AT =: \tilde{A} = \begin{bmatrix} \tilde{A}_{11} & \tilde{A}_{12} \\ 0 & \tilde{A}_{22} \end{bmatrix} \tag{1}$$

where $\Lambda(\tilde{A}_{11}) = \Lambda_I$ and $\Lambda(\tilde{A}_{22}) = \Lambda_{II}$. ($\tilde{A}$ has the same partitioning as A.) Partitioning T and T^{-1} by

$$T = (T_1, T_2), \quad T^{-1} = \begin{bmatrix} S^* \\ \Sigma^* \end{bmatrix} \tag{2}$$

where $T_1 \in \mathbb{C}^{n\times q}$ and $\Sigma^* \in \mathbb{C}^{p\times n}$, then (1) is equivalent to

$$AT_1 - T_1\tilde{A}_{11} = 0, \tag{3}$$

or

$$AT_1 - T_1(S^*AT_1) = 0, \tag{4a}$$

where

$$S^*T_1 = I_q. \tag{4b}$$

Remarks. We call (4) the *decoupling equation*, if we wish to characterize this *decoupling of* A only in terms of the invariant subspace T_1 and S^*. Note that T_2 and Σ^* play no role in the decoupling per se. □

Emphasizing (4) in a framework for interpreting general invariant subspace algorithms seems not to have been done in the past, although it has certainly been used in special cases. When S is fixed, it is closely related to a standard Riccati equation, as we see shortly. It also plays a central role for the case where A is symmetric, e.g., the error for an approximate invariant subspace can be bounded in terms of the corresponding residual for (4) (Parlett, 1980). Equation (4) reflects the relation between the number of equations and unknowns; considering it as equations for T_1 and its left inverse S^* shows that there are pq too many unknowns, although their interrelationship is not entirely straightforward. Directly interpreting (3) as an invariant subspace equation for unknowns T_1 and A_{11} would be misleading because A_{11} cannot be arbitrarily specified.

Three approaches have generally been used for the matrix decoupling, and we shall interpret them as arising from truncating the Taylor expansion for (4). We consider unitary and Gaussian transformations in this section and Riccati transformation in §3.

The Taylor expansion for

$$F(T_1,S^*) := AT_1 - T_1(S^*AT_1) = 0, \tag{5}$$

around the pair X and Z^* satisfying $Z^*X = I_q$ is

$$\begin{aligned} F(T_1,S^*) = {} & [F(X,Z^*)] + [A(T_1-X) - (T_1-X)Z^*AX - X(S^*-Z^*)AX \\ & - XZ^*A(T_1-X)] + [-(T_1-X)Z^*A(T_1-X) - X(S^*-Z^*)A(T_1-X) \\ & - (T_1-X)(S^*-Z^*)AX] + [-(T_1-X)(S^*-Z^*)A(T_1-X)]. \end{aligned} \tag{6}$$

The orthogonality relation (4b) could be considered as a quadratic equation in T_1 and S^* and its Taylor expansion truncated too; however, the standard algorithms satisfy this orthogonality exactly at all steps because this normalization is inherited by the choice made for T_1. Whether or not useful hybrid algorithms could be derived by approximating (4b) also remains to be seen.

1. *UNITARY TRANSFORMATION*

T is unitary, so $T_1^*T_1 = I_q$ and $S^* = T_1^*$ is the pseudoinverse. Thus, (5) is a cubic equation in T_1 with exact Taylor expansion around X given by

$$\begin{aligned}&[AX-X(X^*AX)] + [A(T_1-X) - (T_1-X)X^*AX - X(T_1^*-X^*)AX - XX^*A(T_1-X)] \\ &\quad + [-(T_1-X)X^*A(T_1-X)-X(T_1^*-X^*)A(T_1-X) - (T_1-X)(T_1^*-X^*)AX] \\ &\quad + [-(T_1-X)(T_1^*-X^*)A(T_1-X)] = 0. \qquad (7)\end{aligned}$$

Choosing T_1 unitary still leaves $\frac{q(q-1)}{2}$ degrees of freedom. It is customary to restrict the structure of $T_1^*AT_1$ to being upper triangular, i.e., take the additional conditions

$$H_{ij}(T_1) := (T_1^*AT_1)_{ij} = 0, \quad i > j. \qquad (8)$$

Expanding H_{ij} in its exact Taylor series about X gives

$$H_{ij}(T_1) = H_{ij}(X) + [(T_1^*-X)AX + X^*A(T_1-X) + (T_1^*-X^*)A(T_1-X)]_{ij} = 0, \quad i > j. \qquad (9)$$

Consider now the first order truncated expansion of (7) consisting of the terms in the first bracket, plus the second and third terms in the second bracket, plus the last term in the third bracket: calling the initial guess X_0 and the next one X_1, we have

$$AX_0-X_0X_0^*AX_0-(X_1-X_0)X_0^*AX_0-X_0(X_1^*-X_0^*)AX_0-(X_1-X_0)(X_1^*-X_0^*)AX_0 = 0,$$

or

$$AX_0 = X_1(X_1^*AX_0). \qquad (10)$$

Taking the first order truncated expansion of (9) consisting of $H_{ij}(X)$ and the first term in brackets gives

$$(X_1^* A X_0)_{ij} = 0, \quad i > j. \tag{11}$$

Together, (10) and (11) express the fact that AX_0 is decomposed into the product of a unitary X_1 and an upper triangular $X_1^*AX_0$. This is in fact precisely the QR-factorization, or the *orthogonal iteration method* (following terminology used by Golub and vanLoan (1983), e.g., even for complex A) for computing a dominant invariant subspace: Given unitary X_0,

$$AX_{k-1} = Z_k, \quad Z_k = Q_k R_k, \quad X_k = Q_k, \quad k = 1,2,\ldots .$$

Using similar arguments but instead taking the terms in the first bracket plus the first and last terms in the second bracket in (7), and the $H_{ij}(X)$ term and the second term in the bracket in (9), we obtain the *inverse orthogonal iteration method*: Given unitary X_0,

$$AZ_k = X_{k-1} \quad (\textit{find } Z_k), \quad Z_k = X_k R_k \quad (\text{QR-}\textit{decomposition}) \quad k = 1,2,\ldots ,$$

(where R_k^{-1} is upper triangular).

2. *GAUSSIAN REPRESENTATION*

T is unit lower triangular so $T_1 = \begin{bmatrix} L_{11} \\ L_{21} \end{bmatrix}$ and $S^* = (L_{11}^{-1}, 0)$, where L_{11} is unit lower triangular. There remain as before $\frac{q(q-1)}{2}$ degrees of freedom which can be used to structure S^*AT_1, an obvious choice being $H_{ij}(T_1,S^*) := (S^*AT_1)_{ij} = 0$, $i > j$. Using the first order truncated expansion for (6) as in the first case above and the same correction for the H_{ij}, we obtain *treppeniteration* (or LR *iteration* in the case $q = n$): Given X_0,

$$AX_{k-1} = Z_k, \quad Z_k = X_k R_k \ (\text{LU-}\textit{factorization}), \quad k = 1,2\ldots .$$

As above, an inverse iteration counterpart exists. Of course, in practice such a representation generally requires pivoting for successful implementation (Wilkinson, 1965); so the imposed extra structure is $\hat{H}_{ij}(T_1,S^*) := (S^*P^TAPT_1)_{ij} = 0$, $i > j$, for suitable permutation matrix P. □

Since we have only considered methods for computing invariant subspaces which arise from simple 1st order truncated expansions, the above derivation shows immediately that convergence can generally only be expected to be linear (although this is not to say that a careful look at the structure would not often provide further insight into the methods' convergence properties). It is possible, at least in theory, to produce a variety of different methods by taking other parts of the Taylor expansion (6). However, finding novel methods which satisfy the feasibility constraints in the general case is nontrivial. We have not succeeded in interpreting the popular shift strategies (for the QR algorithm, for instance) directly in terms of truncated Taylor expansions. Nevertheless, there are two situations where the structure of the equations simplifies considerably and we can devise meaningful algorithms and interpret shift strategies as well. One is when S is fixed, the other (not necessarily unrelated) is when the invariant subspace is of dimension 1, i.e., an eigenvector problem. These are discussed in §§3 and 4, respectively.

III. ON SOLVING THE RICCATI EQUATION

We now consider the invariant subspace problem for the case where the chosen representation for T (or T_1) leaves no remaining degrees of freedom. We shall see that this occurs by fixing a normalization for T_1, i.e., $S^*T_1 = I_q$ for a *fixed* S^* (assumed to be full rank).

When $T_1 = \begin{bmatrix} I_q \\ R \end{bmatrix}$ and $S^* = (I_q,0)$, then (1) reduces to the algebraic Riccati equation

$$G(R) := A_{21} + A_{22}R - RA_{11} - RA_{12}R = 0. \tag{12}$$

In fact, for general fixed $S \in \mathbb{C}^{n\times q}$ one still essentially solves a Riccati equation, as we see below.

Theorem 1. *There is a decoupling for a given* S *iff a certain quadratic (Riccati) equation is solvable. Each such solution uniquely determines* T_1.

Proof. For given S, suppose that unitary Q satisfies $QS = \begin{bmatrix} U_S \\ 0 \end{bmatrix}$ with U_S upper triangular. Setting $U = \begin{bmatrix} U_S^{-1} & 0 \\ 0 & I_p \end{bmatrix}$, then $\tilde{S} = UQS = \begin{bmatrix} I_q \\ 0 \end{bmatrix}$ so T_1 satisfies (4b) iff

$T_1^* Q^* U^{-1} UQS = I_q$. Letting $\tilde{T}_1 := (U^*)^{-1} Q T_1 =: \begin{bmatrix} I_q \\ R \end{bmatrix}$, (4a) becomes $B\tilde{T}_1 - \tilde{T}_1(\tilde{S}^* B \tilde{T}_1) = 0$, where $B = (U^*)^{-1} Q A Q^* U^*$. Partitioning B in the usual way, this implies $B_{21} + B_{22}R - RB_{11} - RB_{12}R = 0$. □

Generally, methods for solving (4) consist of giving an initial approximation for T_1, say X_0, and successively refining it to get the desired invariant subspace to sufficient accuracy. The theorem essentially says that if we do not want the successive approximations to T_1 to correspond to iteratively solving a *fixed* Riccati equation for R_k (with $X_k = \begin{bmatrix} I_q \\ R_k \end{bmatrix}$), then we must allow S^* to also change during the iteration, because fixing S^* essentially reduces to solving a Riccati equation. This observation is in Demmel (1987), who also notes that it is often desireable to fix q^2 entries in T_1 (as done computationally by Dongarra, Moler and Wilkinson, (1983)), which again involves solving a Riccati equation.

It follows from Theorem 1 that T is unitary only if S is to begin with, and then the matrix Q reducing S to $\begin{bmatrix} I_q \\ 0 \end{bmatrix}$ must decouple A. Also, if $T = \begin{bmatrix} I_q & 0 \\ R & I_p \end{bmatrix}$ decouples A and if $T = QU$ is the QU-factorization of T, then Q decouples A. Conversely, if unitary $Q = \begin{bmatrix} Q_{11} & Q_{12} \\ Q_{21} & Q_{22} \end{bmatrix}$ decouples A with Q_{11} nonsingular, then so does the associated Riccati transformation $T = \begin{bmatrix} I_q & 0 \\ R & I_p \end{bmatrix}$, $R := Q_{21}Q_{11}^{-1}$. This is useful when a unitary invariant subspace is wanted but the Riccati solution is known instead. In situations where the Riccati solution R is wanted, as in optimal control applications, it can be obtained via a QR-factorization of A. This is the essence of the Schur method to find R for problems in control theory (see Laub (1979), Arnold and Laub (1984)). The nonsingularity of the upper left q×q block of Q can be guaranteed after an appropriate row permutation $\hat{Q} = PQ$ of Q, and then $\hat{R} = \hat{Q}_{21}\hat{Q}_{11}^{-1}$ gives a decoupling for PAP^T. Chatelin (1984) considers Newton's method for solving (4) with S fixed, viz., given X_0 such that $S^*X_0 = I_q$,

$$X_{k+1} = X_k - F'(X_k)^{-1}F(X_k), \quad k = 0,1,..., \tag{13}$$

where now $F(X) := AX - X(S^*AX)$ and $F'(X)\colon Z \to (A - XS^*A)Z - ZS^*AX$. She shows that $S^*X_{k+1} = I_q$ as long as $S^*X_0 = I_q$ and S^*AX_k is invertible. She also proposes some modifications of Newton's method, each based on an approximate expression for the Jacobian $F'(X)$, and each only linearly convergent. In §4 we use the following modification of one of these quasi-Newton methods: Approximate F' by

$$\tilde{F}'(X)\colon Z \to (A - XS^*A)Z - ZS^*AX_0. \tag{14}$$

Given X_0 such that $S^*X_0 = I_q$, the iteration with $\tilde{F}'$ replacing F' in (13) is

$$AX_{k+1} - X_{k+1}S^*AX_0 = X_k(S^*AX_{k+1} - S^*AX_0), \quad k = 0,1,\ldots. \tag{15}$$

Notice that if $S^*X_k = I_q$ and S^*AX_0 is invertible, then multiplying (15) by S^* implies $(S^*X_{k+1} - I_q)S^*AX_0 = 0$, so $S^*X_{k+1} = I_q$.

Now consider the case $S^* = (I_q, 0)$ so that we search for solutions of the Riccati equation (12). The finite Taylor expansion for $G(R)$ around R_0 is

$$G(R) = G(R_0) + G'(R_0)(R - R_0) + \frac{1}{2}\, G''(R_0)(R - R_0)^2,$$

where the 1st and 2nd Fréchet derivatives are

$$Z \to G'(R)Z := A_{22}Z - ZA_{11} - ZA_{12}R - RA_{12}Z$$

and

$$Z \to G''(R)Z^2 := -2ZA_{12}Z,$$

so that

$$\begin{aligned} G(R) = G(R_0) &+ [A_{22}(R - R_0) - (R - R_0)A_{11} - (R - R_0)A_{12}R_0 - R_0A_{12}(R - R_0)] \\ &+ [-(R - R_0)A_{12}(R - R_0)] = 0. \end{aligned} \tag{16}$$

Several iterative methods have been proposed to solve (12), and they can easily be derived by considering different truncated expansions of (16):

(i) Taking the approximation

$$G(R_k) + A_{22}(R_{k+1}-R_k) = 0,$$

with $R_0 = 0$ we obtain

$$R_{k+1} = -A_{22}^{-1}(A_{21}-R_kA_{11}-R_kA_{12}R_k), \quad k = 0,1,..., \tag{17a}$$

and analogously

$$G(R_k) - (R_{k+1}-R_k)A_{11} = 0$$

gives with $R_0 = 0$,

$$R_{k+1} = (A_{21}+A_{22}R_k-R_kA_{12}R_k)A_{11}^{-1}, \quad k = 0,1,\ldots . \tag{17b}$$

The scheme (17a) has been used by Kokotovic (1975), who shows that it converges linearly. This scheme is appealing for singular perturbation problems for which the fast and slow variables force A_{22} to be $O(\frac{1}{\epsilon})$, and the linear convergence constant is then $O(\epsilon)$, $0 < \epsilon \ll 1$.

(ii) Using the truncated expansion

$$G(R_k) + A_{22}(R_{k+1}-R_k) - R_kA_{12}(R_{k+1}-R_k) = 0,$$

we have the scheme, $R_0 = 0$,

$$(A_{22}-R_kA_{12})R_{k+1} = -A_{21} + R_kA_{11}, \quad k = 0,1,..., \tag{18a}$$

and analogously

$$R_{k+1}(A_{11}+A_{12}R_k) = A_{21} + A_{22}R_k, \quad k = 0,1,\ldots . \tag{18b}$$

It is fairly easy, following the line of argument of Stewart (1973), to give sufficient conditions for linear convergence of these schemes. Specifically, for (18a), we can prove the following:

Lemma. *Assume* A_{22} *is nonsingular. Let* $\kappa := \dfrac{||A_{12}||\,||A_{21}||}{\mathrm{sep}^2(A_{22},0)}$ *and* $\alpha := \dfrac{||A_{11}||}{\mathrm{sep}(A_{22},0)}$ *for the zero matrix* $0 \in \mathbb{R}^{q\times q}$. *If* $\kappa < \dfrac{(1-\alpha)^2}{4}$, *then there is a unique solution of the Riccati equation satisfying*

$$0 \le ||R|| \le \frac{(1-\alpha) - \sqrt{(1-\alpha)^2 - 4\kappa}}{2\kappa}\,\frac{||A_{21}||}{\mathrm{sep}(A_{22},0_q)}\,.$$

This is an asymptotically stable solution obtained from (18a), *whose domain of attraction includes the matrices* R *satisfying the above inequality.* □

The lemma is valid for either the 2-norm or the Frobenius norm. An analogous result holds for (18b), so the schemes retain the characteristics of (17a, b) and perform better the farther the spectrum of $A_{22}(A_{11})$ is from zero.

To our knowledge, the schemes (18a, b) have not appeared before in the Riccati equation literature. While they have the obvious disadvantage of requiring a new matrix factorization at each step, they have some interesting features. Theoretically, they are inverse and direct iteration schemes, respectively. Specifically, at step k they look like

$$A\begin{bmatrix} I_q \\ R_{k+1} \end{bmatrix} = \begin{bmatrix} I_q \\ R_k \end{bmatrix} C_{k+1} \quad \text{and} \quad A\begin{bmatrix} I_q \\ R_k \end{bmatrix} = \begin{bmatrix} I_q \\ R_{k+1} \end{bmatrix} C_{k+1},$$

which is seen by solving for C_{k+1}. Computationally, they can be thought of as alternatives to Newton's method and could be very appealing when A is known to have large eigenvalues forced into the lower (upper) diagonal block of $\tilde{A}$.

(iii) Taking

$$G(R_k) + A_{22}(R_{k+1}-R_k) - (R_{k+1}-R_k)A_{11} = 0,$$

we obtain the frequently used scheme $R_0 = 0$,

$$A_{22}R_{k+1} - R_{k+1}A_{11} = -A_{21} + R_kA_{12}R_k, \quad k = 0,1,\dots\,. \tag{19}$$

It converges linearly to a locally unique solution of (12) (Stewart, 1973). Stewart's analysis shows that the scheme can be expected to perform satisfactorily as long as $\Lambda(A_{11})$ and $\Lambda(A_{22})$ are sufficiently separated. Finding R_{k+1} involves solving a Sylvester equation (19) at each step. Basic algorithms, involving a preprocessing step to reduce A_{11} and A_{22}

both to triangular or to Hessenberg and triangular forms, have been proposed by Bartels and Stewart (1972) and by Golub, Nash and vanLoan (1979), respectively. The reduction step is done only once, with a cost of $O(q^3+p^3)$ flops, while the solution step is $O(pq^2+qp^2)$ flops. In many applications, the sought invariant subspace has dimension $q \ll p$, which makes the reduction and solution steps essentially $O(n^3)$ and $O(n^2)$ flops, respectively.

With R_0 given, Newton's method is $R_{k+1} = R_k - [G'(R_k)]^{-1}G(R_k)$, $k = 0,1,\ldots$, or

$$(A_{22}-R_kA_{12})R_{k+1} - R_{k+1}(A_{11}+A_{12}R_k) = -A_{21} - R_kA_{12}R_k, \quad k = 0,1,\ldots \tag{20}$$

When $R_0 = 0$, the guaranteed convergence region of (20) is not much smaller than for (19), and Demmel (1987) shows that the scheme performs satisfactorily if $\Lambda(A_{11})$ and $\Lambda(A_{22})$ are sufficiently separated.

Comparing with (20), we see that (19) is simply a quasi-Newton method

$$R_{k+1} = R_k - [G'(R_0)]^{-1}G(R_k), \quad k = 0,1,\ldots, \tag{21}$$

with $R_0 = 0$. Indeed, from this we see that the linear convergence results for (19) can be derived directly from the more general Kantorovich analysis (e.g., in Kolmogorov and Fomin (1980)).

With Newton's method, we need to solve a Sylvester equation for which the blocks to be reduced change, so each step costs $O(n^3)$ flops. This is potentially expensive, but the quadratic convergence of the scheme can compensate for this cost, and in fact Newton's method is often used to refine an approximation obtained by another scheme (Arnold and Laub (1984)).

We now propose a new modified secant scheme which compromises between the order of convergence and the cost per iteration. It can be introduced in several ways, and we choose the simplest as opposed to the most rigorous. Assuming

$$G'(R_k)(R_k-R_{k-1}) \approx Y_k := G(R_k) - G(R_{k-1}), \tag{22}$$

we have

$$Y_k = A_{22}(R_k - R_{k-1}) - (R_k - R_{k-1})A_{11} - R_kA_{12}R_k + R_{k-1}A_{12}R_{k-1}$$
$$= A_{22}(R_k - R_{k-1}) - (R_k - R_{k-1})A_{11} - R_kA_{12}(R_k - R_{k-1}) - (R_k - R_{k-1})A_{12}R_{k-1},$$

and thus obtain an approximation for the Jacobian

$$Z \to \Delta_1(R_k)Z := A_{22}Z - ZA_{11} - R_kA_{12}Z - ZA_{12}R_{k-1}. \tag{23a}$$

Replacing $G'(R_k)$ in Newton's method with this, we obtain the iteration

$$(A_{22} - R_kA_{12})R_{k+1} - R_{k+1}(A_{11} + A_{12}R_{k-1}) = -A_{21} - R_kA_{12}R_{k-1}. \tag{24a}$$

Alternatively, we can add and subtract $R_{k-1}A_{12}R_k$ to Y_k to obtain another approximation to the Jacobian

$$Z \to \Delta_2(R_k)Z := A_{22}Z - ZA_{11} - R_{k-1}A_{12}Z - ZA_{12}R_k, \tag{23b}$$

and hence another iterative method

$$(A_{22} - R_{k-1}A_{12})R_{k+1} - R_{k+1}(A_{11} + A_{12}R_k) = -A_{21} - R_{k-1}A_{12}R_k. \tag{24b}$$

The two approximations (23) can be interpreted as distinct secant-like methods (although only $\Delta_1(R_k)$ satisfies the secant equation $\Delta(R_k - R_{k-1}) = G(R_k) - G(R_{k-1})$), arising because of the non-commutativity of the matrices in the quadratic term. While solving for R_{k+1} from (24a) or (24b) still involves solving a Sylvester equation in which the blocks to be reduced change at each step, making the cost of each step comparable to that for Newton's method, by instead alternating one step (24a) with one of (24b) the cost is essentially halved. This suggests the following *modified secant iteration*: Given R_0, R_1 (e.g., $R_0 = 0$, R_1 from (19)), let

$$\begin{aligned}(A_{22} - R_{k-1}A_{12})R_{k+1} - R_{k+1}(A_{11} + A_{12}R_k) &= -A_{21} - R_{k-1}A_{12}R_k,\\ (A_{22} - R_{k+1}A_{12})R_{k+2} - R_{k+2}(A_{11} + A_{12}R_k) &= -A_{21} - R_{k+1}A_{12}R_k,\end{aligned} \tag{25}$$
$$k = 1,3,5,\ldots.$$

The convergence properties of this method can be obtained by modifying the arguments of Stewart (1973) and Demmel (1987) for analyzing (19) and (20).

Theorem 2. *If* $\kappa = \dfrac{||A_{12}||\ ||A_{21}||}{\text{sep}^2(A_{11},A_{22})}$ *satisfies* $\kappa < \frac{1}{12}$, *then the modified secant method* (25) *converges to a unique solution inside the ball*

$$||R|| < \frac{2||A_{21}||}{\text{sep}(A_{11},A_{22})}.$$

If $E_k := R_k - R$ *then* $\lim_{j\to\infty} \dfrac{||E_{j+1}||}{||E_j||^{\frac{1+\sqrt{5}}{2}}} =$ *constant.*

Proof. The second equation of (25) gives

$$(A_{22}-R_{k+1}A_{12})E_{k+2} - E_{k+2}(A_{11}+A_{12}R_k) = -E_{k+1}A_{12}E_k,$$

which implies (for either the 2-norm or the Frobenius norm)

$$||E_{k+2}|| \le \frac{||A_{12}||\ ||E_k||\ ||E_{k+1}||}{\text{sep}(A_{11}+A_{12}R_k,\ A_{22}-R_{k+1}A_{12})}, \tag{26}$$

while for the first equation we get

$$||E_{k+1}|| \le \frac{||A_{12}||\ ||E_{k-1}||\ ||E_k||}{\text{sep}(A_{11}+A_{12}R_k,\ A_{22}-R_{k-1}A_{12})}. \tag{27}$$

To derive bounds on the growth of $||R_j||$, from (25) we have

$$||R_{k+1}|| \le \frac{||A_{21}|| + ||A_{12}||\ ||R_{k-1}||\ ||R_k||}{\text{sep}(A_{11},A_{22}) - ||A_{12}||(||R_k|| + ||R_{k-1}||)}$$

$$||R_{k+2}|| \le \frac{||A_{21}|| + ||A_{12}||\ ||R_{k+1}||\ ||R_k||}{\text{sep}(A_{11},A_{22}) - ||A_{12}||(||R_k|| + ||R_{k+1}||)} \qquad k = 1,3,\ldots,$$

where we use the property $\text{sep}(A+B,\ C+D) \ge \text{sep}(A,C) - ||B|| - ||D||$ (Stewart, 1973). If

$$g_j := \frac{\text{sep}(A_{11},A_{22})}{||A_{21}||}\ ||R_j||,$$

we have $g_0 = 0$, $g_{j+1} \le \dfrac{1 + \kappa g_j g_{j-1}}{1 - \kappa(g_j+g_{j-1})}$, $j = 1,2,\ldots$. It is easy to check that $\{g_j\}$ is a monotonically increasing sequence approaching $g^* = \dfrac{1 - \sqrt{1 - 12\kappa}}{6\kappa}$ since $\kappa < \frac{1}{12}$ (the same restriction on κ needed for Newton's method). Therefore,

$$||R_j|| < \frac{1 - \sqrt{1 - 12\kappa}}{6\kappa} \frac{||A_{21}||}{\text{sep}(A_{11},A_{22})}, \tag{28}$$

and from (26) and (27)

$$||E_{j+1}|| \leq ||E_j||\ ||E_{j-1}|| \frac{3||A_{12}||}{\text{sep}(A_{11},A_{22})(2 + \sqrt{1 - 12\kappa})}, \quad j = 1,2,\dots . \tag{29}$$

Setting $C := \dfrac{3||A_{12}||}{\text{sep}(A_{11},A_{22})(2 + \sqrt{1 - 12\kappa})}$ and $d_j := C||E_j||$, (29) reads $d_{j+1} \leq d_j d_{j-1}$. Since $d_0 d_1 \leq d < 1$, we obtain $d_j \leq d^{\gamma_j}$ with $\{\gamma_j\}$ the familiar Fibonacci sequence. Therefore, $\gamma_j = \frac{1}{\sqrt{5}}\left[\left(\frac{1 + \sqrt{5}}{2}\right)^j - \left(\frac{1 - \sqrt{5}}{2}\right)^j\right]$, $\lim_{j\to\infty} \frac{\gamma_{j+1}}{\gamma_j} = \frac{1 + \sqrt{5}}{2}$, and we recover the order of convergence characteristic of the secant method. □

Thus, (25) is an alternative to Newton's method for refining an approximate solution found by another means. It is a superlinearly convergent method requiring half of the number of matrix factorization of Newton's method (when solving the Sylvester equations). It is possible to capitalize on a further (minor) savings in terms of the formation of the right-hand sides of (24), because one matrix multiplication can be avoided every double iteration. The operation counts for (19), (20), and (25) using the Golub-Nash-vanLoan algorithm, assuming that $p \geq q$ and that intermediate matrix products are not stored from one iteration to the next, are shown in Table 1. If for instance $q \ll p$ (or, with modification $q \gg p$), the modified secant method could then be roughly 38% more efficient than Newton's method, with their difference otherwise being more modest, as in the example (taken from Golub and vanLoan (1983)).

Example. We seek a representation $\begin{bmatrix} I_2 \\ R \end{bmatrix} \in \mathbb{R}^{5\times 2}$ for an invariant subspace of

$$A = \begin{vmatrix} 2 & 3 & 4 & 5 & 6 \\ 4 & 4 & 5 & 6 & 7 \\ 0 & 3 & 6 & 7 & 8 \\ 0 & 0 & 2 & 8 & 9 \\ 0 & 0 & 0 & 1 & 10 \end{vmatrix}$$

For this, we iterate with $R_0 = 0$ to find $R \in \mathbb{R}^{3\times 2}$ satisfying (12), using the stopping criterion $||G(R_k)||_F < 10^{-13}$. The iteration (19) overflows, while Newton's method (20) and the modified secant method (25) both converge to the subspace associated with $\{\lambda_1 \approx -0.33541642,\ \lambda_2 \approx 1.5014220\}$ after 8 iterations and require .013 and .011 seconds,

Table 1. *Operation count for the schemes* (19), (20), (25).

Method	k = 1	k ≥ 2
Iteration (19)	$\frac{5}{3}p^3 + 10q^3 + 5p^2q + \frac{9}{2}pq^2$	$5p^2q + \frac{9}{2}pq^2$
Newton (20)	$\frac{5}{3}p^3 + 10q^3 + 6p^2q + \frac{9}{2}pq^2$	$\frac{5}{3}p^3 + 10q^3 + 6p^2q + \frac{9}{2}pq^2$
Modified secant (25)	$\frac{5}{3}p^3 + 10q^3 + 6p^2q + \frac{9}{2}pq^2$	i = even, $10q^3 + 5p^2q + \frac{9}{2}pq^2$ i = odd, $5/3p^3 + 6p^2q + \frac{9}{2}pq^2$

respectively (on an IBM 3081 using double precision). □

In light of Theorem 1, it is easy to extend the modified secant method to the case in which S^* is fixed, but not necessarily equal to $(I_q,0)$. Also, the method has a natural extension to the generalized eigenvalue problem setting, where the potential savings over Newton's method is much greater, as will be investigated elsewhere.

There is an important distinction to be made between using the Riccati transformation to compute an invariant subspace and the closely related use of Gaussian transformations such as with treppeniteration (cf. §2). For the latter, or for QR iteration, there is a chain of similarity transformations, and orthogonal factorizations are generally recommended in order to avoid the possible buildup of roundoff errors. But when we instead directly solve the Riccati equation with iterative methods, each iteration involves the blocks of the original matrix A, so the very nature of the iterative process diminishes the possibility of such error accumulation, at least as long as the Riccati equation itself is well-conditioned. That this often happens is confirmed by the approach taken by Arnold and Laub (1984), as well as our own experience with the various schemes, although it is not necessarily the case (Dieci, 1989).

A potential drawback of an iterative scheme to solve the Riccati equation is the dependence on having a good initial guess. While this is often a restriction in practice, it is also the case that the convergence regions predicted by the theory can be very conservative. If convergence is to the wrong solution corresponding to the wrong splitting of eigenvalues of A, it turns out to be generally straightforward to recover the desired one (Dieci, 1989) (as it is for QR iteration).

An important source of invariant subspace problems is a general differential system $\frac{dx}{dt} = A(t)x$. In some contexts, the matrix invariant subspace problem is to be solved for various values of t (Kreiss, Nichols and Brown, 1986), and the iterative methods are particularly appealing. Alternatively, a dynamic change of variables $y = T^{-1}(t)x$ is sought such that

$$\frac{dy}{dt} = \tilde{A}(t)y =: \begin{bmatrix} \tilde{A}_{11} & \tilde{A}_{12} \\ 0 & \tilde{A}_{22} \end{bmatrix} y.$$

The generalization of the decoupling equations (4) is

$$\frac{dT_1}{dt} = AT_1 - T_1\tilde{A}_{11}, \tag{30a}$$

$$\tilde{A}_{11} = S^*(AT_1 - \frac{dT_1}{dt}). \tag{30b}$$

Decoupling is then important both as an analytic tool, e.g., as a fundamental tool for analyzing methods for solving boundary value problems (Ascher, Mattheij and Russell, 1988), and as a computational tool. For the latter, there are analogues of the 3 approaches for the matrix problem (30a), and much of the method development parallels this case (Dieci and Russell, 1987).

IV. THE EIGENVECTOR PROBLEM

We finish with the familiar one dimensional invariant subspace problem, i.e., the eigenvector problem. In this case there are no extra degrees of freedom to force further structure on the system (on T_1 or $\tilde{A}$), no matter which of the three previous approaches is used.

Perhaps the simplest and most straightforward methods to solve this problem are the direct and inverse power methods, which are extremely powerful when used in conjunction with a shift strategy. In particular, the standard way to refine an eigenvalue estimate is via so-called inverse iteration, which is a shifted inverse power method (see Golub and vanLoan (1983) and references therein). Recall that for inverse iteration, given a unit vector ν_0 and scalar μ such that $A - \mu I$ is nonsingular,

$$(A-\mu I)z_{k+1} = \nu_k, \quad \nu_{k+1} = \alpha_{k+1}z_{k+1}, \quad k = 0,1,2,..., \tag{31}$$

where each α_{k+1} is a normalization factor (such as $\frac{1}{||z_{k+1}||_2}$). If A has a complete set of eigenvectors $\{v_1,...,v_n\}$ with $Av_i = \lambda_i v_i$ and $|\lambda_j-\mu| < |\lambda_i-\mu|$, $i \neq j$, the procedure converges to a multiple of v_j provided ν_0 is not deficient in the v_j direction, and the convergence is linear with ratio of successive errors $O\left(\min_{i\neq j} \frac{|\lambda_j-\mu|}{|\lambda_i-\mu|}\right)$. The possibility that ν_0 is deficient in the v_j direction is generally precluded by the presence of roundoff errors, and the process in practice converges for virtually any μ. In fact, if the eigenvalues are real, the only values of μ to be avoided in theory are the points midway between two adjacent eigenvalues (as in the example in (Peters and Wilkinson (1979) where things go awry), and in the complex case they are a set of points lying midway between a pair of eigenvalues. Convergence, while very slow if μ is not a good approximate eigenvalue, is essentially global.

Using the previous framework, inverse iteration can easily be seen to correspond to a modified Newton iteration. To see this, note that the Newton method (13) for finding an eigenvector is, given x_0,

$$(A-(s^*Ax_k)I)x_{k+1} = (s^*Ax_{k+1} - s^*Ax_k)x_k, \quad k = 0,1,\ldots \tag{32}$$

Chatelin (1984) points out that the generated sequence is equivalent (actually, to within a normalization) to a sequence of right-Rayleigh quotient iteration vectors (i.e., using inverse iteration with updated shifts):

$$\nu_0 = \frac{x_0}{||x_0||_2}, \quad \mu_0 = s^*Ax_0,$$

$$(A-\mu_k I)z_{k+1} = \nu_k, \quad \nu_{k+1} = \frac{z_{k+1}}{||z_{k+1}||_2}, \quad \mu_{k+1} = \frac{s^*A\nu_{k+1}}{s^*\nu_{k+1}}, \quad k = 0,1\ldots \tag{33}$$

Inverse iteration in (31) corresponds to keeping the μ_k's equal to μ_0. Moreover, the modified Newton method (15) is the inverse iteration (31) with $\mu = s^*Ax_0$, because (15) in this one dimensional case is

$$(A-\mu I)x_{k+1} = (s^*Ax_{k+1} - s^*Ax_0)x_k, \quad k = 0,1,\ldots, \tag{34}$$

so that the sequence $\{x_k\}$ is equivalent to the sequence $\{\nu_k\}$ generated from (31). The choice $\mu = s^*Ax_0$ is natural since μ is an approximate eigenvalue if x_0 is an approximate eigenvector (with s^* the left inverse for x_0).

The first step of (34) is the same as with Newton's method, which, as observed by Peters and Wilkinson (1979), explains the initial rapid improvement from inverse iteration observed in practice. This equivalence also shows that the modified Newton scheme (34) essentially converges for all μ since (31) does. Deviating slightly from the previous structure by fixing $s^* = (0,\ldots,0,1)$ and choosing $x_0 = (0,\ldots,0,1)$, then $x_k^* = (r_k^*,1)$. For the resulting "inverse" Riccati equation, $\mu = a_{nn}$, which corresponds to the shift used for Hessenberg (or tridiagonal) A.

Finally, consider the unitary representation case. Then from (7) Newton's method can be written as, given x_0 with $x_0^*x_0 = 1$,

$$(A - (x_k^* A x_k) I) x_{k+1} = [(x_{k+1}^* - x_k^*) A x_k + x_k^* A (x_{k+1} - x_k)] x_k. \tag{35}$$

By additionally requiring $x_{k+1}^* x_{k+1} = 1$, $k = 0,1,\ldots$ (i.e., taking the exact expansion for the equation $x^* x - 1 = 0$ as in §2), this is easily seen to be equivalent to the *Rayleigh quotient iteration*: given q_0 with $q_0^* q_0 = 1$,

$$(A - \mu_k I) z_{k+1} = q_k, \quad (\mu_k = q_k^* A q_k)$$

$$q_{k+1} = \frac{z_{k+1}}{\|z_{k+1}\|_2}, \quad k = 0,1,\ldots .$$

As for the higher dimensional problem, it is interesting to ask if efficient higher order methods (and shift strategies) can be developed using this framework.

V. REFERENCES

1. Arnold, W. F., Laub, A. J.: Generalized eigenproblem algorithms and software for algebraic Riccati equation, *Proc. IEEE*, 72, 1746-1754 (1984).

2. Ascher, U. M., Mattheij, R. M., Russell, R. D.: *Numerical Solution of Boundary Value ODEs*, Englewood Cliffs, Prentice-Hall (1988).

3. Bartels, R. H., Stewart, G. W.: Algorithm 432. Solution of the matrix equation AX+XB = C, *Comm. ACM*, 15, 820-826 (1972).

4. Chatelin, F.: Simultaneous Newton's iteration for the eigenproblem, *Computing Suppl.*, 5, 67-74 (1984).

5. Demmel, J. W.: Three methods for refining estimates of invariant subspaces, *Computing*, 38, 43-57 (1987).

6. Dieci, L.: Some numerical considerations and Newton's method revisited for solving AREs, submitted to *IEEE AC* (1989).

7. Dieci, L., Russell, R. D.: On the computation of invariant subspaces, LCCR Tech. Rep. 87-11, Simon Fraser University (1987).

8. Dongarra, J. J., Moler, C. B., Wilkinson, J. H.: Improving the accuracy of computed eigenvalues and eigenvectors, *SIAM J. Numer. Anal.*, 20, 23-45 (1983).

9. Golub, G. H., Nash, S., vanLoan, C. F.: A Hessenberg-Schur method for the problem AX + XB = C, *IEEE Trans. Autom. Control*, AC-24, 909-913 (1979).

10. Golub, G. H., vanLoan, C. F.: *Matrix Computations.* Baltimore, The Johns Hopkins University Press (1983).

11. Kokotovic, P. V.: A Riccati equation for block-diagonalization of ill-conditioned systems, *IEEE Trans. Autom. Control*, AC-20, 812-814 (1975).

12. Kokotovic, P. V., Khalil, H. K., O'Reilly, J.: *Singular Perturbation Methods in Control: Analysis and Design.* London, Academic Press (1986).

13. Kolmogorov, A. N. and Fomin, S: *Elementy teorii funktsij: Funktsianol'nogo analiza.* Moskva, Nauka (1980).

14. Kreiss, H.-O., Nichols, N., Brown, D.: Numerical methods for stiff TPBVP, *SIAM J. Numer. Anal.*, 23, 325-368 (1986).

15. Laub, A. J.: A Schur method for solving algebraic Riccati equations, *IEEE Trans. Autom. Control*, AC-24, 913-921 (1979).

16. Parlett, B. N.: *The Symmetric Eigenvalue Problem*, Englewood Cliffs, Prentice Hall (1980).

17. Peters, G., Wilkinson, J. H.: Inverse iteration, ill-conditioned equations, and Newton's method, *SIAM Review*, 21, 339-360 (1979).

18. Stewart, G. W.: Error and perturbation bounds for subspaces associated with certain eigenvalue problems, *SIAM Review*, 15, 727-764 (1973).

19. Wilkinson, J. H.: *The Algebraic Eigenvalue Problem*, Oxford, Clarendon Press (1965).

Error Bounds via Complete Monotonicity for a Uniform Asymptotic Expansion of the Legendre Function P_n^{-m} (cosh z)

C. L. Frenzen Professor, Mathematics Department, Southern Methodist University, Dallas, Texas

I. INTRODUCTION

Completely monotonic functions are the asymptotic analyst's ally, for when f is completely monotonic on [0, b],

$$0 \le (-1)^n f^{(n)}(x) \le (-1)^n f^{(n)}(0)$$

for all non-negative n and all x ∈ [0, b], giving bounds on the derivatives of f. Here we shall use complete monotonicity to derive a uniform asymptotic expansion of the Legendre function $P_n^{-m}(\cosh z)$ for fixed $m + \frac{1}{2} > 0$ and large n which is uniformly valid for $z \in [0, \infty)$. We also provide uniform error bounds for our expansion. Like several recent authors ((Ursell, 1984), (Shivakumar and Wong, 1988)) we begin

Research supported in part by NSF Grant DMS-8718941

with an integral representation. An equivalent asymptotic expansion with error bounds, obtained from the asymptotic theory of linear second order ordinary differential equations, can be found in (Olver, 1974, Ch. 12). The derivation given here parallels that of (Shivakumar and Wong, 1988). However by using complete monotonicity we are able to shorten the calculation and sharpen the error bounds. A comparison with the bounds of (Shivakumar and Wong, 1988) is given at the end of this paper. Complete monotonicity has also been used to obtain error bounds for gamma function ratios in (Frenzen, 1987).

We begin with the integral representation, valid for $m+\frac{1}{2} > 0$,

$$P_n^{-m}(\cosh z) = \left(\frac{1}{2\pi}\right)^{\frac{1}{2}} \frac{(\sinh z)^{-m}}{\Gamma(m+\frac{1}{2})} \int_{-z}^{z} e^{-ut}(\cosh z - \cosh t)^{m-\frac{1}{2}} dt, \tag{1}$$

where $u = n + \frac{1}{2}$. From (Shivakumar and Wong, 1988, eqs. 2.4, 2.5), we have the series expansion

$$\cosh z - \cosh t = \frac{\sinh z}{2z} (z^2 - t^2) \sum_{\nu=0}^{\infty} (-1)^\nu \phi_\nu(z)\, (z^2 - t^2)^\nu\ , \tag{2}$$

where $\phi_0(z) = 1$ and

$$\phi_\nu(z) = \frac{1}{2^\nu(\nu+1)!} \frac{I_{\nu+\frac{1}{2}}(z)}{z^\nu I_{\frac{1}{2}}(z)}\ , (\nu \geq 1). \tag{3}$$

The expansion

$$(\cosh z - \cosh t)^{m-\frac{1}{2}} = \left(\frac{\sinh z}{2z}\right)^{m-\frac{1}{2}} (z^2 - t^2)^{m-\frac{1}{2}} \sum_{\nu=0}^{\infty} (-1)^\nu \psi_\nu(z)(z^2-t^2)^\nu \tag{4}$$

follows immediately, where the coefficients $\psi_\nu(z)$, analytic functions of z, satisfy the recurrence relation

$$\psi_{\nu+1}(z) = \sum_{j=0}^{\nu} \left[(m-\tfrac{1}{2}) - \frac{j}{\nu+1}(m+\tfrac{1}{2})\right] \phi_{\nu+1-j}(z)\, \psi_j\,(z), \qquad (5)$$

$\nu = 0, 1, 2, \ldots$ with $\psi_0(z) = 1$; see (Shivakumar and Wong, 1988) for the first three. Introducing the remainder $\Delta_p(z, t)$ defined by

$$\left\{\frac{\cosh\, z - \cosh\, t}{z^2 - t^2}\, \frac{2z}{\sinh\, z}\right\}^{m-\frac{1}{2}} = \sum_{\nu=0}^{p-1} (-1)^\nu \psi_\nu(z)\, (z^2 - t^2)^\nu$$

$$+ (-1)^p (z^2 - t^2)^p\, \Delta_p(z, t) \qquad (6)$$

and recalling that

$$I_\nu\,(uz) = \frac{(2z)^{-\nu} u^\nu}{\pi^{1\backslash 2}\Gamma(\nu+\frac{1}{2})} \int_{-z}^{z} (z^2 - t^2)^{\nu-\frac{1}{2}}\, e^{-ut}\, dt\ , \qquad (7)$$

we insert (6) into (1) and use (7) to obtain

$$P_n^{-m}\,(\cosh\, z) = \left(\frac{z}{\sinh\, z}\right)^{\frac{1}{2}} \left[\sum_{\nu=0}^{p-1} c_\nu(z)\, \frac{I_{m+\nu}(z)}{u^{m+\nu}} + \epsilon_p(z, u)\right]\ , \qquad (8)$$

where

$$c_\nu(z) = (-1)^\nu\, \frac{\Gamma(m+\nu+\frac{1}{2})}{\Gamma(m+\frac{1}{2})}\, (2z)^\nu\, \psi_\nu(z)\ , \qquad (9)$$

and

$$\epsilon_p(z, u) = \frac{(-1)^p}{\pi^{1/2}(2z)^m\Gamma(m+\frac{1}{2})} \int_{-z}^{z} e^{-ut}(z^2-t^2)^{p+m-\frac{1}{2}}\, \Delta_p(z, t)\, dt\ . \qquad (10)$$

Note from (3), (5) and (9) that $c_\nu(z) = O(1)$, $(z \to \infty)$.

We shall soon show that

$$\left|\Delta_p(z,\,t)\right| \le \frac{M_p^{(1)}}{(z+2)^p} + \frac{M_p^{(2)}}{(z+4)^p}, (-z \le t \le z) \;, \tag{11}$$

for explicit constants $M_p^{(1)}$ and $M_p^{(2)}$ (given in (39) and (43)). Combining (11), (10) and (7) then yields the uniform error bound

$$\left|\epsilon_p(z,\,u)\right| \le \frac{\Gamma(p+m+\frac{1}{2})}{\Gamma(m+\frac{1}{2})}\;\frac{I_{m+p}(uz)}{u^{p+m}}\left[\frac{M_p^{(1)}(2z)^p}{(z+2)^p} + \frac{M_p^{(2)}(2z)^p}{(z\;+\;4)^p}\right] \tag{12}$$

for the asymptotic expansion in (8).

2. PRELIMINARY RESULTS

From (2), cosh z- cosh t is a function of z^2 - t^2. Let $x = z^2 - t^2$ so that $0 \le x \le z^2$ when $-z \le t \le z$ and define

$$G(x) := \frac{\cosh z \;-\; \cosh\sqrt{(z^2-x)}}{x}\;\;\frac{2z}{\sinh z} \tag{13}$$

on $[0,\,z^2]$ with $G(0) = 1$. From (2),

$$G(x) = \sum_{\nu=0}^{\infty} (-1)^{\nu}\;\phi_{\nu}(z)x^{\nu}. \tag{14}$$

Lemma 0. G is completely monotonic on $[0,\,z^2]$; i.e. for all $0 \le x \le z^2$ and $k = 0, 1, 2, \ldots,$

$$0 < (-1)^k\,G^{(k)}\,(x) \le (-1)^k\,G^{(k)}(0).$$

proof. From (13) it follows that

$$G(x) = \frac{2z}{\sinh z}\sum_{p=1}^{\infty}\frac{1}{(2p)!}\;\frac{(z^2)^p - (z^2-x)^p}{z^2-(z^2-x)}$$

$$= \frac{2z}{\sinh\,z}\sum_{p=1}^{\infty}\frac{1}{(2p)!}\sum_{j=0}^{p-1}(z^2-\;x)^j\,(z^2)^{p-1-j}$$

$$= \frac{2z}{\sinh z} \sum_{k=0}^{\infty} \left[\sum_{j=o}^{\infty} \frac{z^{2j}}{(2j+2k+2)!} \right] (z^2 - x)^k \tag{15}$$

Differentiating the series in (15) with respect to x then yields the above inequalities. □

We remark that Lemma 0 also shows that G^r, where r is a non-negative integer, is also completely monotonic on $[0, z^2]$ and therefore is enveloped by its Maclaurin series.

As in (Shivakumar and Wong, 1988), we now study the function σ, defined by

$$\sigma(z, t) := 1 - \frac{2z}{\sinh z} \frac{\cosh z - \cosh t}{z^2 - t^2} . \tag{16}$$

σ is a function of x; denoting it by g and using (13) and (14),

$$\sigma(z, t) = g(x) := 1 - G(x) = \sum_{\nu=1}^{\infty} (-1)^{\nu+1} \phi_\nu(z)\, x^\nu , \tag{17}$$

where G is given in (13). We give $\frac{g(x)}{x}$ the value $\phi_1(z)$ at the removable singularity $x = 0$.

Lemma 1. For $0 \le x \le z^2$, $0 < \frac{g(x)}{x} \le \phi_1(z)$.

Proof. By the mean value theorem there exists a ς between 0 and x such that

$$\frac{g(x)}{x} = -\frac{(G(x) - G(0))}{x} = -G^{(1)}(\varsigma)$$

From Lemma 0, $0 < -G^{(1)}(\varsigma) \le -G^{(1)}(0)$, and $-G^{(1)}(0) = \phi_1(z)$. □

Lemma 2. For any positive integer n and $0 \le x \le z^2$,

$$|g^{(n)}(x)| \leq n!\,\phi_n(z) \tag{18}$$

Proof. By Lemma 0 and (17),

$$|g^{(n)}(x)| = |G^{(n)}(x)| \leq |G^{(n)}(0)| = n!\,\phi_n(z)\ . \ \square$$

From Lemma 0, $g^{(1)}(x) > 0$ on $[0, z^2]$, so that

$$g(x) \leq g(z^2) = 1 - \frac{2}{z}\left(\frac{\cosh z - 1}{\sinh z}\right) = 1 - \frac{\tanh \frac{z}{2}}{\frac{z}{2}} < 1\ . \tag{19}$$

Lemma 3. The function H, defined by

$$H(x) = \frac{g(x)}{x(1 - g(x))}\ , \quad (0 \leq x \leq z^2)\ , \tag{20}$$

with $H(0) = \phi_1(z)$, is increasing, and

$$\phi_1(z) \leq H(x) \leq \frac{1}{2z}\ , \ (0 \leq x \leq z^2)\ .$$

proof. From (13) and the identity

$$\frac{\cosh a - \cosh b}{a^2 - b^2} = \frac{1}{2}\ \frac{\sinh\left(\frac{a+b}{2}\right)}{\left(\frac{a+b}{2}\right)}\ \frac{\sinh\left(\frac{a-b}{2}\right)}{\left(\frac{a-b}{2}\right)} \tag{21}$$

it follows that

$$H^{(1)}(x) = \frac{1}{x^2}[1 - W(a, b)]\ , \tag{22}$$

where

$$W(a, b) = \frac{\frac{\sinh a}{a}\ \frac{\sinh b}{b}}{\left(\frac{\sinh\left(\frac{a+b}{2}\right)}{\left(\frac{a+b}{2}\right)}\right)^2 \left(\frac{\sinh\left(\frac{a-b}{2}\right)}{\left(\frac{a-b}{2}\right)}\right)^2}\ .$$

and $a = z$, $b = \sqrt{(z^2 - x)}$. Since $\frac{\sinh a}{a} = \prod_{n=1}^{\infty}\left(1 + \frac{a^2}{n^2\pi^2}\right)$.

$$W(a, b) = \prod_{n=1}^{\infty} \frac{\left(1+\frac{a^2+b^2}{2n^2\pi^2}\right)^2 - \left(\frac{a^2-b^2}{2n^2\pi^2}\right)^2}{\left(1 + \frac{a^2+b^2}{2n^2\pi^2} + \left(\frac{a^2-b^2}{4n^2\pi^2}\right)^2\right)^2} \leq 1 \quad ,$$

with equality holding only when $a = b$ (i.e., $x = 0$). Consequently, by (22), $H^{(1)}(x) > 0$ on $(0, z^2]$. With $H(0) = \phi_1(z)$, it follows that $\phi_1(z) \leq H(x) \leq H(z^2)$. But

$$H(z^2) = \frac{1}{2z}\left[\coth \frac{z}{2} - \frac{2}{z}\right]$$

and $\coth \frac{z}{2} - \frac{2}{z}$ is non-negative and increases on $[0, \infty)$ with limit 1 as $z \to \infty$. □

Finally we obtain some inequalities for $\phi_\nu(z)$. From (3) and the fact that $I_{\nu+\frac{1}{2}}(z) \leq I_{\frac{1}{2}}(z)$, we have

$$\phi_\nu(z) \leq \frac{1}{(2z)^\nu(\nu+1)!} \tag{23}$$

From (Olver, 1974, p. 60) ,

$$I_\nu(z) = \left(\frac{1}{2} z\right)^\nu \sum_{s=0}^{\infty} \frac{\left(\frac{1}{4} z^2\right)^s}{s\,!\,\Gamma(\nu+s+1)} \tag{24}$$

and the inequality

$$\frac{\Gamma(\nu + 1)}{\Gamma(\nu+s+\frac{3}{2})} \leq \frac{1}{\Gamma(s+\frac{3}{2})} \quad , (\nu = 0, 1, 2,\ldots, s = 0, 1, 2,\ldots),$$

it follows that

$$\phi_\nu(z) \leq \frac{1}{4^\nu\ \nu!\ (\nu+1)!} \tag{25}$$

For $0 \leq z \leq 2,\ z+2 \leq 4$ and (25) implies $\phi_\nu(z) \leq \frac{1}{(z+2)^\nu\ \nu!\ (\nu+1)!}$.

For $z \geq 2,\ 2z \geq z+2$ and (23) implies $\phi_\nu(z) \leq \frac{1}{(z+2)^\nu(\nu+1)!}$.

Combining the last two inequalities then gives

$$\phi_\nu(z) \leq \frac{1}{(z+2)^\nu(\nu+1)!} \quad , \tag{26}$$

for $\nu = 0, 1, 2, \ldots,$ and $0 \leq z < \infty$.

3. DERIVATION OF ERROR BOUNDS

We now give a proof of the result in (11) which leads to the error bounds in (12). By the binomial theorem

$$\left[1 - g(x)\right]^{m-\frac{1}{2}} = \sum_{r=0}^{p-1} \frac{\Gamma(\frac{1}{2}-m+r)}{r!\Gamma(\frac{1}{2}-m)} (g(x))^r + (g(x))^p R_p^{(2)} \tag{27}$$

where

$$R_p^{(2)} = \frac{\Gamma(\frac{1}{2}-m+p)}{(p-1)!\ \Gamma(\frac{1}{2}-m)} \int_0^1 (1-u)^{p-1}(1-g(x)u)^{m-\frac{1}{2}-p} du \ . \tag{28}$$

Replacing each $(g(x))^r$, $r = 0, \ldots. p - 1$ on the right hand side of (27) by its Taylor polynomial plus remainder gives

$$(1-g(x))^{m-\frac{1}{2}} = \sum_{j=0}^{p-1} \sum_{r=0}^{p-1} \left[\frac{\Gamma(\frac{1}{2}-m+r)}{r!\ \Gamma(\frac{1}{2}-m)} g_r^{(j)}(0)\right] \frac{x^j}{j!} + \frac{x^p}{p!} R_p^{(1)} + (g(x))^p R_p^{(2)} \tag{29}$$

where $g_r(x) = (g(x))^r$ and

$$R_p^{(1)} = \sum_{r=0}^{p-1} \frac{\Gamma(\frac{1}{2}-m+r)}{r!\Gamma(\frac{1}{2} - m)} g_r^{(p)}(\varsigma_r) \quad , \tag{30}$$

with ς_r lying between 0 and x, $r = 0, \ldots, p - 1$. Note that because $p \geq 1$, the sum in (30) actually starts from $r = 1$ since $g_0^{(p)}(\zeta_0) = 0$. Because $(1-g(x))^{m-\frac{1}{2}}$ also has the expansion given in (6), a comparison of (29) and (6) implies that

$$\left|\Delta_p(z,\, t)\right| \le \frac{1}{p!}\left|R_p^{(1)}\right| + \left(\frac{g(x)}{x}\right)^p \left|R_p^{(2)}\right| . \tag{31}$$

To estimate $R_p^{(1)}$ we shall apply di Bruno's formula for the n-th derivative of a composite function (see, for example (Riordan, 1980)) to $g_r^{(p)}(\zeta_r)$, obtaining

$$g_r^{(p)}(\zeta_r) = \sum_{k=1}^{p} \frac{\Gamma(r+1)}{\Gamma(r-k+1)} (g(\zeta_r))^{r-k} A_{pk} (g^{(1)}(\zeta_r),\ldots, g^{(p)}(\zeta_r)) \tag{32}$$

where $p \ge 1$ and

$$A_{pk}\left[g^{(1)}(\zeta_r),\, \ldots,\, g^{(p)}(\zeta_r)\right] = \sum \frac{p!}{m_1!m_2!\ldots m_p!} \left(\frac{g^{(1)}(\zeta_r)}{1!}\right)^{m_1} \left(\frac{g^{(p)}(\zeta_r)}{p!}\right)^{m_p} \tag{33}$$

The summation in (33) extends over all non-negative integers $m_1, m_2, \ldots m_p$ satisfying

$$\begin{aligned} m_1 + m_2 + \ldots + m_p &= k \\ m_1 + 2m_2 + \ldots + pm_p &= p \end{aligned} \tag{34}$$

By Lemma 2, $\left| g^{(j)}(\zeta_r)\right| \le j\,!\,\phi_j(z)$ for $j \ge 1$, so that (33), (26) and the second equation in (34) imply

$$\frac{\left|A_{pk}\left[g^{(1)}(\zeta_r),\, \ldots,\, g^{(p)}(\zeta_r)\right]\right|}{p!} \le \frac{B_{pk}}{(z+2)^p} , \tag{35}$$

where the constant B_{pk} is defined by

$$B_{pk} := \sum \frac{1}{m_1!m_2! \ldots m_p!} \frac{1}{(2!)^{m_1}} \cdots \frac{1}{((p+1)!)^{m_p}} , \tag{36}$$

and the sum in (36) again extends over solutions of (34). For example,

$$B_{21} = \frac{1}{6} , \quad B_{31} = \frac{1}{24} , \quad B_{32} = \frac{1}{12} .$$

Since $g_0^{(p)}(\varsigma_r) = 0$, $(\Gamma(r - k + 1))^{-1} = 0$ for $k \geq (r + 1)$ and $|g(\varsigma_r)| \leq 1$ (see (19)), (32) together with (35) yields the inequality

$$\frac{|g_r^{(p)}(\varsigma_r)|}{p!} \leq \frac{1}{(z+2)^p} \sum_{k=1}^{r} \frac{\Gamma(r+1)}{\Gamma(r-k+1)} B_{pk} . \tag{37}$$

Finally, using (37) in (30) gives

$$\frac{|R_p^{(1)}|}{p!} \leq \frac{M_p^{(1)}}{(z+2)^p} , \tag{38}$$

where

$$M_p^{(1)} = \sum_{r=1}^{p-1} \left(\left| \frac{\Gamma(\frac{1}{2}-m+r)}{\Gamma(\frac{1}{2}-m)} \right| \left\{ \sum_{k=1}^{r} \frac{B_{pk}}{\Gamma(r-k+1)} \right\} \right) , \tag{39}$$

and B_{pk} is given by (36). The first few $M_p^{(1)}$ are readily determined:

$$M_1^{(1)} = 0, \quad M_2^{(1)} = \frac{|m-\frac{1}{2}|}{6} , \quad M_3^{(1)} = \frac{|m - \frac{1}{2}|}{24} + \frac{|m-\frac{1}{2}|\,|m-\frac{3}{2}|}{8} \tag{40}$$

To obtain a bound for $R_p^{(2)}$, we note that since $0 \leq g(x) < 1$ by (19), $1-u \leq 1-g(x)u$ and

$$\int_0^1 (1-u)^{p-1} (1- g(x))^{m-\frac{1}{2}-p} du \leq \int_0^1 (1-g(x)u)^{m+\frac{1}{2}} (1-g(x)u)^{-2} du$$

$$\leq \int_0^1 (1 - g(x)u)^{-2}\, du$$

$$= \frac{1}{1-g(x)} \quad , \tag{41}$$

where we have used $m+\frac{1}{2} > 0$. Since $(1 - g(x))^p \leq (1-g(x))$, combining (28) and (41) gives

$$\left(\frac{g(x)}{x}\right)^p |R_p^{(2)}| \leq M_p^{(2)} \left(\frac{g(x)}{x(1-g(x))}\right)^p . \tag{42}$$

with

$$M_p^{(2)} = \frac{\Gamma(\frac{1}{2}-m+p)}{(p-1)!\ \Gamma(\frac{1}{2}-m)} \tag{43}$$

Using Lemma 3, (42) then implies

$$\left(\frac{g(x)}{x}\right)^p |R_p^{(2)}| \leq \frac{M_p^{(2)}}{(2z)^p} \tag{44}$$

Also, since g is increasing, $1 - g(x) \geq 1 - g(z^2)$, so that from Lemma 1

$$\frac{1}{1-g(x)}\frac{g(x)}{x} \leq \frac{1}{1-g(z^2)}\phi_1(z)$$

$$= \frac{1}{8}\left(\frac{\coth z - \frac{1}{z}}{\coth z - \frac{1}{\sinh z}}\right) \leq \frac{1}{8} \quad . \tag{45}$$

Consequently, from (42) and (45)

$$\left(\frac{g(x)}{x}\right)^p |R_p^{(2)}| \leq \frac{M_p^{(2)}}{8^p} \tag{46}$$

For $0 \leq z \leq 4$, $z + 4 \leq 8$, so by (46)

$$\left(\frac{g(x)}{x}\right)^p \left|R_p^{(2)}\right| \leq \frac{M_p^{(2)}}{(z+4)^p} \tag{47}$$

For $z \geq 4$, $2z \geq z+4$ so that (44) implies that (47) also holds in this case. Consequently (47) holds for $0 \leq z < \infty$. It now follows from (31), (38) and (44) that

$$\left|\Delta_p(z, t)\right| \leq \frac{M_p^{(1)}}{(z+2)^p} + \frac{M_p^{(2)}}{(z+4)^p} \quad , \tag{48}$$

where $M_p^{(1)}$ is given in (39) and $M_p^{(2)}$ is given in (43).

(Shivakumar and Wong, 1988) give the result

$$\left|\Delta_p(z,t)\right| \leq \frac{\bar{M}_p^{(1)}}{(z+1)^p} + \frac{\bar{M}_p^{(2)}}{(z+1)^p} \quad , \tag{49}$$

where $\bar{M}_p^{(2)} = M_p^{(2)}$, and

$$\bar{M}_p^{(1)} = 0,\ \bar{M}_2^{(1)} = \frac{7}{3}\left|m - \frac{1}{2}\right|,\ \bar{M}_3^{(1)} = \frac{15}{4}\left|m - \frac{1}{2}\right| + \frac{57}{4}\left|m - \frac{1}{2}\right|\left|m - \frac{3}{2}\right| \tag{50}$$

Comparison of (40) and (50) indicates that our bounds are sharper than those in (Shivakumar and Wong, 1988). The last reference also gives a comparison with the bounds obtained in (Olver, 1974) for an equivalent asymptotic expansion.

To conclude, we note that the bound in (12) mimics the first term neglected in the expansion (8) near both $z=0$ and $z=\infty$. Also, when $m-\frac{1}{2}$ is a positive integer, $G^{m-\frac{1}{2}}$ is completely monotonic and is enveloped by its Maclaurin series on $[0, z^2]$. This implies that when $m-\frac{1}{2}$ is a positive integer the error $\epsilon_p(z,u)$ made by truncating the expansion (8) is numerically less than and has the same sign as the first neglected term.

REFERENCES

1. Frenzen, C. L. (1987). SIAM J. Math Anal., 18:890.

2. Olver, F. W. J. (1974). Asymptotics and Special Functions, Academic Press, New York.

3. Riordan, J. (1980). An Introduction to Combinatorial Analysis, Princeton University Press, New Jersey.

4. Shivakumar, P. N., and Wong, R. (1988). Quart Appl. Math., 46:473.

5. Ursell, F. (1984), Math. Proc. Camb. Phil. Soc., 95:367.

Computational Asymptotics of Fourth-Order Operators

P. A. Carinhas* Professor, Department of Physics, Texas A & M University, College Station, Texas

S. A. Fulling† Professor, Department of Mathematics, Texas A & M University, College Station, Texas

1. Introduction

The short-time asymptotic expansion of the heat (or Schrödinger) kernel of a linear differential operator has many applications in physics and pure mathematics. For second-order operators whose leading term is a Laplacian or Laplace–Beltrami operator, algorithms for calculating the terms in the expansion are well known. The algorithm most popular in the physics literature is that of DeWitt (1965). The quantization of the gravitational field leads to more complicated second-order operators (with nonscalar principal

* current address: Physics Department, Oregon State University, Corvallis, Oregon, 97331

† speaker

symbol, in the language of pseudodifferential operators) and to fourth-order operators (Barth and Christensen, 1983). The technique of DeWitt does not apply to these, for reasons which we shall elucidate below. Attempts in the physics literature to develop methods for treating such operators have been inconclusive and sometimes contradictory. (We mention here only the promising recent work of Lee et al. (1986, 1987).) In the rigorous mathematical literature, Gilkey (1980) has shown that many of the terms in the heat-kernel expansion for an operator of fourth or higher even order (whose leading term is a power of a Laplace–Beltrami operator) can be found quite efficiently, but by special tricks which may not be generalizable to a systematic algorithm for all terms.

It is clear that the problem can be solved in principle by the calculus of pseudodifferential operators (Gilkey, 1975; Widom, 1980; Fulling and Kennedy, 1988), but the calculational complexity is such that computerization is necessary. This approach is being pursued by S. M. Christensen and S. A. Fulling, with priority on the second-order operators with nonscalar principal symbol. (The case of fourth-order operators with scalar principal symbol will presumably be easier, but not trivial.)

The research reported here is an exploration of how far the DeWitt method can be pushed, for operators of the form $(\text{Laplacian})^2 + \text{lower-order}$ terms. We first (Sec. 3) show, by considering the case of a constant-coefficient operator, that the short-time expansion of the heat kernel $K(x, y, t)$ does not have the structure assumed in the DeWitt (and the Minakshisundaram–Pleijel (1949)) method, unless the summation index is extended to negative values. (The terms of negative order vanish on the diagonal, $x = y$, but they leave their traces upon the terms of positive order.) A second difference from the second-order case is that the heat kernel of the unperturbed Laplacian, which stands as a factor in the Minakshisundaram–DeWitt series, is not a simple Gaussian in the present case, but an unknown special function. In Sec. 4 we show how to modify the DeWitt procedure (for calculating the series on the diagonal) so that an explicit form for that function is not necessary. We arrive at a hierarchy of equations which are the counterparts of the recursion relations of DeWitt. However, the presence of the negative-order terms prevents these equations from being solved recursively. Our hope is that they will nevertheless be useful in checking the results of computations done by the pseudodifferential method or the method of Lee et al., or in supplementing results such as Gilkey's. In the final section we briefly describe the pseudodifferential approach and its present status.

The existence and asymptotic nature of the series (4.1) in question are well established (for elliptic operators satisfying suitable technical conditions) (e.g., Greiner, 1971; Widom, 1978). The off-diagonal series (4.9) has been less carefully studied. Here we are not so much concerned with reporting new analytical results as with the algebraic structure of the series and with various algorithms for generating it. Also, we deal only with the local expansion of the heat parametrix, not with effects due to boundaries, closed geodesics, etc.

We thank S. M. Christensen for providing the initial and continuing stimulus to this research, for some preliminary calculations along the line of Sec. 3, and for bringing both of us to the Workshop on Quantum Gravity Theory and Computer Symbolic Manipulation in Urbana–Champaign, 1987. S.A.F. thanks H. Widom and G. Kennedy for a discussion in which the final remark in the appendix was made by someone.

2. Notation

We work in Euclidean space, $\mathbf{R}^d$. Greek letters μ, ν, ... are tensor indices (with the summation index in effect), and Greek letters α, β, ... are multiindices [see (A.3)]. Derivatives $\partial_\nu \partial_\mu A$ are sometimes abbreviated as $A_{;\mu\nu}$.

Let

$$H_0 \equiv (-\Delta)^m \equiv (-\partial^\mu \partial_\mu)^m \tag{2.1}$$

(acting on scalar functions on $\mathbf{R}^d$), and let H equal H_0 plus lower-order derivative terms, whose coefficient tensors may depend on the variable, $x \in \mathbf{R}^d$.

To simplify the formulas in this brief exposition, we shall usually specialize to

$$m = 2, \qquad d = 4, \qquad H = \Delta^2 + A(x)^{\mu\nu} \partial_\mu \partial_\nu \,, \tag{2.2}$$

with occasional looks at the familiar case $m = 1$ for the sake of comparison. (Without loss of generality, the tensor A is taken to be symmetric.) Using the methods of DeWitt (1965), it should be possible to generalize our discussion in Sec. 4 to operators whose leading term is the square of a Laplace–Beltrami operator, but at present this would add more computational complexity than insight.

Let $K(x,y,t)$ be a heat-kernel parametrix of H. Thus

$$\partial_t K = -H_x K \qquad (t > 0) \tag{2.3}$$

and

$$\lim_{t \downarrow 0} K = \delta(x-y). \tag{2.4}$$

$K_0(z,t)$ ($z \equiv x-y$) will denote the heat kernel of H_0. (Since the construction of a parametrix is local, H and K actually may be defined only on a subdomain of $\mathbf{R}^d$, and K is not unique. H_0 and K_0, however, refer to the unique self-adjoint realization of $(-\Delta)^m$ on $\mathbf{R}^d$.)

We write

$$\partial^\alpha \mathbf{K}_0 \equiv \partial^\alpha K_0(0,1). \tag{2.5}$$

It is easy to show (see appendix) that

$$\partial^\alpha K_0(0,t) = t^{-\frac{1}{2m}(d+|\alpha|)} \partial^\alpha \mathbf{K}_0 . \tag{2.6}$$

We let

$$\Xi_\alpha \equiv \frac{\partial^\alpha \mathbf{K}_0}{\mathbf{K}_0} . \tag{2.7}$$

The same notation is used with a string of tensor indices in place of α. The calculation of the numbers $\mathbf{K}_0$ and Ξ_α (which depend only on m, d, and α) is described in the appendix.

The "diagonal value" or "coincidence limit", $a(x,x)$, of a two-point function $a(x,y)$ is denoted by $[a]$ or $[a](x)$.

3. Constant coefficients

Consider the operator (2.2),

$$H = \Delta^2 + A^{\mu\nu}\partial_\mu\partial_\nu ,$$

with A independent of x. In this case the problem (2.3)–(2.4) (on the domain $\mathbf{R}^d$) can be solved by Fourier transformation:

$$K(x,y,t) = \frac{1}{(2\pi)^d} \int e^{-(|p|^4 - A^{\mu\nu} p_\mu p_\nu)t} e^{ip\cdot z} \, d^d p, \tag{3.1}$$

where $z \equiv x - y$ (hence we need not distinguish x- from z-derivatives).

The perturbing factor in the integrand has the Taylor expansion

$$e^{A^{\mu\nu}p_\mu p_\nu t} = 1 + A^{\mu\nu}p_\mu p_\nu t$$
$$+ \frac{1}{2} A^{\mu\nu}A^{\rho\sigma}p_\mu p_\nu p_\rho p_\sigma t^2 + \frac{1}{6} A^{\mu\nu}A^{\rho\sigma}A^{\tau\upsilon}p_\mu p_\nu p_\rho p_\sigma p_\tau p_\upsilon t^3 + \ldots . \qquad (3.2)$$

The corresponding terms in (3.1) are recognized as derivatives of the unperturbed heat kernel [see (A.1)–(A.2)]:

$$K(x,y,t) = K_0(z,t) + i^{-2}A^{\mu\nu}\partial_\mu\partial_\nu K_0(z,t)t$$
$$+ \tfrac{1}{2} i^{-4}A^{\mu\nu}A^{\rho\sigma}\partial_\mu\partial_\nu\partial_\rho\partial_\sigma K_0(z,t)t^2 + \ldots . \qquad (3.3)$$

The t dependence of each term, relative to K_0, depends on the number of differentiations as well as on the explicit power of t; from (2.6) and (2.7) we have

$$[K] \equiv K(x,x,t) = K_0(0,t)(1 - \Xi_{\mu\nu}A^{\mu\nu}t^{\frac{1}{2}} + \tfrac{1}{2}\Xi_{\mu\nu\rho\sigma}A^{\mu\nu}A^{\rho\sigma}t + \ldots). \quad (3.4)$$

The coefficient tensors are evaluated in the appendix. For example, when $d = 4$, (A.13) yields

$$[K] = \frac{1}{32\pi^2 t}\left(1 + \frac{\sqrt{\pi}}{8} A^\mu_\mu t^{\frac{1}{2}} + \frac{1}{48}\left((A^\mu_\mu)^2 + 2A^{\mu\nu}A_{\mu\nu}\right) t \right.$$
$$\left. + \frac{\pi}{1536}\left((A^\mu_\mu)^3 + 6A^\mu_\mu A^{\rho\sigma}A_{\rho\sigma} + 8A^\mu_\nu A^\nu_\rho A^\rho_\mu\right) t^{\frac{3}{2}} + \ldots\right). \qquad (3.5)$$

So far, the results are very routine. The series (3.4) or (3.5) has the expected form

$$[K] = K_0(0,t)\sum_n [a_n]t^{\frac{n}{2}}, \qquad (3.6)$$

with the summation range $n = 0, 1, \ldots$ [cf. (4.1)]. Indeed, we have done the calculation for general d, with terms $B^\mu\partial_\mu + C$ included in H, getting results consistent with Gilkey (1980).

The interesting phenomenon appears when one expands the diagonal value of a sufficiently high derivative of K. From (3.1) and (3.2) we have

$$[\partial^\alpha K] = \frac{i^{|\alpha|}}{(2\pi)^d}\int e^{-|p|^4}\, p^\alpha\, (1 + A^{\mu\nu}p_\mu p_\nu t + \ldots)\, d^d p. \qquad (3.7)$$

When $|\alpha|$ is odd, all the integrals are 0 (but not if odd-order derivatives are added to H). When $\alpha = 2\beta$, we find (using (2.6), etc.)

$$[\partial^\alpha K] = t^{-|\beta|/2}K_0(0,t)(\Xi_\alpha - \Xi_{\alpha\mu\nu}A^{\mu\nu}t^{\frac{1}{2}} + \tfrac{1}{2}\Xi_{\alpha\mu\nu\rho\sigma}A^{\mu\nu}A^{\rho\sigma}t + \ldots). \quad (3.8)$$

(The mixture of multiindices and tensor indices is not ambiguous, since the Ξ tensors are totally symmetric.) In particular, when $|\beta| = 2$ (four derivatives), we get

$$[K_{;\theta\iota\kappa\lambda}] = t^{-\frac{d}{4}}\mathbf{K}_0\big(\Xi_{\theta\iota\kappa\lambda}t^{-1} - \Xi_{\theta\iota\kappa\lambda\mu\nu}A^{\mu\nu}t^{-\frac{1}{2}} + \tfrac{1}{2}\,\Xi_{\theta\iota\kappa\lambda\mu\nu\rho\sigma}A^{\mu\nu}A^{\rho\sigma} + \ldots\big). \tag{3.9}$$

The point is that this result is not consistent with the off-diagonal generalization of (3.6),

$$K(x,y,t) = K_0(x-y,t)\sum_n a_n(x,y)t^{\frac{n}{2}}, \tag{3.10}$$

unless the sum includes a term corresponding to $n = -1$. To see this, take the fourth derivative of (3.10) and set $x = y$:

$$\begin{aligned}
[K_{;\theta\iota\kappa\lambda}] &= [K_{0;\theta\iota\kappa\lambda}]\sum_n [a_n]t^{\frac{n}{2}} \\
&\qquad + \Big([K_{0;\theta\iota}]\sum_n [a_{n;\kappa\lambda}]t^{\frac{n}{2}} + 5\text{ perms.}\Big) + [K_0]\sum_n [a_{n;\theta\iota\kappa\lambda}]t^{\frac{n}{2}} \\
&= t^{-\frac{d}{4}}\Big\{\mathbf{K}_{0;\theta\iota\kappa\lambda}\sum_n [a_n]t^{\frac{n-2}{2}} \\
&\qquad + \Big(\mathbf{K}_{0;\theta\iota}\sum_n [a_{n;\kappa\lambda}]t^{\frac{n-1}{2}} + 5\text{ perms.}\Big) + \mathbf{K}_0\sum_n [a_{n;\theta\iota\kappa\lambda}]t^{\frac{n}{2}}\Big\} \\
&= t^{-\frac{d}{4}}\sum_n t^{\frac{n}{2}}\big\{\mathbf{K}_{0;\theta\iota\kappa\lambda}[a_{n+2}] \\
&\qquad + \big(\mathbf{K}_{0;\theta\iota}[a_{n+1;\kappa\lambda}] + 5\text{ perms.}\big) + \mathbf{K}_0[a_{n;\theta\iota\kappa\lambda}]\big\}.
\end{aligned} \tag{3.11}$$

Identify the second term of (3.9) with the $n = -1$ term of (3.11):

$$-\Xi_{\theta\iota\kappa\lambda\mu\nu}A^{\mu\nu} = \Xi_{\theta\iota\kappa\lambda}[a_1] + \big(\Xi_{\theta\iota}[a_{0;\kappa\lambda}] + 5\text{ perms.}\big) + [a_{-1;\theta\iota\kappa\lambda}]. \tag{3.12}$$

A previous, similar calculation (omitted) with the second derivative of (3.10) shows that

$$[a_{0;\kappa\lambda}] = -\Xi_{\kappa\lambda\mu\nu}A^{\mu\nu} + \Xi_{\kappa\lambda}\Xi_{\mu\nu}A^{\mu\nu}, \tag{3.13}$$

and comparison of (3.4) and (3.6) yields $[a_1] = -A^{\mu\nu}\Xi_{\mu\nu}$. Thus (3.12) tells us that

$$\begin{aligned}
[a_{-1;\theta\iota\kappa\lambda}] &= -\Xi_{\theta\iota\kappa\lambda\mu\nu}A^{\mu\nu} + \Xi_{\theta\iota\kappa\lambda}\Xi_{\mu\nu}A^{\mu\nu} \\
&\qquad + \big(\Xi_{\theta\iota}(\Xi_{\kappa\lambda\mu\nu}A^{\mu\nu} - \Xi_{\kappa\lambda}\Xi_{\mu\nu}A^{\mu\nu}) + 5\text{ perms.}\big)
\end{aligned} \tag{3.14}$$

This object does not vanish: When $d = 4$ it simplifies to

$$\frac{\sqrt{\pi}(\pi-3)}{256} A^{\mu}_{\mu}(\delta_{\theta\iota}\delta_{\kappa\lambda} + 2 \text{ perms.}) - \frac{\sqrt{\pi}}{384}(A_{\theta\iota}\delta_{\kappa\lambda} + 5 \text{ perms.}). \tag{3.15}$$

It is worth noting that a_{-1} does not make its presence felt until the fourth derivative of K, nor does such a term exist at any order of differentiation for the operator $H = \Delta^2 + C$ (C = constant). Nor, of course, does it exist for the standard second-order operators.

Incidentally, if we compare the *leading* term of (3.9) with the $n = -2$ term of (3.11), the conclusion is that $[a_{-2;\theta\iota\kappa\lambda}] = 0$. Furthermore, $[a_{-l;\theta\iota\kappa\lambda}] = 0$ for all larger l as well. Similar reasoning shows the following, by induction on $|\alpha|$:

Theorem. *In the expansion (3.10) for the operator (2.2) (in arbitrary dimension), the coincidence limits satisfy*

$$[\partial^\alpha a_{-l}] = 0 \quad \textit{if } |\alpha| \le 2l > 0 \tag{3.16}$$

and

$$[\partial^\alpha a_n] = 0 \quad \textit{for all } n \textit{ if } |\alpha| \textit{ is odd.} \tag{3.17}$$

We expect that (3.16) is "best possible" — that is,

$$[\partial^\alpha a_{-l}] \neq 0 \quad \text{if } |\alpha| = 2l + 2.$$

The theorem implies that the natural expansion parameter for K/K_0 is the variable $zt^{-\frac{1}{4}}$ (rather than z), which is essentially the observation of Lee and Pak (1986). In Sec. 5 we indicate informally how this theorem extends to the variable-coefficient case.

4. Would-be recursion relations

We now return to the case of variable coefficients. Ideally, our goal is to calculate the terms in the series

$$[K](x,t) \sim K_0(0,t) \sum_{n=0}^{\infty} [a_n](x)\, t^{\frac{n}{m}}. \tag{4.1}$$

(a) *Second-order operator*

When $m = 1$ and $H \equiv -\Delta + V(x)$, the standard procedures (Minakshisundaram and Pleijel, 1949; DeWitt, 1965) exploit the fact that

$$K_0(z,t) = (4\pi t)^{-\frac{d}{2}} e^{-|z|^2/4t}. \tag{4.2}$$

Here we present a variant of those methods that makes no use of the explicit formula (4.2), because the latter will not be available when $m > 1$.

One knows from Minakshisundaram and Pleijel (1949) that

$$K(x,y,t) \sim K_0(x-y,t) \sum_{n=0}^{\infty} a_n(x,y) t^n \tag{4.3}$$

in the second-order case. Substitute this into (2.3), obtaining

$$0 = \sum_{n=0}^{\infty} t^n \{(n+1)K_0 a_{n+1} - 2K_{0;\mu} a_{n;\mu} - K_0 \Delta a_n + K_0 V a_n\}. \tag{4.4}$$

This is the "master equation" from which all other results will follow.

Passing to the diagonal, we find immediately (since odd-order derivatives of K_0 vanish there)

$$0 = (n+1)\big[a_{n+1}\big] - \big[\Delta a_n\big] + V\big[a_n\big]. \tag{4.5}$$

We seek to use (4.5) to determine $\big[a_{n+1}\big]$. From (2.4) one can show that $\big[a_0\big] = 1$. To find $\big[\Delta a_0\big]$ we will need the second derivative of the master equation:

$$\begin{aligned} 0 = \sum_{n=0}^{\infty} t^n \{ & (n+1)(K_{0;\mu\nu} a_{n+1} + 2K_{0;(\mu} a_{n+1;\nu)} + K_0 a_{n+1;\mu\nu}) \\ & - 2(K_{0;\lambda\mu\nu} a_{n;\lambda} + 2K_{0;\lambda(\mu} a_{n;\nu)\lambda} + K_{0;\lambda} a_{n;\lambda\mu\nu}) \\ & - K_{0;\mu\nu} \Delta a_n - 2K_{0;(\mu} \Delta a_{n;\nu)} - K_0 \Delta a_{n;\mu\nu} \\ & + K_{0;\mu\nu} V a_n + 2K_{0;(\mu} V_{;\nu)} a_n + 2K_{0;(\mu} V a_{n;\nu)} \\ & + K_0 V_{;\mu\nu} a_n + 2K_0 V_{;(\mu} a_{n;\nu)} + K_0 V a_{n;\mu\nu} \}. \end{aligned} \tag{4.6}$$

(The parentheses on indices denote symmetrization.)

From (2.6) we note that each differentiation acting on K_0 produces an extra factor $t^{-\frac{1}{2}}$ in the coincidence limit. Therefore, when we set $y = x$ in (4.6) we must shift the summation index in the terms involving $K_{0;\mu\nu}$ before separating the powers of t. Doing so, dividing by $\mathbf{K}_0$, and tracing over the index pair $(\mu\nu)$, we get

$$\begin{aligned} 0 = (n+2)\Xi^{\mu}_{\mu}\big[a_{n+2}\big] + (n+1)\big[\Delta a_{n+1}\big] - 4\Xi^{\mu}_{\lambda}\big[a_{n+1;\lambda\mu}\big] - \Xi^{\mu}_{\mu}\big[\Delta a_{n+1}\big] \\ - \big[\Delta^2 a_n\big] + \Xi^{\mu}_{\mu} V\big[a_{n+1}\big] + \Delta V\big[a_n\big] + 2V_{;\mu}\big[a_{n;\mu}\big] + V\big[\Delta a_n\big]. \end{aligned} \tag{4.7}$$

From (A.8) we have $\Xi_{\mu\nu} = -\frac{1}{2}\delta_{\mu\nu}$, hence $\Xi^{\mu}_{\mu} = -\frac{1}{2}d$ and $-4\Xi^{\mu}_{\lambda}\big[a_{n+1;\lambda\mu}\big] = 2\big[\Delta a_{n+1}\big]$. Replace n by $n+1$ in (4.5), multiply by $\frac{1}{2}d$, and add to (4.7):

$$0 = (n+3)\big[\Delta a_{n+1}\big] - \big[\Delta^2 a_n\big] + \Delta V\big[a_n\big] + 2V_{;\mu}\big[a_{n;\mu}\big] + V\big[\Delta a_n\big]. \tag{4.8}$$

This is the formula to determine $\big[\Delta a_{n+1}\big]$. (There is also an untraced version of this formula, which yields individual second derivatives of a_{n+1}.)

In particular, $\big[\Delta a_0\big] = 0$, since a_{-1} does not exist in the second-order case. Thus we now have enough information to determine $\big[a_1\big]$ as $-V(x)$.

To get $\big[a_2\big]$ out of (4.5) requires $\big[\Delta a_1\big]$, and to get that out of (4.8) requires $\big[\Delta^2 a_0\big]$ (as well as some first and third derivatives of a_0). To get a recursion relation for $\big[\Delta^2 a_{n+1}\big]$, take the *fourth* derivative of the master equation, decompose its diagonal value into powers of t as above, and subtract appropriate multiples of (4.5) and (4.8) (with shifted indices) to eliminate terms involving a_{n+3} and a_{n+2}. (Odd-order derivatives and untraced even-order ones are obtained by similar procedures.) This process may be continued indefinitely, to obtain the diagonal values of all the a_n and all their derivatives.

Incidentally, the standard DeWitt algorithm produces equations such as (4.8) directly, without passing through the stage of (4.7). The difference stems from the fact that the master equation of the DeWitt method is (4.4) *divided by* K_0.

(b) *Fourth-order operator*

Now consider the fourth-order operator (2.2), with $A^{\mu\nu}$ a C^∞ function of x. We make the ansatz

$$K(x,y,t) \sim K_0(x-y,t) \sum_{n=-\infty}^{\infty} a_n(x,y)t^{\frac{n}{2}} \tag{4.9}$$

(expecting the series to be asymptotic in powers of t for positive n and in powers of $(x-y)t^{-\frac{1}{4}}$ for negative n). Substituting (4.9) into (2.3), we get the master equation

$$0=\sum_{n=0}^{\infty}t^{\frac{n}{2}}\left(\frac{n+2}{2}K_0a_{n+2}+4\Delta K_{0;\mu}a_{n;\mu}+4K_{0;\mu\nu}a_{n;\mu\nu}\right.$$
$$+2\Delta K_0\Delta a_n+4K_{0;\mu}\Delta a_{n;\mu}+K_0\Delta^2a_n$$
$$\left.+A^{\mu\nu}K_{0;\mu\nu}a_n+2A^{\mu\nu}K_{0;\mu}a_{n;\nu}+A^{\mu\nu}K_0a_{n;\mu\nu}\right). \quad (4.10)$$

Each derivative of K_0 contains a hidden factor of $t^{-\frac{1}{4}}$ in the coincidence limit, $y\to x$. We take these into account by shifting the index n appropriately when that limit is taken. Each layer in the hierarchy of would-be recursion relations will be obtained by taking derivatives of (4.10) *before* setting y equal to x or extracting factors of $t^{\frac{n}{2}}$. The first layer comes from (4.10) itself, and is (after division by $\mathbf{K}_0$)

$$0=\frac{n+2}{2}\left[a_{n+2}\right]+4\Xi_{\mu\nu}\left[a_{n+1;\mu\nu}\right]+2\Xi^{\lambda}_{\lambda}\left[\Delta a_{n+1}\right]$$
$$+\left[\Delta^2a_n\right]+A^{\mu\nu}\Xi_{\mu\nu}\left[a_{n+1}\right]+A^{\mu\nu}\left[a_{n;\mu\nu}\right], \quad (4.11)$$

where the Ξ tensors are now those appropriate to $m=2$, given by (A.8) or (A.13). It is easily checked that the case $n=-1$ of (4.11) is satisfied by the constant-coefficient results in Sec. 3.

Taking one derivative of (4.10), one gets the second layer of relations:

$$0=\frac{n+2}{2}\left[a_{n+2;\sigma}\right]+4\Xi^{\lambda}_{\lambda\mu\sigma}\left[a_{n+2;\mu}\right]$$
$$+4\Xi_{\mu\nu}\left[a_{n+1;\mu\nu\sigma}\right]+2\Xi^{\lambda}_{\lambda}\left[\Delta a_{n+1;\sigma}\right]+4\Xi_{\mu\sigma}\left[\Delta a_{n+1;,\mu}\right]$$
$$+\left[\Delta^2a_{n;\sigma}\right]+A^{\mu\nu}_{;\sigma}\Xi_{\mu\nu}\left[a_{n+1}\right]+A^{\mu\nu}\Xi_{\mu\nu}\left[a_{n+1;\sigma}\right]$$
$$+2A^{\mu\nu}\Xi_{\mu\sigma}\left[a_{n+1;\nu}\right]+A^{\mu\nu}_{;\sigma}\left[a_{n;\mu\nu}\right]+A^{\mu\nu}\left[a_{n;\mu\nu\sigma}\right]. \quad (4.12)$$

Treating the second derivative of the master equation similarly, one gets an equation involving $[a_{n+3}]$ as well as the expected $[a_{n+2;\sigma\tau}]$. In analogy to the passage from (4.7) to (4.8), one can subtract a multiple of (4.11) to eliminate a_{n+3}. The result is

$$0=\frac{n+2}{2}\left[a_{n+2;\sigma\tau}\right]+8\Xi^{\lambda}_{\lambda\mu(\sigma}\left[a_{n+2;\tau)\mu}\right]+4\Xi_{\mu\nu\sigma\tau}\left[a_{n+2;\mu\nu}\right]$$
$$+4\Xi_{\mu\nu}\left[a_{n+1;\mu\nu\sigma\tau}\right]+8\Xi_{\mu(\sigma}\left[\Delta a_{n+1;\tau)\mu}\right]+2\Xi^{\lambda}_{\lambda\sigma\tau}\left[\Delta a_{n+2}\right]$$

$$
\begin{aligned}
&+2\Xi^{\lambda}_{\lambda}\left[\Delta a_{n+1;\sigma\tau}\right]-4\Xi_{\sigma\tau}\Xi_{\mu\nu}\left[a_{n+2;\mu\nu}\right]-2\Xi_{\sigma\tau}\Xi^{\lambda}_{\lambda}\left[\Delta a_{n+2}\right]\\
&+\left[\Delta^{2}a_{n;\sigma\tau}\right]-A^{\mu\nu}\Xi_{\mu\nu}\Xi_{\sigma\tau}\left[a_{n+2}\right]+A^{\mu\nu}_{;\sigma\tau}\Xi_{\mu\nu}\left[a_{n+1}\right]\\
&+2\Xi_{\mu\nu}A^{\mu\nu}_{;(\sigma}\left[a_{n+1;\tau)}\right]+A^{\mu\nu}\Xi_{\mu\nu\sigma\tau}\left[a_{n+2}\right]+A^{\mu\nu}\Xi_{\mu\nu}\left[a_{n+1;\sigma\tau}\right]\\
&+4A^{\mu\nu}_{;(\sigma}\Xi_{\tau)\mu}\left[a_{n+1;\nu}\right]+4A^{\mu\nu}\Xi_{\mu(\sigma}\left[a_{n+1;\tau)\nu}\right]+A^{\mu\nu}_{;\sigma\tau}\left[a_{n;\mu\nu}\right]\\
&+2A^{\mu\nu}_{;(\sigma}\left[a_{n;\tau)\mu\nu}\right]+A^{\mu\nu}\left[a_{n;\mu\nu\sigma\tau}\right]. \qquad (4.13)
\end{aligned}
$$

This can be further reduced by means of the explicit forms for the Ξ tensors in the appendix. For the case $d = 4$ one obtains (after tracing over σ and τ)

$$
\begin{aligned}
0 = &\left(\frac{n}{2}+6-\frac{3\pi}{4}\right)\left[\Delta a_{n+2}\right]+\frac{\sqrt{\pi}}{2}\left[\Delta^{2}a_{n+1}\right]+\left[\Delta^{3}a_{n}\right]\\
&+\left(\frac{1}{4}-\frac{\pi}{16}\right)A^{\mu}_{\mu}\left[a_{n+2}\right]-\frac{\sqrt{\pi}}{8}A^{\mu}_{\mu}\left[\Delta a_{n+1}\right]-\frac{\sqrt{\pi}}{2}A^{\mu\nu}\left[a_{n+1;\mu\nu}\right]\\
&+A^{\mu\nu}\left[\Delta a_{n;\mu\nu}\right]-\frac{\sqrt{\pi}}{4}A^{\mu}_{\mu;\lambda}\left[a_{n+1;\lambda}\right]-\frac{\sqrt{\pi}}{2}A^{\nu\mu}_{;\mu}\left[a_{n+1;\nu}\right]\\
&+2A^{\mu\nu}_{;\lambda}\left[a_{n;\mu\nu\lambda}\right]-\frac{\sqrt{\pi}}{8}\Delta A^{\mu}_{\mu}\left[a_{n+1}\right]+\Delta A^{\mu\nu}\left[a_{n;\mu\nu}\right]. \qquad (4.14)
\end{aligned}
$$

This process can obviously be carried to any order, but the final step of our treatment of the second-order case is not possible in this case. Equation (4.11) with $n = -1$, as a formula for $[a_1]$, requires $[\Delta^2 a_{-1}]$, which we know to be nonzero from the special case studied in Sec. 3. Similarly, (4.14) (which one would like to solve for $[\Delta a_0]$) involves a sixth derivative of a_{-2}, which is presumably nonzero. Thus the "recursion" has no bottom, and the relations (4.11), (4.12), (4.13), ... can't be solved (at least not without some additional information).

The root of the distinction between the second- and the higher-order cases is that K_0 in the latter is a not a Gaussian (or other simple exponential function). For the Gaussian kernel (4.2) the ratios $\partial^\alpha K_0/K_0$ are *polynomials* in t^{-1}, of degree proportional to $|\alpha|$. When K_0 is a more general function, its derivatives bear no particular resemblance to K_0 itself, and hence $\partial^\alpha K_0/K_0$ retains an essential singularity as $t \downarrow 0$ with $z \neq 0$. In the DeWitt calculational approach it is clear that this property carries over to K/K_0; that is, the series in (4.9) won't terminate at the bottom end.

5. Pseudodifferential operators

Since the speaker has discussed the pseudodifferential symbolic calculus in other conferences (Fulling and Kennedy, 1987a,b), we'll be very brief here.

The Fourier method of Sec. 3 does not extend, of course, to variable-coefficient operators. Nevertheless, the introduction of a "local" Fourier transform enables one to obtain asymptotic series for various integral kernels associated with an elliptic operator H. Such kernels are essentially Fourier transforms of "symbols", which possess asymptotic expansions in $|p|^{-1}$ (as $|p| \to \infty$), where p is the Fourier variable.

In particular, the symbol of the kernel of the resolvent operator $(H-\lambda)^{-1}$ has the asymptotic form

$$\sum_{k=0}^{\infty} \sum_{j=1}^{2k+1} b_{k,j}(x,p)\big(|p|^4 - \lambda\big)^{-j}, \tag{5.1}$$

where k determines the term's homogeneity degree in p. A contour integration over λ followed by a Fourier transformation in p yields a formal expansion for the heat kernel $K(x,y,t)$ in powers of $t^{\frac{1}{2m}}$, starting with $t^{-\frac{d}{2m}}$ as in the free kernel [see (2.6)]. The usual proof concerning $[K]$ in the second-order case (e.g., Gilkey, 1974, Chap. 3) extends to the off-diagonal K and the fourth-order case to establish that this series is asymptotic as $t \downarrow 0$, the coefficients being analytic functions of $zt^{-\frac{1}{4}}$ and C^∞ functions of x. (When $z \equiv x - y = 0$ they reduce to the familiar invariants $[a_n]$, constructed polynomially from derivatives of the tensor A.)

When the coefficients are expanded as Taylor series in z, powers of $t^{-\frac{1}{4}}$ appear — but always accompanied by corresponding powers of z. The same is true of K_0 and hence $K_0{}^{-1}$. This establishes that K *has a* (formal, at least) *series of the type (4.9), and the nonzero terms in that series obey the constraint (3.16).* (Terms with $|\alpha| = 2l$ are accounted for by the factor K_0 in (4.9).) This establishes that the formal calculations in Sec. 4 have something to do with reality.

There is a formula for the coefficient $b_{k,j}$ in (5.1) (Fulling and Kennedy, 1988), but applying the formula concretely is a lengthy calculational task. A computer program has been written to list the terms (for any elliptic linear differential operator in the role of H). For operators of the type studied in the present work, all integrals necessary to compute the heat-kernel expansion coefficients from the b's can be evaluated; they are the same integrals discussed in our appendix. Some additional programming is needed to obtain the $[a_n]$ efficiently.

The ultimate aim of this work is to create general tools for solving all such recursive asymptotic calculations routinely. The increasing availability of symbolic computation software on relatively accessible computers will greatly facilitate applications.

Since occasional programming or calculational errors are inevitable, the would-be recursion relations in Sec. 4 will provide a much-needed check on the results of these computations.

APPENDIX

Here we calculate the diagonal values of the free heat kernel, K_0, and its spatial derivatives. The integrals which arise here are needed in all the approaches to asymptotic calculation of the perturbed kernel, K, that involve Fourier analysis.

$K_0(z,t)$ satisfies (2.3) in the form

$$\partial_t K = -(-\partial^\mu \partial_\mu)^m K,$$

along with the initial condition (2.4). This problem is easily solved by Fourier transforms:

$$K_0(z,t) = \frac{1}{(2\pi)^d} \int e^{ip\cdot z} e^{-|p|^{2m} t}\, d^d p. \tag{A.1}$$

(The integral is over $\mathbf{R}^d$, as are all similar later ones.) Take an arbitrary derivative of this formula and set $z = 0$:

$$\partial^\alpha K_0(0,t) = \frac{i^{|\alpha|}}{(2\pi)^d} \int p^\alpha e^{-|p|^{2m} t}\, d^d p, \tag{A.2}$$

in the multiindex notation,

$$|\alpha| \equiv \sum_{\mu=1}^{d} \alpha_\mu, \qquad p^\alpha \equiv \prod_{\mu=1}^{d} {p_\mu}^{\alpha_\mu}, \qquad \alpha! \equiv \prod_{\mu=1}^{d} \alpha_\mu!. \tag{A.3}$$

Letting $p = t^{-1/2m} q$ in (A.2) leads to (2.5) and (2.6) with

$$\partial^\alpha \mathbf{K}_0 = \frac{i^{|\alpha|}}{(2\pi)^d} \int q^\alpha e^{-|q|^{2m}}\, d^d q. \tag{A.4}$$

By symmetry, $\partial^\alpha \mathbf{K}_0 = 0$ if any of the exponents α_μ is odd. Therefore, we may write $\alpha \equiv 2\beta$ in all the nontrivial cases.

When $m = 1$, (A.1) factors into one-dimensional integrals, and the well known formula (4.2) results. For larger m, it does not factor and cannot (to our knowledge) be evaluated in closed form when $z \neq 0$. However, the integrals (A.4) can be evaluated. This can be done by two methods, which we call the "multiindex" and "tensor" methods because of the notation in which each produces an answer most directly. (Of course, the results of either method can be translated into the other notation.)

The multiindex method. Introduce hyperspherical coordinates (Vilenkin, 1968, pp. 435–437)

$$\begin{aligned} q_1 &= q \sin\theta_{d-1} \cdots \sin\theta_2 \sin\theta_1 \,, \\ q_2 &= q \sin\theta_{d-1} \cdots \sin\theta_2 \cos\theta_1 \,, \\ q_3 &= q \sin\theta_{d-1} \cdots \cos\theta_2 \,, \\ &\vdots \\ q_d &= q \cos\theta_{d-1} \,. \end{aligned} \tag{A.5}$$

The ranges of the variables are

$$0 \le q < \infty, \qquad 0 \le \theta_1 < 2\pi, \qquad 0 \le \theta_j < \pi \quad (j > 1).$$

Then (A.4) separates into one-dimensional integrals which can be evaluated.

After considerable labor one arrives at

$$\mathbf{K}_0 = (4\pi)^{-\frac{d}{2}} \frac{\Gamma\left(\frac{d}{2m}\right)}{m\Gamma\left(\frac{d}{2}\right)} \tag{A.6}$$

and

$$\partial^{2\beta}\mathbf{K}_0 = \frac{(-1)^{|\beta|}}{(4\pi)^{d/2}} \frac{(2\beta)!}{4^{|\beta|}\beta!} \frac{\Gamma\left(\frac{d+2|\beta|}{2m}\right)}{m\Gamma\left(\frac{d+2|\beta|}{2}\right)} \,. \tag{A.7}$$

(The labor could have been reduced by applying a trick, described below.) Most useful to us is the ratio $\Xi_{2\beta}$ [see (2.7)], which is

$$\Xi_{2\beta} = (-1)^{|\beta|} \frac{(2\beta)!}{4^{|\beta|}\beta!} \frac{\Gamma\left(\frac{d}{2m} + \frac{|\beta|}{m}\right)}{\Gamma\left(\frac{d}{2m}\right)} \frac{\Gamma\left(\frac{d}{2}\right)}{\Gamma\left(\frac{d}{2} + |\beta|\right)} \,. \tag{A.8}$$

The tensor method. A multiindex α is equivalent to a string of tensor indices in arbitrary order, $(\mu_1, \ldots, \mu_{|\alpha|})$, where α_1 of the indices are equal to 1, etc. An integral of the form $\int F(|q|)\, q^\alpha\, d^d q$ is one component of a symmetric tensor of rank $|\alpha|$.

Lemma. *Let 2β be an even multiindex and F be any function on $\mathbf{R}_+$ that is adequate to cause convergence of the integrals in question. Let*

$$I_F \equiv \int F(|q|)\, {q_\mu}^{2|\beta|}\, d^d q, \tag{A.9}$$

(which is independent of μ and is the value of the tensorial integral when all the tensor indices are equal). Then

$$\begin{aligned}\int F(|q|)\, q^{2\beta}\, d^d q &= I_F\, \mathbf{S}\left[\delta_{\mu_1\mu_2}\delta_{\mu_3\mu_4}\cdots\delta_{\mu_{2|\beta|-1}\mu_{2|\beta|}}\right] \\ &= \frac{2^{|\beta|}|\beta|!}{(2|\beta|)!}\, I_F\left(\delta_{\mu_1\mu_2}\delta_{\mu_3\mu_4}\cdots + \ldots\right).\end{aligned} \tag{A.10}$$

In the first form, $\mathbf{S}$ indicates total symmetrization over the $(2|\beta|)!$ permutations of the indices. In the second form, the summation is over a complete set of permutations which remain essentially distinct *after consideration of the $|\beta|!$ trivial reorderings of the Kronecker symbols and the $2^{|\beta|}$ pair symmetries.*

For example, when $|\beta| = 2$, the second form of (A.10) is

$$\int F(|q|)\, q_\mu q_\nu q_\rho q_\sigma\, d^d q = \frac{1}{3}\, I_F\left(\delta_{\mu\nu}\delta_{\rho\sigma} + \delta_{\mu\rho}\delta_{\nu\sigma} + \delta_{\mu\sigma}\delta_{\nu\rho}\right). \tag{A.11}$$

PROOF: The equivalence of the two versions of (A.10) is clear from the combinatorial facts cited in the lemma itself, as is the numerical consistency of (A.10) and (A.9). Therefore, it suffices to prove the first version of (A.10) with an undetermined constant in the role of I_F. Such assertions are often justified by the claim that no other intrinsically defined tensor with the proper symmetries exists, but a more convincing argument can be based on the inverse Fourier transform

$$\check{F}(x) \equiv \int F(|q|)\, e^{ix\cdot q}\, d^d q.$$

One has

$$\int F(|q|)\, q^{2\beta}\, d^d q = (-i\partial)^{2\beta}\check{F}(0).$$

One can write $\check{F}(x) \equiv G(|x|^2)$, and hence

$$\frac{\partial^2}{\partial x_\nu \partial x_\mu} F = 2\delta_{\mu\nu} \frac{\partial G}{\partial z} + 4x_\mu x_\nu \frac{\partial^2 G}{\partial z^2} ,$$

for instance. (If necessary, differentiability of $G(\zeta)$ can be assured by cutting off $F(|q|)$ at infinity temporarily.) It is easy to see that to produce in any partial derivative a term which survives when x is set to 0, the differentiations must act in pairs, one to create a factor x_{μ_j} , the other to destroy it. The result must be a linear combination of the terms $\delta_{\mu_1\mu_2} \cdots \delta_{\mu_{2|\beta|-1}\mu_{2|\beta|}}$; by symmetry, all the coefficients are equal when the result is stated in tensorial form.

To calculate $\partial^{2\beta}\mathbf{K}_0$ we now need to evaluate I_F when $F = (-1)^{2|\beta|} \times (2\pi)^{-d} e^{-|q|^{2m}}$. This is easily done in analogy to (A.7). Here, we can read the answer off as the special case of (A.7) with $\beta = (|\beta|, 0, \ldots, 0)$:

$$I_F \equiv \partial_\mu{}^{2|\beta|}\mathbf{K}_0 = \frac{(-1)^{|\beta|}}{(4\pi)^{d/2}} \frac{(2|\beta|)!}{4^{|\beta|}|\beta|!} \frac{\Gamma\left(\frac{d+2|\beta|}{2m}\right)}{m\Gamma\left(\frac{d+2|\beta|}{2}\right)} . \tag{A.12}$$

To find the I_F appropriate for calculating $\Xi_{2\beta}$ in tensor form, make the obvious similar modification in (A.8). Note that the combinatorial factor in these equations now cancels most of the one in the second form of (A.10), leaving just $2^{-|\beta|}$.

Specializing to the case (2.2) of a fourth-order operator in dimension 4, we have

$$\mathbf{K}_0 = \frac{1}{32\pi^2} , \tag{A.13a}$$

$$\Xi_{\mu\nu} = -\frac{\sqrt{\pi}}{8} \delta_{\mu\nu} , \tag{A.13b}$$

$$\Xi_{\mu\nu\rho\sigma} = \frac{1}{24} (\delta_{\mu\nu}\delta_{\rho\sigma} + \delta_{\mu\rho}\delta_{\nu\sigma} + \delta_{\mu\sigma}\delta_{\nu\rho}) , \tag{A.13c}$$

$$\Xi_{\mu\nu\rho\sigma\tau\upsilon} = -\frac{\pi}{256} (\delta_{\mu\nu}\delta_{\rho\sigma}\delta_{\tau\upsilon} + 14 \text{ permutations}) . \tag{A.13d}$$

REMARK: In either approach, each integral reduces to a radial integral, times an integral over the unit sphere which is independent of F (i.e., of m). But the integrals for the case $m = 1$ can be evaluated by an independent method, since they factor already in Cartesian coordinates. This allows the integrals over the sphere to be found once and for all without introducing the angular variables in (A.5).

REFERENCES

1. Barth, N. H., and Christensen, S. M. (1983). *Phys. Rev. D* **28**, 1876–1893.
2. DeWitt, B. S. (1965). *The Dynamical Theory of Groups and Fields*, Gordon and Breach, New York.
3. Fulling, S. A., and Kennedy, G. (1987a). in *Differential Equations and Mathematical Physics* (Lecture Notes in Math. No. 1285), ed. by I. W. Knowles and Y. Saito (Springer, Berlin), pp. 126–133.
4. Fulling, S. A., and Kennedy, G. (1987b). in *The Physics of Phase Space* (Lecture Notes in Physics No. 278), ed. by Y. S. Kim and W. W. Zachary (Springer, Berlin), pp. 407–409.
5. Fulling, S. A., and Kennedy, G. (1988). *Transac. Am. Math. Soc.* **310**, 583–617.
6. Gilkey, P. B. (1974). *The Index Theorem and the Heat Equation*, Publish or Perish, Boston.
7. Gilkey, P. B. (1975). *J. Diff. Geom.* **10**, 601–618.
8. Gilkey, P. B. (1980). *Duke Math. J.* **47**, 511–528.
9. Greiner, P. (1971). *Arch. Rat. Mech. Anal.* **41**, 163–218.
10. Lee, H. W., and Pac, P. Y. (1986). *Phys. Rev. D* **33**, 1012–1017.
11. Lee, H. W., Pac, P. Y., and Shin, H. K. (1987). *Phys. Rev. D* **35**, 2440–2447.
12. Minakshisundaram, S., and Pleijel, Å. (1949). *Canad. J. Math.* **1**, 242–256.
13. Vilenkin, N. Ja. (1968). *Special Functions and the Theory of Group Representations*, American Mathematical Society, Providence.
14. Widom, H. (1978). in *Topics in Functional Analysis*, ed. by I. Gohberg and M. Kac, Academic, New York, pp. 345–395.
15. Widom, H. (1980). *Bull. Sci. Math.* **104**, 19–63.

NOTE ADDED

V. P. Gusynin ["Seeley–Gilkey coefficients for the fourth-order operators on a Riemannian manifold," preprint ITP-88-156E, Institute for Theoretical Physics, Kiev] has used the pseudodifferential symbolic calculus to find $[a_2]$ for the general fourth-order operator with scalar principal symbol (in particular, operators of the form (2.2) with Δ replaced by a Laplace–Beltrami operator).

Universal Asymptotic Distribution Functions mod 1

William M. Y. Goh Professor, Drexel University, Department of Mathematics and Computer Science, Philadelphia, Pennsylvania

Eric Schmutz* Professor, Drexel University, Department of Mathematics and Computer Science, Philadelphia, Pennsylvania

I. INTRODUCTION.

The question considered is roughly this: what happens to the asymptotic distribution mod 1 (defined below) of a sequence when one applies high iterates of a transformation T?

Let $\{x_k\}$ denote the fractional part of the real number x_k, i.e. $\{x_k\} := x_k \bmod 1$ We say that g is an a.d.f. (asymptotic distribution function mod 1) for the sequece of real numbers $\langle x_k \rangle$ iff

$$g(x) := \lim_{N\to\infty} \frac{1}{N} \# \Big\{ k \le N : \{x_k\} \in [0,x) \Big\}$$

for each $x \in [0,1]$. Now, given a sequence $\langle x_k \rangle_{k=1}^{\infty}$ and a map $T : [0,1] \to [0,1]$, define $\langle x_k^{(n)} \rangle_{k=1}^{\infty}$ recursively by

$$x_k^{(0)} := \{x_k\} = x_k \bmod 1$$

$$x_k^{(n)} := T(x_k^{(n-1)})$$

* Supported by a Drexel University Faculty Development Minigrant

Finally g_n will denote an a.d.f. for $\langle x_k^{(n)} \rangle_{k=1}^{\infty}$. We shall always assume that g_0 exists,is C^2, and that g_0' is bounded away from zero.

II. TENT FUNCTIONS

We consider first the case where T is one of the family of "tent functions" T defined below. Define h by

$$h(x) := \begin{cases} 2x, & 0 \le x \le 1/2; \\ 2 - 2x, & 1/2 \le x \le 1 \end{cases}$$

Our main result is:

Theorem 1 *Let $k \in C^2([0,1])$ with $k(0) = 0, k(1) = 1$, and $k'(x) > \delta > 0$ $\forall x$. Let $T = k^{-1} \circ h \circ k$. Then $g_n(x) = k(x) + O(e^{-\lambda\sqrt{n}})$*

Remark: Both λ and the constant implicit in the big-Oh notaion depend on k and g_0. As usual, we assume that g_0 is C^2, and that $g_0'(x)$ is bounded away from zero.

The proof contains many technical details, so we begin with a brief outline. Readers familiar with Khintchine(1974) will recognize the influence of Kuzmin. We shall find an increasing sequence $\langle A_r \rangle$, and a decreasing sequence $\langle B_r \rangle$, such that

$$(*) \qquad A_r k'(x) < g_{rn}'(x) < B_r k'(x)$$

and

$$1 - c_1 e^{-c_2 n} < A_r < B_r < 1 + c_1 e^{-c_2 n}$$

(for all r, where c_1 and c_2 are constants depending only on g_0 and k.) This implies that

$$\left| g_{n^2}'(x) - k'(x) \right| = O(e^{-c_2 n})$$

The theorem then follows, by an easy argument, from the lemma 2:

Lemma 2 *If α and β are constants,and*

$$\alpha k'(x) < g_m'(x) < \beta k'(x) \ \ \forall x$$

then

$$\alpha k'(x) < g_{m+1}'(x) < \beta k'(x) \ \ \forall x$$

The proofs of $(*)$ and lemma 2 depend heavily on the fact that $d\mu := k'(x)dx$ is invariant under T. We therefore begin by verifying this fact. Let L and R denote restrictions of T to $[0, k^{-1}(1/2)]$ and $[k^{-1}(1/2), 1]$ respectively. Then

Lemma 1 *$d\mu := k'(x)dx$ is $T-$invariant.*

Pf. Let P_T be the Frobenius-Perron operator:

$$P_T f(x) := \frac{d}{dx} \int_{T^{-1}([0,x])} f(s)ds$$

It suffices to show that $P_T k' = k'$. But

$$P_T k'(x) = \frac{d}{dx}\left(\int_0^{L^{-1}(x)} k'(s)ds + \int_{R^{-1}(x)}^{1} k'(s)ds \right)$$

$$= \frac{d}{dx} \int_{k(0)}^{k \circ L^{-1}(x)} dt + \frac{d}{dx} \int_{k \circ R^{-1}(x)}^{1} dt$$

$$= \frac{1}{2}k'(x) + \frac{1}{2}k'(x) = k'(x)$$

Corollary

$$k'(x) = \frac{k'(L^{-1}(x))}{|L'(L^{-1}(x))|} + \frac{k'(R^{-1}(x))}{|R'(R^{-1}(x))|}$$

One can verify that g_m satisfies an identity similar to that in the preceeding corollary:

$$(**) \qquad g'_{m+1}(x) = \frac{g'_m(L^{-1}(x))}{|L(L^{-1}(x))|} + \frac{g'_m(R^{-1}(x))}{|R(R^{-1}(x))|}$$

With these two identities, it is easy to prove lemma 2. Repeated application of (**) also yields:

Lemma 3

$$g'_n(x) = \sum_{\langle f_i \rangle} g'_0(f_1^{-1} \circ f_2^{-1} \circ \ldots \circ f_n^{-1}) \left| \frac{d}{dx}\left(f_1^{-1} \circ f_2^{-1} \circ \ldots \circ f_n^{-1}\right) \right|$$

where the sum is over all 2^n *sequences* $\langle f_i \rangle_{i=1}^{n}$ *with* $f_i = L$ or R.

We shall need a few more technical lemmas. Their proofs are omitted because they are tedious and straight forward.

Lemma 4 *Let* $\Pi := \left\{ f_1^{-1} \circ f_2^{-1} \circ \ldots \circ f_n^{-1}(x) \ : f_i = L \text{ or } R \right\}$. *For* $f \in \Pi$, *let* $I_f = f([0,1])$ *be the range of* f. *Then*

(a) *Each* $f \in \Pi$ *is monotonic on* $[0,1]$

(b) $\bigcup_{f \in \Pi} I_f = [0,1]$

(c) *For* $f \neq g$, $I_f \cap I_g$ *is empty or a single point.*

Observation: The I_f's form a partition of the unit interval into 2^n subintervals.

Now for any g, let $M(g) := \max_{x \in [0,1]} (g(x))$, and let $m(g) := \min_{x \in [0,1]} (g(x))$. Then we have

Lemma 5 *If $f = f_1 \circ f_2^{-1} \circ \dots f_n^{-1} \in \Pi$, then*

(a) $\frac{m(k')}{M(k')}2^{-n} \le |f'(x)| \le \frac{M(k')}{m(k')}2^{-n}$

(b) $|f''(x)| \le 2^{-n}\left(\frac{M(k'')}{m(k')} + \frac{M^2(k')M(k'')}{m^3(k')}\right)$

(c) *Let $|I_f| = |f(1) - f(0)|$ be the length of the subinterval corresponding to f. Then*

$$|I_f| \le 2^{-n}\left(\frac{M(k')}{m(k')}\right)$$

Finally we have:

Lemma 6

$$|g_n''(x)| \le 2^{-n}\frac{M(g_0')M^2(k')}{m^2(k')} + M(g_0)\left(\frac{M(k'')}{m(k')} + \frac{M^2(k')M(k'')}{M^3(k')}\right)$$

Now at last we can carry out the proof of theorem 1. Choose constants A_0 and B_0 so that

$$A_0k'(x) < g_0'(x) < B_0k'(x) \quad \forall x$$

Let $\Phi_n(x) := g_n'(x) - A_0k'(x)$. As in lemma 3, we have

$$\Phi_n(x) := \sum_{f\in\Pi} \Phi_0(f(x))|f'(x)|$$

It follows from lemma 5 that

$$\Phi_n(x) > \frac{m^2(k')}{M^2(k')}\sum_f \Phi_0(f(x))|I_f|$$

On the other hand, the mean value theorem implies that

$$\int_0^1 \Phi_0(\xi)d\xi = \sum_f \Phi_0(c_f)|I_f|$$

where $c_f \in I_f$. Thus

$$\Phi_n(x) > \frac{m^2(k')}{M^2(k')}\int_0^1 \Phi_0(\xi)d\xi + \frac{m^2(k')}{M^2(k')}\sum_f(\Phi_0(f(x)) - \Phi_0(c_f))|I_f|$$

Since $|\Phi_0(f(x)) - \Phi_0(c_f)| < (M(g_0'') + A_0M(k''))|I_f|$, it follows that

$$g_n'(x) > A_0k'(x) + \frac{m^2(k')}{M^2(k')}\int_0^1 \Phi_0(\xi)d\xi - \frac{m^2(k')}{M^2(k')}\left(M(g_0'') + A_0M(k'')\right)\sum_f |I_f|^2$$

But by lemma 5, $|I_f| < \frac{M(k')}{m(k')}2^{-n}$, so we have

$$g_n'(x) > A_0k'(x) + \frac{m^2(k')}{M^2(k')}\int_0^1 \Phi_0(\xi)d\xi - 2^{-n}\frac{m(k')}{M(k')}\left(M(g_0'') + A_0M(k'')\right)\sum_f |I_f|$$

If we let $c := \frac{m^2(k')}{M^3(k')}\int_0^1 \Phi_0(\xi)d\xi$, let $b := \frac{1}{M^2(k')}(M(g_0') + A_0M(k''))$, and set $A_1 := A_0 + c - b2^{-n}$, then we have already proved that

$$g_n'(x) > A_1k'(x)$$

In the same way, one can define $\Psi_n(x) = B_0k'(x) - g_n'(x)$. One then finds that

$$g_n'(x) < (B_0 - c' + b2^{-n})k'(x) := B_1k'(x)$$

where $c' = \frac{m^2(k')}{M^3(k')}\int_0^1 \Psi_0(t)dt$. Furthermore, for n sufficiently large, we have $A_0 < A_1 < B_1 < B_0$, and $B_1 - A_1 = B_0 - A_0 - (c' + c) + 2b2^{-n}$, where $(c + c') = \frac{m^2(k')}{M^3(k')}\int_0^1(\Phi_0(t) + \Psi_0(t))dt = \frac{m^2(k')}{M^3(k')}(B_0 - A_0)$.

Now begin again with $\langle x_k^{(n)}\rangle$ as the initial sequence, and repeat all the same arguments. Thus we begin with g_n instead of g_0, and by similar reasoning arrive at the inequality $A_2k'(x) < g_{2n}'(x) < B_2k'(x)$ where $A_1 < A_2 < B_2 < B_1$ and $B_2 - A_2 \leq \delta(B_1 - A_1) + 2b_12^{-n}$. Here $b_1 := \frac{M(g_n') + A_1M(k')}{M(k')}$ and $\delta := 1 - \frac{m^2(k')}{M^3(k')}$ is a fixed constant in $(0, 1)$.

Continuing in the same way, we obtain recursively

$$A_rk'(x) < g_{2r}'(x) < B_rk'(x)$$

$$A_{r-1} < A_r < B_r < B_{r-1}$$

and

$$B_r - A_r \leq \delta(B_{r-1} - A_{r-1}) + 2b_{r-1}2^{-n}$$

where

$$b_{r-1} := \frac{M(g_{n(r-1)}') + A_{r-1}M(k')}{M(k')}$$

Furthermore, lemma 6 implies that b_r is bounded above by a fixed (independent of r) constant T. Then

$$B_n - A_n \leq (B - A)\delta^n + 2T2^{-n}(\delta^{n-1} + \delta^{n-2} + \ldots \delta + 1)$$

Thus

$$0 \leq B_n - A_n \ll e^{-\lambda n}$$

where λ is a constant depending only on g_0 and k.

The limit $\lim_{r\to\infty} A_r = \lim_{r\to\infty} B_r = L$ exists. We claim that $L = 1$:

$$L = L\int_0^1 k'(x) = \int_0^1 g'_{n^2}(x)dx + O(e^{-\lambda n})$$

$$= 1 + O(e^{-\lambda n})$$

Thus

$$|g_N - k'(x)| \ll e^{-\lambda\sqrt{N}}$$

as $N = n^2 \to \infty$. The theorem then follows, by an easy argument, from lemma 1.∎

III. OTHER TRANSFORMATIONS

One important feature of theorem 1 is that the limit distribution g is essentially independent of the initial distribution g_0. This is not a special property of tent functions; it seems to be true for many different choices of T. To emphasize this point, we consider two more examples. The first is the map $T : x \mapsto \{\frac{1}{\{x\}}\}$ Assume that $x_n \notin Z$ and that g_0 is C^2. It is well known (see Kuipers and Niederreiter(1974)) that

$$g_1(x) = \sum_{k=1}^{\infty}\left(g_0(\frac{1}{k}) - g_0(\frac{1}{k+x})\right)$$

Then

$$g'_1(x) = \sum_{k=1}^{\infty} g'_0(\frac{1}{k+x})\frac{1}{(k+x)^2}$$

Furthermore g_1 is C^2, so we can repeat. In this way one obtains the identity

$$g'_{n+1}(x) = \sum_{k=1}^{\infty} g'_n(\frac{1}{k+x})\frac{1}{(k+x)^2}$$

This functional equation was considered by Kuzmin in his treatment of Gauss' problem in the metric theory of continued fractions (see Khintchine(1974)). His results imply the following

Theorem 2 $\quad g_n(x) = \frac{\log(1+x)}{\log(2)} + O(e^{-c\sqrt{n}})$

uniformly for $x \in [0,1]$ (here c is an absolute constant)

Finally we consider the classical Bernoulli shift $x \mapsto \{2x\}$. We show that $g_n(x) = x + O(2^{-n})$. More precisely, we prove

Theorem 2 *Let $M := |g''_0|_\infty$ Then for all n and x*

$$|g_n(x) - x| \le \frac{M}{2^n}$$

Proof: We begin by establishing the identity

$$g_{n+1}(x) = g_n(\frac{x}{2}) + g_n(\frac{x+1}{2}) - g_n(\frac{1}{2}) \qquad *(1)*$$

For $n = 0$ we have

$$g_1(x) := \lim_{N\to\infty} \frac{1}{N}\#\Big\{n \le N : x_n^1 \in [0,x)\Big\}$$

$$= \lim_{N\to\infty} \frac{1}{N}\Big(\#\Big\{n \le N : x_n^0 \in [0,\frac{x}{2})\Big\} + \#\Big\{n \le N : x_n^0 \in [\frac{1}{2},\frac{x+1}{2})\Big\}\Big)$$

$$= g_0(\frac{x}{2}) - g_0(0) + g_0(\frac{x+1}{2}) - g_0(\frac{1}{2})$$

$$= g_0(\frac{x}{2}) + g_0(\frac{x+1}{2}) - g_0(\frac{1}{2})$$

This identity, and the fact that g_0 is C^2, together imply that g_1 is C^2. Hence we can repeat the argument for $n = 1$, and recursively for $n > 1$. In this way one obtains the identity $*(1)*$.

Next let $\Delta_n(x) := g_n(x) - x$. We shall prove that Δ_n is small. By equation $*(1)*$ we have

$$\Delta_{n+1}(x) = \Delta_n(\frac{x}{2}) + \Delta_n(\frac{x+1}{2}) - \Delta_n(\frac{1}{2})$$

Differentiating yields

$$\Delta_n'(x) = \frac{1}{2}\Big(\Delta_n'(\frac{x}{2}) + \Delta_n'(\frac{x+1}{2})\Big)$$

Using this identity and mathematical induction, one obtains (for $n \ge 1$):

$$\Delta_n'(x) = \sum_{\epsilon_i = 0 \text{ or } 1} \frac{1}{2^n}\Delta_0'\Big(\frac{x + \epsilon_0 2^0 + \epsilon_1 2^1 + \ldots + \epsilon_{n-1}2^{n-1}}{2^n}\Big)$$

On the other hand, by the mean value theorem, one has

$$\int_0^1 \Delta_0'(t)dt = \sum_{\vec{\epsilon}} \frac{1}{2^n}\Delta_n'\big(\xi_{\vec{\epsilon}}\big)$$

where $\xi_{\vec{\epsilon}}$ is some point in the interval of length 2^{-n} whose left endpoint is $\frac{\sum \epsilon_i 2^i}{2^n}$ Thus we have

$$|\Delta_n'(x) - \int_0^1 \Delta_0'(t)dt| =$$

$$\Big|\sum_{\vec{\epsilon}} \frac{1}{2^n}\Delta_0'(\frac{x + \epsilon_0 2^0 + \epsilon_1 2^1 + \ldots + \epsilon_{n-1}2^{n-1}}{2^n}) - \sum_{\vec{\epsilon}} \frac{1}{2^n}\Delta_n'(\xi_{\vec{\epsilon}})\Big|$$

$$\le \frac{1}{2^n}\sum_{\vec{\epsilon}}\Big|\Delta_0'(\frac{x + \epsilon_0 2^0 + \epsilon_1 2^1 + \ldots + \epsilon_{n-1}2^{n-1}}{2^n}) - \Delta_n'(\xi_{\vec{\epsilon}})\Big|$$

Hence we have

$$|\Delta_n'(x) - \int_0^1 \Delta_0'(t)dt| \leq \frac{M}{2^n}$$

where $M = |g_0''|_\infty$ Observe that $\int_0^1 \Delta_0'(t)dt = \Delta(1) - \Delta(0) = 0$. Thus

$$|\Delta_n'(x)| \leq \frac{M}{2^n}$$

and consequently

$$\left|g_n(x) - x\right| = |\Delta_n(x)| \leq \frac{M}{2^n}.$$

■

CONCLUSION

For several choices of T, we proved that $g_n \to g$ uniformly, and that g is universal in the sense that it is essentially independent of the initial distribution g_0. To the best of our knowledge, no one has studied this universal feature of the invariant measure associated with a transformation T.

REFERENCES

1. Halmos,P., (1956). Lectures on Ergodic Theory, Chelsea.
2. Jetschke and Stiewe,(1987). J.Phys.A : Math.Gen **20**.
3. Khintschine,(1974). ContinuedFractions, Univ of Chicago Press
4. Kuipers and Niederreiter,(1974). Uniform Distribution of Sequences, Wiley-Interscience.
5. Lasota and Yorke,(1973). Trans.A.M.S. **186** 1973.
6. Ryll and Nardzewsky,(1951). StudiaMath. **12** 74.
7. Schuster, Deterministic Chaos, Physik Verlag 1984
8. Waterman, M. Monat.furMathematik **79**:157
9. Wirsing(1974).Acta Arithmetica **24**: 507.

On the *n*-Variable Saddle Point and Steepest Descent Methods

D. Kaminski Professor, Weapons Effectiveness Group, Defence Research Establishment, Valcartier, Courcelette, Quebec, Canada

I. INTRODUCTION

The saddle point method is a well-known technique used to obtain the large λ behaviour of integrals of the form

$$I(\lambda) = \int_\gamma f(z) e^{\lambda S(z)}\, dz. \tag{1}$$

We shall take $\lambda > 0$ and $\operatorname{Re} S \to -\infty$ as $|z| \to \infty$ along γ. The method proceeds by first deforming γ into a new contour passing through one or more *saddle points* (i.e., points where $S' = 0$) such that the new integration contour lies in valleys relative to these saddles (i.e., regions where $\operatorname{Re} S < \operatorname{Re} S(z_0)$ for z_0 a saddle point). Under moderate restrictions on S, f and the initial integration contour γ, it can be shown that the asymptotic behaviour of $I(\lambda)$ is governed by the saddle point z_0 for which $\operatorname{Re} S(z_0) > \operatorname{Re} S(z_0')$ for any other z_0' along this new contour, which we continue to denote by γ.

We assume that only one such saddle point is maximal in this sense, and that all saddle points are non-degenerate, i.e., $S''(z_0) \neq 0$.

As a rule, the saddle point method approximates $I(\lambda)$ by

$$I_1(\lambda) = \int_{\gamma \cap |z-z_0| \leq \epsilon} f(z) e^{\lambda S(z)} \, dz \tag{2}$$

for some (small) $\epsilon > 0$, and subsequently replaces that portion of γ in $|z - z_0| < \epsilon$ by a convenient integration path $\tau \mapsto \gamma_1(\tau)$ (usually a line segment) subject to the restriction that the second derivative of $(S \circ \gamma_1)(\tau)$, evaluated at τ for which $\gamma_1(\tau) = z_0$, have non-positive real part.

In this setting, the integral $I_1(\lambda)$, and hence $I(\lambda)$, has the principal asymptotic behaviour

$$I_1(\lambda) \sim f(z_0) e^{\lambda S(z_0)} \sqrt{\frac{-2\pi}{\lambda S''(z_0)}}, \tag{3}$$

$\lambda \to \infty$, the square root being suitably interpreted; see (Copson, 1965, Ch. 8).

The method of steepest descent comes into the picture by noting that if γ_1 is chosen so that $(S \circ \gamma_1)''(\tau)$, evaluated at τ for which $z_0 = \gamma_1(\tau)$, is negative, then γ_1 is tangent to the steepest descent curve passing through z_0. Steepest descent curves in the plane through z_0 are characterized by:

- $\text{Re}\,[S(z) - S(z_0)]$ is negative and monotonically decreasing along a steepest descent curve as we move away from the saddle point z_0, and
- $\text{Im}\,[S(z) - S(z_0)] = 0$ for any z on the steepest descent curve.

Of course, if we deform the integration path γ for $I(\lambda)$ into a sum of steepest descent contours, we obviate the need for the saddle point method and obtain precision estimates of the errors committed in the asymptotic approximation; see (Olver, 1970).

However, recent successes in the extension of the saddle point method to $\mathbf{C}^n$, $n \geq 2$, raise the question: can the method of steepest descent be similarly extended? As the two methods use similar computational methods once manipulations of the integration contour are complete, it is this manipulation that will serve as our focus.

II. THE n-DIMENSIONAL SADDLE POINT METHOD

The higher dimensional analog of the Laplace integral (1) in the Introduction takes the same form, only now f and S are holomorphic functions of $z = (z_1, z_2, \ldots, z_n) \in \mathbf{C}^n$, $dz = dz_1 \cdots dz_n$ and γ is a smooth manifold of (real) dimension n. We allow for the possibility that γ may have boundary $\partial\gamma$, and we assume that all saddle points (zeroes of, dS, the first derivative of S) are *non-degenerate*, i.e., if z^0 is a saddle point of S, then $\det \operatorname{Hess}_{z^0}(S) = \det(d^2S)(z^0) \neq 0$.

One of the first difficulties that we encounter concerns the problem of the deformation process. While it is clear that any deformation of the manifold γ must "push" the manifold in directions where $\operatorname{Re} S$ is negative, there seems, at first sight, to be no preferred direction, especially since the usual steepest descent requirement $\operatorname{Im}[S(z) - S(z^0)] = 0$ yields an analytic variety of real dimension $2n - 1$.

If we regard S as a function of $2n$ real variables $(x_1, y_1, \ldots, x_n, y_n)$, $z_j = x_j + iy_j$, then it is a straightforward process to find the gradient of $\operatorname{Re} S$. Taking the negative of this vector therefore yields the direction of steepest descent of $\operatorname{Re} S$. By using the Cauchy-Riemann equations, this result can be re-expressed in terms of the holomorphic function S. For an integral curve $t \mapsto z(t)$ of these directions of steepest descent, we have

$$\frac{dz}{dt} = \overline{-(dS)(z)}, \tag{4}$$

where the overline denotes complex conjugation. We choose the real parameter t so that $\operatorname{Re} S$ decreases with increasing t. We note that $\operatorname{Im} S$ is constant along such curves.

With this observation, we can state the fundamental result concerning the deformation process needed for the application of the saddle point method in $\mathbf{C}^n$:

Theorem (Fedoryuk, 1977) Let γ be a compact smooth n-manifold (possibly with boundary) and let f and S be functions holomorphic at $z^0 \in \gamma$. Among the points at which $\max\{\operatorname{Re} S(z) | z \in \gamma\}$ is attained, suppose there are no saddle points of S and no boundary points of γ. Then there exists a manifold γ_1 such that

1. $\displaystyle\int_\gamma f(z) e^{\lambda S(z)}\,dz = \int_{\gamma_1} f(z) e^{\lambda S(z)}\,dz,$
2. $\displaystyle\max_{z\in\gamma_1} \operatorname{Re} S(z) < \max_{z\in\gamma} \operatorname{Re} S(z).$

Proof. The conditions on γ guarantee that if $t \mapsto z(t)$ is an integral descent curve through the point z^0 such that $z(0) = z^0$, denoted γ_{z^0}, then the tangent vector $z'(0)$ is not tangent to γ at z^0 (i.e., the steepest descent curve is *transversal* to γ at z^0). To see this, note first that

$$\left.\frac{d}{dt}(\operatorname{Re} S \circ z)(t)\right|_{t=0} = \sum_{j=1}^{n} \left[(\operatorname{Re} S)_{x_j}(z^0)\dot{x}_j(0) + (\operatorname{Re} S)_{y_j}(z^0)\dot{y}_j(0)\right].$$

Here, the identification $(z_1, \ldots, z_n) = (x_1, y_1, \ldots, x_n, y_n)$ has been made by setting $z_j = x_j + iy_j \equiv (x_j, y_j)$, where $z_j \in \mathbf{C}$, x_j, $y_j \in \mathbf{R}$. From $z'(0) = \overline{-(dS)(z^0)}$ and the Cauchy-Riemann equations, we have $\dot{x}_j(0) = -(\mathrm{Re}\, S)_{x_j}(z^0)$, $\dot{y}_j(0) = -(\mathrm{Re}\, S)_{y_j}(z^0)$, whence

$$\begin{aligned} \frac{d}{dt}(\mathrm{Re}\, S \circ z)(t)\bigg|_{t=0} &= -\sum_{j=1}^{n} \left[(\mathrm{Re}\, S)^2_{x_j}(z^0)\dot{x}_j(0) + (\mathrm{Re}\, S)^2_{y_j}(z^0)\dot{x}_j(0)\right] \\ &= -\|(dS)(z^0)\|^2 \neq 0. \end{aligned}$$

The notation $\|\cdot\|$ denotes Euclidean length. Now if $z'(0)$ were tangent to γ at $z(0)$, then since $\mathrm{Re}\, S(z)$ is smooth and real-valued on γ, with a maximum at $z(0)$, we would have $\nabla \mathrm{Re}\, S = 0$ at z^0, a contradiction to $(dS)(z^0) \neq 0$. Thus, $z'(0)$ is not tangent to γ at z^0.

By continuity of the map $z \mapsto \gamma_z$ (continuous in the sense that steepest descent curves passing through z, γ_z, depend smoothly on z), we know that in a neighborhood U of z^0, $U \subset \gamma$, all steepest curves through points of U are transversal to γ. Thus, we have the setting portrayed in Fig. 1.

The set U is chosen so that $\overline{U} \cap \partial\gamma = \emptyset$, where the overbar here denotes closure. For each $z \in \overline{U}$, choose $t = t(z) \in \mathbf{R}_+$ and "transport" z along the curve γ_z to the point $\gamma_z(t(z))$. We will require $t(z) > 0$ for $z \in U \setminus \partial U$, $t = 0$ on ∂U. Choose $t(z)$ sufficiently smooth on $\overline{U}$ so that the set $\{\gamma_z(t(z)) \mid z \in U\}$ forms a piecewise C^1 manifold (intuitively easy to do, but rife with technicalities which we do not pursue here). Further, $t(z)$ can be chosen sufficiently small so that all points along each curve $\gamma_z(t(z))$, $z \in U$, are contained in a common domain of holomorphy of f and S.

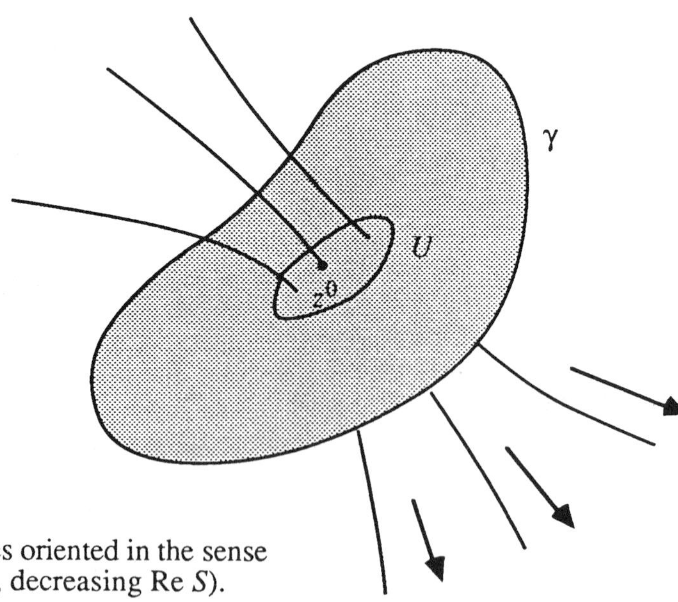

Figure 1: Descent curves oriented in the sense of increasing t (i.e., decreasing Re S).

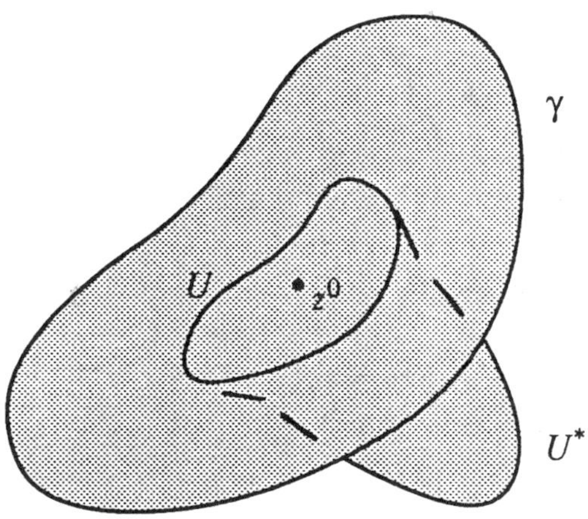

Figure 2: The dimpling effect of the deformation process applied to U (a subset of γ).

Call the result of transporting U in this way U^*. We get a "dimpled" version of γ, displayed in Fig. 2 for the $n = 2$ case. Define a new (possibly piecewise) smooth manifold γ_1 by excising U from γ, and replacing it by U^*. The deformed manifold γ_1 now satisfies the conclusions of the theorem, the second of these following from the orientation of the integral curves as defined in eqn. (4).

To see the first, if we consider the integration chain $\gamma - \gamma_1$ ($= \gamma +$ 'γ_1 with the opposite orientation'), then the pieces common to γ and γ_1 cancel, leaving U and the 'dimple' U^*. These two pieces together bound an $(n+1)$-dimensional object Σ having $U - U^*$ as its boundary manifold. By the Cauchy-Poincaré integral theorem,

$$\int_{\partial\Sigma} fe^{\lambda S} = \int_{\Sigma} d(fe^{\lambda S}) = 0,$$

so that

$$\int_U fe^{\lambda S} = \int_{U^*} fe^{\lambda S}$$

since

$$\int_{\partial\Sigma} fe^{\lambda S} = \int_{U-U^*} fe^{\lambda S} = \int_U fe^{\lambda S} - \int_{U^*} fe^{\lambda S}.$$

Thus, we obtain the first conclusion of the theorem. □

This gives a means of deforming the manifold γ so that the value of the integral is left undisturbed, and so that the saddles and boundaries (if any) retain their privileged role.

We now turn to selecting the ϵ-neighborhoods of saddles, paralleling the small disks that appear in eqn. (2). To accomplish this, Fedoryuk and others, most notably Pham (Pham, 1983), move to homology theory. Fedoryuk, in particular, considers integration over chains drawn from relative homology groups. In brief, let $X \subset \mathbf{C}^n$. A formal linear combination of orientable k-dimensional manifolds (a k-chain), γ, is a *relative k-cycle* modulo a subset $A \subseteq X$ if $\partial\gamma \subset A$. Further, if this relative k-cycle can be expressed as $\gamma = \gamma' + \partial\tilde{\gamma}$ for a k-chain γ' in A and a $(k+1)$-chain $\tilde{\gamma}$ in X, then we say that γ is *homologous to zero* mod A, written $\gamma \approx 0$ mod A. Two k-chains γ_1 and γ_2 are homologous in X mod A, written $\gamma_1 \approx \gamma_2$ mod A, if their difference is homologous to 0 mod A.

In all cases, chains are assumed to be compactly supported; see (Fedoryuk, 1977) and (Cohn, 1967) (the latter gives a relatively painless introduction to homology theory in the context of function theory on the plane).

The use of homologous integration chains permits the easy application of the Cauchy-Poincaré integral theorem. To illustrate, let us take Ω to be a bounded domain of holomorphy of f and S, $A = \Omega \cap \{z | \operatorname{Re} S(z) = a\} \neq \emptyset$, and $X = \Omega \cap \{z | \operatorname{Re} S(z) \geq a\}$. If two n-dimensional chains σ and τ are homologous in X mod A, which is to say that $\sigma, \tau \subset X$, with $\sigma - \tau = v + \partial\varphi$ for an n-chain $v \subset A$ and an $(n+1)$-chain $\varphi \subset X$, then we have

$$\begin{aligned}
\int_{\sigma-\tau} f(z)e^{\lambda S(z)}dz &= \int_{v+\partial\varphi} f(z)e^{\lambda S(z)}dz = \int_{v} f(z)e^{\lambda S(z)}dz + \int_{\partial\varphi} f(z)e^{\lambda S(z)}dz \\
&= \int_{v} f(z)e^{\lambda S(z)}dz + \int_{\varphi} d(f(z)e^{\lambda S(z)}dz) \\
&= \int_{v} f(z)e^{\lambda S(z)}dz \quad \text{(by the Cauchy-Poincaré theorem)} \\
&= O(e^{a\lambda}).
\end{aligned}$$

Thus, integrals over σ and τ differ by a term of order $O(e^{a\lambda})$.

Evidently, this technique, combined with the deformation process outlined in the theorem, permits us to reduce the integral (1) to one over an appropriately small domain, while accumulating errors which are of exponentially small order (for $a < 0$). Ultimately, however, we must perform an explicit evaluation of an integral over an n-chain. To this end, Fedoryuk considers *vanishing cycles* which serve as generators of the relative homology groups, over whose elements he performs his integrations (see (Fedoryuk, 1977) for more information). As a more complete discussion of this would lead us too far astray from our study of the deformation process, we close this section by citing, for the convenience of the reader, two results of Fedoryuk which follow from our discussion, an examination of integration on relative chains, and the use of vanishing cycles as generators of certain homology groups.

Theorem(Fedoryuk, 1977) Let $\max_{z \in \gamma} \operatorname{Re} S(z)$ be attained only at the point $z^0 \in \gamma$, a non-degenerate saddle point in the interior of γ. Then, as $\lambda \to +\infty$,

$$I(\lambda) \sim \frac{1}{\lambda^{n/2}} e^{\lambda S(z^0)} \sum_{k=0}^{\infty} \frac{a_k}{\lambda^k}.$$

This expansion may be differentiated with respect to λ an arbitrary number of times.

Proposition(Fedoryuk, 1977) The principal term of the above expansion is given by

$$I(\lambda) = \left(\frac{2\pi}{\lambda}\right)^{n/2} e^{\lambda S(z^0)} \left[\det(-d^2 S)(z^0)\right]^{-1/2} \left[f(z^0) + O\left(\frac{1}{\lambda}\right)\right].$$

Compare the result of the proposition with the usual (planar) result given in eqn. (3).

Additional information regarding these results can be found in (Fedoryuk, 1977, Ch. 5), or in a translation of §1 of that chapter available (for personal use only) directly from myself.

III. THE NON-UNIQUENESS OF STEEPEST DESCENT SURFACES

It was in (Ursell, 1980) where difficulties with an extension of the steepest descent method to $\mathbf{C}^n$ for $n \geq 2$ began to surface. Ursell examined a multivariate integral in $\mathbf{C}^2$ serving as a canonical diffraction integral (canonical in the sense of catastrophe theory), obtaining the asymptotic behaviour of an integral having a cubic polynomial in two variables as a phase function. In the process, he was compelled to use 'descent surfaces' rather than 'steepest descent surfaces.' Some discussion will reveal why.

If we use the deformation process characterized by (4), applied to the initial integration surface γ, then after deforming γ as much as possible, we obtain a surface, say γ^*. However, as ably pointed out in an example of Ursell, the result is somewhat surprising. Because it occupies a central role in the ensuing discussion, it is worth recounting his illustration.

We begin with the phase function $S(w) = S(w_1, w_2) = -w_1^2 - w_2^2$, and the surface $\gamma = \{(w_1, w_2) | \operatorname{Re} w_1 = 0, \operatorname{Re} w_2 = 0\}$. Note that the initial integration surface qualifies as a 'steepest descent surface' since the exponential in the integrand of (1) is already real, exponentially decreasing and has a negative-definite quadratic phase function along this surface.

Applying the deformation process (4) results in the system of equations

$$(\operatorname{Re} w_1)(t) = (\operatorname{Re} w_1)(0)e^{2t}, \quad (\operatorname{Im} w_1)(t) = (\operatorname{Im} w_1)(0)e^{-2t},$$
$$(\operatorname{Re} w_2)(t) = (\operatorname{Re} w_2)(0)e^{2t}, \quad (\operatorname{Im} w_2)(t) = (\operatorname{Im} w_2)(0)e^{-2t}.$$

Initial conditions—arbitrary $(\operatorname{Re} w_1)(0)$ and $(\operatorname{Re} w_2)(0)$, and zero $(\operatorname{Im} w_1)(0)$ and $(\operatorname{Im} w_2)(0)$—force the resulting deformation process to move points on the $\operatorname{Re} w_1 \operatorname{Re} w_2$-plane to other points on the same plane. Thus, the initial surface is indeed a 'steepest surface' according to (4) in the sense that no further deformation is possible. Thus, we may set $\gamma = \gamma^*$.

If we were to make the biholomorphic change of variables

$$\begin{aligned} w &= (w_1, w_2) = b(z) = (b_1(z_1, z_2), b_2(z_1, z_2)) \\ &= (z_1 \cosh\alpha + iz_2 \sinh\alpha, -iz_1 \sinh\alpha + z_2 \cosh\alpha), \ \alpha \in \mathbf{R}, \end{aligned} \tag{5}$$

we would find that S retains its form as a negative sum of squares, while the integration surface switches to a plane transversal to the initial real plane γ (in the event that b is not the identity map). Evidently, steepest descent surfaces in $\mathbf{C}^n$, $n \geq 2$ are not preserved under holomorphic (or biholomorphic) mappings, contrary to our experience in the plane.

The situation is even worse than that suggested by this example. If $w_1^2 + w_2^2 = b_1^2(z_1, z_2) + b_2^2(z_1, z_2) = z_1^2 + z_2^2$ for some biholomorphism $b = (b_1, b_2)$, then what can be said about b?

Locally, denoting terms of order 2 or more as $O(2)$, we have

$$\begin{aligned} b_1(z_1, z_2) &= \alpha_1 z_1 + \alpha_2 z_2 + O(2), \\ b_2(z_1, z_2) &= \beta_1 z_1 + \beta_2 z_2 + O(2), \end{aligned}$$

so that $\sum b_i^2 = \sum z_i^2$ implies that

$$\alpha_1^2 + \beta_1^2 = 1, \ \alpha_2^2 + \beta_2^2 = 1, \ \alpha_1\alpha_2 + \beta_1\beta_2 = 0.$$

Furthermore, in looking at the preservation of the quadratic form under the linearization of b, we not only require that the Jacobian of b be non-zero at the origin (which yields $\alpha_1\beta_2 - \alpha_2\beta_1 \neq 0$), but that for the linearization

$$\begin{bmatrix} w_1 \\ w_2 \end{bmatrix} = \begin{bmatrix} \alpha_1 & \alpha_2 \\ \beta_1 & \beta_2 \end{bmatrix} \begin{bmatrix} z_1 \\ z_2 \end{bmatrix},$$

we also have

$$\alpha_1\beta_2 - \alpha_2\beta_1 = \pm 1. \tag{6}$$

This follows from consideration of the quadratic forms $\sum w_i^2 = \sum z_i^2$.

The substitutions $\alpha_1 = \cos\varphi$, $\beta_1 = \sin\varphi$, $\alpha_2 = \cos\psi$ and $\beta_2 = \sin\psi$ imply that, for the '+' case in (6), φ and ψ differ by $\pi/2 \bmod 2\pi$, or that we may set $\psi = \varphi + \frac{\pi}{2}$. The '−' case similarly implies that we may take ψ

$= \varphi - \frac{\pi}{2}$. Hence, any biholomorphism b which fixes the origin must have the linearization

$$(db)(0)z^T = \begin{bmatrix} \cos\varphi & \mp\sin\varphi \\ \sin\varphi & \pm\cos\varphi \end{bmatrix} \begin{bmatrix} z_1 \\ z_2 \end{bmatrix}. \tag{7}$$

The set of such biholomorphisms evidently forms a group, their linearizations forming a 1-(complex-)parameter group under the usual matrix multiplication. In this setting, Ursell's example appears as a special case (that of the upper sign choice, along with the evaluation $\varphi = -i\alpha$).

The situation does not improve in considering additional higher order terms. If we examine Maclaurin expansions of b to terms of order 2, we find an underdetermined system of four equations in six unknowns (beyond the characterization of $(db)(0)$ in terms of φ).

Thus, steepest descent cannot be performed in any fashion which requires a unique steepest descent surface constructed using the deformation process of eqn. (4).

IV. COORDINATE-WISE DESCENT SURFACE CONSTRUCTION

A different approach to steepest descent surfaces results from choosing surfaces (or, manifolds in sufficiently high dimensional cases) in order that the second derivative of S exhibit a certain behaviour. Recall the discussion following eqn. (3) concerning the expression $(S \circ \gamma_1)''(\tau)$. There we remarked that if this was negative at the saddle z_0 (for the planar case), then γ_1 was tangent to the steepest descent curve through z_0. By analogy, let $\alpha = e^{i\theta}$ and $\beta = e^{i\phi}$, and consider the function $\tilde{S}(s,t) = S(z^0 + (s\alpha, t\beta)) - S(z^0)$ for real s and t. This function of s and t is a quadratic form near (0,0) which is negative (semi-)definite only if $\tilde{S}_{ss} \leq 0$ and $\det \text{Hess}_{(0,0)}(\tilde{S}) > 0$. Thus, by paralleling the planar situation, we can obtain directions of 'steepest descent' α and β by selecting θ and ϕ so that these two inequalities are satisfied.

Now, $\tilde{S}_{ss}(0,0) = \alpha^2 S_{z_1 z_1}(z^0)$ and $\det \text{Hess}_{(0,0)}(\tilde{S}) = \alpha^2\beta^2 \det \text{Hess}_{z^0}(S)$. Requiring $\tilde{S}_{ss}(0,0) < 0$ and $\det \text{Hess}_{(0,0)}(\tilde{S}) > 0$ implies

$$\begin{aligned} \theta &= \tfrac{\pi}{2} - \tfrac{1}{2}\arg S_{z_1 z_1}(z^0), \\ \phi &= -\tfrac{\pi}{2} + \tfrac{1}{2}\arg S_{z_1 z_1}(z^0) - \tfrac{1}{2}\arg \det \text{Hess}_{z^0}(S). \end{aligned}$$

These two results can be used to determine a surface $(u,v) \mapsto \gamma(u,v) = (\gamma_1(u,v),\ \gamma_2(u,v))$, by setting

$$\gamma(0,0) = z^0,$$
$$\left.\frac{\partial \gamma_1}{\partial u}\right|_{(0,0)} = \alpha, \quad \left.\frac{\partial \gamma_1}{\partial v}\right|_{(0,0)} = 0,$$
$$\left.\frac{\partial \gamma_2}{\partial u}\right|_{(0,0)} = 0, \quad \left.\frac{\partial \gamma_1}{\partial v}\right|_{(0,0)} = \beta,$$

or $(d\gamma_1)(0,0) = (\alpha, 0)$, $(d\gamma_2)(0,0) = (0, \beta)$. Away from the origin in the uv-plane, i.e., away from the saddle z^0, we can invoke the requirement

$$(S \circ \gamma)(u,v) - S(z^0) = (S \circ \gamma)(u,v) - (S \circ \gamma)(0,0) = -u^2 - v^2$$

for real u and v.

In effect, this approach gives the construction of descent surfaces formed in such a manner that $\gamma(u,0)$ is the steepest descent curve of $S(z_1, z_2^0)$ through $z_1 = z_1^0$, and that of $\gamma(0,v)$ is the steepest descent surface of $S(z_1^0, z_2)$ through $z_2 = z_2^0$. Since, at each stage of the descent surface construction, we have all but one coordinate in the parameter space (the uv-plane in this case) fixed, we can effectively compute the steepest descent manifold (surface) by computing its coordinate plane traces.

V. CLOSING REMARKS

As a result of our discussion, we see that while the saddle point method is achieving a certain theoretical maturity through the use of techniques from algebraic topology, the steepest descent method still faces formidable difficulties in its generalization. Certainly the greatest obstacle is the lack of uniqueness of steepest descent manifolds, although this may prove to be a strength since the analyst can then choose a convenient descent manifold, while retaining salient features of the method.

Furthermore, it is not clear if the deformation process described in section II always results in a surface constructed according to the process outlined in the previous section. In any event, Ursell's observation still applies, yielding the same lack of uniqueness of steepest surfaces.

Further work is needed.

REFERENCES

1. Cohn, H. (1967). *Conformal Mapping on Riemann Surfaces*, Dover Publications, Inc., New York.

2. Copson, E.T. (1965). *Asymptotic Expansions*, Cambridge University Press, Cambridge.

3. Fedoryuk, M.V. (1977). *Metod Perevala* (in Russian), Nauka, Moscow.

4. Olver, F.W.J. (1970). *SIAM Review*, Vol. 12, No. 2: 228-247.

5. Pham, F. (1983). *Proc. Symposia Pure Math.*, Vol. 40, Part 2: 319-333.

6. Ursell, F. (1980). *Math. Proc. Camb. Phil. Soc.*, Vol. 87: 249-273.

Applications of the Method of Steepest Descents in Wave-Propagation Problems

Francesco Mainardi Professor, Department of Physics, University of Bologna, Bologna, Italy

Giuliano Vitali Professor, Institute of Agronomy, University of Bologna, Bologna, Italy

1. INTRODUCTION

The study of wave propagation in linear dispersive systems is usually carried out by using Fourier or Laplace transforms (see *e.g.* Brillouin, 1960; Thau, 1974; Whitham, 1974; Bleistein, 1984).

Referring to uniaxial waves in a semi-infinite medium ($x \geq 0$), initially quiescent, and denoting by r the "response" of the system chosen among the field variables, the pulse $r(x,t)$ can easily be expressed at any time $t > 0$ and position $x > 0$ by a Laplace transform. In fact we get

$$r(x,t) = \frac{1}{2\pi i} \int_{Br} ds\, \bar{r}_0(s)\, e^{st-\gamma(s)x} \tag{1.1}$$

in which s is the complex Laplace variable and Br denotes the Bromwich path, *i.e.* a vertical line lying to the right of all singularities of $\bar{r}_0(s)$ and of $\gamma(s)$.

This work was supported in part by the Italian Ministry of Education (MPI: 60 % grants), and by the National Research Council (CNR: FISBAT, Bologna).

The function $\bar{r}_0(s)$ is the Laplace transform of the source pulse,

$$r(0,t) = r_0(t) \quad t \geq 0\,, \tag{1.2}$$

while the function $\gamma(s)$ can be found from the dispersion relation for the given medium (Thau, 1974),

$$\mathcal{D}(\omega, k) = 0\,, \tag{1.3}$$

where ω is the frequency and k is the (complex) wavenumber. Setting $\omega = is$ and solving for $k = k(is)$, we define

$$\gamma(s) = \pm ik(is)\,, \tag{1.4}$$

where the choice of sign is dictated by the condition that the real part of γ should be positive when s is real positive. Here we assume that

$$\frac{\gamma(s)}{s} \to \frac{1}{c} > 0 \quad \text{as} \quad Re\,[s] \to +\infty\,, \tag{1.5}$$

so that, setting

$$\gamma(s) = \frac{s}{c}\, n(s)\,, \tag{1.6}$$

equation (1.1) reads

$$r(x,t) = \frac{1}{2\pi i}\int_{Br} ds\, \bar{r}_0(s)\, e^{s[t-n(s)x/c]}\,. \tag{1.7}$$

initially quiescent,

It is easy to recognize (see *e.g.* Mainardi, 1983a) that c represents the wave front velocity and that $n(s)$ provides, for $s = \pm i\omega$, the complex index of refraction for the given medium.

The exponential function $\bar{\mathcal{P}}(x,s) = \exp[sn(s)x/c]$ can be referred to as the "propagator", since it allows the propagation through the medium of the input pulse $r_0(t)$, with finite speed but with dispersion and attenuation.

Although the Laplace method provides a formal solution to any signalling problem here considered through the Bromwich formula (1.7), it usually cannot provide exact analytical solutions. When the integration can analytically be performed, it often leads to mathematically cumbersome solutions even for the simplest forms of $n(s)$ because of the presence of essential singularities and branch cuts in the propagator.

In his classical paper of 1914 about the propagation of light in dispersive media, Brillouin (1960) used the saddle-point method of integration in the complex frequency plane to compute the waveform of the signal and to define the arrival time of its bulk. This analysis may be extended to any signalling problem where a Fourier-Laplace representation of the solution is available, in particular to the propagation of transient stress waves in a linear viscoelastic material.

From now on we will deal with viscoelastic waves pointing out, however, that our analysis could be applied to other kinds of linear dispersive waves. The corresponding problem will be stated in the next section, where we will provide, for a class of viscoelastic models, the mathematical structure of the function $n(s)$.

Several authors have treated problems of transient waves in linear viscoelastic media. Many of them, including Jeffreys (1931, 1932), Rubin (1954), Chu (1962), Caputo (1969), Mainardi (1972), Chin (1980), Pound (1985), have used the saddle-point method to get an asymptotic approximation for the integral (1.1) or (1.7) for a variety of viscoelastic models; their solutions turn out to be valid for sufficiently large values of x and t.

Alternative methods to get the solution, working directly in the space-time domain (see *e.g.* Achenbach & Reddy, 1967; Brun, 1974; Buchen, 1974) or working in the complex Laplace domain (see *e.g.* Fabrizio, 1968; Longman, 1972; Buchen & Mainardi, 1975; Mainardi & Turchetti, 1975) have been proposed in the literature of viscoelastic waves. These methods, however, are expected to provide an adequate description of the pulse evolution only near to the wave-front and at moderate distances so that they are usually adopted as complementary techniques to the saddle-point method.

For a detailed illustration of the saddle-point method and related matter we refer to books on asymptotics (see *e.g.* Olver, 1974; Bleistein & Handelsman, 1975). Here we would like to recall the principal features of this method with respect to our aim of evaluating the pulse response in a large space - time domain, without using complementary techniques. As it is well known, the method is based on the possibility of deforming the original path of integration (the Bromwich path) into another (the steepest descent path) through the saddle points of the function

$$F_\mu(s) = s[1 - \mu n(s)]\,, \quad \mu = \frac{x}{ct} \quad (0 \le \mu \le 1)\,. \tag{1.8}$$

Such a path depends on the evolution parameter μ; on it the imaginary part of F_μ is constant while its real part attains a finite number of maxima and minima.

In order to get the leading term of the asymptotic series, the ordinary saddle-point method is known not to require a detailed knowledge of this path. Usually the location of the saddle points with respect to the singularities of $\bar{r}_0(s)$ is necessary to get a uniform approximation for large times. However, we note that the steepest descent path could succesfully be used for numerical integration (in any space – time domain), since it allows us to isolate the oscillations of the propagator. The price to be paid is, of course, represented by the detailed evaluation of this path that may be a difficult task because it is generally cumbersome and it depends on the parameter μ.

In the section 3 of this paper we will introduce our specific routine based on Padé Approximants, that allows us to follow with continuity the required path starting from the saddle points. The numerical integration along the path is achieved with a good accuracy so that the wave form can be obtained in the required space – time domain. The analytical result of the saddle-point method is of course recovered for large times. We provide an instructive example of evaluation of transient waves propagating in a linear viscoelastic medium (the Maxwell model), for which the exact solution is analytically known. This allows us to check the good level of accuracy of our computational scheme.

2. STATEMENT OF THE PROBLEM

Problems of transient viscoelastic waves essentially concern the response of a long viscoelastic rod of uniform small cross-section to dynamical (uniaxial) loading conditions.

According to elementary theory (Hunter, 1960), the rod is taken to be homogeneous (of density ρ), semi-infinite in extent ($x \geq 0$), and undisturbed for $t < 0$. For $t \geq 0$ the end of the rod (at $x = 0$) is subjected to a disturbance (input) denoted by $r_0(t)$. The response variable $r(x,t)$ may be either the displacement $u(x,t)$, or the particle velocity $v(x,t) = u_t(x,t)$, or the stress $\sigma(x,t)$, or the strain $\epsilon(x,t) = u_x(x,t)$, where subscripts denote partial derivatives. The general problem consists of solving the equation of motion

$$\sigma_x = \rho u_{tt} \tag{2.1}$$

with the aid of the stress – strain relation, under the boundary condition

$$r(0,t) = r_0(t) \quad t \geq 0\,, \tag{2.2}$$

and the initial conditions

$$r(x,0) = r_t(x,0) = 0 \quad x \geq 0\,. \tag{2.3}$$

The stress – strain relation is the equation describing the mechanical properties of the rod, namely it is the constitutive equation for the assumed viscoelastic model. It is most conveniently taken in the *Laplace representation* that reads (see *e.g.* Christensen, 1982)

$$\bar{\sigma}(x,s) = s\bar{G}(s)\,\bar{\epsilon}(x,s)\,, \quad \text{or} \quad \bar{\epsilon}(x,s) = s\bar{J}(s)\,\bar{\sigma}(x,s)\,, \tag{2.4}$$

where $\bar{G}(s)$ and $\bar{J}(s)$ denote the Laplace transform of the *relaxation modulus* $G(t)$ and of the *creep compliance* $J(t)$, respectively. These functions are called the *material functions* for the given viscoelastic model; they are expressed in terms of appropriate discrete or continuous spectra of characteristic times, referred to as the *relaxation spectrum* for $G(t)$, the *retardation spectrum* for $J(t)$.

Applying the Laplace transform to equations (2.1) – (2.3) and using (2.4), the transform of the solution can be obtained (see *e.g.* Christensen, 1982), and reads

$$\bar{r}(x,s) = \bar{r}_0(s)\, e^{-\gamma(s)x}\,, \tag{2.5}$$

where

$$\gamma(s) = s\left[\rho\, s\bar{J}(s)\right]^{1/2} = s\left[\rho/s\bar{G}(s)\right]^{1/2}, \tag{2.6}$$

with $\gamma(s) \geq 0$ for $s \geq 0$.

The solution $r(x,t)$ is therefore given by the Bromwich formula (1.1), where $\gamma(s)$ turns out to be an analytic function of s over the entire s - plane cut along the negative real axis. The cut extends from $s = -a$ to $s = -b$ $(0 \leq a < b \leq \infty)$ where $1/a$ is the maximum retardation time and $1/b$ is the minimum relaxation time.

We will consider viscoelastic waves propagating with a finite wave-front velocity c; this requires that the given viscoelastic model exhibits an instantaneous elasticity, namely $0 < G(0^+) = 1/J(0^+) < \infty$. In fact, setting $G_0 = G(0^+)$ and $J_0 = J(0^+)$, it is easy to verify that the conditions (1.5), (1.6), that ensure a finite wave-front velocity, are satisfied if

$$c = \sqrt{G_0/\rho} = \sqrt{1/\rho J_0}\,, \tag{2.7}$$

and

$$n(s) = \left[s\bar{J}(s)/J_0\right]^{1/2} = \left[G_0/s\bar{G}(s)\right]^{1/2}. \tag{2.8}$$

A typical and simple viscoelastic model was introduced by Zener (1948) with the denomination of "Standar Linear Solid" (SLS). Denoting by $\Theta(t)$ the step Heaviside function, the material functions for the SLS are

$$G(t) = \left[G_\infty + (G_0 - G_\infty)\, e^{-t/\tau_\epsilon}\right]\Theta(t)\,, \tag{2.9}$$

$$J(t) = \left[J_0 + (J_\infty - J_0)\left(1 - e^{-t/\tau_\sigma}\right)\right]\Theta(t)\,, \tag{2.10}$$

with $\tau_\sigma > \tau_\epsilon$. The SLS is a three-parameter model being

$$G_0 J_0 = G_\infty J_\infty = 1\,, \tag{2.11}$$

$$G_0/G_\infty = J_\infty/J_0 = \tau_\sigma/\tau_\epsilon > 1\,. \tag{2.12}$$

Such a model exhibits a single relaxation time (τ_ϵ) and a single retardation time (τ_σ); for it the function $n(s)$ entering the Bromwich formula (1.7) turns out to be

$$n(s) = \sqrt{\frac{s+b}{s+a}}\,, \quad a = \frac{1}{\tau_\sigma}\,,\; b = \frac{1}{\tau_\epsilon}\,. \tag{2.13}$$

The SLS provides, as a limiting case for $\tau_\sigma \to \infty$, the simplest two-parameter model, known as Maxwell model. For it the material functions read

$$G(t) = G_0\, e^{-t/\tau_\epsilon}\, \Theta(t)\,, \tag{2.14}$$

$$J(t) = \frac{1}{G_0}\,\frac{t}{\tau_\epsilon}\,\Theta(t)\,, \tag{2.15}$$

and the function $n(s)$, that is obtained from (2.13) with $a = 0$, reads

$$n(s) = \sqrt{1 + \frac{1}{s\tau}}\,, \tag{2.16}$$

where τ is denoting the relaxation time.

After having examined the structure of the complex function $n(s)$ we devote our attention to the source function $\bar{r}_0(s)$. In this respect we recall the relations among the Laplace transforms of the response variables $\sigma\,,\,\epsilon\,,\,u\,,\,v$ for $x = 0$ (see *e.g.* Hunter, 1960):

$$\bar{\sigma}_0(s) = \frac{\rho s^2}{\gamma^2(s)}\,\bar{\epsilon}_0(s)\,, \quad \bar{\epsilon}_0(s) = -\gamma(s)\,\bar{u}_0(s)\,, \quad \bar{u}_0(s) = \frac{\bar{v}_0(s)}{s}\,. \tag{2.17}$$

As an example we consider the problem of evaluating the stress $\sigma(x,t)$ when the rod is subjected to impact with constant velocity, *i.e.* $v_0(t) = V\Theta(t)$. From (2.17) and (1.6) we get

$$\bar{\sigma}_0(s) = -\frac{\rho V}{\gamma(s)} = -\frac{\rho c V}{s\, n(s)} \,. \tag{2.18}$$

For this problem the solution in the Bromwich form (1.7) is therefore

$$\sigma(x,t) = \frac{\sigma^*}{2\pi i} \int_{Br} \frac{ds}{s\, n(s)}\, e^{s[t-n(s)x/c]} \,, \tag{2.19}$$

where, for convenience, we have introduced the constant $\sigma^* = -\rho c V$.

The above problem is interesting since for the Maxwell model a simple analytical solution is available in the space – time domain (Lee & Kanter, 1953). in fact, inserting (2.16) into (2.19), the Bromwich integral can now be carried out analytically to yield

$$\sigma(x,t) = \sigma^*\, e^{-t/2\tau}\, I_0\left[\frac{1}{2\tau}\sqrt{t^2 - x^2/c^2}\right] \Theta(t - x/c) \,, \tag{2.20}$$

where I_0 is the modified Bessel function of order 0.

3. SOLUTION OF THE PROBLEM

In order to provide a general (numerical) technique to obtain the solution in any space – time domain, we propose to carry out the Bromwich integral entering equation (1.7) according to the method of steepest descents, as anticipated in the introduction.

In this method we are concerned with the points in the complex $s-$ plane where the argument of the exponential in (1.7) has a zero derivative, *i.e.* the saddle points of the function

$$s\left[t - n(s)\frac{x}{c}\right] = t\, F_\mu(s) = \frac{x}{c}\, F_\theta(s) \,, \tag{3.1}$$

where

$$F_\mu(s) = s[1 - \mu n(s)] \,, \quad \mu = \frac{x}{ct} \,, \tag{3.2}$$

and

$$F_\theta(s) = s[\theta - n(s)] \,, \quad \theta = \frac{ct}{x} \,. \tag{3.3}$$

We adopt the representation with μ or θ according we analyse the response pulse at fixed times or at fixed distances; in the former case we have

$$0 \leq x \leq ct \Rightarrow 0 \leq \mu \leq 1, \tag{3.4}$$

in the latter case

$$t \geq x/c \Rightarrow \theta \geq 1. \tag{3.5}$$

According to the method of steepest descents (see Thau, 1974; Mainardi, 1983a) the original Bromwich path Br is to be substituted by a new path L, that depends on the chosen evolution parameter (say μ), with the following properties:

(i) L passes through the saddle point s_0 of $F_\mu(s) = \Phi(s) + i\,\Psi(s)$, *i.e.*

$$n(s) + s\frac{dn}{ds} = \frac{1}{\mu} \quad \text{at } s = s_0\,; \tag{3.6}$$

(ii) the imaginary part of $F_\mu(s)$ is everywhere constant on L, *i.e.*

$$\Psi(s) = \Psi(s_0)\,, \quad s \in L\,; \tag{3.7}$$

(iii) the real part of $F_\mu(s)$ attains an absolute maximum at s_0 along L, *i.e.*

$$\Phi(s) \leq \Phi(s_0)\,, \quad s \in L\,; \tag{3.8}$$

(iv) the integral on L is equivalent to the original one or differs from it by a finite residue contribution.

For large x and t we can argue that the dominant contribution to the integral along L comes from a neighborhood of s_0: this provides the leading term in the asymptotic expansion of the wave form, that can easily be found by the so called saddle-point method. For this method a detailed knowledge of the path L is usually not required, while the location of the relevant saddle point with respect to the singularities of $\bar{r}_0(s)$ is important to get a uniform approximation.

When the integration is carried out on the *entire* curve L, the exact result is expected to be found for any x and t, and without any trouble related to an oscillating integrand, since on L the imaginary part of F_μ is constant.

For viscoelastic waves the mathematical structure of $n(s)$ (see Mainardi, 1983b) allows us to state that the relevant saddle point s_0 is located on the real axis, moving from $+\infty$ (at the wave-front, $\mu = 1$) up to the largest branch point of $n(s)$ (for $\mu = 0$), and that the path L, of steepest descent through s_0, is a curve that encloses the branch cut of $n(s)$.

We point out, however, that the analytical determination of the path L and of its evolution appears a very difficult task even for simple models of viscoelasticity as the SLS (see Rubin, 1954; Mainardi, 1972; Chin, 1980).

From above one can argue that any numerical technique of integration on L can succesfully be used, provided that the path itself may be determined with a sufficient accuracy, starting from the relevant saddle point s_0. In this respect we find it convenient to use our specific routine that enables us to evaluate the path L in the complex $s-$ plane by solving numerically the equation (3.7). It is based on rational approximations (Padé Approximants of type II), according to a simple algorithm that provides an exponential convergence (Zinn-Justin, 1971).

We have checked our numerical scheme for the *Lee - Kanter problem*, that is illustrated in equations (2.18) - (2.20). Here we have used nondimensional variables

$$\sigma' = \sigma/\sigma^* , \quad x' = x/c\tau , \quad t' = t/\tau , \quad s' = s\tau , \tag{3.9}$$

so that the exact solution (2.20) reads

$$\sigma'(x',t') = e^{-t'/2}\, I_0\left[\frac{1}{2}\sqrt{t'^2 - x'^2}\right] \Theta(t' - x') , \tag{3.10}$$

and the complex function $n(s)$, given by (2.16), becomes

$$n(s') = \left(1 + 1/s'\right)^{1/2} . \tag{3.11}$$

From now on, we agree to drop the primes in the variables.

The equation (3.6) for the saddle points reads

$$\left(1 + 1/s\right)^{1/2} \left[1 - \frac{1}{2(1+s)}\right] = \frac{1}{\mu} , \tag{3.12}$$

that, when rationalized, is a quadratic equation in s with two real solutions. These solutions are the required saddle points; in the μ representation they read

$$s^{\pm} = -\frac{1}{2}\left[1 \mp \frac{1}{\sqrt{1-\mu^2}}\right] . \tag{3.13}$$

These saddle points move on the real axis from $\pm\infty$ (for $\mu = 1$) to the branch points of $n(s)$: $s_*{}^+ = 0$, $s_*{}^- = -1$ (for $\mu = 0$).

The path of steepest descent through $s^{\pm}$ is defined by the condition (3.7), that now reads

$$Im\{F_{\mu}(s)\} = Im\{s - \mu[s(s+1)]^{1/2}\} = 0\,. \tag{3.14}$$

Setting $s = \xi + i\eta$, this condition leads to the following algebraic equation

$$\frac{(\xi + 1/2)^2}{a^2} + \frac{\eta^2}{b^2} = 1\,, \tag{3.15}$$

with

$$a = a(\mu) = \frac{1}{2\sqrt{1-\mu^2}}\,, \qquad b = b(\mu) = \frac{\mu}{2\sqrt{1-\mu^2}}\,. \tag{3.16}$$

Thus the curve L represented by (3.15) – (3.16) is an ellipse, with axis $2a$ and $2b$, foci in the branch points $s_*{}^{\pm}$, that intersects the real axis in the saddle points $s^{\pm}$.

It is easy to prove that s^+ is the relevant saddle point s_0, being a maximum of $Re\{F_{\mu}(s)\}$ on L, and that the original Bromwich path Br can be deformed into the ellipse L. In FIG. 1 the curve L is shown in the complex $s-$ plane for different values of the parameter μ; from the figure we easily understand the evolution of the ellipse from a big circle at infinity (for $\mu = 1$) to the segment of the branch cut (for $\mu = 0$).

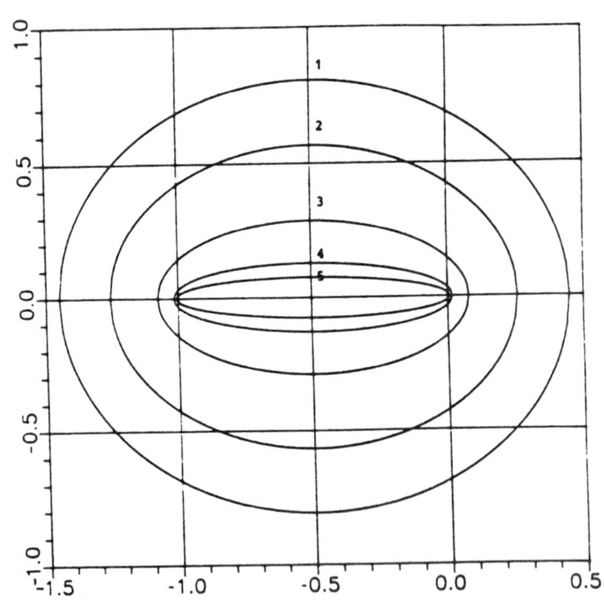

FIG. 1. *The paths of steepest descents in the complex s−plane for the following values of the parameter* μ : $\mu_1 = 0.85$, $\mu_2 = 0.75$, $\mu_3 = 0.50$, $\mu_4 = 0.25$, $\mu_5 = 0.15$.

We have been able to determine such a curve with high numerical accuracy by using our routine; starting from the point s_0 on the real axis, the routine has provided, step by step, a number of points on L, adequate to represent the curve. This provides a check, that enables us to use the routine in other cases where the analytical determination of the path L turns out to be cumbersome or prohibitive.

Now we consider the problem of finding the solution for the response pulse and comparing it with the exact formula (3.10). The saddle-point method provides the following approximate representation (that is expected to be good for large values of time)

$$\sigma(\mu,t) \simeq \sqrt{\frac{2}{\pi\mu t}}\,[s_0(s_0+1)]^{1/4}\, e^{t\{s_0-\mu[s_0(s_0+1)]^{1/2}\}}\,. \tag{3.16}$$

Such an approximation can be proved to be regular for $\mu \to 0$; for $\mu = 0$ (*i.e.* at $x = 0$) it yelds

$$\sigma(0,t) \simeq \frac{1}{\sqrt{\pi t}} \leq e^{-t/2}\, I_0\left(\frac{t}{2}\right); \tag{3.17}$$

where in the R.H.S. the exact value is quoted from (3.10).

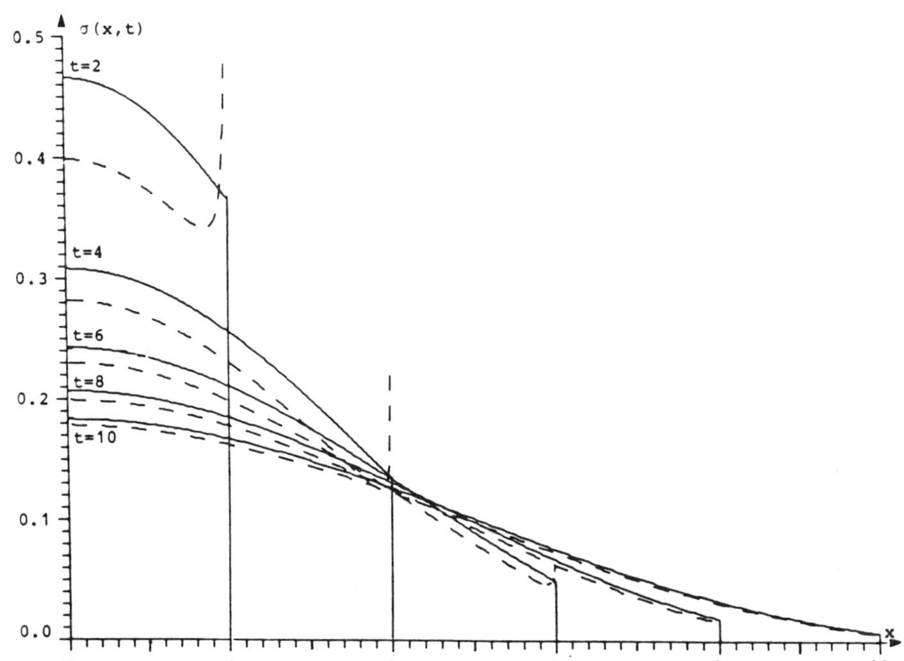

FIG. 2. *The pulse response for the Lee - Kanter problem depicted versus x at the following values of time : t = 2, 4, 6, 8, 10; the continuous lines show the exact solution (3.10), while the dashed lines show the saddle-point approximation (3.16).*

In FIG. 2 we compare the exact solution given by (3.10) with the approximate solution given by (3.16) for some fixed values of time. We recognize that the saddle-point method provides a satisfactory approximation (*i.e.* a discrepancy less than one per cent) for any x only if $t \geq 8$.

For smaller times the contribution to the integral from a neighborhood of the relevant saddle point s_0 is thus inadequate to represent the solution; as a consequence the integration along the entire path of steepest descent is requested. We have verified that the numerical integration along the path $L(\mu)$ has provided, for any μ ($0 < \mu < 1$) and t ($2 \leq t \leq 10$), an accurate representation of the exact solution (3.10).

REFERENCES

- J. Achenbach and D.P. Reddy, *Z. Angew. Math. Phys. 18*: 141 (1967).
- N. Bleistein and R.A. Handelsman, *Asymptotic Expansions of Integrals*, Holt, Rinehart & Wintson, New York (1975), reprinted by Dover, N.Y. (1986).
- N. Bleistein, *Mathematical Methods for Wave Phenomena*, Academic Press, New York (1984).
- L. Brillouin, *Wave Propagation and Group Velocity*, Academic Press, New York (1960).
- L. Brun, *J. Méc. 13*: 449 (1975).
- P.W. Buchen, *Pure and Appl. Geophys. 112*: 1011 (1974).
- P.W. Buchen and F. Mainardi, *J. Méc. 14*: 597 (1975).
- M. Caputo, *Elasticità e Dissipazione*, Zanichelli, Bologna (1969).
- R.C.Y. Chin, *Wave Propagation in Viscoelastic Media* in *Physics of the Earth's Interior* (A.M. Dziewonski & E. Boschi eds): 213-246, International School of Physics *E. Fermi*, North-Holland, Amsterdam (1980).
- R. M. Christensen, *Theory of Viscoelasticity*, 2-nd ed., Academic Press, New York (1982).
- B.T. Chu, *J. Méc. 1*: 439 (1962).
- M. Fabrizio, *Istituto Lombardo (Rend. Sc.) A102*: 437 (1968).
- S.C. Hunter, *Viscoelastic Waves* in *Progress in Solid Mechanics 1* (I.N. Sneddon & R. Hill eds): 3-60, North-Holland, Amsterdam (1960).
- Jeffreys, Sir H., *Mon. Not. R. Atr. Soc., Geophys. Suppl., 2*: 318 (1931).

- Jeffreys, Sir H., *Proc. R. Soc. London A138*: 283 (1932).
- E.H. Lee and I. Kanter, *J. Appl. Phys. 24*: 1115 (1953).
- I.M. Longman, *J. Comp. Phys. 10*: 224 (1972).
- F. Mainardi, *Pure and Appl. Geophys. 99*: 72 (1972).
- F. Mainardi and G. Turchetti, *Mech. Res. Comm. 2*: 107 (1975).
- F. Mainardi, *Wave Motion 5*: 33 (1983a).
- F. Mainardi, *Nuovo Cimento 74 B*: 52 (1983b).
- F. Mainardi, *On Linear Dispersive Waves with Dissipation* in *Wave Phenomena* (C. Rogers & T.B. Moodie eds): 307-317, North-Holland, Amsterdam (1984).
- F.W.J. Olver, *Asymptotic and Special Functions*, Academic Press, New York (1974).
- M.J. Pound, *Wave Motion 7*: 271 (1985).
- R.J. Rubin, *J. Appl. Phys. 25*: 528 (1954).
- S.A. Thau, *Linear Dispersive Waves* in *Non Linear Waves* (S. Leibovich & A.R. Seebass eds): 44-81, Cornell Univ. Press, Ithaca (1974).
- G.B. Whitham, *Linear and Nonlinear Waves*, Wiley, New York (1974).
- C.M. Zener, *Elasticity and Anelasticity of Metals*, Chicago Univ. Press, Chicago (1948).
- J. Zinn-Justin, *Physics Report 1C*: 55 (1971).

The Asymptotics of Pearcey's Integral for Complex Variables

R. B. Paris Professor, Department of Mathematical and Computer Sciences, Dundee Institute of Technology, Dundee, Scotland

1. INTRODUCTION

The canonical form of the oscillatory integral describing the cusp diffraction catastrophe defined by

$$P(X,Y) = \int_{-\infty}^{\infty} \exp i(u^4 + Xu^2 + Yu)\,du \tag{1}$$

is fundamental to the investigation of many short wavelength scattering problems. In most physical applications the variables X and Y are real though here we shall consider them to be arbitrary complex variables. The integral (1) was first evaluated numerically by Pearcey (1946) in his investigation of the electromagnetic field near a cusp and is generally known as Pearcey's integral, although it appears to have been first studied in detail by Brillouin (1916).

Previous work dealing with the asymptotics of this integral for large real values of the variables X and Y has depended upon the detailed structure of the stationary or saddle points of the integrand. If we denote the phase function in (1) by $\phi(u;X,Y) = u^4 + Xu^2 + Yu$ and consider real values of X and Y, then $\phi_u = 0$ gives three real distinct stationary points for (X, Y) situated inside (i.e. to the left) of the cusp $Y^2 + (2X/3)^3 = 0$ in the X, Y-plane, while on the cusp itself (away from the origin) two of these points coalesce to form a double stationary point. For (X,Y) situated outside the cusp only one stationary point is real with the other two forming a complex conjugate pair. By examining the topography of the paths of steepest descent for real X and Y, Wright (1980) has shown that for (X, Y) lying inside the cusp all three stationary points contribute to the asymptotics of $P(X,Y)$, while outside only two stationary points make a contribution. In addition to the above cusp corresponding to the appearance of a double stationary point, Wright has shown that in the interior of the second cusp (the Stokes set) centred about the positive X-axis, given by $Y^2 - (X/3)^3(5 + \sqrt{27}) = 0$, only the real stationary point contributes to $P(X,Y)$.

The situation when the variables X and Y are allowed to take on complex values will be even more complicated and discussion of the topography of the corresponding paths of steepest descent is likely to be difficult. By means of a new integral representation involving a parabolic cylinder function, we show in this paper how the asymptotics of $P(X,Y)$ may be obtained *without* reference to these stationary points for complex X and Y when one of the variables becomes large. The analysis of the case $|X| \to \infty$, Y finite can be carried through to yield the full asymptotic expansion of $P(X,Y)$. The case $|Y| \to \infty$, X finite is considerably more delicate due to the presence of exponentially small terms in certain sectors of the Y-plane, and accordingly only the leading terms in the expansion are given. In this paper we state the results and present only an outline of their derivation; full details, including the proof of the asymptotic nature of the expansions in the case $|X| \to \infty$, will be published elsewhere (Paris, 1989).

The behaviour of $P(X,Y)$ when X and Y both become large is considerably more difficult as it must take into account the behaviour in the neighbourhood of the cusps. This case is not discussed here, except to briefly mention how, when X and Y are real, the new integral representation may be used to obtain the leading terms in the expansion of $P(X,Y)$ as $-X$ and $Y \to +\infty$ on the cusp $Y^2 + (2X/3)^3 = 0$.

To derive the new integral representation we first rotate the path of integration in (1) through an angle $\pi/8$ and define the new variables $t = u\,e^{-\pi i/8}$ and

$$x = Xe^{-\pi i/4}, \qquad y = Ye^{\pi i/8}. \tag{2}$$

The asymptotics of $P(X,Y)$ will appear in a more symmetrical form if expressed in terms of the variables x and y rather than X and Y. Application of Jordan's lemma shows that the contribution from the arcs $\pm\infty$ to $\pm\infty\exp(\pi i/8)$ vanishes, so that (1) may be written alternatively in terms of x and y as

$$P(x,y) = 2e^{\pi i/8} \int_0^\infty \exp(-t^4 - xt^2)\cos yt\, dt. \tag{3}$$

It is readily seen that $P(x,y)$ satisfies the symmetry and conjugacy relations

$$P(x,y) = P(x,-y), \quad e^{-\pi i/8}P(x,y^*) = [e^{\pi i/8}P(x^*,y)]^*, \tag{4}$$

where the asterisk denotes the complex conjugate.

We now apply the result

$$\cos z = \frac{1}{2\pi i}\int_C \Gamma(s)\cos\tfrac{1}{2}\pi s\; z^{-s}\, ds, \tag{5}$$

where the path of integration C denotes the loop which starts from $-\infty$, encircles the origin in the positive sense and returns to $-\infty$ enclosing all the poles of the integrand. This representation is readily deduced from the corresponding integral for e^{-z} [given by (5) with the omission of the cosine term], and is valid without restriction on arg z. If we now temporarily restrict C to lie in the left-hand plane $Re(s)<1$, we obtain from (3)

$$P(x,y) = 2e^{\pi i/8}\int_C \Gamma(s)y^{-s}\cos\tfrac{1}{2}\pi s\,\{\int_0^\infty \exp(-t^4 - xt^2)t^{-s}dt\}ds,$$

where the reversal of the order of integration is justified provided $Re(s)<1$. With the substitution $\tau = t^2$, the inner integral may be evaluated in terms of a parabolic cylinder function making use of the relation

$$D_\nu(z) = \frac{e^{-z^2/4}}{\Gamma(-\nu)} \int_0^\infty \exp\left(-\frac{\tau^2}{2} - z\tau\right) \tau^{-\nu-1} d\tau, \quad Re(\nu) < 0.$$

We then find that

$$e^{-\pi i/8}P(x,y) = 2^{-\frac{1}{4}}e^{x^2/8} \frac{1}{2\pi i} \int_C \Gamma(s)\Gamma(\tfrac{1}{2}-\tfrac{1}{2}s)\, 2^{s/4}y^{-s}\cos\tfrac{1}{2}\pi s\, D_{\frac{1}{2}s-\frac{1}{2}}(x/\sqrt{2})\,ds$$

$$= 2^{-5/4}\pi^{\frac{1}{2}}e^{x^2/8} \frac{1}{2\pi i} \int_C \Gamma(\tfrac{1}{2}s)\, D_{\frac{1}{2}s-\frac{1}{2}}(x/\sqrt{2})\, (y^2/4\sqrt{2})^{-s/2}\, ds$$

upon using the duplication and reflection formulas for the gamma function. Replacing s by $2s$ in the last integral, we finally obtain the desired representation

$$P(x,y) = 2^{-\frac{1}{4}}\pi^{\frac{1}{2}}e^{x^2/8+\pi i/8} \frac{1}{2\pi i} \int_C \Gamma(s) D_{s-\frac{1}{2}}(x/\sqrt{2})\, (y^2/4\sqrt{2})^{-s} ds. \quad (6)$$

From the definition of the parabolic cylinder function in terms of confluent hypergeometric functions, it is seen that $D_{s-\frac{1}{2}}(x/\sqrt{2})$ is an analytic function of s, and the restriction imposed on $Re(s)$ in the derivation of (6) may therefore be removed by analytic continutation. The path of integration in (6) encircles in the positive sense the poles of $\Gamma(s)$ situated at $s = 0, -1, -2, \ldots$. From the asymptotic behaviour of $D_{s-\frac{1}{2}}(x/\sqrt{2})$ as $|s| \to \infty$ in $|arg(-s)|<\pi$ [see (18)], the convergence of the integral in (6) is seen to be controlled by the ratio of gamma functions $\Gamma(s)/\Gamma(\frac{3}{4} - \frac{1}{2}s)$. Application of Rule 2 given in Paris and Wood (1986, p.21) then shows that (6) converges for all finite values of x and y.

The above integral representation for $P(x,y)$ may appear at first sight to be more complicated than the original definition in (1) or (3) since, instead of a relatively simple exponential dependence, the integrand now involves more involved transcendental functions. However, although the x-dependence of the integrand is contained in the parabolic cylinder function whose order is the variable of integration, the y-dependence now appears as a simple algebraic power. This fact, combined with the flexibility associated with the loop integral in the complex plane, enables the asymptotics of $P(x,y)$ to be determined, when one of the variables becomes large, without recourse to knowledge of stationary points.

One immediate deduction which can be made from (6), combined with the symmetry property in (4), is that $P(x,y)$ is exponentially small as $y \to \infty$ in the sectors $|arg(\pm y)| < \pi/8$. This may be seen by first swinging back the contour C to coincide with the imaginary s-axis suitably indented at $s = 0$. The resulting Mellin-Barnes integral

$$P(x,y) = 2^{-\frac{1}{4}}\pi^{\frac{1}{2}}e^{x^2/8 + \pi i/8}\frac{1}{2\pi i}\int_{-\infty i}^{\infty i} \Gamma(s)D_{s-\frac{1}{2}}(x/\sqrt{2})(y^2/4\sqrt{2})^{-s}ds,$$

then defines $P(x,y)$ only in the sector $|arg y| < \pi/8$ [see Rule 1 in Paris and Wood (1986), p.21]. Since the only poles of the integrand are those of $\Gamma(s)$, application of the standard method for asymptotically estimating Mellin-Barnes integrals, by displacing the path of integration to the right parallel to itself, shows that $P(x,y)$ is exponentially small as $y \to \infty$ in $|arg y| < \pi/8$.

2. ASYMPTOTICS OF P(x,y) FOR $|x| \to \infty$

The asymptotics of $P(x,y)$ for $|x| \to \infty$, y finite will be determined by consideration of the integral in (6). This case is more straightforward than the case $|y| \to \infty$,x finite and it is possible to derive the full asymptotic expansion. The analysis we shall present here is formal as no attempt is made to discuss the asymptotic nature of the various expansions; full details will be given in Paris (1989).

We start with the result, valid in $|arg x| < \pi/2$, which expresses the parabolic cylinder function in the form of the first n+1 terms of its asymptotic expansion for large argument together with an exact remainder term

$$D_{s-\frac{1}{2}}(x/\sqrt{2}) = e^{-x^2/8}(x/\sqrt{2})^{s-\frac{1}{2}}\{1 - \frac{(s-\frac{1}{2})(s-\frac{3}{2})}{1!x^2} + \frac{(s-\frac{1}{2})(s-\frac{3}{2})\ (s-\frac{5}{2})(s-\frac{7}{2})}{2!x^4} -$$

$$\ldots + (-)^n \frac{(s-\frac{1}{2})(s-\frac{3}{2}) \ldots (s-2n+\frac{3}{2})(s-2n+\frac{1}{2})}{n!x^{2n}} + E_n(x,s) \} \quad (7)$$

where

$$E_n(x,s) = \frac{1}{\Gamma(\frac{1}{4} - \frac{1}{2}s)} \int_0^\infty t^{-\frac{s}{2}-\frac{3}{4}} R_n(t,x)e^{-t}dt$$

$$R_n(t,x) = \frac{\lambda(\lambda-1)\ldots(\lambda-n)}{n!} (\frac{4t}{x^2})^{n+1} (1 + \frac{4t}{x^2})^{\lambda} \int_0^1 u^n (1 + \frac{4ut}{x^2})^{-\lambda-1} du,$$

with $\lambda = \frac{1}{2}s-\frac{3}{4}$, *and* n is chosen sufficiently large so that $Re(n-\frac{1}{2}s-\frac{3}{4}) > 0$. This representation is valid for arbitary complex s and x provided $1+4t/x^2$ is not a negative real number, that is provided $|\arg x|<\pi/2$. This result may be obtained from the corresponding expression for the Whittaker function given in Whittaker and Watson (1965, p.342) and the fact that $D_{s-\frac{1}{2}}(x/\sqrt{2}) = 2^{\frac{1}{2}s+\frac{1}{4}}x^{-\frac{1}{2}}W_{\frac{1}{2}s,-\frac{1}{4}}(x^2/4)$.

Insertion of (7) into (6) then yields

$$P(x,y) = \sqrt{\frac{\pi}{x}}\, e^{\pi i/8} \frac{1}{2\pi i} \int_C \Gamma(s)(y^2/4x)^{-s} \{ 1 - \frac{(s-\frac{1}{2})(s-\frac{3}{2})}{1!x^2} + \ldots$$

$$+ (-)^n \frac{(s-\frac{1}{2})\ldots(s-2n+\frac{1}{2})}{n!x^{2n}} + E_n(x,s) \} \, ds.$$

Employing the result analogous to that in (5) for e^{-z} and integrating term-by-term, we then find that

$$P(x,y) = \sqrt{\frac{\pi}{x}}\, e^{\pi i/8} \{e^{-y^2/4x} [1 - \frac{a_1(\chi)}{1!x^2} + \frac{a_2(\chi)}{2!x^4} - \ldots + \frac{(-)^n a_n(\chi)}{n!x^{2n}}]$$

$$+ \frac{1}{2\pi i} \int_C \Gamma(s)E_n(x,s)(y^2/4x)^{-s} ds \}, \quad (8)$$

where the coefficients $a_n(\chi)$, with $\chi = y^2/4x$, are given by

$$a_n(\chi) = e^\chi \sum_{m=0}^\infty (m+\tfrac{1}{2})(m+\tfrac{3}{2})\ldots(m+2n-\tfrac{3}{2})(m+2n-\tfrac{1}{2})\frac{(-\chi)^m}{m!}$$

$$= e^\chi \sum_{m=0}^\infty \frac{(-\chi)^m\Gamma(m+2n+\frac{1}{2})}{m!\ \Gamma(m+\frac{1}{2})}$$

$$= \frac{\Gamma(2n+\frac{1}{2})}{\sqrt{\pi}} e^\chi {}_1F_1(2n+\tfrac{1}{2};\tfrac{1}{2};-\chi) = \frac{\Gamma(2n+\frac{1}{2})}{\sqrt{\pi}} {}_1F_1(-2n;\tfrac{1}{2};\chi) \quad (9)$$

upon use of Kummer's theorem for the confluent hypergeometric function. The coefficients $a_n(\chi)$ are seen to be polynomials in χ

of degree $2n$. Alternatively, $a_n(\chi)$ may be expressed in terms of a $2n$th derivative of $e^{-\chi}$ as

$$a_n(\chi) = \chi^{\frac{1}{2}}e^{\chi}\frac{d^{2n}}{d\chi^{2n}}(\chi^{2n-\frac{1}{2}}e^{-\chi}), \qquad n = 0,1,2,\ldots$$

so that

$$a_0(\chi) = 1, \qquad a_1(\chi) = \chi^2 - 3\chi + \tfrac{3}{4},$$

$$a_2(\chi) = \frac{105}{16} - \frac{105}{2}\chi + \frac{105}{2}\chi^2 - 14\chi^3 + \chi^4, \ldots .$$

Discussion of the remainder term in (8) shows that this term is $O(|x|^{-2n-2})$ as $|x| \to \infty$ in $|\arg x| < \frac{1}{2}\pi$. Hence, we have the asymptotic expansion of $P(x,y)$ for $|x| \to \infty$ in the sector $|\arg x| < \frac{1}{2}\pi$

$$P(x,y) \sim \sqrt{\frac{\pi}{x}}\, e^{\pi i/8}\, S_1(x,y), \tag{10}$$

where $S_1(x,y)$ denotes the formal asymptotic sum

$$S_1(x,y) = e^{-y^2/4x}\sum_{n=0}^{\infty}\frac{(-)^n a_n(\chi)}{n!\,x^{2n}}, \qquad \chi = y^2/4x. \tag{11}$$

To obtain the expansion in the remainder of the complex x-plane, we employ the relations

$$D_\nu(-z) = e^{\pm\pi i\nu}D_\nu(z) \pm \frac{i\sqrt{2\pi}}{\Gamma(-\nu)}\, e^{\pm\frac{1}{2}\pi i\nu}D_{-\nu-1}(\pm iz) \tag{12}$$

and write $x = \bar{x}e^{\pi i}$. In order to employ the expansion in (7), the signs above must be chosen in such a way that both arguments of the parabolic cylinder functions lie in the right-hand half plane. This means that we select the upper sign in (12) when $\pi/2 < \arg x < \pi$ and the lower sign when $\pi < \arg x < 3\pi/2$. Substitution of (12) in (6) then yields

$$2^{\frac{1}{4}}\pi^{-\frac{1}{2}}e^{-x^2/8-\pi i/8}\, P(x,y) = I_1 + I_2 \tag{13}$$

where, according to the range of $\arg x$,

$$I_1 = \mp\frac{1}{2\pi}\int_C \Gamma(s)D_{s-\frac{1}{2}}(\bar{x}/\sqrt{2})(e^{\mp\pi i}y^2/4\sqrt{2})^{-s}ds,$$

$$I_2 = \pm(2\pi)^{-\frac{1}{2}}\int_C\frac{\Gamma(s)}{\Gamma(\frac{1}{2}-s)}\, e^{\pm\frac{1}{2}\pi i(s-\frac{1}{2})}D_{-s-\frac{1}{2}}(\pm i\bar{x}/\sqrt{2})(y^2/4\sqrt{2})^{-s}ds.$$

The first integral I_1 is of the same form as that in (6), so that application of (7) in $|arg\bar{x}| < \pi/2$ yields the result

$$I_1 \sim \pm \sqrt{(\pi/x)}\, e^{\pi i/8}\, S_1(x,y),$$

where the upper and lower signs are chosen according as $arg(-x)$ is negative or positive respectively.

The second integral I_2 becomes

$$I_2 = -\frac{2^{\frac{1}{4}} i e^{x^2/8}}{\sqrt{2\pi\bar{x}}} \int_C \frac{\Gamma(s)}{\Gamma(\frac{1}{2}-s)} (\bar{x}y^2/8)^{-s} \Big[1 + \frac{(s+\frac{1}{2})(s+\frac{3}{2})}{1!\bar{x}^2} + \dots + \frac{(s+\frac{1}{2})(s+\frac{3}{2})\dots(s+2n-\frac{3}{2})(s+2n-\frac{1}{2})}{n!\bar{x}^{2n}} + E_n(\pm i\bar{x},s)\Big]ds$$

$$= -\frac{2^{\frac{1}{4}} i e^{x^2/8}}{\sqrt{2\pi\bar{x}}} \int_C \Gamma(s)(\bar{x}y^2/8)^{-s} \Big[\frac{1}{\Gamma(\frac{1}{2}-s)} + \frac{1}{1!\bar{x}^2\Gamma(-\frac{3}{2}-s)} + \dots + \frac{1}{n!\bar{x}^{2n}\Gamma(-2n+\frac{1}{2}-s)} + \frac{E_n(\pm i\bar{x},s)}{\Gamma(\frac{1}{2}-s)}\Big]ds.$$

We now integrate term-by-term making use of the integral for the Bessel function, valid without restriction on z and ν [cf. Watson, 1944 p.191],

$$(\tfrac{1}{2}z)^{-\nu} J_\nu(z) = \frac{1}{2\pi i} \int_C \frac{\Gamma(s)}{\Gamma(1+\nu-s)} (\tfrac{1}{2}z)^{-2s} ds$$

where C is the loop encircling the origin which starts from and returns to $-\infty$ as in (6). Then we find

$$I_2 = 2^{\frac{1}{4}} e^{x^2/8} \sqrt{\frac{2\pi}{\bar{x}}} \Big\{ \sum_{m=0}^{n} \frac{(\xi/2)^{2m+\frac{1}{2}}}{m!\bar{x}^{2m}} J_{-2m-\frac{1}{2}}(\xi) + \frac{1}{2\pi i} \int_C \frac{\Gamma(s)}{\Gamma(\frac{1}{2}-s)} (\bar{x}y^2/8)^{-s} E_n(\pm ix,s)\, ds \Big\} \tag{14}$$

with $\xi = y(\bar{x}/2)^{\frac{1}{2}}$. From the properties of the spherical Bessel functions (Abramowitz and Stegun, 1965 p.437)

$$J_{-2m-\frac{1}{2}}(\xi) = (-)^m \sqrt{(2/\pi\xi)}\, \{P(2m,\xi)\cos\xi - Q(2m,\xi)\sin\xi\},$$

where (after a slight modification in the notation) the finite sums $P(2m,\xi)$ and $Q(2m,\xi)$ in inverse powers of ξ^2 are given by

$$P(2m,\xi) = \sum_{k=0}^{m} \frac{(-)^k (2m+2k)!(2\xi)^{-2k}}{(2k)!\Gamma(2m-2k+1)}$$

$$Q(2m,\xi) = \sum_{k=0}^{m-1} \frac{(-)^k (2m+2k+1)!(2\xi)^{-2k-1}}{(2k+1)!\Gamma(2m-2k)}$$

If we now define the formal asymptotic sum

$$S_2(x,y) = \sum_{m=0}^{\infty} \frac{(y^2/8x)^m}{m!} \{P(2m,\xi)\cos\xi - Q(2m,\xi)\sin\xi\}, \quad \xi = y(-x/2)^{\frac{1}{2}}$$

then analysis of the remainder term in (14) shows that as $x \to \infty$ in $|arg(-x)| < \pi/2$ the integral I_2 has the asymptotic expansion

$$I_2 \sim 2^{\frac{3}{4}} i x^{-\frac{1}{2}} e^{x^2/8} S_2(x,y). \tag{15}$$

It then follows that, for $|x| \to \infty$ in $|arg(-x)| < \pi/2$,

$$P(x,y) \sim \sqrt{(\pi/x)}e^{\pi i/8}\{\pm S_1(x,y) + i\sqrt{2}e^{x^2/4}S_2(x,y)\}$$

where the upper and lower signs are taken according as $arg(-x)$ is negative or positive respectively.

The form of the expansions $S_1(x,y)$ and $S_2(x,y)$ shows that the variable y only appears in the groups $y^2/4x$ and $xy^2/2$ in the right and left-hand half planes respectively. This fact suggests the possibility of transforming the integrals in (6) and (13) so as to make this specific dependence apparent. One way of achieving this is to use the Mellin-Barnes representation of $D_\nu(z)$ in the form

$$D_\nu(z) = \frac{z^\nu e^{-z^2/4}}{\Gamma(-\nu)} \frac{1}{2\pi i} \int_{-\infty i}^{\infty i} \Gamma(s)\Gamma(-2s-\nu)(2z^2)^s ds, \quad |\arg z| < \tfrac{3}{4}\pi$$

where, provided ν is not equal to a nonnegative integer, the path of integration is suitably indented to separate the poles of $\Gamma(s)$ and $\Gamma(-2s-\nu)$. Substitution of this representation into (6), followed by reversal of the order of integration, then shows that, in the sector $|\arg x| < 3\pi/4$,

$$P(x,y) = \sqrt{\frac{\pi}{x}}\, e^{\pi i/8} e^{-y^2/4x} \frac{1}{2\pi i} \int_{-\infty i}^{\infty i} \Gamma(s)\frac{\Gamma(\frac{1}{2}-2s)}{\Gamma(\frac{1}{2})}\, {}_1F_1(2s;\tfrac{1}{2};y^2/4x)x^{-2s} ds,$$

where ${}_1F_1$ denotes the confluent hypergeometric function. This Mellin-Barnes representation of $P(x,y)$ exhibits the desired dependence on the group of variables $y^2/4x$. Application of the standard method for asymptotically evaluating Mellin-Barnes integrals then yields the expansion given in (10), but now valid (in the sense of Poincaré) in the wider sector $|\arg x| < 3\pi/4$. By means of similar reasoning, the second integral I_2 in (13) may be shown to possess the Mellin-Barnes representation in terms of the Bessel function

$$I_2 = 2^{\frac{1}{4}}\pi^{\frac{1}{2}}e^{x^2/8}\ \frac{1}{2\pi i}\int_{-\infty i}^{\infty i} e^{\pm\pi i s}\frac{\Gamma(s)}{\cos 2\pi s}\,(\xi/2x)^{-2s+\frac{1}{2}}\,J_{2s-\frac{1}{2}}(\xi)\,ds,$$

$$\xi = y(\bar{x}/2)^{\frac{1}{2}} \qquad |\arg(\pm i\bar{x})| < \tfrac{3}{4}\pi$$

from which the expansion (15) follows. In Paris (1989) full details of the derivation of these results together with the proof of the asymptotic nature of the expansions $S_1(x,y)$ and $S_2(x,y)$ will be based on analysis of the above Mellin-Barnes integrals, rather than on the forms given in (6) and (13).

To conclude this section, the asymptotic expansion of $P(x,y)$ for $|x| \to \infty$, y finite is given by

$$\begin{aligned} P(x,y) &\sim \sqrt{\frac{\pi}{x}}\ e^{\pi i/8}\,S_1(x,y) && |\arg x| < \tfrac{3}{4}\pi \\ &\sim \sqrt{\frac{\pi}{x}}\ e^{\pi i/8}\{\pm S_1(x,y) + i\sqrt{2}e^{x^2/4}\,S_2(x,y)\}, && |\arg(-x)| < \tfrac{1}{2}\pi \end{aligned} \tag{16}$$

where the upper and lower signs are chosen according as $\pi/2 < \arg x < \pi$ or $-\pi < \arg x < -\pi/2$ respectively. Thus $P(x,y)$ is exponentially large in the sector $|\arg(-x)| < \pi/4$, possessing a dominant algebraic behaviour in the remainder of the x-plane. The choice of signs in the subdominant algebraic expansion $S_1(x,y)$ on $\arg x = \pi$ corresponds to the Stokes phenomenon. Similarly, on $\arg x = \pm\pi/2$, the exponential expansion is of maximum degree of subdominancy and its coefficient changes to become zero in $Re(x) > 0$. The rays $\arg x = \pi$, $\pm\pi/2$ are consequently Stokes lines for $P(x,y)$, with the rays $\arg x = \pm 3\pi/4$

being anti-Stokes lines upon which the exponential and algebraic expansions are equally significant.

In terms of the original variables X and Y in (2), we see that negative real X corresponds to the upper boundary $\arg x = 3\pi/4$ of the exponentially large sector. Thus, for $X \to -\infty$, $P(X,Y)$ will consist of an algebraic and an oscillatory expansion. From (16) with the upper sign, the leading behaviour of $P(X,Y)$ for large real X (but finite complex Y) is then given by

$$P(X,Y) \sim \sqrt{\frac{\pi}{X}}\, e^{\pi i/4} \qquad X \to +\infty$$

$$\sim \sqrt{\frac{\pi}{-X}}\, e^{-\pi i/4}\, [1 + i\sqrt{2}e^{-iX^2/4} \cos(Y\sqrt{-X/2})]. \qquad X \to -\infty \tag{17}$$

This leading behaviour for real X and Y has been given previously by Connor and Farrelly (1981) when $Y = 0$ and Berry et al. (1979) using the method of stationary phase when Y is considered only as a small perturbation. Note that the exponential factor $\exp(-y^2/4x)$ appearing in $S_1(x,y)$ has to be put equal to unity in the leading behaviour. Finally, we remark that the above leading behaviour for $X \to \pm\infty$ can be seen to correspond respectively to the single and three saddle point contributions discussed in Section 1.

3. ASYMPTOTICS OF $P(x,y)$ FOR $|y| \to \infty$

From the symmetry property in (4) it is sufficient to confine our attention to $|\arg y| \leqslant \pi/2$. The asymptotics of $P(x,y)$ for $|y| \to \infty$, x finite can be found by substituting the identity

$$D_\nu(z) = \frac{\Gamma(1+\nu)}{\sqrt{2\pi}} \{ e^{\pi\nu i/2} D_{-\nu-1}(iz) + e^{-\pi\nu i/2} D_{-\nu-1}(-iz)\}$$

into (6), to find

$$P(x,y) = e^{x^2/8+\pi i/8} \{ e^{-\pi i/4} J_+ + e^{\pi i/4} J_- \}$$

where

$$J_\pm = \frac{1}{2\pi i} \int_C 2^{-\frac{3}{4}}\, \Gamma(s)\,\Gamma(\tfrac{1}{2}+s)\, D_{-s-\frac{1}{2}}(\pm ix/\sqrt{2})\, (\mp iy^2/4\sqrt{2})^{-s} ds.$$

The loop C around the origin may now be deformed such that $|s|$ is large everywhere on C. For finite x we may then employ the asymptotic behaviour of the parabolic cylinder function for $|s| \to \infty$

$$D_{-s-\frac{1}{2}}(\pm ix/\sqrt{2}) \sim \frac{\pi^{\frac{1}{2}}\exp(\mp ix\sqrt{s/2})}{2^{\frac{s}{2}+\frac{1}{4}}\;\Gamma(\frac{s}{2}+\frac{3}{4})} \tag{18}$$

uniformly in $|\arg s| \leqslant \pi - \varepsilon, \varepsilon > 0$. Using Stirling's formula for the gamma function, we can show that for large $|s|$ on C the integrals $J_\pm$ to leading order reduce to the form

$$J_\pm \sim 2^{1/6}\sqrt{\frac{\pi}{3}}\,y^{-1/3}\,e^{\pm\pi i/12}\,\frac{1}{2\pi i}\int_{C'}\Gamma(t)\,Z_\pm^{-t}\exp(\mp ix\sqrt{\tfrac{1}{3}(t+\tfrac{1}{4})}\,)\,dt,$$

$$Z_\pm = \frac{3y^{4/3}}{4^{4/3}}\,e^{\mp\pi i/3},$$

where t is the new integration variable given by $t = \frac{3}{4}s-\frac{1}{4}$ and C' denotes the corresponding loop in the t-plane.

For large $|y|$, the integrals $J_\pm$ may be estimated by the saddle point method to find

$$J_\pm \sim \exp(-Z_\pm \mp ix\sqrt{Z_\pm/3})\,e^{x^2/24}, \qquad |\arg y| \leqslant \pi/2.$$

It then follows that, for $|y| \to \infty$ in $|\arg y| \leqslant \pi/2$,

$$P(x,y) \sim 2^{1/6}\sqrt{\frac{\pi}{3}}\,y^{-1/3}\,e^{x^2/6}\,[\exp(-\frac{3y^{4/3}}{4^{4/3}}\,e^{-\pi i/3} - \frac{ixy^{2/3}}{4^{2/3}}\,e^{-\pi i/6} - \frac{\pi i}{6})$$

$$+ \exp(-\frac{3y^{4/3}}{4^{4/3}}\,e^{\pi i/3} + \frac{ixy^{2/3}}{4^{2/3}}\,e^{\pi i/6} + \frac{\pi i}{6})\,]\exp(\pi i/8). \tag{19}$$

The above leading asymptotic behaviour can be seen to satisfy the conjugacy property in (4). Both exponential terms inside the square brackets in (19) vanish as $|y| \to \infty$ in $|\arg y| < \pi/8$, so that $P(x,y)$ is consequently exponentially small in the sectors $|\arg(\pm y)| < \pi/8$.

In terms of the original real variables X and Y we have the oscillatory behaviour

$$P(X,Y) \sim 2^{1/6} \sqrt{\frac{\pi}{3}}\, Y^{-1/3}\, e^{-iX^2/6} \left[\exp i\left(\frac{\pi}{4} - \frac{3Y^{4/3}}{4^{4/3}} + \frac{XY^{2/3}}{4^{2/3}} \right) \right.$$

$$\left. + \exp\left(-\frac{\pi i}{12} - \frac{3Y^{4/3}}{4^{4/3}} e^{-\pi i/6} - \frac{XY^{2/3}}{4^{2/3}} e^{\pi i/6} \right) \right], \quad Y \to +\infty. \tag{20}$$

The second exponential term is subdominant for large positive values of Y. The behaviour as $Y \to \pm\infty$ can be seen to correspond to the two saddle point contributions discussed in Section 1.

Note the presence of the crucial factor $exp(x^2/6)$ in (19). Brillouin (1916, p.54) gave the first terms of an expansion for this case by the method of steepest descents. However, he failed to recognise that his expansion is not asymptotic since his series contains terms corresponding to the series expansion of the factor $exp(x^2/6)$. Thus, Brillouin did not detect this important factor and consequently his 'expansion' would be incorrect even for the leading approximation to $P(x,y)$.

4. CONCLUDING REMARKS

A new integral representation for Pearcey's integral $P(x,y)$, defined in (1), has been given in terms of a loop integral involving a parabolic cylinder function. From this integral it is possible to determine the full asymptotic expansion of $P(x,y)$ for the case $x \to \infty$, y finite and the leading behaviour when $y \to \infty$, x finite. In contrast to previous work dealing with the asymptotics of $P(x,y)$ for real x and y this has been achieved for complex values of the variables without knowledge of the stationary points of (1). The relevant results are summarised in (16) and (19).

For real values of X and Y on the cusp $Y^2 + (2X/3)^3 = 0$, it is possible to employ the representation (6) to determine the asymptotic behaviour of $P(X,Y)$ on the cusp. In this case, as both $-X$ and $Y \to +\infty$ we require the uniformly valid expansion of $D_{s-\frac{1}{2}}(Xe^{-\frac{1}{4}\pi i}/\sqrt{2})$ for large order and argument. This then reduces the problem to an examination of an integral involving exponential functions, as in (1), with a simple and double saddle points. It is found that on the cusp

$$P(X,Y) \sim \frac{\pi}{\sqrt{3|X|}} e^{-2iX^2/3 + \pi i/4} + \frac{2^{-\frac{1}{2}}}{3^{1/3}|X|^{1/6}} \left\{ \Gamma(\tfrac{1}{3}) - i\frac{\Gamma(\frac{2}{3})}{2.3^{1/3}|X|^{2/3}} + \dots \right\} e^{iX^2/12} \qquad |X| \to +\infty \tag{21}$$

in complete agreement (when due account is taken of the slightly different notation) with recent results obtained by Kaminski (1989) using the saddle point method applied to (1).

REFERENCES

M. Abramowitz and I. Stegun (ed.) (1965) *Handbook of Mathematical Functions*. Dover (New York).

M.V. Berry, J.F. Nye and F.J. Wright (1979) "The elliptic umbilic diffraction catastrophe". Phil. Trans. Roy. Soc. Lond. 291, pp.453-484.

L. Brillouin (1916) "Sur une méthode de calcul approchée de certaines intégrales, dite méthode de col". Ann. Sci. Ecole Norm. Sup. (Paris), 3rd series 33, pp.17-69.

J.N.L. Connor and D. Farrelly (1981) "Theory of cusped rainbows in elastic scattering: uniform semiclassical calculations using Pearcey's integral". J. Chem. Phys. 75, pp.2831-2846.

D. Kaminski (1989) "Asymptotic expansion of the Pearcey integral near the caustic". SIAM J. Math. Anal. 20, pp. 987-1005.

R.B. Paris (1989) in preparation.

R.B. Paris and A.D. Wood (1986) *Asymptotics of High Order Differential Equations*. Pitman Research Notes in Mathematics Series, Vol. 129, Longman.

T. Pearcey (1946) "The structure of an electromagnetic field in the neighbourhood of a cusp of a caustic". Phil. Mag. 37, pp.311-317.

G.N. Watson (1944) *Theory of Bessel Functions*. Cambridge University Press.

E.T. Whittaker and G.N. Watson (1965) *Modern Analysis*. Cambridge University Press.

F.J. Wright (1980) "The Stokes set of the cusp diffraction catastrophe". J. Phys. A: Math. Gen. 13, pp.2913-2928.

Asymptotic Methods in Magnetoconvection

N. Rudraiah
O. P. Chandna
R. M. Barron

Department of Mathematics and Statistics, and Fluid Dynamics Research Institute, University of Windsor, Windsor, Ontario, Canada

ABSTRACT

The effects of Lorentz force and non-uniform temperature gradient on the onset of magnetoconvection in an electrically conducting horizontal Boussinesq fluid layer permeated by a uniform transverse magnetic field are studied analytically using linear stability analysis by specifying constant heat flux at the boundaries. It is shown that when the Chandrasekhar number $Q \to \infty$ the correct asymptotic value of the critical Rayleigh number, R_c, can be obtained from the non-viscous MHD equations using a single-term Galerkin expansion. The criterion for the onset of magnetoconvection is determined using a regular perturbation technique with wave number as perturbation parameter. The method of matched asymptotics is used to predict explicitly the effect of the Hartman boundary layer (that exists at the rigid boundary for large values of Q) on the onset of magnetoconvection. It is shown that the effect of the Hartman boundary layer is to increase the asymptotic value of R_c by an amount proportional to the value of the Hartman number M. We find that the ratios R_{ci}/C_{c1}, where R_{ci} (i=1 to 6) are the asymptotic values of R_c for different non-linear temperature profiles, are independent of Q but depend on the thermal depth ε. It is also shown that the power law for asymptotic values of R_{ci} depends crucially on the nature of heating and not on the nature of the boundaries.

1. INTRODUCTION

Linear and nonlinear natural convection in an electrically conducting fluid in the presence of a magnetic field (known as magnetoconvection) caused by buoyancy force ([1] - [5]) or by surface tension variation with temperature ([6], [7]) has been of considerable interest during the last three decades owing to its astrophysical, geophysical and engineering applications. So far, the theoretical and experimental studies on the problem have been concerned with a conducting fluid layer bounded by isothermal boundaries exposed to uniform temperature gradients [1]. In some geophysical, astrophysical and industrial problems, particularly those concerned with providing effective insulation, specifying a uniform temperature at the free surface is more restrictive than specifying a uniform heat flux. Further, in geophysical problems, non-uniform heating is imported by the sun and applying a uniform temperature gradient has limited applications. It is shown ([8], [9]) that the kind and rate of heating significantly influence the value of the critical Rayleigh number R_c and can be effectively used to control convection. These aspects have not been given much attention in problems on magnetoconvection in spite of their relevance in practical problems as discussed above.

The object of this paper is, therefore, to study the effects of non-uniform temperature gradient and magnetic field on the onset of magnetoconvection with specified constant heat flux at the boundaries. These conditions generally pose the following mathematical difficulties:

(i) A steady state basic temperature profile that satisfies constant heat flux at the boundaries cannot be determined uniquely;

(ii) The basic nonlinear temperature profile leads to a stability equation with variable coefficients.

To overcome the first difficulty, we will solve the coupled momentum and energy equation for temperature as a function of position and time. In the case of marginal state, discussed in this paper, we consider a series of temperature profiles ([8], [9]) based on a simplification in the form of quasi-static approximation [10] that consists of freezing the temperature distribution at a given instant of time. This assumption may be justified using two different time scales, one the thermal time scale d^2/κ and the other the viscous time scale d^2/ν where κ is the thermal diffusivity, ν is the kinematic viscosity and d is the width of the layer. The second mathematical difficulty may be overcome by employing a single-term Galerkin expansion technique as in [11] which eliminates the use of elaborate numerical computation. But we find that in the case of rigid-rigid boundaries with constant heat flux, this method is not suitable to predict the required criterion for the onset of magnetoconvection. This is due to the fact that a polynomial type of trial function used in a single-term Galerkin expansion does not take into account the Hartman boundary layer that exists at the rigid-rigid boundaries when $Q \to \infty$. One way to take the boundary layer into account is to consider the higher order trial functions in the Galerkin technique. This naturally requires elaborate numerical computations and also does not predict explicitly the effect of the Hartman boundary layer on the onset of magnetoconvection. Another way to deal with the Hartman boundary layer is to use the method of matched asymptotic expansions. In this paper we follow the latter method.

The plan of this paper is as follows: In Section 2 the basic equations and the relevant boundary conditions are summarized. Section 3 is devoted to finding the asymptotic value of R_c as $Q \to \infty$ from the non-viscous MHD equations employing a single-term Galerkin expansion procedure. A suitable power law for asymptotic values of R_c is also established in this section. Exact solutions of the stability equations are obtained in section 4 using a regular perturbation technique with wave number as perturbation parameter. A singular perturbation technique is used in section 5 to find explicitly the effect of the Hartman boundary layer on the onset of magnetoconvection. Some important conclusions are drawn in the last section.

2. FORMULATION OF THE PROBLEM

We consider an electrically conducting Boussinesq fluid in an infinite horizontal layer of depth d bounded by rigid surfaces at $z=0$ and $z=d$. Suppose that the layer is parallel to the xy-plane and that it is permeated by a uniform magnetic field B_0 in the z-direction. Originally the fluid is at rest in an isothermal state with temperature T_0. At time $t=0^+$, heat is applied in an arbitrary manner with respect to time but uniformly with respect to the lower bounding surface. Let Δt be the temperature difference between the lower and upper bounding surfaces with the lower surface at a greater temperature than the upper surface. The resulting conduction temperature profile T_b will depend on time and the vertical coordinate z. This has to be determined by solving the energy equation

$$\frac{\partial T_b}{\partial t} = \kappa \frac{\partial^2 T_b}{\partial z^2}$$

with suitable initial and boundary conditions. In the quasi-static approximation, this may be written as

$$\frac{dT_b}{dz} = -\frac{\Delta T}{d} f(z). \tag{2.1}$$

Here f(z) represents a non-uniform temperature gradient such that $\int_0^1 f(z)dz=1$. The basic MHD equations for such a system are the conservation of mass, momentum, energy, and magnetic lines of force. In the Boussinesq approximation, these equations, after subtracting the hydrostatic contribution, may be written as

$$\nabla\cdot\vec{q} = 0 \tag{2.2}$$

$$\rho_0 \frac{D\vec{q}}{Dt} = -\nabla p' - \rho' g\hat{k} + \frac{1}{\mu}(\nabla X\vec{B})X\vec{B} + \rho_0 \nu \nabla^2\vec{q} \tag{2.3}$$

$$\rho' = -\rho_0 \alpha T \tag{2.4}$$

$$\frac{DT'}{Dt} = \frac{\Delta T}{d} f(z)\vec{q}.\hat{k} + \kappa\nabla^2 T' \tag{2.5}$$

$$\frac{D\vec{B}}{Dt} = (\vec{B}\cdot\nabla)\vec{q} + \nu_m \nabla^2 \vec{B} \tag{2.6}$$

$$\nabla.\vec{B} = 0. \tag{2.7}$$

Here $\vec{q} = (u,v,w)$ is the fluid velocity, $\vec{B}$ is the magnetic field strength p',ρ', and T' are the departure of the pressure p, density ρ, and temperature T from their values in equilibrium, ρ_0 is the value of ρ at $T=T_0$, T_0 is the ambient temperature, ν_m is the magnetic viscosity, μ is the magnetic permeability, α is the volumetric expansivity of the fluid and $\hat{k}$ is the unit vector in the z-direction. The equation of state (2.4) is true only when the magnetic pressure is always much less than the fluid pressure.

It is convenient to non-dimensionalize these equations, so we shall rescale lengths with d, times with d^2/ν, velocities with ν/d, temperatures with $\frac{\nu}{ad}\left(\frac{\nu\Delta T}{\alpha g d\kappa}\right)^{\frac{1}{2}}$ and magnetic field with B_0. Since we are interested in finding the conditions for the onset of magnetoconvection, which is the realm of linear theory, it is permissible to linearize the equations and seek horizontally periodic solutions. The velocity and magnetic field may be decomposed into poloidal and toroidal parts, the latter vanishing and hence $\vec{q} = \nabla X \nabla X(w\hat{k})$ and $\vec{B} = \hat{k} + \nabla X \nabla X(B\hat{k})$. The resulting equations are separable in z and t , so we may write

$$(w,B,T') = (W(z),\ B(z),\ T(z))F(x,y)\, e^{\sigma t} \tag{2.8}$$

where σ is the frequency of perturbations, $\left(\frac{\partial^2}{\partial x^2} + \frac{\partial^2}{\partial y^2}\right)F(x,y) = -a^2 F$ and a is the horizontal wave number scaled with d. On substituting (2.8) into the linearized equations and assuming the marginal state is valid (see Chandrasekhar [1]), we obtain

$$(D^2-a^2)W = a^2 RT - (Q/P_m)\ a\ D(D^2-a^2)B \tag{2.9}$$

$$(D^2-a^2)T + f(z)W = 0 \tag{2.10}$$

$$(D^2-a^2)B + P_m DW = 0 . \tag{2.11}$$

Here $D = d/dz$, $R = \alpha g \Delta T d^3/\nu\kappa$ is the Rayleigh number, $Q = \mu H_0^2 d^2/\nu\nu_m$ is the Chandrasekhar number, $P_m = \nu/\nu_m$ is the magnetic Prandtl number, ν is the kinematic viscosity and $H_0 (B_0 = \mu H_0)$ is the magnetic field.

We assume that both the boundaries bounding the layer are rigid and that constant heat flux (i.e., adiabatic) conditions hold. Then the relevant boundary conditions are

$$W = DW = B = DT = 0 \qquad \text{at } z = 0 \text{ and } 1 . \tag{2.12}$$

When the layer is heated from below at a constant rate, the basic temperature gradient f(z) in (2.1) is not only non-negative but also decreases monotonically. Our object here is to find a suitable temperature profile that increases or decreases (optimizes) the critical Rayleigh number R_c. For this purpose, we consider different basic temperature profiles [8] for which

$$\begin{aligned}
&\text{(i)} && f(z) = 1, \\
&\text{(ii)} && f(z) = \varepsilon^{-1}, \text{ for } 0 \le z < \varepsilon \text{ and } 0 \text{ for } \varepsilon < z \le 1, \\
&\text{(iii)} && f(z) = 0 \text{ for } 0 \le z < (1-\varepsilon) \text{ and } \varepsilon^{-1} \text{ for } (1-\varepsilon) < z \le 1, \\
&\text{(iv)} && f(z) = \delta(z-\varepsilon), \\
&\text{(v)} && f(z) = 2(1-z), \\
&\text{(vi)} && f(z) = 2z.
\end{aligned} \tag{2.13}$$

Here ε is the thermal depth and f(z) given by (i) to (vi) represents respectively the linear temperature profile, piecewise linear temperature profile with heating from above, piecewise linear temperature profile with cooling from above, step-function temperature profile, inverted parabolic temperature profile and parabolic temperature profile. The critical Rayleigh number for these profiles are respectively denoted by R_{ci} (i = 1 to 6).

3. ASYMPTOTIC VALUE OF R_c AS $Q \to \infty$

In the case of uniform temperature gradient (i.e., f(z) = 1) with isothermal rigid boundaries, Chandrasekhar [1] has shown that R_c obeys the power law $R_c \simeq \pi^2 Q$ for $Q \to \infty$. Recently, Rudraiah et al [11] have shown that this power law $\pi^2 Q$ is true only for stress-free boundaries and that for rigid isothermal boundaries this power law should be

$R_c \approx 12Q$ for $Q \to \infty$. In this section we show that the same power law is true even in the case of adiabatic rigid boundaries. This is proved through the following theorem.

Theorem: The asymptotic value of R_c for $Q \to \infty$ depends on the nature of heating but not on the nature of the bounding surface.

Proof: To prove this theorem, we follow the Galerkin procedure as in [11]. Multiplying (2.9) by W, (2.10) by T and (2.11) by B, and integrating the resulting equations with respect to z from 0 to 1, with the trial functions $W = a_0 W_1$, $T = b_0 T_1$ and $B = c_0 B_1$ we obtain [9]

$$R = \langle (D^2W)^2 + (2a^2 + Q(DW)^2 + a^4W^2 \rangle \langle (DT)^2 + a^2T^2 \rangle / a^2 \langle WT \rangle \langle f(z)WT \rangle \quad (3.1)$$

where $\langle \ldots \rangle = \int_0^1 (..)dz$. This shows that R_c exists at a=0. Substituting suitable trial functions $W = z(1-z)(1+z-z^2)$, T=1, and f(z) given by (2.6), Rudraiah et al [11] obtained R_{ci} (i = 1 to 6). For example for f(z) = 1, they [11] obtained

$$R_{c1} = 720 + 17.14Q. \quad (3.2)$$

When $Q \to \infty$, this gives $R_c \approx 17.14Q$. Substituting the definition of R_c and Q, we obtain

$$\alpha\, g\, \Delta T_c \approx 17.14 \mu H_0^2 d^2 \kappa / \rho_0 \nu_m \quad (3.3)$$

where ΔT_c is the critical temperature difference across the layer. This shows that ΔT_c is independent of kinematic viscosity as in the case of isothermal boundaries [1]. This implies that the dissipation of energy by Joule heating dominates over the dissipation of energy by viscosity because the effect of viscosity on the system becomes negligible when $Q \to \infty$. This suggests that a correct value of R_c and $Q \to \infty$ may be obtained by neglecting the viscous term in (2.9). Applying the single-term Galerkin expansion procedure to the inviscid MHD equations and using the trial functions

$$W = z^2(1-z)^2 \quad \text{and} \quad T = 1 \quad (3.4)$$

we obtain

$$R_c = 17Q/[7\langle f(z)(z-2z^3+z^4)\rangle] \quad (3.5)$$

From this we obtain R_{ci} for different f(z) given by (2.6) in the form

$$R_{c1} = R_{c5} = R_{c6} \approx 12.11Q,\ R_{c2} = R_{c3} \approx 10.65Q,\ R_{c4} \approx 7.777Q. \quad (3.6)$$

Here the equations $R_{c1} = R_{c5} = R_{c6}$ and $R_{c2} = R_{c3}$ are due to symmetry in WT about the midline.

We see that for f(z) = 1, the power law is 12Q which is the same as that obtained in [11] for isothermal rigid boundaries. For other f(z) Rudraiah et al [11] have obtained $R_{c2} = R_{c3} \approx 10Q$ and $R_{c4} \approx 7Q$ for isothermal boundaries. These values are close to those given in (3.6) for adiabatic boundaries. Hence the power law depends on the nature of heating (or cooling). This is true for both adiabatic and isothermal boundaries. Hence the theorem.

The discrepancy between 17.14Q and 12Q is far too large to be acceptable. This large discrepancy is due to the use of a polynomial type of trial function in the single-term Galerkin expansion. One may use the higher order Galerkin expansion to obtain the required R_{ci}. This naturally requires elaborate numerical computations. But, our aim in this paper is to determine the required eigenvalues R_c analytically. In the next section we, therefore, use an exact solution obtained from a regular perturbation technique to compute the required values of R_{ci}.

4. SOLUTIONS OBTAINED USING REGULAR PERTURBATION TECHNIQUE

In the case of constant heat flux at the boundaries, the cell patterns are uni-cellular because of small values of the wave number 'a'. We, therefore, use 'a' as a perturbation parameter and expand the dependent variables in the form

$$(W,T) = (W_0, T_0) + a^2(W_1,T_1) + \dots \tag{4.1}$$

Substituting (4.1) in the combined equations of (2.9) and (2.11) and equating like powers of a, we get the first order equations in the form

$$D^4W_0 = QD^2W_0, \quad D^2T_0 + f(z)W_0 = 0. \tag{4.2}$$

The solutions of (4.2) satisfying the conditions (2.12), are

$$W_0 = 0 \quad \text{and} \quad T_0 = 1. \tag{4.3}$$

The second order equations are

$$D^4W_1 - QD^2W_1 = RT_0 + 2D^2W_0 \tag{4.4}$$

$$D^2T_1 = T_0 - f(z)W_1. \tag{4.5}$$

The general solution of (4.4) is

$$W_1 = RW_1' \tag{4.6}$$

where

$$W_1' = -\frac{1}{2}Q^{-1} z^2 + C_0 + C_1 z + C_2 e^{Mz} + C_3 e^{-Mz} \tag{4.7}$$

where $M = Q^{\frac{1}{2}}$ is the Hartman number and $C_i(1 = 0 \text{ to } 3)$ are the constants obtained using the boundary conditions (2.12)

$$\begin{aligned} C_0 &= (M \cosh M - 2 \sinh M + M)/MQ\Delta) \\ C_1 &= (2 \cosh M - M \sinh M - 2)/MQ\Delta) \\ C_2 &= (2 - M - Me^{M} - 2e^{-M})/(2MQ\Delta) \\ C_3 &= (2e^{M} - M - Me^{M} - 2)/(2MQ\Delta) \\ \Delta &= 2(2 + M \sinh M - 2 \cosh M). \end{aligned} \tag{4.8}$$

The effect of the Hartman boundary layer is also reflected in these constants.

Integrating (4.5) and using (2.12) we obtain the integrability condition

$$R^{-1} = \int_0^1 f(z) W_1' \, dz . \tag{4.9}$$

Substituting f(z) and W_1', given by (2.13) and (4.7), into (4.9) and integrating we get R_{ci}(i = 1 to 6). For example, for f(z) = 1, we obtain

$$R_{c1} = 6Q^2 (4+2M\sinh M-4\cosh M)/[24-4Q+(QM+24M)\sinh M-(8Q+24)\cosh M]. \tag{4.10}$$

When $Q \to \infty$, this gives $R_{c1}=720$ which is the known exact value in the absence of magnetic field. Similarly the asymptotic value of R_c is

$$R_{c1} \approx 12Q \quad \text{when } Q \to \infty \tag{4.11}$$

in contrast to the value 17.14Q given by (3.2). In computing the asymptotic value of R_{ci} for large Q, we found that $QC_0 \neq 0$ when $Q \to \infty$. In that case the expression for W_1 is

$$W_1 \sim \frac{1}{2} R_m[z-z^2 + \frac{1}{M}\{e^{M(z-1)} + e^{-Mz} - 1\}] \tag{4.12}$$

where $R_m = R/Q$. Similarly, the asymptotic values of $R_{ci}(i = 2 \text{ to } 6)$ as $Q \to \infty$ are $R_{c2} = R_{c3} \approx 10.7Q$ attained at $\varepsilon = 0.75$, $R_{c4} \approx 8.0\,Q$ attained at $\varepsilon = 0.578$ and $R_{c5} = R_{c6} \approx 12Q$. These results also confirm the theorem stated in Section 3.

The values of $R_{ci}(i = 1 \text{ to } 6)$ obtained from (4.9) are computed for different values of Q and the results are tabulated in table 1. The values of R_{c2} and R_{c4} reported in this table are computed at $\varepsilon = 0.75$ and $\varepsilon = 0.578$ respectively. We also observe that the asymptotic values of the ratios R_{ci}/R_{c1} are independent of Q but depend on the thermal depth ε. The influence of various temperature profiles on R_{ci}/R_{c1} is shown in Fig. 1. In this figure ε_2, ε_3 and ε_4 represent the critical value of ε for respective profiles. This figure and table 1 reveal that $R_{c2} = R_{c3}$, $R_{c1} = R_{c5} = R_{c6}$ because of the symmetry of W_1T_1 about the midline. Further, for all f(z)'s chosen such that $\langle f(z)\rangle = 1$, the thermal profile effect is most effective for the step-function temperature profile, with jump concentrated at the maximum of W_1T_1 to augment convection. Similarly, the linear temperature profile is most effective to suppress convection.

5. SOLUTIONS USING SINGULAR PERTURBATION TECHNIQUE

The regular perturbation method used in the previous section does not predict explicitly the effect of the Hartman boundary layer on R_{ci}. The best way to deal with the boundary layer is to use the method of matched asymptotic expansions. The analysis of this technique is now well known (for example see Cole [12]) and is extensively used to find a uniformly valid solution in the entire region. But, not much attention has been given to the use of the singular perturbation technique (i.e., the method of matched asymptotic expansions) to study convection problems. Rudraiah and Masuoka [13] used this technique to study convection in porous media. Their [13] analysis does not predict the criterion for the onset of convection. To our knowledge Nield [14] was the first to use the singular perturbation technique to determine the condition for the onset of convection in porous media. In this section we extend the analysis of [14] and [15] to determine the criterion for the onset of magnetoconvection.

For this purpose we rewrite (4.4) and (4.5) in the form

$$\frac{1}{Q}D^4W_1 - D^2W_1 = R_m \tag{5.1}$$

$$D^2T_1 = 1 - f(z)\,W_1. \tag{5.2}$$

The outer solution, $W^{(0)}$, of (5.1) for $Q \to \infty$ satisfies the equation

$$D^2W^{(0)} = -\,R_m. \tag{5.3}$$

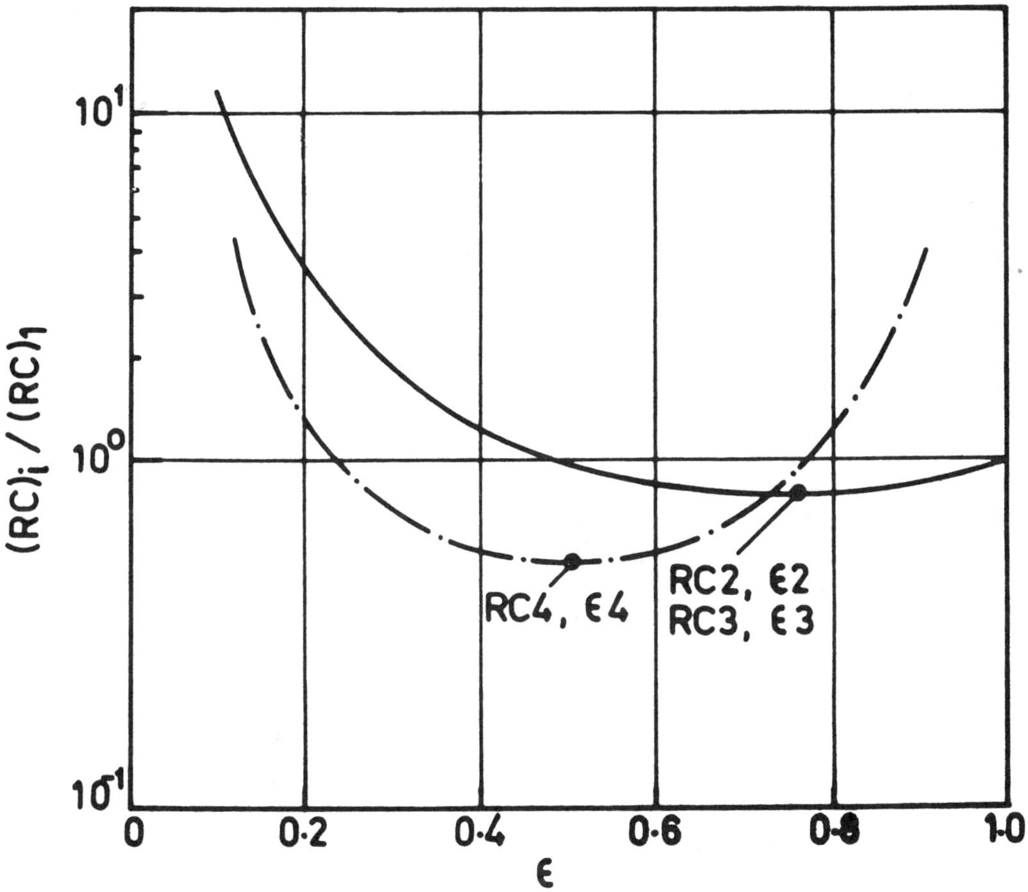

FIG. 1 – CRITICAL RAYLEIGH NUMBER AS A FUNCTION OF THERMAL DEPTH FOR RIGID-RIGID ADIABATIC BOUNDARIES

Table 1 *Critical Rayleigh number for rigid-rigid adiabatic boundaries (using regular perturbation technique)*

Q	R_{c1}	R_{c2}	R_{c3}	R_{c4}	R_{c5}	R_{c6}
0	720.00	600.10	600.10	384.00	720.00	720.00
10^{-1}	721.7140	602.5651	602.5651	385.3403	721.7140	721.7140
10^{0}	737.1225	615.5916	615.5916	393.9846	737.1225	737.1225
10^{1}	889.5114	744.7540	744.7540	480.1778	889.5114	889.5114
10^{2}	2307.9341	1961.2864	1961.2864	1322.5463	2307.9341	2307.9341
10^{3}	1.45938×10^{4}	1.27219×10^{4}	1.27219×10^{4}	9163.6444	1.45938×10^{4}	1.45938×10^{4}
10^{4}	1.274968×10^{5}	1.1259128×10^{5}	1.1259128×10^{5}	8.33763×10^{4}	1.274968×10^{5}	1.274968×10^{5}
10^{5}	1.223059×10^{6}	1.084889×10^{6}	1.084889×10^{6}	8.106543×10^{5}	1.223059×10^{6}	1.223059×10^{6}

Its solution, satisfying the symmetry condition

$$DW^{(0)} = 0 \qquad \text{at } z = \tfrac{1}{2} \tag{5.4}$$

is

$$W^{(0)} = (R_m/2)\,(z-z^2) + d_0. \tag{5.5}$$

To find the inner solution we use the transformation

$$W^{(0)} = Y(M)g(\xi) \tag{5.6}$$

where

$$\xi = Mz \tag{5.7}$$

is the stretching variable introduced to make the terms D^4W_1 and D^2W_1 in (5.1) of the same order. The function $g(\xi)$ satisfies the differential equation

$$d^4g/d\xi^4 - d^2g/d\xi^2 = 0 \tag{5.8}$$

and the boundary conditions

$$g = dg/d\xi = 0 \text{ at } \xi = 0 \quad \text{and} \quad \xi = 1. \tag{5.9}$$

We find the inner solution, valid in the Hartman boundary layer of thickness M^{-1}, separately for z=0 and z=1. The solution of (5.8), satisfying (5.9) near ξ=0 and using the fact that g cannot grow exponentially large when $\xi \to \infty$, is

$$g(\xi) = a_0(-1+e^{-\xi}). \tag{5.10}$$

The matching condition, that

$$(W^{(0)})_{z\to 0} \quad \text{matches } (W^{(i)})_{z\to\infty} \tag{5.11}$$

requires that

$$d_0 = 0,\ a_0 = \tfrac{1}{2} R_m,\ Y(M) = \frac{1}{M}. \tag{5.12}$$

Then the complete solution, valid near z=0 is

$$W_1 = (R_m/2)\,[z - z^2 + \frac{1}{M}(e^{-Mz} - 1)]. \tag{5.13}$$

Adding the similar solution at z=1, we get the uniformly valid solution as

$$W_1 = (R_m/2)\,[z-z^2 + \frac{1}{M}(e^{M(z-1)} + e^{-Mz} - 1)]. \tag{5.14}$$

Note that this coincides with (4.12), implying perfect matching. Since the solution obtained by employing the regular perturbation technique in section 4 is exact, the solution obtained in this section using a singular perturbation technique is also exact.

Substituting (5.14) and (2.13) into (4.9), we obtain expressions for R_{ci} (i = 1 to 6) involving explicitly the effect of the Hartman boundary layer. For example, for f(z) = 1, we obtain

$$R_{c1} = 12Q^2/[Q - 6M + 12(1 - \cosh M + \sinh M)]. \tag{5.15}$$

When $Q \to \infty$, this gives

$$R_{c1} \approx 12Q + 72M. \tag{5.16}$$

Similarly, for other f(z)'s, when $Q \to \infty$ we obtain the expressions

$$R_{c2} = R_{c3} \approx 10.7Q + 56.8M, \quad R_{c4} \approx 8Q + 32M, \quad R_{c5} = R_{c6} = R_{c1}. \tag{5.17}$$

The second term in each of R_{ci} (i = 1 to 6) given by (5.16) and (5.17) represents the contribution from the Hartman boundary layer. Its effect is, therefore, to raise the value of R_{ci} by an amount proportional to M when compared to those obtained from the regular perturbation technique discussed in Section 4 .

8. CONCLUSIONS

The effects of Lorentz force and nonlinear basic temperature profile on the criterion for the onset of marginal convection have been established analytically. Some of the new results of the present work as compared with those of Chandrasekhar [1] are given below.

(1) The present work deals with the effects of non-uniform temperature gradient rather than the case of uniform temperature gradient considered in [1]. Here, it is shown that a suitable non-uniform temperature gradient and the transverse magnetic field suppress or augment convection; this is important in material processing in the laboratory or in the transfer of heat from one region to the other.

(2) In the present analysis, the rigid boundaries are assumed to be maintained at constant heat flux rather than at constant temperature as in [1].

(3) In the case of rigid-rigid boundaries the analysis reported in [1] is based on a variational principle and requires elaborate numerical computations. In the present paper exact solutions based on regular and singular perturbation techniques are reported. These solutions provide a quick and easy method for obtaining R_{ci} and give results closer to the known values than those

obtained from a single-term Galerkin expansion. For example, in the case of rigid boundaries with constant heat flux, the single-term Galerkin expansion used in [11] gives $R_c \simeq 17.14Q$ for $f(z) = 1$ and $Q \to \infty$. This value is much greater than the known value 12Q. This discrepancy is due to the fact that a polynomial type of trial function used in the Galerkin expansion does not take into account the effect of the Hartman boundary layer that exists at the rigid boundaries for large values of Q.

(4) In contrast to the power law $R_{cl} \simeq \pi^2 Q$ predicted in [1], we have obtained the law $R_{cl} \simeq 12Q$ for $f(z) = 1$. Further, we conclude that this power law crucially depends on the nature of heating and not on the nature of the bounding surface.

REFERENCES

1. S. Chandrasekhar, *Hydrodynamic and hydromagnetic stability*, Oxford: Clarendon Press (1961).
2. N. Rudraiah, *Publ. Astron. Soc. Japan*, *33*: 721 1981).
3. N.O. Weiss, *J. Fluid Mech.*, *108*: 247 (1981).
4. N. Rudraiah, V. Kumudini and W. Unno, *Publ. Astron. Soc. Japan*, *37*: 183 (1985).
5. N. Rudraiah, I.S. Shivakumara, P. Geetavani and W. Unno, Unno's 60th birthday commemoration international conference on "Hydrodynamic and Magnetohydrogynamic problems in the sun and stars", University of Tokyo, Japan (1986).
6. D.A. Nield, *ZAMP*, *17*: 131 (1966).
7. N. Rudraiah, O.P. Chandna and M.R. Garg, *Indian J. Tech.*, *24*: 279 (1986).
8. D.A. Nield, *J. Fluid Mech.*, *71*: 441 (1975).
9. N. Rudraiah and O.P. Chandna, *Canadian Journal of Physics*, *64*: 1 (1985).
10. I.G. Currie, *J. Fluid Mech.*, *29*: 237 (1967).
11. N. Rudraiah, O.P. Chandna and R.M. Barron, (Submitted for publication) (1989)
12. J.D. Cole, *Perturbation methods in applied mathematics*, Blaisdell Publishing Company (1968).
13. N. Rudraiah and T. Masuoka, *Int. J. Engg. Sc.*, *20*: 27 (1982).
14. D.A.Nield, Further Results Preprint (1985).
15. N. Rudraiah, Natural convection in porous media (V. Prasad and N.A. Hussain, eds.) *ASME*, *HTD56*, 33 (1986).

The Interior Layer Structure for a Linear Parbolic Problem with Discontinuous Data

Shagi-Di Shih Professor, Department of Mathematics,
University of Wyoming, Laramie, Wyoming

I Introduction

We study the initial value problem for a linear parabolic partial differential equation of the form

$$L_\epsilon [u_\epsilon] \equiv \frac{\partial u_\epsilon}{\partial t} - \epsilon \frac{\partial^2 u_\epsilon}{\partial x^2} + p(x,t)\frac{\partial u_\epsilon}{\partial x} + q(x,t)u_\epsilon = f(x,t), \qquad (1.1)$$

Supported in part by a grant from the NSF, No. INT–8514247.

in the domain $\Omega = \{(x,t) : -\infty < x < \infty,\ 0 < t < \infty\}$ subject to the initial condition

$$u_\epsilon(x,0) = g(x), \qquad -\infty < x < \infty; \tag{1.2}$$

where ϵ is a small positive parameter $0 < \epsilon \ll 1$; $p(x,t)$, $q(x,t)$, and $f(x,t)$ are assumed to be smooth functions; and $g(x)$ is a smooth function everywhere except at $x = 0$, where g is discontinuous, i.e., $g(0+) \neq g(0-)$. Since the early 1950's this type of equations has been studied extensively, see, for instance, Kamenomostskaya (1952), Aronson (1956), Ladyženskaja (1968), Bobisud (1969), Besjes (1974), Holland (1977), Howes (1986).

The aim of this investigation is to generalize the techniques of Shih and Kellogg (1987) to analyze the "interior layer" structure of the exact solution to the problem (1.1), (1.2) due to the propagation of discontinuity of initial data along the characteristic curve of the reduced problem (when $\epsilon = 0$) by explicitly constructing interior layer functions along the characteristics through imposing some initial–boundary value problems for them in a systematic way. Here by interior layer functions we mean some functions which are significantly large only near certain curve in the *interior* of the domain under consideration, and this shows differences to the boundary layer functions occurring for other phenomena. Moreover, this constructive procedure can be carried out to arbitrary order with respect to the small parameter. Finally, we provide a proof of its uniform validity in the closure of the domain by using the maximum principle and some exponential estimates of interior layer functions. For this type of interior layer problems, Bobisud (1969) proved an asymptotic expansion of order 1/2 in ϵ.

II Outer approximation

In order to obtain the first rough approximation to the solution $u_\epsilon(x,t)$ for small values of the parameter ϵ, we consider a function $u_0(x,t)$, which is defined by the reduced equation obtained from (1.1) by putting $\epsilon = 0$,

$$L_0\left[u_0\right] \equiv \frac{\partial u_0}{\partial t} + p(x,t)\frac{\partial u_0}{\partial x} + q(x,t)u_0 = f(x,t), \tag{2.1}$$

with the initial condition (1.2). It is well-known from the theory of hyperbolic partial differential equations that $u_0(x,t)$ is discontinuous across the characteristic curve $x = P(t)$, defined by the following initial value problem

$$\frac{dx}{dt} = p(x,t), \qquad x(0) = 0; \tag{2.2}$$

moreover, away from the curve $x = P(t)$, $u_0(x,t)$ is a smooth function. Since the solution $u_\epsilon(x,t)$ is a smooth function in the region Ω and the difference $u_\epsilon - u_0$ satisfies the differential equation

$$L_\epsilon\left[u_\epsilon - u_0\right] = \epsilon\frac{\partial^2 u_0}{\partial x^2}, \tag{2.3}$$

in a subregion Ω_1 of Ω excluding a neighborhood Ω_0 of the characteristic curve $x = P(t)$, it is quite evident that the approximation u_0 for u_ϵ is not valid in the neighborhood Ω_0 of the curve $x = P(t)$ and is valid in the remaining subregion Ω_1 up to the order $\mathcal{O}(\epsilon)$.

First of all, in terms of the powers of ϵ in the right hand side of the differential operator L_ϵ, a better approximation of the solution $u_\epsilon(x,t)$ in the subregion Ω_1 is to be obtained by adding to $u_0(x,t)$ a sum $\sum_{k=1}^{n} \epsilon^k u_k(x,t)$ which equals zero at the initial time $t = 0$ and has the property that $\sum_{k=0}^{n} \epsilon^k u_k(x,t)$ satisfies (1.1) in Ω_1 up to the order $\mathcal{O}(\epsilon^{n+1})$ for some positive integer n. The functions $u_k(x,t)$ are determined iteratively by the differential equations in Ω_1

$$L_0\left[u_k\right] = \frac{\partial^2 u_{k-1}}{\partial x^2}, \tag{2.4}$$

with the initial conditions

$$u_k(x,0) = 0, \tag{2.5}$$

for $k = 1, 2, \ldots, n$. Notice that nonsmoothness of the function $u_k(x,t)$ across the curve $x = P(t)$ is due to that of the right hand side $\dfrac{\partial^2 u_{k-1}}{\partial x^2}$ of the equation (2.4) across the curve $x = P(t)$ for $k = 1, 2, \ldots, n$.

We will call the series

$$U_n(x,t;\epsilon) = \sum_{k=0}^{n} \epsilon^k u_k(x,t), \tag{2.6}$$

the *outer asymptotic approximation* (or in brief, *outer approximation*). Applying the differential operator L_ϵ to the function $u^*(x,t;\epsilon)$, defined by $u^* = u_\epsilon - U_n$, yields the differential equation in Ω_1

$$L_\epsilon [u^*] = \epsilon^{n+1} \frac{\partial^2 u_n}{\partial x^2}, \tag{2.7}$$

with the initial condition $u^*(x,0;\epsilon) = 0$.

Now the outer approximation $U_n(x,t;\epsilon)$ satisfies the initial condition (1.2) and as in the case of $u_0(x,t)$, it is not smooth in the neighborhood Ω_0 of the characteristic curve $x = P(t)$. To obtain a "uniform" approximation of the solution $u_\epsilon(x,t)$ in Ω in terms of regularity, we eliminate this discrepancy along the curve $x = P(t)$ by introducing "interior layer" functions along the two sides of the curve $x = P(t)$. These functions will have the property that when acted on by the differential operator L_ϵ, the results will be of order $\mathcal{O}(\epsilon^{n+1})$ uniformly in one side of the curve $x = P(t)$. Moreover, the interior layer functions have the property of being asymptotically equal to zero everywhere in Ω except for a small neighborhood of one side of the curve $x = P(t)$. These functions along the characteristic curve $x = P(t)$ satisfy parabolic partial differential equations along one side of the curve $x = P(t)$ and may be called as the *parabolic interior layer functions.* Furthermore, in terms of boundary conditions for parabolic interior layer functions, there are two classes of these functions. One class of functions is designed to remove the C^0–discontinuity of the outer approximation across the characteristic curve $x = P(t)$ while the other class of functions is employed to remove the C^1–discontinuity of the outer approximation across the curve $x = P(t)$. Consequently, the outer approximation supplemented by these interior layer functions will be at least continuously differentiable with respect to both space and time variables across the characteristic curve $x = P(t)$.

Next, the sum of the outer approximation and the interior layer functions may not be twice continuously differentiable with respect to space variable across the characteristic curve $x = P(t)$. Hence, when the differential operator L_ϵ is applied to this sum, the resulting expression may not be continuous with respect to the space variable across the curve $x = P(t)$. To remedy this deficiency, we will construct a function to supplement this sum so that all of the desired properties mentioned above can be obtained.

Finally, in order to establish the uniform validity of the constructed asymptotic expression for the solution $u_\epsilon(x,t)$, the maximum principle will be used in the usual manner.

III Parabolic interior layers along the characteristic curve $x = P(t)$

Let us study separate parabolic interior layers on each side of the characteristic curve $x = P(t)$. For the case of $x > P(t)$, the stretched variable ξ along the curve $x = P(t)$ is defined by $x = P(t) + \sqrt{\epsilon}\,\xi$. The parabolic interior layer along the characteristic curve $x = P(t)$ is defined by the series

$$V^+_{2n+1}(\xi, t; \epsilon) = \sum_{k=0}^{2n+1} \epsilon^{k/2} v^+_k(\xi, t), \tag{3.1}$$

where the functions $v^+_k(\xi,t)$ are defined iteratively by the parabolic partial differential equations of the form

$$M\left[v^+_k\right](\xi,t) \equiv \frac{\partial v^+_k}{\partial t} - \frac{\partial^2 v^+_k}{\partial \xi^2} + \xi\frac{\partial p}{\partial x}(P(t),t)\frac{\partial v^+_k}{\partial \xi} + q(P(t),t)v^+_k = \Pi^+_k(\xi,t), \tag{3.2}$$

over the quarter plane $0 < \xi < \infty$ and $0 < t < \infty$ and the functions $\Pi^+_k(\xi,t)$ have the expressions

$$\Pi^+_k(\xi,t) = \begin{cases} 0, & k = 0; \\ -\sum_{i=1}^{k} \frac{\xi^{i+1}}{(i+1)!}\frac{\partial^{i+1} p}{\partial x^{i+1}}(P(t),t)\frac{\partial v^+_{k-i}}{\partial \xi} - \sum_{i=1}^{k} \frac{\xi^i}{i!}\frac{\partial^i q}{\partial x^i}(P(t),t)v^+_{k-i}, & 1 \le k \le 2n+1. \end{cases} \tag{3.3}$$

Now we are in a position to specify the initial-boundary conditions for each function $v_k^+(\xi,t)$. First, both the function $u_\epsilon(x,t)$ and the outer approximation $U_n(x,t;\epsilon)$ have the same initial condition and thus the homogeneous initial condition for $v_k^+(\xi,t)$ is imposed for each k. Next, since these parabolic interior layer functions are significant only in a neighborhood of the curve $x = P(t)$, the function $v_k^+(\xi,t)$ should approach zero as $x \neq P(t)$ and ϵ tends to zero for each k. Finally, for each k, the boundary condition imposed on the function $v_{2k}^+(\xi,t)$ at $\xi = 0$ is such that the C^0–discontinuity of $u_k(x,t)$ along the characteristic curve $x = P(t)$ disappears while the boundary condition imposed on the function $v_{2k+1}^+(\xi,t)$ at $\xi = 0$ is such that C^1–discontinuity of $u_k(x,t)$ along the curve $x = P(t)$ disappears. In other words, $v_k^+(\xi,t)$ satisfies the following initial-boundary conditions for $0 \leq k \leq 2n+1$,

$$v_k^+(\xi,0) = 0, \tag{3.4}$$

$$\lim_{\xi\to\infty} v_k^+(\xi,t) = 0, \tag{3.5}$$

and for $0 \leq k \leq n$,

$$v_{2k}^+(0,t) = \frac{1}{2}\left\{u_k^- - u_k^+\right\}(P(t),t), \tag{3.6}$$

$$\frac{\partial v_{2k+1}^+}{\partial \xi}(0,t) = \frac{1}{2}\left\{\frac{\partial u_k^-}{\partial x} - \frac{\partial u_k^+}{\partial x}\right\}(P(t),t). \tag{3.7}$$

Here we have used the notations $u_k^+(x,t)$ and $u_k^-(x,t)$ to denote the function $u_k(x,t)$ with $x > P(t)$ and $x < P(t)$, respectively.

For the case of $x < P(t)$, the stretched variable η along the characteristic curve $x = P(t)$ is defined by $x = P(t) - \sqrt{\epsilon}\,\eta$ and the parabolic interior layer along the curve $x = P(t)$ is defined by the series

$$V_{2n+1}^-(\eta,t;\epsilon) = \sum_{k=0}^{2n+1} \epsilon^{k/2} v_k^-(\eta,t), \tag{3.8}$$

where the functions $v_k^-(\eta,t)$ are defined iteratively by the parabolic partial differential equations

$$M\left[v_k^-\right](\eta,t) = \Pi_k^-(\eta,t), \tag{3.9}$$

over the quarter plane $0 < \eta < \infty$ and $0 < t < \infty$ and the functions $\Pi_k^-(\eta,t)$ have the expressions

$$\Pi_k^-(\eta,t) = \Pi_k^+(-\eta,t), \tag{3.10}$$

for $0 \leq k \leq 2n+1$. With the same argument as in the case of $x > P(t)$, we impose the following initial–boundary conditions for $v_k^-(\eta,t)$

$$v_k^-(\eta,0) \;=\; 0, \tag{3.11}$$

$$\lim_{\eta\to\infty} v_k^-(\eta,t) \;=\; 0, \tag{3.12}$$

for $0 \leq k \leq 2n+1$, and

$$v_{2k}^-(0,t) \;=\; \frac{1}{2}\left\{u_k^+ - u_k^-\right\}(P(t),t), \tag{3.13}$$

$$\frac{\partial v_{2k+1}^-}{\partial \eta}(0,t) \;=\; \frac{1}{2}\left\{\frac{\partial u_k^-}{\partial x} - \frac{\partial u_k^+}{\partial x}\right\}(P(t),t), \tag{3.14}$$

for $0 \leq k \leq n$.

For the sake of convenience, the symbols $v_k(\xi,t)$ and $V_{2n+1}(\xi,t;\epsilon)$ are used to denote the parabolic interior layer functions defined on both sides of the characteristic curve $x = P(t)$ with $-\infty < \xi < \infty$.

First we obtain the following result for the parabolic interior layer functions defined on both sides of the characteristics $x = P(t)$.

Theorem III.1 *The solutions v_k^+ and v_k^- for the initial-boundary value problems* (3.2), (3.4), (3.5), (3.6), (3.7), *and* (3.9), (3.11), (3.12), (3.13), (3.14), *respectively, satisfy the following relation*

$$v_k^+(\xi,t) = (-1)^{k+1}v_k^-(\xi,t), \tag{3.15}$$

for $0 < \xi < \infty$, $0 < t < \infty$, *and* $0 \leq k \leq 2n+1$.

Based on Theorem III.1, we could prove the following important relations between outer approximation and parabolic interior layers at the characteristics $x = P(t)$.

Theorem III.2 *For $0 \le k \le n$, the functions $u_k^+(x,t)$, $v_{2k}^+(\xi,t)$, $v_{2k+1}^+(\xi,t)$, $u_k^-(x,t)$, $v_{2k}^-(\eta,t)$, and $v_{2k+1}^-(\eta,t)$ constructed above have the following properties:*

$$u_k^+(P(t),t) + v_{2k}^+(0,t) = u_k^-(P(t),t) + v_{2k}^-(0,t), \tag{3.16}$$

$$\begin{aligned} & u_k^+(P(t),t) + v_{2k}^+(0,t) + \sqrt{\epsilon}\, v_{2k+1}^+(0,t) \\ = \; & u_k^-(P(t),t) + v_{2k}^-(0,t) + \sqrt{\epsilon}\, v_{2k+1}^-(0,t), \end{aligned} \tag{3.17}$$

$$\begin{aligned} & \frac{\partial}{\partial x}\left\{u_k^+(P(t),t) + v_{2k}^+(0,t) + \sqrt{\epsilon}\, v_{2k+1}^+(0,t)\right\} \\ = \; & \frac{\partial}{\partial x}\left\{u_k^-(P(t),t) + v_{2k}^-(0,t) + \sqrt{\epsilon}\, v_{2k+1}^-(0,t)\right\}, \end{aligned} \tag{3.18}$$

$$\begin{aligned} & \frac{\partial}{\partial t}\left\{u_k^+(P(t),t) + v_{2k}^+(0,t) + \sqrt{\epsilon}\, v_{2k+1}^+(0,t)\right\} \\ = \; & \frac{\partial}{\partial t}\left\{u_k^-(P(t),t) + v_{2k}^-(0,t) + \sqrt{\epsilon}\, v_{2k+1}^-(0,t)\right\}. \end{aligned} \tag{3.19}$$

In short, Theorem III.2 gives a justification of imposing the parabolic interior layer functions in such a way that

1. the sum $u_k + v_{2k}$ is continuous across the characteristic curve $x = P(t)$, and
2. the sum $u_k + v_{2k} + \sqrt{\epsilon}\, v_{2k+1}$ is not only continuous but also continuously differentiable with respect to both space and time variables across

the characteristic curve $x = P(t)$. Moreover, the sum of the outer approximation and the parabolic interior layers

$$U_n(x,t;\epsilon) + V_{2n+1}(\xi,t;\epsilon) = \sum_{k=0}^{n} \epsilon^k \left\{u_k + v_{2k} + \sqrt{\epsilon}\, v_{2k+1}\right\}, \tag{3.20}$$

its first x-derivative, and its first t-derivative are continuous across the characteristic curve $x = P(t)$.

Furthermore, we have the following *a priori* estimates for the parabolic interior layer functions and their first ξ-derivatives by using the techniques in Ladyženskaja, Solonnikov, and Ural'ce-va (1968).

Theorem III.3 *There exist two positive constants C and c, independent of ϵ, so that the inequalities*

$$\left|\frac{\partial^j v_k(\xi,t)}{\partial \xi^j}\right| \quad \le \quad C \exp\left[-\frac{c\xi^2}{t}\right], \tag{3.21}$$

holds for $-\infty < \xi < \infty$, $0 < t < \infty$, $0 \le k \le 2n+1$, and $j = 0, 1$.

Now applying the differential operator L_ϵ to the parabolic interior layer V_{2n+1}^+ yields the following result

$$L_\epsilon\left[V_{2n+1}^+\right] = \mathcal{O}(\epsilon^{n+1}), \tag{3.22}$$

for $x \ge P(t)$ and $0 < t < \infty$ by using Theorem III.3. Similarly it follows that

$$L_\epsilon\left[V_{2n+1}^-\right] = \mathcal{O}(\epsilon^{n+1}), \tag{3.23}$$

holds for $x \le P(t)$ and $0 < t < \infty$.

IV Asymptotic representation of the solution

Consider the sum

$$U_n(x,t;\epsilon) + V_{2n+1}(\xi,t;\epsilon) = \sum_{k=0}^{n} \epsilon^k \left\{u_k(x,t) + v_{2k}(\xi,t) + \sqrt{\epsilon}\, v_{2k+1}(\xi,t)\right\} \tag{3.20}$$

to be an approximation to the solution $u_\epsilon(x,t)$ of the initial value problem (1.1), (1.2); the outer approximation U_n is defined in (2.6) while the parabolic interior layer V_{2n+1} is defined in (3.1) and (3.8). Here we have used the stretched variable $\xi = (x - P(t))/\sqrt{\epsilon}$. Then it follows from the analysis of the preceding sections that the difference $u_\epsilon - (U_n + V_{2n+1})$ satisfies the parabolic differential equation

$$\begin{aligned} & L_\epsilon\left[u_\epsilon - (U_n + V_{2n+1})\right] \\ = \; & \epsilon^{n+1}\left\{\frac{\partial^2 u_n}{\partial x^2} - \sum_{k=0}^{2n+1}\left[\frac{1}{(2n+3-k)!}\frac{\partial^{2n+3-k}p}{\partial x^{2n+3-k}}(\tilde{x}_k,t)\,\xi^{2n+3-k}\frac{\partial v_k}{\partial \xi}\right.\right. \\ & \left.\left. + \frac{1}{(2n+2-k)!}\frac{\partial^{2n+2-k}q}{\partial x^{2n+2-k}}(\bar{x}_k,t)\,\xi^{2n+2-k}v_k\right]\right\}, \end{aligned} \tag{4.1}$$

for some $\tilde{x}_k$ and $\bar{x}_k$ between x and $P(t)$. The right hand side of equation (4.1) is continuous across the characteristic curve $x = P(t)$ except its first term $\dfrac{\partial^2 u_n}{\partial x^2}$. In order to overcome this difficulty, first let us consider the function $\hat{v}^+_{2n+2}$ defined by the parabolic differential equation for $0 < \xi < \infty$ and $0 < t < \infty$,

$$M\left[\hat{v}^+_{2n+2}\right](\xi,t) = \frac{1}{2}\left\{\frac{\partial^2 u_n^+}{\partial x^2} - \frac{\partial^2 u_n^-}{\partial x^2}\right\}(P(t),t)v(\xi,t), \tag{4.2}$$

subject to the initial condition

$$\hat{v}^+_{2n+2}(\xi,0) = 0, \tag{4.3}$$

and the boundary conditions

$$\hat{v}^+_{2n+2}(0,t) \;=\; 0, \tag{4.4}$$

$$\lim_{\xi\to\infty}\hat{v}^+_{2n+2}(\xi,t) \;=\; 0. \tag{4.5}$$

Here the function v appearing in the right hand side of equation (4.2) satisfies the following initial-boundary value problem for $0 < \xi < \infty$ and $0 < t < \infty$,

$$M[v](\xi,t) \;=\; 0, \tag{4.6}$$

$$v(\xi,0) = 0, \tag{4.7}$$

$$v(0,t) = 1, \tag{4.8}$$

$$\lim_{\xi\to\infty} v(\xi,t) = 0. \tag{4.9}$$

It then follows that

$$M\left[\hat{v}^+_{2n+2}\right](0,t) = \frac{1}{2}\left\{\frac{\partial^2 u_n^+}{\partial x^2} - \frac{\partial^2 u_n^-}{\partial x^2}\right\}(P(t),t), \tag{4.10}$$

and along the characteristic curve $x = P(t)$, we have

$$\begin{aligned} &L_\epsilon\left[u_\epsilon - (U_n^+ + V^+_{2n+1} + \epsilon^{n+1}\hat{v}^+_{2n+2})\right] \\ &= \frac{1}{2}\epsilon^{n+1}\left\{\frac{\partial^2 u_n^+}{\partial x^2} + \frac{\partial^2 u_n^-}{\partial x^2}\right\}(P(t),t). \end{aligned} \tag{4.11}$$

Similarly we can construct $\hat{v}^-_{2n+2}$ so that

$$\begin{aligned} &L_\epsilon\left[u_\epsilon - (U_n^- + V^-_{2n+1} + \epsilon^{n+1}\hat{v}^-_{2n+2})\right] \\ &= \frac{1}{2}\epsilon^{n+1}\left\{\frac{\partial^2 u_n^+}{\partial x^2} + \frac{\partial^2 u_n^-}{\partial x^2}\right\}(P(t),t), \end{aligned} \tag{4.12}$$

along the characteristic curve $x = P(t)$. Using the maximum principle, see, for example, Protter and Weinberger (1967), we then have

Theorem IV.1 *The solution $u_\epsilon(x,t)$ of the initial value problem* (1.1), (1.2) *has a uniform approximation $\check{u}(x,t;\epsilon)$ in $\overline{\Omega}$ with error $O(\epsilon^{n+1})$, where $\check{u}$ is defined by the series*

$$\check{u}(x,t;\epsilon) = \sum_{k=0}^{n}\epsilon^k\left\{u_k(x,t) + v_{2k}(\xi,t) + \sqrt{\epsilon}\,v_{2k+1}(\xi,t)\right\}. \tag{4.13}$$

REFERENCE

1. D.G. Aronson, *Linear parabolic differential equations containing a small parameter*, J. Rational Mech. Anal., 5 (1956), pp. 1003 – 1014.

2. J.G. Besjes, *Singular perturbation problems for linear parabolic differential operators of arbitrary order*, J. Math. Anal. Appl., 48 (1974), pp. 594 – 605.

3. L.E. Bobisud, *Parabolic equations with a small parameter and discontinuous data*, J. Math. Anal. Appl., 26 (1969), pp. 208 – 220.

4. C.J. Holland, *Singular perturbations in the first boundary value problem for parabolic equations*, SIAM J. Math. Anal., 8 (1977), pp. 368 – 374.

5. F.A. Howes, *Multi-dimensional initial-boundary value problems with strong nonlinearities*, Arch. Rational Mech. Anal., 91 (1986), pp. 153 – 168.

6. S.L. Kamenomostskaya, *On equations of elliptic and parabolic type with a small parameter in the highest derivative*, Mat. Sbornik N.S., 31(73) (1952), pp. 703 – 708 (Russian).

7. O.A. Ladyženskaya, *On equations with small parameter in the higher derivatives in linear partial differential equations*, Vestnik Lenin-grad. Univ. Ser. Mat. Meh. Astr., 12(7) (1957), pp. 102 – 120 (Russian).

8. O.A. Ladyženskaja, V.A. Solonnikov, and N.N. Ural'ce-va, *Linear and quasilinear equations of parabolic type*, Transl. Math. Monographs, vol. 23, American Mathematical Society, Providence, Rhode Island, 1968.

9. M.H. Protter and H.F. Weinberger, *Maximum principle in differential equations*, Prentice-Hall, Inc., Englewood Cliffs, New Jersey, 1967.

10. S.-D. Shih and R.B. Kellogg, *Asymptotic analysis of a singular perturbation*, SIAM J. Math. Anal., 18 (1987), pp. 1467 – 1511.

Asymptotic Expansions of Solutions to $\Delta u - u = f$ at Infinity

D. Siegel Professor, Department of Applied Mathematics, University of Waterloo, Waterloo, Ontario, Canada

I. INTRODUCTION

Consider solutions to the equation $\Delta u - u = f$ in a neighborhood of infinity, $B_R^e = \{(x,y) : r > R\}$, where r is the usual radial polar coordinate. Meyers and Serrin (1960) have studied such exterior problems for linear elliptic partial differential equations. It follows from Theorem 14 of their paper that for f bounded and Hölder continuous, there exists a unique bounded solution satisfying a Dirichlet condition, u a prescribed continuous function on $r = R$.

Research partially supported under Natural Sciences and Engineering Research Council of Canada Grant A9345.

The present paper is concerned with establishing asymptotic expansions for bounded solutions in the case that f is rapidly decreasing at infinity. We will obtain expansions of the form

$$u \sim K_0(r)[c(\theta) + \frac{c_1(\theta)}{r} + \frac{c_2(\theta)}{r^2} + \cdots] \text{ as } r \to \infty,$$

where r and θ are the usual polar coordinates and $K_0(r)$ is a modified Bessel function. These expansions have a "structure": for $n \geq 1$, $c_n(\theta)$ is expressible as a linear combination of even order derivatives of $c(\theta)$, from order 2 up to order $2n$.

A similar "structured" asymptotic expansion appears in the study of solutions to the Helmholtz equation (see Sommerfeld, 1949). More recent treatments of asymptotic expansions for solutions to linear elliptic partial differential equations in a neighborhood to infinity were carried out by Meyers (1963) and Pazy (1967). Our results are not included in either of these works. Whereas Meyers used integral representations and Pazy used semi-group theory, our approach is by means of differential inequalities.

The paper is organized as follows: in Section II two lemmas are proved; in Section III the main results are stated and proved; in Section IV an application is made to exterior capillary surfaces.

We will use the usual asymptotic symbols $(o, O, \sim)$ where all statements are taken to be as $r \to \infty$, uniformly in θ. Also, $I = I_0(r)$ and $K = K_0(r)$ will denote the usual modified Bessel functions (see Lebedev, 1972). They are two independent solutions of the equation $u_{rr} + \frac{1}{r} u_r - u = 0$, with $I(r)$ being regular at $r = 0$ ($I(0) = 1$) and $K(r)$ having a logarithmic singularity at $r = 0$. Asymptotic expansions of I and K are well-known, in particular $I \sim \frac{e^r}{\sqrt{\pi r}}$ and $K \sim \sqrt{\pi/2r}\, e^{-r}$.

II. TWO LEMMAS

In this section we prove two lemmas needed in the proofs of the theorems in Section III. We will be using the following well-known maximum principle for the operator $L[u] = \Delta u - u$ (see Protter and Weinberger 1984 or Gilbarg and Trudinger 1983).

Maximum Principle If $u \in C^2(\Omega) \cap C(\overline{\Omega})$, where Ω is a bounded domain, and $L[u] \geq 0$ in Ω then $u \leq \max\{\max_{\partial\Omega} u, 0\}$.

Lemma 1 Let $\Delta u - u = f$ in B_R^e and $u = o(I)$.

(a) If $f = O(1)$ then $u = O(1)$.

(b) If $f = o(1)$ then $u = o(1)$

(c) If $f = O(\frac{K}{r^2})$ then $u = O(K)$.

Proof Each statement will be proved in turn, making use of the previous part.

(a) We will show that $|u| \leq C = \max\{\max_{r=R_1} |u|, \max_{r\geq R_1} |f|\}$ for $r \geq R_1$, for any $R_1 > R$. For $\epsilon > 0$, let $V = u - C - \epsilon I$, then $L[V] = f + C \geq 0$ for $r \geq R_1$. Also, $v < 0$ on $r = R_1$. For an arbitrary point (x_0, y_0) with $r_0 > R_1$, choose R_2 so large that $R_2 > r_0$ and $v < 0$ on $r = R_2$ (this is possible in view of the hypothesis $u = o(I)$). By the maximum principle, $v \leq 0$ on the annulus $R_1 < r < R_2$. Hence $v(x_0, y_0) \leq 0$, which gives $u(x_0, y_0) \leq C + \epsilon I(r_0)$. Letting $\epsilon \rightarrow 0$, we obtain $u(x_0, y_0) \leq C$. Replacing u by $-u$ in the above derivation gives $-u(x_0, y_0) \leq C$, thus $|u(x_0, y_0)| \leq C$. Since (x_0, y_0) is arbitrary, we have established that $|u| \leq C$ for $r \geq R_1$.

(b) Fix R_1 and a such that $R_1 > R$ and $0 < a < 1$.

Let $A(R') = \sup_{r\geq R_1} |u|/I(aR')$ for $R' > 0$. Note that $L[AI(ar')] = (a^2 - 1)AI(ar')$, where r' is the distance from (x,y) to (x',y'). Let $B^*_{R'}(x',y')$ denote the disk of radius R' centered at (x',y'). By the hypothesis on $f(f = o(1))$, there exists an $R_2 > R_1$ such that $|f| \leq A(1-a^2)$ for $r \geq R_2$ (R_2 depends on R'). Let $v = u - AI(ar')$, then $L[v] = f + (1 - a^2)AI(ar') \geq f + (1-a^2)A \geq 0$ on $B^*_{R'}$ provided $B^*_{R'} \subset B^e_{R_2}$. For $r' = R'$, $v = u - \sup_{r\geq R_1} |u| \leq 0$, hence by the maximum principle $v \leq 0$ in $B^*_{R'}$, or $u \leq AI(ar')$ in $B^*_{R'}$. It follows that

$u(x',y') \leq A$. Replacing u by $-u$ in the above reasoning yields $|u(x',y')| \leq A$. Thus $|u| \leq A(R')$ for $r \geq R_2 + R'$. Letting $R' \to \infty$, we see that $u = o(1)$.

(c) Fix $R_1 > R$ and $a > 0$, with $R_1 > a$. Let $v = u - AK(r)(1 - \frac{a}{r})$ where

$$A = \max\{\max_{r=R_1} |u|/(K(R_1)(1 - \frac{a}{R_1}), \sup_{r \geq R_1} \frac{|f| r^2}{aK(\frac{1}{r} - \frac{K'}{K})}\}.$$

A is a finite positive number since $K' < 0$, $\frac{K'}{K} \sim -1$, and $f = O(\frac{K}{r^2})$. A calculation shows that $L[v] = f + \frac{aAK}{r^2}[\frac{K'}{K} - \frac{1}{r}] \geq 0$ for $r > R_1$. Also, by the choice of A, $v \leq 0$ on $r = R_1$. Applying the maximum principle to the annulus $R_1 < r < R_2$ yields $v \leq \max\{0, \max_{r=R_2} v\}$. Letting $R_2 \to \infty$ and making use of $u = o(1)$ (hence $v = o(1)$) gives $v \leq 0$ on $B^e_{R_1}$, or $u \leq AK(1 - \frac{a}{r})$ on $B^e_{R_1}$. Replacing u by $-u$ in the above reasoning gives $|u| \leq AK(1 - \frac{a}{r})$ on $B^e_{R_1}$, which implies that $u = O(K)$.

Lemma 2 Let $g \in C^2(B^e_R)$ satisfy $P[g] = g_{rr} + g_r(\frac{1}{r} + \frac{2K'}{K}) = O(\frac{1}{r^n})$ for an integer n, $n \geq 2$, and $g_r = o(e^{2r})$, then there exists a continuous function $c(\theta)$ such that $g(r,\theta) = c(\theta) + O(\frac{1}{r^{n-1}})$. If in addition $g = o(1)$, then $g = O(\frac{1}{r^{n-1}})$.

Proof Multiplying by the integrating factor rK^2, we have $(rK^2 g_r)_r = O(\frac{K^2}{r^{n-1}})$. Integrating from r to ∞ gives $rK^2 g_r = O(\int_r^\infty \frac{K^2(s)}{s^{n-1}} ds)$. Since $\int_r^\infty \frac{K^2(s)}{s^{n-1}} ds \sim -\frac{1}{2} \frac{K^2(r)}{r^{n-1}}$ (by L'Hôpital's rule), we have $g_r = O(\frac{1}{r^n})$. Integrating from r_1 to r_2 gives $|g(r_2,\theta) - g(r_1,\theta)| \leq C|\frac{1}{r_2^{n-1}} - \frac{1}{r_1^{n-1}}|$ for some constant C. Thus,

$g(r,\theta)$ is a uniformly Cauchy sequence of functions of θ, as $r \to \infty$. Therefore, $c(\theta) = \lim_{r \to \infty} g(r,\theta)$ exists and is continuous. Taking the limit as $r_2 \to \infty$ in the preceding expression gives

$$|c(\theta) - g(r_1,\theta)| \leq \frac{C}{r_1^{n-1}} \text{ for } r_1 > R$$

which is the desired result. The second statement of the lemma follows directly from the first.

III. MAIN RESULTS

Our main results are presented in two theorems.

Theorem 1 Suppose that $u \in C^4(B_R^e)$, $\Delta u - u = f$ in B_R^e, $u = o(I)$, $D_\theta^j f = O(\frac{K}{r^2})$ for $j = 0,2$, and $\|f\|_{\alpha,B_2(x,y)} = o(\frac{I}{r^2})$, where $\|f\|_{\alpha,B_2(x,y)}$ is the Hölder norm of f taken over $B_2(x,y)$, the ball of radius 2 centered at (x,y), and $0 < \alpha < 1$. Then there exists a continuous function $c(\theta)$ such that

$$u = K(r)[c(\theta) + 0(\frac{1}{r})] \,.$$

Proof By Lemma 1, $u = O(K)$. The interior Schauder estimate, (see Gilbarg and Trudinger, 1983)

$$\|u\|_{2+\alpha,B_1(x,y)} \leq C(\|f\|_{\alpha,B_2(x,y)} + \max_{B_2(x,y)} |u|) \,,$$

implies that u_x, u_y, u_{xx}, u_{xy}, $u_{yy} = o(\frac{I}{r^2})$, from which we deduce that $u_{\theta\theta} = o(I)$. Let $v = u_{\theta\theta}$, then v satisfies $\Delta v - v = f_{\theta\theta} = O(\frac{K}{r^2})$ and $v = o(I)$. It follows from Lemma 1 that $v = O(K)$. Let $u = K(r)w$, then

$$M[w] = w_{rr} + (\frac{1}{r} + \frac{2K'}{K})\, w_r + \frac{1}{r^2} w_{\theta\theta} = \frac{f}{K} \,.$$

Since $w_{\theta\theta} = \frac{u_{\theta\theta}}{K} = O(1)$ and $\frac{f}{K} = O(\frac{1}{r^2})$ we have $P[w] = w_{rr} + (\frac{1}{r} + 2\frac{K'}{K})w_r = O(\frac{1}{r^2})$. As $w_r = \frac{u_r}{K} - \frac{K'}{K}u = o(\frac{e^{2r}}{r^2})$ (again using the above consequence of the interior Schauder estimate), Lemma 2 implies that there exists a continuous function $c(\theta)$ with $w = c(\theta) + O(\frac{1}{r})$. It follows immediately that $u = K(r)[c(\theta) + O(\frac{1}{r})]$.

Theorem 2 Let N be an integer $N \geq 1$. Suppose that $u \in C^{2N+6}(B^e_R)$, $\Delta u - u = f$ in B^e_R, $u = o(I)$, $D^\ell_\theta f = O(\frac{K}{r^{j+2}})$ for $0 \leq \ell \leq 2N - 2j + 2$, $0 \leq j \leq N$, and $\|D^\ell_\theta f\|_{\alpha, B_2(x,y)} = o(\frac{I(r)}{r^2})$ for $0 \leq \ell \leq 2N+4$. Then there exists a function $c(\theta)$ which is $2N+2$ times continuously differentiable and

$$u = K(r)[c(\theta) + \frac{c_1(\theta)}{r} + \cdots + \frac{c_N(\theta)}{r^N} + O(\frac{1}{r^{N+1}})]$$

where $c_j(\theta)$ is a linear combination of even derivatives of $c(\theta)$ from order 2 up to order $2j$, for $1 \leq j \leq N$.

Proof We first establish that $c(\theta)$ has continuous derivatives up to order $2N+2$. Let $v = u_\theta$, then $\Delta v - v = f_\theta$. Theorem 1 implies that there exists a continuous function $\tilde{c}(\theta)$ such that $v = K(r)[\tilde{c}(\theta) + O(\frac{1}{r})]$. Since $\frac{u}{K} = c(\theta) + O(\frac{1}{r})$ and $\frac{u_\theta}{K} = \tilde{c}(\theta) + O(\frac{1}{r})$, uniformly in θ, we see that $c'(\theta)$ exists and $c'(\theta) = \tilde{c}(\theta)$. Repeating this argument gives that $c^{(j)}(\theta)$ exists and is continuous for $1 \leq j \leq 2N+2$.

Next, we explain how the coefficients $c_1(\theta), \ldots, c_N(\theta)$ are to be chosen. Let $u = Kw$ as above. We have $M[w] = \frac{f}{K} = O(\frac{1}{r^{N+2}})$. Let $w_j = c(\theta) + \frac{c_1(\theta)}{r} + \cdots + \frac{c_j(\theta)}{r^j}$ for $1 \leq j \leq N$. Using the asymptotic expansion for K, we can give the asymptotic expansion $\frac{1}{r} + \frac{2K'}{K} = -2 + \frac{1}{4r^2} + \cdots$, an expansion in powers of $\frac{1}{r}$. Putting this into the expression for $M[w_j]$ we can choose $c_i(\theta)$, $1 \le i \le j$, so that the terms with factors $\frac{1}{r^2}, \ldots, \frac{1}{r^{j+1}}$ cancel out. We therefore have $M[w_j] = O(\frac{1}{r^{j+2}})$. As the coefficients of M do not depend on θ, we also have $M[D_\theta^\ell w_j] = O(\frac{1}{r^{j+2}})$ for $0 \le \ell \le 2N - 2j + 2$. The first few coefficients are

$$c_1(\theta) = -\frac{1}{2}c^{(2)}(\theta),\ c_2(\theta) = \frac{1}{8}c^{(4)}(\theta),\ c_3(\theta) = \frac{1}{48}(c^{(2)}(\theta) + c^{(6)}(\theta))\ .$$

In general we see that $c_i(\theta)$ is a linear combination of even derivatives of $c(\theta)$ from order 2 up to order $2i$.

Next, we prove the statements $D_\theta^\ell(w - w_j) = O(\frac{1}{r^{j+1}})$ for $0 \le j \le N$, $0 \le \ell \le 2N - 2j$, by induction on j. Consider $z = w - c(\theta)$ (define $w_0 = c(\theta)$). By the first paragraph of the present proof $D_\theta^\ell z = D_\theta^\ell w - c^{(\ell)}(\theta) = O(\frac{1}{r})$ for $0 \le \ell \le 2N + 2$. Since $M[D_\theta^\ell z] = O(\frac{1}{r^3})$ then $P[D_\theta^\ell z] = O(\frac{1}{r^3})$ for $0 \le \ell \le 2N$. As $D_\theta^\ell z = o(1)$ and $D_r D_\theta^\ell z = o(\frac{e^{2r}}{r^2})$ (arguing as in the proof of Theorem 1), Lemma 2 implies that $D_\theta^\ell z = O(\frac{1}{r^2})$ for $0 \le \ell \le 2N$. We now assume the result is true for j and prove it for $j + 1$, $0 \le j \le N-1$. Let $z = w - w_{j+1}$, then $D_\theta^\ell z = D_\theta^\ell(w - w_j) - \frac{C_{j+1}^{(\ell)}(\theta)}{r^{j+1}} = O(\frac{1}{r^{j+1}})$ for $0 \le \ell \le 2N - 2j$. Since $M[D_\theta^\ell z] = O(\frac{1}{r^{j+3}})$ then $P[D_\theta^\ell z] = O(\frac{1}{r^{j+3}})$ for $0 \le \ell \le 2N - 2j - 2$, hence Lemma 2 implies that $D_\theta^\ell z = O(\frac{1}{r^{j+2}})$. This completes the induction step.

Putting $j = N$, $\ell = 0$, in the preceding result gives $w = w_N + O(\frac{1}{r^{N+1}})$, which then yields the conclusion

$$u = K[c(\theta) + \cdots + \frac{c_N(\theta)}{r^N} + O(\frac{1}{r^{N+1}})] .$$

Comments The condition $\|D_\theta^\ell f\|_{\alpha,B_2(x,y)} = o(\frac{I(r)}{r^2})$ can be replaced by $D_r D_\theta^\ell f = o(\frac{I(r)}{r^2})$ for $0 \leq \ell \leq 2N+4$ and $D_\theta^\ell f = o(\frac{I(r)}{r^2})$ for $2N+3 \leq \ell \leq 2N+5$. The other condition on f is equivalent to $D_\theta^\ell f = O(\frac{K}{r^{\ell'+2}})$ where $\ell' = \min\{N, [N - \frac{\ell}{2} + 1]\}$ for $0 \leq \ell \leq 2N+2$ (where $[a]$ indicates the largest integer $\leq a$).

IV. EXTERIOR CAPILLARY SURFACES

If a vertical cylindrical rod with cross-section Ω is immersed in an infinite resevoir of fluid, a capillary surface is formed. The height $u(x_1,x_2)$ above the undisturbed level of the fluid satisfies the equation

$$\frac{\Delta u}{(1+|\nabla u|^2)^{1/2}} - \sum_{i,j=1}^{2} \frac{u_{x_i} u_{x_j} u_{x_i x_j}}{(1+|\nabla u|^2)^{3/2}} = \kappa u \text{ in } \Omega^e ,$$

where κ is a positive constant and $\Omega^e = \{(x_1,x_2) : (x_1,x_2) \in \mathbb{R}^2 - \overline{\Omega}\}$. By the change of variables $\tilde{u} = \frac{1}{\sqrt{\kappa}} u(\sqrt{\kappa}\, x_1, \sqrt{\kappa}\, x_2)$ the constant can be taken to be one. There is also a nonlinear boundary condition on $\partial\Omega$ (for background see Finn, 1986). In (Siegel, 1980) it was shown that any solution u (with $\kappa = 1$) in an exterior domain Ω^e (i.e. Ω is a bounded domain) satisfies $D_{x_1}^i D_{x_2}^j u = O(K_0(r))$ for all i, j. Because of these bounds, we know that u satisfies

$$\Delta u - u = f \text{ with } D_{x_1}^i D_{x_2}^j f = O(K_0^3(r)) \text{ for all } i,\ j .$$

Theorem 2 of Section III in this case yields a complete asymptotic expansion

$$u \sim K_0(r)[c(\theta) + \frac{c_1(\theta)}{r} + \cdots] \text{ as } r \to \infty .$$

It is not known if the coefficient $c(\theta)$ is a constant. If it were, we would have

$$u = K_0(r)[c + O(\frac{1}{r^N})] \text{ as } r \to \infty$$

for any N.

For the homogeneous equation $\Delta u - u = 0$ there are solutions of the form $u = K_n(r)[a\cos n\theta + b\sin n\theta]$, in which case $c(\theta)$ in the expansion is not a constant.

REFERENCES

1. R. Finn (1986). *Equilibrium Capillary Surfaces*, Springer-Verlag, New York.
2. D. Gilbarg and N.S. Trudinger (1983). *Elliptic Partial Differential Equations of Second Order*, Second Edition, Springer-Verlag, Berlin, p. 33 and p. 90.
3. N.N. Lebedev (1972). *Special Functions and Their Applications*, Dover, New York.
4. N. Meyers (1963). *J. Math. and Mech.* **12**: 247.
5. N. Meyers and J. Serrin (1960). *J. Math. and Mech.* **9**: 513.
6. A. Pazy (1967). *Archive Rat. Mech. Anal.* **24**: 193.
7. M.H. Protter and H.F. Weinberger (1984). *Maximum Principles in Differential Equations*, Springer-Verlag, New York, p. 64.
8. D. Siegel (1980). *Pacific J. Math* **88**: 471.
9. A. Sommerfeld (1949). *Partial Differential Equations in Physics*, Academic Press, New York, p. 191.

Approximating Zeros of Solutions of Second-Order Linear ODEs by "Phase Function" Methods

R. Spigler Professor, Department of Mathematical Methods and Models for Applied Science, University of Padova, Padova, Italy

M. Vianello Professor, Department of Pure and Applied Mathematics, University of Padova, Padova, Italy

I. INTRODUCTION

Among the methods that can be adopted to approximate asymptotically and compute zeros of solutions of second-order linear oscillatory differential equations,

$$y''+q(x)y=0 \quad , \tag{1}$$

This work was supported, in part, by the Italian Ministry of Education (M.P.I. Funds).

a few can be based on the derivation of a "phase function", $\alpha(x)$, relatively to a given basis. Such a function is given by any C^3 solution to equation

$$\operatorname{tg} \alpha(x) = \frac{u(x)}{v(x)} , \tag{2}$$

$u(x)$ and $v(x)$ being any two linearly independent solutions to (1), cf. [4]. A phase function has the property that

$$\left| \alpha(x_k) - \alpha(x_{k-1}) \right| = \pi , \quad k = 1,2,3,\ldots \tag{3}$$

for *any* two consecutive zeros of *any* particular solution to (1), although it depends on the special choice of the basis $(u(x),v(x))$. By inverting relation (3) one can thus compute the zeros, x_k. On the other hand, we obtain from (2)

$$\alpha'(x) = -\frac{W}{u^2(x)+v^2(x)} , \tag{4}$$

where W is the (constant) Wronskian of $u(x),v(x)$.

Equation (4) itself can be used to obtain $\alpha(x)$ whenever u and v are at hand, cf., e.g.,[10]. Moreover, it is possible to construct explicitly a third-order linear differential equation satisfied by $u^2(x)+v^2(x)$, cf. [2,3,11]. Asymptotic and/or numerical approximation of the solutions to such an equation can be then used in order to approximate $\alpha(x)$ from (4). However, it is possible to eliminate u,v in (4) thus obtaining a close-form *nonlinear* third-order differential equation (that is a special case of the so-called Kummer equation) for $\alpha(x)$ alone:

$$\alpha'(x)^2 = q(x) - \frac{1}{2}\{\alpha(x),x\} , \tag{5}$$

where

$$\{\alpha,x\} \equiv \frac{\alpha'''}{\alpha'} - \frac{3}{2}\left(\frac{\alpha''}{\alpha'}\right)^2 \tag{6}$$

denotes the so called "Schwarzian derivative". Equation (5) appears in the theory of the transformations of second-order linear differential equations, cf. [4]. Note that, in this approach, one avoids computing the solutions u,v to equation (1) as well as their combination u^2+v^2.

As it is well known from Boruvka's theories [4], the concept of phase plays a central role in describing solutions and properties of second-order linear differential equations like (1). This provides a key to study any particular solution to (1) as well as its zeros. Approximating a certain phase according to our procedure below will enable us to compute globally the zeros of *any* particular solution.

The last procedure was essentially introduced by F.W.J. Olver in 1950 [6] to compute zeros of cylinder functions. Olver actually used the inverse function $r=r(\alpha)$ that yields directly the generic zero as a solution to

$$r'^2 = \frac{1}{q(r)}\left[1-\frac{1}{2}\{r,\alpha\}\right] \tag{7}$$

(it is easily checked that equations (5) and (7) change one into each other by the transformations $r=r(\alpha), \alpha=\alpha(r)$). In view of equation (7) and of the fact that $\{r,\alpha\}=O(r^{-3})$ as $r \longrightarrow +\infty$ as well as observing tables of differences of the zeros of Bessel functions [7], Olver used the following iterative scheme

$$\begin{aligned} r_0'^2 &= \frac{1}{q(r_0)}, \\ r_{n+1}'^2 &= \frac{1}{q(r_{n+1})}\left[1-\frac{1}{2}\{r_n,\alpha\}\right], \quad n = 0,1,2,\ldots \end{aligned} \tag{8}$$

We found more convenient, instead, basing our approach on equation (5) rather than (7). In [8], therefore, we introduced the iterative scheme

$$\begin{aligned} \alpha_0'^2 &= q(x) \\ \alpha_{n+1}'^2 &= q(x) - \frac{1}{2}\{\alpha_n,x\} \end{aligned} \tag{9}$$

Note that Olver's method requires solving a differential equation at each step (for $r_{n+1}(\alpha)$), while ours merely consists of quadratures. We must however obtain the zeros from $\alpha_n(\cdot)$, while Olver's approach yields them directly. Moreover, Olver's procedure was not rigorously justified in that, for example, no proof of convergence of the algorithm (8) was given at that time. In [8], we proved that the function α_n' defined by (9), with $q(\cdot)$ in some class, converges (in a suitable sense) to a solution, α', of (5), possessing a certain asymptotic behavior. This solution, α', is such that relation (4) holds, (u,v) being a Liouville-Green basis (cf. [5]), and W the (constant) Wronskian of u and v. The initialization itself in (9) has been suggested from the fact that we want to approximate *such a* phase function. We can call this phase function a "Liouville-Green phase".

There are two classes of coefficients q(x) to which our results apply. The first is such that

(A) $q(x) = a+Q_1(x)$

where $a>0$ and Q_1 has a holomorphic extension to a certain (*annular*) sector, S, of the complex plane that contains a real positive half-line, cf. [8]. Moreover, it is required that $Q_1(z) = O(z^{-p})$ as $z\longrightarrow\infty$ in S, $p>1$. The constants implied by the O-symbols that will appear below, depend, in general, on all the parameters entering the problem. Sometimes we shall display their dependence on *a*. The second class consists of the functions of the form

(B) $q(x) = ax^m+Q_2(x)$,

where $a>0$, $m>0$ and Q_2 can be holomorphically continued into a certain sector, S, of the complex plane containing a real positive half-line, cf. [9]. Besides, $Q_2(z) = O(z^\delta)$ as $z\longrightarrow\infty$ in S, $\delta < m/2-1$. The gap between m and δ might be unnecessary and perhaps is needed just for technical reasons.

The two classes are genuinely different, and we must make a distinction in discussing them as we have to treat in a different way the cases m=0 and m>0. Below, we shall refer to these two classes as to "class A" and "class B".

II. THE APPROXIMATING SCHEME.

In (9), we set

$$\Phi_n \equiv(\alpha_n')^2 \quad , \quad n = 0,1,2,\dots \quad , \tag{10}$$

thus obtaining the scheme

$$\begin{aligned} \Phi_0 &= q(x) \\ \Phi_{n+1} &= q(x)+[\Phi_n,x] \quad , \quad n = 0,1,2,\dots \quad , \end{aligned} \tag{11}$$

where we introduced the symbol

$$[\Phi,x] \equiv -\frac{1}{4}\frac{\Phi''}{\Phi}+\frac{5}{16}\left(\frac{\Phi'}{\Phi}\right)^2 \tag{12}$$

In this approach Φ_n will approximate Φ, the solution to

$$\Phi = q(x) + [\Phi,x], \tag{13}$$

into which equation (5) is transformed by setting

$$\Phi \equiv (\alpha')^2 \tag{14}$$

This procedure has the advantage of avoiding handling square roots (which gives some complications when working in the complex plane). On the other hand, there is no ambiguity in deciding the sign of α' knowing Φ, as it turns out that $W = 1$ (cf. (4)) for the real Liouville-Green basis (see (18) below).

As in proving the convergence of Φ_n to Φ, estimates are needed on the derivatives of $\Phi_n-\Phi$ (cf. $[\Phi_n,x]$ in (11)), we "complexify" all quantities involved, extending them from the real half-line into a sector S. We start with a *complex* Liouville-Green basis in S. Following Olver [5], we split the coefficient $q(z)$ into two parts, $f(z)$ and $g(z)$,

$$q(z) = f(z) + g(z) \quad , \tag{15}$$

where we chose

$$f(z) = a, \quad g(z) = Q_1(z) = O(z^{-p}) \quad , \tag{16'}$$

for class A, and

$$f(z) = az^m, \ g(z) = Q_2(z) = O(z^s) \quad , \tag{16''}$$

for class B. Theorem 11.1 in [5,Ch.6] yields the basis

$$U(z) = (f(z))^{-1/4} \exp\{i\int^z (f(t))^{1/2}dt\}(1+\varepsilon_1(z))$$
$$V(z) = (f(z))^{-1/4} \exp\{-i\int^z (f(t))^{1/2}dt\}(1+\varepsilon_2(z)) \quad , \tag{17}$$

and the splitting of q(z) as above guarantees that $\varepsilon_j(z) = o(1)$, $f^{-1/2}\varepsilon_j'(z) = o(1)$, $j = 1,2$, as $z \to \infty$ in S. More precisely we got $\varepsilon_j(z) = O(a^{-1/2}z^{1-p})$ for class A, and $\varepsilon_j(z) = O(z^{-b})$, where $b = \min\{1+m/2,-1-\delta+m/2\}$, for class B. With the splitting adopted it was very simple obtaining the so-called "ξ-progressive" paths, as required by Theorem 11.1 of [5,Ch.6].

We then derived from (17) a *real* Liouville-Green basis as

$$u(x) \equiv \frac{U(x)+V(x)}{2}$$
$$v(x) \equiv \frac{U(x)-V(x)}{2i} \quad ,x \in S\cap R \tag{18}$$

(in fact, $V(x) = \overline{U(x)}$) and

$$\Phi \equiv \frac{W^2}{(u^2(x)+v^2(x))^2} = \frac{1}{(U(x)V(x))^2} \tag{19}$$

Then we extended $\Phi(x)$ into the complex plane, in a subsector S' of S where $U(z)V(z) \neq 0$. At the same time, equation (13) can be considered for x in S' rather than for real values of x only.

Now we are ready to get an asymptotic approximation for Φ, preliminary to the numerical work. From (19) and the asymptotic behavior of $\varepsilon_j(z)$, we first obtain

$$\Phi(z) = a+O(z^{1-p}), \quad z \to \infty \text{ in } S' \tag{20'}$$

correspondingly to class A, and

$$\Phi(z) = az^m+O(z^{m-b}), \quad z \to \infty \text{ in } S' \tag{20''}$$

for the other class. Introducing these relations in the right hand side of equation

$$\Phi = q(z) + [\Phi(z),z], \tag{21}$$

we get the refinements

$$\Phi(z) = q(z) + O(a^{-1/2}z^{-1-p}), \quad z \longrightarrow \infty \text{ in } S'' \quad , \tag{22'}$$

and

$$\Phi(z) = q(z) + O(z^{-2}), \quad z \longrightarrow \infty \text{ in } S'' \quad , \tag{22''}$$

respectively. Here S" is a suitable subsector of S',

$S'' \equiv \{z \in C: |z| \geq r > 0, \ |\arg(z)| < \Theta \leq \pi/2\}$.

Restriction of S' to S" is due to the differentiation of the O-symbols, cf. [5,Ch.1,Th.4.2].

Denote by $S(0,\Theta)$ the sector with its vertex in the origin and semiangle Θ. Then consider the sequence of decreasing sectors

$$S_n = x_n + S(0,\Theta) \quad , \tag{23}$$

x_n being the increasing sequence of their vertices. Here are the main results.

THEOREM II.1

Suppose that equation (1) is given, with q(x) in class A.

Then there exists $x_0 \in S''\cap\mathbf{R}$, that depends, in general, on all parameters entering the problem, such that the quantity $\Phi_n(z)$ given by the complexified scheme (11) "converges" to the function $\Phi(z)$ related to the Liouville-Green basis as in (19).

The convergence is intended in the sense that, for any $\beta > a^{-1/2}$,

$$|\Phi_n(z)-\Phi(z)| = a^{-1/2} \ (\beta a^{1/2})^{-2n} \ O((|z|-n\beta 2^{1/2})^{-1-p}), \tag{24}$$

n=0,1,2,...,for $z \in S_n$, where

$$x_n = x_0 + \frac{n\beta 2^{1/2}}{\sin\Theta} \quad , \tag{25}$$

(cf. (23)).

THEOREM II.2

If q(x) in (1) belongs to class B, then there exists a quantity x_0, similarly to Theorem II.1, so that the approximation of $\Phi_n(z)$ to $\Phi(z)$

is intended as

$$|\Phi_n(z)-\Phi(z)| = h_n(z)\ O((|z|-n2^{1/2})^{-2}) \quad , \tag{24'}$$

n=0,1,2,..., for $z \in S_n$, where

$$x_n = x_0 + \frac{n2^{1/2}}{\sin\Theta} \quad , \tag{25'}$$

and

$$\begin{aligned} h_0(z) &\equiv 1, \\ h_n(z) &= \prod_{j=0}^{n-1} O((|z|-j2^{1/2})^{-m}) \quad ,n=1,2,3,\ldots \end{aligned} \tag{26}$$

We do not give here the details of the proofs of Theorems II.1, II.2, that were obtained by induction on *n*. We only notice that a particular care has been paid to estimating the Lipschitz constant of the function

$$G(w_1,w_2,w_3) \equiv \frac{1}{4}\frac{w_3}{w_1} - \frac{5}{16}\frac{w_2{}^2}{w_1{}^2} \tag{27}$$

of the three *complex* variables w_1,w_2,w_3, defined in a certain polydisc. In fact,

$$|\Phi_{n+1}(z)-\Phi(z)| = |G(\Phi_n,\Phi_n',\Phi_n'')-G(\Phi,\Phi',\Phi'')| \tag{28}$$

The polydisc referred to above is determined by the asymptotic structure of Φ, [8,9].

As a final remark, we observe that the "geometric structure" of the sequence of sectors S_n was suggested by the need of controlling the differentiation operators in the algorithms.

III. NUMERICAL EXAMPLES.

If $\alpha(x)$ denotes *a* Liouville-Green phase, relation (3) yields

$$\alpha(x_k)-\alpha(x_{k-1}) = \int_{x_{k-1}}^{x_k}(\Phi(t))^{1/2}dt = \pi \tag{29}$$

for *any* two consecutive zeros, x_{k-1},x_k, of *any* particular solution of (1). Having the n-th approximation to Φ, Φ_n (see Theorems II.1, II.2), we can solve numerically equation (29) for x_{k-1}, given x_k. In practice, we start from *a known zero*, say σ, and come down to compute the zeros smaller than σ.

It is interesting, at this point, to comment on the *rate of convergence* of our algorithms. We first re-write the estimates (24), (24') given by Theorems II.1, II.2, *uniformly* in z, for $z \in S_n$. In fact, as $|z| - n\beta 2^{1/2} > x_0$ and $|z| - n2^{1/2} > x_0$, respectively, in S_n, we get

$$|\Phi_n(z)-\Phi(z)| = a^{-1/2} \ (\beta a^{1/2})^{-2n} \ O(x_0^{-1-p}) \quad , \tag{30}$$

$$|\Phi_n(z)-\Phi(z)| = \{\prod_{j=0}^{n-1} O((x_{n-j})^{-m})\} \ O(x_0^{-2}) \quad , \tag{30'}$$

cf. [8,9]. Using the fact that $x_{n-j}=x_0+(n-j)2^{1/2}/\sin\Theta$, the latter equation yields the estimate

$$|\Phi_n(z)-\Phi(z)| = \{\prod_{j=0}^{n-1} O([(n-j)2^{1/2}/\sin\Theta]^{-m})\} \ O(x_0^{-2})$$

$$\leq K^n \ (\sin\Theta/2^{1/2})^{mn} \ (n!)^{-m} \ O(x_0^{-2}) \quad , \tag{30''}$$

where K estimates the constants implied by the O-symbols appearing in the products in (30'). Notice that

$$K^n(\sin\Theta/2^{1/2})^{mn}(n!)^{-m} \equiv (A^n/n!)^m \sim \ [(Ae/n)^n(2n\pi)^{-1/2}]^m \quad , \tag{31}$$

where we set

$$A \equiv K^{1/m} \ \sin\Theta/2^{1/2} \quad , \tag{32}$$

for short, and Stirling approximation formula (valid as $n \longrightarrow \infty$) has been used. This result makes it clear that the error in (30') is *more than* exponentially small as $n \longrightarrow \infty$ (while in (30) it is exponentially small). The sequence x_n of vertices of sectors S_n grows, instead, just linearly with *n*.

Here we give *two* examples, one in class A and one in class B. The first is about a well known special function. As its zeros are tabulated with ten digits in [1], this example provides a way to check our results. Other examples in this class can be found in [8]. The second example considers a nontrivial case that can be thought of as a perturbed generalized Airy equation. Further examples in this second class are given elsewhere (cf. [9]). In practice, as in such examples q(x) is an elementary function, we took advantage of using symbolic manipulations (with MACSYMA), to obtain $\Phi_n(x)$ according to our procedure.

A. *EXAMPLE 1.*

We consider the case of the zeros of Bessel function $J_0(x)$. We tested our method by evaluating the discrepancy

$$E_{k,n} \equiv \left| \int_{x_{k-1}}^{x_k} (\Phi_n(t))^{1/2} dt - \pi \right| \tag{33}$$

between the integrals in (29) and π, when Φ is approximated by Φ_n, for several values of n and k (cf. (29)). These quantities are reported in Table 1. In Table 2, we give approximations for some zeros of $J_0(x)$ computed by solving for x equation

$$\int_{x}^{x_k} (\Phi_n(t))^{1/2} dt = \pi \quad , \tag{34}$$

with x_k given. In parentheses, we show the differences (in absolute value) between the zeros we computed (with 20 digits) and those tabulated in [1]. The x_k's, k=20,15,10,5,2 , have also been taken from [1]. The number appearing in the square brackets represents the number of iterations needed to solve equation (34) by Newton-Raphson method.

B. *EXAMPLE 2.*

As an example pertaining to class B, we consider equation (1) with

$$q(x) = x^2 - \frac{3}{4x^2} \tag{35}$$

This coefficient has been constructed from the phase function

$$\alpha(x) = x^2/2 \tag{36}$$

TABLE 1 $E_{k,n}$, defined in (33), for several values of k and the iteration number, n, in Example 1.

n \ k	0	1	2
2	$3.0\ 10^{-3}$	$1.0\ 10^{-3}$	$8.4\ 10^{-4}$
5	$1.9\ 10^{-5}$	$5.3\ 10^{-7}$	$3.2\ 10^{-8}$
10	$8.3\ 10^{-7}$	$4.9\ 10^{-9}$	$7.4\ 10^{-11}$
15	$1.5\ 10^{-7}$	$3.5\ 10^{-10}$	$1.7\ 10^{-11}$
20	$4.4\ 10^{-8}$	$1.0\ 10^{-10}$	$4.6\ 10^{-11}$

TABLE 2 *The computed zeros for Example 1, for several k's and n's. The departures from the values tabulated in Ref. [1] are given in parentheses. The number in square brackets is the Newton-Raphson iteration number.*

k \ n	0	1	2
1	2.407906271 (2.8 10^{-3}) [2]	2.4037934169 (1.0 10^{-3}) [2]	2.4056580109 (8.3 10^{-4}) [2]
4	11.7915531688 (1.9 10^{-5}) [2]	11.7915339077 (5.3 10^{-7}) [2]	11.7915344714 (3.2 10^{-8}) [2]
9	27.4934799598 (8.2 10^{-7}) [2]	27.4934791271 (4.9 10^{-9}) [2]	27.4934791321 (1.0 10^{-10}) [2]
14	43.1997918600 (1.5 10^{-7}) [2]	43.1997917128 (4.0 10^{-10}) [2]	43.1997917132 (0.0) [2]
19	58.9069839701 (4.4 10^{-8}) [2]	58.9069839260 (1.0 10^{-10}) [2]	58.9069839261 (0.0) [2]

TABLE 3 *$E_{k,n}$, defined in (33), for several values of k and the iteration number, n, in Example 2.*

n	0	1	2
k			
10	$2.7\ 10^{-4}$	$3.8\ 10^{-7}$	$1.7\ 10^{-9}$
20	$7.1\ 10^{-5}$	$2.6\ 10^{-8}$	$3.1\ 10^{-11}$
30	$3.2\ 10^{-5}$	$5.2\ 10^{-9}$	$2.8\ 10^{-12}$
50	$1.2\ 10^{-5}$	$6.9\ 10^{-10}$	$1.4\ 10^{-13}$
100	$2.9\ 10^{-6}$	$4.4\ 10^{-11}$	$2.2\ 10^{-15}$
200	$7.4\ 10^{-7}$	$2.8\ 10^{-12}$	$3.6\ 10^{-17}$

and (5). On the other hand, the solutions to equation $\sin\alpha(x)=0$, i.e. $x_k \equiv (2k\pi)^{1/2}$, represent the zeros of a particular solution of equation (1) (with $q(x)$ given by (35)), that we computed by our method. As in Example 1, we give in Table 3 the quantities $E_{k,n}$ (see (33)), and in Table 4 the computed zeros along with the differences between them and the exact zeros $(2k\pi)^{1/2}$, as well as the number of Newton-Raphson iterations.

TABLE 4 *The computed zeros for Example 2, for several k's and n's. The departures from the "exact" values are given in parentheses. The number in square brackets is the Newton-Raphson iteration number.*

k \ n	0	1	2
10	7.9266254792284643654 ($2.9\ 10^{-5}$) [2]	7.9266546426586107448 ($4.7\ 10^{-8}$) [3]	7.9266545949937568359 ($2.2\ 10^{-10}$) [3]
20	11.209976590712628801 ($5.8\ 10^{-6}$) [2]	11.209982435091436927 ($2.3\ 10^{-9}$) [3]	11.209982432793094807 ($2.8\ 10^{-12}$) [3]
30	13.729366279764809413 ($2.2\ 10^{-6}$) [2]	13.729368493338627878 ($3.8\ 10^{-10}$) [3]	13.729368492956327162 ($2.1\ 10^{-13}$) [3]
50	17.724537848804636876 ($6.4\ 10^{-7}$) [2]	17.724538509094517135 ($3.9\ 10^{-11}$) [3]	17.724538509055152465 ($7.8\ 10^{-15}$) [3]
100	25.066282630357686889 ($1.2\ 10^{-7}$) [2]	25.066282746311778789 ($1.7\ 10^{-12}$) [3]	25.066282746310004935 ($8.9\ 10^{-17}$) [3]
200	35.449076997341023294 ($2.1\ 10^{-8}$) [2]	35.449077018110399716 ($7.9\ 10^{-14}$) [3]	35.449077018110320545 ($9.7\ 10^{-19}$) [3]

IV. FINAL REMARKS.

The approximations for Φ, Φ_n, derived in Theorems II.1, II.2, are of asymptotic nature. They can be used, as described in Sec.III, as a starting point to obtain better (numerical) approximations for the zeros of any given differential equation of type (1), with q(x) in class A or class B. This last step is beset by the *numerical instability* arising from the approximate evaluation of certain derivatives appearing in our algorithms.

In many important cases when q(x) is explicitly known in terms of elementary functions, it is possible to take advantage of *symbolic manipulation*, i.e. the so called, computer algebra techniques. Though, in principle, *only roundoff errors* will now affect our method, in practice we have to cope with an *exponential computational complexity*. Convergence is, however, rather fast (cf. (24) and (24')), so that *satisfactory results* are usually attained *within few iterations*.

REFERENCES

1. M. Abramowitz and I. A. Stegun, Ed.s, *Handbook of Mathematical Functions*, Dover, New York, 1968.
2. P. Appel, *Sur la transformation des équations différentielles linéaires, C.R. Acad. Sci. Paris 91:* 211 (1880).
3. R. Bellman, *On the linear differential equations whose solutions are the products of the solutions of two given linear differential equations, Boll. Un. Mat. Ital. (3) 12:* 12 (1957).
4. O. Borůvka, *Linear Differential Transformations of the Second Order*, The English Universities Press, London, 1971.
5. F. W. J. Olver, *Asymptotics and Special Functions*, Academic Press, New York, 1974.
6. F. W. J. Olver, *A new method for the evaluation of zeros of Bessel functions and of other solutions of second-order differential equations, Proc. Cambridge Philos. Soc. 46:* 570 (1950).
7. F. W. J. Olver, *private communication*.

8. R. Spigler and M. Vianello, *A numerical method for evaluating zeros of solutions of second-order linear differential equations,* to appear.
9. R. Spigler and M. Vianello, *The phase function method for approximating zeros of solutions of certain second-order linear differential equations,* to appear.
10. R. Spigler, *Alcuni risultati sugli zeri delle funzioni cilindriche e delle loro derivate, Rend. Sem. Mat. Univers. Politec. Torino 38:* 67 (1980).
11. R. Spigler, *The linear differential equations whose solutions are the products of solutions of two given differential equations, J. Math. Anal. Appl. 98:* 130 (1984).

Asymptotic Expansions of Integrals of Two Bessel Functions

B. J. Stoyanov
R. A. Farrell
J. F. Bird

The Johns Hopkins University, Applied
Physics Laboratory, Laurel, Maryland

I. INTRODUCTION

A number of problems in applied mathematics and physics lead to results represented by definite integrals containing products of special functions, such as Bessel functions, whose argument involves a parameter of certain physical content. Asymptotic behavior of the integrals for large positive values of the parameter is usually of particular interest in these problems. However, the evaluation of the integrals in this limit is, as a rule, not a trivial task. Direct numerical methods for accurate estimations of such integrals can, at

---

This work was supported in part by the U.S. Navy and the National Institutes of Health.

best, provide sets of numbers and are of little use when the parameter is large enough. Infinite series representations of the integrals in ascending powers of the parameter also are of little value since poor convergence means that many terms are required to attain the necessary accuracy. In general, these series are not readily transformable into representative asymptotic series in inverse powers of the large parameter. Thus, direct asymptotic expansions of the integrals are required to obtain convenient analytic expressions that are useful for large values of the parameter.

The literature on methods of asymptotic expansions of integrals is quite extensive; see, for example, publications by Erdelyi (1956), Olver (1974), Bleistein and Handelsman (1986), and references therein. However, the bulk of the literature is devoted to classical methods specifically designed for integrals of what may be called canonical types, such as Laplace, Fourier, Airy, and probability integrals among others. These are often further subdivided into subclasses of integrals with exponential, algebraic, trigonometric or oscillatory, etc., kernels. In practice, such a classification of integrals involving products of Bessel functions is usually not clear-cut, or may even be impossible, without preliminary, not always obvious, transformations of the integrals. Then it may not be at all obvious, at first, what asymptotic method or methods will work on a particular integral. Therefore it is often desirable and reassuring from the standpoint of an applied researcher to have available, in addition to more sophisticated and systematically developed methods, a straightforward, unsophisticated asymptotic approach.

Two such different methods, which are complementary in character, will be compared and illustrated here. The straightforward method is a heuristic approach that readily produces asymptotic formulae and leading correction terms for integrals of two Bessel functions of arbitrary (generally different) orders but of the same argument involving a large parameter. It proceeds by introducing auxiliary parameters, by means of which the integrals are partitioned into parts that can be either easily evaluated or reduced to known results. The other method, employing Mellin transforms in formally abstruse but well-established recipes, is better suited than the straightforward method for routinely generating very high order terms.

II. SPECIFICATION OF THE PROBLEM AND RESULTS

The two methods briefly described above will be used to obtain explicit asymptotic expansions of the following integrals:

$$I_1(\lambda) = \int_0^{\pi/2} dx\ J_\mu(\lambda \sin x) J_\nu(\lambda \sin x) , \tag{1}$$

$$I_2(\lambda) = \int_0^\infty \frac{dt}{1+t^2} J_\mu(\lambda t) J_\nu(\lambda t) , \tag{2}$$

where J_μ, J_ν are Bessel functions of first kind of orders μ, ν, and λ is a large positive parameter. These integrals arise in various practical contexts; specifically, the first integral is of a type encountered in a variational formulation of diffraction theory by Stoyanov and Farrell (1987,1988) and in crystallography (see Wong (1988)), while the second occurs in variational grating theory of Bird and Gray (1989). The first integral, with an obvious change of variables, can be put into the form of the second integral, so that the two may be written as the general h-transform of Bleistein and Handelsman (1986),

$$I_i(\lambda) = \int_0^\infty dt\ f_i(t) h(\lambda t) , \quad i = 1, 2 , \tag{3}$$

where the kernel h(t) is defined by

$$h(t) \equiv h_{\mu\nu} = J_\mu(t) J_\nu(t) , \tag{4}$$

while

$$f_1(t) = \begin{cases} \dfrac{1}{\sqrt{1-t^2}} , & 0 \le t \le 1 , \\ 0 , & t > 1 , \end{cases} \tag{5}$$

and

$$f_2(t) = \frac{1}{(1+t^2)} , \quad 0 \le t < \infty . \tag{6}$$

Note that, formally, the first integral has now an infinite range of integration, but its integrand function $f_1(t)$ is singular and discontinuous at t=1. Detailed calculations were presented for this integral by Stoyanov and Farrell (1987) for the particular case of $\mu=\nu=0$ using the straightforward method. A more general case of $\mu=\nu>-\frac{1}{2}$ was considered by Wong (1988) via clever application of Mellin transform techniques systematically developed in Bleistein and Handelsman (1986) and also in Wong (1983). In the following sections we shall consider the case of general μ,ν by both methods (with certain simplifications and modifications) for direct comparison. The results will include the earlier special $\mu=\nu$ formulae.

The second integral of Eq. (3) has a regular integrand function $f_2(t)$ and is one of Hankel's infinite integrals, which can be evaluated in closed form in great generality (including all λ) for $\mu+\nu$ = *odd* integer (see Watson (1944)). However for all other cases, including the scattering application where $\mu+\nu$ is an *even* integer, not even an asymptotic formula is available. Neither of the above methods, when applied to this integral, imposes any severe restraint on μ,ν, so that general μ,ν will be considered. In particular, the results include all integer μ,ν, so that in addition to cross-checks between the results of the two methods we have also a check against established results for $\mu+\nu$ odd.

Detailed applications of the two methods will be given in Sections III and IV to obtain asymptotic formulae through terms of order λ^{-3}. Here, certain general properties are discussed and the final formulae are given. It will prove advantageous to write the kernel h(t) as the sum of two terms, one of which, $h^a(t)$, is algebraic in the large argument limit, while the other, $h^o(t)$, is oscillatory, i.e.,

$$h(t) = h^a(t) + h^o(t) \ , \tag{7a}$$

where

$$h^a(t) = [H_\mu^{(1)}(t)\ H_\nu^{(2)}(t) + H_\mu^{(2)}(t)\ H_\nu^{(1)}(t)]/4 \ , \tag{7b}$$

$$h^o(t) = [H_\mu^{(1)}(t)\ H_\nu^{(1)}(t) + H_\mu^{(2)}(t)\ H_\nu^{(2)}(t)]/4, \tag{7c}$$

with $H_v^{(\ell)} = J_v - (-1)^\ell iY_v$, $\ell = 1,2$, the v-th order Hankel functions of the first and second kind, respectively, Y_v being the v-th order Neumann function. Hankel's asymptotic expansions (see, e.g., Olver (1974)) can be used to show that for large arguments:

$$h^a(t) \sim \sum_{\ell=0}^{\infty} \left[c_{2\ell} t^{-(2\ell+1)} \cos\zeta + c_{2\ell+1} t^{-(2\ell+2)} \sin\zeta\right] , \tag{8a}$$

where $\zeta = (\mu-\nu)\pi/2$ and the first few coefficients are

$$c_0 = \frac{1}{\pi},\ c_1 = \frac{1}{2\pi}(\mu^2-\nu^2),\ c_2 = -\frac{1}{8\pi}\left[(\mu+\nu)^2-1\right]\left[(\mu-\nu)^2-1\right] ; \tag{8b}$$

and

$$h^o(t) \sim \sum_{\ell=0}^{\infty} \left[u_{2\ell} t^{-(2\ell+1)} \sin w(t) + u_{2\ell+1} t^{-(2\ell+2)} \cos w(t)\right] , \tag{9a}$$

where $w(t) \equiv 2t - (\mu+\nu)\pi/2$ and the first few coefficients are

$$u_0 = \frac{1}{\pi},\ u_1 = \frac{1}{2\pi}(\mu^2+\nu^2-\tfrac{1}{2});\ u_2 = \frac{1}{16\pi}\left[1-2(\mu^2+\nu^2)\right]\left[(\mu^2+\nu^2)-5/2\right]. \tag{9b}$$

These display the respective algebraic and oscillatory natures of h^a and h^o for large t. It is also necessary to use the well-known Maclaurin series expansions of the functions $f_1(t)$ and $f_2(t)$ for small values of the argument:

$$f_i(t) = \sum_{\ell=0}^{\infty} p_{i\ell}\, t^{2\ell} , \quad i = 1,2 , \tag{10a}$$

with the first two coefficients being

$$p_{10} = p_{20} = 1;\ p_{i1} = \begin{cases} 1/2, & i=1 , \\ -1, & i=2 . \end{cases} \tag{10b}$$

We shall demonstrate that for $\lambda \to \infty$ the integrals in Eqs. (1),(2) have asymptotic expansions which can be compactly combined together as

$$\begin{aligned} I_i(\lambda) &= \lambda^{-1}\{[\ln(\lambda/2|p_{i1}|) - \psi_0]C_0\cos\zeta + (1/2)\sin\zeta\} \\ &+ \lambda^{-3/2}(\delta_{i1}/2\sqrt{\pi})\sin\phi(\lambda) - \lambda^{-2}(\delta_{i2}\pi C_1/2)\sin\zeta \\ &+ \lambda^{-5/2}[\delta_{i1}(\mu^2 + \nu^2 - 9/8)/4\sqrt{\pi}]\cos\phi(\lambda) \qquad (11) \\ &+ \lambda^{-3}p_{i1}\{[C_2[\ln(\lambda/2|p_{i1}|) - \psi_0 + (3-\delta_{i1})/2] - (\mu^2+\nu^2-1)/2\pi]\cos\zeta \\ &+ (\pi C_2/2)\sin\zeta\} + (\delta_{i1})O(\lambda^{-7/2}) + (\delta_{i2})O(\lambda^{-4}) , \end{aligned}$$

where δ_{ij} is the Kronecker delta function, $\phi(\lambda) \equiv w(\lambda) - \pi/4$, and $\psi_0 \equiv \gamma + \psi((1+\mu+\nu)/2) + \psi((1+\mu-\nu)/2)$, with γ Euler's constant and ψ the digamma function [4, pp. 255,258]. The δ_{ij} were introduced to enable a single expression to hold for both integrals. The results for the two integrals differ otherwise only through the value of the Maclaurin coefficient p_{i1}. The terms that contain $\phi(\lambda)$ are contributions from the upper limit of the integral in Eq. (1); they are absent in $I_2(\lambda)$.

The singular behavior of the Bessel functions of Eqs. (1), (2) at $t = 0$ for negative values of the indices (μ,ν) restricts the validity of Eq. (11) to $\mathrm{Re}(\mu+\nu) > -1$. Negative integer values of the indices are an important exception, for which the well-known relationship $J_m = [\mathrm{sgn}(m)]^m J_{|m|}$ is used to restrict attention to positive values for integer indices. The poles of the digamma functions, $\psi((1+\mu+\nu)/2)$ and $\psi((1+\mu-\nu)/2)$, contained in ψ_0 might appear to invalidate Eq. (11) when the argument of one or the other of these functions is zero or a negative integer. However, as just mentioned, the values of (μ,ν) considered do not include the poles of $\psi((1+\mu+\nu)/2)$. Also, using the μ,ν symmetry in Eqs. (1),(2) one can define μ and ν such that $\mu \geq \nu$ and avoid the poles of the other digamma function. Alternatively, the reflection property $\psi(z) = \psi(1-z) - \pi\cot(\pi z)$ could have been used to show that $\psi((1+\mu-\nu)/2)\cos\zeta = \pi\sin\zeta$ for $(\mu-\nu)$ negative odd integer. More generally, for any (μ,ν), the reflection formula can be used to show that the combination $[\psi((1+\mu-\nu)/2)\cos\zeta-(\pi/2)\sin\zeta]$ that appears in Eq. (11) is symmetric in (μ,ν). This makes it clear that Eq. (11) possesses this symmetry which is manifest in the original integrals. It is useful to point out that, although negative integer values of (μ,ν) are avoidable, Eq. (11) as displayed is in fact valid for *all* integer (μ,ν), as can be demonstrated by similar application of the reflection formula to singular digamma functions.

III. STRAIGHTFORWARD METHOD

Starting with the heuristic method, we note that, except for the neighborhood of t=0, the kernel in Eq. (3) can be approximated by its asymptotic value for $\lambda \to \infty$ and μ,ν fixed. Thus, introducing an auxiliary parameter B such that $1 \ll B \ll \lambda$, the integration is split into two parts

$$I_i(\lambda) = \left(\int_0^{B/\lambda} + \int_{B/\lambda}^{\infty}\right)dt\; f_i(t)h(\lambda t) \equiv I_{i1}(\lambda) + I_{i2}(\lambda) \; . \tag{12}$$

The integrals I_{i1} and I_{i2} will be treated separately. They each depend on the auxiliary parameter B. However, since B simply partitions the integration interval into two parts, the combination in Eq. (12) must be independent of B. In particular, the coefficients of various powers of λ that appear in the final asymptotic expansion of $I_i(\lambda)$ are independent of B. Thus terms that appear in I_{i1} and are not independent of B must be cancelled by corresponding terms in I_{i2} when the sum $I_{i1} + I_{i2}$ is formed. This cancellation could be demonstrated explicitly. However, we choose to avoid the tedious algebra involved in this demonstration and simply retain only those terms that are independent of B in the separate integrals, discarding all others. The explicit cancellation is illustrated for terms which depend on $\ell n B$ appearing in the coefficient of λ^{-1}.

In $I_{i1}(\lambda)$, $t \leq B/\lambda \ll 1$ so we can expand $f_i(t)$ in the Maclaurin series of Eq. (10a). Then, letting $t = x/\lambda$, we obtain

$$I_{i1}(\lambda) = \sum_{\ell=0}^{\infty} \left[p_{i\ell} \int_0^B dx\; x^{2\ell}h(x)\right] \lambda^{-2\ell-1} \equiv \sum_{\ell=0}^{\infty} p_{i\ell}\eta_\ell\lambda^{-2\ell-1} \; . \tag{13}$$

The coefficients η_ℓ in Eq. (13) do not depend on λ so that this equation provides a convenient expansion for large λ. In this paper, the coefficients η_0 and η_1 will be obtained, with the remaining terms being coefficients of λ^{-5} or higher.

The coefficient η_0 is

$$\eta_0 = \int_0^B dx\ J_\mu(x) J_\nu(x) \ , \tag{14}$$

where the upper limit is large compared to unity but small compared to λ. This suggests that we make use of the corresponding infinite integral; but it diverges. Therefore we call on the shifted-argument technique introduced by Stoyanov and Farrell (1987) to circumvent this divergence. Thus we write

$$\eta_0 = \lim_{\delta \to 0_+} \int_0^B dx\ J_\mu(x) J_\nu\big(x(1+\delta)\big) \ , \tag{15}$$

where δ is *arbitrarily* small compared to B but need not be explicitly related to B as in Stoyanov and Farrell (1987). This shifts J_μ and J_ν out of phase asymptotically so that the infinite integral will converge. Hence, we can now write

$$\eta_0 = \lim_{\delta \to 0_+} \Big(\int_0^\infty - \int_B^\infty\Big) dx\ J_\mu(x) J_\nu\big(x(1+\delta)\big) \ , \tag{16}$$

where the first integral on the right-hand side is of known Weber-Schafheitlin type [4, p. 487], while asymptotic expansions of J_μ, J_ν can be used to reduce the second integral to elementary integrations. Thus, from [4, Eqs. (11.4.34) and (15.3.10)], we find as $\delta \to 0_+$

$$\int_0^\infty dx\ J_\mu(x) J_\nu\big(x(1+\delta)\big) \to \frac{-[\ln(2\delta)+\gamma+\psi_0]}{\pi} \cos\zeta \ , \tag{17}$$

where ζ, γ, and ψ_0 are defined below Eq. (8a) and Eq. (11). A sufficient condition for result (17) is $\mathrm{Re}(\mu+\nu) > -1$ as indicated in [4, Eq. (11.4.34)], imposed to secure convergence at the lower limit.

The last integral of Eq. (16) can be written as

$$\int_B^\infty dx\ J_\mu(x)J_\nu\big(x(1+\delta)\big) = \int_B^\infty dx\Big[J_\mu(x)J_\nu\big(x(1+\delta)\big) - \frac{\cos(\delta x+\zeta)}{\pi x}\Big] \tag{18a}$$

$$+ \int_B^\infty dx\ \frac{\cos(\delta x+\zeta)}{\pi x}$$

$$= \int_B^\infty dx\Big[h(x) - \frac{\cos\zeta}{\pi x}\Big] - \big[\mathrm{Ci}(B\delta)\cos\zeta - \mathrm{si}(B\delta)\sin\zeta\big]/\pi\ , \tag{18b}$$

where the lead term in the asymptotic expansion of $J_\mu(x)J_\nu(x(1+\delta))$ was subtracted and added in Eq. (18a). In obtaining Eq. (18b), the small δ-limit was taken in the brackets of Eq. (18a), and $\mathrm{Ci}(B\delta)$ and $\mathrm{si}(B\delta)$ denote the cosine- and sine-integral of argument $B\delta$, respectively [4, p. 232]. Thus

$$\int_B^\infty dx\ J_\mu(x)J_\nu\big(x(1+\delta)\big) \rightarrow -\frac{1}{\pi}\ \{[\ln(B\delta) + \gamma]\cos\zeta + (\pi/2)\sin\zeta\}\ , \tag{19}$$

where the integral appearing in Eq. (18b) has been discarded since it does not lead to any B-independent terms for large B, as can be easily verified using Eqs. (8a),(9a). The cosine- and sine-integrals have been replaced by their small argument values. The $\ln B$ term is retained in order to show explicitly the cancellation discussed above. Subtracting Eq. (19) from Eq. (17) in accord with Eq. (16), one sees the shift parameter δ disappear, leaving

$$\eta_0 = \frac{1}{\pi}\{[\ln B - \ln 2 - \psi_0]\cos\zeta + (\pi/2)\sin\zeta\}\ . \tag{20}$$

The μ,ν symmetry of this result (see discussion after Eq. (11)) reassures that it makes no difference to which of the two Bessel functions in Eq. (15) the shifted-argument technique is applied.

Evaluating the integral apearing in η_1, the coefficient of λ^{-3} in Eq. (13), is somewhat trickier. The presence of x^2 in this integral means it cannot be evaluated in the same way as the first integral (cf. Eqs. (15)-(17)) because the corresponding infinite integral does not exist even with shifted arguments. To avoid this trouble, we would like to reduce the integral appearing in η_1 to that represented by η_0, already evaluated above. The best way to accomplish this is to exploit an extension (see below) of Schafheitlin's reduction formula (see Watson (1944)) used by Stoyanov and Farrell (1987) for the special case $\mu=\nu=0$. Indeed, another alternative would be to invoke the argument-shift technique again and to apply repeatedly the basic indefinite-integral formula (see, e.g., Watson (1944) or [4, Eq. (11.3.29)]) to reduce the integral to known results. But this generates such onerous algebra that it becomes unattractive. The commonly known Schafheitlin formula cited above is only good for the special case of two Bessel functions of the same kind, order, and argument. Fortunately, it turns out that this formula is capable of great generalization via a straightforward extension of the derivation outlined in Watson (1944) (see also Straubel (1941) where this generalization was obtained as a by-product among other results). In particular, the generalized Schafheitlin formula yields the integral η_1 in terms of the integral η_0 (Eq. (14)), plus integrated terms, viz.,

$$\eta_1 \equiv \int_0^B dx\ x^2 h(x) = \pi C_2 \int_0^B dx\ h(x) + [J_\mu J_\nu + J'_\mu J'_\nu]B^3/4$$
$$- [(\mu^2-\nu^2+1)J_\mu J'_\nu + (\nu^2-\mu^2+1)J'_\mu J_\nu]B^2/8 - (\mu^2+\nu^2-1)J_\mu J_\nu B/8 \ , \qquad (21)$$

where the unspecified arguments of Bessel functions are all B, and C_2 is defined in Eq. (8b). The B-independent contribution from the integrated terms in Eq. (21) is $[3C_2/2 - (\mu^2+\nu^2-1)/2\pi]\cos\zeta$, as can be verified by replacing derivatives by $J_\nu{}' = (\nu/B)J_\nu - J_{\nu+1}$ and using Hankel's expansions (Eqs. (8a),(9a)). The integral on the right-hand side of Eq. (21) is simply η_0, whose B-independent contribution is obtained by discarding the $\ell n B$ term in Eq. (20). With these results, the B independent contribution to the coefficient η_1 of Eq. (13) is

$$\eta_1 = -[C_2(\ell n2 + \psi_0 - 3/2) + (\mu^2+\nu^2-1)/2\pi]\cos\zeta + (\pi C_2/2)\sin\zeta. \qquad (22)$$

Higher-order coefficients, η_ℓ, in Eq. (13) could be similarly derived via repeated applications of the generalized Schafheitlin formula of

Straubel (1941).

Turning now to the integral $I_{i2}(\lambda)$ in Eq. (12), we can again use the expansion of $h(\lambda t)$ represented by Eqs. (8a),(9a). The algebraic contributions furnished by $h^a(\lambda t)$ are readily evaluated for both $f_i(t)$ by elementary integrations. Then, retaining the term with $\ln B/\lambda$ but discarding all other B-dependent terms, we arrive at

$$\int_{B/\lambda}^{\infty} dt f_i(t) h^a(\lambda t) = (\pi\lambda)^{-1}[\ln(\lambda/|p_{i1}|) - \ln B]\cos\zeta + T_i(\lambda) \tag{23a}$$

with

$$T_i(\lambda) \equiv \begin{cases} \lambda^{-3}[\ln(2\lambda) - 1/2](C_2/2)\cos\zeta + O(\lambda^{-5}\ln\lambda), & i=1 , \\ -\lambda^{-2}(\pi/2)C_1\sin\zeta - \lambda^{-3}(\ln\lambda)C_2\cos\zeta + O(\lambda^{-4}), & i=2 , \end{cases} \tag{23b}$$

where further terms could be explicitly written down if desired. (When Eqs. (20) and (23) are combined according to Eqs. (12),(13), the $\ln B$ terms in the coefficient of $1/\lambda$ cancel out. Q.E.D.)

The oscillatory contributions to $I_{i2}(\lambda)$ of Eq. (12), due to $h^o(\lambda)$, can be found by parts integration. For i=2, the elementary integration-by-parts formula readily shows (cf. Olver (1974)) that the $h^o(\lambda t)$ terms do not produce any B-independent contributions, and thus are discarded. For i=1, however, the singular $f_1(t)$ and arbitrary μ,ν in $h^o(\lambda t)$ require an extension of the standard integration-by-parts procedure. The straightforward integrations used by Stoyanov and Farrell (1987) are effective for evaluating the $1/\lambda$ contribution. However, the extension of the parts method based on neutralizer techniques developed by Erdelyi (1956) (see also Sec. 3.4 of Bleistein and Handelsman (1986)) is more convenient when higher order terms are sought. Specifically, we wish to evaluate

$$\int_{B/\lambda}^{1} dt \frac{h^o(\lambda t)}{\sqrt{1-t^2}} \sim \sum_{\ell=0}^{\infty} \int_{B/\lambda}^{1} \frac{dt}{\sqrt{1-t^2}} \left[\frac{u_{2\ell}\sin w(\lambda t)}{\lambda^{2\ell+1}\, t^{2\ell+1}} + \frac{u_{2\ell+1}\cos w(\lambda t)}{\lambda^{2\ell+2}\, t^{2\ell+2}}\right], \tag{24}$$

in accordance with Eqs. (12), (5), and (9a). For this purpose, Erdelyi's formulae (see, e.g., Bleistein and Handelsman (1986), Eqs.

(3.4.12),(3.4.17)) are employed with $a=B/\lambda$, $b=1$, $\lambda\to 2\lambda$, $\alpha=1$, $\beta=\frac{1}{2}$, and with $F(t)=(1+t)^{-\frac{1}{2}}t^{-(2\ell+1)}$ for the $u_{2\ell}$ terms and $F(t) = (1+t)^{-\frac{1}{2}}t^{-(2\ell+2)}$ for the $u_{2\ell+1}$ terms. Also, the cosine and sine are represented as complex exponentials. Then the result is that the B-independent contributions arise entirely from the upper endpoint of integration in Eq. (24). Thus, the oscillatory contributions are

$$\int_{B/\lambda}^{\infty} dt\ f_i(t)h^o(\lambda t) = \begin{cases} \dfrac{\sin\phi(\lambda)}{\lambda^{3/2}\ 2\sqrt{\pi}} + \dfrac{(\mu^2+\nu^2-9/8)}{\lambda^{5/2}\ 4\sqrt{\pi}}\cos\phi(\lambda)+O(\lambda^{-7/2}), i=1, \\ 0,\ i=2, \end{cases} \tag{25}$$

where $\phi(\lambda)$ is defined after Eq. (11). This result will be also useful in the Mellin transform method.

It was remarked by Stoyanov and Farrell (1987) that the corresponding term of order $\lambda^{-3/2}$ in Eq. (25) could be alternatively obtained as the stationary phase point (at $x=\pi/2$) contribution to the original integral of Eq. (1). However, the simple stationary phase formula would be inadequate in providing correct higher terms, i.e., of order $\lambda^{-5/2}$, etc., of Eq. (25), since it supplies only the lead order contribution from each term in the expansion (9a). On the other hand, the correct contribution of order $\lambda^{-5/2}$, e.g., is a combination of the leading and next order contributions from respectively the u_1 and u_0 terms in Eq. (9a). Thus, an adequate alternative to the Erdelyi formulae would be a similar extension of the standard method of stationary phase, as discussed in Sec. 6.1 of Bleistein and Handelsman (1986).

Finally, a combination of Eqs. (20), (22), (23), and (25), in accord with Eqs. (13) and (12), yields the desired results of Eq. (11).

IV. MELLIN TRANSFORM METHOD

The formal Mellin transform technique, systematically developed in Bleistein and Handelsman (1986) using the conventional analytic continuation approach, will be applied to obtain the asymptotic expansions of the integrals in Eq. (3). (For an alternative version of this technique, exploiting the regularization approach developed in the distribution theory, see Wong (1983).) Certain modifications and simplifications will be introduced here while utilizing the Mellin transform technique for our integrals.

The Mellin transform method is based on the Parseval formula (see, e.g., Eq. (4.4.3) of Bleistein and Handelsman (1986)):

$$I(\lambda) = (1/2\pi i) \int_{r-i\infty}^{r+i\infty} dz\, \lambda^{-z}\, G(z) \,, \tag{26}$$

where

$$G(z) \equiv M[h;z]M[f;1-z] \,, \tag{27}$$

with M[f;s] denoting the Mellin transform of f(t), i.e.,

$$M[f;s] = \int_0^\infty dt\, t^{s-1}\, f(t) \,. \tag{28}$$

The constant r is such that Re(z) = r lies in the common strip of analyticity of the Mellin transforms M[h;z] and M[f;1-z]. For $h(t) = J_\mu(t)J_\nu(t)$, the Mellin transform (see Bleistein and Handelsman (1986), p. 415, Eq. (25)) is

$$M[h;z] = \frac{\pi 2^{z-1}\Gamma[(z+\mu+\nu)/2][\Gamma(z)\sin(\pi z)]^{-1}}{\Gamma[1+(\mu+\nu-z)/2]\Gamma[1+(\nu-\mu-z)/2]\Gamma[1+(\mu-\nu-z)/2]} \,, \tag{29}$$

where $-\mathrm{Re}(\mu+\nu) < \mathrm{Re}(z) < 1$. Therefore, $\mathrm{Re}(\mu+\nu)$ must be greater than -1 in order for Eq. (29) to hold.

The simpler integral $I_2(\lambda)$ is considered first. The Mellin transform for $f_2(t) = 1/(1+t^2)$ as defined in Eq. (6) is (see Bleistein and Handelsman (1986), p. 413, Eqs. (3),(10)):

$$M[f_2;1-z] = \pi/[2\cos(\pi z/2)] \tag{30}$$

for $-1 < \mathrm{Re}(z) < 1$. Thus, there is a common strip of analyticity to provide a value of r on the condition that $\mathrm{Re}(\mu+\nu) > -1$. Although

Eqs. (29),(30) only define $G(z) \equiv G_2(z)$ for Re(z) in the common strip of analyticity, it is clear that a *meromorphic* function $G_2(z)$ can be defined in the region beyond this strip, viz., for $\text{Re}z \geq 1$. Indeed, upon multiplying the right-hand sides of Eqs. (29) and (30) in accord with Eq. (27), we have

$$G_2(z) = \frac{\pi^2 2^{z-3}\Gamma[(z+\mu+\nu)/2][\Gamma(z)\sin(\pi z/2)\cos^2(\pi z/2)]^{-1}}{\Gamma[1+(\mu+\nu-z)/2]\Gamma[1+(\nu-\mu-z)/2]\Gamma[1+(\mu-\nu-z)/2]}, \tag{31}$$

which explicitly shows that, for $\text{Re}(z) > -\text{Re}(\mu+\nu)$, the only singularities of $G_2(z)$ are first- and second-order poles at $z = 2,4,\ldots$ and $z = 1,3\ldots$, respectively.

Furthermore, if z is written as x + iy, then Eq. (3.2.41) of Bleistein and Handelsman (1986) shows that for x finite and $|y| \to \infty$

$$\Gamma(x+iy) = O(|y|^{x-1/2}e^{-(\pi/2)|y|}) . \tag{32}$$

This, along with Eq. (31), can be used to show that $G_2(x+iy)$ is $O(|y|^{x-3/2}e^{-(\pi/2)|y|})$ for arbitrary $x \geq r$ and $|y| \to \infty$, i.e., $G_2(z)$ vanishes in this limit. Thus to evaluate $I_2(\lambda)$ via Eq. (26), we utilize Cauchy's integral theorem with a closed contour generated by displacing the vertical line of integration in the Parseval formula to the right, say to some $\text{Re}(z) = R > r$. Then, by the residue theorem, this standard procedure leads to

$$I_2(\lambda) = -\sum_{r<\text{Re}(z)<R} \text{res}[\lambda^{-z}G_2(z)] + (1/2\pi i)\int_{R-i\infty}^{R+i\infty} dz\, \lambda^{-z}G_2(z) , \tag{33}$$

where the summation is over the poles of $G_2(z)$ for which $\text{Re}(z) < R$, and the remainder integral is $O(\lambda^{-R})$. Since $G_2(z)$ decays quite rapidly along vertical lines, R can be allowed to go to ∞, so that an infinite asymptotic expansion would be obtained as an infinite sum of residues. In particular, applying the standard residue formula to evaluate contributions from the first three poles located at $z = 1,2$, and 3, one readily arrives, after routine but somewhat lengthy calculations, at the result given in Eq. (11) for i=2. Note that the terms with $\ell n\lambda$ arise from the second-order pole residues due to the derivative of the factor λ^{-z} premultiplying $G_2(z)$.

The above simple procedure used to derive the asymptotic expansion of $I_2(\lambda)$ is not allowed, however, when dealing with the integral $I_1(\lambda)$ via Eq. (26) unless its integrand is properly modified. This is due to the behavior of $G(z) \equiv G_1(z)$ for $|y| \to \infty$. The Mellin transform of $f_1(t)$ defined in Eq. (5) is represented by the particular Beta function $B(1/2,z/2)/2$ (see [4, p. 258]; Bleistein and Handelsman (1986), p. 413, Eq. (8)), so that

$$M[f_1;1-z] = \frac{\sqrt{\pi}}{2}\,\frac{\Gamma(z/2)\,\sin(\pi z/2)}{\Gamma[(1+z)/2]\,\cos(\pi z/2)}\,, \tag{34}$$

for $\mathrm{Re}(z) < 1$. This, together with Eq. (29) and in accord with Eq. (27), leads to

$$G_1(z) = \frac{\pi^2\Gamma[(z+\mu+\nu)/2]\{\Gamma[(1+z)/2]\cos(\pi z/2)\}^{-2}}{4\Gamma[1+(\mu+\nu-z)/2]\Gamma[1+(\nu-\mu-z)/2]\Gamma[1+(\mu-\nu-z)/2]}\,, \tag{35}$$

whose only singularities, for $\mathrm{Re}(z) > -\mathrm{Re}(\mu+\nu)$, are second-order poles at $z = 1,3,\ldots$. However, using Eq. (32) one finds that $G_1(x+iy)$ is $O(|y|^{x-2})$ as $|y| \to \infty$. Thus, $G_1(z)$ decays only for $x < 2$ and hence Cauchy's integral theorem becomes nugatory, while the integral $I_1(\lambda)$ itself is absolutely convergent only for $x < 1$.

Comparison of this behavior of $G_1(\lambda)$ for $|y| \to \infty$ with that of $G_2(\lambda)$ uncovers the dual reason for the failure of the standard procedure in this case. Whereas for $f_2(t)$ the Mellin transform in Eq. (30) is $O(e^{-(\pi/2)|y|})$, the Mellin transform of $f_1(t)$ in Eq. (34) is $O(|y|^{-1/2})$ in the limit $|y| \to \infty$. Also, the kernel $h(t)$ being identical for both cases, $i = 1,2$, contains the oscillatory part $h^o(t)$ (see Eqs. (7)) whose Mellin transform $M[h^o;z] = O(|y|^{x-3/2})$ as $|y| \to \infty$ (see Marichev (1983)). Since, on the other hand, for the algebraic part of the kernel $M[h^a;z] = O(|y|^{x-3/2}e^{-\pi|y|})$ in this limit, an immediate remedy (used by Wong (1988)) to circumvent this difficulty would be to treat the algebraic and oscillatory kernels separately all the way through.

A different (but closely related) approach will be taken here to make the standard procedure work in this case as well. We observe that the behavior of $M[f_1;1-z]$ is distinct from that of $M[f_2;1-z]$ when $|y| \to \infty$ due primarily to the fact that $f_1(t)$ is singular and discontinuous at the upper limit of integration, t=1, in $I_1(\lambda)$. Therefore we isolate this critical point by the neutralization process (see Secs. 3.3 and 6.3 of Bleistein and Handelsman (1986)), but without separating the kernel into its algebraic and oscillatory parts. Thus, by using neutralizers $\nu_1(t)$ and $\nu_2(t)$, we write

$$I_1(\lambda) = \int_0^1 dt\ h(\lambda t) f_1(t) \nu_1(t) + \int_0^1 dt\ h(\lambda t) f_1(t) \nu_2(t)$$
$$\equiv I_{11}(\lambda) + I_{12}(\lambda) \ . \tag{36}$$

At t=0, $\nu_1(t)=1$ but all of its derivatives vanish, while at t=1 both $\nu_1(t)$ and all of its derivatives vanish. Thus $\nu_1(t)$ is essentially unity for $0 \le t \le a$ with $0 < a < 1$ and goes continuously to zero at some point $b > a$ such that $0 < a < b < 1$. The other neutralizer $\nu_2(t)$ is simply $1-\nu_1(t)$, as the form of Eq. (36) implies. The separation of $I_1(\lambda)$ in this way is reminiscent of that achieved by introducing the parameter B in the straightforward method described earlier in Section III. Indeed, focusing attention on $I_{12}(\lambda)$, we note that since $\nu_2(t)$ is essentially zero for $t \le a$, the kernel $h(\lambda t)$ can be replaced by its asymptotic expansion for large argument (see Eqs. (8),(9)), and then each term integrated. That is, we have

$$I_{12}(\lambda) \sim \sum_{\ell=0}^{\infty} \int_a^1 \frac{dt\ \nu_2(t)}{\sqrt{1-t^2}} \left[\frac{C_{2\ell} \cos\zeta}{\lambda^{2\ell+1}\ t^{2\ell+1}} + \frac{C_{2\ell+1} \sin\zeta}{\lambda^{2\ell+2}\ t^{2\ell+2}}\right]$$
$$+ \sum_{\ell=0}^{\infty} \int_a^1 \frac{dt\ \nu_2(t)}{\sqrt{1-t^2}} \left[\frac{u_{2\ell} \sin w(\lambda t)}{\lambda^{2\ell+1}\ t^{2\ell+1}} + \frac{u_{2\ell+1} \cos w(\lambda t)}{\lambda^{2\ell+2}\ t^{2\ell+1}}\right] . \tag{37}$$

The integrals in the sum on the top line of Eq. (37) can be evaluated by parts integration using the fact that all the derivatives of $\nu_2(t)$

vanish at the limits. Each term is of the form $\nu_2(t)S_\ell(t)\big|_a^1$, where $S_\ell(t)$ is the indefinite integral

$$S_\ell(t) = \frac{C_{2\ell}\cos\zeta}{\lambda^{2\ell+1}} \int^t \frac{dx}{x^{2\ell+1}\sqrt{1-x^2}} + \frac{C_{2\ell+1}\sin\zeta}{\lambda^{2\ell+2}} \int^t \frac{dx}{x^{2\ell+2}\sqrt{1-x^2}}, \tag{38}$$

with $\ell=0,1,\ldots$. Each of these simple table integrals vanishes at t=1 and, since $\nu_2(t)$ vanishes at the lower limit, the sum on the top line of Eq. (37) does not contribute to I_{12}. Similar integration by parts of the terms on the second line of Eq. (37) reduces them to the form of Eq. (24) with B/λ replaced by a. The Erdelyi formula for the upper endpoint of integration already used in Section III shows that $I_{12}(\lambda)$ is given by the contributions of Eq. (25) for i=1.

The other integral of Eq. (36), $I_{11}(\lambda)$, will be evaluated via Mellin transforms. Using the definition of the Mellin transform given in Eq. (28), and integration by parts, one can easily show (see Sec. 6.2 of Bleistein and Handelsman (1986)) that $M[f_1\nu_1;1-z] = o(|y|^{-R})$ as $|y| \to \infty$, i.e., the Mellin transform of $f_1\nu_1$ decreases faster than any power of $|y|^{-1}$ in this limit. Since, on the other hand, the Mellin transform of the (total) kernel h(t) grows algebraically for $|y| \to \infty$, the product (see Eq. (27)) $M[h;z]M[f_1\nu_1;1-z]$ decays along vertical lines for arbitrary $x \geq r$. Therefore we can obtain an infinite expansion of $I_{11}(\lambda)$ as a residue series in the standard way. Furthermore, since the Maclaurin series of $f_1\nu_1$ is the same as that for f_1 (see Eqs. (10)), the residue contributions arising from poles of $M[f_1\nu_1;1-z]$ will be identical with the corresponding contributions from $M[f_1;1-z]$. Specifically, since the poles of the Mellin transforms $M[h;z]$ and $M[f_1\nu_1;1-z]$ occur only at positive integers, z=m, one can use integration by parts to show that $\operatorname{res}_{z=m} M[h;z]\{M[f_1\nu_1;1-z] - M[f_1;1-z]\} = 0$. Thus we can use $G_1(z)$ in Eq. (35) to evaluate the residue contributions from $I_{11}(\lambda)$ via the standard residue formula. In particular, the contributions from the first two second-order poles at z=1 and z=3 yield the terms with λ^{-1} and λ^{-3}, respectively, in Eq. (11) for i=1. This completes the derivation of the asymptotic expansion of $I_1(\lambda)$ by the Mellin transform method.

V. ACKNOWLEDGEMENT

The authors would like to thank Dr. James R. Kuttler for helpful discussions of Mellin transforms.

VI. REFERENCES

1. Bird, J. F. and Gray, E. P. (1989), to be published.
2. Bleistein, N. and Handelsman, R. A. (1986), *Asymptotic Expansions of Integrals*, Dover, New York.
3. Erdelyi, A. (1956), *Asymptotic Expansions*, Dover, New York, Sec. 2.8.
4. *Handbook of Mathematical Functions* (1972), AMS 55 (M. Abramowitz and I. A. Stegun, eds.), Dover, New York.
5. Marichev, O. I. (1983), *Handbook of Integral Transforms of Higher Transcendental Functions, Theory and Algorithmic Tables,* Ellis Horwood Ltd., West Sussex, England, p. 210, Eq. 48(1).
6. Olver, F. W. J. (1974), *Asymptotic and Special Functions*, Academic Press, New York, pp. 238, 75.
7. Stoyanov, B. J. and Farrell, R. A. (1987), *Math. Comp., 49*: 275.
8. Stoyanov, B. J. and Farrell, R. A. (1988), "Effects of Simple Shadowing on Variational Calculations," *Proceedings of the 1987 Scientific Conference on Obscuration and Aerosol Research*, Edgewood, Maryland, pp. 347-356.
9. Straubel, R. (1941), *Ing.-Arch., 12*: 325.
10. Watson, G. N. (1944), *Theory of Bessel Functions*, Cambridge Univ. Press, Cambridge, 2nd ed., pp. 430, 137-138, 134.
11. Wong, R. (1983), *Congress. Numer., 37*: 145.
12. Wong, R. (1988), *Math. Comp., 50*: 229.

On Eigenvalues with Exponentially Small Imaginary Part

A. D. Wood Professor, School of Mathematical Sciences, Dublin City University, Glasnevin, Dublin, Ireland

R. B. Paris Professor, Department of Mathematical and Computer Sciences, Dundee Institute of Technology, Dundee, Scotland

1 INTRODUCTION

We shall be concerned with the behaviour of eigenvalues of a boundary value problem in ordinary differential equation when a perturbation of form ε times a function of the independent variable x is introduced into one of the coefficients. As ε increases from 0, the perturbed eigenvalue $\lambda(\varepsilon)$ describes a locus in the complex plane. In the case under consideration, the unperturbed problem is self-adjoint, with the result that the unperturbed eigenvalues lie on the real-axis. The effect of the perturbation is to destroy this self-adjointness, thus permitting the eigenvalues to move off the real-axis. There are several applications where this imaginary part of λ, although perhaps small compared with the perturbation in the real part, is of physical importance. Typically we might have $\text{Im } \lambda(\varepsilon) = O(e^{-1/\varepsilon})$ while

$$\lambda(\varepsilon) = \lambda_0 + \varepsilon\lambda_1 + \varepsilon^2\lambda_2 + \cdots, \lambda_k \in \mathbf{R}, \quad \text{as} \quad \varepsilon \to 0+$$

The fact that $e^{-1/\varepsilon} \ll \varepsilon^n$ for any integer n as $\varepsilon \to 0+$ makes $\text{Im } \lambda$ very hard to compute by standard perturbation methods or numerical schemes. We are thus in the area known as exponential asymptotics (R. E. Meyer, 1980) or asymptotics beyond all orders (M. D. Kruskal and H. Segur, 1989), although the problem of computing exponentially small amplitudes for above-barrier reflection of high energy particles from a one-dimensional potential barrier was first identified by Povrovskii and Khalatnikov (1961).

Even when dealing with real quantities, exponentially small terms can still be of importance. F. W. J. Olver in his book (1974: p. 76) points out the danger of neglecting terms which are exponentially small compared with other terms in the asymptotic expansion. Although such terms are negligible in Poincare's sense, their inclusion can significantly improve the numerical computation. As an example (which we have modified in terms of small ε) he cites the evaluation of

$$I(\varepsilon) = \int_0^\pi \frac{\cos(t/\varepsilon)}{1+t^2}\,dt.$$

The method of integration by parts yields an asymptotic expansion as a power series in ε^2. For $\varepsilon = 0.1$ the first three terms contribute -0.0005271. Direct numerical quadrature gives an answer of -0.0004558 correct to seven decimal places. The large relative error arises from the neglect of an exponentially small error term. It is a standard exercise in contour integration to show that

$$\int_0^\infty \frac{\cos(t/\varepsilon)}{1+t^2}\,dt = \frac{1}{2}\pi e^{-1/\varepsilon}$$

and hence that

$$I(\varepsilon) = \frac{1}{2}\pi e^{-1/\varepsilon} - \int_\pi^\infty \frac{\cos(t/\varepsilon)}{1+t^2}\,dt.$$

The method of integration by parts gives exactly the same asymptotic series representation in powers of ε^2 for the integral (albeit with a much smaller error bound). This new representation differs only from the original by the exponentially small term, which can be evaluated to be 0.0000713 when $\varepsilon = 0.1$. This is almost exactly the difference between our original estimate and the correct answer obtained by quadrature.

We shall see that the perturbed differential equations where this exponentially small behaviour arises are of the type known as singular perturbations. A perturbation is singular if the fundamental nature of the problem changes abruptly as $\varepsilon \to 0$, otherwise the perturbation is regular. Boundary value problems can always be set up as linear operators in a Hilbert space and as such are rigorously treated in Chapters 5-7 of the book (Kato, 1984) where he describes holomorphic families of operators. Because of the analyticity in ε, both eigenvalues and eigenfunctions can be expanded as power series in ε with non-zero radius of convergence. Although the results on asymptotic perturbation theory in Chapter 8 roughly correspond to singular perturbation theory, they are abstract in nature and not immediately applicable to the differential equation problem.

It is sometimes claimed that a singular perturbation problem is one where the highest derivative vanishes as $\varepsilon \to 0$. For instance, the second order problem

$$\varepsilon y'' + y' + y = 0$$

$$y(0) = 0, \quad y'(0) = 1$$

becomes a first-order differential equation in the limit as $\varepsilon \to 0$ and it is not possible for its solution to satisfy two boundary conditions. This problem is clearly singular, but not all singular perturbations are of this type. The problem

$$y'' + 2\varepsilon y' + y = 0, \quad x \in \mathrm{I}$$

$$y(0) = 0, \quad y'(0) = 1$$

has exact solution

$$\begin{aligned} y(x,\varepsilon) &= \frac{e^{-\varepsilon x}}{\sqrt{1-\varepsilon^2}} \sin(\sqrt{1-\varepsilon^2}x) \\ &= \sin x - \varepsilon x \sin x + O(\varepsilon^2 x^2) \text{ as } \varepsilon \to 0. \end{aligned}$$

If I is bounded, say I= $[0, X]$ for fixed X, we may replace $O(\varepsilon^2 x^2)$ by $O(\varepsilon^2)$ and the perturbation is regular. If I is unbounded, say I= $(0, \infty)$, we cannot make this substitution, and the perturbation is singular. The model problem which we consider is of this second type.

2 STOKES PHENOMENON

The occurence of eigenvalues with exponentially small imaginary part is intimately connected to Stokes phenomenon for the solutions of the differential equation. We illustrate this by the Airy equation

$$u'' - zu = 0$$

which will form the basis for our model problem. The origin is a regular point of this equation and we can find two linearly independent Taylor series solutions about $z = 0$ in the form of the Airy functions

$$Ai(z) = 3^{-2/3} \sum_{n=0}^{\infty} \frac{z^{3n}}{9^n n! \Gamma(n+\frac{2}{3})} - 3^{-4/3} \sum_{n=0}^{\infty} \frac{z^{3n+1}}{9^n n! \Gamma(n+\frac{4}{3})},$$

$$Bi(z) = 3^{-1/6} \sum_{n=0}^{\infty} \frac{z^{3n}}{9^n n! \Gamma(n+\frac{2}{3})} + 3^{-5/6} \sum_{n=0}^{\infty} \frac{z^{3n+1}}{9^n n! \Gamma(n+\frac{4}{3})}.$$

The point at infinity, on the other hand, is an irregular singular point. A series solution is not possible and we construct an approximate solution via the Liouville-Green substitution $u(z) = \exp[S(z)]$. Under the assumption $S'' \ll (S')^2$ as $z \to \infty$ (which we can later justify) this leads to the asymptotic differential equation

$$(S')^2 \sim z^{\beta}$$

with solutions $S_{1,2}(z) \sim \pm\frac{2}{3}z^{3/2}$. This gives the controlling behaviours at infinity. To obtain the next term, we substitute

$$u(z) = \exp\left[\pm\frac{2}{3}z^{3/2} + T(z)\right]$$

which gives $T(z) \sim -\frac{1}{4}\log z$. Continuing in this way, we may construct two linearly independent expansions

$$E_1(z) = z^{-1/4} e^{\zeta} \sum_{m=0}^{\infty} (-1)^m \frac{(\frac{1}{3}, m)}{(2\zeta)^m}$$

$$E_2(z) = z^{-1/4} e^{-\zeta} \sum_{m=0}^{\infty} \frac{(\frac{1}{3}, m)}{(2\zeta)^m}$$

where we have written

$$\zeta = \frac{2}{3} z^{3/2}$$

and

$$(\nu, m) = \frac{\Gamma(\nu + m + \frac{1}{2})}{m! \Gamma(\nu - m + \frac{1}{2})}.$$

Thus the general solution of Airy's equation has expansion

$$u(z) \sim c_1 E_1(z) + c_2 E_2(z) \quad \text{as} \quad z \to \infty.$$

It is now a question of connecting the behaviour at infinity to that at the origin by finding the values of c_1, c_2 appropriate to $Ai(z)$ and $Bi(z)$ respectively. These values are dependent on $\arg z$ as follows. In the sector $|\arg z| < \pi/3$, $E_2(z) \ll E_1(z)$ as $z \to \infty$, the maximum degree of subdominancy occuring on $\arg z = 0$; this is known as a Stokes line. It is characterised as an asymptote as $|z| \to \infty$ of the curve $\mathrm{Im}[S_1(z) - S_2(z)] = 0$. On the rays bounding the sector, $\arg z = \pm\frac{\pi}{3}, E_1(z)$ and $E_2(z)$ are of the same order of magnitude, this is an anti-Stokes line, characterised as an asymptote to $\mathrm{Re}[S_1(z) - S_2(z)] = 0$. Note that we have adopted the definition of Stokes line given by F. W. J. Olver (1974): in some texts, the definitions of Stokes and anti-Stokes lines are reversed. It follows that the values of the Stokes multipliers c_1, c_2 will change from sector to sector, $c_{1,2}$ jumping in value on the Stokes lines corresponding to maximum subdominancy of $E_{1,2}(z)$ compared to $E_{2,1}(z)$.

Careful derivation of the full asymptotic expansion of $Bi(z)$ from the known results for the modified Bessel functions via the identity

$$Bi(z) = 2(z/3)^{1/2}[I_{1/3}(\zeta) + \pi^{-1}\sin(\pi/3)K_{1/3}(\zeta)],$$

valid for all z (Abramowitz and Stegun, 1970: p. 447) yields the Stokes multipliers

$$\begin{aligned} c_1 &= \pi^{-1/2}, \quad c_2 = -\frac{1}{2}i\pi^{-1/2}, \quad -\frac{2}{3}\pi < \arg z < 0 \\ c_1 &= \pi^{-1/2}, \quad c_2 = \frac{1}{2}i\pi^{-1/2}, \quad 0 < \arg z < \frac{2}{3}\pi \\ c_1 &= \frac{1}{2}\pi^{-1/2}, \quad c_2 = \frac{1}{2}i\pi^{-1/2}, \quad \frac{2}{3}\pi < \arg z < \frac{4}{3}\pi. \end{aligned}$$

The sectors of validity are shown in Fig. 1. When we apply a similar analysis to $Ai(z)$, we find that $\arg z = 0$ is not a Stokes line since $Ai(z)$ is exponentially small as $z \to +\infty$, the discontinuous change in the coefficients occuring only on $\arg z = \pm\frac{2}{3}\pi$.

We therefore have as $|z| \to \infty$ in $|\arg z| < \frac{2}{3}\pi$

$$\begin{aligned} Bi(z) &\sim \pi^{-1/2}z^{-1/4}e^{\zeta} \pm \frac{1}{2}i\pi^{-1/2}z^{-1/4}e^{-\zeta} \\ Bi'(z) &\sim \pi^{-1/2}z^{1/4}e^{\zeta} \mp \frac{1}{2}i\pi^{-1/2}z^{1/4}e^{-\zeta}, \end{aligned} \tag{1}$$

recalling that ζ denotes $\frac{2}{3}z^{3/2}$, where the upper sign is taken for $\arg z > 0$, and the lower sign for $\arg z < 0$. Because $\arg z = 0$ is not a Stokes line for $Ai(z)$ we have as $|z| \to \infty$ in $|\arg z| < \pi$

$$\begin{aligned} Ai(z) &\sim \frac{1}{2}\pi^{-1/2}z^{-1/4}e^{-\zeta} \\ Ai'(z) &\sim -\frac{1}{2}\pi^{-1/2}z^{1/4}e^{-\zeta} \end{aligned} \tag{2}$$

3 THE MODEL PROBLEM

We now illustrate some of the difficulties associated with exponential asymptotics by considering the model problem

$$i\phi_t = -\phi_{xx} - \varepsilon x\phi \tag{3}$$

with the general linear homogeneous boundary condition

$$\phi_x(0,t) + h\phi(0,t) = 0$$

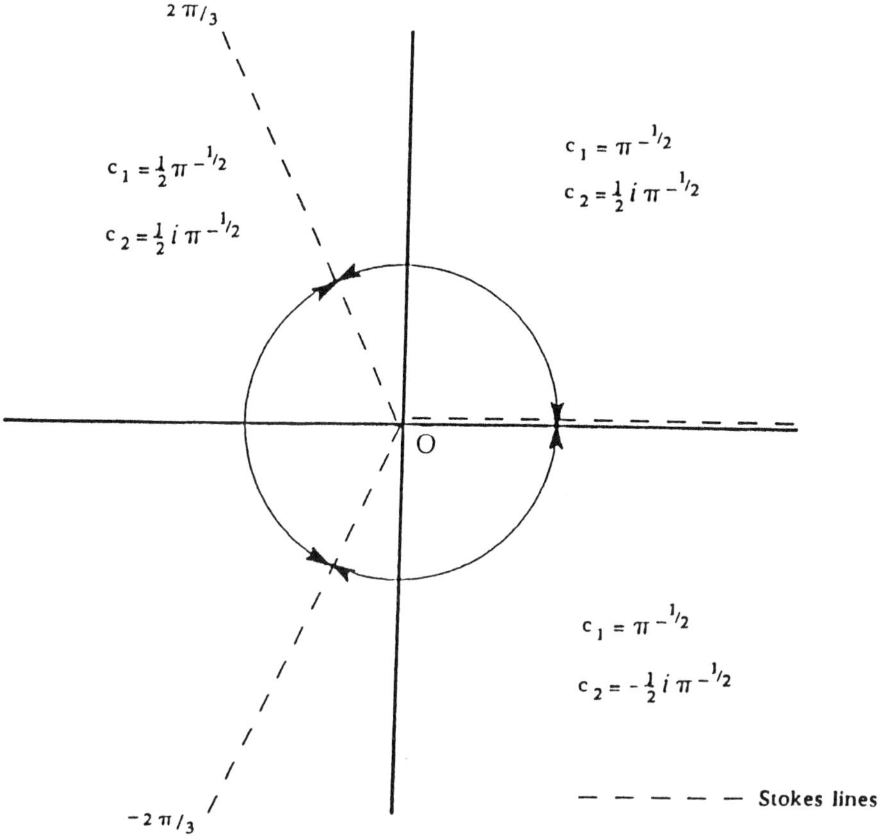

Fig 1. Stokes Multipliers and Stokes Lines for Bi(z)

for some real constant h. We separate variables

$$\phi(x,t) = e^{-i\lambda t}y(x), \quad \text{with } \operatorname{Im} \lambda < 0. \tag{4}$$

to obtain the eigenvalue problem

$$y'' + (\lambda + \varepsilon x)y = 0 \quad \text{on} \quad (0, \infty) \tag{5}$$

$$y'(0) + hy(0) = 0 \tag{6}$$

This equation lacks a boundary condition at infinity. A more difficult equation of the same type is used by Kath and Kriegsmann (1989) to describe radiation losses in bent fibre optic waveguides. A curved waveguide does not trap light perfectly and there is a slow leakage of energy out of the core to radiate into the cladding. This effect is known as optical tunnelling. In this application, the imaginary part of the eigenvalue corresponds to the rate of leakage. For physical reasons, we therefore require that the solution $y(x)$ be an outgoing wave at infinity, beyond the turning point $x = -\lambda/\varepsilon$ of (5). Solutions which may have been evanescent to the left of the turning point become oscillatory in nature to the right. This will imply that, for some positive function $p(x)$,

$$y(x) \text{ has controlling behaviour } e^{ip(x)} \text{ as } x \to +\infty. \tag{7}$$

The unperturbed problem obtained by setting $\varepsilon = 0$ in (5), together with boundary conditions (6, 7) can be shown to be self-adjoint with a single negative real eigenvalue $\lambda = -h^2$.

The problem (5, 6, 7) appears at first glance to be self-adjoint, in which case the spectrum would be real, but a careful analysis shows that the conditions for self-adjointness are broken at infinity. To see this set $Lu \equiv u'' - \varepsilon x u$ and denote by $<,>$ the usual inner product in the Hilbert space $L^2(0,\infty)$. Then for any functions $u, v \in L^2(0,\infty)$ satisfying the boundary conditions, an integration by parts shows that

$$\langle Lu, v\rangle = -[u'(x)\overline{v(x)}]_0^\infty + \int_0^\infty u'(x)\overline{v'(x)}dx - \varepsilon \int_0^\infty xu(x)\overline{v(x)}dx$$

and

$$\langle u, Lv\rangle = -[u(x)\overline{v'(x)}]_0^\infty + \int_0^\infty u'(x)\overline{v'(x)}dx - \varepsilon \int_0^\infty xu(x)\overline{v(x)}dx.$$

The self-adjointness condition $\langle Lu, v\rangle = \langle u, Lv\rangle$ holds if and only if the integrated terms are equal. Since they are clearly equal at the origin, this condition is equivalent to

$$\lim_{x\to\infty} u'(x)\overline{v(x)} = \lim_{x\to\infty} u(x)\overline{v'(x)}.$$

When we insert the proposed outgoing wave behaviour $\exp[ip(x)]$, we find that $u'(x)\overline{v(x)} \sim ip'(x)$, but $u(x)\overline{v'(x)} \sim -ip'(x)$ as $x \to +\infty$. The problem is thus non self-adjoint and non-real eigenvalues may occur.

A straightforward regular perturbation expansion about $\varepsilon = 0$ shows that the eigenvalue of (5, 6, 7) is given by the asymptotic series

$$\lambda = -h^2 - \frac{\varepsilon}{2h} - \frac{\varepsilon^2}{8h^4} - \frac{5\varepsilon^3}{32h^7} - \frac{11\varepsilon^4}{32h^{10}} + 0(\varepsilon^5). \tag{8}$$

This expression contains no information on the imaginary part of λ. Subsequent analysis will show that this is hardly surprising, since $\mathrm{Im}(\lambda) = o(\varepsilon^n)$ for any positive integer n, but (8) is certainly a valid estimate of $\mathrm{Re}(\lambda)$.

The model problem (3) was chosen because an exact solution is possible in terms of either Airy or Hankel functions whose asymptotic properties are well-known. We shall find the outgoing wave solution of (5) and substitute it into the condition (6) at the origin to obtain the eigenvalue relation. We introduce the scaling $x = \varepsilon^{-1/3}x^*$ and $\lambda^* = \varepsilon^{-2/3}\lambda$, setting $z = \lambda^* + x^*$ to obtain the differential equation

$$y'' + zy = 0, \tag{9}$$

where $'$ now denotes differentiation with respect to z. This is, of course, Airy's equation, with linearly independent solutions $Ai(-z)$ and $Bi(-z)$. Using the well-known relation between Airy and Hankel functions, these may be combined to yield an outgoing wave solution

$$\begin{aligned} y(z) &= iAi(-z) + Bi(-z) \\ &= ie^{\pi i/6}(z/3)^{1/2}H_{1/3}^{(1)}(\zeta) \quad \text{where} \quad \zeta = \frac{2}{3}z^{3/2}, \\ &\sim \pi^{-1/2}z^{-1/4}\exp\left[\frac{2}{3}iz^{3/2} + \frac{\pi i}{4}\right] \quad \text{as} \quad z \to \infty \end{aligned} \tag{10}$$

as required. Written in the z variable, the boundary condition (6) at the origin becomes

$$\varepsilon^{1/3}y'(\lambda^*) + hy(\lambda^*) = 0. \tag{11}$$

Substituting the outgoing wave solution, we find the eigenvalue relation

$$-\varepsilon^{1/3}[Ai'(-\lambda^*) - iBi'(-\lambda^*)] + h[Ai(-\lambda^*) - iBi(-\lambda^*)] = 0. \tag{12}$$

Now let $H = -\lambda^*$ and put $\tau = \frac{2}{3}H^{3/2}$. Since $\arg\lambda < 0$, we will have $\arg H > 0$. The eigenvalue relation then becomes

$$-\varepsilon^{1/3}[Ai'(H) - iBi'(H)] + h[Ai(H) - iBi(H)] = 0. \tag{13}$$

Substituting the leading terms given by (1, 2) in (13) gives

$$\varepsilon^{1/3}H^{1/2}\left[\frac{1}{2}e^{-2\tau} + i \pm \frac{1}{2}e^{-2\tau}\right] + h\left[\frac{1}{2}e^{-2\tau} - i \pm \frac{1}{2}e^{-2\tau}\right] \simeq 0.$$

It is easily seen that taking lower signs ($\arg z < 0$) causes the $e^{-2\tau}$ terms to cancel, given the result $\text{Im}(\lambda) = 0$. In $\arg z > 0$ we take upper signs to find

$$\text{Im}(\lambda) \sim -\frac{4h^2}{e}\exp\left[-\frac{4}{3\varepsilon}h^3\right].$$

It is shown by the authors (Paris and Wood, 1989) by different methods, based on either expanding the solution in the small imaginary part of the eigenvalue or using a variational method, that the correct result is

$$\text{Im}(\lambda) \sim -\frac{2h^2}{e}\exp\left[-\frac{4}{3\varepsilon}h^3\right].$$

Averaging the above results would give the correct answer. A heuristic justification of this process in a quantum mechanics setting is given by Bender and Wu (1973, p. 1627). Recently M. V. Berry (1989) has offered a more rigorous justification of this averaging by controlling exponentially small terms in the dominant series using Dingle's method of Borel summation of late terms, starting with the least term. The net result is that, if the change in the Stokes multiplier across the Stokes line is not discontinuous but occurs smoothly from one to the other, the functional dependence is that of the error function. The Airy function $Bi(z)$ is given as an example in Section 3 of Berry's paper. This permits us to obtain the correct estimate of the imaginary part of the eigenvalue.

Recent work by N. Brazel has shown that if we replace x by x^2 in the potential in the differential equation (5) the exponentially small behaviour of $\text{Im}(\lambda)$ persists, but is now $0(e^{-1/\sqrt{\varepsilon}})$. The extension to general x^n is currently under consideration.

4 OTHER PHYSICAL EXAMPLES

Most of the known examples of exponentially small eigenvalues arise in sub-particle physics. Outside of this, apart from the optical fibre application given in Section 3 above and the application to the formation of needle-like crystals (Kruskal and Segur, 1989), the only example known to the authors is in the paper of Lozano and Meyer (1976) on leakage and response of waves trapped by round islands. Here too the problem is non self-adjoint and complex eigenfrequencies are possible. Physically speaking, a very small leakage gives a large response to incident waves. When the leakage is exponentially small the waves are resonant in character. Regular perturbation methods and a "brilliant numerical attack" failed because the frequency bandwidth of response was too narrow. The structure of the strongly resonant wave modes was characterised by an eigenvalue problem for an ordinary differential equation whose imaginary parts gave an exponentially small leakage.

In subnuclear physics, apart from the work of Bender and Wu (1973) cited in Section 3, S.

Coleman (1987) in a paper on quantum field theory discusses estimates of imaginary parts of eigenvalues which are exponentially small, but incorrect by factor $\frac{1}{2}$. Although his methods are totally different, this appears to be related to the averaging process across Stokes lines discussed above. In quantum mechanics, this exponentially small behaviour of eigenvalues arises in the Stark effect in hydrogen (for a recent treatment of these resonances see Hinton and Shaw (1985)). Quasi-resonance of long life is considered by R. E. Meyer (1986) in treating Schrödinger's equation for two body scattering in the tunnelling range of energy, where the WKB method fails. This paper contains an interesting description of turning points and Stokes lines for non-real energies and again eigenvalues of exponential proximity to the real axis are obtained. In their paper on nonexistence of small-amplitude breather solutions in ϕ^4 theory, Segur and Kruskal (1987) calculate a radiation rate which lies beyond all orders of ε in the asymptotic expansion. If we go to non-linear problems, examples outside of subnuclear physics, involving boundary layer problems where the method of matched asymptotic expansions fails because of the appearance of exponential small terms, are given by Kath, Knessl and Matkowsky (1987) and Byatt-Smith (1989) (in studying resonant oscillations of water waves).

It is clear from the above examples that there is need for a unified, rigorous treatment of these exponential asymptotics which takes account of the many mathematical and physical facets of the problem. Recent remarks by M. D. Kruskal indicate that the surreal number system, introduced by D. Knuth and J. H. Conway, may be of assistance in distinguishing different kinds of infinitesimals.

REFERENCES

M. Abramowitz and I. Stegun (ed.) (1970) *Handbook of Mathematical Functions.* Dover (New York).

C. M. Bender and T. T. Wu (1973) "Anharmonic Oscillator II", *Physical Review D,* 7, 6, pp. 1620-1636.

M. V. Berry (1989) "Uniform asymptotic smoothing of Stokes discontinuities" *Proc. Roy. Soc. Lond.*, to appear.

J. G. B. Byatt-Smith (1989) "The asymptotic solution of a connection problem of a second order ordinary differential equation". To be published.

S. Coleman (1979) "The Uses of Instantons" in *The Whys of Subnuclear Physics*, ed. A. Zichichi, pp. 805-826, Subnuclear series 15, Plenum Press.

D. B. Hinton and K. Shaw "Spectral concentration for Hamiltonian systems" *Helvetica Physica Acta 58*, pp. 982-994.

W. L. Kath and G. A. Kriegsmann (1989) "Optical Tunnelling: Radiation losses in bent fibre-optic waveguides". *IMA Journal of Applied Math.*, to appear.

W. L. Kath, C. Knessl and B. J. Matkowsky (1987) "A variational approach to non-linear singularly perturbed boundary value problems". *Studies in Applied Math. 77*, pp. 61-68.

T. Kato (1984) *Perturbation theory for linear operators.* Second Edition. Springer-Verlag (Berlin).

M. D. Kruskal and H. Segur (1989) "Asymptotics beyond all orders in a model of dendrites" to be published.

R. E. Meyer (1986) "Quasiresonance of long life". *J. Math. Phys. 27*, (1) pp. 238-248.

R. E. Meyer (1980) "Exponential Asymptotics", *SIAM Review 22*, 2, pp. 213-224.

F. W. J. Olver (1974) *Asymptotics and Special Functions.* Academic Press, New York.

R. B. Paris and A. D. Wood (1989) "A model equation for optical tunnelling", *IMA Journal of Applied Math.*, to appear.

V. L. Povrovskii and I. M. Khalatnikov (1961) "On the problem of above barrier reflection of high-energy particles" *Soviet Physics JETP 13,* 6, pp. 1207-1210.

H. Segur and M. D. Kruskal (1987) "Nonexistence of small-amplitude breather solutions in ϕ^4 theory". *Physical Review Letters 58*, 8, pp. 747-750.

Index